ANTOINE HENRI BECQUEREL

(1852–1908)

The Discoverer of Radioactivity

One Hundred Years after the Discovery of Radioactivity

Special issue
of
Radiochimica Acta
International journal for chemical aspects of nuclear science and technology

R. Oldenbourg Verlag München 1996

Foto Umschlag: Marie und Pierre Curie (Association Curie et Joliot-Curie)

Preface

The Editors of Radiochimica Acta have dedicated this special issue to the hundredth anniversary of the discovery of radioactivity not only to provide a historical retrospective, but also to present an overview of present day radiochemistry, thus demonstrating its viability one century after the beginning.

The discovery of the natural radioactivity of uranium by Antoine Henri Becquerel in 1896 opened a new era of science that can be considered as the beginning of nuclear sciences. This event resulted from the search for new kinds of radiation after X-rays had been discovered in 1895 by Wilhelm Conrad Röntgen. It was soon realized that radioactivity was a hitherto unknown property of matter. Pierre and Marie Curie established later that the presence of other elements in uranium was mainly responsible for the radiation observed, and this led to the discovery of radium and polonium. Radium was found to be useful as a radiation source and subsequently opened the way to the first applications of radioactivity in medicine. Ernest Rutherford investigated the deflection of α particles and postulated the existence of a heavy positively charged nucleus inside the atom.

The discovery of the neutron in 1932 by James Chadwick provided a new means for the investigation of nuclear reactions on a large scale. In 1934, Frédéric Joliot and Irène Curie synthesized for the first time new radioactive matter using α-particles. Attempts to prepare transuranium elements by the reaction of neutrons with uranium led to the unexpected discovery of nuclear fission by Otto Hahn and Fritz Strassmann in 1938. In the following years, the first transuranium elements were prepared by Glenn T. Seaborg and coworkers and the last gaps in the Periodic Table were filled by the synthesis of the elements technetium, promethium and astatine. Historical milestones of radioactivity are reviewed in Part A.

During the last several decades many new aspects of radioactive transformations and nuclear reactions have been investigated. The mechanisms of spontaneous and particle-induced nuclear fission have been elucidated. New nuclides far from the β-stability line have been obtained and proton emission has been observed as a fourth kind of radioactive transformation. This was followed by the discovery of the emission of clusters such as carbon-14. New types of accelerators, in particular for heavy ions, have enabled the investigation of increasingly complex nuclear reactions. These new developments are covered in Part B, devoted to modern aspects of radioactivity and nuclear reactions.

New heavy elements have been synthesized by reactions of heavy ions, in particular with nuclei having closed nucleon shells. Since their half-lives are very short, down to the millisecond range, the investigation of the chemical properties of these elements is extremely difficult. However, owing to relativistic effects in the electron shells, the determination of these properties is of great interest. The expected magic number of 114 protons will probably be reached in the near future, although the presently attainable number of neutrons still lies appreciably below the corresponding magic number of these nucleons. The question of the existence of relatively stable double-magic superheavy elements is therefore still open. The properties of "quasi-atoms" and "quasi-molecules" formed within very short periods of time in the course of heavy-ion reactions are also a matter of research. Some aspects of new synthetic elements are considered in part C.

Radioactivity has found broad applications in many fields of science and technology. The importance of these applications was emphasized by Glenn T. Seaborg at the first Geneva Conference in 1955. In chemistry, biochemistry and agriculture, radionuclides are applied as tracers in the study of reaction mechanisms and transport processes. In the field of analytical chemistry, activation methods and tracer techniques are often methods of choice owing to their high sensitivity. Radiochemical methods have largely contributed to cosmo- and geochemistry. Dating with radioactive nuclides is an indispensable tool in archaeology, geology and other fields.

Applications in life science are of major importance. Already in 1921 George de Hevesy employed a natural radioisotope of lead to investigate the metabolism of this element in plants. His concept of radioactive indicators became the base of a new interdisciplinary speciality, called nuclear medicine. With the invention of the cyclotron in 1930 by Ernest O. Lawrence, the spectrum of radionuclides for biomedical studies greatly broadened. The availability of nuclear reaction in the late forties led to the application of new long lived β^--emitters, such as tritium and carbon-14 which played a paramount role in the biomedical discoveries of the past 50 years. During the last 20 years a renaissance of the cyclotron took place. Together with the advent of new imaging devices, such as single photon emission tomography (SPET) and particularly positron emission tomography (PET), it became possible to study regional biochemistry in the living man.

New radionuclides became available for in vivo studies. Besides the generator-nuclide technetium-99m, the workhorse in nuclear medicine, iodine-123 substituted iodine-131 in functional imaging with SPET. For PET, short-lived emitters such as carbon-11 and fluorine-18 are used to label biomolecules and drugs. Furthermore, therapeutic radionuclides are being used routinely or are presently evaluated for endotherapy. It is an exciting new challenge to bridge radiochemistry of short-lived no-carrier-added radionuclides with the appropriate organic or coordination chemistry to label the ever increasing number of compounds. Adding a new dimension to the tracer concept of de Hevesy, they are aimed to learn more about how life is organized on a molecular level in a living system and to detect or to cure malfunction.

Amongst the manifold benefits of radioactivity, only a few examples can be given in Part D.

The operation of nuclear reactors leads to the production of radioactive wastes which demand safe storage conditions for long periods of time. Fission products and actinide elements have been dispersed on a large scale into the soil and in the atmosphere as a consequence of nuclear explosions and accidents. The behaviour of natural and artificial radionuclides in the geosphere, biosphere and environment is dealt with in Part E.

J. P. Adloff   K. H. Lieser   G. Stöcklin

Index

A. Historical Notes

B. Modern Aspects of Radioactivity and Nuclear Reactions

C. New Synthetic Elements

D. Applications of Radioactivity

A. Historical Notes

Radiochimica Acta *70/71*, 3–12 (1995)

The Discovery of Uranic Rays: A Short Step for Henri Becquerel but a Giant Step for Science

By Michel Genet

Groupe de Radiochimie, Institut de Physique Nucléaire, Université Paris Sud, B.P. 1, 91406 Orsay, France

(Received June 19, 1995; revised June 28, 1995)

History / Becquerel / Discovery of Radioactivity

Abstract

In March 1896, Antoine Henri Becquerel discovered radioactivity. This success is largely due to a tradition of high-level scientific research which prevailed over three generations in the Becquerel family. The circumstances of the discovery and the controversies which followed are reviewed within the framework of their scientific and historical context.

1. The saga of the Becquerels

The Becquerel family is exceptional and unique. In four successive generations, covering the historical period from Napoleon the First to World War II, each male member of the family dedicated his life to research (s. next page). Few other such examples can be found in the field of science. During the 17th and 18th centuries, the Cassini family was renowned for its contribution to astronomy; three members of the Bernouillis influenced the development of mathematics and three generations of De Jussieu, from 1750 to 1850, contributed to significant progress in botany [1].

The Becquerel progeny exemplifies how tradition and continuity in research can lead to the highest achievements in science.

a. Antoine César Becquerel (1788–1878)

The first famous Becquerel, Antoine César, was born in Châtillon-sur-Loing (now Châtillon-Coligny in the Loiret Département, 150 km south of Paris) on March 7, 1788. His father, Louis-Hector Becquerel de la Chevrottière was Lieutenand du Roy (King's Lieutenant) and mayor of the community. The vicissitudes of his life were greatly marked by the confused periods of the French Revolution and the First Empire. His mother's aunt, Anne Angélique Cordier, married Antoine Florent Girodet; their son was to become the well known painter Antoine Louis Girodet (1767–1824).

In 1806, Antoine César passed the entrance examination for the prestigious Ecole Polytechnique. From 1810 to 1812, he actively participated in the Spanish War initiated by Napoleon. At the age of 27, he left the army and started a very fruitful scientific career which lasted over sixty years. In 1828 he was elected Member of the Academy of Sciences, and ten years later he was appointed to the Chair of Physics created on his behalf at the Muséum d'Histoire Naturelle. Most of his work dealt with electricity, optics, phosphorescence, electrochemistry, chemical fertilizers and meteorology. He was in advance of his time when in 1829 he constructed the first electrochemical cell capable of producing a stable current, known as the pile Daniell in 1836. He entered also in politics, devoting part of his life as mayor and country councillor to the Loiret subdivision. Antoine César died on January 18, 1878.

b. Alexandre Edmond Becquerel (1820–1891)

Alexandre Edmond was the second son of Antoine César Becquerel and Aimée Cécile Darlu (1794–1883). He entered the Ecole Normale Supérieure in 1838 and was received the following year at the Ecole Polytechnique. However, he preferred to accept his father's proposal to become his assistant at the Museum. From then on, and for the next fifty years, he undertook research, both alone and with his father. In joint publications, the authors appeared as Becquerel and Edmond Becquerel; the first names Antoine César were regularly omitted.

Father and son published numerous articles on pyrometry, electrochemistry, magnetism and on the chemical effects of light. This latter topic was chosen as the main subject of Alexandre Edmond's thesis, that was submitted in 1840 and which led him towards the invention of an electrochemical actinometer. Early in his scientific career he was impressed by the photographic process that had recently been improved by Daguerre. Alexandre Edmond spent ten years trying to set up a procedure for a colour photography, but his colour prints remained stable only in the dark! It is possible that during this period he was in touch with Abel Niepce de Saint-Victor, as will be considered in a following section.

In 1852, Alexandre Edmond obtained a professorship for applied physics at the Conservatoire National des Arts et Métiers. He constructed a "phosphoroscope" for measuring the phosphorescence of minerals and synthetic samples. He found that the emission of

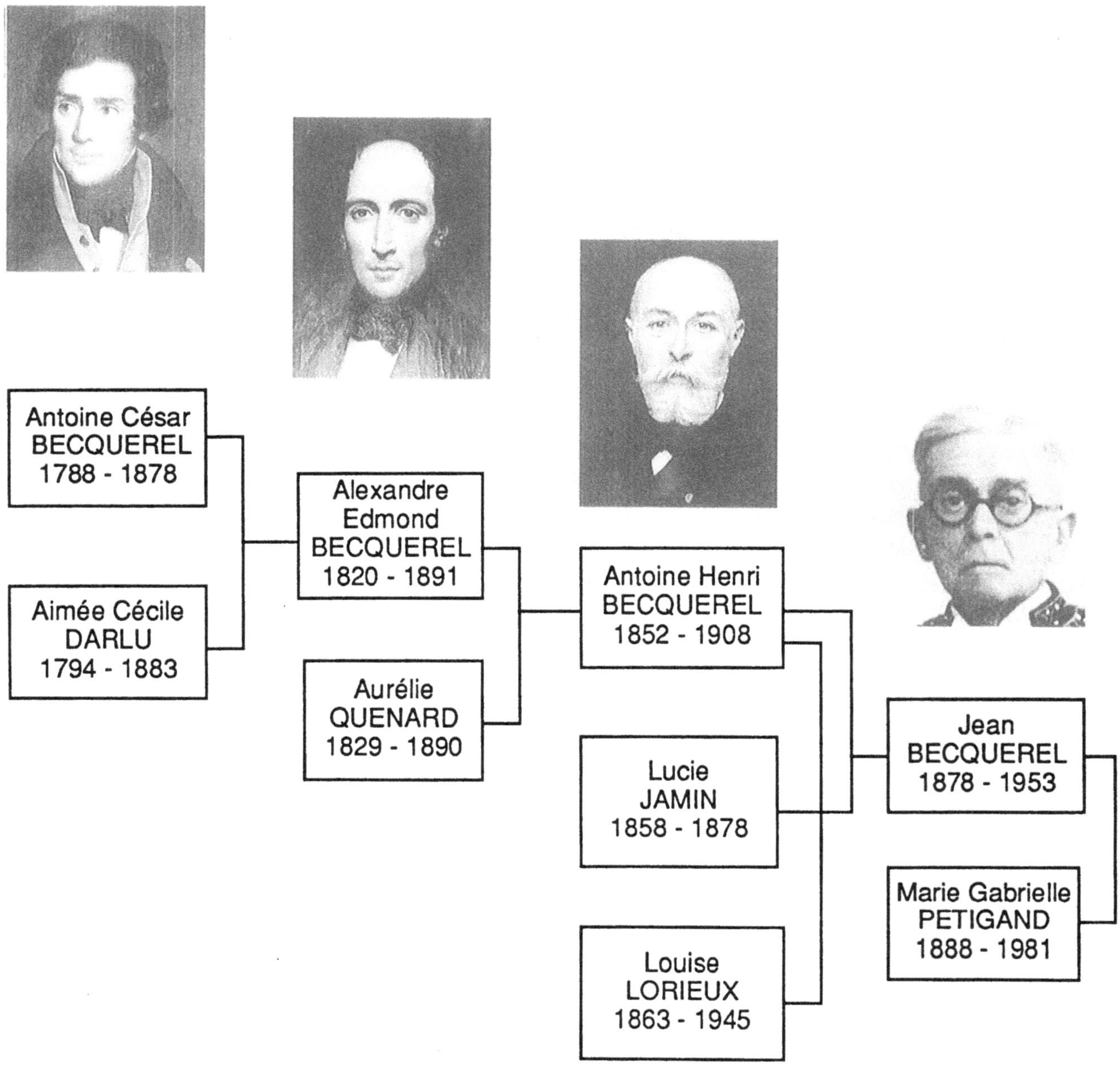

A part of the Becquerels' family tree

light was due to impurities present in the phosphorescent materials and he became a pioneer in the analysis of uranium salts, using a technique based on their phosphorescence. He also discovered new lines in the ultraviolet (in 1842) and infrared (1873) regions of the solar spectrum.

Alexandre Edmond was elected Member of the Academy of Sciences in 1863 and became President of this Institution in 1881. He held a deep veneration for his father, whom he succeeded for a few years as Professor at the Museum.

c. Antoine Henri Becquerel (1852–1908)

The third Becquerel, Antoine Henri was the most famous one; he was born on December 15, 1852 as the first son of Alexandre Edmond and Aurélie Quénard (1829–1890). The parents had a second son, André Paul, five years later. Henri lived a peaceful childhood, playing in the Jardin des Plantes close to his father's laboratory and to the family home, 57 rue Cuvier. He entered the Ecole Polytechnique in 1872, where he met his best friend, Henri Deslandres, who later became a famous astronomer. After a two-year term, he pursued his studies for three more years at the Ecole des Ponts et Chaussées, but before receiving an engineer's diploma, he began with his father a research project which led to the discovery of the magneto-optic field.

After the death of his wife in 1878, he left his apartment in the rue Gay Lussac and returned to live with his father, who had in the meantime replaced Antoine César in the Chair of Physics at the Museum. At this point, Antoine Henri replaced his own father as a naturalist assistant in the same laboratory.

In 1876, Antoine Henri gave his first lecture as "répétiteur" at the Ecole Polytechnique and 5 years later he was nominated "répétiteur adjoint" with an annual salary of 2500 French francs. He was already renowned for the discovery of magnetic properties of gases and for his work on the rotation of polarized light induced by a magnetic field. He was very skillful in the conception and realization of small scale experiments with a tiny budget.

During the years 1883–84, he pursued his father's work on the infrared emission spectra of the sun and of hot metallic vapors. Simultaneously he performed a careful analysis of the annealing of phosphorescence by infrared light. On March 15, 1888 he submitted his thesis "Recherches sur l'absorption de la lumière" [Research on the absorption of light].

Antoine Henri had been interested in the absorption of light by crystals since 1886 and showed the importance of crystal symmetry in the absorption spectra of polarized light. He noticed that tetravalent uranium compounds were not phosphorescent, whereas uranyl salts exhibited a bright luminescence under the same conditions of excitation. Interestingly enough, this was the second experiment performed by a Becquerel on uranium. Like his father, Antoine Henri was fascinated by the phenomenon of phosphorescence, and at the time nobody suspected the secret hidden in the mysterious element. This strange coincidence might be regarded as a premonitory sign of destiny or as the first step towards a major discovery.

continued tirelessly and published in 1890 a long paper on the temperature of soil, thereby completing work he had started in collaboration with his grandfather.

From 1891 on, Becquerel lectured at several places: the Museum, the Conservatoire des Arts et Métiers and at the Ecole Polytechnique. In the latter institute, he replaced in 1895 the ailing Alfred Potier as Professor of Physics until his death. He was elected to this position despite the indignation of Alfred Cornu, Professor at the Ecole and President of the Academy of Sciences, because there were only two candidates and Henri's rival, René Colson did not have the appropriate qualification.

One year after the celebration of the centenary of the Ecole Polytechnique, Henri Becquerel began to teach physics. His lectures at the Ecole and at the Museum were not well received, although he prepared them carefully. According to statements of his students, Henri's lectures were difficult to follow and his pedagogy was poor. However, despite his faulty elocution and obscure reasoning, he compelled respect from the audience.

In their traditional annual feast "Séances des Ombres" (Seances from the Shadows) the students used to joke about Henri Becquerel's teaching. An excerpt of the satirical pamphlet of the academic year 1897/1898 is shown below:

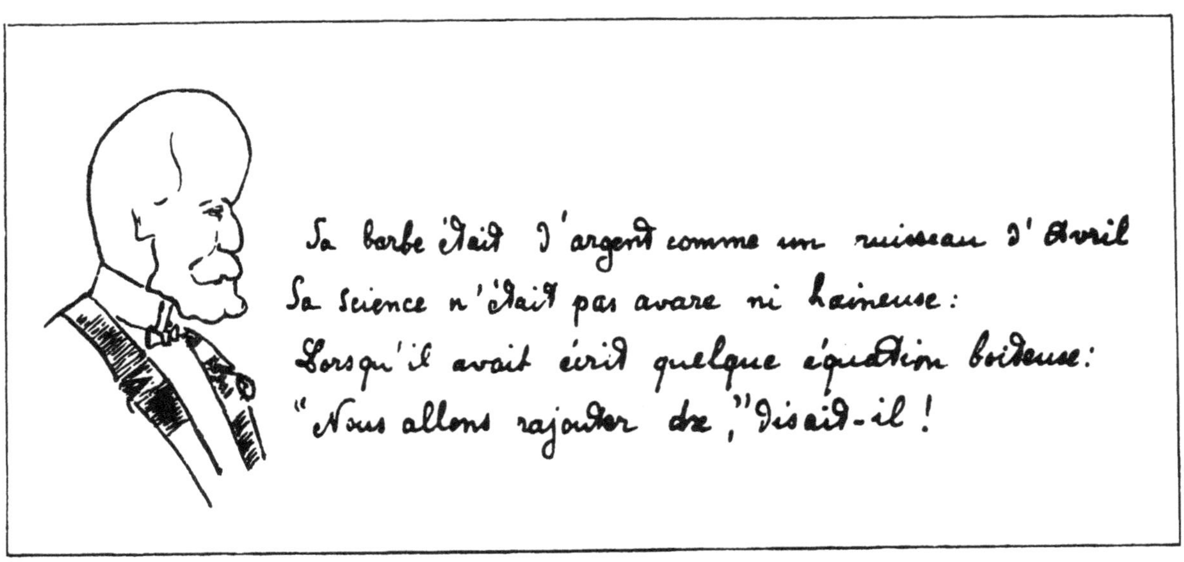

(Reproduced with permission of Archives de l'Ecole Polytechnique)

It is noteworthy that in 1852, G. G. Stokes had stated a definition of fluorescence that was seldom used by Alexandre Edmond Becquerel. The latter never acknowledged the distinction between fluorescence and phosphorescence and Antoine Henri held to this tradition [2].

At the age of 36, on March 27, 1889, Antoine Henri entered in his turn the Academy of Sciences in replacement of the chemist Marcellin Berthelot (1827–1907), who was elected as "secrétaire perpétuel". Despite these honours and the recognition of his fame, he

In a free translation:

His silvery beard flowed like the April stream that fled
His approach to Science was not biased, nor furious,
And when his equations appeared quite dubious
"We're going to add "dx" he said!

The modern terms of optics, electricity, magnetism and radiations commonly used by the family were difficult to understand. Curiously enough, Henri mentioned the existence of uranic rays in his lectures at the Ecole Polytechnique only two years after the discovery and summarized this important event in less than 20 lines [3]:

103. Rayonnement spontané de l'uranium et de diverses autres substances.

On a découvert en 1896 (1) que l'uranium métallique et tous les sels de ce métal, émettaient un rayonnement dont les propriétés sont très voisines de celles des rayons de Röntgen ; il impressionne les plaques photographiques au travers de divers corps opaques pour la lumière, et provoque la décharge des corps électrisés en rendant conducteurs les gaz qui les entourent. Ce rayonnement paraît spontané, car il se maintient avec une intensité sensiblement constante dans des corps conservés depuis trois ans à l'abri de toute excitation connue, sans que l'on ait pu reconnaître à quelle source ces corps empruntent l'énergie qu'ils rayonnent sans cesse. Le même phénomène a été observé avec le Thorium et récemment Mr et Mme Curie en utilisant ces propriétés comme un nouveau moyen d'analyse, ont découvert deux matières considérablement plus actives que l'Uranium, qu'ils ont nommées le Polonium et le Radium. L'intensité du rayonnement de ces matières est suffisante pour exciter la phosphorescence du platino-cyanure de Barium.

(1) H. Becquerel. Comptes-rendus de l'Académie des Sciences (1896)

(Reproduced with permission of Archives de l'Ecole Polytechnique)

Translation:

Spontaneous radiation of uranium and of various others substances
It has been discovered in 1896(1), that metallic uranium and all salts of uranium emit a radiation with properties very similar to those of the Röntgen rays; it acts on photographic plates passing through various bodies opaque to light, discharges electrified bodies and renders surrounding gases conductive. This radiation seems to be spontaneous, because it persists with a fairly constant intensity in bodies that have been stored for three years away from any known excitation; the source from which these bodies acquire the incessant energy which they radiate has not been determined. The same phenomenon has been observed with Thorium, and recently Mr and Mme Curie, using these properties as a new analytical tool, discovered two substances considerably more active than Uranium, which they named Polonium and Radium. The intensity of the radiation of these substances is sufficient to excite the phosphorescence of barium platinocyanide.

(1) H. Becquerel, Comptes-rendus at the Academy of Sciences (1896).

d. Jean Becquerel (1878–1953)

Antoine Henri Becquerel's first wife, Lucie Jasmin (1858–1878), was daughter of Jules Célestin Jamin, Professor of Physics at the Ecole Polytechnique. She had only one son, Jean, born on February 5, 1878. The glorious great grandfather Antoine César had died on January 18, without having seen the last member of the Becquerel dynasty and Jean's unfortunate mother followed him on March 18 at the age of 20. The couple had been married for only fourteen months. Jean was brought up by his grandmother Aurélie and his great grandmother Aimée Cécile.

On August 14, 1890, Antoine Henri Becquerel remarried with Louise Désirée Lorieux, who was 12 years younger. She took care of Jean like a second mother, and this was opportune because the grandmother died at about the same time.

Like his ancestors, Jean passed the entrance examination at the Ecole Polytechnique and succeeded his father at the Ecole des Ponts et Chaussées in 1899. He was appointed the Chair of Applied Physics at the Museum in 1909.

Jean's scientific career started with the 20th century. At that time, fundamental discoveries had revolutionized physics and chemistry. Nevertheless, Jean Becquerel followed the track of his predecessors, Edmond and Henri. He investigated the optical and magnetic properties of crystals and was particularly interested in the Zeeman effect. He used new low-temperature techniques to improve his experimental setup, which allowed him to provide a clear evidence of magnetic rotatory polarization. As of 1924, and for 25 years afterwards, he pursued a family tradition, being first "répétiteur" and later "examinateur" at the Ecole Polytechnique. In 1939, before the outbreak of World War II, he discovered metamagnetism. In January 1946, he was elected at the Academy of Sciences in replacement of Jean Perrin.

In the last part of his scientific career he was particularly interested in modern aspects of physics such as the expanding universe and the cosmological theories in relation with this field, together with radioactivity, nuclear energy and special relativity.

Jean Becquerel died of a heart attack, like his father, on July 4, 1953 in his seaside house in Brittony. He had been married twice, without children. With his death, the name of the Becquerels disappeared from the scientific scene.

The Becquerel dynasty is unique with respect to the repetitive events within four generations: studies at the Ecole Polytechnique (where Henri and Jean taught physics), professorships in Physics at the Museum, Members of the Academy of Sciences.

The discovery of radioactivity by Henri Becquerel at the end of the 19th century must be considered in the light of the traditions introduced by Antoine César and Alexandre Edmond. Without taking into account the continuity of events, one would not be able to understand the final success of the research pursued by the three generations of Becquerels. The revolution in the field of physics, and the circle of great French scientists close to the Becquerels were factors that set the stage for of a major discovery. Gaspard Monge, the founder of the Ecole Polytechnique, was Antoine César's professor of mathematics. This first Becquerel knew Abbé René-Just Haüy (1743–1822), the father of mineralogy, and became a friend of the chemist Louis Joseph Gay-Lussac (1778–1850) when he applied for entry to the Academy of Sciences. At the Museum, Antoine César and Edmond encountered Michel Eugène Chevreul (1786–1889). Edmond was a contemporary of the chemist Eugène Péligot (1811–1890), who was the first to prepare metallic uranium in 1841.

A statue of the founder of the dynasty, Antoine César, was erected in 1882 in his native village Châtillon-Coligny. There also a museum is dedicated to the famous Becquerel family.

2. The discovery of uranic rays

a. Upstream

The 19th century was fertile in new techniques. Nicéphore Niepce (1765–1833) invented photography and communicated his passion for this process to his nephew Abel Niepce de Saint Victor (1805–1870), who spent about 15 years from 1851 working on heliochromy. At that time, fluorescence and phosphorescence were already known, and, as said, Edmond Becquerel tried to fix a colour print. Abel Niepce was particularly interested whether a compound exposed to light or to sun might retain an impression to this exposure in the dark.

The answer to this question was provided in a first series of papers (1857–1858) [4–7] presented to the Academy of Sciences by Chevreul, who had carefully followed and encouraged the work of Abel Niepce. The latter's results are summarized in the statements:

- Some compounds exposed to light exhibit in the dark the same effect as that produced by the direct action of light,
- Cardboards impregnated with both uranium nitrate and tartaric acid are "active", the intensity being stronger for the former compound,
- The "activity" remaining on the exposed cardboard is revealed by its action on a photographic plate,
- The "activity" is persistent and acts at a distance in the dark; it cannot pass through glass,
- The effect is not due to phosphorescence,
- Finally, the activity is attributed to invisible "chemical" rays.

It is amazing how closely these experiments resembled those performed by Henri Becquerel about 30 years later. However, some questions remained unanswered:

- Niepce mentioned that the cardboard must be dyed with a uranium salt solution until the color becomes as yellow as straw. Was the amount of uranium sufficient to darken the photographic plate?
- How long was the impregnated carboard in contact with the plate?
- What was the actual sensitivity of the photographic plate? At that time a wet collodion preparation was still in use, whereas Becquerel already disposed of dry photographic plates.

Today, these experiments cannot be reproduced under their original conditions J. Orcel and F. Kraut [8] have carefully examined the leading role played by Abel Niepce and have concluded that even if he had performed the right experiments, he could not have proposed a correct interpretation. A similar confusion prevailed in the discovery of uranic rays by Becquerel, but the great merit of the latter consisted in his proof that the observed effect was due to uranium.

b. The fruitful mistake of Henri Poincaré

Towards the end of the 19th century the Crookes tube had become a familar instrument in physical laboratories for the production of cathode rays. These rays are emitted when an electric discharge is established between two electrodes in a glass vessel containing a gas at low pressure. The cathode rays which traverse the wall of the tube have the property of exciting a bright fluorescence on a screen coated with barium platinocyanide.

On November 8, 1895, Wilhelm Conrad Röntgen (1845–1923), Professor at the University of Würzburg, tried to stop the rays with cardboard and observed that the screen was still luminescent. The luminescence was extinguished by interposing a dense substance and appeared again when a panel of wood or a thin aluminum foil was used. When the screen was replaced by a photographic plate, the latter was rapidly darkened. Rays behaving like "invisible light" and which traversed opaque and dense matter were unknown. Röntgen therefore called them X-rays because the letter X was the symbol usually used in mathematics for denoting an unknown entity.

News of the discovery spread rapidly over Europe, and was accompanied by the first X-ray radiography. Röntgen had photographed with X-rays the hand of his wife Berta. The developed image clearly showed the bones of her hand as well as the ring she was wearing. This picture was reproduced in the newspapers, with an effect that did not fail to contribute to the weird nature of these rays, which were capable of revealing the inner and invisible part of bodies.

The mathematician Henri Poincaré (1854–1912) and Antoine Henri Becquerel commented on these documents on January 20, 1896 at the weekly session of the Academy of Sciences. Poincaré suggested that, since the Röntgen rays seem to emerge from the fluorescence of the walls of the Crookes tube, other fluorescent substances might emit both visible light and invisible X-rays. (Actually, X-rays were emitted from the interaction of cathode rays with the glass wall, which played the role of the anticathode in X-ray tubes.)

In the course of a few weeks, Charles Henry, G. H. Niewenglowski, L. Troost, and M. d'Arsonval investigated various kinds of fluorescent compounds and the results seemed to confirm Poincaré's hypothesis. Interestingly enough, d'Arsonval detected an effect only when the experiment was designed with a "urane glass" (a glass made of uranium oxide) inserted between the photographic plate and the source of fluorescence!

Henri Becquerel was quite aware that the fluorescence of zinc and calcium sulfides used respectively by C. Henry and G. H. Niewenglowski was due to the presence of impurities in the crystals. The only truly phosphorescent substances were uranium salts, owing to the large amount of uranium present in these compounds. Accordingly, Henri Becquerel thought that the effect should be much more important in the case of a pure uranium salt than with an element at the trace level as in the sulfides.

He chose potassium uranyl sulfate for his tests. The compound was placed on a photographic plate well wrapped in tight black paper and then exposed to sunlight. It was found that the uranium phosphorescence acted on the plate through an aluminum foil. Indeed, when the plate was developed, the position of the uranium salt was clearly shown by the presence of dark spots. This experiment was reported at the Academy of Science on February 24 [9].

One week later, Henri Becquerel described the results of an experiment performed under similar conditions. When the experimental setup was ready, Becquerel left it in a drawer of his desk because the weather was not sunny enough. After two days, on Wednesday 26 and Thursday 27 February, the weather conditions improved and Henri Becquerel decided to start a new experiment. However, before replacing the photographic plate, he developed the one which had been kept in the dark in contact with the uranium. To his astonishment he found the same kind of dark spots, but with an even greater intensity than when the exposure had lasted for a few hours in the sunlight. Accordingly, it was not necessary to insolate the uranium salt in order to produce a darkening of the plate. This unexpected result could only be interpreted by assuming the existence of a penetrating radiation emitted spontaneously by uranium. This key observation was made on Saturday February 29, 1896 (it was a leap year) and was reported at the Academy of Sciences the following Monday [10].

The absorption of the "radiant activity" increased with the thickness of a copper plate interposed between the uranium and the photographic plate [10]. Another most important observation was that the radiation could render the air conducting and discharge an electroscope [11], a property which was subsequently found useful for the conception of a new type of detector.

Further, Henri Becquerel found that the radiation was present in all types of uranium compounds. He also proposed experimental evidence for the reflection and refraction of the penetrating radiation [11], but this was disproved in later experiments.

In a Note of March 23, it was reported that phosphorescent compounds without uranium were without effect and that uranous sulfate, which is not phosphorescent, was efficient in blackening the plate [12]. In his next paper, Becquerel compared the two methods of detecting the "rays", consisting of the photographic plate and the electroscope [13].

About two months later, on May 18, at the same session of the Academy of Sciences, Henri Moissan (1852–1907) described the preparation and properties of metallic uranium [14] and Becquerel showed that the activity was the most intense with the metal. He confirmed that all uranium compounds were active and demonstrated that the phenomenon was independent of the physical state of the uranium present. The same effect was produced, whether the material was in molten, dissolved, or crystallized form [15]. It was now clear to Becquerel that uranium itself produced the effect by the emission of penetrating rays. In his last Note in the Comptes-Rendus for the year 1896, Henri Becquerel used for the first time the terms "uranic radiation" and "uranic rays" [16].

c. The birth of radioactivity

At that time, nobody was aware of the importance of Becquerel's discovery, which was much less spectacular than that of X-rays. Not only did the newspapers pay little attention to the event, but even the scientific community did not realize its implications. Henri Becquerel himself gradually lost interest in the phenomenon. In 1896, between February and November, he published seven short Notes in the Comptes Rendus of the Academy of Sciences [9–13, 15–16], two more in 1897 [17–18], none in 1898 and one in 1899 [19]. Two years after his discovery, he was much more interested in his studies on the Zeeman effect than in uranic rays.

The real nature of the rays was unknown. The source of the energy carried by the radiation was not understood, but apparently nobody cared. In 1898, only, a single paper by G. Sagnac "Luminescence and X-rays" mentioned the existence of uranic rays [20].

The complete oblivion that had settled on Henri Becquerel's discovery was lifted in the first months of 1898 when Marie Sklodowska-Curie (1867–1934) in France [21] and Gerhardt C. Schmidt (1865–1949) in Germany [22] found independently that thorium compounds exhibited an activity similar to that of uranium. At the same time, without any relationship with Marie Curie, Ernest Rutherford (1871–1937) in Cambridge started his own research on Becquerel's rays. The term "radioactivity" was introduced by Marie Curie after she had discovered with Pierre Curie (1859–1906) polonium [23] and radium [24], while Ernest Rutherford and Frederick Soddy (1877–1956) worked on the nature of the rays produced by radioactive decay [25].

After the mediatic successes of Marie Curie, Henri Becquerel returned to the field of radioactivity. He received a sample of radium from the Curies with which he succeeded in separating the α and β particles emitted by the element. Friedrich Oskar Giesel in Germany (1852–1927) was on the point of performing the same experiment. In 1900, Henri Becquerel narrowly missed the discovery of γ rays realized by Paul Villard (1860–1934). In 1903, he published an extended review on his investigations on radioactivity in the "Mémoires" of the Academy of Sciences [26].

d. Downstream

Henri Becquerel's discovery brought him petty annoyances. When the work of Abel Niepce de Saint-Victor was brought to light, many persons thought that Henri had been aware of these publications prior to his research on uranium. Several facts seem to support this belief. Henri's father, Alexandre Edmond, had reported several details from Abel Niepce in a book entitled "Light: its causes and effects" published in 1869 [27]. When Henry Carrington Bolton (1843–1903) reviewed in 1869 the work of Abel Niepce, he mentioned a "remarkable property of uranium nitrate to absorb the actinic rays of light, retaining them in an active condition for a long time" [28]. In 1866 and 1869, J. Jamin, Henri's first stepfather, lectured at the Ecole Polytechnique on the Niepce effects.

The doubts continued after 1896, being sustained by G. H. Niewenglowski (1897) and were evoked in 1905 in a book by Gustave Le Bon "The evolution of matter" [29], in which Henri Becquerel was severely attacked and accused of plagiarism. As a scientist, Henri was certainly grieved by this unwarranted invective. He remained silent in this unpleasant situation, but one can be imagine that the controversy may well have been responsible for his decreased activity in the two years following the discovery.

In 1908, the intrigue still persisted since Lucien Poincaré again mentioned Abel Niepce's former work [30]. This matter has been discussed in details in 1990 in a comprehensive review by Paul Fournier and Josette Fournier [31].

In a paper (1965) entitled "Radioactivity before the Curies", Lawrence Badash relates [32] that an English scientist, Silvanus P. Thompson performed an experiment with uranium salts and a photographic plate at the end of February 1896. The effect observed was called "hyperphosphorescence". However, S.P. Thompson was more interested in the darkening of the plate by X-rays than by uranium salts. On June 6, 1896, he published his results confirming those of Becquerel obtained three months before [33] and gave a series of conferences in London in 1896 [34].

Further remarkable revelations were brought to light by Gabriel Bertrand (1867–1962), a biochemist and contemporary of Henri Becquerel. He was impressed by the celebration of the fiftieth anniversary of the discovery of radioactivity and took the occasion to publish in November 1946 a short note in the Comptes-Rendus about the event [35]. The story is worth mentioning. In 1886, Gabriel Bertrand went to the Museum to attend the School of Chemistry directed by Edmond Fremy (1814–1894). Three years later, he passed the final examination and obtained a research position. His laboratory at the Museum was close to that of Becquerel. At the end of 1893, Henri Becquerel met Bertrand and he asked the young chemist "What is your opinion about the darkening of a photographic plate left a few days in contact with a piece of pitchblende?". G. Bertrand answered that he had already analysed this mineral and had found a certain amount of sulfide. He suggested to Becquerel that hydrogen sulfide might be liberated and could reduce silver salts in the plate in a reaction enhanced by humidity. He proposed then to verify this interpretation by inserting a paper strip impregnated with lead acetate between the plate and the uranium ore in order to detect the presence of hydrogen sulfide vapour. Bertrand prepared the paper and brought it a few days later to Becquerel, who said that the control experiment was now useless, because he had already found the right interpretation of the phenomenon...!

The question remains open if Becquerel really did find the correct explanation two years before March

Henri Becquerel, his wife, and his son Jean in their library

1896. In the affirmative, why did he not mention it? Or was he on the wrong path?

From this anecdote, it can only be concluded that at the end of the 19th century, Henri Becquerel obviously disposed of optimal experimental conditions for discovering radioactivity with uranium compounds and a photographic plate. Some had claimed that he was simply lucky; on the other hand, as stated by Louis Pasteur (1822–1895), "Chance favour only the prepared minds". Henri Becquerel used to say that "the issue of some sixty years of work performed in the same place would inevitably lead to the discovery of radioactivity at the right time".

3. The Nobel Prize award

The Nobel Prizes for physics and chemistry were created in 1901. The first to receive this award in physics was W. C. Röntgen for the "discovery of the remarkable rays"; the prize for 1902 was shared between H. A. Lorentz and P. Zeeman for their research on the "influence of magnetism on the radiation phenomenon". On November 12, 1903, the Royal Academy of Sciences decided to award half the prize to Henri Becquerel for his "discovery of spontaneous radioactivity" and the other half to Pierre Curie and Marie Curie "for their joint research on the radiation phenomena discovered by H. Becquerel". Oddly enough, the word "radioactivity" coined by the Curies, was only associated with the name of Becquerel.

At that time, Pierre and Marie Curie were unable to travel to Stockholm for the ceremony; they did so, however, one and a half year later, on June 6, 1905, to attend the Nobel Conference and deliver their address. The prize was awarded on December 10, 1903, and in his dissertation Antoine Henri Becquerel outlined the details of his discovery. The 100 000 French francs associated with the prize were most welcome to the Curies, who where living in precarious circumstances. Marie was expecting her second daughter, Eve, and Pierre was searching for a position that would improve his income. Part of the prize would be useful in acquiring new equipment for the laboratory or for purchasing a small quantity of radium. In 1903, the cost of one gram of radium was twice the total amount of the Nobel Prize.

In their part, the Becquerels enjoyed a completely different social status. They were wealthy and attended the high Parisian society. In his apartment, rue Durmont d'Urville, Antoine Henry had collected several devices that had been built or used by Alexandre Edmond and Antoine César. Picture corresponds to a rare photograph of Becquerels in company of Henri's son Jean in the library of their apartment.

Conclusion

It may be stated that the discovery of radioactivity occurred at the right period towards the end of the 19th century. It followed immediately that of X-rays in 1895, preceding the discovery of the electron by J. J. Thomson in 1897, and was succeeded in 1898 by the discoveries of polonium and radium. All these events had shaken the contemporary scientific world. Thermodynamic laws and the notion of mass conservation had to be revised with new concepts based on energy conservation. The field of electromagnetism was forced to include new radiations (X and γ rays) and the discovery of new elements foreshadowed the completion and extension of the periodic table.

Nowadays the consequences of the discovery of radioactivity are so important for humanity that it is hard to conceive why the announcement on March 1896 had remained virtually unheeded for two years and had not been recognized by the scientific community as a major breakthrough in science.

This question can be partly answered: first the number of scientists was relatively small, consisting of only a few hundred in France. Secondly, the discovery was not exploited by Antoine Henri, who considered that it was a natural extension of the work carried out by his grandfather. This particular aspect is well summarized in a text published by O. Lodge [36]: "Professor Becquerel claims that one of the causes conducive to the productiveness of his laboratory was the continuity of the work accomplished there. Antoine Becquerel began the study of phosphorescent substances under electric discharge. Edmond Becquerel continued the work and recognized the exceptional properties of the salts of uranium. Henri Becquerel pursued the subject, and effected the climax. He himself definitely recognized the long tradition of pertinacious and successful inquiry on which he had built and he spoke thus: "*It was perfectly appropriate that the discovery of radioactivity should have been made in our laboratory, and if my father had lived in 1896, he it is who would have made it.*"

The author has consulted several publications on Becquerel, of which the following:

L. Badash: Chance favoured the prepared mind: Henri Becquerel and the discovery of Radioactivity, Archives Internationales d'Histoire des Sciences, **18**, 55–66 (1965).

L. Barbo: La radioactivité. Histoire d'une découverte, Centre d'Histoire des Sciences et des Techniques, Université de Nantes, **18** (1990–1991).

J. L. Basdevant: L'enseignement d'Henri Becquerel à l'Ecole Polytechnique (1885–1908), Bulletin de la Société des Amis de la Bibliothèque de l'Ecole Polytechnique (1988) 2–18.

J. Warschnitter: La saga des Becquerel, private manuscript.

Acknowledgment

The author thanks J. P. Adloff and H. J. MacCordick for comments and revision of the manuscript. He is indebted to Mme F. Masson, Conservatoire en Chef de la Bibliothèque de l'Ecole Polytechnique, for permission of reproduction of documents from the "Archives de l'Ecole Polytechnique", Palaiseau, France.

References

1. Rebière, A.: *Les savants modernes.* Librairie Nony, Paris (1899).
2. Harvey, E. N.: *A history of luminescence.* American Physical Society, Philadelphia (1957), p. 390.
3. Becquerel, H.: *Cours de Physique.* Ecole Polytechnique, 2ème Division, 1898–1899. Archives de l'Ecole Polytechnique, Palaiseau, France.
4. Niepce de Saint-Victor, A.: Mémoire sur une nouvelle action de la lumière. Compt. Rend. Acad. Sc. Paris **45**, 811 (1857).
5. Niepce de Saint-Victor, A.: Deuxième mémoire sur une nouvelle action de la lumière. Compt. Rend. Acad. Sc. Paris **46**, 448 (1858).
6. Niepce de Saint-Victor, A.: Cinquième mémoire sur une action de la lumière inconnue jusqu'ici. Compt. Rend. Acad. Sc. Paris **53**, 33 (1861).
7. Niepce de Saint-Victor, A.: Sixième mémoire sur une nouvelle action de la lumière. Compt. Rend. Acad. Sc. Paris **65**, 505 (1867).
8. Orcel, J., Kraut, F.: Niepce de Saint-Victor et la découverte de la radioactivité. Congrès des Sociétes Savantes, Lyon. Section des Sciences **3**, 93 (1964).
9. Becquerel, H.: Sur les radiations émises par phosphorescence. Compt. Rend. Acad. Sc. Paris **122**, 420 (1896).
10. Becquerel, H.: Sur les radiations invisibles émises par les corps phosphorescents. Compt. Rend. Acad. Sc. Paris **122**, 501 (1896).
11. Becquerel, H.: Sur quelques propriétés nouvelles des radiations invisibles émises par divers corps phosphorescents. Compt. Rend. Acad. Sc. Paris **122**, 559 (1896).
12. Becquerel, H.: Sur les radiations invisibles émises par les sels d'uranium. Compt. Rend. Acad. Sc. Paris **122**, 689 (1896).
13. Becquerel, H.: Sur les propriétés différentes des radiations invisibles émises par les sels d'uranium, et du rayonnement de la paroi anticathodique d'un tube de Crookes. Compt. Rend. Acad. Sc. Paris **122**, 762 (1896).

14. Moissan, H.: Préparation et propriétés de l'uranium. Compt. Rend. Acad. Sc. Paris **122**, 1088 (1896).
15. Becquerel, H.: Emission de radiations nouvelles par l'uranium métallique. Compt. Rend. Acad. Sc. Paris **122**, 1086 (1896).
16. Becquerel, H.: Sur diverses propriétés des rayons uraniques. Compt. Rend. Acad. Sc. Paris **123**, 855 (1896).
17. Becquerel, H.: Recherches sur les rayons uraniques. Compt. Rend. Acad. Sc. Paris **124**, 438 (1897).
18. Becquerel, H.: Sur la loi de décharge dans l'air de l'uranium électrisé. Compt. Rend. Acad. Sc. Paris **124**, 800 (1897).
19. Becquerel, H.: Note sur quelques propriétés du rayonnement de l'uranium et des corps radioactifs. Compt. Rend. Acad. Sc. Paris **128**, 771 (1899).
20. Sagnac, G.: Luminescence et rayons X. Revue Générale des Sciences Pures et Appliquées **9**, 314 (1898).
21. Sklodowska Curie, M.: Rayons émis par les composés de l'uranium et du thorium. Compt. Rend. Acad. Sc. Paris **126**, 1101 (1898).
22. Schmidt, G. C.: Sur les radiations émises par le thorium et ses composés. Compt. Rend. Acad. Sc. Paris **126**, 1264 (1898).
23. Curie, P., Curie, S.: Sur une substance nouvelle radio-active contenue dans la pechblende. Compt. Rend. Acad. Sc. Paris **127**, 175 (1898).
24. Curie, P., Curie, M., Bémont, G.: Sur une nouvelle substance fortement radioactive, contenue dans la pechblende. Compt. Rend. Acad. Sc. Paris **127**, 1215 (1898).
25. Rutherford, E., Soddy, F.: The cause and nature of radioactivity. Phil. Mag. and Journal of Science **6**, IV, 561 and 576 (1903).
26. Becquerel, H.: Recherche sur une nouvelle propriété de la matière. Activité radiante spontanée ou radioactivité de la matière. Mémoires de l'Académie des Sciences, Paris 46 (1903).
27. Becquerel, E.: La lumière: ses causes et ses effets; Tome 2: Effets de la lumière. F. Didot, Paris (1868), p. 50.
28. Bolton, H. C.: On the action of light on uranium. Amer. J. of Sciences and Arts, 2nd Series 48, 206 (1869).
29. Le Bon, G.: Evolution de la matière. Flammarion, Paris (1905), p. 21.
30. Poincaré, L.: La Physique Moderne. Flammarion, Paris (1908).
31. Fournier, P., Fournier, J.: A. Niepce de Saint-Victor (1805–1870), M. E. Chevreul (1786–1889) et la découverte de la radioactivité. New J. Chem. **14**, 785 (1990).
32. Badash, L.: Radioactivity before the Curies. Amer. J. Phys. **33**, 128 (1965).
33. Thompson, S. P.: On hyperphosphorescence. Phil Mag. **42**, 103 (1896).
34. Thompson, S. P.: Radiations visibles et invisibles. Hermann, Paris (1914).
35. Bertrand, G.: Sur l'origine de la découverte de la radioactivité. Compt. Rend. Acad. Sc. Paris **223**, 698 (1946).
36. Lodge, O.: Becquerel Memorial Lecture. J. Chem. Soc. **101**, 2005 (1912).

Radiochimica Acta 70/71, 13–22 (1995)

The Dawn of Radiochemistry

By J. P. Adloff and H. J. MacCordick

Centre de Recherches Nucléaires, Laboratoire de Chimie Nucléaire, B.P. 28, 67037 Strasbourg Cedex 2, France

(Received April 25, 1995)

History / Pierre Curie / Marie Curie / Polonium / Radium / Actinium

Abstract

Chemical research on radioactivity started in 1898. That year, three short key Notes by the Curies reawakened the topic of uranic rays discovered two years earlier by Becquerel. The discoveries of polonium and radium were milestones in the history of nuclear sciences. The exegesis of the Notes reveals the progress and processes which ineluctably led within a few months to the announcement of the existence of elements in amounts invisible at that time.

1. Introduction

The photographic plate by which radioactivity was discovered was the first known radiation detector. It was merely qualitative but nevertheless quite useful in seeking compounds other than those of uranium which could also produce an effect on photosensitive materials. During the years 1896 to 1898 Becquerel's discovery stimulated a search for these strange rays and led to a short flare of false or at least doubtful observations. In an extensively documented review on this period, Lawrence Badash points out "how a proliferation of related phenomena can obscure the relevant facts, and how a sterile approach can effectively kill a newborn science" [1].

The interest of scientists consequently faded rather rapidly after the initial excitement aroused in 1896 by Becquerel's Notes in the *Comptes Rendus*. The next important announcement, just 2 years later, was that thorium and its compounds were capable of blackening a photographic plate. This observation was made by another physicist, Gerhardt S. Schmidt (1865–1949), also a leader in the study of phosphorescence, at Erlangen in Germany [2–4]. The discovery of the radioactivity of thorium was published on February 4, 1898.

It might appear obvious today that the search for substances sharing the peculiar property of uranium and its compounds should have been a logical and systematic process following Becquerel's observation. It is amazing how few *scientists* followed this approach. Apart from Schmidt, a young woman in Paris entered the scene: this was Marie Curie, who missed by two months the discovery of the radioactivity of thorium.

2. Marie Curie's first Note: April 12, 1898 [5]

Marie Curie (née Marya Sklodowska, 1867–1934) came to Paris in 1891 and obtained a "licence" in mathematics and physics in 1894. The following year she married Pierre Curie (1859–1906), a physicist renowned for his work in magnetism and crystal symmetry (Figure 1). Together with his brother Jacques, he had discovered the piezoelectric effect in 1882.

Becquerel's discovery had raised a puzzling problem. The uranium salts appeared to maintain an undiminished ability to blacken a photographic plate in the course of time. What was the origin of this continuous emission of energy? Marie Curie, who was thinking of a subject for her thesis, decided to investigate the new phenomenon.

At this point, little progress would have been made without the genius of Pierre Curie. Besides blackening a photographic plate, the uranic rays also rendered air conducting for electricity. This property was much more amenable to quantitative measurement of the action of the rays, and thus of the intensity of their emission. In his early experiments, Becquerel had already used electroscopes, but the results of measurements were unreliable.

The device invented by Pierre Curie was the prototype of an ionization chamber. The apparatus consisted of two parallel-mounted metallic electrodes between which an electric field was applied (Figure 2). One electrode was loaded with a powder of the material to be tested and the very small electric current produced in the air by the rays and flowing between the electrodes was measured with a highly sensitive and reliable compensation method. The principle of this technique was that a displacement of charges produced by applying a known weight to a leaf of piezoelectric quartz was opposed to the ionization current. The amount of charges produced is proportional to the length and thickness of the foil of the leaf and to the stretching strength.

The compensation of the opposite currents from the ionization chamber and from the quartz was controlled by the motion of the foil of a quadrant electrometer. The progressive accumulation of charges on the collecting electrode tends to deviate the foil from its rest position. This is counteracted by applying progressively a weight to the quartz leaf. Eventually the charges from the chamber are completely compensated and are equal to the amount of charges conveyed by the quartz, as calculated from the weight and from the

Fig. 1. Pierre Curie and Marie Curie, about 1898.

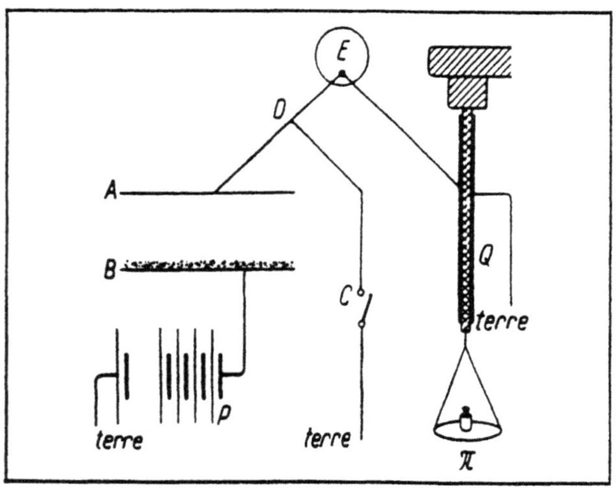

Fig. 2. Apparatus used for the measurement of "the conductivity of air by the influence of active radiations". AB, ionization chamber; the active material was deposited on plate B. CD, connection to ground. E, quadrant electrometer. Q, piezoelectric quartz. Π, weight [12].

time during which the weight is applied. For the first time, the emission of uranic rays could be quantified on the basis of the intensity of the current (in the picoampere range) or of the charge in e.s.u. units.

The handling of this device required considerable skill, but Marie Curie now had an invaluable tool for routine measurements. She did not always calculate values for the current but listed in her notebooks the strength of the sources in terms of the weight in grams and the application time in seconds (Figure 3). The device invented by Pierre Curie was rapidly adopted by most researchers in the field and remained in use at the authors' Laboratory until the early 1950's.

On February 11, 1898, Marie Curie started a systematic search for elements and compounds capable of imparting electrical conductivity to air. She tested all samples at hand or borrowed them from various collections; materials consisted of "a large number of metals, salts, oxides and minerals" [5]. The complete list was later recapitulated in her doctoral dissertation [6]:

"1. All metals and metalloids commonly available and a few rarer and pure compounds from Mr. Etard's collection at the Ecole de Physique et de Chimie Industrielles de la Ville de Paris.
2. The following rare substances: gallium, germanium, neodymium, praseodymium, niobium, scandium, gadolinium, erbium, samarium and rubidium (samples lent by Mr. Demarçay); yttrium, ytterbium together with a new erbium (samples lent by M. Urbain).
3. A large number of rocks and minerals."

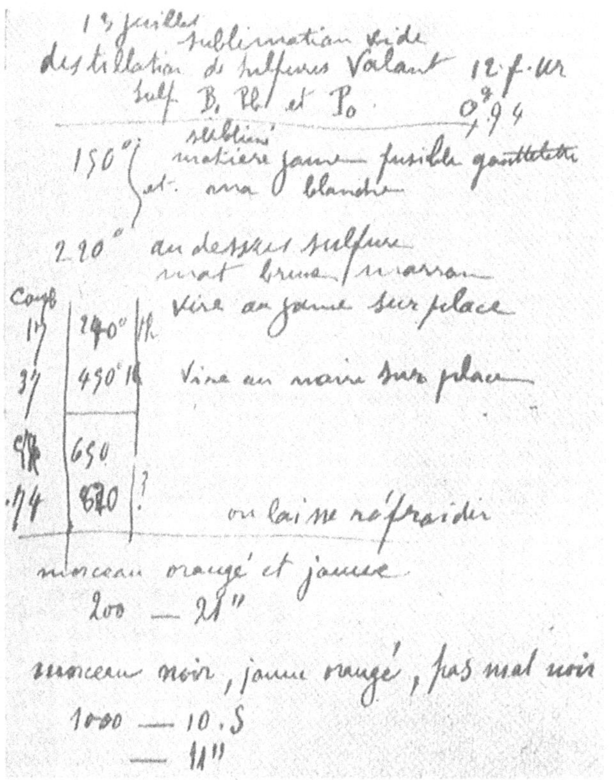

Fig. 3. Excerpt from the laboratory notebook of Pierre and Marie Curie; on this page, dated July 13, 1898, the symbol of polonium appears for the first time. The text refers to the sublimation of sulfides of Bi, Pb and Po with activity 12 times that of uranium.

Table 1. Intensity of the ionization current for various materials [5]

Compound	Intensity in pA
Metallic uranium with slight amounts of carbon	24
Black uranium oxide U_2O_5	27
Green uranium oxide U_3O_8	18
Ammonium, potassium and sodium uranates	about 12
Thorium oxide, 0,25 mm layer	22
Thorium oxide, 6 mm layer	53
Potassium fluoxytantalate	2
Potassium fluoxyniobate	0,3
Pitchblende from Johanngeorgenstadt	83
Pitchblende from Joachimsthal	67
Natural chalcolite	52
Autunite	27
Orangite	20
Samarskite	11

Marie Curie's first Note was presented at the Academy by her former Professor Gabriel Lippman (1845–1921), who was also member of the examination board of her Thesis in 1903 and a Nobel Prize Laureate in 1908 for the invention of a method of colour photography. Curiously, the paper does not bear the name of Pierre Curie, without whom the experiments would not have been possible. Part of the data reported in [5] are shown in Table 1.

With two exceptions, all compounds contained U or Th, and the intensity was roughly proportional to the fraction of the element in the compound. This behaviour had already been observed by Becquerel, who concluded that the property of making air conducting was specific for the element uranium, independently of its state or chemical combination. An important step towards this conclusion was the measurement of uranium metal, which produced the same emission as uranium salts, but with greater intensity owing to the higher concentration of the element. The recognition of the proportionality between the uranium content and the intensity was the première of *radioanalytical chemistry.*

However, the result of the second sample in Table 1 was confusing. It was convenient to take the intensity of metallic uranium as a reference. Even if the metal prepared by Moissan by reduction of uranium oxide with carbon was not quite pure, as indicated, it was difficult to understand why the black U_2O_5 exhibited a 10% higher intensity. This odd behaviour appears also, still unexplained, in Marie Curie's thesis and has been properly clarified by R. L. Wolke [7].

Marie Curie emphasized that all minerals which were active contained the element uranium or, as she had found independently of Schmidt, the element thorium. The most important result was that pitchblende, a massive variety of uraninite (UO_2), is nearly 4 times more active than metallic uranium, and chalcolite (also known as metatonbernite), $Cu(UO_2)_2(PO_4)_2$, 6–8 H_2O, is about twice as active as the metal. The Note's key sentence actually states how Marie Curie will pursue her research: "*This fact is quite remarkable and suggests that these minerals may contain an element much more active than uranium (itself)*".

Initial evidence in favour of this hypothesis appears in the next sentence, since Marie Curie knew how to prepare chalcolite "*I have prepared chalcolite by the Debray method with pure products; this artificial chalcolite is not more active than other uranium salts*". Marie Curie then concluded that the unknown element exists only in the uraniferous minerals which are more active than metallic uranium. A third mineral, autunite, a phosphate of Ca and UO_2, also exhibited an anomalous, although less spectacular, activity that was higher than in uranium.

Interestingly enough, the samples investigated included two potassium compounds with feeble activities. Marie Curie assigned the activity of potassium fluoxytantalate and niobate to the elements Ta and Nb. Neither of these elements has long-lived natural radioisotopes. It may therefore be assumed that Marie Curie unknowingly discovered the radioactivity of ^{40}K. The potassium oxyfluotantalate in question was probably Marignac's salt, $K_4Ta_4O_5F_{14}$, with a specific ^{40}K activity of 4 Bq g^{-1}. On the other hand, Marie Curie did not report a signal from rubidium (^{87}Rb) or from samarium (^{147}Sm), which have much higher specific activities (88 Bq g^{-1} for RbCl and 122 Bq g^{-1} for Sm_2O_3). This may be due to the lower energies of the emitted radiations but, in any case, the activities would be close to the limit of detection.

At this stage, further search for the identity of the supposed element became a matter of paramount importance and urgency. Pierre Curie himself was fascinated by Marie's findings; on March 18 he abandoned his own research projects and joined his wife in the adventure.

When Marie Curie's Note reporting the radioactivity of thorium appeared, Schmidt hastened to announce to the French scientists that *he* was the discoverer of the phenomenon and published the following month in the *Comptes Rendus* a one-page summary of his original paper [3].

3. The second Note: July 18, 1898 [8]

The title of Marie Curie's second publication was signed with Pierre Curie (as first author) and announced that the search for "the element much more active than uranium" was successful: "On a new substance, radio-active, contained in pitchblende". This was also the first appearance of the word *radio-active*. The Curies omitted the hyphen the following year. The Note was presented at the Academy by Henri Becquerel who, surprisingly, was not associated with the two most important publications of the Curies.

It is noteworthy that research on radioactivity now turned from physics to chemistry. Neither Pierre nor Marie Curie were chemists and they had to ask for assistance. The collaboration of Gustave Bémont (1867–1932) is acknowledged in a footnote. Bémont was "Chef de Travaux", i.e. in charge of practical training for the students at the Ecole Municipale de Physique et Chimie Industrielles in Paris. The handwriting of Bémont appeared in the Curies' notebooks on May 5, 1898 [9]. Bémont was associated with the discovery of radium, as co-author of the third Note of the Curies [10]. This paper suggests that he was a close collaborator during the period May to December 1898. Apart from that, however, he remained the "forgotten man of radioactivity".

The chemical analysis of pitchblende was neither particularly difficult nor innovative. The method followed the classical scheme of analysis given by Fresenius and could be carried out in cookbook fashion. Besides, pitchblende had already been analysed long before and the material was known to contain several metals such as Cu, As, Sb, Bi and Pb.

However, the Curies were now able to follow the procedure by a new and highly sensitive method, involving the measurement of radioactivity of the element searched. In a biography of Pierre Curie published in 1924 [11], Marie Curie explained: "The method we have used is a new one for *chemical research based on radioactivity*. It consists in separations performed with the ordinary procedures of analytical chemistry, and in the measurement of the radioactivity of all compounds separated. In this way, one can recognize the chemical character of the radioactive element sought; the latter is concentrated in fractions

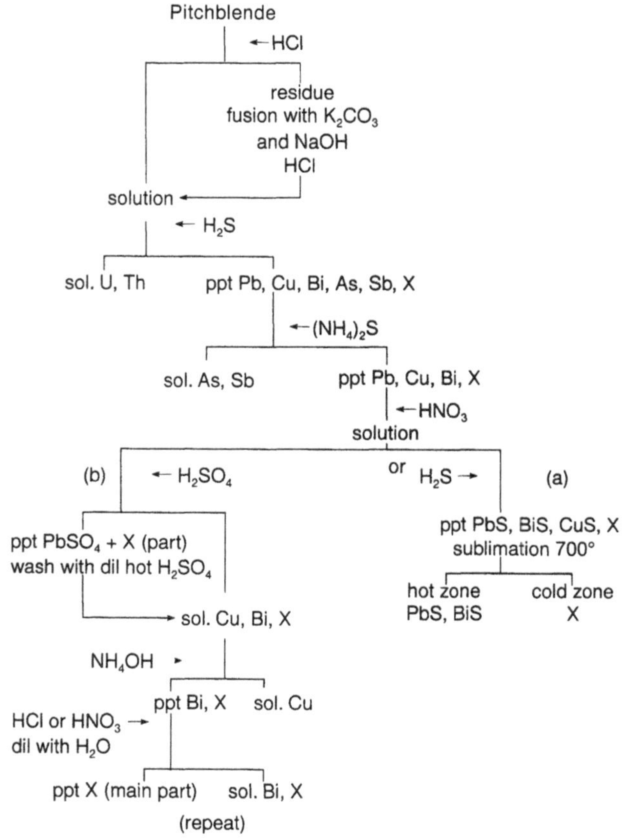

Fig. 4. Flow chart of the chemical separations which led to the discovery of the element X, polonium. The polonium was partially separated from Bi by sublimation [path (a)] or by a wet method based on fractional precipitation of hydroxides [path (b)].

which become increasingly radioactive in the course of the separation".

More details of the procedure than those found in the Note were given in a 10-page report by Marie Curie published on January 30, 1899 [12], as well as in subsequent publications on polonium [13]. It was known that various elements in a mineral could sometimes be separated by sublimation at an appropriate temperature. The first observation made by the Curies was that very active products, 20-fold more active than uranium, could be separated by sublimation of pitchblende in vacuum, but with a very low yield. Hence they decided to attack the minerals with HCl.

The treatment began on April 14, whereby a 100 g sample of pitchblende was first ground and then subjected to the action of HCl. The mineral was "two and a half times more active than uranium", and thus probably from Joachimsthal (Table 1) [14]. Insoluble residues remained which were generally very active, and these were solubilized in acids after fusion with a mixture of potassium carbonate and sodium hydroxide.

The Curies and Bémont used several procedures in parallel runs, although only one is described in the Note. The flow chart shown in Figure 4 has been drawn up in accordance with the procedures described in references [8] and [12].

The treatment of the acidic solution with H_2S was already a significant step, since the precipitate of sulfides was more active than that of the residual solution. The latter contained U and Th, although the occurrence of Th in the mineral was not previously mentioned. The activity contained in the sulfides was insoluble in ammonium sulfide, and thus could be separated from As and Sb. The residual insoluble sulfides were dissolved in HNO_3; after addition of H_2SO_4, the activity was partially carried by lead sulfate. Eventually, the activity was found in the last fraction, that contained only bismuth. At each step the activity increased, and the inactive fractions were discarded.

The separation of the active substance from Bi by wet methods was found to be difficult. The Curies used fractional precipitation by addition of water to the acidic solution. The fractions which precipitated first were those that carried the greater part of the activity.

On June 6, Marie Curie precipitated sulfides from a nitric acid solution containing Pb, Bi and the active substance. She underlined the result in her notebook: the solid was *150 times more active* than uranium. The same day, Pierre Curie heated the precipitate in an evacuated tube. In the course of heating at several hundred degrees, the bismuth and lead sulfides remained in the hottest part of the tube, while a thin black layer of powder carrying the activity condensed in the cooler part at 250–300 °C. That day, the Curies succeeded in preparing a sample 330 times more active than uranium.

By repeating the various steps, fractions with continually increasing activities were isolated. Finally, a batch was obtained that was about *400 times more active* than uranium itself.

The Curies carefully verified that "compounds of nearly all elements ... including those of the rarest substances" were not active and that the property of radioactivity was limited to U and Th and "perhaps Ta".

The Note ends with the words *"We ... believe that the substance we recovered from pitchblende contains a heretofore unknown metal, similar to bismuth in its analytical properties. If the existence of this new metal is confirmed, we propose that it be named polonium in honor of the native land of one of us".*

The name polonium with the symbol Po appeared in the laboratory notebook for the first time on July 13, in the handwriting of Pierre Curie (Figure 3). Prior to polonium, four elements were given names referring to countries: germanium, ruthenium, gallium and scandium. But the designation "polonium" had a provocative significance, since as a state Poland had virtually disappeared in 1795, being parcelled out between Prussia, the Austrian Empire and Russia. At the time when Marie Curie had left her homeland, a Grand Dukedom of Warszaw had been constituted, but politically it was a mere dependence of Russia.

The announcement of the existence of a new element that remained invisible and could be identified solely on the basis of its emission of "uranic rays" was unique in the history of chemistry. It had become customary that no such claim was considered valid until a pure substance had been isolated, the atomic weight of the element determined and its spectral lines measured. Eugène Demarçay (1852–1904), who was a recognized expert in spectroscopy, examined the spectrum of the new element but, to the disappointment of the Curies, could not distinguish any new characteristic line apart from those of impurities. The authors admitted that "... this fact does not favour the idea of the existence of a new metal".

The mass equilibrium ratio ^{238}U to ^{210}Po is 1.34×10^{10}. Pitchblende may contain as much as 80% of uranium [14]. Accordingly, the 100 g sample used by the Curies may have contained of the order of 6 ng of Po. Even with a 100% separation yield, such amounts were naturally invisible and far below the sensitivity limits of spectroscopical detection attainable at that time. One of the largest preparations of polonium in the early history of radioactivity was carried out by Marie Curie and André Debierne (1874–1949), who started with several tons of the residue from the sulfuric acid extraction of uranium ore. Their final product weighed about 2 mg and probably contained about 0.1 mg (0.45 Ci) of polonium. The spark spectrum of this material revealed for the first time a few lines characteristic of the element, 12 years after its discovery [13]. The largest reported amount of natural ^{210}Po is about 40 Ci. It was extracted from a residue of 37 tons of lead oxide from the Port Hope radium refinery [15].

The isolation of polonium from uranium had thus been accomplished, although the Curies were unaware of the genetic relationship between the two elements. They considered the whole material as a mixture. They also knew nothing at the time of the phenomenon of radioactive decay. In this sense, it was a matter of chance since the separations were performed within 3 months, which was a short time with respect to the 138-day half-life of polonium. It was only a few years later that the Curies noticed with astonishment and great perplexity that polonium was progressively disappearing. Moreover, the Curies could not imagine that they were dealing with *tracer* amounts of the new element. By simple reasoning, the substance in the ore that was more active than uranium itself must obviously be present in very small amount, since it had not yet been found. However, it was not evident at first that this amount was below the limit of any chemical means of detection.

Without knowing it, the Curies had invented the use of carriers; in the final step the polonium was effectively *carried* by bismuth. The words "tracers", "chemistry at the tracer level", "carriers", are terms that arose from the pioneer work of the Curies and that introduced the classical vocabulary of radiochemistry.

The assignment of polonium in the Periodic Table of the elements does not seem to have preoccupied the discoverers. At the time, the terminal part of the Table was unoccupied beyond $_{83}Bi$, with the exceptions of

$_{90}$Th and $_{92}$U. In principle, the new element could have fitted into any of the vacant positions.

It appeared remarkable that the polonium was concentrated in the bismuth fraction. This affinity led Marie Curie to consider that the two elements had similar chemical properties. Radionuclides are, however, often carried physically by any kind of solid. As pointed out by Wolke [7], polonium hydroxide, $PoO(OH)_2$, and polonium sulfide, PoS, are co-precipitated at trace concentrations with $Bi(OH)_3$ and insoluble sulfides, but this behaviour does not necessarily imply a strong chemical resemblance.

Nevertheless, it seemed reasonable to assume that polonium should be situated not too far from Bi in the Periodic Table. The insolubility of the sulfide as well as other properties of the elements in groups 1, 2, 17 and 18 [in the IUPAC version of the Periodic Table] preclude for polonium the positions 87 Eka-Cs, 88 Eka-Ba, 85 Eka-I and 86 Eka-Xe. Rare earths are not carried by sulfides, and this further excludes the 89 Eka-La position.

In a Faraday Lecture given before the Chemical Society of London in 1889, Mendeleev expounded on still undiscovered elements, and predicted that "in the series which contains Hg = 204, Pb = 206 and Bi = 208 ... there exists an element analogous to tellurium which ... can be described as *dvi-tellurium*" [16]. He anticipated the properties of the element Z = 84 that were later confirmed when larger amounts of polonium became available. Nevertheless, Mendeleev had made a mistake: the element which he anticipated was *dvi-selenium* or *eka-tellurium*.

As will be pointed out in the Epilogue, the persistence of Marie Curie in considering that the chemistry of Po should be closely similar to that of Bi turned out to be detrimental to the discoverers of the element.

The second Note of the Curies also ended the short story of polonium for several years. Now the Curies were on the verge of another, much more important, discovery. All the same, Marie Curie conserved a sentiment of ownership for polonium, which she defended with emotion, vehemence, and even with aggressiveness. In one sense, she was right, owing to the role of her polonium in the discoveries of the atomic nucleus, the neutron, artificial radioactivity and fission.

4. The third Note: December 26, 1898 [10]

The title of the third Note, like the second one, announced the discovery of a radioactive element and bore one additional word: "On a new *strongly* radioactive substance contained in pitchblende".

The notebook entries of the Curies were interrupted from July to November 11th 1898. Polonium had been found in the sulfide fraction precipitated from the soluble part of the pitchblende. Even before their announcement of the discovery of polonium, the Curies probably pursued the analysis of the residual solution, suspecting that the latter might contain still another radioactive substance. They focused their attention on the sulfides because it was easier to search for the activity concentrated in this material. Bismuth and other metals whose sulfides were insoluble were the carriers of Po. This role was now ensured by barium, which fortunately was present in sufficient amount in the mineral.

Few details on the procedure appear in the Note. The Curies began with a recapitulation of the properties of Po in order to emphasize the strikingly different nature of the "new radio-active substance". It behaved like "nearly pure barium": it was not precipitated by hydrogen sulfide or ammonium sulfide, nor by ammonia. On the other hand, the sulfate and the carbonate were insoluble in water. The high solubility of barium chloride in water, and its insolubility in concentrated HCl and in alcohol proved to be of particular importance in the concentration of the radioactive substance. Finally, the spectra showed the characteristic pattern of pure barium.

The authors concluded that the new radioactive substance must be contained in the barium. They advanced three reasons in favour of this hypothesis. First, Marie Curie verified that "normal" barium was inactive. For this purpose, she undertook fractional crystallizations on 50 kg of commercial barium chloride until a final amount of 10 g was obtained and which showed no activity. Accordingly, the radioactive substance was *not* barium.

The second proof was the progressive concentration of the radioactive substance in the course of fractional crystallization of barium chloride. The first hydrated chloride was 60 times more active than uranium. Upon dissolution in water, followed by partial precipitation with alcohol, the solid was much more active than the solution. The Curies pursued the progressive concentration until the activity of the chlorides was *900 times* higher than that of uranium. At this point they had to cease the experiment because of lack of material. They concluded that the solubility of the chloride of the radioactive element must be lower than that of barium chloride. Later, the precipitation of chlorides from water-alcohol mixtures was replaced by the more efficient procedure of crystallization of chlorides or bromides from water alone.

The third and last argument was the decisive one. This time, the spectroscopic test was successful. Demarçay observed in the spectrum of the radioactive barium chloride several lines which could not be assigned to any known element [17]. The wavelength of the most intense line was 3814.8 Å (a recent value is 3814.42 Å). The intensity of the line increased with the radioactivity of the sample from "very weak" with the first sample up to "notable" for the sample with a 900-fold activity of that of uranium. The authors conclude "We think this is a very serious reason to attribute it [the line] to the radio-active part of our substance. The various reasons which we have enumerated lead us to think that the new radioactive substance contains a new element, to which we propose to give the name *radium*".

This name, followed by a question mark, appeared in the notebook for the first time on December 18. The chemical behaviour of the new substance was much easier to determine than that of polonium. Clearly, it belonged to the family of the alkaline earths, for which the best fingerprint is the insolubility of the sulfates. Indeed, the statement "radium sulfate more soluble in concentrated H_2SO_4 than barium sulfate" also with a question mark, was written by Pierre Curie.

Besides the spectroscopic analysis, a second "official" proof for the existence of radium would have been the determination of the atomic mass. Marie Curie was aware of this exigence. She determined the amount of chlorine present in the radiferous anhydrous barium chloride and compared it to that of a batch of ordinary barium chloride. The calculated atomic masses of the radioactive material were *always slightly higher*, but still within limits of errors. Obviously, the amount of radium was too low to change the apparent atomic mass of Ba, 137.4 (now 137.327).

From the equilibrium mass ratio of ^{238}U to ^{226}Ra, and assuming a quantitative separation yield, it is calculated that the amount of radium present was, at the most, of the order of 30 μg.

The determination of the atomic mass of radium became an obsession for Marie Curie that lasted for years. In successive trials, as the radiferous barium became more and more enriched in radium, the measured atomic mass increased. Eventually, on July 21, 1902 she published the value of 225±1 (now 226.0254). This astonishingly close value was obtained with a self-luminous sample of 0.120 g of $RaCl_2$ with a Ra:Ba ratio of 10^6, which was *one million times* more active than uranium [18]. Now radium could be definitely placed in the 88 Eka-Ba position in the Periodic Table.

The first preparation of radium, as noted by the Curies, "contains very likely a high proportion of barium; nevertheless, the radio-activity is considerable. Hence the radio-activity of radium must be *enormous*". The authors now compared the radioactivity of the four known radioelements: U, Th, Po and Ra. The last two left an image on a photographic plate within half a minute, U and Th only after exposure times of several hours.

When the Curies ran out of material, they were aware that the concentration procedure of radium in radiferous barium should be pursued. They were also aware that vast amounts of raw material would be necessary in order to prepare "visible" or at least much larger quantities of Po and Ra. In a footnote of the publication reporting the discovery of radium, the authors thanked Professor M. Suess of the University of Vienna, also Corresponding Member of the French Academy, for his intervention with the Austrian Government. The latter offered to the Curies free of charge 100 kg of a uranium-free residue from the treatment of the Joachimsthal pitchblende [14]. The residue still contained Po and Ra. "This shipment will greatly facilitate our research" concluded the authors.

With the foregoing discovery of polonium, the Curies had oddly enough begun with the most difficult part of the work. In its own right, radium had outstanding advantages; its concentration in the ore was about 5000 times greater than that of polonium; it is a true chemical analogue of barium, from which it can be separated and it could be readily assigned to its correct place in the Periodic Table.

The Nobel Prize for Physics was attributed in 1903 to "H. A. Becquerel for his discovery of spontaneous radioactivity and to P. and M. Curie for their joint researches on the radiation phenomena". At this time, no mention was made of the discoveries of polonium and radium. Thus it is perhaps significant that Becquerel travelled alone to Stockholm; in the meantime, the Curies were allegedly occupied with their teaching duties.

In 1911, Marie Curie received the Nobel Prize in chemistry "for her services to the advancement of chemistry by the discovery of the elements radium and polonium, by the isolation of radium, and for the study of the nature and compounds of this remarkable element".

5. Epilogue

L. Badash stated that the "great discoveries of 1898 [of Po and Ra] ... served to renew interest in radioactivity. The moribund subject was resurrected and interest now never flagged until radioactivity merged into atomic physics and evolved into nuclear physics" [19].

During the year 1898, the discovery of uranic rays also stimulated research outside of France. Julius Elster and Hans Geitel in Wolfenbüttel (Germany) published a first paper on November 11, 1898 [20]. Their work was mainly directed toward verifying the first (and erroneous) interpretations of the phenomenon of radioactivity, but they also confirmed the "very interesting discovery" of the Curies, since they succeeded in the chemical isolation of a strongly active substance from Joachimsthal minerals. Stefan Meyer (1872–1949) in Vienna began a study of the rays of radium a few months after the discovery of the element. Like Suess, he aided the Curies in obtaining the shipment of residues of uranium ore.

When Ernest Rutherford (1871–1937) left the Cavendish Laboratory in Cambridge for McGill University in Montreal in August 1898, he had already shown that the uranic rays had two components, which he named for "convenience" the α and β rays. This discovery was published in January 1899 [21].

In America, the Becquerel experiments were reviewed in April 1898 [22]. Interestingly enough, the discovery of radium was announced in the January 28, 1899 issue of *Scientific American*, scarcely one month after the publication had appeared in the *Comptes*

Rendus [23]. The translation came to the attention of Bertram Borden Boltwood (1870–1927), one of the first Americans interested in radioactivity, as early as 1898 [19].

It was, however, in Germany that the discovery of the radioelements had the strongest impact. In List, near Hannover, the chemical Company E. de Haën produced uranium for the glass industry, using extracts from the Joachimsthal ore. On July 14, 1899, the proprietor of the plant, E. de Haën himself, stated in a short note that he had succeeded in separating radium; indeed, a batch produced by him was active enough to exhibit self-luminosity, a property observed for the first time [24]. The interest of de Haën was obviously commercial, since the radium preparations were offered for sale. Owing to the facilities of his factory, he was able to prepare strongly active sources more rapidly than the Curies, who treated 20 kg batches with their own hands.

Friedrich Oskar Giesel (1852–1927), chief chemist at a quinine factory in Braunschweig, was the "scientist" occupied with the treatment of the residues of the de Haën factory. Shortly after the publication of the discovery of Po by the Curies, he found that the residues of the treatment of the uranium minerals, rather than the ore itself, were also a convenient source for the extraction of radioactive substances. He followed the Curies' procedure and separated a substance that did not behave like Po but had properties *closer to those of barium*. Hence, Giesel had nearly anticipated the discovery of radium by Pierre and Marie Curie.

The material available to Giesel was much more active than that of the Paris team. Within a few years, he was producing radium and offering it at a nominal price to anyone who wanted it. He confirmed de Haën's observation of the self-luminosity of radiferous Ba, in particular of the non-hydrated salts, which was so intense that one could read by the bluish light emitted by radiferous barium bromide. In his first publications on August 9 and 24, 1899, Giesel described original observations, such as the colour change of active barium salts due to self-irradiation, the increase in the activity of radium during the first weeks following recristallization (i.e., the in-growth of active deposit), and noted the differences in the absorption of the rays emitted by Ra, "which traverse considerably a silver Thaler" and of those of Po that were completely absorbed by thin metallic foils [25–26]. He also observed the disappearance of the activity of the precipitated lead sulfides within the course of a few months.

Meanwhile, the Curies had begun the extraction of radium from several tons of residues from the Joachimsthal ores. The Vienna Academy of Sciences had intervened with the Austrian Government (the proprietor of the mine) to sell the material to the Curies at a low price, which Pierre and Marie paid from their own money. The tedious processing of the residues under primitive conditions has been popularized in various biographies, novels and movies. The problems encountered, the reasonings and the procedure are described in Marie Curie's thesis. This document has been analysed didactically by Wolke [7]. The sentence which summarized the success of the endeavour stated "we succeeded ... in extracting from ... thousands of kilograms of starting material a few decigrams of products".

André Debierne, a former student of Pierre Curie, had agreed to take over the large scale processing of the pitchblende residues. Advised by the Curies, he began looking for other active substances in the tailings. Logically, he selected the substances which did not precipitate with H_2S, but were precipitated completely with ammonia. This choice ought to exclude Po and Ra. In October 1899, he announced that he had discovered a new fraction which had properties of titanium but with an activity 100000 times higher than that of uranium [27].

Later he found that the new substance more closely resembled thorium, but it was also much more active than this element. Since it behaved in a manner different from that of the known radioelements, Debierne claimed the discovery of *actinium* [28].

In fact, Debierne's assertion was false. To our knowledge, it appears that his substance was a mixture of two thorium isotopes, ionium (^{230}Th) and radioactinium (^{227}Th) with little, if any, of the element now known as actinium [29]. The error was quite understandable, however, owing to the extreme complexity of the mixture of natural radionuclides. The real discoverer of actinium was probably Giesel.

Marie Curie was aware of the importance of proving that Ra and Po were *elements*. She recognized that the demonstration would be much easier for Ra. For that reason, she worked untiringly to obtain radium in ever higher purity and in visible amounts in order to determine the atomic mass. It is understandable that, initially, the scientific community included sceptics, among whom were leading authorities. Even in 1906, Lord Kelvin argued that radium was a molecular combination of lead and helium, because both elements appeared spontaneously in radium.

As for Po, doubts and controversy followed the euphoria of the discovery. Apart from the activity of the "pseudo-bismuth" considered by the Curies as polonium, no other argument could be advanced in favour of its existence. Nearly four years after the discovery, the Curies stated that "experiments over several years show that the activity of U, Th, Ra and probably Ac ... does not change with time". On the other hand, they recognized that polonium progressively lost its activity. Hence they complemented the previous assertion by a footnote [30] indicating that "this substance [Po] is a kind of active bismuth; *it has not yet been proven that it contains a new element,* ... it does not emit radiations deviated by a magnetic field [which was wrong] and does not produce induced radioactivity [which was correct]". The doubts of the Curies could not have been expressed at a worse moment: Willy Marckwald in Berlin had claimed the discovery of a new radioelement in the bismuth fraction from a pitchblende residue, which *he was able to separate from bismuth*. One of the harshest controversies in the early history of radiochemistry followed.

Marckwald had received from the chemical firm Richard Stahmer in Hamburg several kilograms of residues from the treatment of uranium ores. Upon being asked to search for radioactive substances, he extracted about one per cent by weight of a very active bismuth oxide chloride, BiOCl. After several abortive trials to precipitate the active constituent, he succeeded with the help of electrolysis [31]. The first metal which plated out was quite active, leaving most of the Bi in solution. In a still simpler experiment, he dipped a Bi foil into the solution of the active BiOCl; a thin film deposited spontaneously and after a few days the entire activity was separated from Bi. This represented the goal which had been unsuccessfully pursued by the Curies.

In a following publication, on October 10, 1899, Marckwald showed that the radioactive substance was more electronegative than antimony, and suggested that it might belong to the S-Te group, rather than to the Sb-Bi group [32]. This proposal was upheld by the vacant Eka-Te position in the Periodic Table. The resemblance of the radioactive substance with Te was further confirmed by the reaction of $SnCl_2$ with the active BiOCl, which produced brownish flakes that carried most of the activity. Furthermore, the latter did not appear to decrease with time. Therefore *it could not be polonium*. Encouraged by these observations, Marckwald named "provisionally" the new substance *radiotellurium* [33].

That was too much for Marie Curie, who felt outraged by the new name given to *her* polonium. On December 12, "Frau" Curie reacted vigorously to Marckwald's assertions [34]. Obviously, she regretted the misfortunate (she wrote *misinterpreted*) words of the footnote to which "Mr. Marckwald gave a significance that they did not bear". She said that the Curies had never intended to suggest that polonium was anything more than activated bismuth, but only that they had not yet been able to prove that it was a new element. She was convinced that Marckwald's radiotellurium, which was carried by bismuth and emitted easily absorbed rays, was identical with polonium. The only difference would be that "Mr. Marckwald's polonium did not suffer a decrease of activity". She warned that only long-term measurement over months had shown the decay of the activity of polonium, and she defied Marckwald to prove the constancy of the activity.

Neither of the two authors was completely right. The chemical properties of polonium were undoubtedly those of Eka-Te. On the other hand, Frederick Soddy (1877−1956) showed that Marckwald's statement on the constancy of the activity of his radiotellurium "seems to be physically irreconcilable with our present knowledge of the nature of α rays..." [35]. He concluded that "most men of science will agree with Mme Curie in protesting against a new name given to it [to polonium]". Soddy added that "the practice of rechristening well-known bodies and sending them back to the country of their origin with new names and as new discoveries, ... seems to be prevalent among German organic chemists. If adopted in the case of the radioactive bodies, [it] would lead to the recognized number being exactly doubled". At the time, the situation was already sufficiently confused as a result of the ever-increasing number of new radioactive entities found in uranium and thorium ores.

The controversy subsided when Marckwald reported in January 1905 that radiotellurium decayed with a half-life of 139.8 days [36]. One year later, Marie Curie announced with great satisfaction that she had found a half-live value of 140 days with sources of polonium prepared by her usual method of fractional hydrolysis of a solution of bismuth sub-nitrate, as well as with samples obtained by the "very convenient" method used by Mr. Marckwald [37]. ... "This provides definitive proof that the body studied by Mr. Marckwald under the name *radiotellurium* is identical with polonium."

Eventually, in a final publication dated June 1st 1906, Marckwald surrendered, proposing to replace in the future the name of radiotellurium by polonium [38]. However, before the sentence announcing the decision, he insisted once again that the chemical properties of the radioelement were identical with those of Eka-Te. Marckwald concluded by quoting from Shakespeare's "Romeo and Juliet":

> What's in a name? that which we call a rose
> By any other name would smell as sweet.

Nonetheless, Marckwald was vexed and remained unconvinced that the original polonium of the Curies was not a mixture of several radioelements. For her part, Marie Curie was still reluctant to acknowledge the conclusions of Marckwald that polonium was a homologue of tellurium: "By no means can one pretend that polonium exhibits the reactions of tellurium rather than those of bismuth ... To know its properties, polonium should be available in weighable amounts ..." [37]. Five more years of work were necessary to reach this goal [13].

The discoveries of 1898 were the premises of a long story which ended by the mid-1930's, when the first generation of radiochemists eventually unravelled the natural radioactive families.

References

1. Badash, L.: Radioactivity before the Curies. Am. J. Phys. **33**, 128 (1965).
2. Schmidt, G. C.: Über die vom Thorium und den Thorverbindungen ausgehende Strahlung. Verh. Dtsch. Phys. Ges. Berlin **17**, 14 (1898).
3. Schmidt, G. C.: Sur les radiations émises par le thorium et ses composés. Compt. Rend. Acad. Sc. Paris **126**, 1264 (1898).
4. Badash, L.: The discovery of thorium's radioactivity. J. Chem. Educ. **43**, 219 (1966).

5. Sklodowska Curie, M.: Rayons émis par les composés de l'uranium et du thorium. Compt. Rend. Acad. Sc. Paris **126**, 1101 (1898).
6. Sklodowska Curie, M., Thesis: *Recherches sur les substances radioactives,* Faculté des Sciences de Paris, June 12, 1903, 155 p.; published by Gauthier-Villars, Paris (1904).
7. Wolke, R. L.: Marie Curie's doctoral thesis: prelude to a Nobel prize. J. Chem. Educ. **65**, 561 (1988).
8. Curie, P., Curie, S.: Sur une substance nouvelle radio-active, contenue dans la pechblende. Compt. Rend. Acad. Sc. Paris **127**, 175 (1898).
9. Joliot, I.: Les carnets de laboratoire de la découverte du polonium et du radium. Appendix to "Pierre Curie" by Marie Curie', Editions de Noël, Paris (1955).
10. Curie, P., Curie, M., Bémont, G.: Sur une nouvelle substance fortement radioactive, contenue dans la pechblende. Compt. Rend. Acad. Sc. Paris **127**, 1215 (1898).
11. Curie, M.: *Pierre Curie,* Payot, Paris (1924).
12. Curie, S.: Les rayons de Becquerel et le polonium. Rev. Gén. Sci. **10**, 41 (1899).
13. Curie, M., Debierne, A.: Sur le polonium. Compt. Rend. Acad. Sc. Paris **150**, 386 (1910).
14. Zeman, A., Beneš, P.: Radiochim. Acta **70/71**, 23–29 (1995).
15. Moyer, H. V., in *Polonium*, USAEC, TID-5221, Oak Ridge, July 1956, p. 4.
16. Mendeleev, D.: The periodic law of the chemical elements. J. Chem. Soc. **55**, 634 (1889); also Goldwhite, H.: Mendeleev's other prediction. J. Chem. Educ. **56**, 35 (1979).
17. Demarçay, Eug.: Sur le spectre d'une substance radio-active. Compt. Rend. Acad. Sc. Paris **127**, 1218 (1898).
18. Curie, M.: Sur le poids atomique du radium. Compt. Rend. Acad. Sc. Paris **135**, 161 (1902).
19. Badash, L.: *Radioactivity in America. Growth and Decay of a Science.* The John Hopkins University Press, Baltimore and London (1979), 327 pages.
20. Elster, J., Geitel, H.: Versuche an Becquerelstrahlen. Ann. Phys. **66**, 735 (1898).
21. Rutherford, E.: Uranium radiation and the electrical conduction produced by it. Philos. Mag. **47**, 109 (1899).
22. Stewart, O. M.: A résumé of the experiments dealing with the properties of Becquerel rays. Phys. Rev. **6**, 239 (1898).
23. Curie, P., Curie, M., Bémont, G.: Radium, A new body, strongly radio-active, contained in pitchblende". Sci. Am. **80**, 60 (1899).
24. de Haën, E.: Über eine radioaktive Substanz. Ann. Phys. Leipzig **69**, ser. 3, 902 (1899).
25. Giesel, F.: Einiges über das Verhalten des radioaktiven Baryts and über Polonium. Ann. Phys. Leipzig **69**, ser. 3, 91 (1899).
26. Giesel, F.: Über Radium und Polonium. Phys. Z. **1**, 16 (1899).
27. Debierne, A.: Sur une nouvelle matière radio-active. Compt. Rend. Acad. Sc. Paris **129**, 593 (1899).
28. Debierne, A.: Sur un nouvel élément radioactif: l'actinium. Compt. Rend. Acad. Sc. Paris **130**, 906 (1900).
29. Kirby, H. W.: in *Gmelin Handbook of Inorganic Chemistry,* 8th Edition, Actinium, Supplement volume 1, Springer Verlag, Berlin (1981).
30. Curie, P., Curie, S.: Sur les corps radioactifs. Compt. Rend. Acad. Sc. Paris **134**, 85 (1902).
31. Marckwald, W.: Über das radioaktive Wismuth (Polonium). Ber. Dtsch. Chem. Ges. **35**, 2285 (1902).
32. Marckwald, W.: Das radioaktive Wismut (Polonium). Phys. Z. **4**, 51 (1902/3).
33. Marckwald, W.: Über den radioaktiven Bestandteil des Wismuths aus Joachimsthaler Pechblende. Ber. Dtsch. Chem. Ges. **35**, 4239 (1902).
34. Frau Curie, M.: Über den radioaktiven Stoff "Polonium". Phys. Z. **4**, 234 (1902/3).
35. Soddy, F.: Radio-tellurium. Nature **69**, 347 (1904).
36. Marckwald, W.: Über das Radiotellur IV. Ber. Dtsch. Chem. Ges. **38**, 591 (1905).
37. Curie, M.: Sur la diminution de la radioactivité du polonium avec le temps. Compt. Rend. Acad. Sc. Paris **142**, 2273 (1906). This Note was presented at the Academy by Pierre Curie on January 29, 1906. He died in an accident on April 19.
38. Marckwald, W.: Über Polonium und Radiotellur. Phys. Z. **7**, 369 (1906).

Radiochimica Acta 70/71, 23–29 (1995)

St. Joachimsthal Mines and their Importance in the Early History of Radioactivity

By A. Zeman and P. Beneš

Department of Nuclear Chemistry, Faculty of Nuclear Science and Physical Engineering, Czech Technical University, 11519 Prague 1, Brehová 7, Czech Republic

(Received April 7, 1995)

St. Joachimsthal mines / Uranium mining / Uranium production / Radium production / Radiotherapy / History

Abstract

A short discussion of the history of the ore mining in St. Joachimsthal is presented, emphasizing its role in the discovery and production of radium, production of uranium dyes and in the development of radiotherapy. The historical procedures used in the production of uranium and radium in St. Joachimsthal are also briefly outlined.

1. Introduction

For the general public the town of Jáchymov (St. Joachimsthal or Joachimstal in German), situated in a valley at the foothill of the Czech part of the Ore Mountains, about 20 km north of Karlovy Vary (Carlsbad), is connected with the discovery of the element radium by the Curies in 1898 and with the extensive mining of uranium ores towards the middle of this century. However, the history of this town is much older and it is characterized by frequent changes of ups and downs of its prosperity. It can be divided into several periods, some of which overlap and include mining of silver, mining of pitchblende for manufacturing of uranium dyes, production of radium, mining of uranium ores for military purposes and for power production after World War II, and therapeutical use of St. Joachimsthal waters.

2. Period of silver mining [1–5]

In spite of the fact that already in prehistoric times tin for bronze production was obtained in the area of the present locality of Jáchymov, the town itself was founded much later, with the onset of silver mining, at the beginning of the 16th century. Konradsgrün, originally a small village, was located at its present site; the inhabitants earned their living mainly from forestry and farming and only to a small extent from mining. This community became the property of the Schlick Counts in 1454, when Emperor Sigismund bestoved Caspar Schlick, Chancellor of Bohemia, with the fief of the towns Loket (Elbogen) and Ostrov (Schlackenwerth) together with the neighbouring country in recognition of his outstanding merits. The rapid growth and the onset of the first illustrious period of St. Joachimsthal history is intimately connected with the family of Schlicks.

This history began in 1512, when rich veins of silver ore were found that in several localities protruded up to the surface. Silver was mined on the Saxon slopes of the Ore Mountains already in the 12th century, whereby the town of Freiburg played an important part. The discovery of new rich ores on the Czech territory resulted in a precipitate migration of Saxon miners. A colony with the name of "Tal" (Valley) was formed and a saying "Ins Tal, ins Tal, mit Mutter, mit All" is still remembered, characterizing the hectic atmosphere of the time. In 1519, the increasing output of mining incited King Ludwig to promote this community with 1200 homes to the status of a free royal town with the name St. Joachimsthal. This name was chosen on the basis of the Saxon mining towns Annaberg and Marienberg (St. Joachim was the husband of St. Anne and father of the Holy Virgin). The same year, the King gave Count Stephan Schlick the right to mint his own silver coins, known by the name "Joachimsthaler Groschen", "Talersgroschen", "Joachimstaler", "Schlickentaler" or, in later years, simply as "Taler" (tolar in Czech), from which originated the name of the American dollar. At its time, this coin, originally with the Czech heraldic lion on one side and St. Joachim's and Schlick's coat of arms on the other, had been used as a means of payment throughout the whole of Central Europe.

Progress became rapid, and the population increased concurrently. In 1516 the population of St. Joachimsthal was 1050; by 1520 it had attained 5000 and in 1534 18200. At that time, there were 13 silver smelting plants and St. Joachimsthal was the second largest town in Bohemia after Prague. However, the rapid growth was also a latent cause of rapid decline, a situation that occurred repeatedly in the history of Jáchymov. In 1527, the new Emperor Ferdinand I reserved for himself both the right of silver purchase and the mintage of coins and in 1545 the rule of the Schlick family in St. Joachimsthal came to an end.

In the second half of the 16th century and at the beginning of the 17th century all mining activities declined, not only owing to the shortage of silver ore but also as a result of the Thirty Years War. The town suffered from several military campaigns, heavy requisitions and war taxes. Further, it was devastated by the Swedish army, and by Saxons and Croats. As a result, the town had a population of only 2000 in 1621 and in 1635 only 144 houses were inhabited. Another disaster came in the form of the plague in 1645 (and again in 1771 and 1772), so that finally, in 1651, the imperial mint was moved to Prague.

The next two centuries witnessed an intermittent rise and decline in silver mining but as in previous years, other metals were also produced, such as Pb, Cu, Bi, Ni, As and, in particular, Co. In 1716, a metallurgical school was established in St. Joachimsthal, probably as the first of its kind in the world [5]. In the mid-19th century the St. Joachimsthal smelting factory was reconstructed and, at the same time, Adolf Patera, metallurgical chemist and Imperial Councillor, introduced the wet treatment of the ore that decreased production costs and that was practised until 1867.

3. Period of the uranium dyestuffs production

It was pitchblende (Pechblende, Uranpecherz, smolinec in Czech) that saved St. Joachimsthal in the critical period when the output of silver ore, mined predominantly in the state-owned mines (Concorde Mine, Kaiser Josef, Elias, and Werner that were drained by the mine galleries Barbara and Daniel) steadily declined [1, 2]. Its occurrence in the form of thin black veins and isolated lenses brought "bad luck" (Pech in German, smula in Czech) to the miners at the time of the silver rush because in the deeper formations, where pitchblende appeared, silver could no longer be found. This situation was a consequence of the geological conditions of the St. Joachimsthal locality [3–7], where the basic formation is determined by the presence of gneiss and the crystalline schists rich in biotite, muscovite, quartz and feldspar. These formations are pervaded with veins of various kinds: porphyry from the Perm period, ore veins, and tertiary veins of basalt and phonolite. The mineralogic richness of the Jáchymov veins is enormously diversified and abundant. More than 100 minerals were found, and some of them were described here for the first time. The silver ores are the oldest and are located in the uppermost layers. They contain mainly argentite, proustite, pyrargyrite, stephanite, polybasite, chlorargyrite and pure metallic silver. The Co-, Ni- and Bi-containing ores are found in the upper layers that include smaltine, chloantite, nickeline, millerite and metallic Bi. Uranium ores are the youngest and are found at depths exceeding 300 m, where they contain pitchblende in its amorphous form. A number of secondary minerals were formed by weathering of this material. The deads of the uranium and of most silver ores were formed by schists, quartz, pyrite, dolomite and limestone.

Pitchblende from St. Joachimsthal was an ore very rich in uranium. For instance, the contents of U_3O_8 in pitchblende from the Concorde Mine amounted to more than 80%, and in the Werner Mine to 76–77%. Other components such as Fe_2O_3, PbO, SiO_2 and CaO were present in amounts of several %, and Bi_2O_3, As_2O_5, Na_2O and S in amounts of approximately 1%. Concentrations of ZnO, MnO, MgO, K_2O and of rare earths amounted to tenths of % [8–9].

In contrast to former times, when pitchblende was deposited on dumps as a useless waste, it suddenly became a much sought-after raw material. This event is ascribed to A. Patera, who in 1852 prepared uranium yellow in his laboratory and demonstrated its usefulness for coloration of glass and china. The production of sparkling and highly estimated uranium dyes based on uranates was introduced in 1856, following a procedure developed by A. Patera. Since 1871 the uranium dyes were produced in the so-called K. K. Uranfarbenfabrik that consisted of the extended old Renaissance works for smelting silver as at the time of the Schlicks. The production of uranium dyes, prepared in six various yellow colours and in one black colour, as well as of uranium oxide and uranium protoxide (a mixture of UO_3 and UO_2) launched the second period of the rise of St. Joachimsthal and its environs [1–2].

Following exploitation of the old dumps remaining after the mining of silver, the extraction of pitchblende proceeded in the old Jáchymov mines Concorde, Werner and Gallery of Saxon Noblemen (in the fifties of this century the last two mines were renamed "Rovnost" [Equality] and "Bratrstvi" [Brotherhood]); the annual output was 15–30 tons. Gradually the mining operations reached deeper and deeper layers when suddenly, at a depth of 500 m, a strong spring of water was struck which flooded the Concorde Mine [2, 10]. The mine was pumped out in 1867 but it flooded again in 1901.

4. Discovery and production of radium

When the mining excavation reached greater depths, it became less profitable and was therefore progressively abandoned. In this difficult situation, Jáchymov was rescued by the discovery in 1898 of radium (and polonium) by Pierre and Marie Curie in a sample of St. Joachimsthal pitchblende [11–16]. One year later, A. Debierne discovered another element, actinium, in the fraction of rare earths separated from the St. Joachimsthal uranium tailings.

The Curies studied the behaviour of individual components during the chemical separation procedures by measuring the radioactivity of the fractions: this approach can be considered as the historical beginning of radiochemical analysis. It was also proved that the new elements can be separated from their natural carriers and suitable procedures were proposed [12].

The results obtained indicated the presence of very minute amounts of the newly discovered elements in pitchblende. Preparation of weighable amounts obviously required treatment of ton amounts of the raw material. Instead of the expensive pitchblende, the Curies used for this purpose St. Joachimsthal's uranium tailings, at that time worthless waste, that remained after leaching of uranium with H_2SO_4 in the production of uranium dyes. The violet-grey tailings contained virtually all Ra originally present in the ore; they had the consistence of a slurry and had 3–5 times higher activity per dry weight than pitchblende itself. Besides a number of other elements, the tailings contained mainly sulfates of Pb, Al and Ca, hydrated silicon dioxides, clay and iron oxides [5, 12]. Radium was present in the form of the least soluble sulfate. The Austrian Government granted the Curies initially 100 kg of the tailings in 1898 and then another 1000 kg gratis [16]. The government also reserved for the Curies another ton of this material, whose price began to rise rapidly.

The Curies used the following procedure for the separation and concentration of Ra [12]:

a. The uranium tailings were boiled with soda to decompose silicates and Pb, Al and Ca sulfates.

b. The insoluble part was washed with water and treated with HCl. Ra remained in the residue, and from the filtrate it was possible to extract Po and Ac.

c. The residue insoluble in HCl was boiled again in soda to convert the still undecomposed sulfates into carbonates. The carbonates were washed with water, treated with HCl, and from the filtrate the raw Ba sulfate containing Ra and small amounts of Ca, Pb, Fe and Ac were precipitated with H_2SO_4. From 1 ton of uranium tailings, 10–20 kg of the raw sulfate were obtained.

d. Further purification of the raw sulfates consisted of boiling with soda, conversion of carbonates into chlorides, precipitation (with H_2S) and separation of sulfides containing Po, and precipitation and removal of hydroxides contaminated with Ac. The next steps included precipitation of carbonates, their dissolution in HCl and evaporation of the solution to dryness. The residue was washed with concentrated HCl to dissolve $CaCl_2$, whereby $BaCl_2$ remained undissolved. About 8 kg of this raw chloride were obtained from 1 ton of uranium tailings.

e. The separation of Ra from the large excess of Ba was accomplished by fractional crystallization of the chlorides (see below, Part 5).

By the procedure described above, Marie Curie obtained nearly 0.1 g of spectroscopically pure $RaCl_2$, an amount that was sufficient for determination of the atomic weight of Ra, which she found to be 225 [18]. The Curies' achievements aroused widespread interest in the scientific community and many other researchers began to study radioactivity. Since uranium tailings were provided also for them at that time, most of the studies were performed using material from St. Joachimsthal.

As early as 1906, the industrial production of Ra started in St. Joachimsthal's uranium factory [2], which worked up the old uranium tailings till the end of World War I. Up to 1916, the production, amounting to about 2 g per year, had no competition throughout the world. In the following years, a new producer of Ra appeared in the United States, exploiting the carnotite strata in Colorado and Utah, that produced up to 30 g of Ra per year until 1924. The American competition was surpassed by the Ra production in Oolen, Belgium (from 1922), where pitchblende from the rich layers in Katanga, accessible from the surface, was processed.

However, even the Belgian production fell behind when rich pitchblende deposits were found at the Great Bear Lake in Canada, where 60 g of Ra were produced in 1937 [10].

In spite of this situation, the annual production of 1–5 g of Ra in Jáchymov continued until World War II, together with the production of uranium dyes and other uranium compounds. After World War I, the Czechoslovak Government decided to reconstruct and modernize the mines and the formerly flooded mine of Concorde was drained. Work was undertaken to establish the underground interconnection of the mines Concorde, Werner and the Gallery of Saxon Noblemen and prospection of the extent of uranium veins was carried out [3].

5. Methods for production of uranium and radium in the twenties and thirties of the 20th century [3, 8, 19, 20]

The production was based on the original procedures of A. Patera and on those applied by the Curies and other research workers. It took place in the old factory shown in Figure 1, which was demolished in 1941. The procedures used, nowadays of only historical significance, are described briefly below (some of them were repeated several times in order to decrease the losses of Ra). The ore with a U_3O_8 content ranging from 40 to 70% was used for the treatment in the factory.

Production of uranium compounds

a. Calcination of the ground ore to red heat to expell As, S etc.

b. Calcination of the mixture with soda and salpeter to convert all U into the form of uranates.

c. Washing with hot water to remove excess soda and other soluble salts, e.g. of As, V and W.

d. Dissolution of the remaining ore in concentrated H_2SO_4, followed by the addition of HNO_3 to change the sulfates of U(IV) into the more soluble sulfates of U(VI).

e. The solution of uranyl sulfate that contains practically no Ra, was used for the production of uranium dyes, while the insoluble remains, the so-called "ra-

Fig. 1. View of St. Joachimsthal/Jáchymov from about 1920: Left, Radium Palace Hotel; Right, State Factory for Uranium Dyes and Radium.

dioactive residues", were used for the production of Ra.

Altogether, about 10 tons of uranium dyes of various shades were produced annually. These included uranium yellow I, uranium yellow II, uranium orange, potassium uranium orange etc. For instance, uranium yellow I, that was used as the raw material for the production of all other dyes, was produced such that the solution of uranyl sulfate was oversaturated with soda, permitting separation of the soluble uranyl carbonate complex from the precipitate of insoluble hydroxides and carbonates. The uranium solution was transferred to vats, in which it was neutralized by H_2SO_4. In this way, a precipitate of the yellow sodium diuranate was obtained. Other compounds such as ammonium diuranate, uranium protoxide and uranyl nitrate were also produced.

Production of radium

a. The insoluble radioactive residues, after leaching with H_2SO_4 (point e above) contained approximately 40 mg of Ra per 100 kg and from 1 ton of the Jáchymov ore of average richness, about 370 kg of such residues were obtained. The latter, in 100 kg portions, were first boiled with a 15% solution of NaOH in boilers of cast-iron, with the result that the silicic acid and some of the sulfates were brought into solution. The remaining insoluble part, containing Ra, was extracted in stoneware dishes on a water bath with HCl. By this procedure, the "prepurified residues" were obtained, that contained the main part of Ra; nevertheless, the HCl extract also contained a certain portion of Ra, that was precipitated in the form of sulfate and also utilized.

b. Boiling of the prepurified residues for several hours with the 15% soda solution in the cast-iron boilers converted Ra into its carbonate, which was decanted with HCl. From the "radium liquor I" obtained, which contained about 60% of the original Ra, the element was precipitated together with Ba by H_2SO_4. "Crude sulfate" was thus obtained. The undissolved part of the prepurified residues, consisting of the "depleted residues" was then treated twice as described above in order to convert the material first to carbonates and then to sulfates. This led to "crude sulfates II and III".

c. By boiling with soda all sulfates were then transformed into the carbonates which were washed and extracted with HCl. From the resulting solution containing Ra, the remaining Pb was removed by precipitation with H_2S and, after thickening, oversaturated with HCl at low temperature, which resulted in an almost complete separation of Ba and Ra chlorides. The maximum content of Ra in this raw chloride was about 0.03%. The whole procedure was much more complicated because after the dissolution of carbonates in HCl, there remained an insoluble residue consisting of undecomposed sulfates and a considerable amount of $PbCl_2$, that contained about 20% of the total amount of Ra. The insoluble residues were therefore repeatedly treated for removal of Pb by extraction with hot water, turned into carbonates and dissolved in HCl in order to recuperate all remaining traces of Ra.

d. The next step consisted in the separation of Ra from Ba by fractional crystallization, based on the lower solubility of $RaCl_2$. The crystallization was carried out in porcelain dishes from solutions saturated at 80 °C and cooled to 20 °C. Under these conditions, a

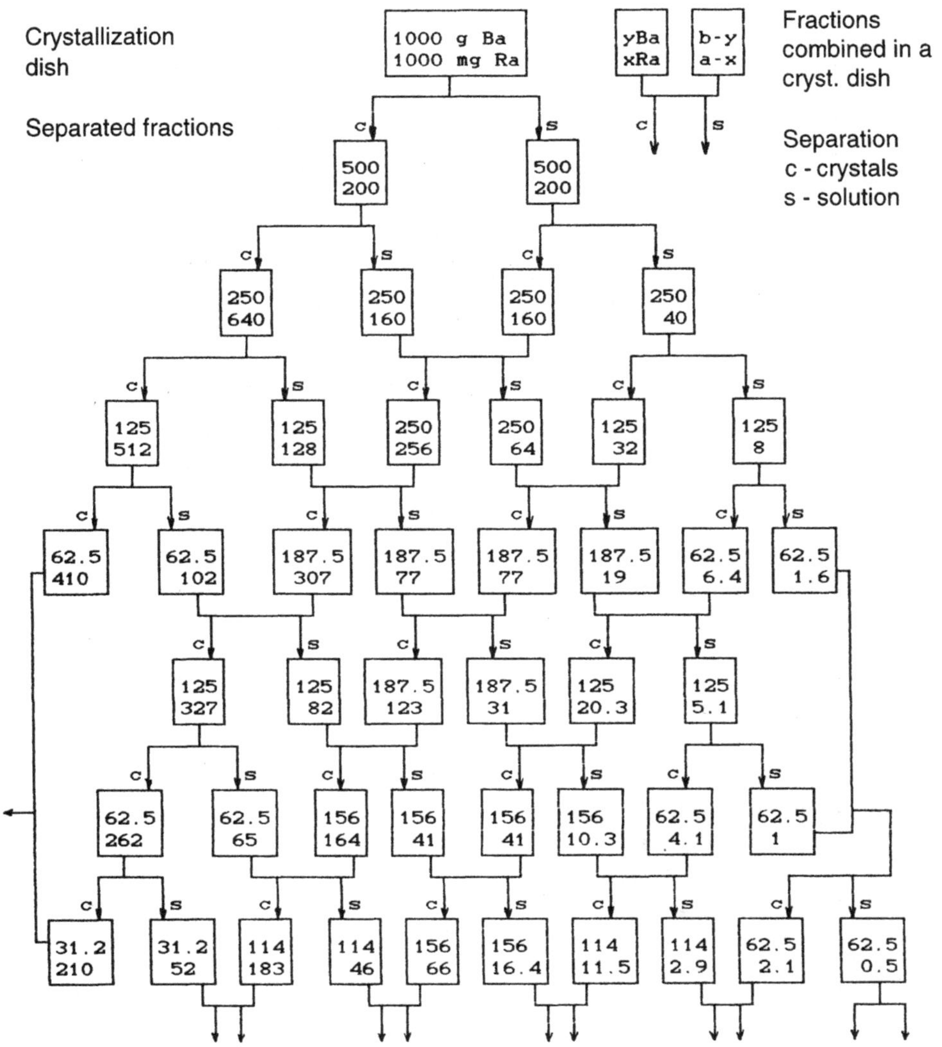

Fig. 2. Scheme of the separation by fractional crystallization used in the production of radium in the St. Joachimsthal radium factory [19–20]. Further details are given in the text.

near equilibrium distribution of Ra in the crystals is achieved, which can be described by the separation factor

$$K_{x,y} = \frac{x(b-y)}{y(a-x)}$$

where a, b are the respective total amounts of Ra and Ba in the system, and x, y are the corresponding amounts in the crystals. Since $K_{\mathrm{Ra,BaCl_2}} = 4$ at 20 °C, 80% of the Ra are concentrated in the crystals if 50% of the $BaCl_2$ are crystallized.

By repeated crystallizations (up to 2000) carried out according to a well-designed scheme, whereby the intermediate fractions with the same contents of Ra were combined (Figure 2), it was possible to obtain practically all Ra present in the mixture of chlorides of several kg in weight, in the form of almost 100% preparations of $RaCl_2$. The number of crystallizations dishes used during this procedure usually did not exceed 20. The gradual enrichment of crystals with Ra could be followed visually by the increasing development of a brown coloration in the crystals owing to radiation effects.

e. The $RaCl_2$ produced was carefully dried at 120 °C and delivered in double glass vials, fused-in and sealed with Pb. After four weeks, i.e., following establishment of radioactive equilibrium with the short-lived deposit, the sealed preparation was certified by comparing its γ radiation intensity with that of a standard.

6. Uranium mining after World War II [21]

By the end of the War the three historical mines Werner, Gallery of Saxon Noblemen and Concorde with several hundred employees were still operating. In September 1945 the mines were taken over by the then Soviet Army and the conclusion of the Czechoslovak-Soviet Bilateral Agreement, signed on September 23, initiated increased exploitation of the uranium ores. An extensive geological survey led to the discovery of a number of new localities; in 1950, 16 new pits were opened. By the end of 1951 altogether 56 pits or galleries were already in operation. Whereas in 1945 the annual output amounted to only 0.9 ton, in 1950 it reached 281 tons and increased to 1200 tons of uranium ore in 1953. It should be recalled that in the 1920–1930's the output was of the order of 200 to 300 tons.

The development of the mining activity required an ever-increasing number of miners and various financial, catering and housing benefits offered to the miners launched a new variant of the "silver rush". Accordingly, in the second half of the 1950's more than 25% of all employees of the Carlsbad district worked in the Jáchymov mines. However, a rather dark page of Jáchymov history should be recorded in this connection, since hard labour of prisoners was extensively used for uranium mining for the whole period up to 1960, at which time amnesty was granted. In the early years of this period, particularly up to 1951, about two thirds of the prisoners consisted of persons imprisoned for political "crimes"; later their number decreased.

The geological survey revealed new deposits of uranium not only in the Jáchymov region, but also in many other localities in the Czech Republic. However, complex interrelations between mining costs, the degree of the deposit's exploitation, the existence of sufficiently high strategic reserves, as well as the then declining importance of the military use of uranium ore by the end of the 1950's, led first to a gradual reduction of the mining activities in the Jáchymov locality, and then to their extinction in 1964. At present, Jáchymov is primarily a spa with a rather limited population of 2700 (in 1991). The Concorde Mine is still kept in operation since water for the spa treatment is pumped from it, and the Gallery of Saxon Noblemen is used as a depository for radioactive wastes.

7. Therapeutical use of Jáchymov waters [2, 22, 23]

Since early times, the miners of Jáchymov suffered from the so-called "Jáchymov miners'" disease, that was also called "Schneeberg disease" on the other side of the Ore mountain [10]. Detailed studies of this disease started only in the second decade of this century, when the deleterious effects of Ra in the human organism were recognized [24]. This discovery was preceded by the well-known case of 17 women whose decease was caused by the licking of brushes used for the application of radioactive luminescent pigments containing Ra. The Jáchymov miners' disease is a cancer of the lungs caused by long-term breathing of the mine air contaminated with radioactive decay products of radon which are deposited in the human lungs.

On the other hand, well before the discovery of radioactivity, the therapeutical effect of Jáchymov mine waters was well known. Heinrich Mache and Stephan Meyer of the Physical Institute in Vienna, who measured the radioactivity of mineral waters from Carlsbad and other spas in Western Bohemia, studied in 1905 the radioactivity of Jáchymov mine springs. They discovered that their radioactivity reaches very high values and that it increases with the depth. The values, expressed by them in the now obsolete Mache unit, M.u. (1 M.u. = 13.4 $Bq\,L^{-1}$ = 362 $pCi\,L^{-1}$) amounted to 33, 50 and 185 Mache units for the depths of 60, 266 and 375 m, respectively. They also obtained abnormally high values as, for example, 1200 M.u. for a spring in the Daniel Gallery.

An outstanding geologist, Josef Štěp, Mining Councillor and Head of the Jáchymov Mines management, acted as a guide for the two scientists. He became very interested in the methods of radioactivity measurements, which he carried out himself, and after discovering several other springs in the Werner Mine he became an enthusiastic pioneer of the utilization of mine waters [10]. The one time Ministry of Land Cultivation appointed Gottlieb, the District Physician, to perform a more detailed investigation of the therapeutical effects of Jáchymov waters. After confirming very good results in the treatment of arthritis, rheumatism and nervous disorders, the spa treatments were authorized in the home of Kuhn, a baker, as early as 1906. This spa consisted of two tubs, the water for which was brought by a miner in two buckets from the Werner Mine. In 1908, a pipeline of 4325 m in length was laid, which supplied 4 tubs in the Uranium Factory. In 1911, the State Institute for Radium Therapy was established and a building with 40 tubs was built. In 1912, a luxurious Radium Palace Hotel (shown in Figure 1) was erected.

Owing to the ever-increasing number of spa patients (e.g., in 1922 there were 3000 guests and 10000 in 1930) the amounts of water, a mixture from several springs in the Werner Mine with the average activity of 8 $kBq\,L^{-1}$ (600 M.u.), became insufficient. After draining of the Concorde Mine in 1923, the springs of this mine were also used for therapeutical purposes. The following Jáchymov springs can be rated as the richest ones that are used for medical treatment: the Curie spring (60 $L\,min^{-1}$, 28.5 °C, 5.85–6.83 $kBq\,L^{-1}$), the Becquerel spring found in its close vicinity in 1928 (50 $L\,min^{-1}$, 26 °C, 683 $Bq\,L^{-1}$) and the Evangelist spring found after 1950 (33 $L\,min^{-1}$, 28.5 °C, 962 $Bq\,L^{-1}$). However, there are also many other springs, especially in the Jáchymov region, and springs and wells with water rich in radon can be found even at the ground level.

One of the most abundant springs, with an activity of 10.2 $kBq\,L^{-1}$, was drilled in the Concorde Mine in 1963 and it was named in honour of the Academician Běhounek.

František Běhounek (1898–1973) was Professor of the Charles University of Prague and Professor of the Faculty of Nuclear Science and Physical Engineering of the Technical University of Prague. In the early twenties, he worked in the Institut du Radium in Paris under the supervision of Marie Curie. It was thanks to him that several Czech institutions of radiological character were established. In the twenties and thirties, as an outstanding specialist in radiobiology and dosimetry, he made many important contributions to the field of activity measurements of rocks, minerals, air and water of the Jáchymov region and thus fostered substantially to the recognition of the natural wealth of the region and to its utilization, particularly in medicine [25].

The therapeutical properties of the Jáchymov springs are primarily determined by their content in radon that is adsorbed in the passage of their waters through rock formations. However, their higher contents in lithium and iron are also beneficial for patients. At present, they are used for the treatment of diseases of the locomotor organs of the human body, for the treatment of diseases of the nervous system and of disorders in body secretion. Moreover, with baths and inhalations, radiation treatment, brachyradiumtherapy and roentgentherapy are also provided.

References

1. Kořan, J.: *450 Years of St. Joachimsthal Mines* (in Czech), Academia, Praha 1967.
2. *St. Joachimsthal,* Verlag der k.k. Berg- und Hüttenverwaltung St. Joachimsthal, Wien 1911.
3. Běhounek, F., Heyrovský, J.: *Introduction to Radioactivity* (in Czech), JČMF, Praha 1931.
4. Knopf, K.: *Die Wunderstadt St. Joachimsthal*, Deutschböhmische Verlagsanstalt Sonnenwirbel, Weipert 1930.
5. Sturm, H.: *Abriss der geschichtlichen Entwicklung von Stadt und Bezirk St. Joachimsthal*, Verlag der Buchhandlung Rudolf Weis, St. Joachimsthal 1932.
6. Ježek, B.: Hornický věstník **9**, 525 (1927).
7. *Czechoslovak uranium deposits* (in Czech), SNTL, Praha 1984.
8. Koblic, O.: Hornický věstník **9**, 289 (1927).
9. Curie, M.: *Le radium et les radioéléments*, Librairie J. B. Bailliere et fils, Paris 1925.
10. Běhounek, F.: *Atom is frightening the World* (in Czech), Nakl. Ing. Mikuta, Praha 1947.
11. Sklodowska Curie, M.: Rayons émis par les composés de l'uranium et du thorium. Compt. Rend. Acad. Sc. Paris **126**, 1101 (1898).
12. Curie, M.: *Die Radioaktivität*, Akademische Verlagsgesellschaft, Leipzig 1911.
13. Curie, P., Curie, M.: Sur une substance nouvelle radioactive, contenue dans la pechblende. Compt. Rend. Acad. Sc. Paris **127**, 175 (1898).
14. Curie, P., Curie, M., Bémont, G.: Sur une nouvelle substance fortement radioactive, contenue dans la pechblende. Compt. Rend. Acad. Sc. Paris **127**, 1215 (1898).
15. Demarçay, E.: Sur le spectre d'une substance radioactive. Compt. Rend. Acad. Sc. Paris **127**, 1218 (1898).
16. Adloff, J. P., MacCordick, H. J.: Preceding article.
17. Seidlerová, I.: Joachimsthal uranium ore in the beginnings of radiology, Čs. čas. fys. (Czech. J. Phys. **A 22**, 198 (1972).
18. Curie, M.: Sur le poids atomique du radium. Compt. Rend. Acad. Sc. Paris **135**, 161 (1902).
19. Majer, V.: *Fundamentals of Nuclear Chemistry* (in Czech), SNTL, Praha 1961.
20. Koblic, O.: Hornický věstník **9**, 305 (1927).
21. Kaplan, K., Pacl, V.: *Secret Space Jáchymov* (in Czech), Actys, České Budějovice 1993.
22. Urban, J., Slánský, J., Mach, M.: *Jáchymov,* MNV Jáchymov 1960.
23. Pirchan, A.: Hornický věstník **9**, 557 (1927).
24. Williams, A. R., Kirchmann, R. J.: Radium – A Historical Introduction. In: *The Behaviour of Radium in the Environment.* IAEA, Vienna 1991.
25. Majer, V.: Chem. Listy **62**, 1273 (1968).

Radiochimica Acta 70/71, 31–37 (1995)

History of the Unraveling of the Natural Decay Series

By G. W. A. Newton
Chemistry Department, The University, Manchester M13 9PL, U.K.

(Received December 20, 1994; revised March 23, 1995)

Natural decay series / History / Isotopes / Emanation / Radioactivity

Abstract

Milestones in the subject of radioactivity, which were important to the unraveling of natural decay series, have been identified and described.

The interplay between all these issues is very complex, and some aspects of the development of the uranium series is discussed as an example, the thorium and actinium series evolved in a similar way from research by many of the same people.

All three series as understood in 1913 are compared with their present-day representation.

Introduction

1996 is the 100th anniversary of the discovery of radioactivity by Becquerel [1], and it is a good time to reflect on how some parts of the subject developed in the early years. Emphasis will be given to the history of the unraveling of the chemistry of the uranium decay series as an example of the early research.

Following the discovery of radioactivity, important milestones in that subject were reached which helped in understanding the natural decay series, some of these were:

1. Discovery of the existence and nature of α, β and γ radioactivity.
2. Methods of measurement of radioactivity.
3. Methods of standardisation of radioactivity.
4. Development of a disintegration Theory.
5. The Geiger-Nuttall rule.
6. Isotopes.
7. The alpha-beta rule.
8. The nuclear atom.
9. Identification of emanation.

Each of these will be considered in turn, followed by the uranium series.

Milestones

1. Types of radioactivity

Rutherford [2] in studying the ionization of gases by the radiation from uranium identified two types: the α-radiation which produces by far the greater part of the ionization, and is adsorbed by a single sheet of ordinary paper, and β-radiation producing much less ionization but capable of penetrating several millimeters of Al, Cu, etc. In 1900 γ-radiation was discovered by Villard [3] who found that Ra emitted an extremely penetrating kind of radiation capable of detection through 20 cm of iron or several cm of lead.

Alpha – Helium found in minerals of thorium and uranium was thought to be produced by radioactivity [4]. The decay from radium was measured spectroscopically and shown to be helium [5]. Rutherford thought that the α particle was a helium atom, and others thought that helium was the end-product of radioactivity. 25 mg radium bromide was sealed in a U-tube, filled with hydrogen and exhausted ten times, and after three months the gas over the radium bromide had the spectrum of helium [6]. The final proof was the passage of α particles through a thin-walled glass tube into an evacuated collecting vessel. The gas in the collecting vessel had the spectrum of helium [7].

In 1906 Rutherford postulated that α particles had a charge of 2+, and it was assumed [8] that the α particle was $^4He^{2+}$.

Beta – Beta rays were assumed to be similar to cathode rays [9], that is they were negatively charged. This was proved by passing the β rays through an earthed screen and letting them impinge on a metal plate which became negatively charged [10].

2. Measurement

The techniques available were the quadrant electrometer, electroscopes, ionisation of gases and scintillation, and these have been described [11]. The spinthariscope developed by Crookes [12] and Elster and Geitel [13] "rendered single events visible to the eye". The cloud chamber [14] was important in visualising decay processes and nuclear reactions but it was not used in natural decay series research.

3. Standardisation

Marie Curie undertook to prepare a 20 mg radium standard [15], which she achieved. For smaller quantities emanation was used and the unit adopted was the curie – the activity of emanation in equilibrium with 1 g of radium [16]. The number of α in equilibrium with 1 g of radium was initially 3.4×10^{10} s^{-1} [17], later 3.57×10^{10} s^{-1} [18], and then 3.72×10^{10} s^{-1} [19].

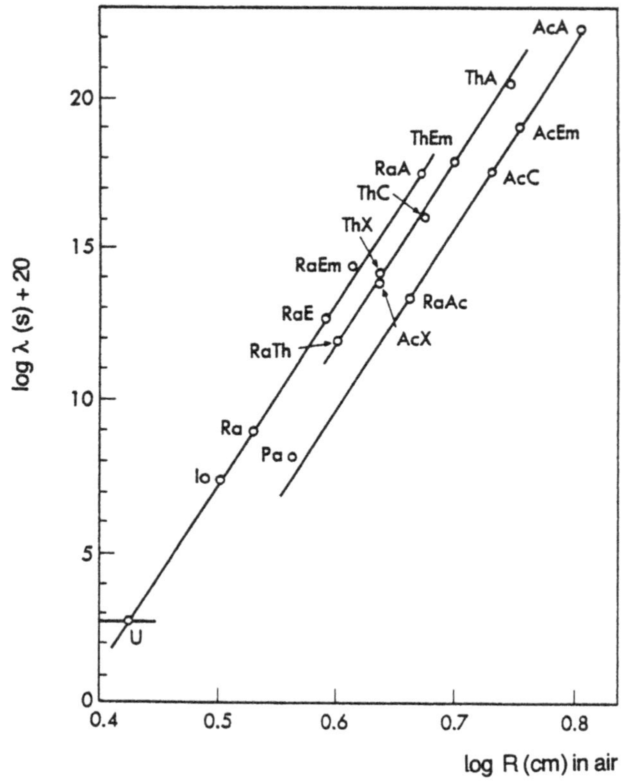

Fig. 1. Geiger-Nuttall relation for naturally occurring α emitters [Adapted from H. Geiger, Z. Physik 8 (1921) 45].

This decay rate became known as the Curie. In 1950, the definition of the Curie was changed to be any radioactive substance decaying at the rate of 3.700×10^{10} disintegrations per second.

4. Disintegration theory

Rutherford and Soddy [20], from measurements on uranium and thorium, came to the conclusion [21] in 1903 that radioelements were undergoing spontaneous transformations from one atom to another by the expulsion of a radiant particle. This was revolutionary; in that it disagreed with the well entrenched Dalton theory that the atom was indivisible. However, it was rapidly accepted as a way to explain the links between the 'new' radioelements, which were about 20 in number.

5. Geiger–Nuttall rule

Rutherford [22] tabulated the period (half-life), energy and range of α particles from 23 substances. He noted an inverse relation between half-life and energy. This was developed by Geiger and Nuttall [23] who plotted $\log \lambda$ against log range (Fig. 1). These were useful in predicting one of the properties of α particles, usually the half-life from the range.

6. Isotopes

The early stages in establishing relationships between radioelements came from noticing that thorium and radiothorium [24], and RaD and lead [25] were chemically similar. Boltwood noticed a radioelement was chemically similar to thorium [26] and suggested the name ionium [27].

It was also shown that thorium X, actinium X, mesothorium I and thorium are chemically identical and are members of the alkaline-earth group of elements [28, 29]. Inseparable pairs were thorium and radiothorium [30], radium and mesothorium, and RaD and lead [31–33]. Ionium, radiothorium, radioactinium and thorium were also chemically identical [34]. It was also well-established at this time that RaD and lead were chemically the same [34]. In fact, 20 different methods failed to separate lead from RaD [35]. It was thought that these are not merely chemical analogues but chemical identities. The atomic weights of the members of these groups were different, but some radioelements from different families had the same atomic weight, e.g. mesothorium I, radiothorium, RaD and polonium.

Electrochemical studies were very important and measurement of deposition voltages showed similarities between:

a) ThC, RaE and bismuth,
b) RaD, ThB and lead.

This led to the introduction of tracers for bismuth and lead [36, 37].

In the following years these ideas were extended and a systematic study [38] showed that the following combinations were chemically identical: [39]

a) radioactinium and thorium
b) mesothorium II and actinium
c) the three B members of the series, RaD and lead
d) the three C members of the series, RaE and bismuth
e) RaA and polonium
f) thorium D, actinium D and thallium.

At this stage all of the known radioelements had been shown to share its chemical nature with at least one other element.

To express this relationship between elements, Soddy [40] introduced the term "**ISOTOPES**". Isotopes all occupy the same place in the Periodic table and are indistinguishable chemically.

7. Alpha and beta rule

By 1913 Soddy, Fleck [41] and Russell [42] had independently announced an important generalisation from the detailed chemistry of the radioactive species. This was that the chemistry of the daughter element was connected with the emitted radiation. Soddy [43] published the then current disintegration series and the analogies between the three of them [40].

"The loss of a helium atom or α-particle appears to cause the change of the element, not into the next family but into the next but one. ... The progress is certainly not so straightforward as this. In several cases the matter appears to *alternate* in its passage, passing through the same family not once but *twice*" [44].

By 1913 [45] the α-ray rule had been shown to be general and the alternating effect was due to β emission which caused the element to shift one place in the Periodic Table in the opposite direction to that caused by the α-ray change. The radioelements could now be incorporated into the Periodic Table [40].

8. Nuclear atom

Geiger and Marsden [46] studied the scattering of α particles by gold foils and were surprised to find that some were scattered back in the direction of incidence. Rutherford was heard to say "... It was quite the most incredible event that has ever happened to me in my life. It was almost as incredible as if you had fired a 15-inch shell at a piece of tissue-paper and it came back and hit you" [47]. This surprising result led Rutherford [48] to postulate the nuclear atom. This is an atom of matter consisting of a central massive nucleus carrying a positive charge which is surrounded, at distances relatively great compared with its diameter, by "planetary" electrons. The central nucleus contains all the positive electricity in the atom and therefore all its mass. The weight of the atom and its radioactive properties are associated with the nucleus; its chemical properties, on the other hand, are properties of its planetary electrons. Detailed scattering experiments [49] proved Rutherford's theory and measured the charge on the nucleus; this was roughly half the atomic weight (measured in units of e).

9. Emanation

It was shown by electrochemical methods that compounds of thorium are different from uranium in that besides α and β "they communicate temporary radioactivity to the surrounding atmosphere" [50]. Actinium was also shown to give a characteristic emanation which imparted it's activity to the neighbouring objects [51–53], and the emanation substance was described as emanium [54–56].

The idea that emanation was a gas came from wrapping thorium compounds in porous material and observing the same effect as unwrapped compounds. Whereas if impervious material was used then no radioactivity escaped from the thorium compounds [50]. Rutherford concluded that emanation could diffuse like a gas. Later work by the same group using physical and chemical methods established that emanations given off by thorium, radium and actinium were high molecular weight radioactive gases. These were present in such minute amounts, of the order of 10^{-11} cm^3, that they could only be detected by their radioactivity. It was also shown that the emanation was positively charged and could be deposited as a thin film. Radium, thorium and actinium were associated with emanation, but uranium and polonium were not.

Leading the radium and thorium emanations into tubes containing powerful reagents, such as platinum black, copper oxide and zinc dust, at high temperatures resulted in no absorption or change in their radioactivity [57]. The conclusion was that emanations are gases similar to the argon family. This was endorsed [58] when sparking emanation with oxygen over potassium hydroxide gave no reaction. Condensation of emanation at −154 C supported the argon family idea and gave a method of preparing pure radium emanation [58, 59].

Uranium series

There was no emanation associated with uranium, but there was temporary radioactive matter needed to be separated chemically. Sir William Crookes [60] carried out several experiments with uranyl nitrate. In one experiment in which ammonium carbonate is added in excess, the precipitate initially formed redissolves leaving a very small residue. This residue contained all the radioactivity, and he called it uranium X(UX). Becquerel [61, 62] agreed with this and added the important observation that on standing one year the uranium regained it's radioactivity and UX lost it. It was further shown [63] that it was only the β activity of uranium that was separated. The sum of the uranium and UX radioactivity was always constant and the half-life for UX decay was 22 days [64]. The whole α radiation of uranium accounted for the transition of uranium to UX, and the whole of the β radiation accounted for the subsequent transitions of UX [65]. Becquerel [66] showed that UX could be separated from boiling solutions of uranium by adding lampblack.

An interesting development was the discovery of a small amount of feebly penetrating radiation with UX [67]. This was thought to be α radiation, and could not be due to the original uranium. The separation procedure was to dissolve uranyl nitrate in various solvents, acetone, alcohols, methyl and ethyl acetate etc., stir in freshly precipitated iron hydroxide and filter. The filtrate contained the uranium free from UX. It was thought [68] that UX decayed to ionium and then to radium, and the α activity was due to the growth of ionium, however, measurements failed to confirm this [69].

Radium was found to grow in uranyl nitrate solutions proportional to the time squared [70]. This means there is only one long-lived (on the timescale of the experiments, 4 years) intermediate, and that is ionium. It had been thought that thorium, and not uranium, had been the parent of radium [70]. As mentioned earlier, Boltwood [27] suggested the name ionium and noted

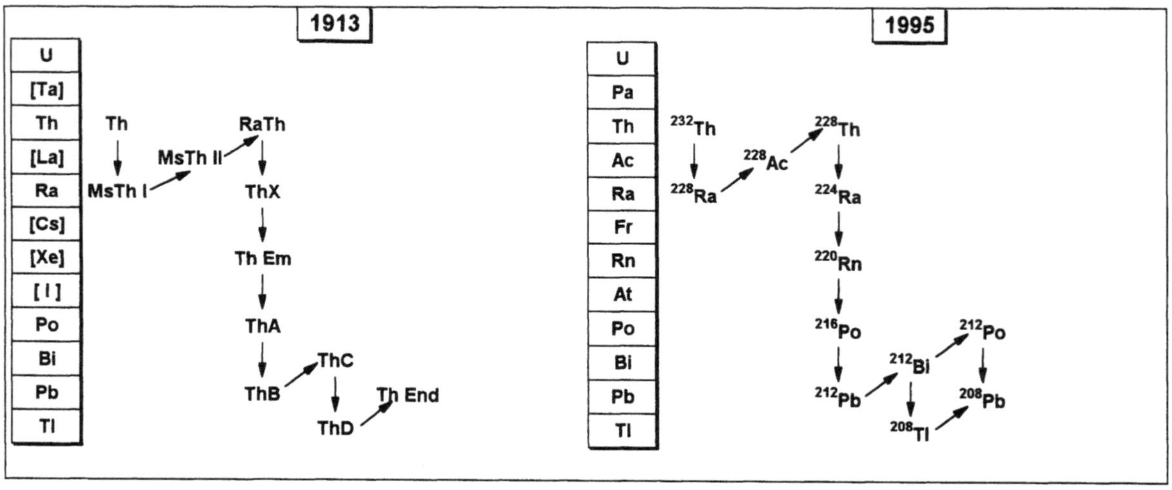

Fig. 2. The 4n radioactive family in 1913 and 1995.

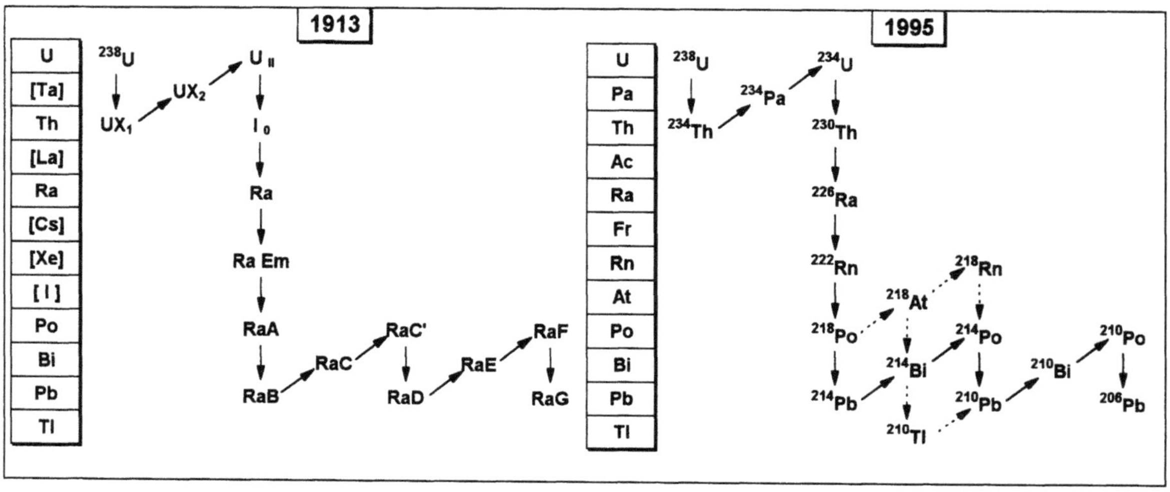

Fig. 3. The 4n + 2 radioactive family in 1913 and 1995. Dotted arrows indicate minor branchings.

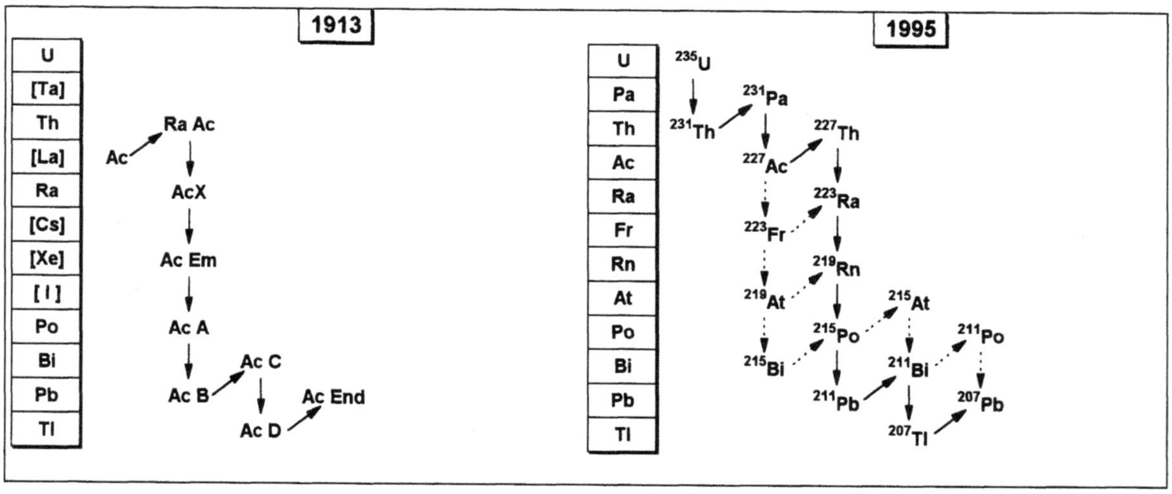

Fig. 4. The 4n + 3 radioactive family in 1913 and 1995. Dotted arrows indiate minor branchings.

that it occupies an intermediate position in the uranium to radium series. Radium and Radium F (Polonium, formerly designated radium E) was discovered by the Curies [71, 72] and is the subject of the preceding article.

Rutherford [73] proposed a convenient nomenclature after studying the decay of radium emanation. He found that after the complete rapid decay of the active deposit, there were two stages, one emitting β particles, the second α only. He suggested a series:

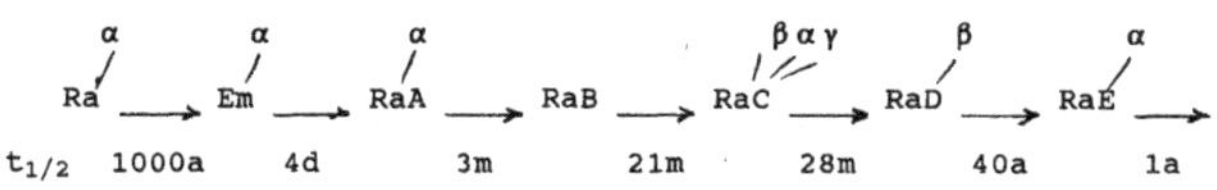

It was probable that the "radio-lead" separated from pitchblende owed it's radioactivity to radium D [74, 75]. RaF corresponded well with the chemistry of polonium. Electrochemical separations [76] of RaB and RaC showed that RaC is "nobler" than RaB. The decay curves in the above series could be explained better [77–79] if it was assumed that RaB to RaC was not a rayless transition but involved very weak β rays, also that RaC should have a shorter half-life than RaB.

Recoil experiments with pure RaC gave new rapidly decaying recoil products [80], RaC_1 and RaC_2, with half-lives from 1 to 2.4 minutes. Very feeble recoil effects indicated that RaC_1 gives β and not α rays. RaC_2 emitted α particles with an exponential decay and had a half-life of about two minutes.

The possibility of branch series in addition to the main path had been considered, but the first experimental evidence of the formation of a branch series came with the later work [81] using much larger quantities of radium. They showed that the half-life of RaC_2 was 1.38 minutes and that it decayed by β and not α rays. Further, the decay products of RaC_2, which were unknown, were not the α emitters either. Therefore, RaC_1 must be the α emitter, and RaC_2 cannot be the product of the α decay of RaC_1. By identifying RaE and RaF as decay products of RaD, it was found that about 25% of the total RaD was collected by recoil onto a plate, hence RaD is the product of the RaC_1 α decay [82]. Using the Geiger-Nuttall rule, it was assumed that the α rays of RaC_1 did not come directly from RaC_1 because the half-life of RaC_1 is 19.5 minutes, but from the range the half-life of the α decay should be 10^{-6} seconds. A complex sequence, based on a similarity with the thorium series [83], was put forward which involved two routes for the decay of RaC: one via RaC_1 to RaD and one via RaC_2 to an unknown product.

The post RaC species were better understood, and it was shown [84] that the polonium examined from four sources:

a) residual activity from radium (RaF)
b) radiolead (RaD, RaE and RaF)
c) "radiotellurium" (now called polonium)
d) radioactive bismuth

all had the same decay rate with a mean half-life of 137 days. Marckwald [85] had shown that "radiotellurium" or polonium was chemically similar to the S–Se–Te family, and Rutherford showed that RaF and "radiotellurium" had the same half-life of 143 days. Marckwald's measurement of the half-life of RaF as 139 days was remarkably close to the currently accepted value of 138.4 days for ^{210}Po (RaF).

RaG was considered to be the end of the uranium series and with an atomic weight between 206 and 207 was thought to be either lead or bismuth.

Before the branching of decay series was understood "new" members of the series were often postulated. For example, RaH was thought to be the result of RaG undergoing β decay to a bismuth isotope which then underwent α decay to a thallium isotope. This was put forward to explain the presence of bismuth and thallium in pitchblende. From the Geiger-Nuttall rule it was estimated that the half-life of the α emitter was between 10^5 and 10^6 years. It is now known that ^{210}Bi has an α branch with a half-life 3×10^6 years to ^{206}Tl.

Conclusion

There were many detailed side-tracks in the unraveling of the uranium series, and the author apologises for any serious omissions in this account.

The thorium and actinium series were developed in a similar way and over the same period with many of the same scientists contributing to their understanding.

By 1913 the radioelements had been placed in the Periodic Table and these are shown in Figures 2–4 together with the 1995 equivalents for comparison.

The 1913 series were remarkably accurate although the concept of lanthanides and actinides was unknown, and Ta, Cs and I were used instead of Pa, Fr and At. (Note: ^{234m}Pa was discovered in 1913 by Fajans and Göhring [86], and it was called brevium, the name protoactinium (shortened to protactinium in 1949) did not appear until 1918 (Hahn and Meitner [87]); Fr was discovered in 1939 (Perey [88]); At was synthesized in 1940 (Corson, McKenzie and Segré [89]).

The actinium series was less well developed, and it was only with the discovery of actino-uranium (^{235}U) by Dempster [90] in 1935 that it was clear that there were two independent uranium families.

The natural decay series became known as thorium (4n), uranium (4n+2) and actinium (4n+3) series, where each member of the family had a mass number divisible by 4 and a remainder shown by the +x in the brackets. Following the work on the Manhattan project during the Second World War and shortly after, the missing (4n+1) neptunium series was reported [91, 92].

References

1. Becquerel, H.: "Sur les radiations invisibles émises par les corps phosphorescence", C. R. Acad. Sci. Paris **122**, 501 (1896).
2. Rutherford, E.: "Uranium Radiation", Philos. Mag. **47**, 109 (1899).
3. Villard, P.: "Sur la réflexion et la réfraction des rayons cathodiques et des rayons déviables du radium", C. R. Acad. Sci. Paris **130**, 1010 (1900), "Sur le rayonnement du radium", ibid. p. 1178.
4. Rutherford, E., Soddy, F.: "Radioactive Change", Philos. Mag. **5**, 579 (1903).
5. Ramsay, W., Soddy, F.: "Experiments in radioactivity, and the production of helium from radium", Proc. Roy. Soc. **72** (A), 204 (1903).

6. Himstedt, K., Meyer, G.: "Über die Bildung von Helium aus der Radium Emanation", Ann. Phys. **15**, 184 (1904).
7. Rutherford, E., Royds, T.: "The Nature of the Alpha Particle from Radioactive Substances", Philos. Mag. **17**, 281 (1909).
8. Soddy, F.: Ann. Rep. Prog. Chem. **6**, 232 (1910).
9. Thompson, J. J.: *Conduction of Electricity Through Gases*, Cambridge University Press, 1906.
10. Strutt, R. J.: "An Experiment to Exhibit the Loss of Negative Electricity by Radium", Philos. Mag. **6**, 588 (1903).
11. Makower, W., Geiger, H.: *Practical Measurements in Radioactivity*. Longmans, Green and Co., London 1912.
12. Crookes, W.: "The emanations of radium", Proc. Roy. Soc. **71**, 405 (1903).
13. Elster, J., Geitel, H.: Phys. Z. **15**, 437 (1903).
14. Wilson, C. T. R.: "A Method of Making Visible the Paths of Ionising Particles Through a Gas", Proc. Roy. Soc. **85**, 285 (1911).
15. International Standards Committee, Le Radium **7**, 65 (1910).
16. Curie, S.: "Mesure de la constante de l'émanation du radium", Le Radium **7**, 32 (1910).
17. Rutherford, E., Geiger, H.: "An Electrical Method of Counting the Number of Alpha Particles from Radio-active Substances", Proc. Roy. Soc. **81A**, 141 (1908).
18. Rutherford, E.: "Radium Constants on the International Standard", Philos. Mag. **28**, 320 (1914).
19. Soddy, F.: Ann. Rep. Prog. Chem. **17**, 235 (1921).
20. Rutherford, E., Soddy, F.: "The Radioactivity of Thorium Compounds II. The Cause and Nature of Radioactivity", J. Chem. Soc. **18**, 837, 860 (1902).
21. Rutherford, E., Soddy, F.: "A Comparative Study of the Radioactivity of Radium and Thorium", Philos. Mag. **5**, 445 (1903).
22. Rutherford, E.: "The Velocity and Energy of the Alpha Particles from Radioactive Substances", Philos. Mag. **13**, 110 (1907).
23. Geiger, H., Nuttall, J. M.: "The ranges of the Alpha Particles from the Thorium and Actinium Products", Philos. Mag. **24**, 647 (1912).
24. Zerban, F.: "Zur Frage nach der Radioaktivität des Thoriums", Ber. Dtsch. Chem. Ges. **38**, 557 (1905).
25. Elster, J., Geitel, H.: "Radiolead", Philos. Mag. **9**, 550 (1905).
26. Boltwood, B. B.: "On Ionium, a New Radioactive Element", Am. J. Sci. **25**, 365 (1908).
27. Boltwood, B. B.: "A New Radioactive Element", Nature **76**, 544, 589 (1907).
28. Stromholm, D., Svedberg, T.: "Untersuchungen über die Chemie der Radioaktiven Grundstoffe", Z. Anorg. Chem. **61**, 338–346 (1909).
29. Stromholm, D., Svedberg, T.: "Untersokningar ofver de Radioaktiva Grundamnenas Kemi", Z. Anorg. Chem. **63**, 197–206 (1909).
30. McCoy, M. N., Ross, W. H.: "The Specific Radioactivity of Thorium and the Variation with Chemical Treatment and with Time", J. Am. Chem. Soc. **29**, 1709–1718 (1917).
31. Marckwald, W.: "Mesothorium", Ber. Dtsch. Chem. Ges. **43**, 3402–3422 (1910).
32. Soddy, F.: "The Chemistry of Mesothorium", J. Chem. Soc. **99**, 72 (1911).
33. Herschfinkel, H.: "Sur le radioplomb", Radium **7**, 198–200 (1910).
34. Hahn, O., Meitner, L.: Phys. Z. **11**, 493 (1911).
35. Paneth, F., von Hevesy, G.: "Attempts to Separate RaD from Lead", Monatsh. Chem. **34**, 1393–1400 (1913).
36. Paneth, F., von Hevesy, G.: "Radio-elements as Indicators in Analytical Chemistry", Monatsh. Chem. **34**, 1401–1407 (1913).
37. Paneth, F., von Hevesy, G.: "The Electrochemical Identity of Radioactive Elements", Monatsh. Chem. **34**, 1593 (1913).
38. Fleck, A.: "The Chemical Nature of UX, Radioactinium and ThB", Chem. News **106**, 128 (1912).
39. Metzener, W.: "The Chemical Properties of ThC and ThD", Ber. Dtsch. Chem. Ges. **46**, 979–986 (1913).
40. Soddy, F.: *The Chemistry of the Radio-elements. Part II. The Radio-elements and the Periodic Law*. Longmans, Green and Co, London 1914.
41. Fleck, A.: "The Chemical Nature of Some Radioactive Disintegration Products", J. Chem. Soc. **103**, 381, 1052 (1913).
42. Russell, A. S.: "The Periodic system and the Radioelements", Chem. News **107**, 49–52 (1913).
43. Soddy, F.: *Chemistry of the Radio-elements*, Longmans, Green and Co. 1911, end of volume.
44. Soddy, F.: *Chemistry of the Radio-elements*, Longmans, Green and Co. 1911, p. 29.
45. Soddy, F.: *The Chemistry of the Radioelements. Part II. The Radio-elements and the Periodic Law*. Longmans, Green and Co, London 1914, p. 2.
46. Geiger, H., Marsden, E.: "On a Diffuse Reflection of the Alpha Particles", Proc. Roy. Soc. A **82**, 495–500 (1909).
47. "Rutherford by Those Who Knew Him", Physical Society, London (The first five Rutherford Lectures).
48. Rutherford, E.: "The Scattering of Alpha and Beta Particles by Matter and the Structure of the Atom", Philos. Mag. **21**, 669 (1911).
49. Geiger, H., Marsden, E.: "The Law of Deflection of Alpha Particles Through Large Angles", Philos. Mag. **25**, 604–623 (1913).
50. Rutherford, E.: "Radioactive Substance Emitted from Thorium Compounds", Philos. Mag. **49**, 1–14 (1900).
51. Debierne, A.: "Sur une nouvelle matière radio-active". C. R. Acad. Sci. Paris **129**, 593 (1899).
52. Debierne, A.: "Sur un nouvel élément radio-actif, l'actinium", C. R. Acad. Sci. Paris **130**, 906–8 (1900).
53. Debierne, A.: "Sur la radioactivité induite provoquée par les sels d'actinium", C. R. Acad. Sci. Paris **136**, 446–449 (1903).
54. Giesel, F.: "Über Radium und radio-aktive Stoffe", Ber. Dtsch. Chem. Ges. **35**, 3608 (1902).
55. Giesel, F.: "Über den Emanationskörper aus Pechblende und über Radium", Ber. Dtsch. Chem. Ges. **36**, 342–347 (1903).
56. Giesel, F.: "Über den Emanationskörper (Emanium)", and "Über Emanium", Ber. Dtsch. Chem. Ges. **37**, 1696–1699, 3963–3966 (1904).
57. Rutherford, E., Soddy, F.: "The Radioactivity of Thorium Compounds I. An Investigation of the Radioactive Emanation", Proc. Roy. Soc. **18**, 120 (1902).
58. Rutherford, E., Soddy, F.: "The Radioactivity of Thorium Compounds II. The Cause and Nature of Radioactivity", J. Chem. Soc. **81**, 321–350 (1902).
59. Rutherford, E., Soddy, F.: "Condensation of the Radioactive Emanations", Philos. Mag. **5**, 561 (1903).
60. Crookes, W.: "Radio-activity of Uranium", Proc. Roy. Soc. **66** (A), 409 (1900).
61. Becquerel, H.: "Sur le rayonnement de l'uranium", C. R. Acad. Sci. Paris **131**, 137–138 (1900).
62. Becquerel, H.: "Sur la radioactivité de l'uranium", C. R. Acad. Sci. Paris **133**, 977–980 (1901).
63. Soddy, F.: "The radioactivity of Uranium", J. Chem. Soc. **81**, 860–865 (1902).
64. Rutherford, E., Soddy, F.: "A Comparative Study of the Radioactivity of Radium and Thorium", Philos. Mag. **5**, 445–447 (1903).
65. Soddy, F.: Ann. Rep. Prog. Chem. **1**, 266 (1905).
66. Becquerel, H.: "Sur quelques propriétés des rayons α du radium", C. R. Acad. Sci. Paris **141**, 485–490 (1905).
67. Moore, R. B., Schlundt, H.: "Some New Methods for Separating Uranium X from Uranium", Philos. Mag. **12**, 393–396 (1906).
68. Soddy, F.: Ann. Rep. Prog. Chem. **6**, 261–262 (1910).
69. Soddy, F.: "The Rays and Product of UX", Philos. Mag. **18**, 858–865 (1909).
70. Hahn, O.: "Über die Muttersubstanz des Radium", Ber. Dtsch. Chem. Ges. **40**, 4415–4420 (1907).

71. Curie, P., Curie, S.: "Sur une substance nouvelle radio-active contenue dans la pechblende", C. R. Acad. Sci. Paris **127**, 175–178 (1898).
72. Curie, P., Curie, S.: "Sur une substance nouvelle fortement radio-active contenue dans la pechblende", C. R. Acad. Sci. Paris **127**, 1215–1217 (1898).
73. Rutherford, E.: "Slow Transformation Products of Uranium", Philos. Mag. **8**, 636 (1904).
74. Hofmann, K. A., Strauss, E.: "Über das radioactive Blei", Ber. Dtsch. Chem. Ges. **34**, 3033–3039 (1901).
75. Giesel, F.: "Radio-active Material", Chem. News **83**, 122–123 (1901).
76. von Lerch, F.: "Trennungen des RaG vom RaB", Ann. Phys. **20**, 345–354 (1906).
77. Schmidt, H. W.: "Über den Zerfall von RaA, RaB und RaC, 1. Mitt.", Phys. Z. **6**, 897–903 (1905).
78. Schmidt, H. W.: "Über die Absorption der Beta-Strahlen des Radiums, 1. Mitt.", Phys. Z. **7**, 764–766 (1906).
79. Schmidt, H. W.: "Über den Zerfall von Radium A, B und C, 2. Mitt.", Ann. Phys. **21**, 609–664 (1906).
80. Hahn, O., Meitner, L.: "Proof of the Complex Nature of Radium C", Phys. Z. **10**, 697–703 (1909).
81. Fajans, K.: "The Complex Nature of Radium C", Phys. Z. **12**, 369–378 (1911).
82. Soddy, F.: Ann. Rep. Prog. Chem. **8**, 288 (1912).
83. Fajans, K.: "Die Verzweigung der Radium Zerfallsreihe", Phys. Z. **13**, 699–705 (1912).
84. Meyers, J., von Schweidler, E. R.: "Mitteilung der Frau Sklodowska Curie. Über die Zeitconstante des Poloniums", Phys. Z. **7**, 257–258 (1906).
85. Marckwald, W.: "Über das Radiotellur. IV", Ber. Dtsch. Chem. Ges. **38**, 591–594 (1905).
86. Fajans, K., Göhring: "Über das Uran X2 – das neue Element der Uranreihe", Phys. Z. **14**, 240–259 (1913).
87. Hahn, O., Meitner, L.: "Parent of Actinium. New Radioactive Element of Long Life", Phys. Z. **19**, 208–218 (1918).
88. Perey, M.: "Sur un élément 87, dérivé de l'actinium", C. R. Acad. Sci. Paris **208**, 97 (1939).
89. Corson, D. R., McKenzie, K. R., Segre, E.: "Possible Production of Radioactive Isotope of Element 85", Phys. Rev. **57**, 459 (1940).
90. Dempster, A. J.: "Isotopic Constitution of Uranium", Nature **136**, 180 (1935).
91. Hageman, F., Katzin, L. I., Studier, M. H., Ghiorso, A., Seaborg, G. T.: "The (4n+1) Radioactive Series: Products of U^{233}", Phys. Rev. **72**, 253 (1947).
92. English, A. C., Cranshaw, T. E., Demers, P., Harvey, J. A., Hincks, E. P., Jelley, J. V., May, A. N.: "The (4n+1) Radioactive Series", Phys. Rev. **72**, 253 (1947).

Radiochimica Acta 70/71, 39–49 (1995)

On the Discovery of Artificial Radioactivity

By Robert Guillaumont and Didier Trubert
Groupe de Radiochimie, Institut de Physique Nucléaire-IN2P3-CNRS, BP 1, 91406 Orsay, France

(Received June 8, 1995)

History / I. Curie / F. Joliot / Discovery / Artificial radioactivity

Abstract

A review is presented of the elegant Joliot-Curie physical experiments which showed for the first time the possibility of producing artificial radioactive nuclei. An attempt has been made to explain the results of their equally elegant chemical experiments that unequivocally demonstrated the chemical change of elements by nuclear processes. These experiments are discussed in the light of present-day radiochemistry, keeping in mind the scientific and historical context in which they were originally performed.

Introduction

The discovery of artificial radioactivity in 1934 by Frédéric Joliot and Irène Curie was a major breakthrough in the field of nuclear sciences because for the first time a radioactive nuclide was synthesized and identified chemically on the scale of a few atoms (Figure 1). At the same time, the topic of "radioactivity" that had emerged 38 years before expanded to the point of dividing into nuclear physics and into the more chemically oriented domains of nuclear chemistry and radiochemistry.

1. The discovery of artificial radioactivity

In contrast to the discovery of natural radioactivity, which resulted from a fortuitous observation and was completed within a few weeks, that of artificial radioactivity is a long story which evolved progressively and logically during a period extending from the late twenties to the mid-thirties. Within this time, two decisive events occurred in the year 1932. J. Chadwick discovered the neutron, which established definitively the atomic model of a nucleus with protons and neutrons, rather than a nucleus formed by protons or α particles, surrounded with electrons [1]. Indeed, at that time, α particles appeared to be particularly stable. For his own part, C. D. Anderson discovered in cosmic rays the positron, the antiparticle of the electron, whose existence had been predicted by P. Dirac [2].

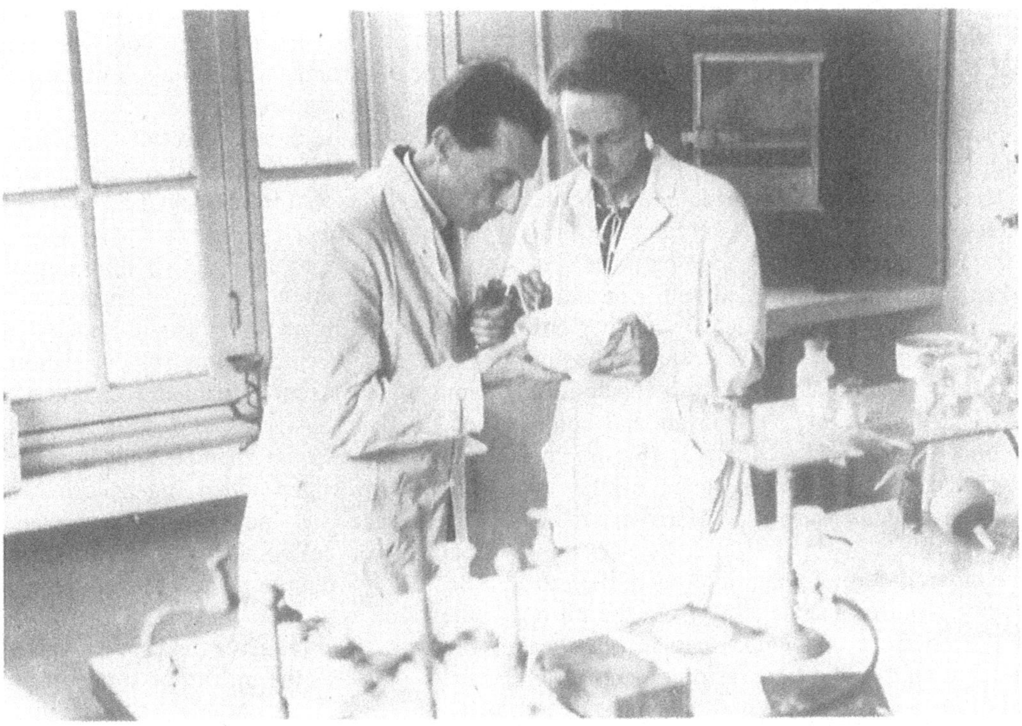

Fig. 1. Frédéric Joliot (1900–1958) and Irène Curie (1897–1956).

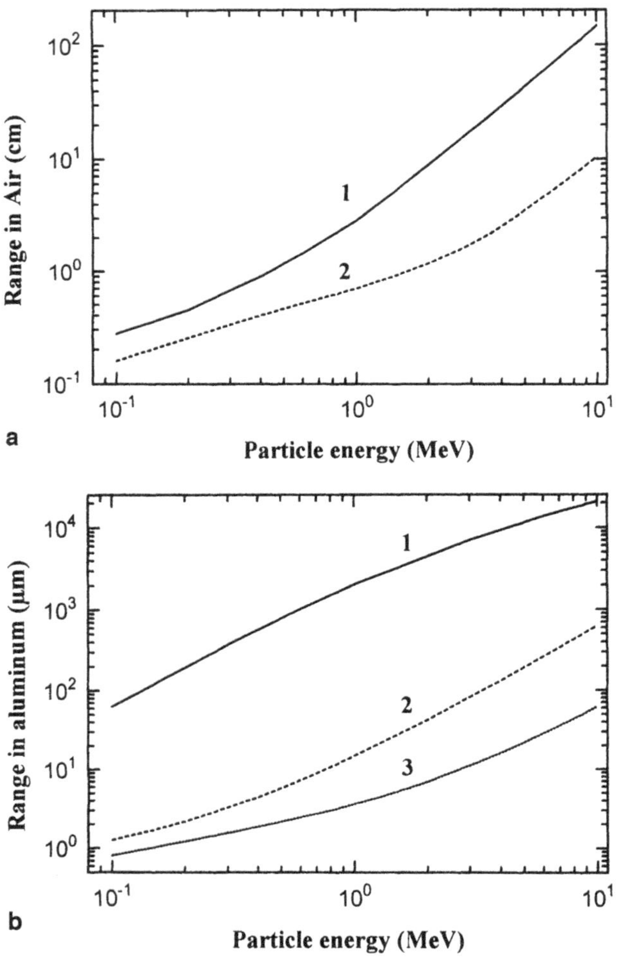

Fig. 2. a. Range of protons (curve 1) and α particles (curve 2) in air. b. Range of protons (curve 1), α particles (curve 2) and electrons (curve 3) in aluminum [3, 42].

a. Towards the discovery of the neutron

As early as 1914, Marsden had noticed the emission of long-range protons when H_2 was bombarded with α rays. The experiments were conducted with the 7.7 MeV α particles (6.5 cm range in air) of a source of RaC/RaC′ [^{214}Bi/^{214}Po] (Figure 2a). From the kinematics of elastic scattering, the mean energy of the ejected protons is about 2 MeV and the range in air is 9 cm. Marsden also observed long range particles similar to recoil protons in air bombarded with α particles. This unexpected effect was elucidated by Rutherford, who repeated the experiment with N_2 as a target using a ZnS detector. He found that the rare protons emitted (1 for 10^6 α) had a range of 1.8 m in air and corresponded to an energy of 11.4 MeV [3] (Figure 2b). This greatly exceeded the 6.5 cm range of the incoming α particles and also surpassed Marsden's recoil protons from H_2.

Since the target did not contain hydrogen, the protons necessarily acquired their energy from the nitrogen "atom". Rutherford concluded that the nitrogen "disintegrates" owing to the intense forces exerted in the collision of the α particle, and that the emitted proton was a constituent of the nitrogen nucleus [4]. Actually, as is now well known, Rutherford had realized the first transmutation, and what he considered as a "disintegration" was the first nuclear reaction, presently written ^{14}N(α,p)^{17}O (Table 1).

At the time, this discovery was considered to have a limited scope. Four years later, using a Wilson chamber, P. M. S. Blackett visualized the long ranges of the α particles and protons, together with the short tracks of eight ^{17}O recoil atoms among 400000 trajectories recorded on the image. Since ^{17}O is stable, its formation could only be inferred from balistics. Assuming a one mbarn cross section for the reaction, a source strength of 1 mCi RaC/RaC′ in 2 π geometry would yield a few hundred ^{17}O atoms per min in a N_2 target at 1 atm pressure.

An anecdote relates that, at a meeting of the Manchester Literary and Philosophical Society in May 1919, Rutherford made the first public announcement of the breakdown of the nitrogen nucleus by an α particle. Asked what happens to the other nitrogen atom of the molecule when its partner is transmuted, Rutherford replied "Oh well, in terms of the size of the atomic nuclei you have to imagine the other nitrogen atom miles and miles away". This story has been reported by Wolfenden [5] who considered this occasion as the "dawn of hot atom chemistry" because it was probably the first time that the chemical consequence of a nuclear transmutation had been evoked.

In 1928, W. Bothe and H. Becker commenced work in the aim of confirming the disintegration theory of Rutherford and attempted to ascertain whether the reaction was accompanied by the emission of γ rays. They bombarded Be and B with α particles and measured with a Geiger-Müller counter the absorption of the radiation in Pb in order to determine the energy. In 1930, these authors reported the emission of a hard radiation of about 14 MeV, which was much more energetic than the α rays and concluded that the origin of the electromagnetic radiation was the disintegration of the target nucleus [6].

During the months of December 1931 and January 1932, the Joliots repeated the experiment of Bothe and Becker with a strong 100 mCi Po source, using an ionization chamber coupled to a very sensitive Hoffman electrometer provided with a thin aluminum entrance window of a few mg cm^{-2} (apparently 5 μm). They indeed confirmed the Bothe-Becker radiation and made a further, extremely important observation. The ionization current remained constant when thin foils of C, Al, Cu, Ag and Pt were interposed between the Po source and the chamber, but it increased strongly when a hydrogen containing material was employed. The effect was most pronounced when a sheet of paraffin or cellophane was used. Thus, the Bothe-Becker "γ" radiation was able to knock out protons in the same way as α particles that collided with light elements. The range of the protons was about 200 μm in Al, whereas the maximum range of protons ejected by the 5 MeV α particle of Po is only 22 μm [3] (Figure 2b). The value found by the authors may be

Table 1. Nuclides and radionuclides formed in (α,n) and (α,p) reactions on some targets [42–43].

	(α,n) Reactions			(α,p) Reactions		
Target isotopic abundance %	Threshold (MeV)	Nuclide (decay)	Half-life	Threshold (MeV)	Nuclide (decay)	Half-life
^{7}Li 92.5	4.40	^{10}B	stable	4.04	^{10}Be (β^-)	1.6x10^{6} y
^{9}Be 100	-	^{12}C	stable	9.95	^{12}B (β^-)	20 ms
^{10}B 19.9	-	^{13}N (β^+)	9.9min	-	^{13}C	stable
^{11}B 80.1	-	^{14}N	stable	-	^{14}C (β^-)	5730 y
^{12}C 98.9	11.35	^{15}O (β^+, EC)	2.04 min	6.63	^{15}N	stable
^{14}N 99.6	6.10	^{17}F (β^+)	1.08 min	1.54	^{17}O	stable
^{16}O 99.8	15.19	^{19}Ne (β^+, EC)	17.2 s	10.15	^{19}F	stable
^{19}F 100	2.36	^{22}Na (β^+)	2.6 y	-	^{22}Ne	stable
^{20}Ne 90.5	8.69	^{23}Mg (β^+)	11.3 s	2.83	^{23}Na	stable
^{22}Ne 9.25	0.60	^{25}Mg	stable	4.17	^{25}Na (β^-)	stable
^{23}Na 100	3.48	^{26}Al (β^+) ^{26m}Al (β^+, EC)	7.3x10^{5} y 6 s	-	^{26}Mg	stable
^{24}Mg 79	8.39	^{27}Si (β^+)	4.14 s	1.87	^{27}Al	stable
^{25}Mg 10	-	^{28}Si	stable	1.40	^{28}Al (β^-)	2.25 min
^{26}Mg 11	0.02	^{29}Si	stable	3.36	^{29}Al (β^-)	6.5 min
^{27}Al 100	3.05	^{30}P (β^+,EC)	2.5 min	-	^{30}Si	stable
^{28}Si 92.2	9.26	^{31}S (β^+)	2.6 s	2.2	^{31}P	stable
^{31}P 100	6.39	^{34}Cl (β^+) ^{34m}Cl (β^+)	1.5 s 32.2 min	-	^{34}S	stable
^{32}S 95	9.71	^{35}Ar (β^+)	1.8 s	2.10	^{35}Cl	stable
^{35}Cl 75.8	6,54	^{38}K (β^+) ^{38m}K (β^+)	7.7 min 0.9s	-	^{38}Ar	stable
^{37}Cl 24.2	4.31	^{40}K (β^-)	1.3x10^{9} y	1.77	^{40}Ar	stable
^{40}Ar 99.6	2.52	^{43}Ca	stable	3.66	^{43}K (β^-)	22.3 h
^{39}K 93.25	7.90	^{42}Sc (β^+) ^{42m}Sc (β^+)	0.6 s 1.01min	0.14	^{42}Ca	stable
^{41}K 6.7	3.72	^{44}Sc (β^+, EC) ^{44m}Sc (IT, EC)	3.9 h 2.4 d	-	^{44}Ca	stable
^{40}Ca 96.9	12.27	^{43}Ti (β^+)	0.5 s	3.90	^{43}Sc (β^+)	3.9 h
^{58}Ni 68.3	9.93	^{61}Zn (β^+)	1.5 min	3.31	^{61}Cu (β^+)	3.35 min
^{60}Ni 26.1	8.43	^{63}Zn (β^+)	38.5 min	4.01	^{63}Cu	stable
^{107}Ag 51.8	8.03	^{110}In (β^+, EC) ^{110m}In (β^+)	4.9 h 1.15 h	3.11	^{110}Cd	stable
^{109}Ag 48.2	6.76	^{112}In (β^+, EC) ^{112m}In (IT)	14.4 min 20.8 min	3.24	^{112}Cd	stable

compared to the 10 MeV and 700 μm range of protons ejected by the neutron from the ^{9}Be(α,n) reaction [7].

The Joliots reported this observation in the January 18, 1932, issue of the "Comptes Rendus de l'Académie des Sciences" [8]. It fortunately happened that the Al window of the ionization chamber was thin enough to permit the recoil protons to traverse the foil. A thicker Al window, as verified by the Joliots, would have stopped the particles. This observation was not in contradiction with the postulated electromagnetic nature of the Bothe-Becker rays, but it implied an inconceivably high γ-ray energy of at least 20 MeV.

Soon afterwards, the Joliots repeated the experiment using a Wilson chamber. The prototype of this first imaging detector had been invented by C. T. R. Wilson in 1911. The tracks of electrons, positrons and recoil atoms could be fixed instantaneously on a photographic plate and the energy of the particles determined from the curvature of the trajectories in a magnetic field. The Wilson chamber was Joliot's favourite detector and, in later work, he constructed his own low-pressure chamber.

The Joliots were able to visualize the recoil protons, as well as He, Be and N atoms with energies that were inversely proportional to the masses, and that were expelled by the mysterious radiation. This was reported in the Comptes Rendus dated February 22, 1932 [9] and confirmed two weeks later [10].

In the meantime, J. Chadwick in Cambridge had undertaken similar experiments with a more sophisticated device consisting of an ionization chamber connected to a proportional amplifier and an oscilloscope. With this equipment the radiations could be counted and the energy determined simultaneously. This device was not available before 1933 at the Institut du Radium, a fact that was highly deplored by F. Joliot.

Chadwick's findings were similar to those of the Joliots, and besides he allowed the radiation to interact not only with H_2, but also with He and N_2. From the energies of the recoiling atoms he concluded that the

Bothe-Becker radiation was a massive neutral particle, the *neutron*. Obviously, the neutron originated from the $^{9}Be(\alpha,n)^{12}C$ nuclear reaction. The discovery of the neutron was reported in a letter to Nature on February 17 [1]. The Joliots narrowly missed this major discovery. The following day, the Physical Review acknowledged a paper from H. Urey announcing the discovery of the deuteron.

After the announcement of Chadwick's discovery, the Joliots pursued their experiments on the irradiation of light elements with α particles [11] and brought further experimental proofs of the existence of the neutron [12].

b. The positron puzzle

C. D. Anderson's discovery of the positive electron (positron) resulted from the observation of cosmic ray tracks in a Wilson chamber with an applied magnetic field; a clue to the existence of the positron was, among others, the direction of curvature of the particle trajectories. However, before Anderson, the Joliots had noticed a similar odd behaviour when the radiation from a Po-Be source entered their Wilson chamber at low pressure. They thought that the phenomenon was due to electrons bouncing towards the source rather than to positrons leaving it [12].

After the announcement of the discovery of the positron, the Joliots immediately realized that their source was emitting positrons [13] (again they had missed an important discovery!) and this was confirmed in a subsequent experiment. The radiation from the Po-Be source, after traversing a foil of Pb, produced simultaneously positive and negative electrons, but not with an Al foil [14]. The same effect occurred with the radiation of ThC″ (^{208}Tl). Thus the creation of positrons was not due to the neutron. What the Joliots had discovered, was the materialization of energetic γ rays in an electron-positron pair.

Two months later, on June 19, 1933, the Joliots made a further important step which in fact opened the way to the discovery of artificial radioactivity [1]. They found that the mere covering of a 20 mCi Po source with a 50 μm Al foil in which the α's are fully stopped (or B but not *Ag, nor paraffin or Li*) produced few protons, but a large number of positrons which were not correlated to an electron as in the materialization process. Furthermore, the positrons were not monokinetic (most of them had an energy below 1.6 MeV, others had 2 to 3 MeV) and were emitted isotropically. This experiment was achieved with the low pressure Wilson chamber constructed two years before by F. Joliot [16]. The trajectories of the particles showed a 75-fold increase in length with respect to those in a normal chamber. The Al-covered Po source was located close to the entrance window (a 100 μm Al foil) of the chamber. Figure 3, drawn from data given in [17] represents the statistics of the number of emitted electrons and positrons as a function of energy.

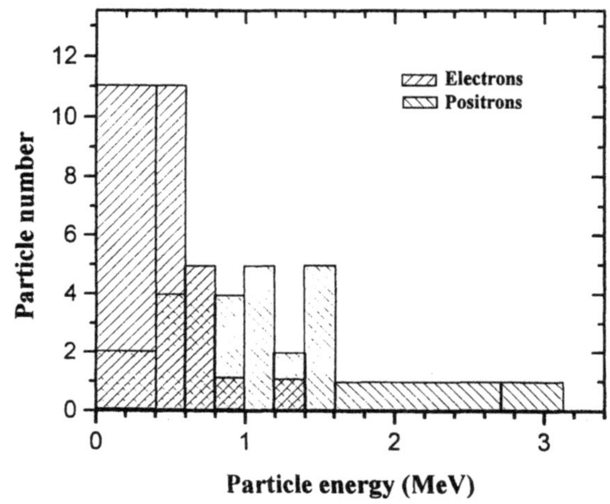

Fig. 3. Histogram of the energy distribution of positrons and electrons emitted in the bombardment of Al with α particles of Po. (After I. Curie and F. Joliot [17].) In 230 experiments L. Meitner observed 11 recoil protons (displaced by neutrons), 170 positrons and 1000 electrons. Blackett reported 10% electrons and 90% positrons [20].

Since Rutherford's first transmutation reaction, it was known that light elements bombarded with α particles emit "transmutation protons" (in fact the (α,p) reaction; formerly the proton emission had also been called "H rays"). Because very few protons were observed, the authors suggested that in some cases the transmutation releases a neutron and a positive electron rather than a proton, i.e. $^{27}Al(\alpha,p)^{30}Si$ could be replaced by $^{27}Al(\alpha, n\, e^{+})^{30}Si$. The Joliots thought for a short time how the formation of an excited $^{30}Si^{*}$ state could possibly occur and decay by neutron-positron emission. Nevertheless, these assumptions supported the hypothesis that the proton was constituted by a neutron and a positive electron. To conclude, the authors noted that the new phenomenon was the most effective means of producing positrons with a yield, in Al, of one positron for 2×10^{6} incident α particles.

It has been reported that the experiment had actually been planned to search for a weak β^{-} emission of ^{210}Po which had been suspected a few years before, and would have led to an isotope of the (at the time) unknown element 85 (At). For this purpose, the Po source had been deliberately covered with an Al foil to absorb all α particles [18].

The Solvay Congress held in Brussels in October 1933 and attended by the most renowned scientists in the field of radioactivity, was the occasion for the Joliots to review their findings, report their latest experiments and confront their data with those obtained by other research workers [19–20]. Since the June publication, the Joliots had identified the origin of the electrons emitted by ^{210}Po as the internal conversion of the 0.8 MeV γ radiation. The present-day tables indicate that this process has a probability of 1.07×10^{-3}% with $\alpha_K = 8.16 \times 10^{-3}$ and $\alpha_{total} = 1.05 \times 10^{-2}$ [21]. Thus a 60 mCi source would emit in 4 π geometry 30 conversion electrons per s. The electrons recorded by the Joliots were probably ejected along the tracks of

α particles in Al. More importantly, the authors had clearly demonstrated the emission of neutrons from Al bombarded with the α particles of Po by recording 0.7 MeV recoil protons. Already in February 1933, they had found that the neutron emission required at least 4 to 5 MeV α particles [22–23].

Other participants at the congress, in particular L. Meitner and P. M. S. Blackett confirmed part of the results presented by the Joliots. L. Meitner stated that she did not observe any neutron emission, but two months later she recognized that she had been wrong and reported even better statistics on electrons and positrons than those of the Joliots, as well as data on recoil protons [20].

The highlight of the discussions was Joliot's hypothesis of a $^{27}Al(\alpha, ne^+)^{30}Si$ reaction, which was received with some scepticism. Among the comments of E. Rutherford, W. Bothe, W. Heisenberg, W. Pauli and others, those of F. Perrin were the most pertinent. The latter suggested that the neutron and the positron were perhaps not emitted simultaneously, but in succession: first a neutron, followed by the formation of a transient unstable nucleus that might decay with emission of a positron. This picture would be consistent with a continuous positron spectrum, similar to that known for β^- radioactivity [20]. The previous month, in Leningrad, F. Perrin had suggested that this process could result from the transformation, inside the nucleus, of a neutron into a proton and an electron; conversely a proton could transform into a neutron and a positron [24].

The ideas were not quite clear and upon their return to the Institut du Radium the Joliots undertook to verify whether the thresholds of neutron and positron emission were the same or not. During the two months needed for the preparation of the experiment, F. Joliot, in another work published on December 18, 1933, gave the first experimental proof of the annihilation of a positron by an electron [25].

The threshold of the neutron emission was remeasured using an ionization chamber with paraffin walls, filled with CH_4 and connected to a proportional amplifier. With this device the number of recoil protons ejected from the paraffin was increased. The ionization chamber was coupled to another chamber with a variable pressure of CO_2. The energy of the α particles emitted by a 60 mCi Po source located at the top of the chamber was monitored by increasing the pressure. Inside this chamber, no recoil protons could be formed. These α particles struck a thin Al foil (probably of 50 µm thickness) at the bottom and the neutrons were recorded with the first chamber. A more precise value than the previous one, i.e., 4 to 4.5 MeV, was obtained for the threshold of the neutron emission.

c. Towards the discovery of artificial radioactivity

The next task was to measure the threshold for the positron emission. The same chamber was used to

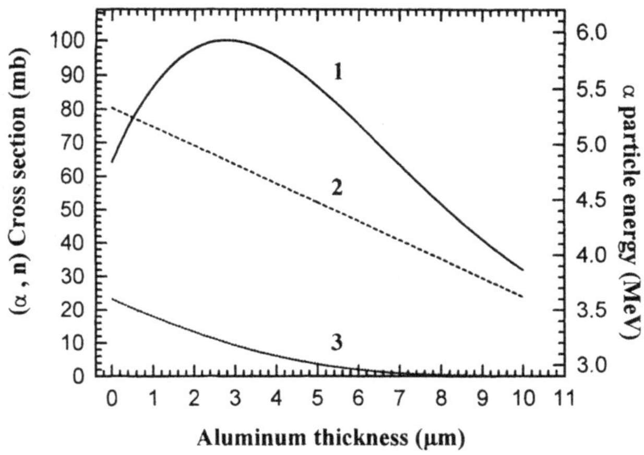

Fig. 4. 1) Range-energy curve of α particles in Al (right scale). 2) Cross section of the $^{27}Al(\alpha,n)^{30}P$ reaction versus Al thickness [31–32]. 3) Cross section of the $^{10}B(\alpha,n)^{13}N$ reaction versus Al thickness [31–32].

monitor the energy of the α particles. Positrons were first detected with the Wilson chamber. However, on January 11, 1934, F. Joliot used a Geiger-Müller detector with a 70 µm-thick Al window, connected to a numerical counter. This new type of detector had recently been introduced at the laboratory; it had been invented by Geiger in 1908 at Rutherford's laboratory and improved by Müller in 1932. The detector had a near 100% efficiency for counting electrons, probably had a poor time resolution and was certainly tricky to handle.

The Al foil was located close to the counter. Indeed, the positron emission appeared for the same α particle energy as for the neutron emission. Furthermore, the emission of positrons continued when the energy of the α particles had decreased below the threshold of the nuclear reaction. Even when the irradiation was stopped, positrons were still emitted with an intensity that decayed exponentially with time.

F. Joliot immediately understood the phenomenon. Besides the $^{27}Al(\alpha,p)^{30}Si$ reaction, two successive transmutations occurred: another nuclear reaction on Al, namely $^{27}Al(\alpha,n)^{30}P$, followed by a spontaneous decay $^{30}P \rightarrow {}^{30}Si + e^+$. Two major discoveries were made in one step: artificial radioactivity and a new type of radioactivity.

The experiment was repeated more simply, by direct irradiation of an Al foil (50 µm thick) placed 0.5 mm above the 100 mCi Po source [19], followed by counting of the activity. The half-life of ^{30}P was found to be 3.4 min, independently of the energy of the α particles. The intensity of the emission, as a function of the irradiation time, followed the same law known for the growth of a short-lived natural nuclide into a long-lived parent, such as ^{222}Rn into ^{226}Ra.

The decisive experiments which led to the discovery had lasted only a few hours. The details were reported much later by F. Joliot himself in his 1948–49 lectures at the Collège de France edited by P. Radvany and M. Bordry [19].

During the following two days, the Joliots irradiated further light elements with α particles (H, Li, Be, B, C, N, O, F, Na, Mg, Ca, Ni, Ag). Among these elements, B and Mg produced two additional new radioactivities, those of the 10 min ^{13}N and ^{27}Si. They wrote the famous paper "Un nouveau type de radioactivité" which was presented by J. Perrin at the weekly session of the Academy of Science on January 15, 1934. This is the official date of the discovery of artificial radioactivity [26]. The event appeared in Nature on February 10, 1934 [27].

The energy of the β^+ particles, measured by absorption, was 2.2 MeV for ^{30}P and 0.7 MeV for ^{13}N and ^{27}Si. However, ^{27}Si could not have been produced. The correct assignment of the corresponding activity was given by the Joliots in March; they had produced the β^- emitter ^{28}Al in the $^{25}Mg(\alpha,p)$ reaction [28].

At the time, the announcement of the discovery of artificial radioactivity was based only on experiments relevant to nuclear physics. Two weeks later, the Joliots provided the chemical proof of the formation of ^{30}P and ^{13}N at the tracer or subtracer level. The elegant radiochemical identification of the two radionuclides has been amply publicized and is discussed in detail in a following section. Eventually, on January 29, 1934, the undeniable proof of artificial radioactivity was definitively established [29]. V. M. Vdovenko in 'Radiochimie' IX, 574 (1967), French translation of Radiokhimija, states that the Russian scientist N. A. Orlov discovered artificial radioactivity in 1903!

Additional results on the half-life and on the energy spectrum of the β^+ radiation from ^{30}P and ^{13}N, as well as of the β^- radiation of ^{28}Al were reported in March [28]. The Joliots suggested that the same radionuclides could be synthesized in reactions induced by other particles, such as protons or deuterons, e.g. $^{12}C(d,n)^{13}N$ and by neutrons, which they used themselves [30]. The number of artificial radionuclides produced increased rapidly; over 200 hundred were known in 1937.

The favourite particle became the neutron. E. Fermi in Rome had started the production of radionuclides as early as March 1934 with a much stronger neutron source than that available in Paris, a Rn-Be source prepared from one gram of Ra. Six months later, Fermi discovered that the reaction yield increased significantly when a paraffin or water screen was interposed between the neutron source and the target. This was immediately followed by the observation of the Szilard-Chalmers effect [31]. Within a few years, thermal neutrons were to lead to the discovery of fission and of the first transuranium elements.

The story of the discovery of artificial radioactivity ended on December 12, 1935 when Irène Curie and Frédéric Joliot were awarded the Nobel Prize in chemistry "in recognition of their synthesis of new radioactive elements". Strictly speaking, the term "radioactive elements" should have been replaced by "radioactive isotopes" or "radioactive nuclides". The designation "radioelement" applies to an element of which all isotopes are radioactive. The first artificial radioelement, technetium, was synthesized by C. Perrier and E. Segrè in 1937.

2. Physical considerations

At the time of these events, there were four laboratories where intense natural radioactive sources could be prepared, namely in Berlin, Vienna, Cambridge and Paris. In this respect, the Curie Laboratory of the Institut du Radium in Paris was the foremost location owing to the long tradition and the experience acquired since Marie Curie's pioneering work. Multicurie sources of ^{226}Ra, as well as large amounts of RaD(^{210}Pb), ^{227}Ac, and MsTh(^{228}Ra) had been accumulated through the years.

Accordingly, the Joliots used a variety of sources: pure RaF (^{210}Po, 138 d, α 5.53 MeV), bulbs of ^{222}Rn (3.8 d) in equilibrium with RaC/RaC′ ($^{214}Bi/^{214}Po$, α 7.7 MeV), active deposits of thoron including the hard γ emitter ThC″ (^{208}Tl, γ 2.65 MeV), electroplated deposits of pure RaE (^{210}Bi, 5 d, β^-, E_{max} 1.2 MeV), as well as Po-Be.

In the present context, Po sources were of paramount importance. Deposits as large as 200 mCi were prepared. However, even with such activities, the yield of radionuclides in (α,n) or (α,p) reactions on light elements was very low owing to the small cross sections, or the reaction could not occur because of an excessively high threshold (Table 1). As an example, with a 100 mCi source emitting 1.85×10^9 α particles s^{-1} in 2π geometry, the saturation yield of ^{30}P in a thick Al target is only 4.5×10^5 atoms and the activity 2080 Bq [32−33] (Figure 7a).

This value is compared to the result of Joliot's experiment: with their 60 mCi Po source, about 10^{-6} to 10^{-7} atom of ^{30}P were produced per α particle and after 15 min irradiation the Geiger-Müller counter registered 150 counts min^{-1}, a low value which explains the poor statistics of the data reported by the authors.

Characteristics of (α,n) and (α,p) reactions on isotopes of light elements up to $Z = 20$, (also for Ni and Ag) are given in Table 1. With a Po source, and the experimental set-up of the Joliots, only the three main radionuclides ^{13}N, ^{30}P and ^{28}Al can be produced − which were found by the authors − and possibly ^{29}Al and ^{61}Cu − which were apparently missed. This shortcoming was probably due to the small cross section for the $^{58}Ni(\alpha,p)^{61}Cu$ reaction and to the threshold value for other reactions. The maximum energy of the α particles emerging from the 5 μm Al foil covering the Po source is only 4.4 MeV. Nuclei at the extreme limit of observation could have been ^{43}Sc from a Ca target and ^{44}Sc from K. It seems that the Joliots, in the course of the irradiations, suspected the formation of a few more short-lived radionuclides, but without lasting positron emission [28].

One may wonder if artificial radionuclides could have been formed unknowingly in the period before

1933 either with natural emitters or with particle beams from accelerators. In the first case, the only choice would be the 1 min ^{17}F β^+ emitter, for instance by (α,n) on ^{14}N. In his historical transmutation reaction in 1919, Rutherford certainly obtained ^{17}F, but he could not detect positrons with ZnS. In further irradiations of B, F, Na, Al and P, Rutherford, as well as G. Kirsh and H. Petterson in their experiments on Be, Mg and Si, reported only the emission of protons. The nuclide ^{17}F was obtained in late 1934 by Wertenstein with the 8.78 MeV α particles from the active deposit of thoron.

The first nuclear reaction with artificially accelerated particles was realized by J. J. Cockroft and E. T. S. Walton in 1932. The accelerator, based on the technology developed by Van de Graaff, produced a 0.7 MeV proton beam with an intensity probably less than one μA. This was sufficient for the $^7Li(p,\alpha)\alpha$ reaction; the two α particles emitted in opposite directions were counted in coincidence with a ZnS detector. A few weeks after the Joliots had reported the discovery of ^{13}N, J. D. Cockroft and E. T. S. Walton obtained the nuclide via the $^{12}C(p,\gamma)^{13}N$ reaction and measured the correct half-life of 10.5 min. Joliot's value of 14 min was overestimated owing to counting losses [34].

Beam intensities as low as a few μA produced intense fluxes of particles focussed on small areas of the target. The emission of particles from radioactive sources was largely surpassed. The same number of α particles can be obtained with a 1.2×10^{-3} μA beam and with a 100 mCi source in 4 π geometry.

The cyclotron invented in 1932 by E. O. Lawrence and M. S. Livingston delivered 1.2 MeV beams of protons first, followed by deuterons. The nuclide ^{13}N was easily obtained by the $^{12}C(d,n)^{13}N$ reaction; at the same time it was observed that parts of the cyclotron as well as various accessories became radioactive. Thus, already in 1932, artificial radionuclides were produced without the knowledge of the experimenters [19].

Radioactive sources were rapidly supplanted by beams of accelerated particles, including the Ra-Be neutron sources. Monoenergetic fast neutron beams could be obtained by fusion reactions such as $^2H(d,n)^3He$.

The discovery of artificial radioactivity closed the "primary" or "natural" era of nuclear sciences. However, renewed interest in natural radioactivity and natural radionuclides flared many years later, for instance with the discovery of the natural nuclear reactors or the cluster radioactivity first observed with AcX(^{223}Ra).

3. Chemical and radiochemical considerations

The success of the Joliots in the various steps which led to the discovery of artificial radioactivity was due in large part to their experience in the preparation and handling of very active Po sources. Procedures for the separation of Po from old Rn bulbs were established first by Irène Curie [35] and later together with Frédéric Joliot in his thesis work on the electrodeposition of the radioelement [36–37]. At that time, the Rn source at the laboratory was provided by 1.5 g of Ra. The Rn bulbs were washed with a mixture of dilute HNO_3 and HCl. The solution contained typically 70 mg of RaD (50 mCi) together with a few mg of Pb from the glass walls.

Po (or Po oxide) was deposited spontaneously on Ag or Ni from 0.2 M HNO_3–0.1 M HCl solutions. The activities were measured using the compensation method invented by Pierre Curie, and indicated in units of electrostatic charges (esu). Typical values were of the order of 1300 esu mm^{-2}, with a 90% yield equivalent to 1 mCi (0.22 μg) Po. Higher activity densities, up to 6.5 mCi, could be obtained with a yield of 25%. When strictly water-free sources were required in order to avoid extraneous proton recoil atoms, Po was volatilized from the Ni support and carried by a stream of CO_2 and N_2.

The most commonly used sources had a diameter of 4 to 5 mm and an activity of about 40 mCi over 1.2 mm^2. Occasionally special sources had an activity as high as 200 mCi spread over 1 cm^2, corresponding to a superficial mass density of 44.5 $\mu g\ cm^{-2}$. The range of α particles in Po being about 22 mg cm^{-2}, self absorption was negligible. The colour of the deposits changed from yellow to black with increasing activity from self-radiolysis with formation of Po(IV) oxide.

Owing to the relatively short half-life of 138 d, the Po sources were periodically recycled, i.e., old sources were dissolved and gathered for the preparation of fresh material. This was the case for the source used in the key experiment in January 1934. It is noteworthy that a large part of the Po was extracted from a solution of RaD offered several years before to Irène Curie by the Belgian "Union Minière du Haut Katanga" and which she brought herself to the Institut du Radium [38]. The 60 mCi source used in the historical experiment was volatilized on a 20 μm Al foil, and the 100 mCi one was deposited on Ni [39].

The activities handled by the Joliots largely exceeded those permitted today without special precautions and the authors must have suffered severe irradiation during the physical experiments and the chemical operations. The equivalent dose rate at one meter from a 100 mCi Rn source is 0.001 mSv h^{-1} (Figure 5) and 0.1 mSv h^{-1} for RaD. Strong α sources emit by sputtering solid aggregates which may be transported as aerosols and enter the body by inhalation.

Details on the procedure used by the Joliots for the chemical proof of the formation of ^{30}P and ^{13}N are scarce, both in their publications and in their notebooks. It seems that, in contrast to the physical demonstration, the chemical part was conducted in a more empirical and qualitative way. Additional information was provided by F. Joliot in his 1948 lectures at the

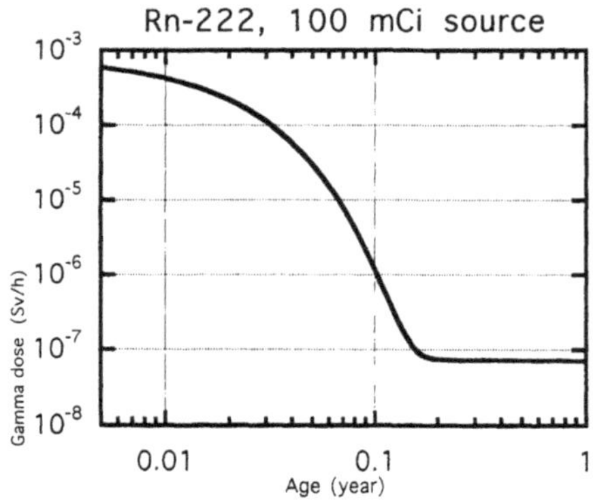

Fig. 5. Gamma dose rate at a distance of one meter from a Rn source as a function of time. (The absorption of X rays is neglected.) From J. P. Shapira, personal communication.

Collège de France [19]. The experiments and the reactions involved are scrutinized below.

a. The formation of ^{30}P

The assignment of the activity induced in the irradiated Al foil to an isotope of phosphorous was straightforward from the balance of the nuclear reaction since the emission of the neutron and of the positron appeared at the same threshold energy. Hence, the chemical proof had to rely on the known chemical properties of the element, although it was clear that the number of ^{30}P atoms was extremely small, which would require the addition of macroscopic amounts of carriers. These were probably the reasonings of the Joliots when they elaborated three tests to prove the chemical identity of the radionuclide.

The first one started with the dissolution of the Al foil in HCl, presumably after addition of a small piece of red phosphorous (a few ten mg?). The gas released in the chemical reaction was collected in a thin-walled tube over water and was found to be radioactive, whereas the $AlCl_3$ deposit after evaporation of the solution was inactive (Figure 6). The presence of the carrier was not specifically mentioned in the publications, but evoked in [19]. It might be thought that the use of carriers in radiochemical separations was so customary that the authors found it superfluous to indicate this detail.

In a further test, Al was dissolved in a mixture of HNO_3 and HCl. In one trial, probably still in the presence of inactive P, the activity remained in the solution. In the ultimate experiment, Na_3PO_4 was added to the solution, followed by $Zr(NO_3)_4$. The activity was carried down with the precipitate of zirconium phosphate, whereas the mother liquor was inactive.

The chemical behaviour of ^{30}P was correctly interpreted, but without comments on the chemical state of the nucleogenic P atoms, nor on the nature of the reactions undergone by extremely small amounts of radioactive matter. Today, in the light of knowledge gained from hot atom chemistry and from recent developments in the kinetics and thermodynamics of reactions

Fig. 6. Facsimile of the device used by the Joliots for the chemical proof of ^{30}P formed in Al bombarded by α particles (Courtesy Institut du Radium, Paris).

between micro- and macrocomponents in a chemical system, better insight is available on the chemistry involved in the Joliot experiments [40].

Neglecting the thin Al_2O_3 film of a few nm thickness which always cover an Al foil, it can be assumed that most of the ^{30}P atoms were formed within the first 5 μm of the target (Figure 7b). The mean recoil energy of the ^{30}P atoms is 2 MeV, and the range in Al 1.5 μm. It is assumed that the target was located at a distance of 0.5 to 1 mm in front of a source with a diameter of 10 mm. This arrangement corresponds to a geometry of 0.135 sr, from which it is calculated that 2×10^5 ^{30}P atoms were present at the end of a 15 min irradiation, i.e., on the average 1 P atom for 2.5×10^{14} Al atoms [41]. When the ^{30}P atoms come to rest, they behave like P atoms implanted in Al and end up in the P(0) oxidation state. Although the β^+ decay is formally a reductive process, the daughter ^{30}Si, as well as the same nuclides formed directly in the concurrent $^{27}Al(n,p)^{30}Si$ reaction (corresponding to about 5×10^4 atoms), stabilize as neutral P(0) atoms in the Al matrix. This statement is well supported by the large scale doping of the semiconductor Si with P(0) in the $^{30}Si(n,\gamma)^{31}Si, \beta^- \rightarrow {}^{31}P$ sequence.

The main reaction between macrocomponents upon dissolution of Al is

$$Al + 3\,H^+ \rightarrow Al^{3+} + 3/2\,H_2 \qquad (1)$$

which is spontaneous ($\Delta G° = -0.5$ eV at open air, Table 2). It has been verified that a fast dissolution requires a HCl concentrations above 5 M.

The standard or apparent reduction potentials of the reactions of P(0) or P compounds in which the

Table 2. Standard and apparent reduction potentials in Volt from: Standard potentials in aqueous solution, Ed. A. J. Bard, R. Pearsons and J. Jordan, IUPAC, Marcel Dekker, New York, 1985

Reaction	Potential
$2\,H^+ + 2\,e^- \rightarrow H_2$	0
$H^+ + e^- \rightarrow H$	−1.05
$Al^{3+} + 3\,e^- \rightarrow Al$	−1.676
$P(red) + 3\,H^+ + 3\,e^- \rightarrow PH_3$	−0.111
$2\,P(red) + 2\,H^+ + 2\,e^- \rightarrow P_2H_4$	−0.633
$H_3PO_4 + 5\,H^+ + 5\,e^- \rightarrow 4\,H_2O + P(red)$	−0.383
$H_3PO_3 + 3\,H^+ + 3\,e^- \rightarrow 3\,H_2O + P(red)$	−0.454
$H_3PO_2 + H^+ + e^- \rightarrow H_2O + P(red)$	−0.365
$NO_3^- + 2\,H^+ + e^- \rightarrow NO_2(gas) + H_2O$	0.775
$H_4SiO_4 + 4\,H^+ + 4\,e^- \rightarrow 4\,H_2O + Si$	−0.848
$Si + 4\,H^+ + 4\,e^- \rightarrow SiH_4$	−0.143

oxidation state of P lies between the limits P(−III) and P(V) are negative with respect to the reference couple H^+/H_2. In an acidic medium P(0) can be either reduced or oxidized. Therefore the P carrier may participate in the formal macrocomponent reactions

$$P + 3/2\,H_2 \rightarrow PH_3 \qquad \Delta G^\circ = 0.33\ \text{eV} \qquad (2)$$

$$P + 4\,H_2O \rightarrow H_3PO_4 + 5/2\,H_2 \qquad \Delta G^\circ = -1.9\ \text{eV} \qquad (3)$$

for which the ΔG° have been calculated with the data in Table 2. In an open system, with escape of PH_3 and production of H_2, reaction (2) is favoured with respect to (3), despite the negative ΔG° of the latter. Furthermore, red phosphorous is quite stable in aqueous medium and (3) is kinetically hindered. In any case, this last reaction will not proceed to a large extent.

In discussing the fate of the ^{30}P atoms released in reaction (1), two extreme cases in the presence or absence of P carrier will be considered. Little information is available on the size of the irradiated Al foils. It can be assumed that the surface was about 1.5 cm² to match that of the Po source, and the thickness 50 μm (quoted in [19]). The initial concentration of ^{30}P after instantaneous and complete dissolution of the target in 2 cm³ of HCl 5 M would be of the order of 2×10^{-16} M, while the HCl concentration decreased to 3.9 M. The dissolution of Al produces 7.5×10^{-4} mole or 25 cm³ of H_2, an amount which can react with 23 mg of P and yield a maximum of 13 cm³ of PH_3.

Owing to the two-phase system involved in reaction (2) and for kinetic reasons, only a small part of P will effectively react and the gas phase is mainly composed of H_2 with traces of phosphine. Hence the amount of carrier P added to HCl was not critical. Nothing prevents the mixing of ^{30}P with ^{31}P nor its reaction with H_2. $^{30}PH_3$ is carried with $^{31}PH_3$ and swept out from the solution with the H_2 stream.

Without P carrier, only $^{30}PH_3$ is formed according to (2). This is now a reaction between a macro- and a microcomponent, which has no kinetic or thermodynamic hindrance [40]. The radioactive phosphine is removed from the solution by the H_2 stream. Today, carrier-free chemical reactions are known to occur frequently [40].

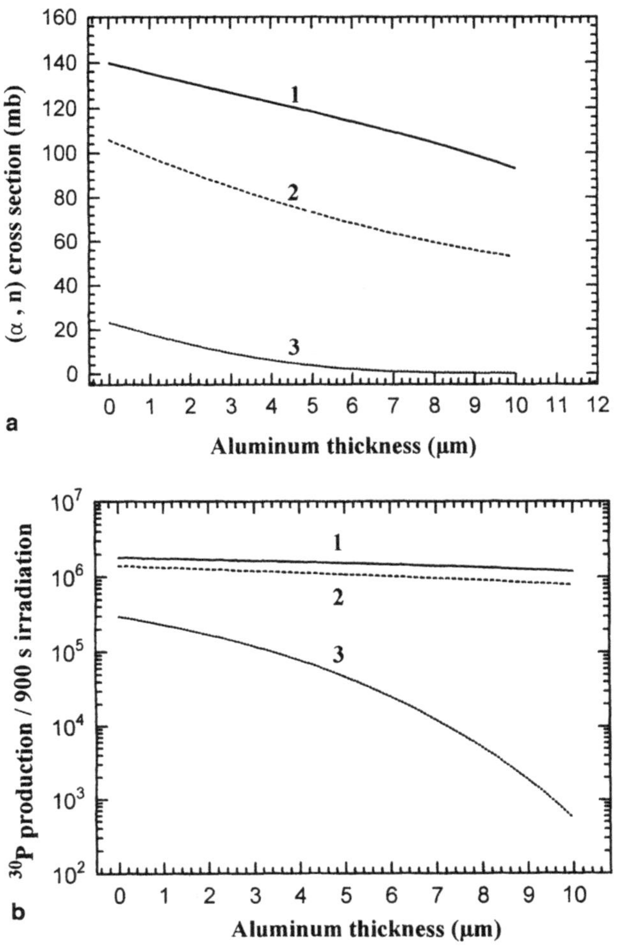

Fig. 7. a. Cross section for the production of ^{30}P with various natural α sources versus Al thickness [3, 31−32, 42]. 1) ^{214}Po (RaC/RaC′), E_α 7.68 MeV. 2) ^{212}Po + ^{212}Bi (thoron source), E_α 8.78 MeV (36%) and 6.07 MeV (64%). 3) ^{210}Po (RaF), E_α 5.3 MeV. b. Production of ^{30}P with various natural sources. Captions as for 7 a. With a ^{210}Po source, 92% of ^{30}P is formed in 5 μm of Al.

In conclusion, the chemical proof of the formation of $^{30}PH_3$ did not require the presence of P carrier, which leaves the question open whether or not the Joliots added red P to the medium.

In both cases, traces of P(V) compounds, e.g. $H_2{}^{31}PO_4^-$, could appear either from P impurities in HCl or from reaction (3). Trace impurities in the reagents would largely exceed the number of ^{30}P atoms. Exchange reactions of $^{31}P/^{30}P$ atoms between $^{30}PH_3/^{31}PH_3$ or between ^{30}P and $^{31}P(V)$ compounds are incompatible with the short half-life of ^{30}P. At the subtracer scale the formation of dimers such as P_2H_6 is excluded [40]. Once formed, $^{30}PH_3$ remains as such.

Upon dissolution of the Al target in the oxidizing mixture of HCl and HNO_3, in the presence or absence of P carrier, reaction (1) is completed by the overall process

$$P + 5\,NO_3^- + 5\,H^+ \rightarrow H_3PO_4 + 5\,NO_2 + H_2O \qquad \Delta G^\circ = -5.77\ \text{eV} \qquad (4)$$

in favourable competition with (2). Most of the activity of ^{30}P remains in solution and the coprecipitation of $^{30}PO_4^{3-}$ with weighable zirconum phosphate is straightforward.

A discussion of the fate of the 5×10^4 ^{30}Si atoms released into the HCl solution together with the ^{30}P atoms might at first appear superfluous because such a small number of stable atoms obviously cannot be detected. As mentioned, however, the Joliots had suggested the formation of ^{30}Si in an excited state, possibly in the combined form $^{30}Si^*H_4$. Accordingly, the following considerations seem appropriate.

Silicon in bulk form is rather inert towards chemical reagents, with the exception of fluorine ions, but may probably become very reactive in an extremely diluted state. In accordance with the redox potentials between Si and Si(IV) species (Table 2), the overall reaction expected is

$$Si + 4\,H_2O \rightarrow H_4SiO_4 + 2\,H_2 \qquad \Delta G^\circ = -3.36\ \text{eV} \tag{5}$$

resulting from the oxidation process

$$Si + 4\,H^+ \rightarrow Si^{4+} + 2\,H_2$$

followed by the hydrolysis of the Si^{4+} ion

$$Si^{4+} + 4\,H_2O \rightarrow H_4SiO_4 + 4\,H^+.$$

Reaction (5) can proceed easily in an open system. It may be noted that the amount of H_4SiO_4 arising from the leaching of SiO_2 in the glass walls by concentrated HCl is certainly much larger than that resulting from reaction (5). The formation of SiH_4, i.e. the formal reduction of Si

$$Si + 2\,H_2 \rightarrow SiH_4$$

is not predominant because the high positive ΔG° (0.56 eV) precludes competition with (5). It follows that the formation of SiH_4 briefly suggested by the Joliots is highly improbable. Even if excited ^{30}Si atoms had been formed, and had decayed by neutron-positron emission, they could not appear as a gaseous compound. On the other hand, the disproportionation of Si

$$2\,Si + 4\,H_2O \rightarrow H_4SiO_4 + SiH_4$$

although thermodynamically favoured at usual concentrations ($\Delta G^\circ = -2.8$ eV) according to reactions (5) and (6), is precluded at the subtracer level.

The dissolution of the irradiated material in HNO_3 instead of HCl should not invalidate the preceding discussion, since the former acid has no particular effect on the chemistry of Si.

b. The formation of ^{13}N

Owing to the insolubility of elemental B, the Joliots irradiated a thin layer of a few mg BN powder, a compound with a graphite-like structure. The target was rapidly dissolved in a watch glass with a hot dilute NaOH solution and evaporated to dryness. The residual salt was inactive. In a second test, the released gas was first trapped at liquid nitrogen temperature and then transferred to a glass bulb which then contained the activity. Finally, a white radioactive deposit appeared when paper soaked with HCl was brought into contact with the gas. These observations are compatible with the formation of radioactive NH_3.

^{13}N is formed in the (α,n) reaction on ^{10}B together with ^{14}N [via $^{11}B(\alpha,n)$], ^{13}C by decay of ^{13}N and ^{14}C [$^{11}B(\alpha,p)$] in the first µm below the surface of the BN grains. Assuming, for simplification, an ionic matrix of N^{3-} and B^{3+} for BN, the hot nucleogenic ^{13}N atoms will adapt as $^{13}N^{3-}$ anions, all the more so because they are embedded in an environment of macro-amounts of isotopic ^{14}N carrier atoms. Accordingly, the behaviour of ^{13}N follows the normal reactions of nitrogen. The formation of $^{13}NH_3$ in the Joliot tests is straightforward in the acid-base reaction

$$B(III)N(-III) + 3\,NaOH \rightarrow Na_3BO_3 + NH_3$$

as well as

$$HCl + NH_3 \rightarrow NH_4Cl.$$

Conclusion

The events which led to the discovery of artificial radioactivity at the Institut du Radium have been reconsidered within the scientific context of the thirties. The first simple and demonstrative nuclear and radiochemical experiments carried out by the Joliots in January 1934 have been presented in the light of our present state of knowledge.

Acknowledgments

The authors thank J. P. Adloff for pertinent highlighting suggestions and P. Radvany, H. Langevin and G. Bouissières for helpful comments.

References

1. Chadwick, J.: The existence of a neutron. Nature **129**, 312 (1932); Proc. Roy. Soc. (London) **A 136**, 692 (1932).
2. Anderson, C. D.: Science **76**, 238 (1932).
3. Northcliffe, L. C., Schilling, R. F.: Range and stopping power for heavy ions. Nuclear Data **A7**, 233 (1970).
4. Rutherford, E.: Collision of α particles with light atoms. IV. An anomalous effect in nitrogen. Philos. Mag. **37**, 581 (1919).
5. Wolfenden, J. H.: The dawn of hot atom chemistry. J. Chem. Ed. **32**, 276 (1955).
6. Bothe, W., Becker, H.: Z. Phys. **66**, 289 (1930).
7. Worwald, M. R., Akgun, A. E., Clayton, C. O.: Int. J. Appl. Rad. Isot. **34**, 363 (1983).
8. Curie, I., Joliot, F.: Emission de protons de grande vitesse par les substances hydrogénées sous l'influence des rayons γ très pénétrants. Compt. Rend. Acad. Sc. Paris **194**, 273 (1932).
9. Curie, I., Joliot, F.: Effet d'absorption de rayons γ de très haute fréquence par projection de noyaux légers. Compt. Rend. Acad. Sc. Paris **194**, 708 (1932).
10. Curie, I., Joliot, F.: Projection d'atomes par les rayonnements pénétrants excités dans les noyaux légers. Compt. Rend. Acad. Sc. Paris **194**, 876 (1932).

11. Curie, I., Joliot, F., Savel, P.: Quelques expériences sur les rayonnements excités par les rayons α dans les corps légers. Compt. Rend. Acad. Sc. Paris **194**, 2208 (1932).
12. Curie, I., Joliot, F.: Preuves expérimentales de l'existence du neutron. J. Phys. Radium **4**, 21 (1933).
13. Curie, I., Joliot, F.: Contribution à l'étude des électrons positifs. Compt. Rend. Acad. Sc. Paris **196**, 1105 (1933).
14. Curie, I., Joliot, F.: Sur l'origine des électrons positifs. Compt. Rend. Acad. Sc. Paris **196**, 1581 (1933).
15. Curie, I., Joliot, F.: Electrons positifs de transmutation. Compt. Rend. Acad. Sc. Paris **196**, 1885 (1933).
16. Joliot, F.: Réalisation d'un appareil Wilson pour pressions variables. J. Phys. Radium **5**, 216 (1934).
17. Curie, I., Joliot, F.: Electrons de matérialisation et de transmutation. J. Phys. Radium **4**, 494 (1933).
18. Curie, I., Joliot, F.: Etude du rayonnement absorbable accompagnant les rayons α du polonium. J. Phys. Radium **2**, 20 (1931).
19. Radvany, P., Bordry, M.: *La radioactivité artificielle et son histoire.* Ed. Seuil and CNRS, Paris (1984), pp. 106–114.
20. Curie, I., Joliot, F.: Rayonnement pénétrant des atomes sous l'action des rayons α. Proceedings of the 7th Solvay Congress in Physics, in *Frédéric et Irène Joliot-Curie, Oeuvres Scientifiques complètes*, Presses Universitaires de France, Paris (1961), p. 475.
21. Internal conversion coefficients for $Z=68$ to $Z=104$. Atomic Data **21**, N° 4–5 (1978).
22. Curie, I., Joliot, F.: Sur les conditions d'émission des neutrons par les éléments légers. Compt. Rend. Acad. Sc. Paris **196**, 397 (1933).
23. Curie, I., Joliot, F.: Nouvelles recherches sur l'émission des neutrons. J. Phys. Radium **4**, 278 (1933).
24. Langevin-Joliot, H.: Historical aspect of the discovery of artificial radioactivity. In *Artificial radioactivity*, Narayama Rao, E. K., Arnikar, H. J., Eds., Tata McGraw-Hill, New Delhi (1985), p. 1; also Amaldi, U., De la radioactivité artificielle aux bosons intermédiaires, in *Cinquantenaire de la radioactivité artificielle*, Laberrigue-Frolow, J., Ersidat, N., Eds., Université P. et M. Curie, Paris, p. 7.
25. Joliot, F.: Preuves expérimentales de l'annihilation des électrons positifs. Compt. Rend. Acad. Sc. Paris **197**, 1622 (1933).
26. Curie, I., Joliot, F.: Un nouveau type de radioactivité. Compt. Rend. Acad. Sc. Paris **198**, 559 (1934).
27. Joliot, F., Curie, I.: Artificial production of a new kind of radio-element. Nature **133**, 201 (1934).
28. Curie, I., Joliot, F.: I. Production artificielle d'éléments radioactifs. II. Preuves chimiques de la transmutation des éléments. J. Phys. Radium **5**, 153 (1934).
29. Curie, I., Joliot, F.: Séparation chimique des nouveaux radioéléments émetteurs d'électrons positifs. Compt. Rend. Acad. Sc. Paris **198**, 559 (1934).
30. Curie, I., Joliot, F., Preiswerk, P.: Radioéléments créés par le bombardment de neutrons. Compt. Rend. Acad. Sc. Paris **198**, 2089 (1934).
31. Szilard, L., Chalmers, T.: Chemical separation of the radioactive element from its bombarded isotope in the Fermi effect. Nature **134**, 462 (1934).
32. Stelson, P. M., McGovan, K.: Cross section for (α,n) reactions for medium weight nuclei. Phys. Rev. **4B**, 133 (1964).
33. Williamson, R. M., Katman, T., Burton, B. S.: ^{19}F, ^{23}Na and ^{27}Al(α,n) reactions. Phys. Rev. **117**, 1325 (1960).
34. Joliot, F., Curie, I.: Radioactivité artificielle, in *Actualités Scientifiques et Techniques*, N° 199, Hermann, Paris (1934) and [20], p. 861.
35. Curie, I.: Extraction et purification du dépôt actif à évolution lente du radium. J. Chim. Phys. **22**, 471 (1925).
36. Curie, I., Joliot, F.: Préparation de sources de polonium de grande intensité d'activité. J. Chim. Phys. **28**, 201 (1931).
37. Joliot, F.: Sur les propriétés électrochimiques du polonium. Compt. Rend. Acad. Sc. Paris **189**, 986 (1929).
38. Radvany, P.: Personal communication.
39. Notebook of I. Curie and F. Joliot, with permission of the Institut du Radium.
40. Adloff, J. P., Guillaumont, R.: *Fundamentals of Radiochemistry*, CRC Press, Boca Raton (1993), p. 195 and 313.
41. Dearnaley, G., Gibbs, W. R., Leachman, R. B., Rogers, C.: Investigation of the ^{27}Al(α,p) reaction by cross section fluctuation studies. Phys. Rev. **139, 5B**, 1170 (1965).
42. Lederer, C. M., Shirley, V. S., Eds.: *Table of Isotopes*, 7th Edition, John Wiley, New York (1978).
43. Howerton, R. J., Braff, D. J., Cahill, W. J., Chazan, N.: Threshold for ^{4}He induced reactions. UCRL, LBL 14005 (1964).

Radiochimica Acta 70/71, 51–67 (1995)

The Discovery of Nuclear Fission – Good Solid Chemistry Got Things on the Right Track

By Günter Herrmann
Institut für Kernchemie, Universität Mainz, D-55099 Mainz, Germany

(Received August 30, 1995)

Fission history / Nuclear fission / Transuranium elements

Abstract

An outline of the history of fission is given covering the period between spring 1934, when for the first time uranium was fissioned by irradiation with neutrons but the fission products were erroneously attributed to transuranium elements, and winter 1938/1939, when fission was discovered by a radiochemical experiment and confirmed world-wide with a variety of methods.

Introduction

"In the end it was good solid chemistry that got things on the right track" – with this remark concludes Otto Robert Frisch, one of the key persons in fission history, his reminiscences [1]. Indeed, after four years of difficult experiments aimed at the identification of transuranium elements expected to be heavy homologs of rhenium and platinum metals, an amazingly simple radiochemical experiment performed by Otto Hahn and Fritz Strassmann on December 17, 1938, leads to one of the great discoveries of this century. The whole procedure is described in eight lines in their laboratory notebook: Radium is coprecipitated with barium chloride from uranium exposed to neutrons, and the sample is separated into three portions by fractional crystallization of barium bromide. For each fraction, one page of counting data collected during five days reveals the totally unforeseen: A much lighter element, barium, is formed when uranium captures neutrons. Thus, fission is discovered in an indirect way by the identification of a fission product, to the astonishment of the discoverers themselves as is evident in their first historical publication [2]: As chemists they feel obliged to draw conclusions that are at variance with what they know as nuclear chemists. Their doubts disappear after a series of further radiochemical experiments in which Hahn and Strassmann [3] demonstrate their mastery. "The experiments described in Hahn and Strassmann's 'second' paper rank among the most careful and unambiguous ever carried out in radiochemistry [4]".

Lise Meitner and Otto Robert Frisch [5], informed in advance by Hahn, immediately arrive at a straightforward explanation of this new type of nuclear reaction on the basis of the already existing liquid-drop model of the nucleus. Numerous researchers, at first Frisch [6], confirm fission by the direct observation of fission fragments with ionization chambers or cloud chambers. The signals are much stronger than those produced by any other nuclear process. "But who can say whether even Fermi would have recognized them for what they were if he had seen them?" [7]. Tragically enough, these events occur in a period of strong political tensions at the eve of the second world war, and thus, very soon fission finds military applications before any peaceful uses.

The history of fission is an outstanding example for the intricate way science often pursues to important discoveries. Although some of the most experienced researchers of that time are involved, more than four years pass until it is realized what happens. For a too long time, two premises are maintained, which seem to be firmly established by theory and experiment but eventually turn out to be wrong: First, in nuclear transmutations – spontaneous as well as induced – atomic numbers change only slightly; thus, only elements in the vicinity of uranium are considered as candidates for the assignment of the new radionuclides. Second, in the Periodic Table of the Elements, Figure 1, thorium, protactinium, and uranium are located in subgroups IV, V and VI; hence, element 93 is expected to belong to subgroup VII below rhenium, and elements 94 to 96 to subgroup VIII below the platinum metals – chemical experiments aimed at the identification of transuranium elements are designed accordingly.

In the following survey, the major steps in fission history will be recalled. A more detailed report with reproductions of further original documents can be found elsewhere [8].

Prologue, Rome 1934: element 93, eka-rhenium – the first transuranium element?

When in spring 1934 Enrico Fermi and his team in Rome irradiate uranium with neutrons, they do not realize at all that they fission uranium into two fragments. Their intention is quite different: By the synthesis of an artificial element, 93, they want to exceed the natural limit of the list of chemical elements set at uranium. This most remarkable perspective is opened in their pioneering work on the use of neutrons in nuclear reactions [9]. Neutrons are found to attach easily

1934

H 1																	He 2
Li 3	Be 4											B 5	C 6	N 7	O 8	F 9	Ne 10
Na 11	Mg 12											Al 13	Si 14	P 15	S 16	Cl 17	Ar 18
K 19	Ca 20	Sc 21	Ti 22	V 23	Cr 24	Mn 25	Fe 26	Co 27	Ni 28	Cu 29	Zn 30	Ga 31	Ge 32	As 33	Se 34	Br 35	Kr 36
Rb 37	Sr 38	Y 39	Zr 40	Nb 41	Mo 42	Ma 43	Ru 44	Rh 45	Pd 46	Ag 47	Cd 48	In 49	Sn 50	Sb 51	Te 52	I 53	Xe 54
Cs 55	Ba 56	La 57 71	Hf 72	Ta 73	W 74	Re 75	Os 76	Ir 77	Pt 78	Au 79	Hg 80	Tl 81	Pb 82	Bi 83	Po 84		Rn 86
	Ra 88	Ac 89	Th 90	Pa 91	U 92												

LANTHANIDES	La 57	Ce 58	Pr 59	Nd 60		Sm 62	Eu 63	Gd 64	Tb 65	Dy 66	Ho 67	Er 68	Tm 69	Yb 70	Lu 71

Fig. 1. The Periodic Table of the Elements in 1934. Shaded boxes indicate elements missing at that time.

to many elements, producing isotopes of the initial element that emit β^--rays and, hence, decay into the element with the next higher atomic number [10]. If the same capture-decay sequence would occur with uranium, one would expect the formation of element 93:

$$^{238}_{92}U + ^{1}_{0}n \rightarrow ^{239}_{92}U \xrightarrow{\beta^-} {}^{239}93.$$

Element 93 should be radioactive and, if it is again a β^--ray emitter, might even decay to another transuranium element, 94. Indeed, at least five β-ray emitters are found by Fermi *et al.* [9] after irradiations of uranium with neutrons, with half lives of 10 s, 40 s, 13 min, 90 min and ≈1 day.

In order to prove the formation of new elements in a more direct way, one has to show that the nuclides ascribed to transuranium elements behave chemically according to their expected location in the Periodic Table. This is a difficult task since the nuclides are produced in tiny amounts that can only be detected by their radioactivity. The situation is quite similar to the identification of natural radioelements in the preceding decades, and the approach followed is the same. The radionuclide supposed to be a new element is mixed with milligram quantities of a homologous stable element, and a chemical operation characteristic for the homolog is performed. If then the radioelement accompanies the inactive carrier, the chemical analogy between both elements is proven.

It is not easy, however, to identify element 93 as being eka-rhenium, because one has to work in the, then, least familiar group of the Periodic Table. The two homologous elements 43, masurium, and 75, rhenium, had been discovered not long before in 1925. Whereas rhenium is already available in small quantities but not yet well investigated, masurium turns out to be a phantom, later replaced by technetium. Fermi and his coworkers rather focus on manganese as model element for subgroup VII and choose as a characteristic reaction the precipitation of manganese dioxide in strong nitric acid. This is not a fortunate choice, because the surface-rich precipitate is known to adsorb a variety of elements. At a first glance their concept seems to be successful, however. The 13-min nuclide, among the new nuclides being most convenient for such chemical studies, is found in the precipitate, but neither uranium nor lighter elements down to lead are present, as is verified with their radioisotopes [11]. As Fermi *et al.* mention incidentally, the 13-min nuclide also coprecipitates with rhenium sulfide in strong hydrochloric acid, a by far more specific reaction. The 90-min nuclide behaves similarly to the 13-min nuclide. Fermi presents the results quite cautiously under the title "Possible production of elements of atomic number higher than 92" [11].

Objections are being raised. Von Grosse and Agruss [12] ascribe Fermi's nuclides to the known element 91, protactinium, because according to their experiments, protactinium coprecipitates completely with manganese dioxide and to a large extent with rhenium sulfide. Otto Hahn and Lise Meitner in Berlin are challenged by this dispute in view of their experience in protactinium chemistry gained as the codiscoverers of this element and, Hahn, as discoverer of nuclear isomerism with protactinium-234. Lise Meitner [13] is from the beginning very attracted by Fermi's neutron-induced reactions. She persuades [14] Otto Hahn to re-establish their collaboration after twelve years of interruption and to focus attention on the transuranics.

In a first note [16] they present support for Fermi's interpretation: The 13-min and the 90-min nuclide can be separated from protactinium added in the form of natural 6.7-h protactinium-234g. The two transuranic nuclides precipitate together with platinum sulfide whereas protactinium remains completely dissolved. Vice versa, the new nuclides remain dissolved if protactinium is coprecipitated with zirconium phosphate. The two transuranium elements may be chemically different, the 13-min nuclide could be element 93, the 90-min nuclide element 94 [16]. Fermi *et al.* [17] arrive at this conclusion, too, and presume the decay of the 13-min into the 90-min nuclide. This is the last

Fig. 2. Otto Hahn (1879–1968) and Lise Meitner (1878–1968) at an institute outing, ca. 1934/35, and Fritz Strassmann (1902–1980), ca. 1936; after [26].

word of the Rome group about the "uranium case"; they leave it to those who are more experienced in radiochemistry [18] and focus on another of their discoveries, the strong interaction of thermalized neutrons with nuclei.

Ida Noddack's criticism [20] is of more conceptual nature. In a comment on two claims for the discovery of element 93 – besides the work of the Rome group an unsubstantiated claim for its natural presence in the Joachimsthal pitchblende – she points to the nonspecifity of the manganese dioxide precipitation and questions also the validity of rhenium sulfide for the characterization of element 93. In this context she writes [20]: "It would be conceivable that in the bombardment of heavy nuclei with neutrons these nuclei disintegrate into several *larger* fragments, which are isotopes of known elements but not necessarily neighbours of the irradiated elements"[1]. Can this statement be taken as the prediction of nuclear fission, as is sometimes advocated [21]? Not really, because Ida Noddack herself does not consider her suggestion of a novel nuclear process to be meaningful enough to test it experimentally. Her objection is hardly noticed in Rome [22] and Berlin [23] and is not relevant to the future development [25].

Progress, Berlin 1935–1938: eka-platinum elements 94 through 96?

From now on the studies of transuranics are pursued by Otto Hahn and Lise Meitner who soon ask the much younger Fritz Strassmann to join them. No photograph exists showing them together at that time; thus, one has to combine two separate ones, Figure 2. The experiments are performed in three rooms on the ground floor of the Kaiser-Wilhelm-Institut für Chemie in Berlin-Dahlem. A room at the end of a wing is used for irradiations with radium-beryllium neutron sources; chemical separations are carried out in Hahn's "private laboratory" about 20 m away, and the final samples are counted in a room across the floor.

The very simple equipment is depicted in Figure 3. Near the front of the table two cylindrical Geiger-Müller counting tubes are placed in their lead shields, the right one being open. Between the shields several counting samples and absorber foils can be seen. In the center two amplifiers are shown with two amplifier tubes each. To their left, mechanical recorders are placed which count the electrical pulses delivered from the amplifiers. The counting tubes are coupled to the amplifiers via high-impedance resistors consisting of a glass tube filled with toluene and embedded in paraffin. Power supplies for the amplifiers are located in the back. Voltages for the counting tubes are taken from anode batteries underneath the table. In the upper right corner a cylindrical paraffin block is shown, and next to it the radium-beryllium neutron sources, six brass capsules of 6 cm length containing about 1.2 g of radium-226. For irradiations with slow neutrons the sources are installed in the paraffin block, and the target sample is placed on top. The flux of thermal neutrons is about 10^4 neutrons per $cm^2 \cdot s$. At the right edge, a filter flask reminds of the crucial role that chemistry plays. The laboratory notebook "Chem. II" can be seen in the lower right corner, opened at page 76 written on December 17, 1938.

In order to produce significant count rates of the artificial nuclides – about a hundred counts per minute – with such weak neutron sources, ten- to hundred-gram samples of uranium have to be irradiated. The natural decay products of uranium, 24-d ^{234}Th–1.17-min ^{234m}Pa with penetrating β^--rays, would cause great problems since they introduce four to five orders of magnitude stronger activities. Therefore, the uranium has to be purified shortly before an irradiation by coprecipitating thorium and protactinium with iron hydroxide in the presence of carbonate ions which keep

[1] „Es wäre denkbar, daß bei der Beschießung schwerer Kerne mit Neutronen diese Kerne in mehrere *größere* Bruchstücke zerfallen, die zwar Isotope bekannter Elemente, aber nicht Nachbarn der bestrahlten Elemente sind" (emphasis as in the original).

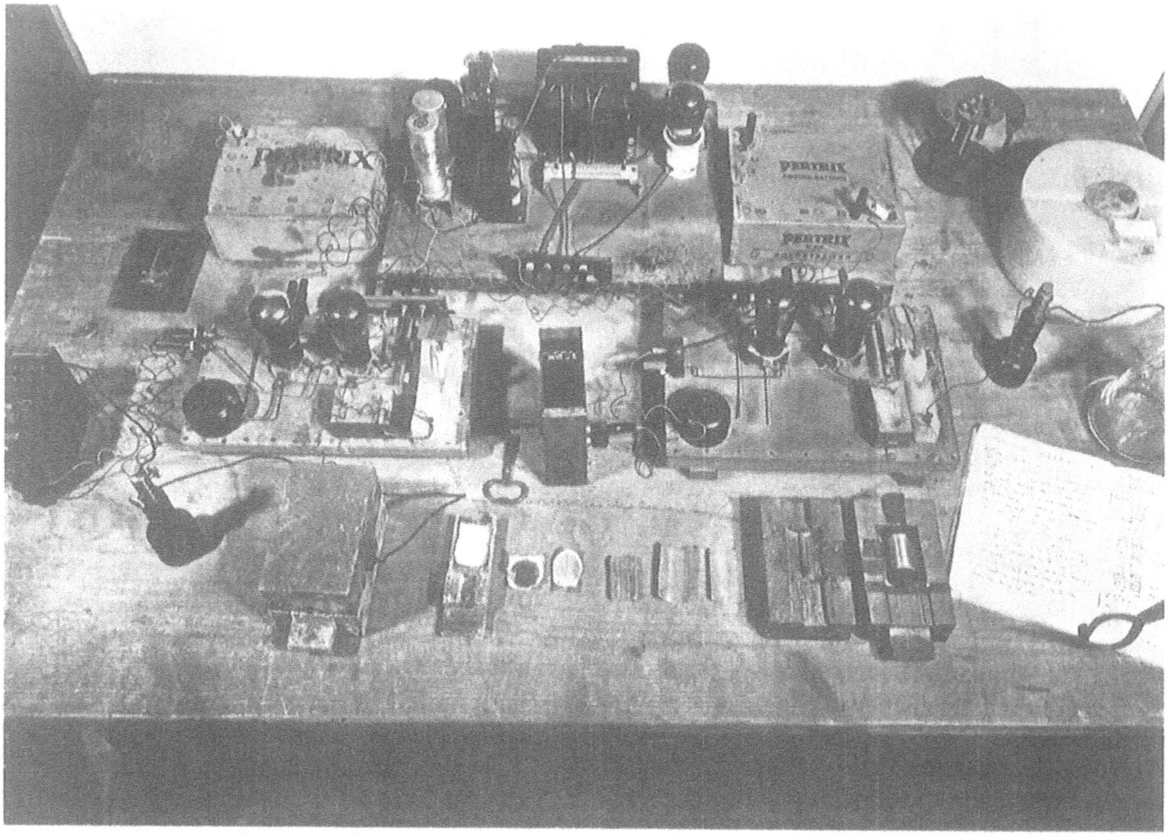

Fig. 3. Instruments at the Kaiser-Wilhelm-Institut für Chemie, Berlin, now in the Deutsches Museum, München: Geiger-Müller counting tubes (front part), pulse amplifiers and mechanical counters (middle part), high voltage supply (rear part), neutron sources and paraffin moderator (upper right), laboratory notebook (lower right). For more details see text.

$_{92}$U	$_{93}$Eka-Re	$_{94}$Eka-Os	$_{95}$Eka-Ir	$_{96}$Eka-Pt	$_{97}$Eka-Au
10 sec	→ 2,2 min	→ 59 min	→ 66 h →	2,5 h →	?
40 sec	→ 16 min	→ 5,7 h	→ 60 d →	?	
23 min →	?				

Fig. 4. Decay chains of uranium and of transuranium elements produced by irradiation of uranium with slow neutrons; situation as of July 1938 according to the last joint publication by O. Hahn, L. Meitner, and F. Strassmann [29].

uranium in solution. From the filtrate, ammonium diuranate is deposited and used for irradiations.

The more Hahn, Meitner and Strassmann progress with the radiochemical analysis of the artificial nuclides produced in uranium, the more complex are the results. Further transuranium elements and nuclides are discovered, their formation reactions and decay properties are studied, and the chemical behaviour of the transuranics is investigated. After several short notes, two detailed publications in May 1937 [27, 28] summarize the results. A last joint note [29] adding a new product, 60-d eka-iridium, is submitted on July 12, 1938, a few days before Lise Meitner's escape from Germany.

The scheme of the transuranics now [29] looks as is shown in Figure 4. Three β^--ray emitters are assigned to uranium. Two of them initiate decay chains extending to element 95, eka-iridium, and even to element 96, eka-platinum. For the third uranium isotope with 23 min half-life, no decay product is found although this nuclide is very clearly identified chemically as belonging to uranium, and in a cloud chamber exposure it is shown to be a β^--ray emitter. Thus, the 23-min uranium certainly decays into element 93, but a too long half-life may prevent the detection of the daughter product. All three uranium nuclides are assigned to mass number 239.

This scheme is based on chemical and physical arguments. The coprecipitation of all transuranics with platinum sulfide in medium-strong hydrochloric acid is considered to be the most striking difference between such elements on the one hand and uranium and cisuranics on the other. In detailed comparisons of individual transuranics with their homologs, chemical similarities prevail, including very specific reactions such as the coprecipitation of eka-rhenium with nitron perrhenate and of eka-platinum with ammonium hexachloroplatinate. If deviations are noted, they appear to be in accord with trends expected within the respective subgroups of the Periodic Table. Physical studies give further support for the assignments. With a cloud chamber it is proven that the transuranium nuclides decay by β^--ray emission. Parent-daughter relationships within the decay chains are established. The production reactions are investigated; the cross-sections for all three chains turn out to be in an order of magnitude that is compatible only with uranium-238 as the initial nucleus, rather than with the rare

$$^{232}\text{Th}\,(n,\alpha) \longrightarrow {}^{229}\text{Ra} \xrightarrow[<1\text{ Min.}]{\beta} {}^{229}\text{Ac} \xrightarrow[\sim 18\text{ Min.}]{\beta} \text{Th?}$$

$$^{232}\text{Th}\,(n,\alpha) \longrightarrow {}^{229}\text{Ra} \xrightarrow[15\text{ Min.}]{\beta} {}^{229}\text{Ac} \xrightarrow[\sim 3{,}5\text{ Std.}]{\beta} \text{Th?}$$

$$^{232}\text{Th}\,(n,\alpha) \longrightarrow {}^{229}\text{Ra} \xrightarrow[\sim 4\text{ Std.}]{\beta} {}^{229}\text{Ac} \xrightarrow[20\text{ bis }30\text{ Std.}]{\beta} \text{Th?}$$

Fig. 5. Isomeric decay chains of radium and actinium produced by irradiation of thorium with fast neutrons; L. Meitner, F. Strassmann, and O. Hahn, May 1938 [33].

uranium isotopes 235 and 234. Thus, a triple nuclear isomerism in uranium-239 is postulated which even is transmitted to the decay products. This remains a puzzle, as well as the stability against α-decay when a region is approached where this decay mode should increasingly dominate.

Interlude, Paris and Berlin 1935–1938: radium from thorium?

Thorium is another heavy element of interest for neutron activation because of the possible access to the missing, long-sought 4n + 1 decay series through reactions like ^{232}Th(n,γ)^{233}Th or ^{232}Th(n,α)^{229}Ra. Although such experiments concern well-known elements, they are difficult since one has to cope with all nine members of the natural 4n decay series of the target thorium-232.

Fermi's team is again the first to start with such experiments [10]. They find two β-ray emitters with 1 min and 24 min half-life, the latter one is assigned to thorium by its precipitation with thorium fluoride and peroxide. Hahn and Meitner [30] confirm these findings and assign the short-lived nuclide to radium-229. They argue that a 11-min nuclide, added by them, may be the actinium daughter product. Irène Curie, Hans von Halban and Pierre Preiswerk in Paris [31] enter the field with two more β-ray emitters with 2.5 min and 3.5 h half-life [32], all assigned to thorium and its neighbours; the 3.5-h nuclide – which plays an important role later – is attributed to actinium because it precipitates together with lanthanum fluoride.

Three years later, the Berlin team adds even more complexity [33]. The 23-min thorium is shown to be formed in a resonance capture of neutrons; hence, it is thorium-233. The daughter product, 27-d protactinium-233, is observed. After irradiations with fast neutrons, three radium isotopes are chemically identified, Figure 5, including one with only 0.8 min half-life. They decay into three actinium isotopes. All three radium isotopes are attributed to the ^{232}Th(n,α)^{229}Ra reaction, although such reactions have up to then been observed only with light elements. Thus – like in uranium – a triple isomerism is assumed, which is inherited from the parent to the daughter element, a conclusion that is difficult to understand [33]. Support comes from outside, however. The nuclides observed in Berlin are confirmed by Nishina's group in Tokyo [34] in irradiations with stronger neutron fluxes at a cyclotron. Furthermore, the observation of prompt, energetic α-particles from a (n,α)-reaction in thorium is reported in a short note by Paul Scherrer *et al.* [35], whereas in Meitner's laboratory a search for such α-particles by von Droste [36] remains unsuccessful.

Turning-point, Paris 1937–1938: a different type of transuranics?

The next important step in the uranium case is taken during summer 1937 by Irène Curie and Paul Savitch [37] in Paris. In order to circumvent the cumbersome purification of uranium prior to irradiation, they cover the irradiated sample with a copper foil which cannot be penetrated by the energetic β-rays of protactinium-234m. In the decay curves recorded under such conditions, they find – in addition to the known 40-s, 2-min, and 16-min components – a new nuclide of 3.5 h half-life. This nuclide does not coprecipitate with sulfides in acid solution, in contrast to the transuranics reported thus far.

The new nuclide seems to accompany natural thorium-234, and an assignment to the ^{238}U(n,α)^{235}Th reaction is proposed [37]. This is, however, at variance with earlier [27] and more recent results [38] in Berlin where thorium is not found after neutron irradiation of uranium. Then, in their following communications, Curie and Savitch offer a very surprising suggestion for the assignment of the 3.5-h nuclide: Its chemical properties are similar to those of the lanthanide elements; either, it is actinium – heavy homolog of the lanthanides – or a new transuranium element with properties completely different from eka-rhenium and eka-platinum elements [39]. As they are able to separate the 3.5-h nuclide from actinium [40], Curie and Savitch favour the hypothesis of a new transuranic with lanthanum-like behaviour, although they admit that this is difficult to understand.

These surprising results are presented in detail in a following publication [41]. The 3.5-h nuclide is chemically isolated from irradiated uranium with lanthanum as carrier element. Several methods are compared; the best is the coprecipitation with the sparingly soluble potassium lanthanum sulfate. For further identification, the 3.5-h nuclide in the lanthanum fraction is mixed with natural actinium, and lanthanum oxalate is precipitated in fractions. The 3.5-h nuclide is found in the head fraction, actinium – as is expected – in the tail. The similarity of the 3.5-h nuclide to the lanthanide elements is summarized as follows [41]: "On the whole, the properties of $R_{3.5\text{-h}}$ are those of lanthanum, from which, it seems so far, it can be separated only by fractionation"[2]. One is tempted to argue that if Irène Curie and Paul Savitch had omitted the second half of that sentence, the puzzle would have been solved. However, the observation of such a frac-

[2] "Dans l'ensemble, les propriétés de $R_{3.5h}$ sont celles du lanthane, dont il semble jusqu'ici qu'on ne puisse le séparer que par fractionnement".

tionation is correct, as will be shown below. In conclusion the researchers in Paris challenge their competitors in Berlin by the statement: "One could imagine, for example, that this body [the 3.5-h nuclide] has the atomic number 93 and that the bodies investigated by Hahn, Meitner, and Strassmann have the atomic numbers 94 to 97"[3]. In Berlin, one is always sceptical and calls the 3.5-h nuclide "curiosum" [42].

On the right track, Berlin, autumn 1938: radium from uranium?

As soon as the detailed publication of I. Curie and P. Savitch [41] arrives in the third week of October 1938, Hahn and Strassmann are able to reproduce the 3.5-h nuclide using the coprecipitation with potassium lanthanum sulfate. Strassmann immediately recognizes [26] that if radium would be present, it would also deposit as sulfate and would appear during the subsequent fractionation in the head fraction, as reported for the 3.5-h nuclide. The presence of radium and actinium would explain the chemical behaviour of the 3.5-h nuclide, and would, hence, trace it back to already known elements. Radium could possibly be formed via a $^{238}U(n,\alpha)^{235}Th$ reaction, if the product thorium-235 would disintegrate by α-particle emission into radium-231, which could further β^--decay into actinium-231. In their previous negative searches for thorium in neutron-irradiated uranium [27, 38], the Berlin team did not check on α-particle emitters.

Radium with its straightforward chemistry seems to be a better candidate for clarifying the situation than the more complex actinium. Strassmann suggests [43, 44] isolation of radium by coprecipitation with barium chloride. This compound crystallizes from concentrated hydrochloric acid in beautiful, colorless, pure needles. The precipitate is soluble in water; thus, the radium fraction is easily available for further chemical treatment. With this elegant method radium can quickly and unambiguously be identified; it is accompanied only by barium. The conventional method, coprecipitation with barium sulfate, would yield a surface-rich compound which is notorious for adsorption effects and is much more difficult to redissolve.

Within a matter of days, Hahn and Strassmann obtain new results, as Hahn notes on October 25, 1938, in a letter to Lise Meitner [45]. The "countdown" for the discovery of fission begins. It is documented in the laboratory notebook "Chem. II" [47]. We are well informed about the following historical events by the frequent letters exchanged between Otto Hahn and Lise Meitner, now in Stockholm; answers often arrive within two days. It is especially informative to read letters and laboratory notes in chronological order [45].

Notebook "Chem. II" opens on November 1, 1938, with an experiment to verify whether the 3.5-h

$$\text{„Ra I"?} \xrightarrow[<1\text{ Min.}]{\beta} \text{Ac I} \xrightarrow[<30\text{ Min.}]{\beta} \text{Th?}$$

$$\text{„Ra II"} \xrightarrow[14 \pm 2\text{ Min.}]{\beta} \text{Ac II} \xrightarrow[\sim 2{,}5\text{ Std.}]{\beta} \text{Th?}$$

$$\text{„Ra III"} \xrightarrow[86 \pm 6\text{ Min.}]{\beta} \text{Ac III} \xrightarrow[\sim\text{mehrere Tage?}]{\beta} \text{Th?}$$

$$\text{„Ra IV"} \xrightarrow[250\text{—}300\text{ Std.}]{\beta} \text{Ac IV} \xrightarrow[<40\text{ Std.}]{\beta} \text{Th?}$$

Fig. 6. "Radium" isomers and their "actinium" daughter products observed in the barium fraction of neutron irradiated uranium; O. Hahn and F. Strassmann, December 1938 [2].

nuclide might be radium or actinium. The irradiated uranium is dissolved in hydrochloric acid. Barium chloride is precipitated, reprecipitated and further purified by scavenger precipitations of platinum sulfide and lanthanum hydroxide. The barium-radium fraction is especially active. Three radium isotopes are observed, with half-lives of 25 min, 110 min, and several days; they seem to be related to three actinium daughter products of 40 min, 4 h, and 60 h half-life, which are separated from radium by coprecipitation with hydroxides. Thus, the sought 3.5-h nuclide is observed in the actinium fraction as decay product of a radium precursor. Hahn and Strassmann very soon submit a note [48]. All three radium isotopes are ascribed to radium-231, in accordance with the presumed production via the $^{238}U(n,\alpha)^{235}Th$ reaction, although strangely enough, the (n,α) reaction occurs with slow neutrons. Figure 6 shows the radium-actinium decay chains, now increased to four as published shortly later [2]. As already in the case of radium from thorium, an inherited multiple isomerism is assumed. What is not recognized is the similarity of the half-lives for radium-actinium from uranium, Figure 6, and from thorium, Figure 5. Of course, in terms of the postulated (n,α) reactions, one would expect different mass numbers for the radium isotopes, viz. 231 from uranium and 229 from thorium.

Breakthrough, Berlin, December 17, 1938: "radium" is barium — fission discovered

On December 22, 1938, Otto Hahn and Fritz Strassmann submit a detailed paper entitled "On the identification and behaviour of alkaline earth metals formed upon irradiation of uranium with neutrons" [2]. Four out of the five printed pages deal with the interesting but not spectacular radium-actinium decay chains. Proof is shown for all reported nuclides by decay curves; Figure 7 gives an example. An improved version of the decay chains is presented, Figure 6. The chemical procedures are discussed, and it is emphasized that the activity precipitated with barium chloride can only stem from radium, "if one does not consider for the present barium itself"[4], as it is pointed

[3] "On pourrait imaginer, par exemple, que ce corps a le nombre atomique 93 et que les corps étudiés par Hahn, Meitner et Strassmann ont des nombres atomiques de 94 à 97".

[4] "[. . .] wenn man das Barium selbst außer Betracht läßt".

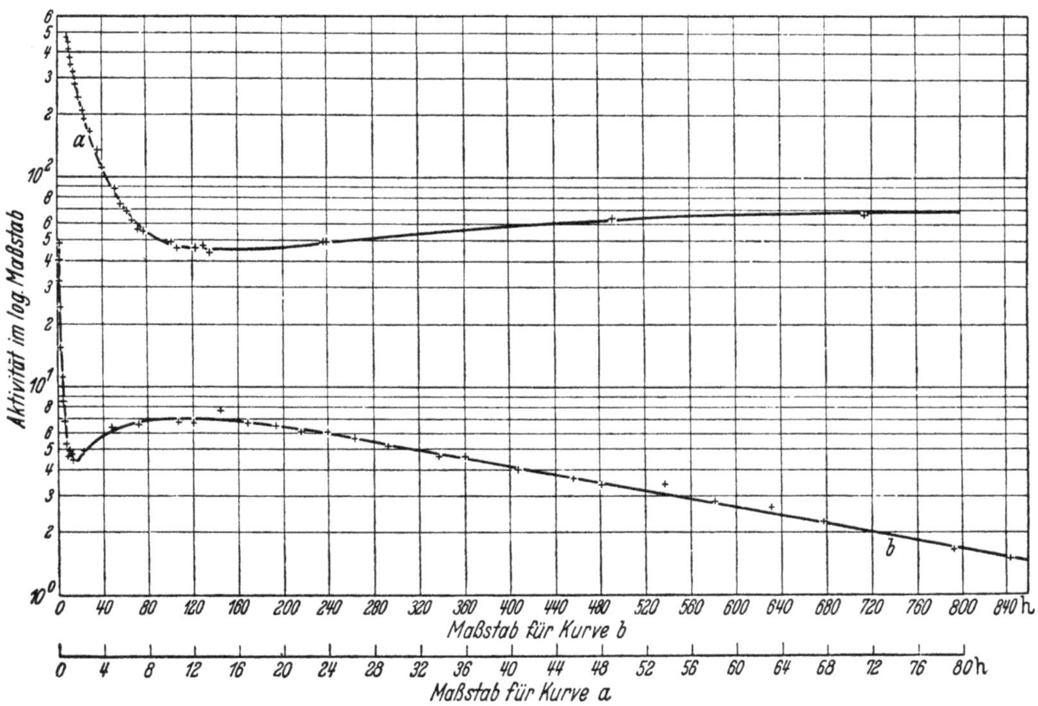

Fig. 7. Decay curve – count rate versus time – of the radium-barium fraction isolated from uranium after four days of irradiation with neutrons. The upper curve (a) shows the initial part on a ten-times expanded time scale than used in the lower curve (b). The initial decrease is due to the 86-min "radium III", the subsequent rise to 40-h "actinium IV" growing from ≈12-d "radium IV" which governs the final decrease. O. Hahn and F. Strassmann, December 1938 [3].

out in the beginning of the text. A careful reader may wonder why in the title the more general wording "alkaline earth metals" replaces "radium" used in the preceding note [48], and why in the text and also the decay chains "radium" is often put in quotation marks.

The big surprise comes at the end of the paper. Hahn and Strassmann report with hesitation some very recent results aimed at a rigorous chemical proof for the presence of radium. Presumably these control experiments are motivated by objections against the inherited isomerism and the unusual production reactions. For Hahn the most substantial doubts are probably the ones he hears [26, 42] on November 13 and 14, 1938, in Copenhagen, where he presents the results in a seminar and meets Niels Bohr, Lise Meitner and her nephew Otto Robert Frisch, also a nuclear physicist. In other statements, however, Hahn [23, 44] gives a quite trivial reason for these experiments, namely to improve the counting efficiency for radium isotopes by removing the bulk of barium carrier.

Anyway, how radium can be enriched with respect to barium is known since the early days of radiochemistry: Upon fractional crystallization of certain barium salts, radium concentrates in the head fractions. The first fractionation is attempted on November 25, 1938 [47]. Barium chloride is used for the separation of radium from uranium and for the fractionation procedure. Three barium chloride fractions are prepared and are counted successively for about 15 min for each of the fractions. The count rates decrease with a half-life of about 40 min, corresponding to a mixture of radium II and III. No sudden jump can be perceived between the first and second, and the second and third sample; hence, there is no evidence for an enrichment of radium in the head fractions. In the next attempt, barium bromide is chosen for the fractionations because it is known to yield stronger enrichments of radium. With the longer-lived radium nuclides now applied, the fractions can be counted alternately and the rates can directly be compared. The results on November 28 with 86-min radium III and on December 6 to 8 with ≈12-d radium IV are the same as before: The activity is distributed equally among the fractions [47].

Should this unexpected behaviour of radium be due to its extremely low concentration, being much lower than during the pioneering studies of radium fractionation with less sensitive detection methods? In the "active" laboratory on the 1st floor Hahn starts to fractionate natural radium in amounts which were common earlier. When he then switches to the small amounts present in their current experiments, no difference is observed – radium is enriched as usual [23, 42].

Now there is only one more logical step left: The chemical behaviour of the artificial and the natural radium isotopes has to be compared directly – the indicator experiment by Otto Hahn and Fritz Strassmann on December 17, 1938. The fractionation of 86-min radium III from uranium is performed in the presence of natural 5.8-a radium-228. This radium isotope, also

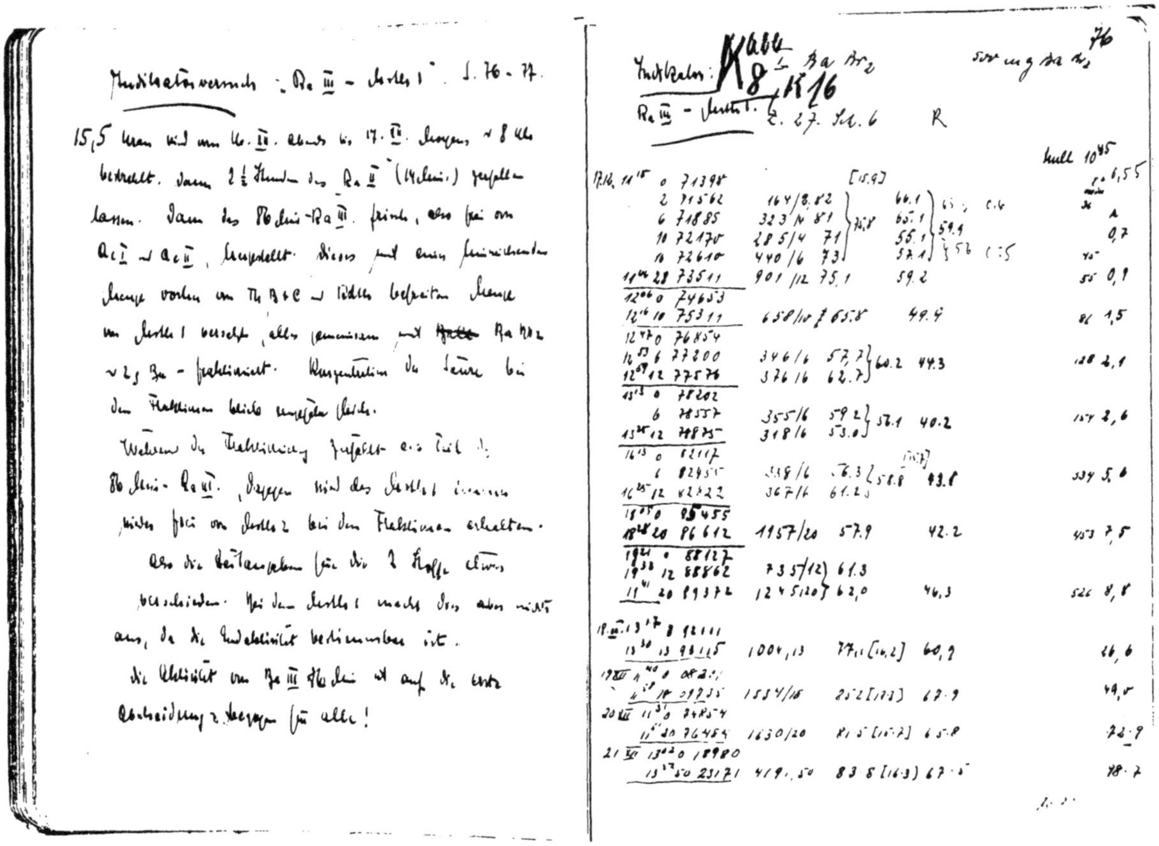

Fig. 8. The first two pages of the laboratory record of the indicator experiment by Otto Hahn and Fritz Strassmann on Saturday, December 17, 1938: Fractional crystallisation of barium bromide in presence of the 86-min "radium III" from uranium and the natural 5.8-a radium-228 (mesothorium 1). Description of the chemistry (left page) [51], and counting data for the first barium bromide fraction (right page) [52]. Translation and explanation in the text.

called mesothorium 1, and its daughter product, 6.1-h actinium-228, mesothorium 2, were Hahn's first important discoveries in Berlin [49, 50]. The experiment is simple in principle; the chemical separation and fractionation is finished in less than one hour. The first paragraph of the protocol [51], Figure 8, gives all the essential information: "15.5 [g] of uranium is irradiated from the evening of Dec. 16 until the morning of Dec. 17, ~8 o'clock. Then 'Ra II' (14 min) [is] allowed to decay for 2 1/2 hours. Then the 86 min-Ra III [is] freshly prepared, viz. free of Ac I and Ac II. This [is] mixed with a sufficiently large amount of Msth 1, purified beforehand from ThB +C [decay products], all that together [is] fractionated with $BaBr_2$[,] ~2 g Ba. The concentration of the acid in the fractions remained approximately the same".

Figure 8 shows the counting data for the first 500 mg fraction of barium bromide [52]. The columns represent from left to right: 1: date, 2: clock time, elapsed time in minutes, 3: meter counts, 4: difference between successive meter counts, respective time interval, 5: count rate in counts/min, 6: count rate after subtraction of background, 15.9 cpm, 7 (at the right edge) elapsed time in minutes and hours, with zero set at 10:45 h. On top, "K8, K16" refer to sheets with decay curves, "Z.27" to the counter and "Sch.6" to the sample. The readings start with 59.9 cpm and decrease to 40.2 cpm within two hours when the 86-min nuclide is decayed. Then the growth of actinium-228 is observed until it reaches equilibrium with the parent radium-228 at 67.6 cpm.

Figure 9 shows the decay curves for all three fractions, taken from the next publication [3]. The plateau values differ appreciably, as can be seen in part A, indicating the enrichment of natural radium in the order fraction I > II > III, as expected. After subtraction of the initial growth according to part B, three practically identical decay curves of the 86-min nuclide result. Hence, this nuclide is distributed equally among the three fractions, in contrast to natural radium: The 86-min nuclide produced in uranium *must be barium.*

Otto Hahn draws the conclusion, Figure 10, at the end of the data sheet for the third barium bromide fraction [53], and even more concisely in his pocket calendar [54]: "Morning at the revenue office about Lise: complained. New application to be made. Indicator experiment Msth 1 + our Ra III. Ra III is not enriched. Msth 1 strongly!!"[5].

[5] "Vormittags auf dem Finanzamt wegen Lise: geschimpft. Neue Eingabe zu machen. Indikatorversuch Msth 1 + unser Ra III. Ra III reichert sich nicht an. Msth 1 stark!!".

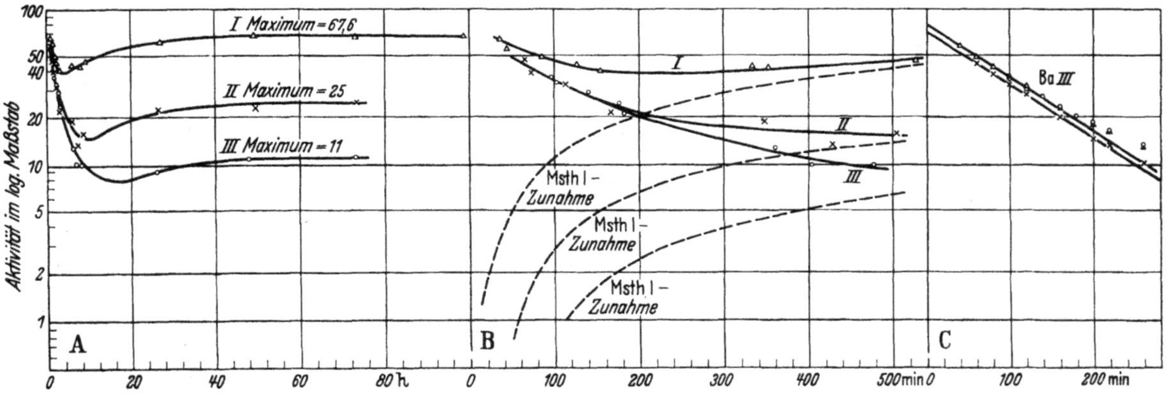

Fig. 9. Decay curves of the three successive barium bromide fractions prepared in the indicator experiment on December 17, 1938 [3]. The raw data, depicted in part A, show the strong enrichment of the indicator radium-228 (Msth 1) in fractions I > II > III, as is evident from the maximum count rates reached by the growing-in daughter 6.1-h actinium-228. Part B explains how this growth is subtracted from the initial part of the curves. Part C gives the resulting three decay curves of 86-min barium III and shows them to be identical.

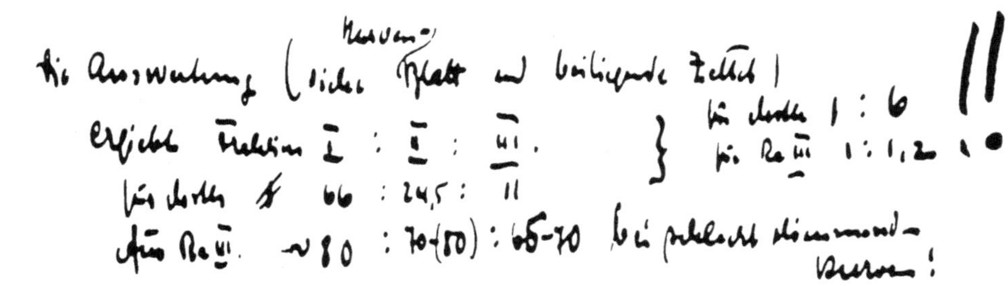

Fig. 10. Otto Hahn's conclusion from the indicator experiment "radium III" – mesothorium 1 written at the end of the counting data of the third barium bromide fraction [53]:

"The evaluation (see decay-curve sheet and attached note) yields

fraction	I : II	: III	} for Msth	1 : 6
for Msth	66 : 24,5	: 11	} for Ra III	1 : 1,2
for Ra III	~80 : 70(–80)	: 65–70	with not well matching curves!"	

The last remark refers to the last data points in Figure 9C which are somewhat too high, probably due to a contribution of ≈12-d Ra IV.

On the evening of December 19, he informs Lise Meitner [56]: "Then, on Saturday, Strassmann + I fractionated one of our "Ra"-isotopes with Msth 1 as indicator. Mesothor was enriched as expected, not so our Ra. All this may still be due to a very peculiar coincidence. But more and more we come to the frightening conclusion: Our Ra-isotopes do not behave like Ra but like Ba. [...] I agreed with Strassmann that we only tell *you* about all this for now. Maybe you can suggest some phantastic explanation. We know ourselves that it *cannot* really break into Ba"[6]. Hahn mentions that they want to check whether the "actinium"-nuclides growing in the "radium"-fraction behave like lanthanum. This experiment is already underway when the letter is being written. The 40-h "actinium IV" is milked from "radium IV" and mixed with natural actinium-228 and lanthanum carrier. Upon fractionation of lanthanum oxalate, actinium-228 becomes enriched in the tail fraction whereas "actinium IV" is uniformly distributed: It is lanthanum.

All these results are presented on the last page of Hahn and Strassmann's first historical publication [2] and are summarized in a famous statement: "As chemists we should, on the basis of the briefly reported experiments, rename the scheme shown above [Figure 6] and, instead of Ra, Ac, Th, insert the symbols Ba, La, Ce. As 'nuclear chemists' who are close to physics in certain ways we cannot yet bring ourselves to take this jump which contradicts all the experiences of nuclear physics to date. Perhaps it is still possible that a number of strange chance occurrences have simulated our results"[7].

[6] "Dann haben Strassmann + ich am Samstag eines unserer "Ra"-Isotope mit Msth 1 als Indikator fraktioniert. Das Mesothor wurde programmgemäß angereichert, unser Ra nicht. Es könnte noch ein höchst merkwürdiger Zufall vorliegen. Aber immer mehr kommen wir zu dem schrecklichen Schluss: Unsere Ra-Isotope verhalten sich nicht wie Ra, sondern wie Ba. [...] Ich habe mit Strassmann verabredet, daß wir vorerst nur *Dir* dies sagen wollen. Vielleicht kannst Du irgendeine phantastische Erkärung vorschlagen. Wir wissen dabei selbst, dass es eigentlich nicht in Ba zerplatzen *kann*" (emphases as in the original).

[7] "Als Chemiker müßten wir aus den kurz dargelegten Versuchen das oben gebrachte Schema [Figure 6] eigentlich umbenennen und statt Ra, Ac, Th die Symbole Ba, La, Ce einsetzen. Als der Physik in gewisser Weise nahestehende 'Kernchemiker' können wir uns zu diesem, allen bisherigen Erfahrungen der Kernphysik widersprechenden Sprung noch nicht entschließen. Es könnten doch noch vielleicht eine Reihe seltsamer Zufälle unsere Ergebnisse vorgetäuscht haben".

It is probably unique that a result of such significance is presented in such a strange way. Hahn decides to publish immediately, without repeating the experiment nor waiting for Lise Meitner's comment on his letter of December 19. The personnel of the Kaiser-Wilhelm-Institut goes on a two-week Christmas vacation on December 21. In view of the competition with the Paris group, it seems to be risky to wait this long. As he explains later [58] he uses a nearly finished manuscript of a more detailed publication on the radium isotopes, and adds, on December 21 and 22 [54], at the end the results of the fractionation experiments. Several inconsistencies remain unnoticed; for example, the printed text says that the long-lived "radium IV" was used in the indicator experiment, whereas protocol and decay data clearly indicate 86-min "radium III". A change of the title from "radium" to "barium" is considered, but then the whole text would have to be rewritten; "alkaline earth metals" is a way out. The typescript is picked up by Paul Rosbaud, advisor to the publisher of "Naturwissenschaften", on the evening of December 22 [54]. He recognizes the significance of the results presented in the final part and manages the publication already in the next issue appearing on January 6, 1939.

Lise Meitner's comment [45] arrives on December 23 [60]. The assumption of such a drastic break-up appears difficult for her to understand but in nuclear physics there had been so many surprises that one cannot really call anything impossible. She also asks for the transuranium elements, and this request may have caused Hahn to insert [61] during proof reading on December 27 [54] a short paragraph indicating, that the transuranics might be lighter rather than heavier homologs of rhenium and platinum metals, formed in a binary break-up as complementary fragments of barium.

The decay curves of the indicator experiment on December 17, Figure 9, are presented in the second historical paper by Otto Hahn and Fritz Strassmann [3]. The title now clearly indicates the main theme "Proof of the formation of active barium isotopes from uranium and thorium by neutron irradiation" and promises even more by continuing with "Identification of further active fragments in uranium fission". Indeed, a wealth of results is obtained between January 4, 1939, when work is resumed at the Berlin institute, and the submission of the paper on January 28. The indicator experiments are now repeated with other barium compounds and several combinations of barium and radium isotopes. Masterpiece is a fractionation of barium chromate where the 3.7-d radium-224 (Th X) is enriched in the first fraction by a factor 19.5 with respect to the second, whereas the 86-min barium III is equally distributed. With the ≈12-d barium IV a cyclic experiment is performed in which the barium carrier circulates through seven compounds without any change of its specific activity. Thorium is fissile by neutrons, too, as is shown by the identification of barium isotopes in a barium-radium indicator experiment.

The second focus is the search for fragments complementary to barium, namely krypton

$$_{92}U \rightarrow {}_{56}Ba + {}_{36}Kr$$

and its decay products rubidium, strontium, and yttrium. Strontium isotopes are discovered by precipitation of the nicely crystallizing strontium nitrate, and are found to decay into yttrium isotopes. A noble gas is identified which decays into an alkali metal; this pair is either krypton–rubidium or xenon–cesium. A search for palladium and silver, products of a break-up into two fragments of equal size, is undertaken but remains without positive results. Thus, evidence for the asymmetric mass split, becoming one of the puzzles of the fission process, is already obtained at that early moment.

Lise Meitner congratulates on February 5 [45]: "Your experiments are a beautifully consistent series of proofs and it is marvellous how much you could achieve during these few weeks"[8]. It is striking how correct these results are, though they are deduced from complex β-decay curves measured at low counting rates. In the chains shown in Figure 6, the "Ra III" is identical with 83.0-min ^{139}Ba decaying into stable ^{139}La, the "Ra–Ac IV" with 12.7-d ^{140}Ba–40.3-h ^{140}La, the "Ra–La II" is a mixture of 18.3-min ^{141}Ba and 10.7-min ^{142}Ba with the daughters 3.93-h ^{141}La and 1.54-h ^{142}La, and the "Ra-La I" can be associated with 0.24-min ^{143}Ba–14-min ^{143}La.

The question of transuranics becomes, with changing views, a major theme in the Hahn-Meitner correspondence [45]. For Lise Meitner this aspect is especially important [62] because if all the transuranium elements were in fact fission products her work during the last years would be lost without her participation in a discovery of great significance. After additional chemical experiments and reflections, it is pointed out that for the time being the transuranics are retained [3].

World-wide echo, early in 1939: fission confirmed

Lise Meitner meets Otto Robert Frisch at Chistmas 1938 in Kungälv on the western shore of Sweden [1]. Through Hahn's letter and an incomplete copy of the typescript of the first paper [2] they are well informed about what has happened. When walking together out in the snow they come upon a straightforward explanation of the novel nuclear process. The emission of large fragments from heavy nuclei should not be viewed – analogous to α-decay – as a quantum mechanical tunnel process, which would be extremely unlikely for such heavy masses. Rather, the already existing liquid-drop model of the nucleus [63] should be ap-

[8] "Eure Versuche stellen eine wunderbar geschlossene Beweiskette dar und es ist fabelhaft, was Ihr in diesen kurzen Wochen alles gemacht habt".

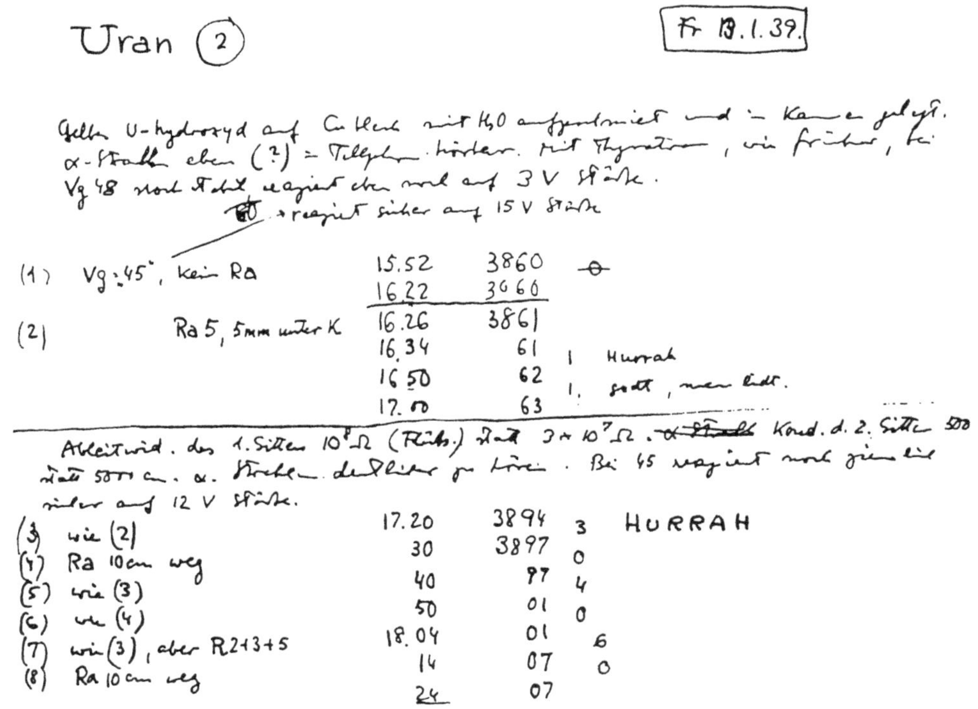

Fig. 11. First physical proof of nuclear fission by the detection of fission fragments with an ionization chamber, performed by Otto Robert Frisch on January 13, 1939, at the University of Copenhagen. Initial part of the protocol sheet "Uran 2", from [67].

plied. Because in very heavy nuclei the attractive nuclear forces between nucleons are nearly compensated by the repulsive forces between protons, "[i]t seems [...] possible that the uranium nucleus has only small stability of form, and may, after neutron capture, divide itself into two nuclei of roughly equal size [...]. These two nuclei will repel each other and should gain a total kinetic energy of c[a]. 200 MeV, as calculated from nuclear radius and charge. This amount of energy may actually be expected to be available from the difference in packing fraction [i.e., binding energy] between uranium and the elements in the middle of the periodic system. The whole 'fission' process can thus be described in an essentially classical way [...]". This is the central statement in their note submitted on January 16 [5]. They coin the term "fission" because of the similarity between the nuclear process and biological cell fission [1].

Meitner and Frisch discuss further aspects of fission [5]. Krypton is thought to be the complementary fragment of barium, and the transuranics are associated with lighter elements; this would make the strange isomeric chains superfluous. The similarity between "radium-actinium" chains from thorium and from uranium is realized, with the conclusion that thorium is fissile, too. The production of 23-min uranium-239 demonstrates that not all nuclear vibrations induced by neutron capture in uranium would lead to fission.

Back in Copenhagen, Frisch informs Niels Bohr on January 3 and on the next day Hahn [64]. Bohr's reaction is drastic: "Oh what fools we have been! We ought to have seen that before [1]". Presumably he had not taken the liquid-drop model that seriously [65] because it was developed to describe static nuclear properties without paying much attention to dynamic aspects. Meitner and Frisch prepare on the telephone their note [5] which is drafted by Frisch on January 6 and modified after a discussion with Bohr. On January 7, Frisch gives him the first two pages of the typescript at the railway station where Bohr is ready to depart for the USA [66]. Bohr promises not to release anything before the note is published [65].

Meanwhile Hahn and Strassmann's first publication [2] has appeared. Frisch discusses the new nuclear process with the theoretician G. Placzek who is very suspicious of it [66]. This motivates Frisch to think about a physical proof of fission [1]. In a gas the energetic fragments should create an unusually large number of ions, by far more than α-particles; this should yield extraordinarily large signals. The simple experiment [1] requires only two days, January 13 and 14, 1939. Figure 11 shows a part of the protocol [67], Figure 12 its author. Frisch has a splendid idea how to suppress the intense background of α-particles during the search for rare signals from fission fragments: He uses electronic pulse-height discrimination, rather than absorption of α-particles in a foil covering the target which would suppress the fission fragments, too.

Frisch places a uranium sample inside a hydrogen-filled ionization chamber of 1 cm depth and 3 cm diameter, and irradiates it with neutrons from a 100 mg radium-beryllium source. The electrical pulses are transmitted to a thyratron serving as the pulse-height discriminator. The upper line of the protocol says: "Yellow U-hydroxide [ammonium diuranate] smeared

Fig. 12. Otto Robert Frisch (1904–1975) around the time of the "Frisch-experiment", from [15].

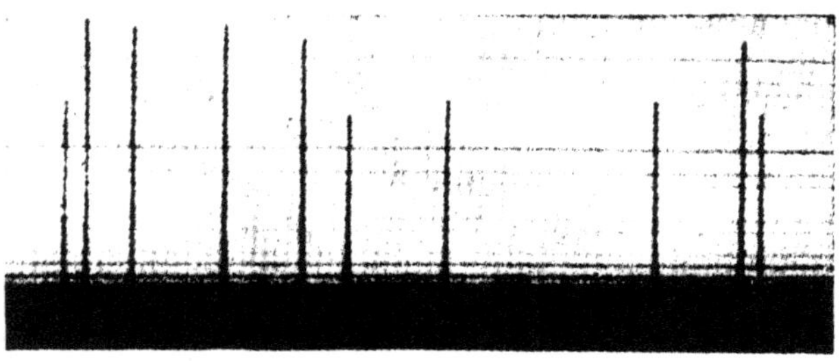

Fig. 13. Oscillographic record of ionization burst caused by fission fragments, taken by Herbert Anderson and John Dunning [68, 69] during the first observation of fission on the American continent on January 25, 1939, at Columbia University, New York; reproduced from [70].

on Cu-plate with H_2O and put into the chamber". The subsequent lines at top and in the middle contain information about the electronic circuit. Readings are alternately taken without (no. 1, 4, 6, 8) and with the neutron source "Ra" (2, 3, 5, 7). Column 3 gives the clock time, 4 the meter readings, and 5 the number of events per counting interval. The first event occurs on Friday, January 13, between 16:34 h and 16:50 h; Frisch welcomes it with "Hurrah". He reports [6]: "By means of a uranium-lined ionization chamber, connected to a linear amplifier, I have succeeded in demonstrating the occurrence of such bursts of ionization". Thorium produces the same effect. A note is submitted on January 16 together with the Meitner-Frisch paper; they appear on February 11 [5] and 18 [6], respectively.

Since Frisch's note does not include an oscillogram of the fission fragment pulses, Figure 13 is taken from another historical experiment [68], the first observation of fission on the American continent on January 25, 1939, [69] at Columbia University, New York. Two quite different groups of fragment signals can be seen which correspond to the light and the heavy fragment of asymmetric fission. In a similar, independent experiment by W. Jentschke and F. Prankl [71] in Vienna, the mean kinetic energies of both fragment groups are already determined surprisingly well to be 98 and 61 MeV, and the total kinetic energy release to about 160 MeV [72]; this should be compared with the present data, 99.8 ± 1.0 MeV and 68.4 ± 0.7 MeV [73].

The employees of Hahn's institute in Berlin learn about the big news when they return from vacations early in January 1939 and the first paper is already in press [74]. Flügge and von Droste [75] publish soon an interpretation of fission on similar lines to the ones of Meitner and Frisch. An experimental proof with an ionization chamber follows somewhat later [76].

At the Collège de France in Paris, Hans von Halban first takes notice of the news and thinks about a physical proof of fission [77]. Hahn and Strassmann's paper arrives around January 16 [78]. Frédéric Joliot now takes the initiative for an experiment performed on January 26 [78]. He infers, independently of Meitner and Frisch, that fission fragments should carry a large kinetic energy, and concludes, that therefore they should recoil out of a thin target layer. Thus, he designs an experiment which does not rely upon pulse height analysis. A thin layer of uranium is spread onto the outer surface of a brass cylinder of 20 mm diameter which contains the radium-beryllium neutron source. Within 3 mm distance from the surface a bakelite cylinder is mounted concentrically. On its inner surface, radioactive fission products are collected during irradiation. With absorber foils placed between target and catcher, the range of fission fragments is estimated to be about 3 cm in air. Thorium shows the same behaviour, with a $\approx$4-h component clearly seen in the decay curve of the collected fission products. Joliot [79] presents his results already on January 30 to the French Academy of Sciences, before Meitner and Frisch's notes [5, 6] appear.

In the same Academy session Irène Curie and Paul Savitch [80] report results of further radiochemical studies which they have resumed a few days earlier [78]. The 3.5-h nuclide is identified in irradiated thorium showing that thorium is fissile, too, and an alkali nuclide is found, observations made in Berlin somewhat earlier [3]. Curie and Savitch comment on their previous work on the 3.5-h nuclide: "At a certain moment, we considered the idea of a break-up of the uranium atom, but we rejected this idea because of the analogy between the formation conditions of the body analogous to rare earths which we found, and the *transuranics* of Hahn, Meitner and Strassmann"[9]. When Georg von Hevesy sees Hahn and Strassmann's first paper he remembers [66] a remark by Irène Curie made to him in autumn 1938, that she might produce very light elements from uranium. Frédéric Joliot pub-

[9] "Nous avions, à un moment donné, envisagé l'idée de la rupture de l'atome d'uranium, mais nous avions rejeté cette idée en raison de l'analogie des conditions de formation du corps analogue aux terres rares que nous avions trouvé et des *transuraniens* de Hahn, Meitner et Strassmann".

Fig. 14. Demonstration of fission with a cloud chamber by Frédéric Joliot [81] at the Collège de France, Paris, in February 1939. The thick, nearly vertical fragment track originates from a uranium target placed at the lower edge of the photograph.

lishes soon [81] one of the first cloud chamber photographs of fission fragments, Figure 14. This is the most direct way of demonstrating fission.

In the USA the news arrives on January 16 [65, 82] through the theoretician Léon Rosenfeld, who accompanied Bohr across the Atlantic but does not know about the obligation to be silent. He reveals it to his companion John Wheeler on the one-hour train trip from New York to Princeton and reports that evening in the regular Physics Journal Club at Princeton University [82]. When Bohr, arriving one or two days later, discovers this, he tells Wheeler more and both start to work on fission theory. As soon as Hahn and Strassmann's first paper arrives in Princeton, Bohr submits a letter [83] commenting on the mechanism of fission with reference to the two notes of Meitner and Frisch. Not until January 22 Frisch writes to Bohr [65] that the two notes are submitted, and reports to him on his experiment performed after Bohr's departure. On the way to a theory conference in Washington, Bohr stops on January 25 in New York, looking for Fermi at Columbia University who has already left. He speaks to Herbert Anderson, one of Fermi's students, who starts that night [69] with John Dunning the experiment furnishing Figure 13. On January 26 the conference is opened with Bohr and Fermi reporting the news, the highlight of the meeting [84]. When the conference is closed on January 28, a local group demonstrates [85] to Bohr and Fermi the huge fragment pulses with an ionization chamber; others communicate this observation by cable [86]. Newspapers report in headlines [88, 89].

A note in a newspaper read by Luis Alvarez on January 27 [90] brings the news to Berkeley where strong, cyclotron-produced neutron fluxes are available. He informs Robert Oppenheimer who instantly pronounces the reaction impossible and proves this by calculations [91]. When the next day Green and Alvarez [92] are able to demonstrate the fission fragment pulses on an oscilloscope, Oppenheimer immediately agrees that the reaction is authentic, and he speculates already about the boil-off of extra neutrons and the possibility of a chain reaction [91]. Most stunned is Philip Abelson [90], a student of Alvarez, who just is attempting to measure the L X-ray energies of 66-h eka-iridium with a bent-crystal spectrometer, in order to prove the assignment to element 95 (Figure 4). A sample isolated after an intense irradiation of uranium is in the spectrometer, but the exposure is not yet completed when the news arrives.

Within a day Abelson [90] switches to the faster, albeit less precise critical absorption technique. He identifies the X-rays to be K X-rays of iodine and confirms by chemical tests, that the 66-h nuclide is tellurium and its 2.4-h daughter is iodine, rather than eka-platinum. Abelson reports this in nine printed lines [93] submitted as a telegram [91]. Thus, two of the "transuranics" of the Berlin researchers are now identified to be lighter elements. This includes the 2.4-h nuclide, one of the cornerstones of their scheme, Figure 4. Not considering light elements as products, the Berlin team did not take into account, that iodide would also be incorporated in ammonium hexachloroplatinate by replacing chloride. One of the first cloud chamber photographs of fission fragments is soon taken in Berkeley [94]. Edwin McMillan [95] independently performs a recoil experiment like the one of Joliot [79] but using a planar geometry, and he measures the penetration depth of fission fragments in aluminium foils and cigarette paper which corresponds to a range of 2.2 cm in air.

Glenn Seaborg [96] vividly describes the reaction of a young nuclear scientist: "I remember very well the Journal Club Meeting the last Monday in January [January 30] in LeConte Hall. You [P. Abelson] were probably there when we were told that the two German chemists had come up with this remarkable discovery of nuclear fission. I remember after that announcement was made it just hit me like a thunderclap, that 'My God, this is obvious', and I can remember after the meeting just walking the streets of Berkeley for hours saying to myself how stupid we had been that we had not been able to figure this out, but, of course, also exhilarated that a couple of fellow chemists had finally straightened out the mess and come, well, come nearly to the right conclusion. I think they were a little reluctant themselves". Literally a flood of studies on fission and fission products is set in motion. About one year after the discovery, L. A. Turner [25] completes the first review article which is based on close to one hundred publications.

How could we understand in retrospective that it took four years to discover the phenomenon, but only a few days to confirm it world-wide? As already pointed out in the Introduction, the two premises made – on the nuclear reactions and the chemical properties – were maintained for a too long time. Only at one earlier occasion in 1936 the Berlin team touched the solution, when Fritz Strassmann in some night-time playful chemistry tried to work out a fast separation for eka-rhenium [97]. Being aware that permanganate gets incorporated into the barium sulfate lattice, he attempted to coprecipitate eka-rhenium – presumed to form a perrhenate-like ion – with barium sulfate. The precipitate was radioactive, but did not show the characteristic 16-min eka-rhenium. Reporting this to Lise Meitner, Strassmann had to admit, that unspecific adsorption effects cannot be ruled out. In turn, she rejected the results, in a friendly manner but quite firmly [97].

The Berlin and Paris teams were confronted with an extremely complex mixture of elements and nuclides. By the end of the second world war, about 100 fission products of 25 elements had been found in Berlin [23], and 170 products of 37 elements were listed in the release [98] of the results kept secret in the USA during the Manhattan-Project [99]. This complexity becomes also obvious in a study [100] with modern high-resolution γ-ray spectroscopy of the whole group of "transuranics" and of individual fractions prepared according to the prescriptions of the Berlin team. The historically most interesting 16-min eka-rhenium (Figure 4) consisted to one half of ^{101}Tc, the remaining half was a mixture of ^{101}Mo, ^{130g}Sb, ^{131}Sb, and ^{131g}Te. The 59-min eka-osmium was mainly ^{134}I with some ^{134}Te and ^{128}Sn, and the 5.7-h eka-osmium was ^{105}Ru with some ^{129}Sb and ^{133}I. The 3.5-h lanthanide-like nuclide of I. Curie and P. Savitch was 3.5-h ^{92}Y – as soon realized [25] – with some contributions of 3.9-h ^{141}La, 1.6-h ^{142}La and, in unprocessed samples, 1.3-h ^{87}Kr. Now it is understandable that with yttrium-92 as the main component, the 3.5-h activity could be fractionated from lanthanum carrier.

Compared to these problems the physical detection of fission fragments with ionization chambers is straightforward. In several leading nuclear physics laboratories, searches were underway for prompt, energetic α-particles from the supposed (n,α)-reactions in uranium and thorium. Such experiments were carried out in Berkeley [91], Berlin [36], Cambridge [101], Heidelberg [102], Rome [7, 18], and Zürich [35]. To suppress during neutron irradiation the strong background of natural α-particles, the targets were often covered with absorber foils. As soon as fission became known, the foils had only to be removed in order to see the fragments. If this happened accidentally, the strong pulses were ignored as being electrical noise [103, 104], or were attributed to the searched-for α-particles [35].

Epilogue, Berkeley, 1939–1940, and Darmstadt, 1981: element 93 and eka-rhenium

After element 93 and eka-rhenium have now disappeared, the search for the first transuranium element begins again. The 23-min uranium-239, left over in the scheme of Figure 4, *must* decay into element 93. The discovery of the daughter product proves to be difficult, however. Edwin McMillan's recoil experiment [95] is a first decisive step. He observes that – in contrast to the complex decay curve of the fission product mixture in the catcher foils – the irradiated thin uranium target shows only two radioactive components: one with 23-min half-life, which he ascribes to uranium-239, and a very strong one with about 2 d half-life. Is this the sought-after element 93? Its recoil energy should be so small that it should stick in the uranium target. From an analysis of the decay curves, Emilio Segrè [105] concludes, however, that the 2.3-d nuclide does not emerge from 23-min uranium-239; at this point he is misled [106] by the low β^--ray energy, 0.47 MeV, of the 2.3-d nuclide. Furthermore, he finds quite unexpected chemical properties. The 2.3-d component does not coprecipitate with rhenium sulfide but rather behaves like a lanthanide element, accompanying lanthanum fluoride, oxalate, and other characteristic precipitates. Furthermore, it can be separated from actinium by fractional crystallization of lanthanum oxalate. Still biased by the eka-rhenium concept, Segrè ascribes the 2.3-d nuclide to a lanthanide fission product and concludes [105] that the 23-min uranium decays into a long-lived element 93 which has not yet been found.

McMillan resumes his recoil experiments after one year [107]. Stronger neutron fluxes allow him the use of thinner uranium layers. Even then, the 2.3-d nuclide remains in the uranium and can, thus, completely be freed from fission products. The 2.3-d nuclide is produced with resonance neutrons just as the 23-min uranium-239; this points to some connection of both nuclides. Philip Abelson, who is now in Washington, observes that the 2.3-d nuclide does not always behave like a lanthanide element. In mid-May 1940 [90] the two start working together at Berkeley. Within five days they find the solution [108]: In a strongly oxidizing medium, bromate, the 2.3-d nuclide does not coprecipitate with cerium fluoride; in a reducing medium, sulfite, it precipitates quantitatively. Based on these results McMillan and Abelson [108] perform their crucial milking experiment: From a reducing solution of 23-min uranium-239, the 2.3-d nuclide is coprecipitated every 20 minutes with cerium fluoride. The initial count rates of the 2.3-d nuclide in these successive samples decrease with a half-life of 23 min. Hence, the 2.3-d nuclide is the decay product of uranium-239: It is element 93, later called neptunium. The chemical behaviour is similar to uranium, in oxidized form to the hexavalent, in reduced form to the tetravalent state. McMillan and Abelson [108] presume a new series of closely related elements starting with uranium.

In Berlin a search for the 2.3-d daughter of the 23-min uranium-239 is undertaken after its description by McMillan [95], but remains without success, as Otto Hahn writes Lise Meitner in May 1939 [45]. Evidently, stronger samples are required. Kurt Starke develops an enrichment procedure based on a Szilard-Chalmers process with uranyl benzoyl acetonate as the irradiated compound; the produced uranium-239 becomes partly converted into an inorganic species and can thus be concentrated from a large bulk of inactive uranium [109]. He observes the 2.3-d element-93 but the publication of his results is delayed [110]. Strassmann and Hahn [111] succeed in finding element 93 by a direct separation from large amounts of uranium, after they have access to a thin-window counter penetrated by the soft β-rays of 2.3-d element 93. In a detailed study they confirm already described chemical properties [105, 108] and add new ones. Everything points at a discontinuity in the Periodic Table: Element 93 has nothing in common with eka-rhenium.

It takes more than four decades until eka-rhenium, now element 107, is eventually discovered by Peter Armbruster, Gottfried Münzenberg and their team at the heavy-ion accelerator in Darmstadt [112]. The isotope 262 with 8.2 ms half-life is synthesized by fusion of chromium-54 projectiles with bismuth-209 targets, and is identified by a long decay chain of known nuclides [113]. Whether it chemically behaves like eka-rhenium is not known yet. At present, element 105 is the heaviest element whose chemistry has been investigated [114, 115].

References

1. Frisch, O. R.: The Discovery of Fission – How It All Began. Phys. Today **20** (11), 43–48 (1967).
2. Hahn, O., Strassmann, F.: Über den Nachweis und das Verhalten der bei der Bestrahlung des Urans mittels Neutronen entstehenden Erdalkalimetalle. Naturwissenschaften **27**, 11–15 (1939) [received Dec. 22, 1938].
3. Hahn, O., Strassmann, F.: Nachweis der Entstehung aktiver Bariumisotope aus Uran und Thorium durch Neutronenbestrahlung; Nachweis weiterer aktiver Bruchstücke bei der Uranspaltung. Naturwissenschaften **27**, 89–95 (1939) [Jan. 28, 1939].
4. Hyde, E. K.: *The Nuclear Properties of the Heavy Elements, Vol. III, Fission Phenomena,* Prentice Hall, Englewood Cliffs 1964, p. 4.
5. Meitner, L., Frisch, O. R.: Disintegration of Uranium by Neutrons: a New Type of Nuclear Reaction. Nature **143**, 239–240 (1939) [Jan. 16, 1939].
6. Frisch, O. R.: Physical Evidence for the Division of Heavy Nuclei under Neutron Bombardment. Nature **143**, 276 (1939) [Jan. 16, 1939].
7. Segrè, E.: Fermi and Neutron Physics. Rev. Mod. Phys. **27**, 257–263 (1955).
8. Herrmann, G.: Vor fünf Jahrzehnten: Von den „Transuranen" zur Kernspaltung. Angew. Chem. **102**, 469–496 (1990); Five Decades Ago: From the "Transuranics" to Nuclear Fission. Angew. Chem. Int. Ed. Engl. **29**, 481–508 (1990).
9. Fermi, E.: Radioactivity Induced by Neutron Bombardment. Nature **133**, 757 (1934).
10. Fermi, E., Amaldi, E., D'Agostino, O., Rasetti, F., Segrè, E.: Artificial Radioactivity Produced by Neutron Bombardment. Proc. R. Soc. London **146**, 483–498 (1934).
11. Fermi, E.: Possible Production of Elements of Atomic Number Higher than 92. Nature **133**, 898–899 (1934).
12. Grosse, A. V., Agruss, M. S.: The Identity of Fermi's Reactions of Element 93 with Element 91. J. Am. Chem. Soc. **57**, 438–439 (1935).
13. Meitner, L.: Über die Umwandlung der Elemente durch Neutronen. Naturwissenschaften **22**, 759 (1934).
14. Meitner, L.: Letter to M. von Laue of Sept. 4, 1944, printed in [15], pp. 128–129, under the erroneous date of Sept. 4, 1941.
15. Lemmerich, J.: *Die Geschichte der Entdeckung der Kernspaltung*. Catalogue for the Exhibition at the Deutsches Museum, München, Bibliothek der Technischen Universität, Berlin 1988.
16. Hahn, O., Meitner, L.: Über die künstliche Umwandlung des Urans durch Neutronen. Naturwissenschaften **23**, 37–38 (1935).
17. Amaldi, E., D'Agostino, O., Fermi, E., Pontecorvo, B., Rasetti, F., Segrè, E.: Artificial Radioactivity Produced by Neutron Bombardment II. Proc. R. Soc. London **149**, 522–558 (1935).
18. Amaldi, E.: in [19], p. 278.
19. Amaldi, E.: From the Discovery of the Neutron to the Discovery of Nuclear Fission. Phys. Rep. **111**, 1–332 (1984).
20. Noddack, I.: Über das Element 93. Angew. Chem. **47**, 653–655 (1934).
21. Van Assche, P. H. M.: The Ignored Discovery of the Element Z = 43. Nucl. Phys. **A 480**, 205–214 (1988).
22. Amaldi, E.: in [19], p. 277.
23. Hahn, O.: Von den natürlichen Umwandlungen des Urans zu seiner künstlichen Zerspaltung. Nobel Lecture on December 13, 1946. In: *Les Prix Nobel en 1946*, Norstedt, Stockholm 1948, pp. 167–183; reprinted in [24], pp. 247–267.
24. Hahn, O.: *Mein Leben*, 1st edit. 1968, new edit. by D. Hahn, Piper, München 1986.
25. Turner, L. A.: Nuclear Fission. Rev. Mod. Phys. **12**, 1–29 (1940) [completed Dec. 6, 1939].
26. Strassmann, F.: *Kernspaltung – Berlin Dezember 1938*, private printing, Mainz 1978.
27. Meitner, L., Hahn, O., Strassmann, F.: Über die Umwandlungsreihen des Urans, die durch Neutronenbestrahlung erzeugt werden. Z. Phys. **106**, 249–270 (1937) [May 14, 1937].
28. Hahn, O., Meitner, L., Strassmann, F.: Über die Transurane und ihr chemisches Verhalten. Ber. Dtsch. Chem. Ges. **70**, 1374–1392 (1937) [May 15, 1937].
29. Hahn, O., Meitner, L., Strassmann, F.: Ein neues langlebiges Umwandlungsprodukt in den Trans-Uranreihen. Naturwissenschaften **26**, 475–476 (1938) [July 12, 1938].
30. Hahn, O., Meitner, L.: Die künstliche Umwandlung des Thoriums durch Neutronen: Bildung der bisher fehlenden radioaktiven 4n + 1 Reihe. Naturwissenschaften **23**, 320 (1935).
31. Curie, I., von Halban, H., Preiswerk, P.: Sur la création artificielle des éléments d'une famille radioactive inconnue, lors de l'irradiation du thorium par les neutrons. C. R. Hebd. Séances Acad. Sci. **200**, 1841–1843 (1935).
32. Curie, I., von Halban, H., Preiswerk, P.: Sur la création artificielle d'éléments appartenant à une famille radioactive inconnue, lors de l'irradiation du thorium par les neutrons. J. Phys. Radium **6**, 361–364 (1935).
33. Meitner, L., Hahn, O., Strassmann, F.: Künstliche Umwandlungsprozesse bei Bestrahlung des Thoriums mit Neutronen; Auftreten isomerer Reihen durch Abspaltung von α-Strahlen. Z. Phys. **109**, 538–552 (1938) [May 16, 1938].
34. Nishina, Y., Yasaki, T., Kimura, K., Ikawa, M.: Artificial Production of Uranium Y from Thorium. Nature **142**, 874 (1938) [Sept. 13, 1938].
35. Braun, A., Preiswerk, P., Scherrer, P.: Detection of α-Particles in the Disintegration of Thorium. Nature **140**, 682 (1937) [Aug. 31, 1937].
36. von Droste, G.: Über Versuche eines Nachweises von α-Strahlen während der Bestrahlung von Thorium und Uran mit Radium + Beryllium-Neutronen. Z. Phys. **110**, 84–94 (1938) [May 25, 1938].
37. Curie, I., Savitch, P.: Sur les radioéléments formés dans l'uranium irradié par les neutrons. J. Phys. Radium **8**, 385–387 (1937) [Aug. 1, 1937].

38. Meitner, L., Hahn, O.: Letter to I. Curie of Jan. 20, 1938, Archives of the Institut Curie, Paris.
39. Curie, I., Savitch, P.: Sur le radioélément de période 3,5 heures formé dans l'uranium irradié par les neutrons. C. R. Hebd. Séances Acad. Sci. **206**, 906–908 (1938) [March 21, 1938].
40. Curie, I., Savitch, P.: Sur la nature du radioélément de période 3,5 heures formé dans l'uranium irradié par les neutrons. C. R. Hebd. Séances Acad. Sci. **206**, 1643–1644 (1938) [May 30, 1938].
41. Curie, I., Savitch, P.: Sur les radioéléments formés dans l'uranium irradié par les neutrons II. J. Phys. Radium **9**, 355–359 (1938) [July 12, 1938].
42. Hahn, O.: *Erinnerungen 1901–1945*, written in June 1945, published in [43], pp. 15–73.
43. Hahn, O.: *Erlebnisse und Erkenntnisse* (D. Hahn, ed.), Econ, Düsseldorf 1975.
44. Hahn, O.: Die Auffindung der Uranspaltung. In: *FIAT Review of German Science 1939–1946, Nuclear Physics and Cosmic Rays, Part 1*, Dieterich, Wiesbaden 1948, pp. 171–178.
45. Krafft, F. (ed.): *Briefwechsel Otto Hahn – Lise Meitner, 11. Aug. 1938–25. März 1945*, printed in [46], pp. 74–135, 228–337.
46. Krafft, F.: *Im Schatten der Sensation – Leben und Wirken von Fritz Strassmann*, Verlag Chemie, Weinheim 1981.
47. Hahn, O., Strassmann, F.: Laboratory notebook "Chem. II", Nov. 1, 1938–Jan. 14, 1939, Deutsches Museum, München; commented transcription (without counting data) in [46], pp. 234–280.
48. Hahn, O., Strassmann, F.: Über die Entstehung von Radiumisotopen aus Uran durch Bestrahlen mit schnellen und verlangsamten Neutronen. Naturwissenschaften **26**, 755–756 (1938) [Nov. 8, 1938].
49. Hahn, O.: Ein neues Zwischenprodukt im Thorium. Ber. Dtsch. Chem. Ges. **40**, 1462–1469 (1907).
50. Hahn, O.: Ein kurzlebiges Zwischenprodukt zwischen Mesothor und Radiothor. Phys. Z. **9**, 246–248 (1908).
51. Hahn, O.: in [47], p. 75v.
52. Strassmann, F., Bohne, I., Lieber, C., Hahn, O.: in [47], p. 76r.
53. Hahn, O.: in [47], p. 77r.
54. Hahn, O.: Taschenkalender 1938, reprinted, partly as facsimile, in [55].
55. Brix, P.: Die folgenreiche Entdeckung der Uranspaltung – und wie es dazu kam. Phys. Bl. **45**, 2–10 (1989); Naturwiss. Rundsch. **42**, 1–9 (1989); The Discovery of Uranium Fission – Its Intricate History and Far-Reaching Consequences. Interdisc. Sci. Rev. **15**, 314–322 (1990).
56. Hahn, O.: Letter to L. Meitner on Dec. 19, 1938, facsimile in [8] and [57], p. 171, transcription in [8] and [46], pp. 263–264.
57. Hahn, D. (ed.): *Otto Hahn, Leben und Werk in Texten und Bildern*, Insel, Frankfurt 1988.
58. Hahn, O.: in [59], pp. 131–132.
59. Hahn, O.: *Vom Radiothor zur Uranspaltung – eine wissenschaftliche Selbstbiographie*, 1st edit. 1962, reprint ed. by D. Hahn, Vieweg, Braunschweig 1989.
60. Hahn, O.: Letter to E. von Bahr-Bergius on Dec. 23, 1938, printed in [46], p. 267.
61. Krafft, F.: in [46], pp. 254, 267.
62. Sime, R. L.: Lise Meitner und die Kernspaltung: "Fallout" der Entdeckung. Angew. Chem. **103**, 956–967 (1991); Lise Meitner and Fission: Fallout from the Discovery. Angew. Chem. Int. Ed. Engl. **30**, 942–953 (1991).
63. Gamov, G.: Mass Defect Curve and Nuclear Constitution. Proc. R. Soc. London A **126**, 632–644 (1930).
64. Frisch, O. R.: Letters to L. Meitner on Jan. 3, 1939, printed in [65], and to O. Hahn on Jan. 4, printed in [46], pp. 271–272.
65. Stuewer, R. H.: Bringing the News of Fission to America. Phys. Today **38** (10), 49–56 (1985).
66. Frisch, O. R.: Letter to L. Meitner on Jan. 8, 1939, printed in [15], pp. 179, 182.
67. Frisch, O. R.: Laboratory protocol of Jan. 13, 1939, facsimile in [15], p. 181.
68. Anderson, H. L., Booth, E. T., Dunning, J. R., Fermi, E., Glasoe, G. N., Slack, F. G.: The Fission of Uranium. Phys. Rev. **55**, 509–510 (1939).
69. Anderson, H. L.: The Legacy of Fermi and Szilard. Bull. At. Sci. **30** (9), 56–62 (1974).
70. Glasstone, S.: *Sourcebook on Atomic Energy*, 3rd. edit., Van Nostrand, New York 1967, p. 477.
71. Jentschke, W., Prankl, F.: Der Nachweis von schweren Kernbruchstücken beim Zerfall von neutronenbestrahltem Uran. Anz. Akad. Wiss. Wien, mathem.-naturw. Kl. **1939** (5), 19–20 [Feb. 9, 1939].
72. Jentschke, W., Prankl, F.: Untersuchung der schweren Kernbruchstücke beim Zerfall von neutronenbestrahltem Uran und Thorium. Naturwissenschaften **27**, 134–135 (1939) [Feb. 14, 1939].
73. Milton, J. C. D., Fraser, J. S.: Time-of-Flight Fission Studies on U^{233}, U^{235}, and Pu^{239}. Canad. J. Phys. **40**, 1626–1663 (1962).
74. Flügge, S.: Zur Entdeckung der Uranspaltung vor zehn Jahren. Z. Naturforschg. **4a**, 82–84 (1949).
75. Flügge, S., von Droste, G.: Energetische Betrachtungen zu der Entstehung von Barium bei der Neutronenbestrahlung von Uran. Z. Phys. Chem. B **42**, 274–280 (1939) [Jan. 22, 1939].
76. von Droste, G.: Über die Energieverteilung der bei der Bestrahlung von Uran mit Neutronen entstehenden Bruchstücke. Naturwissenschaften **27**, 198 (1939) [March 17, 1939].
77. Radvanyi, P., Bordry, M.: *La radioactivité artificielle et son histoire*, Seuil/CNRS, Paris 1984, p. 149.
78. Weart, S. R.: Scientists in Power, Harvard Univ. Press, Cambridge 1979, pp. 63–64.
79. Joliot, F.: Preuve expérimentale de la rupture explosive des noyaux d'uranium et de thorium sous l'action des neutrons. C. R. Hebd. Séances Acad. Sci. **208**, 341–343 (1939) [Jan. 30, 1939].
80. Curie, I., Savitch, P.: Sur les radioéléments formés dans l'uranium et le thorium irradiés par les neutrons. C. R. Hebd. Séances Acad. Sci. **208**, 343–346 (1939) [Jan. 30, 1939].
81. Joliot, F.: Observation par la méthode de Wilson des trajectoires de brouillard des produits de l'explosion des noyaux de l'uranium. C. R. Hebd. Séances Acad. Sci. **208**, 647–649 (1939) [Feb. 27, 1939].
82. Wheeler, J. A.: Fission in 1939: The Puzzle and the Promise. Annu. Rev. Nucl. Part. Sci. **39**, XIII–XXVIII (1989).
83. Bohr, N.: Disintegration of Heavy Nuclei. Nature **143**, 330 (1939) [Jan. 20, 1939].
84. Squire, C. F., Brickwedde, F. G., Teller, E., Tuve, M. A.: The Fifth Washington Conference on Theoretical Physics. Science **89**, 180–182 (1939).
85. Roberts, R. B., Meyer, R. C., Hafstad, L. R.: Droplet Fission of Uranium and Thorium Nuclei. Phys. Rev. **55**, 416–417 (1939) [Feb. 4, 1939].
86. Fowler, R. D., Dodson, R. W.: Communication by cable, Nature **143**, 233 (1939) [Feb. 3, 1939], see also [87].
87. Fowler, R. D., Dodson, R. W.: Intensely Ionizing Particles Produced by Neutron Bombardment of Uranium and Thorium. Phys. Rev. **55**, 417–418 (1939) [Feb. 3, 1939].
88. Anon.: Atom Explosion Frees 200,000,000 Volts; New Physics Phenomenon Credited to Hahn. New York Times, Jan. 28, 1939, reprinted in [59], p. 145.
89. Badash, L., Hodes, E., Tiddens, A.: Nuclear Fission: Reaction to the Discovery in 1939. Proc. Am. Philos. Soc. **130**, 196–231 (1986).
90. Abelson, P. H.: Products of Neutron Irradiation. J. Chem. Educ. **66**, 364–366 (1989).
91. Alvarez, L. W.: *Alvarez – Adventures of a Physicist*, Basic Books, New York 1987, pp. 72–76.
92. Green, G. K., Alvarez, L. W.: Heavily Ionizing Particles from Uranium. Phys. Rev. **55**, 417 (1939) [Jan. 31, 1939].

93. Abelson, P.: Cleavage of the Uranium Nucleus. Phys. Rev. **55**, 418 (1939) [Feb. 3, 1939].
94. Corson, D. R., Thornton, R. L.: Disintegration of Uranium. Phys. Rev. **55**, 509 (1939) [Feb. 15, 1939].
95. McMillan, E.: Radioactive Recoils from Uranium Activated by Neutrons. Phys. Rev. **55**, 510 (1939) [Feb. 17, 1939].
96. Seaborg, G. T.: Discussion remark. In: *The Robert A. Welch Foundation Conference on Chemical Research XXXIV – Fifty Years with Transuranium Elements*. Welch Foundation, Houston 1990, p. 19.
97. Strassmann, F.: Personal communication to the author, ca. 1950, and hand-written note 1969, reproduced in [46]. p. 221.
98. Siegel, J. M.: Nuclei Formed in Fission: Decay Characteristics, Fission Yields, and Chain Relationships. J. Amer. Chem. Soc. **68**, 2411–2442 (1946); Rev. Mod. Phys. **18**, 513–544 (1946).
99. Steinberg, E. P.: Radiochemistry of the Fission Products. J. Chem. Educ. **66**, 367–372 (1989).
100. Menke, H., Herrmann, G.: Was waren die „Transurane" der dreißiger Jahre in Wirklichkeit? Radiochim. Acta **16**, 119–123 (1971).
101. Goldhaber, M.: Reminiscenses from the Cavendish Laboratory in the 1930s. Annu. Rev. Nucl. Part. Sci. **43**, 1–25 (1993).
102. Maier-Leibnitz, H., quoted in A.-L. Edingshaus: *Heinz Maier-Leibnitz – Ein halbes Jahrhundert experimentelle Physik*, Piper, München 1986, p. 66.
103. Curran, S., quoted in G. N. Walton: Nuclear Fission. Q. Rev. Chem. Soc. **15**, 74 (1961).
104. Flügge, S.: How Fission was Discovered. In: *50 Years with Nuclear Fission* (J. W. Behrens, A. D. Carlson, eds.), Amer. Nucl. Soc., La Grange Park 1989, pp. 26–29.
105. Segrè, E.: An Unsuccessful Search for Transuranic Elements. Phys. Rev. **55**, 1104–1105 (1939) [May 10, 1939].
106. Segrè, E. G.: The Discovery of Nuclear Fission. Phys. Today **42** (7), 38–43 (1989).
107. McMillan, E.: The Transuranium Elements; Early History. Nobel Lecture on December 12, 1951. In: *Les Prix Nobel en 1951*, Norstedt, Stockholm 1952, pp. 165–173.
108. McMillan, E., Abelson, P. H.: Radioactive Element 93. Phys. Rev. **57**, 1185–1186 [May 27, 1940].
109. Starke, K.: Abtrennung des Elements 93. Naturwissenschaften **30**, 107–108 (1942) [Jan. 19, 1942].
110. Starke, K.: Zur Frühgeschichte des Neptuniums. Isotopenpraxis **26**, 349–351 (1990).
111. Strassmann, F., Hahn, O.: Über die Isolierung und einige Eigenschaften des Elements 93. Naturwissenschaften **30**, 256–260 (1942) [Apr. 2, 1942].
112. Münzenberg, G., Hofmann, S., Heßberger, F. P., Reisdorf, W., Schmidt, K. H., Schneider, J. H. R., Armbruster, P., Sahm, C. C., Thuma, B.: Identification of Element 107 by α Correlation Chains. Z. Phys. A **300**, 107–108 (1981).
113. Münzenberg, G.: Discovery, Synthesis and Nuclear Properties of the Heaviest Elements. Radiochim. Acta **70/71**, 193–206 (1995).
114. Schädel, M.: Chemistry of the Transactinide Elements. Radiochim. Acta **70/71**, 207–223 (1995).
115. Trautmann, N.: Fast Radiochemical Separations for Heavy Elements. Radiochim. Acta **70/71**, 237–245 (1995).

Radiochimica Acta 70/71, 69–90 (1995)

Transuranium Elements: the Synthetic Actinides

By Glenn T. Seaborg*

Lawrence Berkeley Laboratory, Berkeley, California 94720, USA

(Received April 3, 1995)

Transuranium elements / Atomic number 93–103 / History

Abstract

The era "One Hundred Years of Radioactivity" began with discovery of the radioactivity of uranium in early 1896. For more than half of that span of time we have been investigating the radioactivity of the elements beyond uranium, i.e., the "Transuranium Elements". The first two transuranium elements, neptunium (atomic number 93) and plutonium (94) were synthesized and identified, i.e., discovered, 55 years ago. I became interested in the transuranium field more than 60 years ago, very soon after the discovery of artificial radioactivity. In this account I shall describe my involvement with the first eleven transuranium elements, atomic numbers 93–103, inclusive, the synthetic actinide elements.

Before fission and fission

I began to read the papers coming out of Italy and Germany describing the synthesis and identification of several elements thought to be transuranium elements about five years before the discovery of nuclear fission, as a first year graduate student at Berkeley in 1934. In their original work in 1934, Enrico Fermi, Edoardo Amaldi, Oscar D'Agostino, Franco Rasetti, and Emilio Segrè bombarded uranium with neutrons and obtained a series of beta-particle-emitting radioactivities. On the basis of the periodic table of that day (Fig. 1) they were led to believe that the first transuranium element, with atomic number 93, should be chemically like rhenium (i.e., be eka-rhenium, Eka-Re), element 94 like osmium (Eka-Os) and so forth. Therefore they assigned a 13-minute activity to element 93. I quote from a classical paper [1] written by Fermi, entitled "Possible Production of Elements of Atomic Number Higher than 92", which I remember reading at that time:

> This negative evidence about the identity of the 13 min-activity from a large number of heavy elements suggests the possibility that the atomic number of the element may be greater than 92. If it were an element 93, it would be chemically homologous with manganese and rhenium. This hypothesis is supported to some extent also by the observed fact that the 13 min-activity is carried down by a precipitate of rhenium sulfide insoluble in hydrochloric acid. However, as several elements are easily precipitated in this form, this evidence cannot be considered as very strong.

I recall reading soon thereafter a paper by Ida Noddack [2], entitled "Über das Element 93" [On Element 93], which took issue with this interpretation, suggesting that the radioactivities observed by Fermi *et al.* might be due to elements of medium atomic numbers:

> Es wäre denkbar, daß bei der Beschießung schwerer Kerne mit Neutronen diese Kerne in mehrere größere Bruchstücke zerfallen, die zwar Isotope bekannter Elemente, aber nicht Nachbarn der bestrahlten Elemente sind. [One could think that in the bombardment of heavy nuclei with neutrons these nuclei disintegrate into several larger fragments which, although they are isotopes of known elements, are not neighbors of the irradiated elements.]

However this paper, which intimated the possibility of the nuclear fission reaction, was not taken seriously.

Experiments in Germany during the following years by Otto Hahn, Lise Meitner, and Fritz Strassmann (Fig. 2) appeared to confirm the Italian interpretation and for several years the "transuranium elements" were the subject of much experimental work and discussion. In a typical paper by Hahn, Meitner, and Strassmann [3], which I read, part of a series they published during 1935–1938, they reported a 16-minute ${}_{93}$Eka-Re^{237}, 2.2-minute ${}_{93}$Eka-Re^{239}, 12-hour ${}_{94}$Eka-Os^{237}, 59-minute ${}_{94}$Eka-Os^{239}, 3-day ${}_{95}$Eka-Ir^{239}, 12-hour ${}_{96}$Eka-Pt^{239}.

In 1938 Irene Curie and Pavle Savitch [4], working in Paris, found a product of 3.5 hours half-life that seemed to have the chemical products of a rare earth, but they failed to give an interpretation of this astonishing discovery. Their paper, which I also read at the time, had the title, "Sur la nature du radioélément de période 3,5 heures formé dans l'uranium irradié par les neutrons" [On the Nature of a Radioactive Element with 3.5 Hour Half-Life Produced in the Neutron Irradiation of Uranium] and included the following:

> Nous avons montré qu'il se forme dans l'uranium irradié par les neutrons un radioélément de période 3,5 heures dont les propriétés chimiques sont semblables a celles des terres rares. Nous la désignerons ci-dessous par la notation $R_{3,5\,h}$.

---

* The author is University Professor of Chemistry at the University of California, Berkeley, and Associate Director-at-Large of the Lawrence Berkeley Laboratory. Support for this article was provided by the U.S. Department of Energy.

1 H																	2 He
3 Li	4 Be											5 B	6 C	7 N	8 O	9 F	10 Ne
11 Na	12 Mg											13 Al	14 Si	15 P	16 S	17 Cl	18 Ar
19 K	20 Ca	21 Sc	22 Ti	23 V	24 Cr	25 Mn	26 Fe	27 Co	28 Ni	29 Cu	30 Zn	31 Ga	32 Ge	33 As	34 Se	35 Br	36 Kr
37 Rb	38 Sr	39 Y	40 Zr	41 Nb	42 Mo	(43)	44 Ru	45 Rh	46 Pd	47 Ag	48 Cd	49 In	50 Sn	51 Sb	52 Te	53 I	54 Xe
55 Cs	56 Ba	57–71 La–Lu	72 Hf	73 Ta	74 W	75 Re	76 Os	77 Ir	78 Pt	79 Au	80 Hg	81 Tl	82 Pb	83 Bi	84 Po	(85)	86 Rn
(87)	88 Ra	89 Ac	90 Th	91 Pa	92 U	(93)	(94)	(95)	(96)	(97)	(98)	(99)	(100)				
		57 La	58 Ce	59 Pr	60 Nd	(61)	62 Sm	63 Eu	64 Gd	65 Tb	66 Dy	67 Ho	68 Er	69 Tm	70 Yb	71 Lu	

Fig. 1. Periodic table before World War II. Parentheses indicate undiscovered elements at that time.

Fig. 2. Fritz Strassmann, Lise Meitner, and Otto Hahn, Mainz, Germany 1956.

$R_{3,5\,h}$ se sépare nettement de Ac, allant en tête de fractionnement, alors que Ac va en queue. Il semble donc que ce corps ne puisse être qu'un élément transuranien possédant des propriétés très différente de celles des autre éléments transuraniens connus, hypothèse qui soulève des difficultés d'interprétation.

[We have shown that in the neutron irradiation of uranium a radioactive element with a half-life of 3.5 hours is produced, with chemical properties similar to those of rare earths. In the following we will refer to it as $R_{3.5\,h}$.

$R_{3.5\,h}$ separates cleanly from Ac by going to the 'head' (beginning) of the fractionation while Ac goes to the 'tail' (end). It seems, therefore, that this species cannot be but a transuranic element having properties very different from those of the other known transuranic elements, a hypothesis which raises interpretational difficulties.]

The breakthrough came early in 1939, when Hahn and Strassmann [5], on the basis of experiments performed in December 1938, and with interpretive help from Meitner, who had been forced to leave Germany, described experiments in which they had observed barium isotopes as the result of bombardment of uranium with neutrons. This historic paper, which I also read at the time, had the title, "Über den Nachweis und das Verhalten der bei der Bestrahlung des Urans mittels Neutronen entstehenden Erdalkalimetalle" [On the Identification and the Behavior of Rare Earth Metals Produced in the Neutron Irradiation of Uranium] and contained the following conclusion:

> Als Chemiker müßten wir aus den kurz dargelegten Versuchen das oben gebrachte Schema eigentlich umbenennen und statt Ra, Ac, Th die Symbole Ba, La, Ce einsetzen. Als der Physik in gewisser Weise nahestehende 'Kernchemiker' können wir uns zu diesem, allen bisherigen Erfahrungen der Kernphysik widersprechenden, Sprung noch nicht entschließen. Es könnten doch noch vielleicht eine Reihe seltsamer Zufälle unsere Ergebnisse vorgetäuscht haben.
>
> [We, as chemists, based on the briefly described experiments, should rename the above-mentioned scheme and replace Ra, Ac, Th with the symbols Ba, La, Ce. As nuclear chemists, being in some respects close to physics, we have not yet been able to take this leap which contradicts all previous experiences in nuclear physics. It could be that a series of strange coincidences could have mimicked our results.]

Subsequent work showed that the radioactivities previously ascribed to transuranium elements are actually due to uranium fission products, and hundreds of radioactive fission products of uranium have since been identified.

Thus in early 1939 there were again, as five years earlier, no known transuranium elements. During these five years I developed an increasing interest in the transuranium situation. When as a graduate student I gave my required annual talk at the College of Chemistry weekly Research Conference in 1936, I chose the transuranium elements as my topic, describing the work of Hahn, Meitner, and Strassmann referred to above.

During the two years following my seminar talk in 1936 and before the discovery of fission, my interest in the neutron-induced radioactivities in uranium continued unabated and, in fact, increased. I read and reread every article published on the subject. I was puzzled by the situation, both intrigued by the concept of the transuranium interpretation of the experimental results and disturbed by the apparent inconsistencies in their interpretation. I remember discussing the problem with Joseph W. Kennedy, a colleague in research, by the hour, often in the postmidnight hours of the morning at the old Varsity Coffee Shop on the corner of Telegraph and Bancroft Avenues near the Berkeley campus where we often went for a cup of coffee and a bite to eat after an evening spent in the laboratory.

I first learned of the correct interpretation of these experiments, that neutrons split uranium into two large pieces in the fission reaction, at the weekly Monday night seminar in nuclear physics conducted by Professor Ernest O. Lawrence in Le Conte Hall. On this exciting night in January 1939, we heard the news from Germany of Hahn and Strassmann's beautiful chemical experiments. I recall that at first the fission interpretation was greeted with some skepticism by a number of those present, but, as a chemist with a particular appreciation for Hahn and Strassmann's experiments, I felt that this interpretation just had to be accepted. I remember walking the streets of Berkeley for hours after this seminar in a combined state of exhilaration in appreciation of the beauty of the work and of disgust at my inability to arrive at this interpretation despite my years of contemplation on the subject.

First transuranium elements, neptunium (93) and plutonium (94)

With those radioactivities identified as fission products, there were no longer any transuranium elements left. However, in later investigations by Edwin M. McMillan [6] at Berkeley and others elsewhere, one of the radioactivities behaved differently from the others. The beta radioactivity with a half-life of about 2 days did not separate by recoil from thin layers of uranium, as did the energetic fission products, when uranium was bombarded with slow neutrons. Along toward the spring of 1940, McMillan (Fig. 3) began to come to the conclusion that the 2.3-day activity might actually be due to the daughter of the 23-minute uranium-239 and thus might indeed be an isotope of element 93 with the mass number 239 (93−239). Philip H. Abelson (Fig. 4) joined him in this work in the spring of 1940, and together they were able to chemically separate and identify and thus discover [7] element 93 formed in the following reaction sequences:

$$^{238}_{92}U + ^{1}_{0}n \rightarrow ^{239}_{92}U + \gamma$$

$$^{239}_{92}U \xrightarrow[t_{1/2}=23\ min]{\beta^-} {}^{239}93 \xrightarrow[t_{1/2}=2.3\ d]{\beta^-} .$$

They showed that element 93 has chemical properties similar to those of uranium and not similar to those of rhenium as suggested by the periodic table of that time (Fig. 1).

Immediately thereafter, during the summer and fall of 1940, McMillan started looking for the daughter product of the 2.3-day activity, which obviously would be the isotope of element 94 with mass number 239 (94−239). Not finding anything he could positively

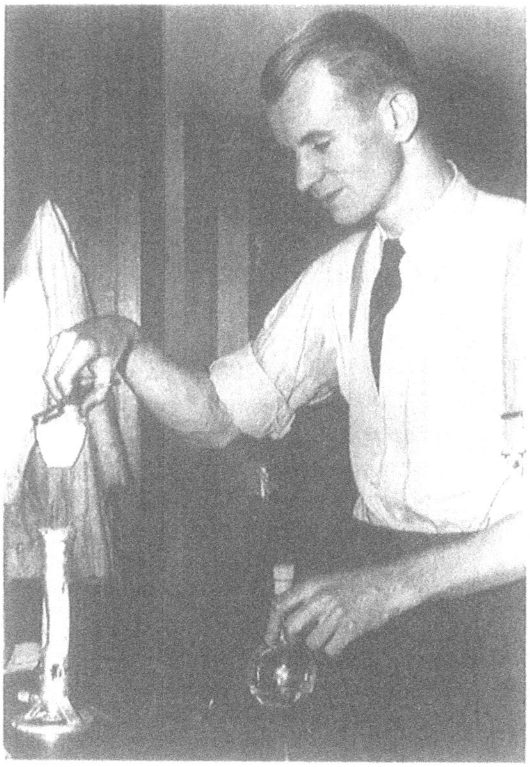

Fig. 3. The discoverers of neptunium: Edwin M. McMillan, Berkeley, California. June 8, 1940.

Fig. 4. The discoverers of neptunium: Philip H. Abelson.

identify as such, he began to bombard uranium with deuterons in the 60-Inch Cyclotron in the hope that he might find a shorter-lived isotope – one of a higher intensity of radioactivity that would be easier to identify as an isotope of element 94. Before he could finish this project, he was called away for war work on radar at MIT.

During this time my interest in the transuranium elements continued. Since McMillan and I lived only a few rooms apart in the Faculty Club, we saw each other quite often, and, as I recall, much of our conversation, whether in the laboratory, at meals, in the hallway, or even going in and out of the shower, had something to do with element 93 and the search for element 94. I must say, therefore, that his sudden departure for MIT came as something of a surprise to me – especially since I did not even know when he had gone.

In the meantime, I had asked Arthur C. Wahl, one of my two graduate students, to begin studying the tracer chemical properties of element 93 with the idea that this might be a good subject for his thesis. My other coworker was Joe Kennedy, now a fellow instructor in the Department of Chemistry and also very interested in the general transuranium problem.

I quote from my diary of Friday, August 30, 1940:

> This afternoon Wahl, Kennedy, and I irradiated our first sample of uranium with neutrons to produce the recently discovered isotope 93^{239} in order to begin a program of study, by tracer technique, of the chemical properties of element 93. This research and its possible expansion into the search for the next transuranium element, element 94, may provide a suitable subject for Wahl's Ph.D. thesis. Today's bombardment used 5.5 g uranyl nitrate hexahydrate [$UO_2(NO_3)_2 6H_2O$ or UNH] placed directly behind a beryllium target bombarded with 16 MeV deuterons in the 60-inch cyclotron. This UNH sample was dissolved in water, and the oxidation-reduction cycle was performed by Wahl in order to isolate the 93^{239} from uranium and fission products in order to characterize its radiation and follow its decay to see if it displays the known half-life of 93^{239}, 2.3 days. Absorption measurements of the radiations in aluminum will be made.
>
> In order to distinguish between this and subsequent bombardments, we intend to use the following nomenclature. Since our primary product of interest is element 93, our designation in each case will start with 93 followed in turn by a number designating the bombardment and a second designating the chemical fraction within that bombardment. Thus, this first bombardment is designated 93-1; the chemical fractions separated are designated 93-1-1, 93-1-2, and so forth. Today the purified element 93 fraction upon which the decay and radiation measurements are being made is being designated 93-1-3.

When I learned that McMillan had gone, I wrote to him asking whether it might not be a good idea if we carried on the work he had started, especially the deuteron bombardment of uranium. He readily assented.

Our first deuteron bombardment of uranium was conducted on December 14, 1940. What we bombarded was a form of uranium oxide, U_3O_8, which was literally plastered onto a copper backing plate. From this bombarded material Wahl isolated a chemical fraction of element 93. The radioactivity of this fraction was measured and studied. We observed that it had different characteristics than the radiation from a sample of pure 93-239. The beta-particles, which in this case were due to a mixture of 93-239 and the new isotope of element 93 with mass number later shown to be 238 (93-238), had a somewhat higher energy than the radiation from pure 93-239 and there was more gamma radiation. But the composite half-life was about the same, namely, 2 days. However, the sample also differed in another very important way from a sample of pure 93-239. Into this sample there grew an alpha-particle-emitting radioactivity. A proportional counter was used to count the alpha-particles to the exclusion of the beta-particles. This work led us to the conclusion that we had a daughter of the new isotope 93-238 – a daughter with a half-life of about 50 years and with the atomic number 94. This is much shorter-lived than the now known half-life of 94-239, which is about 24,000 years. The shorter half-life meant a higher intensity of alpha-particle emission, which explains why it was so much easier to identify what proved to be the isotope of element 94 with the mass number 238 (94-238).

The reactions are:

$$^{238}_{92}U + ^{2}_{1}H \rightarrow {}^{238}93 + 2^{1}_{0}n$$

$$^{238}93 \xrightarrow[t_{1/2}=2\,d]{\beta^-} {}^{238}94\ (t_{1/2} = 50\ y)\,.$$

On January 28, 1941, we sent a short note to Washington describing our initial studies on element 94; this communication also served for later publication in *The Physical Review* under the names of Seaborg, McMillan, Kennedy, and Wahl [8]. We did not consider, however, that we had sufficient proof at that time to say we had discovered a new element and felt that we had to have chemical proof to be positive. So, during the rest of January, and into February, we attempted to identify this alpha activity chemically.

Our attempts proved unsuccessful for some time. We did not find it possible to oxidize the isotope responsible for this alpha radioactivity. I recall that we then asked Professor Wendell M. Latimer, whose office was on the first floor of Gilman Hall, to suggest the strongest oxidizing agent he knew for use in aqueous solution. At his suggestion we used peroxydisulphate with argentic ion as catalyst.

On the stormy night of February 23, 1941, in an experiment that ran well into the next morning, Wahl performed the oxidation with peroxydisulphate plus argentic ion as catalyst which gave us proof that what we had made was chemically different from all other known elements. That experiment, and hence the first chemical identification of element 94, took place in Room 307 of Gilman Hall, the room that was dedicated as a National Historic Landmark, 25 years later. Thus, we showed that the chemical properties of element 94 were similar to those of uranium and not like osmium (as suggested by Fig. 1).

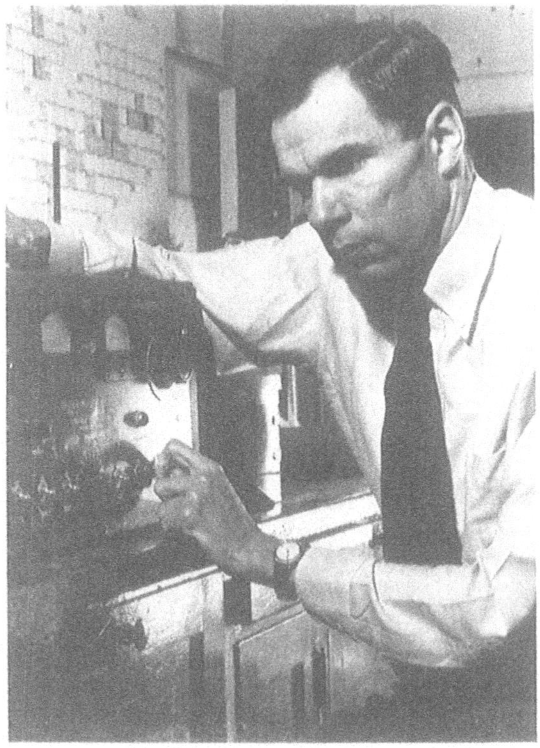

Fig. 5. The discoverers of plutonium: Glenn T. Seaborg at a Geiger counter in Room 303 Gilman Hall, University of California, Berkeley. 1941.

The communication to Washington describing this oxidation experiment, which was critical to the discovery of element 94, was sent on March 7, 1941, and this served for later publication in *The Physical Review* under the authorship of Seaborg (Fig. 5), Wahl (Fig. 6), and Kennedy (Fig. 7) [9].

How element 94 eventually got the name plutonium is an interesting story and one worth telling. This work was carried on under self-imposed secrecy in view of its potential implications for national security. Following the discovery in February 1941, we used only the name "element 94" among ourselves and the few other people who knew of the element's existence. But we needed a code name to be used when we might be overheard. Someone suggested "silver" as a code name for element 93, and we decided to use "copper" for element 94. This worked fine until, for some reason I cannot recall now, it became necessary to use real copper in our work. Since we continued to call element 94 "copper" on occasion we had to refer to the real thing as "honest-to-God-copper".

The first time a true name for element 94 seemed necessary was in writing the report to the Uranium Committee in Washington in March of 1942, which was published later under the authorship of Seaborg and Wahl [10]. I remember very clearly the debates

Fig. 6. The discoverers of plutonium: Arthur C. Wahl and Seaborg with the sample of ^{239}Pu on which fission was demonstrated in 1941. The cigar box belonged to Gilbert N. Lewis. Room 307 Gilman Hall, University of California, Berkeley. February 21, 1966.

Fig. 7. The discoverers of plutonium: Joseph W. Kennedy at Seaborg's parents' home, South Gate, California. December 25, 1940.

within our small group as to what the name should be. It eventually became obvious to us that we should follow the lead of Ed McMillan, who had named element 93 neptunium (chemical symbol Np) because Neptune is the next planet after Uranus, which had served as the basis for the naming of uranium 150 years earlier. Thus we should name element 94 for Pluto, the next planet beyond Neptune. But, and this is a little-known story, it seemed to us that one way of using the base name Pluto was to name the element "plutium". We debated the question of whether the name should be "plutium" or "plutonium", the sound of which we liked much better. We finally decided to take the name that sounded better. I think we made a wise choice, and I believe it also etymologically correct.

There was also the matter of the need for a symbol. Here, too, a great deal of debate was engendered because, although the symbol might have been "Pl", we liked the sound of "Pu" – for the reason you might suspect. We decided on "Pu", and, I might add, we expected a much greater reaction after it was declassified than we ever received.

Fission of plutonium-239

Almost concurrent with this work was the search for, and the demonstration of the fission of, the isotope of major importance – 94-239, the radioactive daughter of 93-239. Emilio Segrè participated in this work together with Kennedy, Wahl, and me. The importance of element 94 stems from its fission properties and its capability of production in large quantities. This work involved, the 60-Inch Cyclotron, the Old Chemistry Building, the Crocker Laboratory, and the 37-Inch Cyclotron, all of which have by now been removed from the Berkeley campus (Fig. 8). The 0.5-microgram sample on which the fission of 94-239 was first demonstrated was produced by transmutation of uranium with neutrons from the 60-Inch Cyclotron; it was chemically isolated in rooms in Old Chemistry Building and Crocker Laboratory and in Room 307 Gilman; and the fission counting was done using the neutrons from the 37-Inch Cyclotron.

A sample of uranyl nitrate weighing 1.2 kilograms was distributed in a large paraffin block (neutron-slowing material) placed directly behind the beryllium target of the 60-Inch Cyclotron in the Crocker Laboratory and was bombarded for two days with neutrons produced by the impact of the full 16 MeV deuteron beam on beryllium. The irradiated uranyl nitrate was placed in a continuously-operating glass extraction apparatus, and the uranyl nitrate was extracted into di-

Fig. 8. University of California at Berkeley, in 1940: 1. LeConte Hall; 2. Gilman Hall; 3. Chemistry Building; 4. Chemistry Annex; 5. Freshman Chemical Laboratory; 6. Radiation Laboratory; 7. Crocker Laboratory; 8. East Hall; 9. Drake's Restaurant and Smorgasbord; 10. Varsity Candy Shop; 11. Alta Vista Building.

ethyl ether. Neptunium-239 was isolated from the aqueous layer by use of the oxidation-reduction principle (described in the next section) with lanthanum and cerium fluoride carrier and was reprecipitated six times in order to remove all uranium impurity. Measurement of the radiation from the neptunium-239 made it possible to calculate that 0.5 microgram was present to yield plutonium-239 upon decay. The resulting alpha activity corresponded to a half-life of 30,000 years for the daughter plutonium-239, in good agreement with the present best value for the half-life of 24,110 years.

The group first demonstrated, on March 28, 1941, with the sample containing 0.5 microgram of plutonium-239, that this isotope undergoes slow neutron-induced fission with a probability of reaction comparable to that of uranium-235. The sample was placed near the screened window of an ionization chamber that could detect the fissions of plutonium-239. Neutrons were then produced near the sample by bombarding a beryllium target with deuterons in the 37-Inch Cyclotron of Berkeley's "Old Radiation Laboratory" (the name applied to the original wooden building, since torn down to make way for modern buildings). Paraffin around the sample slowed the neutrons down so they would be captured more readily by the plutonium. This experiment gave a small but detectable fission rate when a six microampere beam of deuterons was used. To increase the accuracy of the measurement of the fission cross section, this sample, which had about five milligrams of rare-earth carrier materials, was subjected to an oxidation-reduction chemical procedure that reduced the amount of carrier to a few tenths of a milligram. A fission cross section for plutonium-239, some 50 percent greater than that for uranium-235, was found, agreeing remarkably with the accurate values that were determined later. This result was communicated to Washington on May 29, 1941, and this served as the basis for the later publication of an expurgated version by Kennedy, Segrè, Wahl, and me [11].

First isolation and weighing of plutonium

The observation that plutonium-239 is fissionable with slow neutrons provided the information that formed the basis for the U.S. wartime Plutonium Project of the Manhattan Engineer District (MED) centered at the Metallurgical Laboratory of the University of Chicago. Given impetus by the entry of the United States into the war in December 1941, I and some of my colleagues moved to Chicago in the spring of 1942. The mission of the Met Lab was to develop (1) a method for the production of plutonium in quantity, and (2) a method for its chemical separation on a large scale.

The key to solving the first problem was the demonstration by Enrico Fermi and his colleagues of the first sustained nuclear chain reaction in uranium on December 2, 1941.

Important to the solution of the second problem was the determination of the chemical properties of plutonium, an element so new that little was known of its characteristics, and the application of these to the design of a chemical separation process to separate the plutonium from the enormous quantity of fission products and the uranium. I served as leader of the large group of chemists who worked in collaboration with the chemical engineers to solve this problem.

The earlier tracer chemical investigations at Berkeley, continued at Chicago, served to outline the nature of the chemical separation process. The key was the oxidation-reduction cycle in which plutonium is carried in its lower oxidation state(s) by certain precipitates and not carried by these same precipitates when it is present in its higher oxidation state. Thus it is separated from the fission products, which do not exhibit this difference in carrying behavior from oxidizing and reducing solutions. However, the carrying properties of plutonium at tracer (extremely small) concentrations might be different at the macroscopic concentrations that would exist under actual operating conditions in the chemical separation plant.

It occurred to me that central to the achievement of such a separation process would be chemical work on concentrations that would exist in the chemical separation plant. This seemed a very far-out idea, and I can remember a number of people telling me that they thought it was essentially impossible because we had no large source of plutonium. But I thought we could irradiate large amounts of uranium with the neutrons from cyclotrons since the indications were that we probably could produce sufficient plutonium, if we could learn to work on the microgram or smaller-than-microgram scale. That way we could get concentrations as large as those that would exist in the chemical separation plant.

I knew rather vaguely about two schools of ultramicrochemistry – the School of Anton Benedetti-Pichler at Queens College in New York and the School of Paul Kirk in the Department of Biochemistry at the University of California at Berkeley.

I went to New York in May 1942, looked up Benedetti-Pichler, and told him that I needed a good ultramicrochemist. He introduced me to Michael Cefola, and I offered him a job, which he accepted immediately. That he was on the job about three weeks later illustrates the pace at which things moved in those days.

Then, early in June, I took a trip to Berkeley, where I looked up my friend Paul L. Kirk and put the same problem to him. I could not tell any of these people why we wanted to work with microgram amounts or what the material was, but this did not seem to deter their willingness to accept. Paul Kirk introduced me to Burris B. Cunningham. When I asked him if he would come to Chicago, he accepted and was in town by the end of the month. He told me as soon as he arrived that he had a fine student, Louis B. Werner, he would like to invite, and I was, of course, delighted. Werner came along in a few weeks.

These, then, are the people who began the task of isolating plutonium from large amounts of uranium. We brought from Berkeley a little cyclotron-produced sample prepared by Wahl. It contained a few micrograms of plutonium mixed with several milligrams of rare earths. Using that sample, the ultramicrochemists Cunningham and Werner (Fig. 9) and Cefola, isolated the first visible amount – about a microgram – of pure plutonium in the form of the fluoride. It was not weighed, but it could be seen! We were all very excited when we were the first to see a man-made element on August 20, 1942.

In the meantime, hundreds of pounds of uranium were being bombarded with neutrons produced by the cyclotron at Washington University, under the leadership of Alexander S. Langsdorf, and at the 60-Inch Cyclotron at Berkeley, under the leadership of Joseph

Fig. 9. Louis B. Werner and Burris B. Cunningham in Room 404 of the Jones Laboratory at the University of Chicago, August 20, 1942.

G. Hamilton. This highly radioactive material was then shipped to Chicago. Arthur H. Jaffey, Truman Kohman, and Isadore Perlman led a team of chemists who with the help of Daniel E. Koshland, Jr., Elton H. Turk, Bertrand Goldschmidt, and others, put this material through the ether extraction process and the oxidation and reduction cycles to bring it down to a few milligrams of rare earths containing perhaps 100 micrograms of plutonium. This was turned over to Cunningham and Werner who prepared the first sample in pure form by going through oxidation-reduction cycles, the plutonium iodate, and the hydroxide, etc., on to the oxide.

This 2.77-microgram sample was weighed on September 10, 1942 (Fig. 10). The fist aim was to weigh it with a so-called Emich balance, which was somewhat complicated and had electromagnetic compensation features. As it turned out, owing to the heavy load in the shops, this weighing balance would have taken perhaps six months to build.

Cunningham then had the idea of using a simple device consisting of a quartz fiber about 12 centimeters long and $\frac{1}{10}$ of a millimeter in diameter suspended at one end with a weighing pan hung on the other end. Then the depression of that end of the fiber with the pan containing the sample would relate to the weight of the sample. Cunningham measured the depression of the quartz fiber with a telescope. He built this balance himself, although he found out later that an Italian named Salvioni invented it earlier, and so it became known as the Salvioni balance. A description of this first isolation and first weighing of plutonium was published by Cunningham and Werner [12] after World War II.

The chemical separation (extraction) process that finally evolved had three stages: (1) the separation from uranium (extraction) and from the fission products (decontamination) used oxidation-reduction cycles with bismuth phosphate as the carrier precipitate; (2) the concentration (volume reduction) step used an oxidation-reduction cycle with rare earth fluoride as the carrier precipitate; (3) the isolation step consisted of the precipitation of pure (carrier-free) plutonium peroxide from acid solution. There was widespread concern that bismuth(III) phosphate would not carry plutonium(IV) quantitatively at the concentrations that would exist in the chemical separation plant. The critical experiments on the ultramicro-chemical scale showed that plutonium(IV) phosphate is carried completely (>95%) at these concentrations. The so-called Bismuth Phosphate Process operated very successfully in both the plutonium pilot plant at Oak Ridge, Tennessee, and the production plant at Hanford, Washington.

The revised periodic table

At this time we thought that the transuranium elements had the same kind of relationship as the rare earths – a new group of rare earths – and there should be 14 of them, with uranium as the prototype. This we would call the uranide series, just like the lanthanide series. It was on this basis that we predicted that element 95 and element 96 would be chemically like plutonium, neptunium, and uranium – a little different, but more or less the same. When we tried by transmutation reactions to produce elements 95 and 96 by this method and to identify them chemically, we could not do it.

In 1944, I got the idea that maybe these elements were misplaced in the periodic table. Perhaps the new heavy rare earth series should start back at thorium (Fig. 11) with actinium as its prototype – thus dubbing the collection the actinide series. With such an arrangement the position of elements 95 and 96 would suggest that they be chemically similar to europium and gadolinium. When we tried this idea, we found that it was right. We identified elements 95 and 96. A year later, I published the rearrangement of the periodic table in *Chemical and Engineering News* [13]. I remember at the time that when I showed this table to a number of my friends, knowledgeable inorganic chemists, and said that I was contemplating publishing

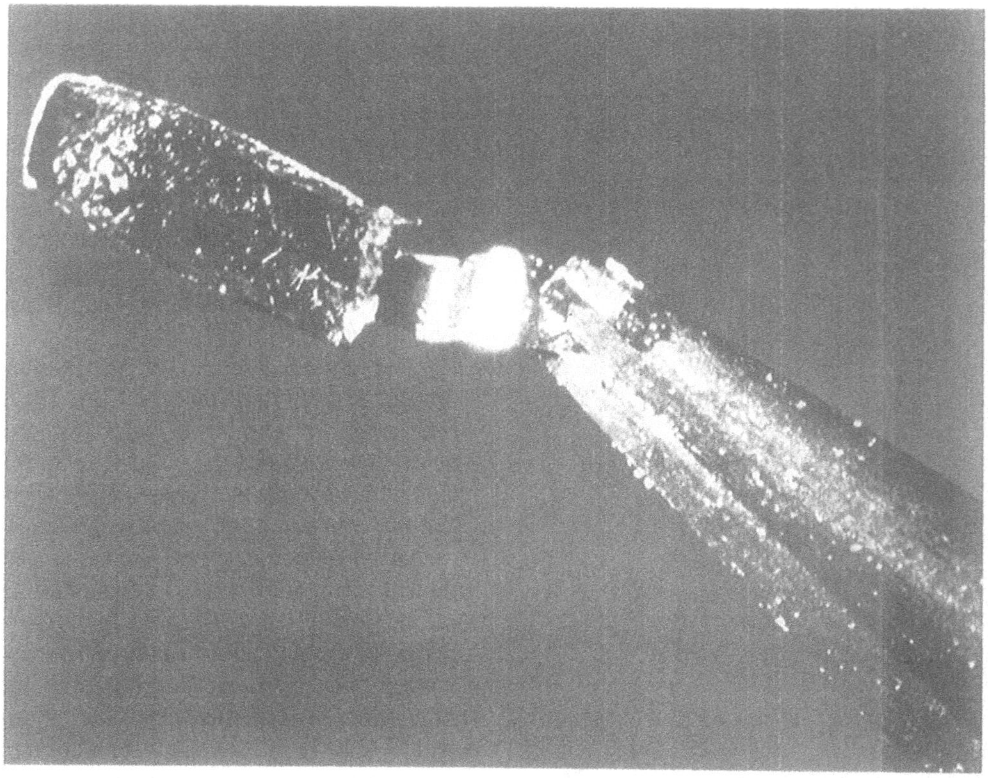

Fig. 10. First weighed sample of plutonium (as an oxide) on a platinum foil held by forceps, University of Chicago, Metallurgical Laboratory. September 10, 1942.

1 H 1.008																1 H 1.008	2 He 4.003
3 Li 6.940	4 Be 9.02											5 B 10.82	6 C 12.010	7 N 14.008	8 O 16.000	9 F 19.00	10 Ne 20.183
11 Na 22.997	12 Mg 24.32	13 Al 26.97										13 Al 26.97	14 Si 28.06	15 P 30.98	16 S 32.06	17 Cl 35.457	18 A 39.944
19 K 39.096	20 Ca 40.08	21 Sc 45.10	22 Ti 47 90	23 V 50 95	24 Cr 52.01	25 Mn 54 93	26 Fe 55.85	27 Co 58 94	28 Ni 5869	29 Cu 63.57	30 Zn 65 38	31 Ga 69.72	32 Ge 72 60	33 As 74.91	34 Se 78.96	35 Br 79.916	36 Kr 83.7
37 Rb 85.48	38 Sr 87.63	39 Y 88.92	40 Zr 91.22	41 Cb 92.91	42 Mo 95.95	43	44 Ru 101.7	45 Rh 102.91	46 Pd 106.7	47 Ag 107.880	48 Cd 112.41	49 In 114.76	50 Sn 118.70	51 Sb 121 76	52 Te 127 61	53 I 126.92	54 Xe 131.3
55 Cs 132.91	56 Ba 137.36	57 LA 138.92 / 58-71 SEE La SERIES	72 Hf 178.6	73 Ta 180.88	74 W 183 92	75 Re 186.31	76 Os 190 2	77 Ir 193.1	78 Pt 195.23	79 Au 197.2	80 Hg 200.61	81 Tl 204 39	82 Pb 207.21	83 Bi 209.00	84 Po	85	86 Rn 222
87	88 Ra	89 AC / SEE Ac SERIES	90 Th	91 Pa	92 U	93 Np	94 Pu	95	96								

LANTHANIDE SERIES	57 La 138.92	58 Ce 140.13	59 Pr 140.92	60 Nd 144 27	61	62 Sm 150.43	63 Eu 152 0	64 Gd 156.9	65 Tb 159.2	66 Dy 162.46	67 Ho 163.5	68 Er 167.2	69 Tm 169.4	70 Yb 173.04	71 Lu 174.99
ACTINIDE SERIES	89 Ac	90 Th 232.12	91 Pa 231	92 U 238 07	93 Np 237	94 Pu	95	96							

Fig. 11. Periodic table published by Seaborg in 1945, showing the heaviest elements as members of an actinide series.

it in *Chemical and Engineering News*, they said, "Don't do it, you'll ruin your scientific reputation". I had a great advantage – I didn't have any scientific reputation at the time – so I went ahead and published it.

This concept had great predictive value, and its success led to the discovery of the remainder of the actinide elements and its acceptance by the scientific community. The modern periodic table contains not

Fig. 12. The discoverers of americium and curium: Ralph A. James. 1945.

only a full lanthanide series, but a full actinide series and transactinide elements as well.

Americium and curium (95 and 96)

At the wartime Metallurgical Laboratory, after the completion of the most essential part of the chemical investigations involved in the production of plutonium, attention was turned to the synthesis and identification of the next transuranium elements. Ralph A. James (Fig. 12), Leon O. Morgan (Fig. 13), Albert Ghiorso (Fig. 14), and I were collaborators in this endeavor.

As indicated above, the first attempts to produce these elements ended in failure. Small amounts of $^{239}_{94}Pu$ were irradiated with neutrons and deuterons but no new alpha-emitting products were found due to the use of insensitive detection techniques and because the experiments were based upon the premise that these elements should behave chemically like plutonium, i.e., they could be oxidized to the VI oxidation state and chemically isolated. It was not until the summer of 1944, when it was first recognized that these elements were a part of an actinide transition series (with stable +3 oxidation states) that any progress was made. Success in their identification followed quickly.

Fig. 13. The discoverers of americium and curium: Leon O. Morgan with Mrs. L. O. (Betty) Morgan. 1944.

Once it was realized that these elements could be oxidized above the III state only with difficulty, the use of a proper chemical procedure led quickly to the identification of an isotope of a transplutonium element. Thus, a new alpha-emitting nuclide, now known to be $^{242}_{96}Cm$ (half-life 162.9 d), was produced in the summer of 1944 [13] by the bombardment of $^{239}_{94}Pu$ with 32-MeV helium ions:

$$^{239}_{94}Pu + ^{4}_{2}He \rightarrow ^{242}_{96}Cm + ^{1}_{0}n .$$

The bombardment took place in the Berkeley 60-inch cyclotron after which the material was shipped to the Metallurgical Laboratory at Chicago for chemical separation and identification. A crucial step in the identification of the alpha-emitting nuclide as an isotope of element 96, $^{242}_{96}Cm$, was the later identification of the known $^{238}_{94}Pu$ as the alpha-decay daughter of the new nuclide.

The identification of an isotope of element 95, in late 1944 and early 1945, followed after the identification of this isotope of element 96 (^{242}Cm) as a result of the bombardment of $^{239}_{94}Pu$ with neutrons in a nuclear reactor; this was published in 1950 [14]. The production reactions, involving multiple neutron capture by plutonium are:

$$^{239}_{94}Pu + ^{1}_{0}n \rightarrow ^{240}_{94}Pu + \gamma$$

$$^{240}_{94}Pu + ^{1}_{0}n \rightarrow ^{241}_{94}Pu + \gamma$$

$$^{241}_{94}Pu \xrightarrow[t_{1/2}=14.4\,y]{\beta^-} ^{241}_{95}Am\ (t_{1/2} = 432.7\ y)$$

Fig. 14. The discoverers of americium and curium: Albert Ghiorso (with Arthur H. Jaffey in background) in the Met Lab counting room, University of Chicago. January 1946.

$$^{241}_{95}\mathrm{Am} + ^{1}_{0}\mathrm{n} \rightarrow ^{242}_{95}\mathrm{Am} + \gamma$$

$$^{242}_{95}\mathrm{Am} \xrightarrow[t_{1/2}=16.0\,\mathrm{h}]{\beta^-} ^{242}_{96}\mathrm{Cm}\,.$$

A confirmation of the identification of the nuclide $^{241}_{95}$Am at the Los Alamos Laboratory involved the physical separation (based upon volatility) of $^{241}_{95}$Am from its parent $^{241}_{94}$Pu in a separated mass 241 sample.

Some comments should be made, at this point, concerning the similarity of these two elements to the rare-earth elements. The hypothesis that elements 95 and 96 should have a stable III oxidation state and greatly resemble the rare-earth elements in their chemical properties proved to be true. In fact, the near identity of their properties greatly hindered the efforts of the discovery team. The better part of a year was spent in trying, without success, to separate chemically the two elements from each other and from the fission product and carrier rare-earth elements. Although the discovery team was confident on the basis of their chemical and radioactive properties and the methods of production, that isotopes of elements 95 and 96 had been produced, the complete chemical proof still was lacking. The elements remained unnamed during this period of futile attempt at separation (although one of the group referred to them as "pandemonium" and "delirium" in recognition of their difficulties). The key to their chemical separation, which occurred later at Berkeley, and the technique which made feasible the separation and identification of subsequent transuranium elements was the ion-exchange technique.

The present names of these new elements were proposed on the basis of their chemical properties. The name "americium" (chemical symbol Am) was suggested for element 95, after the Americas, by analogy with the naming of its rare-earth counterpart or homologue, europium, after Europe; and the name "curium" (chemical symbol Cm) was suggested for element 96, after Pierre and Marie Curie, by analogy with the naming of its homologue, gadolinium, after the Finnish rare-earth chemist Johan Gadolin. These names and symbols were accepted by the International Union of Pure and Applied Chemistry (IUPAC).

By chance, the discovery of these elements was revealed informally on a nationally-broadcast radio program, the "Quiz Kids", on which the author appeared as a guest on November 11, 1945 (Fig. 15). The discovery information had already been declassified (i.e., removed from the "secret" category) for presentation at an American Chemical Society symposium at Northwestern University in Chicago the following Friday. Therefore, when one of the youngsters asked – during a session in which the author was trying to answer their questions – if any additional new elements had been discovered in the course of research on nuclear weapons during the war, he was able to reveal the existence of the elements 95 and 96. Apparently many children in America told their teachers

Fig. 15. "Quiz Kids" Sheila Conlan and Robert Burke with Seaborg when he informally announced the discovery of elements 95 and 96 on a radio show. November 11, 1945.

about it the next day, and, judging from some of the letters which the author subsequently received from such youngsters, they were not entirely successful in convincing their teachers. The formal announcement of the discoveries was, of course, made later in the week, as planned.

Berkelium and californium (97 and 98)

The story of the discovery of berkelium and californium began shortly after the end of World War II. I recall that we began planning for the possible synthesis and identification of transuranium elements as soon as, or even before, we returned to Berkeley from the Chicago Metallurgical Laboratory; i.e., in late 1945 and in 1946. I thought that this would be a good Ph.D. thesis problem for Stanley G. Thompson and it was, of course, natural that Al Ghiorso would participate on the radiation detection end of the problem as he had in the discovery of americium and curium in Chicago a year or two earlier.

On the basis of our confidence in the actinide concept we felt we could make the chemical identification, although we knew we would have to develop better chemical separation methods than were then available to us. And it seemed clear that we would use helium ion bombardments of americium and curium for our production reactions once these elements became available in sufficient quantify through production by prolonged neutron bombardment of plutonium, and we learned how to handle safely their intense radioactivity.

We knew these things, but we didn't anticipate how long it would take to solve these simple problems. Actually, three years went by before we found ourselves ready to make our first realistic experiment.

The most important prerequisite to the process for making the transcurium elements was the manufacture of sufficiently large amounts of americium and curium to serve as target material. Because of the intense radioactivity of americium and curium, even in milligram or submilligram amounts, it was necessary to develop extremely efficient chemical separation methods to isolate the new elements from the target materials. This large degree of separation was necessary to detect the very small amounts of radioactivity due to the new elements produced in the presence of the highly radioactive starting materials. The dangerous radioactivity of the source material also made it necessary to institute complicated remote control methods of operation to keep health hazards at a minimum.

These problems were solved after three years work. Americium for target material was prepared in milligram amounts by intense neutron bombardment of plutonium over a long period of time, and curium target materials were prepared in microgram amounts as the result of the intense neutron bombardment of some of this americium. Both of these neutron bombardments took place in high-flux reactors (i.e., reactors that deliver large concentrations of neutrons that can be used for transmutation purposes).

Element 97 was discovered by Thompson, Ghiorso, and me in December 1949 as the result of the bombardment of milligram quantities of $^{241}_{95}Am$ with 35 MeV helium ions accelerated in the 60-inch cyclotron at Berkeley [15]. The nuclear reaction was

$$^{241}_{95}Am + {}^{4}_{2}He \rightarrow {}^{243}_{97}Bk + 2{}^{0}_{1}n\ .$$

The new nuclide was expected to have a short half-life and thus relatively rapid chemical separation techniques had to be employed. For this purpose cation-exchange was used.

The actual discovery experiments were not as simple as this description would indicate. During the fall of 1949 we made a number of bombardments of americium with helium ions in the 60-inch cyclotron, with emphasis on looking for alpha-particle emitting isotopes of element 97, all with negative results. It was becoming clear that we should look for electron capture decay by detecting the accompanying conversion electrons and X-rays so Ghiorso worked to improve the detection efficiency for such radiations.

The first successful experiment was performed on Monday, December 19, 1949. A target containing 7 milligrams of ^{241}Am was bombarded with helium ions in the 60-inch cyclotron, after which the chemical separation was started at 10:00 a.m. After the removal of the bulk of the americium by two oxidation cycles (utilizing oxidation to the hexapositive, fluoride-soluble, oxidation state of americium, which had just been discovered by Larned B. Asprey, Stephen E. Stephanou, and Robert A. Penneman at Los Alamos), the Bk, Cm, and remaining Am were carried on lanthanum fluoride, dissolved, and subjected to a group separation from fission product lanthanide elements (using a method of elution with concentrated HCl, just discovered by Kenneth E. Street, Jr.), after which the actinide fraction was put through a cation exchange adsorption-elution procedure; this entire process was completed in seven hours. The prediction that element 97 would elute ahead of Cm and Am, in sequence, was of course the key to its successful chemical identification. In this case, and especially in considering the data from following elution experiments, we were somewhat surprised to see the rather large gap between 97 and curium; we shouldn't have been surprised because there is a notably large gap between the elution peaks of the homologous lanthanide elements terbium and gadolinium.

Detected in the samples that eluted at the peak corresponding to element 97 were conversion electrons, X-rays of energy corresponding to decay by electron capture, and alpha particles at very low relative intensity (less than 1%). These radiations were found to decay with a half-life of about 4.5 hours, and it was immediately assumed that the isotope was ^{244}Bk produced by the reaction: ^{241}Am(α,n)^{244}Bk. Soon thereafter it was correctly surmised that the main isotope, that giving rise to the observed alpha particles, was actually ^{243}Bk produced by the reaction ^{241}Am(α,2n)^{243}Bk.

It is interesting to note that experiments as early as the first day, i.e., Monday night, indicated that element 97 has two oxidation states, III and IV. The actinide concept provided the guidance to look for these two oxidation states, by analogy with the homologous element, terbium. In fact, the chemical identification procedure had been devised to accommodate either oxidation state and the large gap in the elution positions of element 97 and the curium was at first erroneously thought to be due to the fact that element 97 was in the IV oxidation state at that stage.

Element 98 was first produced and identified similarly by Thompson, K. Street, Jr., Ghiorso, and me (Fig. 16), soon afterward in February of 1950, again at Berkeley [16]. The first isotope produced is now assigned the mass number 245 and decays by alpha-particle emission and orbital electron capture with a half-life of 44 minutes. This isotope was produced by the bombardment of microgram amounts of $^{242}_{96}$Cm with 35-MeV helium ions accelerated in the 60-inch cyclotron:

$$^{242}_{96}\mathrm{Cm} + {}^{4}_{2}\mathrm{He} \rightarrow {}^{245}_{98}\mathrm{Cf}\ (t_{1/2} = 44\ \mathrm{m}) + {}^{1}_{0}\mathrm{n}\ .$$

It is interesting to note that this identification of element 98 was accomplished with a total of only some 5,000 atoms; someone remarked at the time that this number was substantially smaller than the number of students attending the University of California.

The key to the discovery of element 98 was once again the use of ion-exchange techniques. On the basis of column calibration experiments, element 98 was expected to elute onto collection plate #13 in the 26th and 27th drops of eluant and this is exactly where it was found after a total elapsed chemical separation time of two hours. The half-life and alpha particle energy were also in agreement with predictions.

Element 97 was called berkelium (chemical symbol Bk) after the city of Berkeley, California, where it was discovered, just as its rare-earth analog, terbium, was given a name derived from Ytterby, Sweden, where so many of the early rare-earth minerals were found. Element 98 was named californium (chemical symbol Cf), after the university and state where the work was done. These names and symbols were immediately accepted by the IUPAC. This latter name, chosen for the reason given, does not reflect the observed chemical analogy of element 98 to dysprosium, as "americium", "curium", and "berkelium" signify that these elements are the chemical analogs of europium, gadolinium, and terbium, respectively. In their announcement of the discovery of element 98 in *The Physical Review,* the authors commented, "The best we can do is point out, in recognition of the fact that dysprosium is named on the basis of a Greek word meaning 'difficult to get at', that the searchers for another element (Au) a century ago found it difficult to get to California".

Upon learning about the naming of these elements, the "Talk of the Town" section of the *New Yorker* magazine had the following to say:

> New atoms are turning up with spectacular, if not downright alarming frequency nowadays, and the University of California at Berkeley, whose scientists have discovered elements 97 and 98, has christened them berkelium and californium, respectively. While unarguably suited to their place

Fig. 16. The discoverers of berkelim and californium in Seaborg's office, Lawrence Berkeley Laboratory, on the 25th anniversary of their discovery. January 20, 1975. From left to right: Kenneth E. Street, Jr., Stanley G. Thompson, Seaborg, Albert Ghiorso.

of birth, these names strike us as indicating a surprising lack of public relations foresight on the part of the university, located, as it is, in a state where publicity has flourished to a degree matched perhaps only by evangelism. California's busy scientists will undoubtedly come up with another atom or two one of these days, and the university might well have anticipated that. Now it has lost forever the chance of immortalizing itself in the atomic tables with some such sequence as universitium (97), offium (98), californium (99), berkelium (100).

The discoverers sent the following reply:

"Talk of the Town" has missed the point in their comments on naming of the elements 97 and 98. We may have shown lack of confidence but no lack of foresight in naming these elements "berkelium" and "californium". By using these names first, we have forestalled the appalling possibility that after naming 97 and 98 "universitium" and "offium", some New Yorker might follow with the discovery of 99 and 100 and apply the names "newium" and yorkium".

The answer from the *New Yorker* staff was brief:

We are already at work in our office laboratories on "newium" and "yorkium"! So far we have just the names.

Einsteinium and fermium (99 and 100)

The discoveries of many of the transuranium elements were the result of careful planning, taking into account predictions of chemical and physical properties.

Element 99 and 100, however, were unexpectedly discovered in debris from the "Mike" thermonuclear explosion which took place in the Pacific on November 1, 1952. This was the first large test of a thermonuclear device. Debris from the explosion was collected, first on filter papers attached to airplanes which flew through the clouds (this sampling effort cost the life of First Lieutenant Jimmy Robinson who waited too long before returning to his base, tried to land on Eniwetok and ditched about a mile short of the runway) and, later in more substantial quantity, gathered up as fall-out material from the surface of a neighboring atoll. This debris was brought to the United States for chemical investigation in a number of laboratories to establish the properties of the explosion.

Early analysis of the "Mike" debris by scientists at the Argonne National Laboratory near Chicago and the Los Alamos Scientific Laboratory in New Mexico showed the unexpected presence of new isotopes of plutonium, $^{244}_{94}Pu$ and $^{246}_{94}Pu$. (At the time the heaviest known isotope of plutonium was $^{243}_{94}Pu$.) This observation led to the conclusion that the $^{238}_{92}U$ in the device had been subjected to an enormous neutron flux and

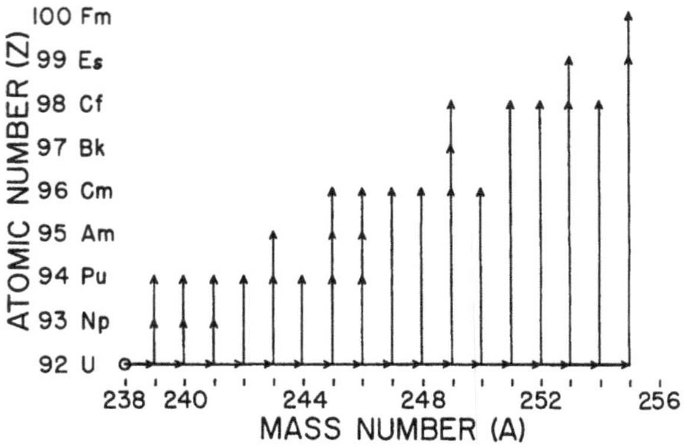

Fig. 17. Production of uranium isotopes in the November 1951 "Mike" thermonuclear device, and their decay to beta-stable nuclei.

had successively captured numerous neutrons. (Later calculations showed an integrated neutron fluence of $1-4\times10^{24}$ neutrons was delivered in a few nanoseconds – a few *moles* of neutrons!!)

Armed with the knowledge of the multineutron capture by ^{238}U, we at the University of California immediately began a search for transcalifornium isotopes in the bomb debris. Ion-exchange experiments of the type previously mentioned in the case of berkelium and californium immediately demonstrated the existence of a new element and within a few weeks, of a second new element. The first identification of element 100 was made with only about 200 atoms. To secure a larger amount of source material, it was necessary later to process many hundreds of pounds of coral from one of the atolls adjoining the explosion area. Eventually, such coral was processed by the ton, using bismuth phosphate as the carrier for the tripositive actinide elements, in a pilot-plant operation which went under the name of "Paydirt".

Without going into the details, it may be pointed out that such experiments involving the groups at the three laboratories led to the positive identification of isotopes of elements 99 and 100. A twenty-day activity emitting alpha particles of 6.6-MeV energy was identified as an isotope of element 99 (with the mass number 253), and a 7.1 MeV alpha activity with a half-life of 22 hours was identified as an isotope of element 100 (with the mass number 255).

The path of successive neutron captures by $^{238}_{92}$U and subsequent β^- decay of the capture products is shown in Fig. 17. The β^- decay chains for each A value ended in the first β stable nuclide. Thus the first isotopes of elements 99 and 100 produced in such a device are those with $A = 253$ and 255, respectively.

The large group of scientists who contributed to the discovery of elements 99 and 100 included Ghiorso, Thompson, Gary H. Higgins, and me from the Radiation Laboratory and Department of Chemistry of the University of California; Martin H. Studier, Paul R. Fields, Sherman M. Fried, Herbert Diamond, Joseph F. Mech, Gray L. Pyle, John R. Huizenga, Albert Hirsch, and Winston M. Manning of the Argonne National Laboratory; and Charles I. Browne, H. Louise Smith, and Roderick W. Spence of the Los Alamos Scientific Laboratory [17] (Fig. 18). These researchers suggested the name einsteinium (symbol E) for element 99 in honor of the great physicist Albert Einstein; and for element 100, the name fermium (symbol Fm) in honor of the father of the atomic age, Enrico Fermi, making these the first in a series of elements named after eminent scientists. These names and the chemical symbols Es and Fm were soon adopted by IUPAC for these elements. The choice of name of fermium for element 100 has proven to be prescient since it is the last element to be synthesized using neutron capture reactions (which were extensively studied by Fermi).

Before removal of the "secret" label from this information and the subsequent announcement of the original discovery experiments could be accomplished, isotopes of elements 99 and 100 were produced by other, more conventional methods. Chief among these was that of successive neutron capture as the result of intense neutron irradiation of plutonium in the high-flux Materials Testing Reactor (MTR) at the National Reactor Testing Station in Idaho (Fig. 19). The difference between this method of production and that of the "Mike" thermonuclear explosion is one of time as well as of starting material. In a reactor, it is necessary to bombard gram quantities of plutonium for two or three years; thus, the short-lived, intermediate isotopes of the various elements have an opportunity to decay. The path of element production proceeds up the valley of β^- stability. In the thermonuclear device larger amounts of uranium were subjected to an extremely high neutron flux for a period of nanoseconds; the subsequent beta decay of the ultraheavy isotopes of uranium led to the nuclides found in the debris.

Mendelevium (101)

The discovery of mendelevium was one of the most dramatic in the sequence of transuranium element syntheses. It marked the first time in which a new element was produced and identified one atom at a time.

By 1955 we at Berkeley had prepared an equilibrium amount of $\sim10^9$ atoms of $^{253}_{99}$Es by neutron irradiation of plutonium in the Materials Testing Reactor. As the result of a "back of the envelope" calculation done by Ghiorso during an airplane flight, we thought it might be possible to prepare element 101 using the reaction

$$^{253}_{99}\text{Es} + {}^{4}_{2}\text{He} \rightarrow {}^{256}101 + {}^{1}_{0}\text{n}\ .$$

The amount of element 101 expected to be produced in an experiment can be calculated using the formula

$$N_{101} = \frac{N_{\text{Es}}\sigma\phi(1-e^{-\lambda t})}{\lambda}$$

where N_{101} and N_{Es} are the number of element 101

Fig. 18. The discoverers of elements einsteinium and fermium at a symposium commemorating the 25th anniversary of their discovery held at Lawrence Berkeley Laboratory. January 23, 1978. Front row, from left to right: H. Louise Smith, Sherman M. Fried, and Gary H. Higgins. Back row, from left to right: Albert Ghiorso, Roderick W. Spence, Seaborg, Paul R. Fields, and John R. Huizenga.

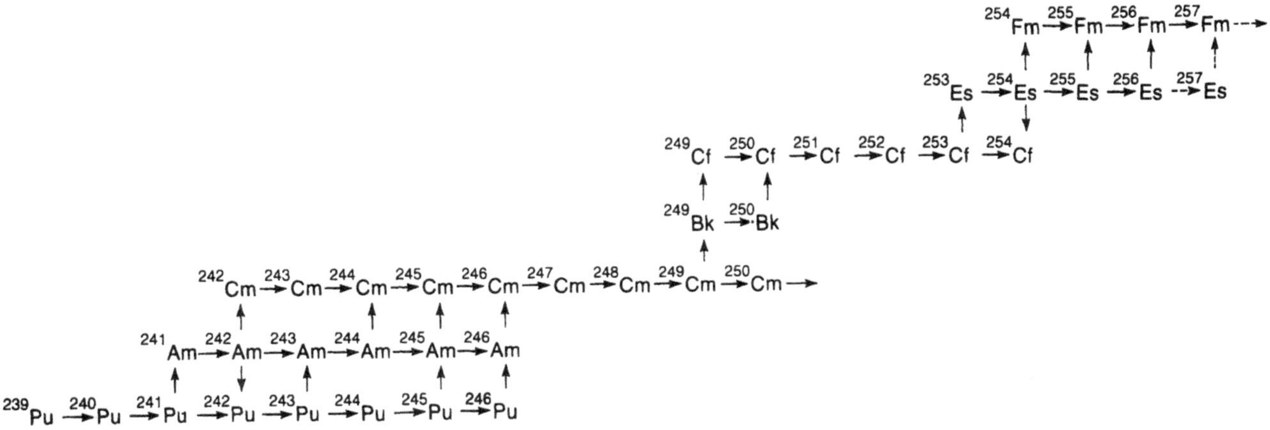

Fig. 19. The sequence of nuclides produced in a high-flux reactor neutron irradiation of a ^{239}Pu target. The horizontal arrows represent neutron capture, vertical arrows up represent β decay, and vertical arrows down represent electron capture decay.

atoms produced and the number of $^{253}_{99}Es$ target atoms, respectively, σ is the reaction cross section (estimated to be $\sim10^{-27}$ cm^2), ϕ the helium ion flux ($\sim10^{14}$ particles/sec − cm^2), λ the decay constant of $^{256}_{101}Md$ (estimated to be $\sim10^{-4}$ sec^{-1}) and t the length of each bombardment ($\sim10^4$ sec).

$$N_{101} \approx \frac{(10^9)(10^{-27})(10^{14})(1-e^{-(10^{-4})(10^{+4})})}{(10^{-4})} \approx 1 \text{ atom}$$

Thus the production of only one atom of element 101 per experiment could be expected!

Adding immeasurably to the complexity of the experiment was the absolute necessity for the chemical separation of the one atom of element 101 from the 10^9 atoms of einsteinium in the target and its ultimate, complete chemical identification by separation with the ion-exchange method. This separation and identification would presumably have to take place in a period of hours, or perhaps even one hour or less, because the expected half-life was of this order of magnitude or less. Furthermore the target material had a 20-day half-life and one needed a non-destructive technique of using the target material over and over again.

These requirements indicated the desperate need for new techniques, together with some luck. Fortunately, both were forthcoming. The first new technique involved separation of the element 101 by *the recoil method* from the einsteinium in the target. The einsteinium was placed on a gold foil in an invisibly thin layer. The helium-ion beam was sent through the back of the foil so that the atoms of element 101, recoiling through a vacuum due to the momentum of the impinging helium ions, could be caught on a second thin gold catcher foil. This second gold foil, which contained recoil atoms and was relatively free of the einsteinium target material, was dissolved and was used for later chemical operations.

The preparation of the ^{253}Es target for this recoil experiment was another technical *tour-de-force*. After approximately five failures to prepare the target by vaporization of ^{253}Es from a hot filament, the essentially weightless deposit of ^{253}Es was electroplated onto the gold foil within a very small area.

An extremely reliable ion exchange separation scheme had to be developed to unambiguously chemically identify atoms of a new element that were made. It took several months, involving hundreds of column elutions to develop the appropriate procedure. The final choice was the use of a Dowex 50 ion exchange column run at an elevated temperature (87 °C) with an alpha-hydroxyisobutyrate eluant. The procedure was so well developed that the discovery team could tell exactly in which drops of eluant the interesting activities would appear. Finally the 60-inch cyclotron $^4He^{2+}$ beam was increased by an order of magnitude in intensity to 100 microamperes per cm^2.

The earliest experiments were confined to a search for short-lived, alpha-emitting isotopes that might be due to element 101. For this purpose it was sufficient to look quickly at the actinide chemical fraction as separated by the ion-exchange method. No alpha activity was observed that could be attributed to element 101, even when the time between the end of bombardment and the beginning of the alpha particle analyses had been reduced to five minutes. The experiments were continued and, in one of the subsequent overnight bombardments, two large pulses in the electronic detection apparatus due to spontaneous fission were observed. With probably unjustified self-confidence, it was thought that this might be a significant result. Although such an attitude might ordinarily have been considered foolish, it must be recalled that rapid decay by spontaneous fission was − up until that time − confined to only a few nuclides, none of which should have been introduced spuriously into the experiment. In addition, background counts due to this mode of decay should be zero in proper equipment.

The major question, of course, was whether the experiment could be repeated. In a number of subsequent bombardments, one or two spontaneous fission events were observed in some, while none was observed in other experiments. This, of course, was to be expected, because of the statistical fluctuation inherent in the production of the order of one atom per bombardment. Furthermore, more advanced chemical experiments seemed to indicate that spontaneous fission counts, when they did appear, came in about the element 100 or 101 chemical fractions.

The definitive experiments were performed in a memorable, all-night session, February 18, 1955. To increase the number of events that might be observed at one time, three successive three-hour bombardments were made, and, in turn, their transmutation products were quickly and completely separated by the ion-exchange method. Some of the nuclide $^{253}_{99}Es$ was present in each case so that together with the $^{246}_{98}Cf$ produced from $^{244}_{96}Cm$ also present in the target (via the $^{244}Cm(^4He,2n)$reaction), it was possible to define the positions in which the elements came off the column used to contain the ion-exchange resin. Five spontaneous fission counters then were used to count simultaneously the corresponding drops of solution from the three runs.

A total of five spontaneous fission counts were observed in the element 101 position, while a total of eight spontaneous fission counts were also observed in the element 100 position. No such counts were observed in any other position.

The rate of spontaneous fission in both the element 101 and 100 fractions decayed with a half-life of about three hours (later determined to be 160 minutes). This and other evidence led to the hypothesis that this isotope of element 101 has the mass number 256 and decays, by electron capture (designated by the symbol EC), with a half-life of the order of one-and-one-half hours, to the isotope $^{256}_{100}Fm$, which is responsible for the spontaneous fission decay. The discovery reactions were:

Fig. 20. The discoverers of mendelevium at the Lawrence Berkeley Laboratory on the 25th anniversary of their discovery. March 28, 1980. From left to right: Gregory R. Choppin, Seaborg, Bernard G. Harvey, and Albert Ghiorso.

$$^{253}_{99}\mathrm{Es} + {}^{4}_{2}\mathrm{He} \rightarrow {}^{256}_{101}\mathrm{Md} + {}^{1}_{0}\mathrm{n}$$

$$^{256}_{101}\mathrm{Md} \xrightarrow[t_{1/2}=1.3\ \mathrm{h}]{\mathrm{EC}} {}^{256}_{100}\mathrm{Fm}$$

$$^{256}_{100}\mathrm{Fm} \xrightarrow[t_{1/2}=2.63\ \mathrm{h}]{} \text{spontaneous fission} .$$

On the basis of this evidence and the experiments which led to the production of a total of 17 atoms of element 101, Ghiorso, Bernard G. Harvey, Gregory R. Choppin, Thompson, and I (Fig. 20) announced the discovery of element 101 [18]. The name mendelevium (symbol Mv) was suggested for the element, in recognition of the role of the great Russian chemist, Dmitri I. Mendeleev, who was the first to use the periodic system of the elements to predict the chemical properties of undiscovered elements, a principle which was used in nearly all the transuranium element discovery experiments. The name mendelevium was immediately accepted by IUPAC, who suggested the chemical symbol, Md.

It is comforting to be able to record that subsequent experiments using larger amounts of einsteinium in the target led to the production of thousands of atoms of mendelevium, lending confirmation to the evidence on which the original conclusions were made. Subsequent experiments have shown that, as expected, mendelevium is a typical tripositive actinide element.

Fig. 21. The discoverers of nobelium at the HILAC building, Lawrence Berkeley Laboratory. 1958. From left to right: Albert Ghiorso, Torbjorn Sikkeland, and John R. Walton (Seaborg is absent).

Nobelium (102)

In 1957 a team of scientists from the Argonne National Laboratory in the United States, the Atomic Energy Research Establishment at Harwell in England, and the Nobel Institute of Physics in Stockholm announced the discovery of element 102 based upon work done at the Nobel Institute [19]. The group reported that in

Fig. 22. The discoverers of lawrencium at the HILAC building, Lawrence Berkeley Laboratory. 1961. From left to right: Torbjorn Sikkeland, Albert Ghiorso, Almon E. Larsh, and Robert M. Latimer.

irradiations of $^{244}_{96}$Cm with ^{13}C ions accelerated at the Nobel Institute Cyclotron, they found an 8.5 MeV alpha-emitter with a half-life of about 10 minutes, presumably due to an isotope of element 102 of mass number 253 or 251 produced in the $^{244}_{96}$Cm($^{13}_{6}$C,4n) or $^{244}_{96}$Cm($^{13}_{6}$C,6n) reactions. They claimed that this activity had been identified as a new element on the basis of ion-exchange chromatography in which the 8.5 MeV activity appeared in the "expected" element 102 position when eluted from a cation exchange column with alpha-hydroxyisobutyrate. The name of nobelium (chemical symbol No) was suggested for the new element in recognition of Alfred Nobel's contributions to the advancement of science and was immediately accepted by the IUPAC.

However, neither experiments at Berkeley [20] nor related experiments at the Kurchatov Institute in Moscow, USSR [21] confirmed this Stockholm work. In fact, subsequent experiments done in Berkeley have shown that the most stable oxidation state of element 102 in solution is +2; thus it would not appear in the "expected" element 102 tripositive position in a cation-exchange column.

In 1958 Ghiorso, Torbjorn Sikkeland, John R. Walton, and I [22] (Fig. 21) announced the positive identification of $^{254}_{102}$No produced using the Berkeley heavy ion linear accelerator (HILAC).

$$^{246}_{96}\text{Cm} + ^{12}_{6}\text{C} \rightarrow ^{254}_{102}\text{No} + 4^{1}_{0}\text{n}$$

$$^{254}_{102}\text{No} \xrightarrow[t_{1/2}\sim 3\text{ sec}]{\alpha} ^{250}_{100}\text{Fm}\ .$$

The $^{250}_{100}$Fm daughter of the new element was collected using recoil techniques, one atom at a time. Eleven atoms of the $^{250}_{100}$Fm daughter were identified by their position in a cation exchange elution curve. A half-life of ~3 sec was assigned to $^{254}_{102}$No on the basis of many recoil experiments in which an apparent alpha-emitting daughter of $^{254}_{102}$No was produced and direct counting of an 8.3 MeV alpha-emitting nuclide with a 3-second half-life. The 8.3 MeV alpha-emitter was found also to decay by spontaneous fission in 30% of its decays (and this has subsequently been shown to be correct). It is now know that the 3-second activity originally assigned to $^{254}_{102}$No in the direct counting experiments was, in fact, $^{252}_{102}$No ($t_{1/2}$ = 2.3 sec, $E\alpha$ = 8.4 MeV, 30% SF) produced by a $^{244}_{96}$Cm(^{12}C,4n) reaction in that the "^{246}Cm target" used by Ghiorso *et al.* had 20 × more ^{244}Cm than ^{246}Cm in it. $^{254}_{102}$No is now known to have a 55-sec half-life.

The experimental claim for discovery of a new element, element 102, however, must be judged upon the observation of the $^{250}_{100}$Fm daughter of $^{254}_{102}$No because this is the only evidence that establishes the atomic number of the new element. The recoil studies which assigned a 3-sec half-life to $^{254}_{102}$No by observation of what was thought to be $^{250}_{100}$Fm were probably erroneous. What was probably being observed was the sequence

$$^{252}_{102}\text{No} \xrightarrow[t_{1/2}=2.3\text{ sec}]{\alpha} ^{248}_{100}\text{Fm} \xrightarrow[t_{1/2}=36\text{ sec}]{\alpha} ^{244}_{98}\text{Cf} \xrightarrow[t_{1/2}=19\text{ min}]{\alpha}$$

in which the 19 min $^{244}_{98}$Cf granddaughter (E_α =

1 H																	2 He
3 Li	4 Be											5 B	6 C	7 N	8 O	9 F	10 Ne
11 Na	12 Mg											13 Al	14 Si	15 P	16 S	17 Cl	18 Ar
19 K	20 Ca	21 Sc	22 Ti	23 V	24 Cr	25 Mn	26 Fe	27 Co	28 Ni	29 Cu	30 Zn	31 Ga	32 Ge	33 As	34 Se	35 Br	36 Kr
37 Rb	38 Sr	39 Y	40 Zr	41 Nb	42 Mo	43 Tc	44 Ru	45 Rh	46 Pd	47 Ag	48 Cd	49 In	50 Sn	51 Sb	52 Te	53 I	54 Xe
55 Cs	56 Ba	57 La	72 Hf	73 Ta	74 W	75 Re	76 Os	77 Ir	78 Pt	79 Au	80 Hg	81 Tl	82 Pb	83 Bi	84 Po	85 At	86 Rn
87 Fr	88 Ra	89 Ac	104 Rf	105 Ha	106 Sg	107 Ns	108 Hs	109 Mt	110	111	(112)	(113)	(114)	(115)	(116)	(117)	(118)

LANTHANIDES	58 Ce	59 Pr	60 Nd	61 Pm	62 Sm	63 Eu	64 Gd	65 Tb	66 Dy	67 Ho	68 Er	69 Tm	70 Yb	71 Lu
ACTINIDES	90 Th	91 Pa	92 U	93 Np	94 Pu	95 Am	96 Cm	97 Bk	98 Cf	99 Es	100 Fm	101 Md	102 No	103 Lr

Fig. 23. Modern periodic table.

7.21 MeV) was mistaken for the 30 min $^{250}_{100}$Fm (E_α = 7.43 MeV). The ion exchange elution curve showing 11 atoms of $^{250}_{100}$Fm to appear in the proper position remains as the first definitive evidence for the production of element 102. The correct determination of decay properties ($t_{1/2}$, $E\alpha$, SF) of the isotope subsequently assigned to ^{252}No provides good supporting evidence for the discovery of element 102. The errors in this experiment indicate the difficulties associated with "one-atom-at-a-time" studies.

A parallel line of research on element 102 was carried out by Georgiy N. Flerov and coworkers in the Kurchatov Institute in the USSR. In an experiment reported in 1958 Flerov *et al.* studied the reaction of $^{239}_{94}$Pu with $^{16}_{8}$O ions, reporting an alpha-emitter with E_α = 8.9 MeV and $2 < t_{1/2} < 40$ sec [23]. In 1964 E. D. Donets, V. A. Schegolev, and V. A. Ermakov [24] of the Dubna laboratory reported the production of the new isotope $^{256}_{102}$No using recoil techniques with *chemical identification* of the alpha-emitting daughter $^{252}_{100}$Fm. The first correct identification of the half-life of $^{254}_{102}$No was in 1966–67 by groups working at Dubna.

In summary one can say that the Berkeley group [22] was the first group to clearly identify the atomic number of element 102 (i.e., to "discover" it) but important contributions to the definitive establishment of the existence of element 102 were made by Donets *et al.* [24]; credit for the discovery should be shared. Since the name nobelium and symbol (No) for this element are in common use, the Berkeley scientists have suggested retention of the name provided in connection with the original, incorrect Stockholm experiment.

Lawrencium (103) – The last actinide element

The first identification of an isotope of element 103 was by Ghiorso, Sikkeland, Almon E. Larsh, and Robert M. Latimer in 1961 [25] (Fig. 22). Three micrograms of a mixture of californium isotopes (A = 249, 250, 251, 252) were bombarded with heavy ion beams of ^{10}B and ^{11}B at the Berkeley HILAC. Atoms recoiling from the target were caught by a long metallized mylar type which was moved past a series of alpha-particle detectors. A new alpha-emitting nuclide with $E\alpha$ = 8.6 MeV and $t_{1/2} \sim 8$ sec was observed and assigned to $^{257}_{103}$Lr. Later experiments indicated that this activity was due to $^{258}_{103}$Lr ($E\alpha$ = 8.6 MeV, $t_{1/2}$ = 4.3 sec).

A subsequent identification of the atomic number of element 103 was made by Donets, Schegolev, and Ermakov at Dubna in 1965 [26]. The nuclear reaction used was:

$$^{243}_{95}\mathrm{Am} + {}^{18}_{8}\mathrm{O} \rightarrow {}^{256}_{103}\mathrm{Lr} + 5\,{}^{1}_{0}\mathrm{n}\,.$$

Using a double recoil technique, they identified the alpha-emitter $^{256}_{103}$Lr ($t_{1/2} \sim 45$ sec) and linked it genetically to its granddaughter, the known $^{252}_{100}$Fm via the decay sequence:

$$^{256}_{103}\mathrm{Lr} \xrightarrow{\alpha} {}^{252}_{101}\mathrm{Md} \xrightarrow{\mathrm{EC}} {}^{252}_{100}\mathrm{Fm} \xrightarrow{\alpha}\,.$$

Credit for the discovery of element 103 should be shared by Ghiorso *et al.* [25] and Donets *et al.* [26]. The relatively long half-life of $^{256}_{103}$Lr (now known to be ~30 sec) enabled Robert J. Silva, Sikkeland, Matti J. Nurmia, and Ghiorso [27] to establish in 1970 that element 103 exhibits a stable +3 oxidation state in solution, as expected by the actinide concept.

In the report of the original experiments of Ghiorso *et al.* [25] they suggested the name lawrencium (immediately accepted by the IUPAC) and the chemical symbol Lw for element 103 in honor of E.O. Lawrence, the inventor of the cyclotron and founder of the Radiation Laboratory at Berkeley where so much of the transuranium research has been carried out. The IUPAC adopted the chemical symbol Lr.

The modern periodic table

The modern periodic table (Fig. 23) shows the elements with atomic numbers 90–103, inclusive, as members of an "actinide series" with the known and predicted members (atomic numbers in parentheses) of the "transactinide" elements placed in the last row of the main body of the table.

References

1. Fermi, E.: Possible Production of Elements of Atomic Number Higher Than 92. Nature **133**, 898 (1934).
2. Noddack, I.: Über das Element 93. Angew. Chem. **47**, 653–655 (1934); English translation: On Element 93.
3. Hahn, O., Meitner, L., Strassmann, F.: Elemente jenseits Uran. Berichte **69**, 905–919 (1936); English translation: Elements Beyond Uranium.
4. Curie, I., Savitch, P.: Sur la nature du radioélément de période 3,5 heures formé dans l'uranium irradié par les neutrons. Comptes Rendus **206**, 1643–1644 (1938); English translation: On the Nature of a Radioactive Element with 3.5 Hour Half-Life Produced in the Neutron Irradiation of Uranium.
5. Hahn, O., Strassmann, F.: Über den Nachweis und das Verhalten der bei der Bestrahlung des Urans mittels Neutronen entstehenden Erdalkalimetalle. Naturwissenschaften **27**, 11–15 (1939); English translation: On the Identification and the Behavior of Rare Earth Metals Produced in the Neutron Irradiation of Uranium. Am. J. Phys. **32**, 10–15 (1964).
6. McMillan, E. M.: Radioactive Recoils from Uranium Activated by Neutrons. Phys. Rev. **55**, 510 (1939).
7. McMillan, E. M., Abelson, P. H.: Radioactive Element 93. Phys. Rev. **57**, 1185–1186 (1940).
8. Seaborg, G. T., McMillan, E. M., Kennedy, J. W., Wahl, A. C.: Radioactive Element 94 from Deuterons on Uranium. Phys. Rev. **69**, 366–367 (1946).
9. Seaborg, G. T., Wahl, A. C., Kennedy, J. W.: Radioactive Element 94 from Deuterons on Uranium. Phys. Rev. **69**, 367 (1946).
10. Seaborg, G. T., Wahl, A. C.: The Chemical Properties of Elements 94 and 93. J. Am. Chem. Soc. **70**, 1128–1134 (1948).
11. Kennedy, J. W., Seaborg, G. T., Segrè, E., Wahl, A. C.: Properties of 94 (239). Phys. Rev. **70**, 555–556 (1946).
12. Cunningham, B. B., Werner, L. B.: The First Isolation of Plutonium. J. Am. Chem. Soc. **71**, 1521–1528 (1949).
13. Seaborg, G. T.: The Chemical and Radioactive Properties of the Heavy Elements. Chem. Eng. News **23**, 2190–2193 (1945).
14. Ghiorso, A., James, R. A., Morgan, L. O., Seaborg, G. T.: Preparation of Transplutonium Isotopes by Neutron Irradiation. Phys. Rev. **78**, 472 (1950).
15. Thompson, S. G., Ghiorso, A., Seaborg, G. T.: Element 97. Phys. Rev. **77**, 838–839 (1950).
16. Thompson, S. G., Street, K., Jr., Ghiorso, A., Seaborg, G. T.: Element 98. Phys. Rev. **78**, 298–299 (1950).
17. Ghiorso, A., Thompson, S. G., Higgins, G. H., Seaborg, G. T., Studier, M. H., Fields, P. R., Fried, S. M., Diamond, H., Mech, J. F., Pyle, G. L., Huizenga, J. R., Hirsch, A., Manning, W. M., Browne, C. I., Smith, H. L., Spence, R. W.: New Elements Einsteinium and Fermium, Atomic Numbers 99 and 100. Phys. Rev. **99**, 1048–1049 (1955).
18. Ghiorso, A., Harvey, B. G., Choppin, G. R., Thompson, S. G., Seaborg, G. T.: New Element Mendelevium, Atomic Number 101. Phys. Rev. **98**, 1518–1519 (1955).
19. Fields, P. R., Friedman, A. M., Milsted, J., Atterling, H., Forsling, W., Holm, L. W., Åström, B.: Production of the New Element 102. Phys. Rev. **107**, 1460–1462 (1957).
20. Ghiorso, A., Sikkeland, T., Walton, J. R., Seaborg, G. T.: Attempts to Confirm the Existence of the 10-Minute Isotope of 102. Phys. Rev. Lett. **1**, 17–18 (1958).
21. Flerov, G. N., Polikanov, S. M., Karamian, A. S., Pasiuk, A. S., Parfanovich, D. M., Tarantin, N. I., Karnaukhov, V. A., Druin, V. A., Volkov, V. V., Semchinova, A. M., Oganesian, Iu. Ts., Khalizev, V. I., Khlebnikov, G. I.: Sov. Phys. Dokl. **3**, 546–548 (1958); English translation: Experiments to Produce Element 102.
22. Ghiorso, A., Sikkeland, T., Walton, J. R., Seaborg, G. T.: Element 102. Phys. Rev. Lett. **1**, 17–18 (1958). See also: Ghiorso, A., Sikkeland, T., Walton, J. R., Seaborg, G. T.: At. Energ. (USSR) **7**, 338 (1959); English translation: Sov. J. At. Energy **7**, 819 (1961).
23. Flerov, G. N.: Heavy Ion Reactions. *Proceedings of the Second United Nations International Conference on the Peaceful Uses of Atomic Energy,* Vol. 14, P/2299, Geneva 1958, pp. 151–157.
24. Donets, E. D., Schegolev, V. A., Ermakov, V. A.: At. Energ. **16**, 195–207 (1964); English translation: Synthesis of Element 102 of Mass Number 256. Sov. J. At. Energy **16**, 233–245 (1964).
25. Ghiorso, A., Sikkeland, T., Larsh, A. E., Latimer, R. M.: New Element Lawrencium, Atomic Number 103. Phys. Rev. Lett. **6**, 473–475 (1961).
26. Donets, E. D., Schegolev, V. A., Ermakov, V. A.: At. Energy **19**, 109–113 (1965); English translation: Synthesis of the Isotope of Element 103 (Lawrencium) with Mass Number 256. Sov. J. At. Energy **19**, 995–999 (1965).
27. Silva, R. S., Sikkeland, T., Nurmia, M., Ghiorso, A.: Tracer Chemical Studies of Lawrencium. Inorg. Nucl. Chem. Lett. **6**, 733–739 (1970).

B. Modern Aspects of Radioactivity and Nuclear Reactions

Radiochimica Acta 70/71, 93–105 (1995)

Proton Radioactivity

By Sigurd Hofmann

GSI, Planckstrasse 1, D-64291 Darmstadt, Germany

(Received July 17, 1995)

Proton radioactivity / Proton drip-line / Proton spectroscopy / Proton binding energies / Neutron-deficient nuclei

Abstract

The stability line for neutron deficient nuclei will be determined by radioactive emission of protons. The presently known cases are reviewed, and experimental techniques for production and investigation of proton emitters are described. The emission probabilities are discussed within a semiclassical model.

1. Introduction

The borderlines of nuclear stability could be estimated since nuclear models allowed to calculate the binding energies of nuclei over a wide range of elements and mass numbers. It was expected that a new type of radioactivity, the spontaneous emission of protons, will determine the limit to existence for neutron-deficient nuclei. The process itself presents a unique opportunity to obtain spectroscopic information from a decaying state in a nucleus situated beyond the drip-line. This limiting character and the simplicity of the decay process made search experiments for proton radioactivity an attractive aim in nuclear physics.

Up to the year 1965, the phenomenon of radioactive emission of one or two protons was subject of a number of theoretical papers reviewed by Goldansky [1]. In 1970 the first case of radioactive proton emission was observed [2]. It could be assigned to an isomeric state in ^{53}Co, a nucleus that is proton bound in its ground-state. A first report on the observation of a proton radioactivity from the ground-state was presented in 1981 at the Helsingør conference [3]. The activity could be assigned to the ground-state decay of ^{151}Lu [4]. Within a short time range of three years, three more proton emitters have been observed in the region of nuclei near ^{151}Lu at GSI (^{147}Tm, ^{147m}Tm, ^{150}Lu [5, 6]) and two in the region near ^{100}Sn in Munich by T. Faestermann *et al.* (^{113}Cs and ^{109}I [7, 8]). In recent time the sensitivity of detector systems has been improved and longer beam times at accelerators could be assigned to this field of research. In particular, successful search experiments for new proton emitters have been performed at the Daresbury Laboratory by P. Woods *et al.* [9] and at the Argonne National Laboratory by C. Davids *et al.* [10]. Nine new proton transitions were measured, four of them could be assigned to nuclei near $N = 82$ (^{146}Tm, ^{146m}Tm, ^{156}Ta, ^{160}Re [11–13]) and five to heavier nuclei up to element $Z = 83$ (^{165,166,167}Ir, ^{171}Au, ^{185}Bi [10]). Recently, also proton emission of ^{105}Sb could be observed [14]. This is the second of the important cases of activities of elements 1znjust beyond closed shells ($Z = 50$ and 82). It was searched for a long time. Partly, previously known data, could be improved, and tentative assignments could be confirmed.

In the region of light elements, fragmentation reactions using high energy projectiles have been very successful in mapping the proton drip line. New nuclei at the limit of stability up to astrophysically interesting cases of neutron deficient nuclei in the region of rubidium have been produced and identified recently [15–18].

The present review describes the experimental investigations on proton radioactivity carried out so far. After a chapter on basic properties of proton radioactivity, the experimental techniques are described and the decay data of the presently known proton emitters are given. Simple shell-model considerations help to understand the observed phenomena and allow to make predictions on the decay properties of so far unknown nuclei at the limits of stability.

2. Fundamentals

A. Proton-decay energy

Various radioactive decay channels are energetically open for nuclei on the proton rich slope of the stability valley: β^+ decay or electron capture, β-delayed 1p-, 2p-, 3p-, and α-decay, direct p-, 2p-, and α-decay, decay by emission of heavy clusters, and fission. Nuclear models predict that radioactive proton-decay half-lives decrease fastest with increasing proton to neutron ratio and, therefore, proton radioactivity will determine the limits to nuclear stability for proton-rich nuclei.

A necessary condition for proton radioactivity is a positive Q_p value:

$$Q_p = (M_{Z+1} - M_Z - m_p - m_e)\, c^2 > 0\,. \qquad (1)$$

M_i are the atomic masses including electrons, m_i are the masses of the elementary particles. If we use binding energies, the Q_p value is simply:

$$Q_p = (B_Z - B_{Z+1}); \qquad B_Z = (ZM_H + Nm_n - M_Z)c^2\,. \qquad (2, 3)$$

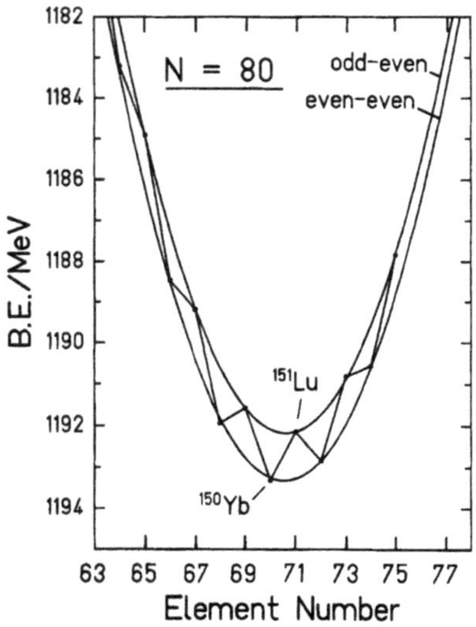

Fig. 1. Binding energies of neutron deficient $N=80$ isotones. The values are calculated using mass predictions given by Möller and Nix [19], which reproduce the measured Q_p value of 1.24 MeV for the proton transition ^{151}Lu$\rightarrow$^{150}Yb + p ($Q_p = B_Z - B_{Z+1}$). Isotonic binding-energy parabolas, not to be mixed with isobaric mass parabolas, connect the data points of odd-even and even-even isotones separated by the proton-pairing energy Δp. Position and slope of the parabolas are mainly determined by the Coulomb and asymmetry term, both quadratic in Z, of the macroscopic liquid-drop part of the mass formulae. Microscopic shell effects lead to significant deviations for isotones just above closed proton shells.

and for two-proton radioactivity for completeness, neglecting a small binding energy of the diproton:

$$Q_{2p} = (B_Z - B_{Z+2}) \, . \tag{4}$$

The energy requirements for proton radioactivity are shown in Fig. 1 for $N=80$ isotones including the known case of the ^{151}Lu proton decay. According to Equation 2, the Q_p values can be read off as differences of binding energies of neighbouring isotones. The binding energies are theoretical values taken from the mass tables of Möller and Nix [19], that well reproduce the measured ^{151}Lu Q_p-value. Parabolas connect the odd-A and even-even isotones, respectively. They are separated by the proton-pairing energy. As a function of the atomic number, the binding energies of isotones exhibit maximum values, $\partial B/\partial Z = 0$, and the change of the slopes allows for positive Q_p values near these maxima. The quadratic Z dependence and the positions of the maxima are determined by the macroscopic liquid-drop behaviour of nuclei. The quadratic terms come from the increasing Coulomb energy and the asymmetry term. Microscopic energies locally lead to deviations from the parabolic dependence. This is especially significant for odd elements just above closed proton shells.

In the example shown in Fig. 1 the first proton unbound isotone is ^{149}Tm and indeed the first candidate with a Q_p value big enough for a measurable decay branch is ^{151}Lu.

Binding energies of neutron-deficient isotones were used by Alkhazov *et al.* [20] and Vogel *et al.* [21] to calculate proton-pairing energies Δp. In Fig. 1 Δp can be read off as distance between the two parabolas. It was found that Δp increases remarkably with increasing proton number, and an empirical expression was proposed for Δp, that is more accurate than the value $12\,A^{-1/2}$ MeV:

$$\Delta p = \{7.55[1-6.07(N-Z)^2]A^{-2}\}\,A^{-1/3}\,\mathrm{MeV}\,. \tag{5}$$

Equation (5) was derived by Jensen *et al.* [22] who extended the investigations to four-nucleon correlations. Results of an extensive investigation of various nuclear pairing models were recently published by Möller and Nix [23].

B. Proton-emission probability

In a first approach to reproduce measured proton decay half-lives, expression (6) can be used:

$$\lambda = \ln 2/T_{1/2} = \nu \times P. \tag{6}$$

It was developed by Gamow [24] and Gurney and Condon [25] to describe the radioactive decay constant λ in a semiclassical way as the product of a frequency factor ν and a transmission coefficient P. For ν the expression:

$$\nu = \sqrt{2}\,\pi^2\,\hbar^2[m^{3/2}R_c^3(zZe^2/R_c - Q_{\mathrm{p,nucl}})^{1/2}]^{-1} \tag{7}$$

was used in analogy to the expression obtained by Bethe [26] and Rasmussen [27] for the α-decay rates of spherical nuclei ($\nu = 6 \cdot 10^{21}$ Hz for the ^{151}Lu p-decay). Applying the WKB approximation the transmission coefficient is:

$$P = P_{j\ell} = \mathrm{e}^{-2G_{j\ell}}. \tag{8}$$

The Gamow factor G_{jl} is an integral over the forbidden region of the potential between the classical turning points R_{in} and R_{out}:

$$G_{jl} = \sqrt{2m/\hbar^2} \int_{R_{\mathrm{in}}}^{R_{\mathrm{out}}} [V_{j\ell}(r) + V_{\mathrm{Coul}}(r) + V_\ell(r) - Q_{\mathrm{p,\,nucl}}]^{1/2} \mathrm{d}r\,. \tag{9}$$

The potential is a superposition of the nuclear potential $V_{j\ell}(r)$, the Coulomb potential $V_{\mathrm{Coul}}(r)$, and the centrifugal potential $V_\ell(r)$. For the nuclear potential the real part of an optical model potential can be used as given, e.g. by Becchetti and Greenlees [28].

The potential in Equation (9) is shown in Fig. 2 for the case of the proton decay of ^{151}Lu. The figure exhibits the long tunnel length for the proton ($R_{\mathrm{out}} \simeq 80$ fm) at decay energies of about 1 MeV. Variations of proton energy and angular momentum will have the greatest influence on the transmission probability.

Conservation of angular momentum and parity selects the allowed transitions:

$$I_{\mathrm{i}} = I_{\mathrm{f}} + \ell + s; \qquad \pi_{\mathrm{i}} = \pi_{\mathrm{f}}(-1)^l. \tag{10, 11}$$

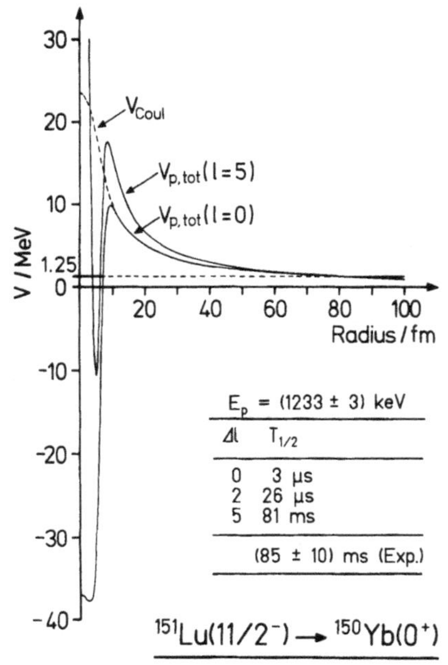

Fig. 2. Proton-nucleus potential for the semiclassical calculation of the ^{151}Lu partial proton half-life.

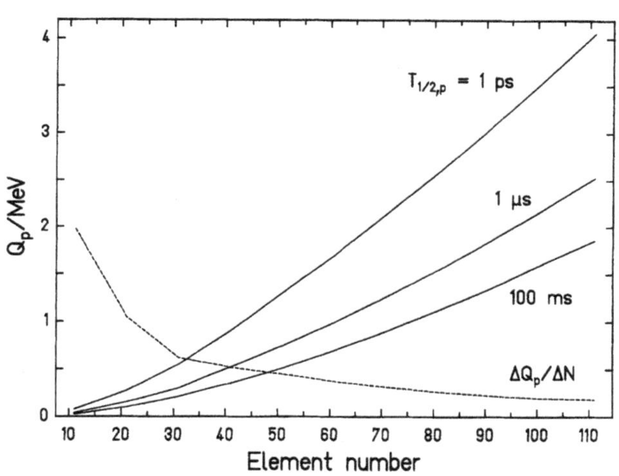

Fig. 3. Semiclassical calculation of proton-decay half-lives. The dashed line is the slope $\Delta Q_p/\Delta N$ for neutron deficient isotopes averaged over pairing effects. For a certain element the ratio between Q_p for a certain half-life and the value of the dashed curve is an estimate for the number of nuclei beyond the drip-line up to the isotope with that half-life. The drip-line is identical to the abscissa.

The good agreement of calculated half-lives with the measured values for ^{147}Tm and ^{151}Lu gave some confidence into the semiclassical description, although nuclear structure effects are not included, except the change of angular momentum. To consider nuclear structure, a spectroscopic factor can be defined:

$$S_{exp} = T_{1/2,\,p,\,WKB}/T_{1/2,\,p,\,exp}\,. \qquad (12)$$

This factor is the ratio of calculated and measured partial proton half-life.

C. The limits for proton-rich nuclei

For light elements the failure of neutron deficient isotopes in reaction experiments indicate the boundary of nuclear existence. The fifteen, presently known proton emitters beyond the drip-line have been observed in regions near ^{100}Sn, near the closed neutron shell $N=82$, and one just beyond $Z=82$. The challenge of experimental work being in progress is to extend the number of proton emitters in the known regions and to find proton radioactivity also in other element regions. To estimate the number of possible proton emitters the partial proton half-lives can be calculated for each element as function of the Q_p value. Fig. 3 shows the result for the three half-life values of 100 msec, 1 μsec, and 1 psec; the first two values are important with respect to competitive β decay and separation time by recoil separators, respectively. The dashed line shows the slope $\Delta Q_p/\Delta N$, which is nearly constant as function of the neutron number for neutron deficient isotopes, if pairing effects are averaged. The slope is steepest for light elements and decreases smoothly to about 0.2 MeV/neutron for the heaviest elements. For each element the ratio between Q_p for a certain half-life and the value of the dashed curve is an estimate for the number of proton unbound isotopes up to that half-life; the abscissa represents the proton drip-line. The ratio is smaller than unity for half-lives of 1 μsec below zirconium, reflecting the difficulties to find proton emitters with $T_{1/2}>1$ μsec in that region. Integration results in the total number of isotopes within a certain half-life window, e.g. between $Z=50$ and 70 are 35 isotopes with partial proton half-lives longer than 100 msec, 15 with half-lives between 100 msec and 1 μsec, and 36 with half-lives between 1 μsec and 1 psec. For heavy elements competition of α decay has to be considered for the longer living isotopes. In case of the heaviest elements ($Z>92$) the limits of stability are determined by spontaneous fission.

3. Experimental techniques

Production rates for neutron-deficient proton emitters are extremely low, half-lives are short, and proton energies are relatively small. Therefore, a preselection of neutron-deficient reaction products from the beam and other reaction products becomes necessary before the decay can be studied using sensitive detectors.

The properties of the detector systems are determined not only by the characteristics of a certain proton decay, like energy and half-life but also by the characteristics of the separators used, like separation time, background suppression, and image size in the focal plane of the separated beam of radioactive nuclei. Therefore, in the following examples the combined systems of separators and detectors are presented, which have been used in major search experiments for proton radioactivity or were used in successful studies of this process.

A. Helium-jet and recoil catching techniques

Already in 1963 very neutron deficient nuclei in the $N=82$ region were identified with the He-jet technique by Macfarlane [32]. The activity escaping from the target is stopped in a helium filled chamber and transported through a capillary to the detectors that are placed in a shielded area. Although the sensitivity is high, the limitations of the He-jet technique using up to several meters long transport capillaries is obvious. Half-life limits for investigation of short lived nuclei are at about 30 msec.

For investigation of the ^{53m}Co proton emitter, Cerny *et al.* [33] produced the activity in reactions with a 32-MeV proton beam. The ^{53m}Co nuclei are stopped in the ^{54}Fe target. A silicon-detector telescope was directed to the target. It was mounted behind a slotted, rotating wheel which controlled the duration of the beam pulse and protected the counters during the beam-on intervals.

A He-jet system was successfully used for investigation of the ^{105}Sb proton decay by Tighe *et al.* [14]. A 0.75 m long capillary was used to transport the activity on a slowly moving tape in the center of a low-energy proton detector ball. It consists of six individual gas-ΔE, gas-ΔE, Si-E triple telescopes that are capable of detecting protons with energies down to $\approx$200 keV.

B. On-line isotope separators

The high volatility of elements around xenon made possible an investigation of their short lived neutron-deficient isotopes at ISOLDE-type mass-separators. Concerning proton radioactivity it could be expected that the protons of odd-element isotopes above the $Z=50$ closed-shell isotopes of tin are loosely bound, in particular the isotopes of antimony, and, therefore, the proton drip-line is closer to the valley of stability, which would make a production of these isotopes easier.

Two different types of reactions have been successful to produce neutron-deficient isotopes above tin: Spallation reactions with high energetic proton beams [34] and heavy ion fusion reactions of neutron deficient projectile and target nuclei [35, 36]. High mass resolution and preparation of thin source layers for high resolution spectroscopy are the outstanding properties of on-line mass separators. New developments on the field of ion-source techniques improved the element specific efficiency and reduced the separation times [37, 38].

In IGISOL systems (ion-guide isotope separator on-line) the recoil ions are thermalized in helium and transported in ionic form into the acceleration chamber of an isotope separator [39]. Like in case of He-jets this technique is well suited for separation of light ion induced reaction products.

C. Gas filled separators

At the U-300 cyclotron at Dubna the gas-filled mass separator BEMS [40] was built for fast separation (within 1 μsec) of the reaction products from the beam (BEMS is a Russian acronym for "fast electromagnetic mass separator"). The recoil nuclei collide with the gas molecules within the magnetic system, which is filled with 1.5 Torr helium. They change statistically the ionic charge. After averaging over the changing curvatures of the trajectories the entire paths of the particles are close to those with an equilibrium charge.

Modern versions of gas-filled separators which in principal can be used to search for proton radioactivity have been built in Argonne [41], Berkeley [42], Darmstadt [43], Dubna [44], Jyväskylä [45], and Saitama [46].

D. Recoil separators and position-sensitive detector systems

Recoil-separating techniques use in contrary to on-line separators the momentum and the ionic charge of the recoiling fusion products received in the reaction process. Therefore, the separation time is fast, and the efficiency independent from element properties.

The GSI separator for heavy ion reaction products (SHIP [47]) is a combination of two electric and four magnetic dipole fields together with two quadrupole triplets. It was upgraded recently in order to increase the sensitivity for detection of reaction products produced at low cross-sections [48–50]. A schematic drawing of the upgraded SHIP is shown in Fig. 4. Main improvements were achieved by a new target wheel, a target position closer to the quadrupole lenses and the addition of another dipol magnet behind SHIP, that further reduced the background by allowing to position the newly developed detectors out of the straight beam line. The separation times depend on the reaction and are in the range of (0.5–2.0) μsec. Also, the suppression factor of the primary beam depends on the reaction and is between 10^8 and 10^{16}. It is high for reactions with great velocity differences between compound nuclei and projectiles.

The background rate behind SHIP is sufficiently small to stop the separated beam of fusion products in the focal plane behind the separator by implantation directly into silicon detectors. Using position sensitive detectors the obtained information is twofold: 1) The absolute positions of the implanted nuclei are measured. The local distribution allows to optimize and to control the field settings of the separator. 2) If the implanted nuclei are radioactive, then the position signals from the decays can be determined relative to the position signals from the implantations or the previous decays. This is so, because recoil effects are small compared to the range of implanted nuclei, emitted protons or α particles and compared to the detector resolution. This method was introduced for investigation of neutron-deficient nuclei near $N=82$ [51]. An

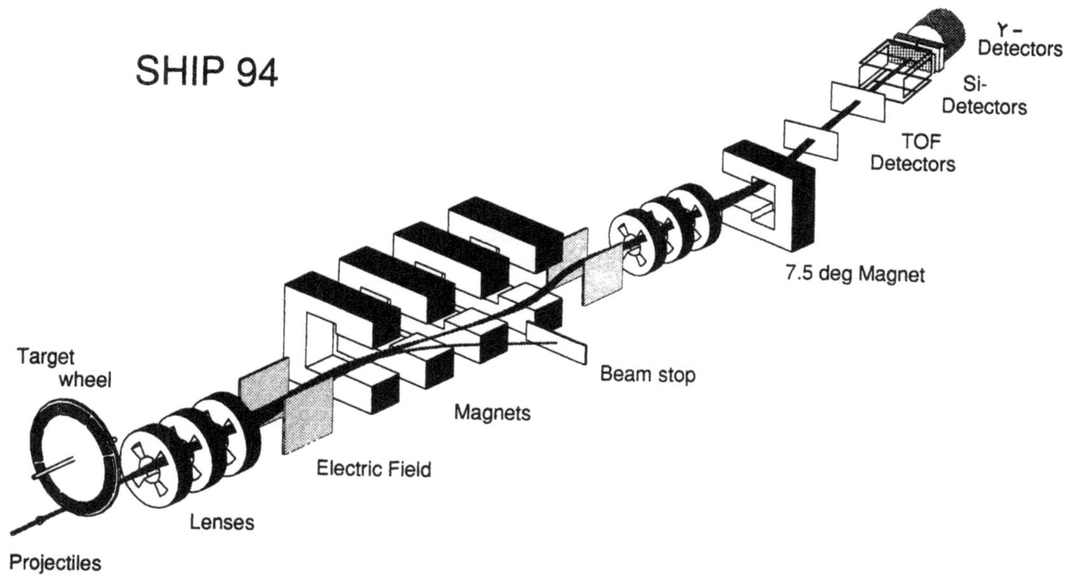

Fig. 4. The velocity filter SHIP, target wheel and detector system, an example for an experimental set-up using recoil separators, high beam currents and ion implantation techniques for identification of the radioactive proton or alpha decay. (From Ref. [50].)

enlarged, more sensitive system has been installed recently for investigation of superheavy elements [52].

Similar to SHIP was the Daresbury recoil mass spectrometer primarily a velocity filter that has been extended by a 50° sector magnet to improve mass resolution [53]. For investigation of proton and alpha radioactivity the Edinburgh group has developed a double-sided silicon strip detector system [54, 55]. The proton emitters ^{146}Tm, ^{156}Ta, and ^{160}Re could be identified recently [11, 13]. After shut-down of the Daresbury accelerator, the detector system was used at the Fragment Mass Analyzer (FMA) at the ATLAS accelerator in Argonne to search for proton emitters. The FMA is a recoil mass spectrometer, in principal similar to SHIP, 8 m in length, with a mass-over-charge resolution $\Delta(M/Q)/(M/Q)$ of 1/340. There, recently, the heaviest proton emitters up to ^{185}Bi were discovered [10].

At the Munich Tandem Linac combination an electrostatic separator has been built and was used for decay studies of neutron deficient evaporation residues [56–58]. These are implanted into an array of 100 PIN diodes arranged in a 10×10 matrix. With the new system decay data of the earlier found proton emitters ^{109}I and ^{113}Cs were remeasured and new proton emitters were searched for.

E. A fast catcher detector system

Gillitzer *et al.* [8] developed a method for investigation of proton radioactivities with extremely short half-lives. A target at the end of an annular large area gas detector system was mounted in a shielded position not visible by the detector. The evaporation residues recoiling from the target were collected in a thin catcher foil, while the direct beam passed the foil through a hole. The subsequent decay of particle unstable nuclei was analyzed during the pauses of the pulsed beam in backward geometry by the gas detector system. Because of the close geometry half-lives down to 10 nsec were accessible for detection. The proton emitters ^{109}I and ^{113}Cs have been discovered using this technique.

F. Fragment separators

In relativistic heavy ion collisions very neutron deficient nuclei are produced in fragmentation processes. Especially light neutron deficient nuclei are produced as projectile fragments with high rates as consequence of relatively high cross-sections, use of thick targets and high separator efficiencies. Pioneering work on investigation of fragmentation for production of new neutron deficient nuclei has been done at the GANIL with projectile energies up to 100 MeV/u on the LISE spectrometer [59]. Further improvement of the spatial separation of projectile fragments could be achieved by an additional velocity separation made by crossed electric and magnetic fields, LISE3 [60]. Typical flight time through the extended separator is 500 nsec. Direct mass and element identification is achieved by TOF and ΔE measurements thus mapping proton-rich nuclei with half-lives longer than few hundred nanoseconds. Nuclei with longer half-lives are finally implanted into silicon detector telescopes where there radioactive decays can be measured. Other major zero-degree spectrometers are in use at MSU [61], RIKEN [62], and GSI [63].

4. Proton emitters: Spectroscopic aspects and predictions

Till today 21 radioactive proton transitions have been observed and could be assigned definitively. In the following we will see that also regions of different nu-

Table 1. Proton-decay energies and energy corrections

Isotope	$E_{p,cl}$ [a,b]	E_R [c]	$E_R-\Delta E_R$ [d]	ΔE_p [e]	E_{SC} [f]	E_p [g]	Q_p [h]	$Q_{p,nucl}$ [i]	E_α [j]	Ref.
^{53m}Co	1564	30	6.0±3.0	1.8	3.3	1560 ±30	1590 ±30	1593 ±30		2,33,64
^{105}Sb	478	4.6	0.9±0.5	0.9	8.3	478 ±15	483 ±15	491 ±15	2001	14
^{109}I	812.9±3.2	7.6	1.5±0.8	1.2	8.8	812.6±4.0	820.2±4.0	829.0±4.0	3757	6–8,12,58,75
^{112}Cs	807	7.3	1.5±0.8	1.2	9.2	807 ±7	814 ±7	823 ±7	3838	65
^{113}Cs	959.7±2.8	8.6	1.7±0.9	1.3	9.2	959.3±3.7	967.9±3.7	977.1±3.7	3360±15	6–8,58,75
^{146m}Tm	1119	7.7	1.6±0.8	1.5	13.1	1119 ±5	1127 ±5	1140 ±5	3482	11
^{146}Tm	1189	8.2	1.6±0.8	1.5	13.1	1189 ±5	1197 ±5	1210 ±5	3482	11
(^{147m1}Tm)	947	6.5	1.3±0.7	1.3	13.1	947.4±5.0	953.9±5.0	967.0±5.0	2821	12
^{147}Tm	1051.1±2.6	7.3	1.5±0.7	1.4	13.1	1051.0±3.3	1058.3±3.3	1071.4±3.3	2821	4,5,12,68
^{147m2}Tm	1118.5±4.5	7.7	1.5±0.8	1.5	13.1	1118.5±5.3	1126.2±5.3	1139.3±5.3	2821	6,12,75
^{150}Lu	1262.8±2.7	8.5	1.7±0.9	1.6	13.7	1262.7±3.6	1271.2±3.6	1284.9±3.6	2978	6,12,75
^{151}Lu	1232.9±2.0	8.3	1.7±0.8	1.6	13.7	1232.8±2.8	1241.1±2.8	1254.8±2.8	2823	3,4,12
^{156}Ta	1022	6.6	1.3±0.7	1.4	14.3	1022 ±13	1029 ±13	1043 ±13	4618	13
^{156m}Ta	1103	7.1	1.4±0.7	1.5	14.3	1103 ±12	1110 ±12	1124 ±12	4618	66
^{160}Re	1261	7.9	1.6±0.8	1.6	15.0	1261 ±6	1269 ±6	1284 ±6	6537±16	13

[a] All energies are in keV.
[b] $E_{p,cl}$ = center of line for decays of detector implanted nuclei.
Data with error bars are calibrated using the ^{151}Lu-proton line and alpha lines of ^{152}Er, ^{151}Ho, and ^{150}Dy.
The ^{151}Lu line was calibrated with alpha lines and conversion electron lines of ^{209}Bi and ^{137}Cs.
[c] E_R = energy of recoil nuclei.
[d] ΔE_R = energy deficit of recoil nuclei in silicon.
[e] ΔE_p = energy deficit of protons in silicon.
[f] E_{SC} = screening correction.
[g] $E_p = E_{p,cl} + \Delta E_p - (E_R - \Delta E_R)$.
[h] $Q_p = E_p + E_R$.
[i] $Q_{p,nucl} = Q_p + E_{SC}$.
[j] Theoretical α-energy values [67]; experimental values for ^{113}Cs [65] and ^{160}Re [13].

Table 2. Partial proton-decay half-lives and branching ratios

Isotope	E_p/keV	$T_{1/2,exp}$	$T_{1/2,p,exp}$ [a]	$T_{1/2,\alpha}$ [b]	$T_{1/2,\beta}$ [c]	$b_p/\%$	$b_\alpha/\%$	$b_\beta/\%$
^{53m}Co	1560	247±12 msec	≃17 sec		200 msec	1.5		98.5
^{105}Sb	478		50 sec	9E12 sec	500 msec	1		99
^{109}I	813	103±5 μsec	103±5 μsec	85 msec	250 msec	99.84	0.12	0.04
^{112}Cs	807	500±100 μsec	500±100 μsec	540 msec	100 msec	99.6	0.1	0.5
^{113}Cs	959	28±7 μsec	28±7 μsec	67 sec	200 msec	99.99		0.01
^{146m}Tm	1119	235±27 msec	1.08 sec	3E10 sec	300 msec	22		78
^{146}Tm	1189	72±23 msec	95 msec	3E10 sec	300 msec	76		24
(^{147m1}Tm)	947	0.1–10 msec	0.1–10 msec	7E16 sec	300 msec	>96.7		0.03–3.3
^{147}Tm	1051	560±40 msec	$2.7^{+2.4}_{-0.9}$ sec	7E16 sec	300 msec	21		79
^{147m2}Tm	1118	360±40 μsec	360±40 μsec	7E16 sec	300 msec	99.9		0.1
^{150}Lu	1263	35±10 msec	46±10 msec	4E16 sec	150 msec	77		23
^{151}Lu	1233	85±10 msec	120^{+120}_{-40} msec	2E18 sec	250 msec	70		30
^{156}Ta	1022	165^{+165}_{-55} msec	198^{+165}_{-55} msec	7.4E4 sec	1.0 sec	84		16
^{156m}Ta	1103	320±80 msec		7.4E4 sec	≃1 sec			
^{160}Re	1261	790±160 μsec	1.31±0.16 msec	2.0 msec	600 msec	60.4	39.5	0.1

[a] $1/T_{1/2,p,exp} = 1/T_{1/2,exp} - 1/T_{1/2,\alpha} - 1/T_{1/2,\beta}$; in case of ^{53m}Co, ^{105}Sb, ^{147}Tm, and ^{151}Lu measured cross-sections were used to determine $T_{1/2,p,exp}$ and b_p. $b_p(^{147}Tm) = 0.15 \pm 0.05$ was measured recently by β-decay studies [68].
[b] $T_{1/2,\alpha}$ calculated with WKB method and α energies from Table 1.
[c] $T_{1/2,\beta}$ for ^{53m}Co from Ref. [64], all others from Ref. [69].

clear structure are covered by the known proton emitters. First, the emission of protons from isomeric states formed by multi nucleon configurations (^{53m}Co), second, a region where pure shell model configurations dominate (around ^{151}Lu) and, third, a transitional region of high complexity (around ^{113}Cs). The decay data are listed in Tables 1 and 2. The experimental half-lives are compared in Table 3 with calculated values. The analysis of the recently discovered heavy proton radioactivities ^{165,166,167}Ir, ^{171}Au and ^{185}Bi [10] is still in progress and the data are preliminary. Therefore, these isotopes are not yet included in the tables, but some of the data are discussed in the text. An energy spectrum, which shows the proton-decay line of the first identified ground-state proton emitter, is shown in Fig. 5.

Table 3. Experimental and calculated proton-decay half-lives

Isotope	$Q_{p,nucl}$/keV[a]	$T_{1/2,p,exp}$[b]	$T_{1/2,p,\Delta l=0}$[c]	$T_{1/2,p,l,j}$[d]	l,j	$S_{l,j,exp}$[e]	$T_{1/2,p,theory}$[f]
^{53m}Co	1593	≃17 sec	3E-20 sec	60 ns	f19/2	3.5E-9	15.7 sec
^{105}Sb	491	50 sec	790 msec	11 sec	d5/2	0.2	
^{109}I	829.0	103±5 μsec	0.44 μsec	5.6 μsec	d5/2	0.05	30−200 μsec
^{112}Cs	824	500±100 μsec	3.3 μsec	39 μsec	d5/2	0.08	
^{113}Cs	977.1	17±2 μsec	27 nsec	0.31 μsec	d5/2	0.018	29−41 μsec
^{146m}Tm	1140	1.08 sec	14 μsec	0.46 sec	h11/2	0.42	
^{146}Tm	1210	95 msec	2.2 μsec	69 msec	h11/2	0.73	
(^{147m1}Tm)	967.0	0.1−10 msec	2.9 msec	2.9 msec	s1/2	29−0.3	
^{147}Tm	1071.4	$2.7^{+2.4}_{-0.9}$ sec	97 μsec	3.3 sec	h11/2	1.2	1.9−5.1 sec
^{147m2}Tm	1139.3	360±40 μsec	14 μsec	130 μsec	d3/2	0.36	210−340 μsec
^{150}Lu	1284.9	46±10 msec	1.5 μsec	38 msec	h11/2	0.83	19−47 msec
^{151}Lu	1254.8	120^{+120}_{-40} msec	2.9 μsec	81 msec	h11/2	0.67	40−110 msec
^{156}Ta	1043	198^{+165}_{-55} msec	5.3 msec	47 msec	d3/2	0.24	
^{156m}Ta	1125		410 μsec	10 sec	h11/2		
^{160}Re	1284	1.31±0.16 msec	23 μsec	0.19 msec	d3/2	0.14	

[a] $Q_{p,nucl}$ see Table 1.
[b] $T_{1/2,p,exp}$ from Table 2.
[c] $T_{1/2,p}$calculated with WKB method for $\Delta l = 0$ transition.
[d] $T_{1/2,p}$ calculated with WKB method and l,j values as given in Column 6.
[e] $S_{l,j,exp} = T_{1/2,p,l,j}/T_{1/2,p,exp}$.
[f] Many-particle theory for proton decay of spherical and deformed nuclei [70, 71].

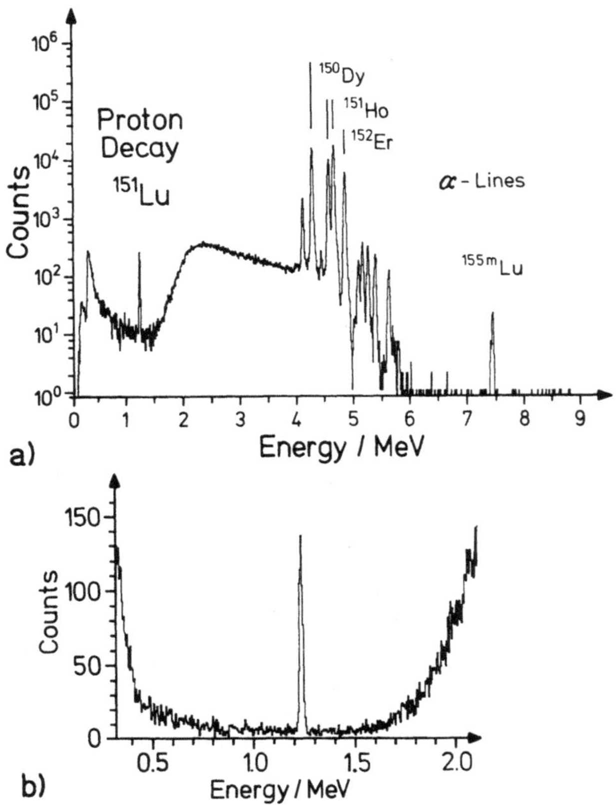

Fig. 5. (a) Energy spectrum obtained during the irradiation of a ^{96}Ru target (97.9% enrichment) with 261-MeV ^{58}Ni projectiles at SHIP. The ^{151}Lu proton-decay line was observed at 1.23 MeV far below the α-decay lines of other reaction products. The α line from the ^{155m}Lu decay originates from target impurities of heavier ruthenium isotopes. The impurities were avoided in an irradiation using mass separated targets. (b) Expanded part of the spectrum showing the ^{151}Lu proton-decay line. The resolution is 19-keV FWHM measured with a 45.5-μm thick silicon surface barrier detector. (From Ref. [4].)

A. Proton radioactivity from isomeric states

The general existence of proton radioactive isomers arising from one or two particle configurations has been discussed by Goldansky [1], many particle configurations were considered by Flerov *et al.* [72]. After the discovery of the proton decay branch of an isomer in ^{53}Co, Peker *et al.* [73] investigated the proton decay of this complex three-particle transition theoretically and predicted a number of three- and four-particle isomeric states with spin ≥ 19/2 in nuclei with $A < 100$ to be *proton radioactive.* They also discussed the similar so far not observed phenomenon of *neutron radioactivity* from isomeric states. Predictions for proton unstable isomeric states in the tin region are given by Ogawa [74]. The possibility of proton decay from isomeric states in nuclei near $N = 82$ and near $Z = 82$ was discussed in Refs. [6, 75, 76, 77]. In this region, isomeric states are in energy close to the ground-states which are already proton unbound. The observed transitions up-to-date are presented in the following.

B. Neutron-deficient light nuclei ($Z < 50$)

Although neutron deficient nuclei of lighter elements ($Z<50$) form the 'classical' region to search for ground-state proton radioactivity, this phenomenon could not be found there so far. For several reasons the region of light elements was favored to search for ground-state proton radioactivity: (1) The proton dripline is relatively close to the valley of stability, especially for elements $Z<28$, and production of neutron deficient nuclei in reactions with p, ^{3}He, and other light ions was possible since the first accelerators started to operate. (2) The Coulomb barriers are low and protons can be emitted already at low Q_p-values. (3) The nuclear structure at low excitation energies is relatively simple. Spherical shell model configurations are adequate to describe many of the neutron deficient nuclei. Proton transition probabilities can be calculated assuming Coulomb plus centrifugal barriers and spectroscopic factors close to one. (4) The charge independence of nuclear forces leads to simple and reliable predictions for masses of mirror nuclei on the neutron deficient side of the $N=Z$ line using known masses of the corresponding partners on the neutron rich side.

The low Coulomb barrier has three severe consequences for search experiments for proton radioactivity: (1) The evaporation of charged particles is almost unhindered and due to low binding energies the production cross-section of neutron deficient nuclei is small. (2) The half-life window, which is determined by the experimental set-up for short half-lives and by the competition from β decay for longer half-lives is very small. Goldansky [78] determined half-life windows from 10 to 10^{-4} sec for proton energies of 0.1 – 0.15 MeV at $Z=20$ and 0.35 – 0.5 MeV at $Z=40$. (3) Weak, low energy lines are difficult to detect in a background of β's and β-delayed particles.

Although ground-state proton radioactivity could not be observed so far in the light element region, all known β-delayed charged particle activities were found there. Bayley and Crane [79] observed delayed alpha's following the reaction ^{7}Li(d,p)^{8}Li. In an attempt to observe delayed proton emission, Alvarez [80] found the three β-delayed α-emitters ^{8}B, ^{12}N, and ^{20}Na. Beta-delayed proton emission was observed by Karnaukhov *et al.* [81] in the reaction ^{20}Ne + Ni and Barton *et al.* [82] identified ^{23}Si as a delayed proton emitter. Beta-delayed two-proton radioactivity of ^{22}Al was discovered by Cable *et al.* [83]. Very recent reviews on the decay properties of neutron-deficient light nuclei were published by Äystö and Cerny [84] and Detraz and Vieira [85].

Proton separation energies derived from reaction data were compiled and adjusted by Wapstra and Audi [86]. One and two proton separation energies of less than minus several hundred keV are known for the lightest proton unbound nuclei of elements below $Z=20$.

Among the heavier elements ^{39}Sc, ^{42}V, and ^{55}Cu are predicted to be proton unbound by 590, 190, and 290 keV [86], respectively. Their shell model configurations should be simply $f^{1}_{7/2}, f^{3}_{7/2}$, and $p^{1}_{3/2}$ and proton transition probabilities should be directly comparable to theory.

Information on the stability lines were obtained for a number of elements in recent experiments at GANIL in Caen. Langevin *et al.* [87] report on the direct identification by time-of-flight energy measurements of the $T_z = -5/2$ nuclei ^{23}Si, ^{27}S, ^{31}Ar, and ^{35}Ca, the latter nucleus already identified by β-delayed 2 proton activity by Äystö *et al.* [88]. No events have been observed for the nuclei ^{42}V, ^{44}Mn, ^{45}Mn, and ^{45}Fe [15]. It is concluded that the three nuclei ^{42}V, ^{44}Mn, and ^{45}Mn are proton unbound.

Blank *et al.* [17] investigated the ^{58}Ni fragmentation at the FRS/SIS of GSI in Darmstadt. The even-Z isotope ^{50}Ni could be observed for the first time demonstrating its particle stability. The odd-Z isotopes ^{49}Co and ^{54}Cu were not observed. According to mass predictions and WKB half-life calculations, these isotopes should decay by proton emission with half-lives shorter than 10^{-10} sec.

Investigation of neutron-deficient isotopes that are important for the astrophysical rapid-proton-capture process (rp) was the aim of a work by Mohar *et al.* [16] at the Cyclotron Laboratory at Michigan State University. The six new isotopes ^{61}Ga, ^{62}Ge, ^{63}Ge, ^{65}As, ^{69}Br, and ^{75}Sr could be identified. Evidence was found for proton instability of ^{73}Rb. This is consistent with the previous result of D'Auria *et al.* [89] at ISOLDE.

Blank *et al.* [18] continued the investigation of the astrophysical rp process in an experiment at the SISSI/LISE facility of GANIL. The five new isotopes ^{60}Ga, ^{64}As, ^{70}Kr, ^{69}Kr and ^{74}Sr were found, but the isotope ^{69}Br was not observed. It was concluded that ^{69}Br is a proton unbound nucleus with a half-life shorter than ≈ 100 nsec. Fig. 6 shows the proton drip-line in the region of elements copper to strontium together with the five new isotopes and a possible path of the rp process.

C. The transitional region near ^{100}Sn

The results on ε_2 deformations in the region $50 \leq (N,Z) \leq 82$ obtained by Möller and Nix [30] are plotted as contour lines in Fig. 7. Marked are the positions of the proton emitters ^{105}Sb, ^{109}I, ^{112,113}Cs, and ^{146,147}Tm, ^{150,151}Lu, ^{156}Ta, ^{160}Re, ^{165,166,167}Ir. It is obvious that the *island* of the light proton emitters (^{105}Sb, ^{109}I, ^{112,113}Cs) is in a region of steep increase of deformation with increasing proton number. Qualitatively we can expect a high hindrance factor for these type of proton transitions between nuclei of different deformation.

No hindrance by change of deformation is expected for proton transitions of nuclei near $N=82$ and for so far unknown proton emitters from holmium to lanthanum. Ground-state proton transitions of these nuclei are either parallel to the contourlines of deformation in the (Z,N)–plane, or they occur between nuclei with similar, though large deformations, e.g. ^{132}Eu.

Bugrov and Kadmenskii [71] calculated the proton-decay widths for ^{113}Cs and ^{109}I. They applied their theory developed for proton decay of deformed nuclei taking many-particle effects into account. They found excellent agreement with the experimental half-lives in case of ^{113}Cs assuming a $3/2^+$ [422] ground-state and reasonable deformations of $\beta_2 \approx 0.21$ ($\beta_2 \approx 1.06\,\varepsilon_2$) and in case of ^{109}I assuming a $1/2^+$ [420] ground-state and $\beta_2 \approx 0.05 - 0.10$. The calculated half-lives are (29 – 41) µsec and (30 – 200) µsec for ^{113}Cs and ^{109}I, respectively.

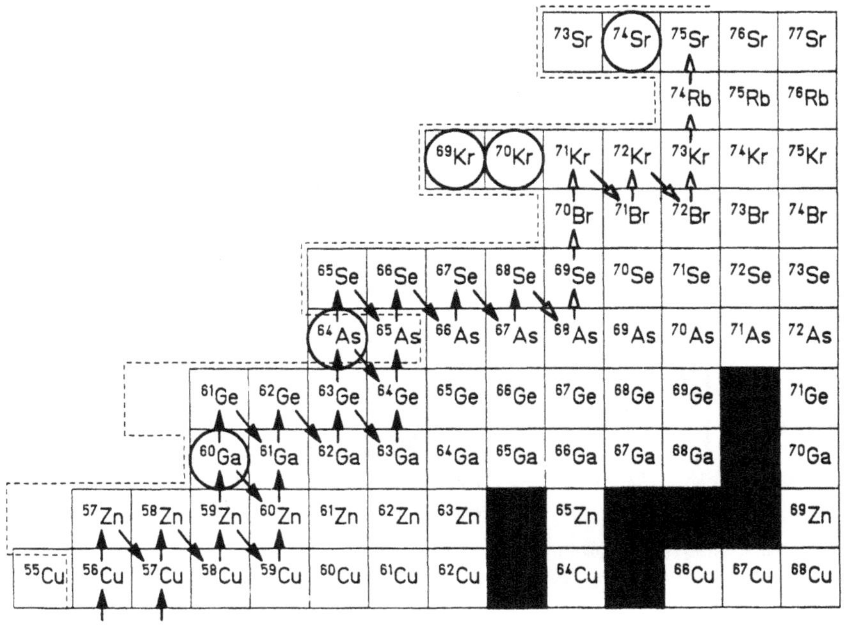

Fig. 6. Section of the chart of nuclides in the region of interest for the astrophysical rp process. Stable nuclei are indicated by filled squares. Five new nuclei marked by circles have been observed in an experiment at the SISSI/LISE facility of GANIL in fragmentation reactions of a ^{78}Kr beam with a nickel target. Nuclei to the left of the dashed line are predicted to be unstable by the mass model of Jänecke and Masson [29]. The arrows indicate a possible flow of the rp process. (From Ref. [18].)

The systematics of Q_p-values for the elements Sb, I, and Cs was discussed by Page *et al.* [65]. Striking is 1) the Q_p-value staggering for odd-odd and odd-even nuclei at the drip line, which is well reproduced by the mass predictions of Wapstra *et al.* [90], but not by the mass calculations of Möller and Nix [30]; and 2) the overestimate of Q_p-values for neutron deficient antimony isotopes by the mass calculations. In both cases the attractive interaction of $d_{5/2}$ protons with $d_{5/2}$ neutrons is responsible for an increased proton-binding energy.

D. The region of deformed neutron-deficient rare-earth nuclei

An estimate for the location in the (Z,N)−plane of possible proton emitters can be obtained by fitting a straight line through the positions of the known proton emitters from antimony to rhenium (see Fig. 7). With $Z = 0.743 \times N + 11.6$, we obtain as most neutron rich candidates for proton radioactivity the nuclei ^{118}La, ^{123}Pr, ^{128}Pm, ^{132}Eu, ^{137}Tb, and ^{142}Ho and their neighbouring isotopes with $\Delta N = \pm 1$. Except the β-delayed proton emitter ^{124}Pr [91] none of these eighteen nuclei have been identified so far. Early results obtained by Karnaukhov *et al.* [92] need further confirmation.

The reason for the experimental difficulties in this region are, that the expected proton emitters cannot be produced in evaporation channels, e.g. p2n, with cross-sections of about 100 µb with stable projectile and target nuclei. Nevertheless, search experiments will be continued, and with higher sensitivity of the experiments one can hope that ground-state proton radioactivity will be established also in this region in the near future.

E. The pure shell-model nuclei near $N = 82$

The structure of nuclei in the quadrant $Z \geq 64$ and $N \leq 82$ is determined by protons in the orbitals 1 $h_{11/2}$, 3 $s_{1/2}$, and 2 $d_{3/2}$, and by neutron holes in the orbitals 3 $s_{1/2}$, 2 $d_{3/2}$, and 1 $h_{11/2}$.

The low energy level schemes of the nuclei in the region of interest show some common properties that can be summarized as shown schematically in Fig. 8. The assigned proton transitions are indicated by arrows. Taking into account the uncertainties of estimated proton energies, β and γ half-lives we conclude that the assigned proton transitions fit well into the presented level schemes.

The two proton transitions from ^{147}Tm to the ground state of ^{146}Er allow to determine the energy of the low spin isomer and establish the $11/2^-$ state to be the ground-state. The energy of the low spin isomer is then at (70 ± 7) keV above the ground-state.

Beyond $N = 82$ and $Z = 64$ the relevant neutron orbitals are $\nu f_{7/2}$, $\nu h_{9/2}$ and the relevant proton orbitals $\pi h_{11/2}$, $\pi d_{3/2}$, $\pi s_{1/2}$. In recent experimental work using α, β, and γ spectroscopy, decay data were measured for isotones up to ^{156}Ta, ^{158}W, ^{161}Re, and ^{162}Os [93–98]. Schematically, with respect to proton radioactivity the measured data on low lying levels can be summarized as following: (1) The even-even nuclei have levels 0^+, 2^+, 4^+, and 6^+ by alignment of two $f_{7/2}$ neutrons up to about 2 MeV. A $[\nu h_{9/2} \nu f_{7/2}]8^+$ level decreases from 2.7 MeV with increasing proton number below 2 MeV and becomes an α-emitting isomer in ^{156}Hf and ^{158}W [93]. (2) Coupling an $h_{11/2}$ proton to these states results for odd-Z isotones in a level sequence $11/2^-$, $15/2^-$, $19/2^-$, $23/2^-$, and an isomeric state $25/2^-$ at about 2 MeV. The configuration of the isomer is domi-

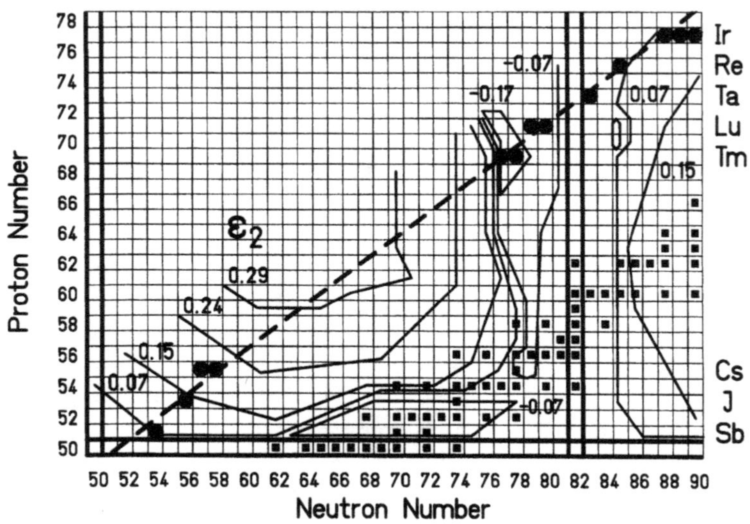

Fig. 7. Contour map of deformation ε_2 versus proton number Z and neutron number N as calculated by Möller and Nix [30]. Stable nuclei are indicated by filled squares, the known proton emitters by dots. The dashed line is a fit through the positions of the known proton emitters ^{105}Sb, ^{109}I, ^{112,113}Cs, ^{146,147}Tm, ^{150,151}Lu, ^{156}Ta, and ^{160}Re. For odd elements the dashed line indicates candidates for proton radioactive nuclei. The figure exhibits three regions with different influence on proton transitions. In region I around ^{109}I and ^{113}Cs the deformation changes rapidly and the contour lines are crossed by the proton transitions $Z+1, N \rightarrow Z, N$. In region II near $N = 82$ the deformations are small (except a region near $Z=70$, $N=75$ with negative ε_2 values) and proton transitions are parallel to the contour lines. In region III between I and II proton transitions are between nuclei of similar but large deformations.

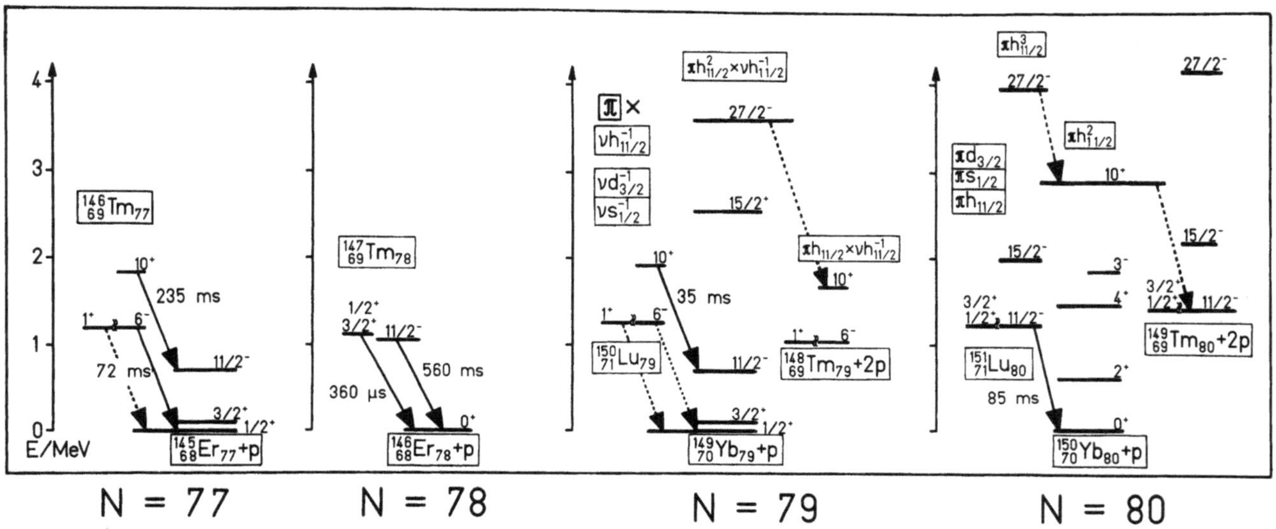

Fig. 8. Partial level schemes of neutron deficient $N = 77$, 78, 79, and 80 isotones obtained by extrapolation from known schemes of nuclei in the region $Z \geq 64$, $N \simeq 82$. The arrows (full lines) mark identified proton transitions, the measured total half-lives are given. Other energetically allowed but so far not observed proton transitions are marked by dashed arrows.

nantly $[\pi h_{11/2} \nu h_{9/2} \nu f_{7/2}]25/2^-$ [96] and it becomes an α emitter in ^{155}Lu [3]. Close to the $11/2^-$ band head a low spin state $[\pi s_{1/2}]$ or $[\pi d_{3/2}]$ is known. (3) In odd neutron, even proton nuclei an $f_{7/2}$ neutron is coupled to the $\pi h^2_{11/2}$ level sequence ending at about 2.5 MeV with an isomeric state of spin and parity $27/2^-$. (4) Odd-odd nuclei have two states $[\pi d_{3/2} \nu f_{7/2}]2^-$ and $[\pi h_{11/2} \nu f_{7/2}]9^+$ close in energy. Until now it is not known which one is the ground-state.

The observed proton decays of ^{160}Re, ^{156}Ta [13] and ^{156m}Ta [66] can be well assigned to levels within the frame of this scheme. Both proton emitters are odd-odd nuclei and the proton-decay widths are in agreement with emission from $d_{3/2}$ orbitals. Therefore, in case of ^{156}Ta the 1022-keV line can be assigned to a transition from a $[\pi d_{3/2} \nu f_{7/2}]2^-$ state to the $\nu f_{7/2}$ ground-state of ^{155}Hf. The $[\pi h_{11/2} \nu f_{7/2}]9^+$ state decays dominantly by a favored Gamow-Teller transition with $T_{1/2} = 320$ msec into the 8^+ isomeric state of ^{156}Hf and by emission of a highly hindered $h_{11/2}$ proton of 1103 keV into the ground-state of ^{155}Hf [66, 93]. Also the 1261-keV proton decay of ^{160}Re can be assigned to the decay of a $[\pi d_{3/2} \nu f_{7/2}]$ state.

F. Neutron-deficient heavy nuclei ($Z > 76$)

Neutron deficient nuclei in the element range from iridium to bismuth have high probabilities for α decay.

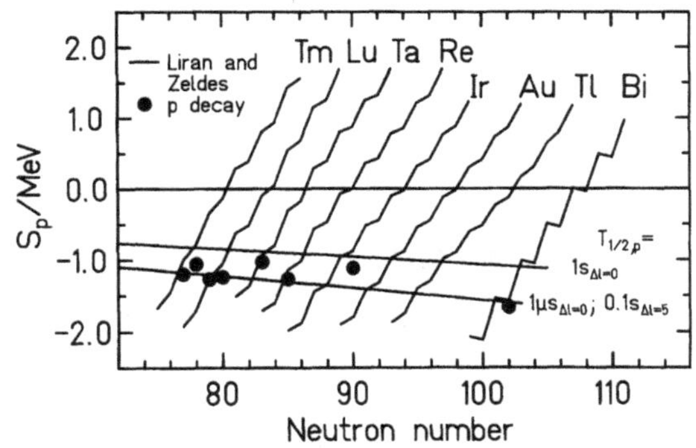

Fig. 9. Proton separation energies ($S_p = -Q_p$) of neutron deficient odd elements from thulium to bismuth. The values are taken from Ref. [99] (Liran and Zeldes). The dots mark measured S_p values of ^{146,147}Tm, ^{150,151}Lu, ^{156}Ta, ^{160}Re, ^{167}Ir and ^{185}Bi. Estimates for partial proton half-lives (WKB-method, $\Delta \ell = 0$) of 1 sec and 1 µsec are indicated by straight lines. Assuming $\Delta \ell = 5$ increases the half-lives from 1 µsec to 0.1 sec.

New isotopes could be unambiguously identified by position-time correlation methods [9, 51, 98]. Great progress was made recently in experiments at the Argonne fragment mass analyzer by Davids *et al.* [10], who reported proton radioactivity of ^{165,166,167}Ir, ^{171}Au and ^{185}Bi. In Fig. 9 we compare the measured proton-separation energies with predictions by Liran and Zeldes [99]. The theoretical values are in good agreement with the experimental data and the extrapolations based on systematics obtained by Wapstra and Audi [86].

The lightest isotopes have partial α half-lives in the range from 5 to 160 msec. These short lived activities offer optimum conditions for experiments at recoil separators with implantation techniques. After the implantation of an evaporation residue, a proton of about 1.5 MeV is emitted followed by a sequence of short lived α decays. The high sensitivity of the correlation technique helped to overcome the difficulty of decreasing cross-sections and small proton-branching ratios. The identification of the new proton radioactivities ^{165}Ir, ^{166}Ir, ^{167}Ir, ^{171}Au and ^{185}Bi [10] demonstrates impressively the great power of the correlation technique.

Predictions for proton radioactivity of light bismuth isotopes were made [6, 75] taking into account the phenomenon of low spin intruder states reviewed by Hyde *et al.* [100]. These $s_{1/2}$ proton hole states form isomeric states known from α and γ spectroscopy for isotopes heavier than ^{187}Bi which decrease in energy with decreasing neutron number. A minimum energy is expected near the middle of the $N = 82$ and 126 shells, i.e. near ^{187}Bi. In ^{185}Bi the predicted proton separation energy is -1.5 MeV, which is high enough, so that proton emission can compete with α decay. The observed proton decay of ≈ 1.65 MeV was correlated with the known α decay of ^{184}Pb. The measured half-life was ≈ 50 µsec [10]. The transition was interpreted as originating from the $s_{1/2}$ intruder state in ^{185}Bi, and proceeding to the ^{184}Pb ground-state [10].

The heaviest odd-Z nucleus has the mass number 272 of element 111 [49]. The proton Q-value is -0.5 MeV according to Liran and Zeldes [99]. The first proton unbound isotope of element 111 has the mass number $A = 269$ ($Q_p = 0.3$ MeV), and the first isotope with a remarkable short calculated partial proton half-life ($T_{1/2} = 100$ msec, $Q_p = 1.9$ MeV) is ^{261}111. In this neutron deficient heavy element region α decay and fission are likely the faster decay modes [101, 102].

G. Two-proton radioactivity

Properties of two-proton radioactivity have been discussed by Goldansky in a series of papers [103–106], by Galitsky and Cheltsov [107], and recently by Brown [108]. This phenomenon would be a new type of radioactive decay of nuclei, but has not been observed so far.

A reason to search for one-proton radioactivity in the region of neutron deficient $N = 82$ nuclei was that high Q_α values allow sensitive measurements on production rates of neutron deficient nuclei by investigation of α decay for nuclei $N \geq 84$ and search for proton emission in absence of α decay for $N < 84$ nuclei. With similar arguments we may predict neutron deficient tellurium isotopes to be candidates for 2p radioactivity. High Q_{2p} values between 1.5 and 3.5 MeV (due to the shell closure at $Z = 50$) can be expected if we extrapolate the Q_{2p} values given by Bos *et al.* [109] beyond ^{106}Te. Due to the $N = 50$ shell effect the predicted Q_α values [67] decrease to 3.0 and 1.2 MeV for ^{103}Te and ^{102}Te, respectively, resulting in $T_{1/2,\alpha} >$ 100 sec. Predicted β half-lives are in the range from (20 – 200) msec [69] and estimates for $T_{1/2,2p}$ are about 1 sec for $Q_{2p} = 2.5$ MeV decreasing to 1 msec for $Q_{2p} = 3.0$ MeV. Difficult to predict with sufficiently high precision is the 1 p emission which can become a serious competition.

In the region of light elements the nuclei ^{39}Ti, ^{42}Cr, ^{45}Fe, and ^{59}Ge are probably ground-state 2p emitters as discussed in Refs. [84, 85]. A recent search for 2p decay of ^{31}Ar was negative [110]. These nuclei can in principle be produced with relatively high cross-sections in fragmentation reactions. The experimental difficulties to identify 2p radioactivity in this region of light elements are low Q_{2p} values and rapid decrease of half-lives with increasing Q_{2p} values within few hundred keV. The resonance like 2p decay of ^{12}O was investigated by Kryger *et al.* [111] resulting in an upper limit of 7% for ^{2}He emission. The decay of the ^{12}O ground-state is dominated by sequential proton emission.

In the heavy element region 2p radioactivity is energetically allowed for ^{152}Hf, ^{154}W, and ^{153}Ta, ^{155}Re (see Fig. 1) but the Coulomb barriers in this element range are too high that 2p activity with the predicted Q_{2p} values can compete with β-decay. But the examples shown in Fig. 1, a plot of binding energies versus proton number, illustrate that a one-step 2p-emission may happen in those cases only, where the first 1p transition is hindered or even forbidden by low energy (or nuclear structure effects) as in case of ^{154}W or ^{152}Hf. Similar arguments let one believe that three-proton emission will always be a two- or three-step process.

References

1. Goldansky, V. I.: Ann. Rev. Nucl. Sci. **16**, 1 (1966).
2. Jackson, K. P., *et al.*: Phys. Lett. **33B**, 281 (1970).
3. Hofmann, S., *et al.*: in Proc. 4th Int. Conf. on Nuclei Far from Stability, CERN 81-09, Geneva, 190 (1981).
4. Hofmann, S., *et al.*: Z. Phys. **A305**, 111 (1982).
5. Klepper, O., *et al.*: Z. Phys. **A305**, 125 (1982).
6. Hofmann, S., *et al.*: in Proc. 7th Int. Conf. on Atomic Masses and Fundamental Constants, Klepper, O., Ed., THD-Schriftenreihe Wissenschaft und Technik **26**, 184 (1984).
7. Faestermann, T., *et al.*: Phys. Lett. **137B**, 23 (1984).
8. Gillitzer, A., *et al.*: Z. Phys. **A326**, 107 (1987).
9. Woods, P. J., *et al.*: Nucl. Phys. **A553**, 485c (1993).
10. Davids, C. N., *et al.*: Proc. of the Int. Conf. on Exotic Nuclei and Atomic Masses, Arles, France, June 19–23, 1995, in press.
11. Woods, P. J., *et al.*: in Proc. 6th Int. Conf. on Nuclei far from Stability + 9th Int. Conf. on Atomic Masses and Fundamental Constants, Neugart, R. and Wöhr, A., Eds., Bernkastel-Kues, 1992, Inst. Phys. Conf. Ser. No 132, 323 (1993).
12. Sellin, P. J., *et al.*: Phys. Rev. **C47**, 1933 (1993).
13. Page, R. D., *et al.*: Phys. Rev. Lett. **68**, 1287 (1992).
14. Tighe, R. J., *et al.*: Phys. Rev. **C49**, R2871 (1994).
15. Borrel, V., *et al.*: Z. Phys. **A344**, 135 (1992).
16. Mohar, M. F., *et al.*: Phys. Rev. Lett. **66**, 1571 (1991).
17. Blank, B., *et al.*: Phys. Rev. **C50**, 2398 (1994).
18. Blank, B., *et al.*: Report Nouvelles du GANIL, 54, 3, February 1995; Phys. Rev. Lett. **74**, 4611 (1995).
19. Möller, P., Nix, J. R.: At. Data Nucl. Data Tables **26**, 165 (1981).
20. Alkhazov, G. D., *et al.*: Z. Phys. **A311**, 245 (1983).
21. Vogel, P., *et al.*: Phys. Lett. **139B**, 227 (1984).
22. Jensen, A. S., *et al.*: Nucl. Phys. **A431**, 393 (1984).
23. Möller, P., Nix, J. R.: Nucl. Phys. **A536**, 20 (1992).
24. Gamow, G.: Z. Phys. **51**, 204 (1928).
25. Gurney, R. W., Condon, E. U.: Phys. Rev. **33**, 127 (1929).
26. Bethe, H. A.: Rev. Mod. Phys. **9**, 69 (1937).
27. Rasmussen, J. O.: *Alpha-, Beta- and Gamma-Ray Spectroscopy,* Siegbahn, K., Ed., North-Holland, Amsterdam, 1966, p. 701.
28. Becchetti, F. D., Greenlees, G. W.: Phys. Rev. **182**, 1190 (1969).
29. Jänecke, J., Masson, P. J.: At. Data Nucl. Data Tables **39**, 265 (1988).
30. Möller, P., Nix, J. R.: At. Data Nucl. Data Tables **39**, 213 (1988).
31. Möller, P.: private communication, 1992.
32. Macfarlane, R. D.: Phys. Rev. **137**, 1448 (1965).
33. Cerny, J., *et al.*: Phys. Lett. **33B**, 284 (1970).
34. D'Auria, J. M., *et al.*: Nucl. Phys. **A301**, 397 (1978).
35. Bruske, C., *et al.*: Nucl. Instrum. Methods **186**, 61 (1981).
36. Nitschke, J. M.: Nucl. Instr. and Meth. **206**, 341 (1983).
37. Kirchner, R., *et al.*: Nucl. Instrum. Methods **B70**, 56 (1992).
38. Kirchner, R.: Nucl. Instrum. Methods **B70**, 186 (1992).
39. Ärje, J., *et al.*: Nucl. Instrum. Methods **247**, 431 (1986).
40. Petrov, L. A., *et al.*: Zh. Eksp. Teor. Fiz. **59**, 1926 (1970); Sov. Phys. JETP **32**, 1042 (1971).
41. Paul, M., *et al.*: Nucl. Instrum. Methods **A277**, 418 (1989).
42. Ghiorso, A., *et al.*: Nucl. Instrum. Methods **A269**, 192 (1988).
43. Ninov, V., *et al.*: Nucl. Instrum. Methods **A357**, 486 (1995).
44. Lazarev, Yu. A., *et al.*: Proc. of the Int. School Seminar on Heavy Ion Physics, Oganessian, Yu. Ts., Penionzhkevich, Yu. E., and Kalpakchieva, R., Eds., Dubna, Russia, May 10–15, 1993, Vol II, p. 497, and Dubna report D7-93-274.
45. Leino, M., *et al.*: Z. Phys. **A348**, 151 (1994).
46. Miyatake, H., *et al.*: Nucl. Instrum. Methods **B26**, 309 (1987).
47. Münzenberg, G., *et al.*: Nucl. Instrum. Methods **161**, 65 (1979).
48. Hofmann, S., *et al.*: Z. Phys. **A350**, 277 (1995).
49. Hofmann, S., *et al.*: Z. Phys. **A350**, 281 (1995).
50. Hofmann, S.: Proc. of the XV. Nuclear Physics Divisional Conf. on Low Energy Nuclear Dynamics, St. Petersburg, Russia, April 18–22, 1995, in press; GSI-Preprint 95-25, p. 1–8, 1995.
51. Hofmann, S., *et al.*: Z. Phys. **A291**, 53 (1979).
52. Hofmann, S.: Proc. of the Int. Conf. Actinides-93, Santa Fe, New Mexico, USA, 19–24 September, 1993; J. Alloys Comp. **213/214**, 74–80 (1994).
53. James, A. N., *et al.*: Nucl. Instrum. Methods **A267**, 144 (1988).
54. Woods, P. J., *et al.*: Nucl. Instrum. Methods **A276**, 195 (1989).
55. Sellin, P. J., *et al.*: Nucl. Instrum. Methods **A311**, 217 (1992).
56. Gillitzer, A., *et al.*: in Proc. Int. Conf. on the Spectroscopy of Heavy Nuclei, Agia Pelagia, Crete, 1989, Sharpey-Schafer, J. F. and Skouras, L. D., Eds., Inst. Phys. Conf. Ser. No. 105, 1989, 191.
57. Heine, F., *et al.*: Z. Phys. **A340**, 225 (1991).
58. Heine, F., *et al.*: in Proc. 6th Int. Conf. on Nuclei far from Stability + 9th Int. Conf. on Atomic Masses and Fundamental Constants, Neugart, R., and Wöhr, A., Eds., Bernkastel-Kues, 1992, Inst. Phys. Conf. Ser. No 132, 331, 1993.
59. Anne, R., *et al.*: Nucl. Instrum. Methods **A257**, 215 (1987).
60. Mueller, A. C., Anne, R.: Nucl. Instrum. Methods **B56**, 559 (1991).
61. Sherill, B. M., *et al.*: Proc. of the 1. Int. Conf. on Radioactive Nuclear Beams, Berkeley 1989, Myers, W. D., Nitschke, J. M., and Norman, E. B., Eds., World Scientific, Singapore, 1990, p. 72.
62. Ishihara, M.: Nucl. Phys. **A538**, 309c (1992).
63. Geissel, H., *et al.*: Nucl. Instrum. Methods **B70**, 286 (1992).
64. Cerny, J., *et al.*: Nucl. Phys. **A188**, 666 (1972).
65. Page, R. D., *et al.*: Phys. Rev. Lett. **72**, 1798 (1994).
66. Livingston, K., *et al.*: Phys. Rev. **C48**, R2151 (1993).
67. Myers, W. D.: *Droplet Model of Atomic Nuclei,* Plenum, New York 1977.
68. Toth, K. S., *et al.*: Phys. Rev. **C47**, 1804–1806 (1993).
69. Takahashi, K., *et al.*: At. Data Nucl. Data Tables **12**, 101 (1973).
70. Bugrov, V. P., *et al.*: Yad. Phys. **42**, 57 (1985); Sov. J. Nucl. Phys. **42**, 34 (1985).
71. Bugrov, V. P., Kadmensky, S. G.: Yad. Phys. **49**, 1562 (1989); Sov. J. Nucl. Phys. **49**, 967 (1989).
72. Flerov, G. N., *et al.*: Zh. Eksp. Teor. Fiz. **47**, 419 (1964); Sov. Phys. JETP **20**, 278 (1965).
73. Peker, L. K., *et al.*: Phys. Lett. **36B**, 547 (1971).
74. Ogawa, K.: in Proc. 7th Int. Conf. on Atomic Masses and Fundamental Constants, Klepper, O., Ed., THD-Schriftenreihe Wissenschaft und Technik **26**, 530 (1984).
75. Hofmann, S., *et al.*: in Proc. Int. School-Seminar on Heavy Ion Physics, Gangrsky, Yu. P., Ed., Dubna, USSR, D7-87-68, 378, 1986.
76. Hofmann, S., *et al.*: in Proc. XI Int. Workshop on Gross Properties of Nuclei and Nuclear Excitations, Hirschegg, Kleinwalsertal, Austria, 1983, Feldmeier, H., Ed., TH-Darmstadt und GSI, Darmstadt, 45, 1983.

77. Hofmann, S., *et al.*: in Proc. Int. School Seminar on Heavy Ion Physics, Gangrsky, Yu. P. and Kholnov, Yu. V., Eds., Alushta, Dubna, USSR, D7-83-644, 114, 1983.
78. Goldansky, V. I.: Nucl. Phys. **19**, 482 (1960).
79. Bayley, D. S., Crane, H. R.: Phys. Rev. **52**, 604 (1937).
80. Alvarez, L. W.: Phys. Rev. **80**, 519 (1950).
81. Karnaukhov, V. A., *et al.*: Proc. of the Third Conf. on Reactions between Complex Nuclei, Ghiorso, A., Diamond, R. M., and Conzett, H. E., Eds., Asilomar, Pacific Grove, California, USA, April 14–18, 1963, p. 434.
82. Barton, R., *et al.*: Can. J. Phys. **41**, 2007 (1963).
83. Cable, M. D., *et al.*: Phys. Rev. Lett. **50**, 404 (1983).
84. Äystö, J., Cerny, J.: *Treatise on heavy-ion science, Vol. 8, Proton-Rich Light Nuclei,* Bromley, D. A., Ed., Plenum Publishing Corp., 207, 1989.
85. Detraz, D., Vieira, D. J.: Ann. Rev. Nucl. Part. Sci. **39**, 407 (1989).
86. Wapstra, A. H., Audi, G.: Nucl. Phys. **A432**, 1 (1985) and Nucl. Phys. **A432**, 55 (1985).
87. Langevin, M., *et al.*: Nucl. Phys. **A455**, 149 (1986).
88. Äystö, J., *et al.*: Phys. Rev. Lett. **55**, 1384 (1985).
89. D'Auria, J. M., *et al.*: Phys. Lett. **66B**, 233 (1977).
90. Wapstra, A. H., *et al.*: At. Data Nucl. Data Tables **39**, 281 (1988).
91. Wilmarth, P. A., *et al.*: Z. Phys. **A325**, 485 (1986).
92. Bogdanov, D. D., *et al.*: Yad. Fiz. **16**, 890 (1972); Sov. J. Nucl. Phys. **16**, 491 (1973).
93. Hofmann, S., *et al.*: Z. Phys. **A333**, 107 (1989).
94. Schardt, D., *et al.*: in Proc. 5th Int. Conf. on Nuclei far from Stability, Towner, I. S., Ed., Rosseau Lake, Ontario, Canada 1987, American Institute of Physics, New York, 477, 1988.
95. Broda, R., *et al.*: Z. Phys. **A327**, 403 (1987).
96. Rathke, G. E.: Gamma-spektroskopische Untersuchung von protonenreichen Kernen der Seltenen Erden in der Gegend um N = 84, Diploma work, Technische Universität München, München, 1984.
97. Rathke, G. E.: in Proc. XXII. Int. Winter Meeting on Nuclear Physics, Bormio, Italy, 1984, 53.
98. Hofmann, S., *et al.*: Nucl. Instrum. Methods **223**, 312 (1984).
99. Liran, S., Zeldes, N.: At. Data Nucl. Data Tables **17**, 431 (1976).
100. Heyde, K., *et al.*: Phys. Rep. **102**, 291 (1983).
101. Moeller, P., Nix, J. R.: J. Phys. G: Nucl. Part. Phys. **20**, 1681 (1994).
102. Smolanczuk, R., Sobiczewski, A.: Proc. of the XV. Nuclear Physics Divisional Conf. on Low Energy Nuclear Dynamics, St. Petersburg, Russia, April 18–22, 1995.
103. Goldansky, V. I.: Nucl. Phys. **27**, 648 (1961).
104. Goldansky, V. I.: Phys. Lett. **14**, 233 (1965).
105. Goldansky, V. I.: Nucl. Phys. **78**, 233 (1966).
106. Goldansky, V. I.: Phys. Lett. **B121**, 11 (1988).
107. Galitsky, V. M., Cheltsov, V. F.: Nucl. Phys. **56**, 86 (1964).
108. Brown, B. A.: Phys. Rev. **C43**, 1513–1517 (1991).
109. Bos, K., *et al.*: Nucl. Phys. **A432**, 140 (1985).
110. Borrel, V., *et al.*: Nucl. Phys. **A531**, 353 (1991).
111. Kryger, R. A., *et al.*: Phys. Rev. Lett. **74**, 860 (1995).

Radiochimica Acta *70/71*, 107–121 (1995)

Alpha Radioactivity

By E. Roeckl

Gesellschaft für Schwerionenforschung mbH, Postfach 110552, D-64220 Darmstadt, Germany

(Received August 18, 1995)

Alpha decay / Nuclear structure

Summary

A review of α-decay experiments is given with emphasis being put on nuclear structure studies performed by means of online mass separators or recoil spectrometers.

1. Introduction

Ever since its discovery by Becquerel almost exactly 100 years ago, which was also the very first observation of a nuclear decay mode, α-decay has always been and is still being used as a probe for investigating nuclear structure. The attractive feature of α-decays is that the experimental observables such as decay energy or decay-width directly relate to nuclear structure parameters such as binding energy or spectroscopic (or hindrance) factor. Moreover, α-particles can be detected rather easily under favourable conditions, i.e. high efficiency, low background and good energy resolution. The resulting nuclear structure data concern: ground state decay energies and mass excess values; energies of isomeric and other excited levels; assignment of spin and of other nuclear structure properties to parent and daughter states; isotopic assignment based on parent-daughter relations, α X-ray coincidences or Q-value systematics.

In order to show how such quantities are deduced from measurements, I shall follow the "traditional" α-decay theory which will be discussed, together with other model descriptions, in Section 2. Subsequently, examples will be given for developments related to source preparation, detectors, energy calibration, and coincidence or correlation techniques (Section 3). In Section 4 selected experimental results will be presented for different areas of the chart of nuclides, namely the island of α-decay above tin, the lanthanide-to-lead region and the area around lead. A few applications of α-decay beyond the research of nuclear structure will be given in Section 5.

This report represents an updated and shortened version of an article [1] that has recently been submitted for publication. It briefly describes recent results on direct (in contrast to β-delayed) α-radioactivity of nuclear ground-states or long-lived isomers, and is focused on experimental developments. A more complete discussion of experimental and theoretical results obtained on α-decay up to 1979, is found in Refs. [2–4]. The related theoretical development during the past two decades is presented in two chapters [5, 6] of the book mentioned in [1] and in a recent review [7].

2. From alpha-decay measurements to nuclear structure data

2.1 Introductory remarks

The measured α-particle energy E_α for a given decay, corrected for recoil effects, yields the total α-decay energy Q_α that represents a link between atomic establishes masses M according to

$$M(Z,N)c^2 = M(Z-2, N-2)c^2 + M(^4\mathrm{He})c^2 + Q_\alpha \,. \tag{1}$$

For far unstable nuclei, Q_α values often represent the only information available on (systematics of) atomic-mass differences and also allow, in some cases, to determine new atomic-mass values by linking them to known ones closer to stability (see Sections 4.1, 4.2 and 4.3). The new masses of ground and isomeric states far from stability can then be compared to predictions based on mass formulae.

If, in addition, the total half-life $T_{1/2}$ of the parent nucleus (Z,N) and the branching ratio b_α for α-emission are determined, one is able to deduce the partial α-decay probability $\lambda^{(\alpha)}$ from

$$\lambda^{(\alpha)} = b_\alpha \cdot (\ln 2)/T_{1/2} \,. \tag{2}$$

The measured $\lambda^{(\alpha)}$ value can be compared to model calculations, which allows one to deduce information about the configuration of the parent and daughter state.

In a simple Gamov picture [8] for a spherical nucleus the decay constant $\lambda_G^{(\alpha)}$ is determined according to

$$\lambda_G^{(\alpha)} = (v_\alpha/2R_i) \cdot P \,. \tag{3}$$

Here one assumes a "pre-formed" α particle inside the nucleus moving with a velocity v_α according to

$$v_\alpha = (2(U_0 + Q_\alpha + \Delta E_{\mathrm{scr}})/M_\alpha)^{1/2} \tag{4}$$

where U_0 is the depth of the nuclear potential $U(R)$. The screening correction

$$\Delta E_{\mathrm{scr}} = 65.3 \cdot Z^{7/5} - 80 \cdot Z^{2/5} \text{ in eV} \tag{5}$$

which accounts for the influence of the atomic electron cloud, amounts to about 16 keV in the tellurium and 41 keV in the fermium region.

In this simplified picture, the decay probability is the product of the "knocking frequency" $v_\alpha/2R_i$ and the barrier penetrability P, which, for s-wave emission, is obtained from the WKB approximation of the one-dimensional problem as

$$P=\exp\{-2\int_{R_\alpha}^{R_i}[(2M_\alpha/\hbar^2)(U(R)-Q_\alpha-\Delta E_{\mathrm{scr}})]^{1/2}dR\}. \tag{6}$$

R_i and R_α are the inner and outer turning points for a classical α-particle trajectory. This treatment has so far been based on assuming the existence of a pre-formed α-cluster inside the parent nucleus. In order to verify this assumption, one has to calculate the pre-formation probability (reduced width or spectroscopic factor) and to compare the results of such calculations to experimental data.

Absolute α-decay rates were first calculated by Mang [9, 10] on the basis of a microscopic theory using shell model wave functions. Later on, Fliessbach and Mang [11] improved these calculations and achieved good agreement with measured α-decay probabilities for ^{212}Po $\rightarrow$ ^{208}Pb$+\alpha$ and ^{211}Po $\rightarrow$ ^{207}Pb$+\alpha$. It is interesting to note that they calculated the spectroscopic factor for ^{212}Po $\rightarrow$ ^{208}Pb$+\alpha$ to be only 4.4×10^{-3}. Quantitative calculations based on such microscopic models have so far been controversial [12–14] and have mainly been restricted to the ^{212}Po case. The importance of pairing and superfluid correlations was recently discussed by Kadmenski [15] and Dodig-Crnkovic *et al.* [16].

Instead of performing microscopic calculations, information on the reduced α-decay width can be gained by determining

$$W_\alpha \sim \lambda^{(\alpha)}/P \tag{7}$$

for a given α-decay, and by considering such W_α values relative to the one for the ^{212}Po ground state α-decay. The reduced width, expressed either as the dimensionless (relative) quantity W_α defined in Eq. (7) or as the parameter δ^2 in MeV [17], describes the preformation probability of the α particle. Following the earlier works of Rasmussen [17] and Hornshøj *et al.* [18], a W_α systematics for s-wave ($0^+\rightarrow0^+$) α-emission is shown in Figure 1. $\lambda^{(\alpha)}$ values have been deduced from the *measured* quantities b_α and $T_{1/2}$, with the b_α determination being rather inaccurate in some cases. The barrier penetrability P has been *calculated* according to Rasmussen [17]. This method consists in a numerical integration of the WKB-integral (Eq. (6)), using the real part of the α-nuclear potential derived by Igo [26] from scattering data. This procedure avoids the introduction of an arbitrary "effective nuclear radius" in contrast to the condition imposed by using sharp cut-off nuclear potentials. The W_α systematics will be further discussed in Section 2.2 and will also be briefly refered to in Section 4.4.

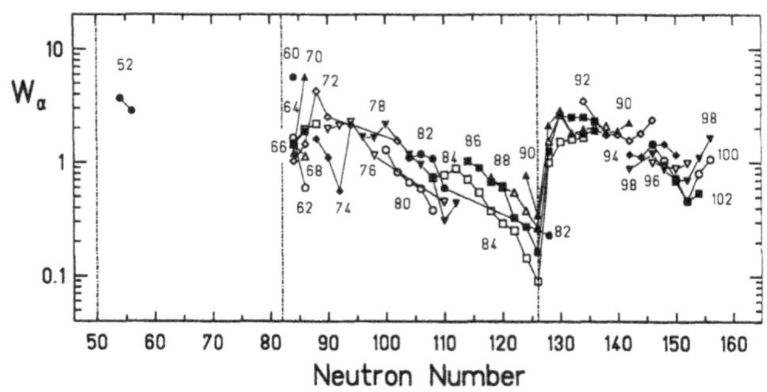

Fig. 1. Reduced width W_α for s-wave α-decay of even nuclei as a function of the neutron number of the emitter. Chains of emitter isotopes of the same element, with the proton number indicated, are joined by straight lines to guide the eye. W_α values are obtained as a ratio between measured transition probabilities and calculated barrier penetrabilities (see text) and are normalized to the ground-state α-decay of ^{212}Po. Experimental data are from Ref. [19] but include the more recent results on the α-emitters ^{146}Sm [20], ^{166}W [21], ^{168}W [22] and ^{218}Ra [23]. The experimental uncertainties have been omitted for reasons of clarity. The experimental results compiled in Ref. [19] for ^{154}Dy [24] and ^{180}W [25] are not shown; they lead to W_α values far off the trend emerging from emitters with neighbouring Z and N, and therefore ought to be verified experimentally. The recent results obtained for platinum-to-uranium isotopes have not been taken into account. These data are listed in Section 3.1, some of them will be discussed to more detail in Section 2.2 and 4.3.

A description of the conceptual and computational difficulties involved with various microscopic calculations of absolute α-decay rates is outside the scope of this report (see Refs. [5–7, 27] for a discussion of the present status of the microscopic theory of α and cluster emission). I shall, however, give an example for a microscopic description of α-decay to intruder 0_2^+ in lead, polonium, mercury and platinum isotopes (Section 4.3). Simpler (cluster) relations for the calculation of α-decay half-lives have recently been proposed [28–31]. These attempts consist in expressing $\lambda^{(\alpha)}$ as a function of Z and Q_α, following the earliest law for the systematic of α-decay half-lives as formulated by Geiger and Nuttall [32]. I shall return to one of these simple parametrisations in Section 2.2.

2.2 The N and Z dependence of the reduced alpha widths for $0^+\rightarrow0^+$ transitions

At first glance, the following two observations can be made from Figure 1:

– The relative α-widths between ^{106}Te and ^{256}Fm are found within a band that deviates at most by a factor of 10 from the reference value for ^{212}Po. Considering the 30 orders of magnitude between the half-lives $T_{1/2}$ of ^{148}Sm (7×10^{15}y) and ^{216}Ra (0.18 μs), this is an impressive proof of the usefulness of the approach based on reduced widths.

– The remaining gross nuclear-structure feature displays a smoothly increasing trend with increasing neutron number between shell closures; this systematical trend is well documented in the region $N=82$ to $N=126$. Along an isotopic chain, sharp discontinuities occur at the shell closure $N=126$ and $N=152$. The effect of the latter is less pronounced.

Additional structures of the W_α systematic are difficult to deduce, because the analysis underlying Figure 1 is based on a *global* spherical potential and since some of the W_α values have rather large experimental uncertainties, which are not shown in Figure 1. Examples of such structures being discussed here are the $Z=82$ shell closure and the influence of other nuclear structures effects. In analogy to the above-mentioned shell effects at $N=126$ and $N=152$, one would expect the reduced widths for lead isotopes ($Z=82$) to be systematically lower than those for Po ($Z=84$), Rn ($Z=86$) and Hg ($Z=80$), Pt ($Z=78$), respectively. However, conflicting experimental data exist for these nuclei. The "lead anomaly", i.e. the strikingly small W_α values of light lead isotopes as deduced by Hornshøj *et al.* [18], was not confirmed in later experiments by Toth *et al.* [33]. The latter authors had found a somewhat irregular behaviour of the lead α-widths as compared to those of neighbouring elements and concluded that the $Z=82$ shell gap disappears around $N\simeq104$. This conclusion as well as the lead anomaly as a whole have to be doubted in view of the recent work of Wauters *et al.* [34, 35]. They studied α decay of $^{188-192}$Pb, $^{196-220}$Po and $^{200-206}$Rn and showed that the discrepancy in the α-branching ratios reported by Hornshøj *et al.* and Toth *et al.* is probably due to problems in determining the α-detection efficiency [34].

It is interesting to compare the updated information on reduced widths of isotopes with $Z\geq78$ [35] to predictions taken from a recent theoretical analysis of ground state to ground state α-decay, performed by Buck *et al.* [30]. They calculated partial α-decay half-lives using a square-well potential and correcting for neutron shell-closures. The latter procedure consisted of introducing adequate principal quantum numbers and exludes a pre-formation probability in contrast to the above-mentioned model of Rasmussen [17]. By fitting the parameters of their model to the experimental data available for even-even nuclei, Buck *et al.* have been able to reproduce the partial α-decay half-lives of almost all of these nuclei within a factor of two. However, their conjecture that no structural information can be extracted from α-decay rates is maybe premature. On the one hand, it was already noticed by Hornshøj *et al.* [18] that a change in shape between the α emitter and the daughter nucleus has indeed no influence the reduced width. This observation has been confirmed by a recent study [36] of the ^{189l}Bi→ ^{187}Tl→^{183}Au α-chain which is characterized by strong shape changes from sphericity to deformation and from oblate to prolate deformation. On the other hand, Wauters *et al.* [35] pointed out that structural effects, in addition to the overall trend emerging from Figure 1, can in fact be deduced: For the case of neutron-deficient polonium isotopes with $N<112$, the decrease of the α-widths with decreasing neutron number is probably related to the mixing of intruder configurations with the ground state (see Section 4.3).

Due to the exponential dependence of the barrier penetrability on the α-particle energy (see Eq. (6)), α decay is generally dominated by the transition to the ground state of the daughter nucleus. However, "fine structure" α-lines can occur because of feeding into low-lying excited states of the daughter nucleus. Here the branching ratios are governed by the energy and the orbital angular momentum of the α-emission (see, e.g., the discussion of $0^+\rightarrow2^+$ d-wave systematics in Ref. [18]). The regular pattern emerging for s-wave α-decays (see Figure 1) can be used as a reference for α-decays with $l\neq0$, where an additional angular momentum barrier has to be taken into account in Eq. (6), as well as for α-decay between nonspherical nuclei. In Sections 4.3 I shall return to the hindrance of α-decay due to $l\neq0$ and due to the influence of the nuclear shape.

3. Experimental techniques

3.1 Source preparation

The standard method of performing α-spectroscopy is to prepare a thin source of the nuclide of interest and to detect the emitted α-particles by means of a suitable detector which is usually the silicon surface barrier type (see Section 3.2). For preparing sources, which are generally positioned close to the detector or on its (cleanable) surface, gas-jet or chemical-separation techniques [37−46] as well as on-line mass separators are applied. As an example, Figure 2 shows data for neutron-deficient tin-to-xenon isotopes (see Section 4.1) obtained at the GSI on-line mass separator. The advantages of preparing thin point-like mass-separated (or in favourable cases even monoisotopic) sources by this method have recently also been used for further α-decay studies at the GSI on-line separator [34, 36, 49, 50, 96] as well as at IRIS [51], ISOCELE [52−54], ISOLDE [35, 49, 55], LISOL [34−36, 49], OASIS [56, 57], UNISOR [58−63] and YASNAPP-2 [64]. These experiments concern the decays of ^{151,152}Ho, ^{155}Tm, ^{155}Yb, ^{155,157}Lu, $^{181-185}$Pt, $^{181-186}$Au, $^{180-184}$Hg, $^{180-185}$Tl, $^{186-192}$Pb, ^{191}Bi, $^{194-198}$Po and ^{200,202}Rn.

As an alternative to the external-source geometry, the α-emitting nuclides can be implanted into a detector by using higher-energy beams, e.g. from recoil separators. This technique (see Figure 3) has been applied very successfully in studies of the heaviest elements [67], and has recently also been used for various other α-decay investigations at Argonne [61, 69], Berkeley [70−72], Daresbury [73, 74], Darmstadt [75], Dubna [76, 77] and Jyväskylä [78−80] and RIKEN [81, 82]. These measurements concern the decays of $^{170-173}$Pt, ^{172}Au, ^{186,188,190}Hg, $^{193-195}$At, ^{196,197}Rn, ^{200,214,217}Fr, ^{203,204,219,220}Ra, ^{207,208,221}Ac, $^{210,211,221-224}$Th, ^{213,214,223,224}Pa, $^{218,223-225}$U, $^{225-227}$Np and $^{228-230}$Pu.

3.2 Detectors

While magnetic spectrometers and gas counters are used for α spectroscopy only in exceptional cases, sili-

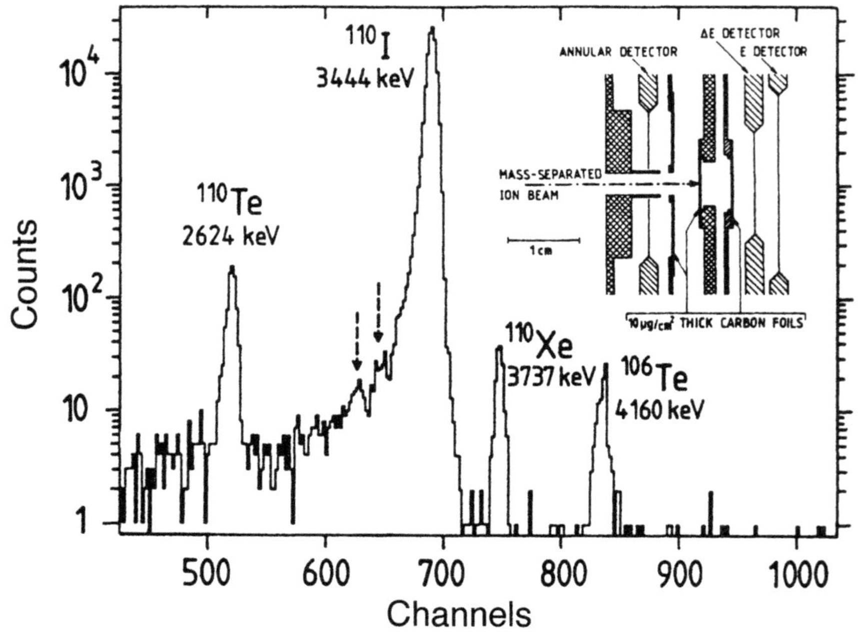

Fig. 2. Example of an α-decay measurement performed at the GSI on-line mass separator [47] using ^{58}Ni (^{58}Ni,xpyn) reactions for production of very neutron-deficient mass-110 isotopes [48]. As an inset, the collector-detector array is shown. The mass separated beam is implanted into a carbon collector after passing through a 3-mm diameter hole of an annular detector (50 μm thick, 300 mm^2) which is protected by a carbon foil against damage by sputtering due to the mass-110 beam of stable isotopes or compounds. An additional carbon foil is mounted in front of the telescope (ΔE: 29 μm, 150 mm^2; E; 676 μm, 450 mm^2) in order to shield the ΔE detector in case of damage of the collector foil. In regular time intervals the carbon collectors are removed from the irradiation position by a remotely controlled wheel which also carries α-sources for calibrating the ΔE and the annular detector. A corresponding wheel with standard sources for calibration of the E detector is not shown. The displayed α-spectrum was obtained at mass 110 from the ΔE detector in anticoincidence with the E detector of the telescope. The anticoincidence condition is necessary for discriminating against β-delayed proton (and α) emission. The counting time amounted to approximately 1 day with an average ^{58}Ni beam intensity of 25 particle-nA. The high-energy α-lines at 3737 ± 30 keV and 4160 ± 30 keV are ascribed to the decays of ^{110}Xe and ^{106}Te, respectively, based on the mass assignment from the magnetic separation and on α-energy systematics. This assignment is confirmed by results from a time correlation analysis, which includes also events detected in the annular detector. The positions of possible fine structure are indicated by arrows; the origin of the low-energy structure due to contamination from the neighbouring mass, i.e. the 3152 keV ^{111}I α-line, can be excluded, as the corresponding contamination of 3×10^{-3} would lie considerably above the level guaranteed by the mass separator.

con surface-barrier detectors are generally preferable due to the following attractive features: (i) large detection efficiency (large area); (ii) good energy resolution; (iii) surface (optionally) cleanable, light-tight, hardened; (iv) selection of low-background material; (v) shape adjustable to experimental requirement with respect to area (e.g. "micro strip" geometry) or (uniform) thickness; (vi) possibility to "bake" up to 200 °C or to cool down to −200 °C and even down to 4 mK (see Section 4.4); (vii) commercial availability.

One of the latest developments in silicon detectors is the "planar process" [83] for the fabrication of passivated ion-implanted silicon detectors with low noise, which allows an energy resolution below 10 keV FWHM for 5 to 6 MeV α-particles to be reached.

Another interesting development is detection via thermal pulses. With a composite diamond-germanium bolometer, for instance, Coron *et al.* [84] measured an energy resolution of 36 keV FWHM for 5 to 6 MeV α-particles at a temperature of 1.3 K and estimated that a resolution of 3 keV FWHM should be possible at 1 K.

3.3 Energy calibration

Precise α-particle energy and intensity values, to be used, e.g., for calibration purposes, can be taken from data compilations such as Ref. [19]. This catalog contains publications received before July 1985 and offers a more complete survey than the recent tables of Rytz [85] or Nichols [86], which include a selected number of α-emitters characterized by large α-branching ratios and very precise α-particle energies. The most accurately known (absolute) E_α values, generally obtained by means of magnetic spectrographs, cannot always be used as references for energy calibration, since measurements with identical source-detector geometry, both for such reference nuclides and the (new) α-emitter(s) of interest, are not always possible. In this context the work of Hofmann *et al.* [87, 88] is of interest. They used the accurately known E_α values from the helium-jet transported activity of ^{211}Bi and ^{215}Po for calibration and determined α-particle energies for a series of emitters between ^{149}Tb and ^{157}Hf with errors of 2 to 3 keV. These results, which were obtained by using implanted radioactivity and included a correc-

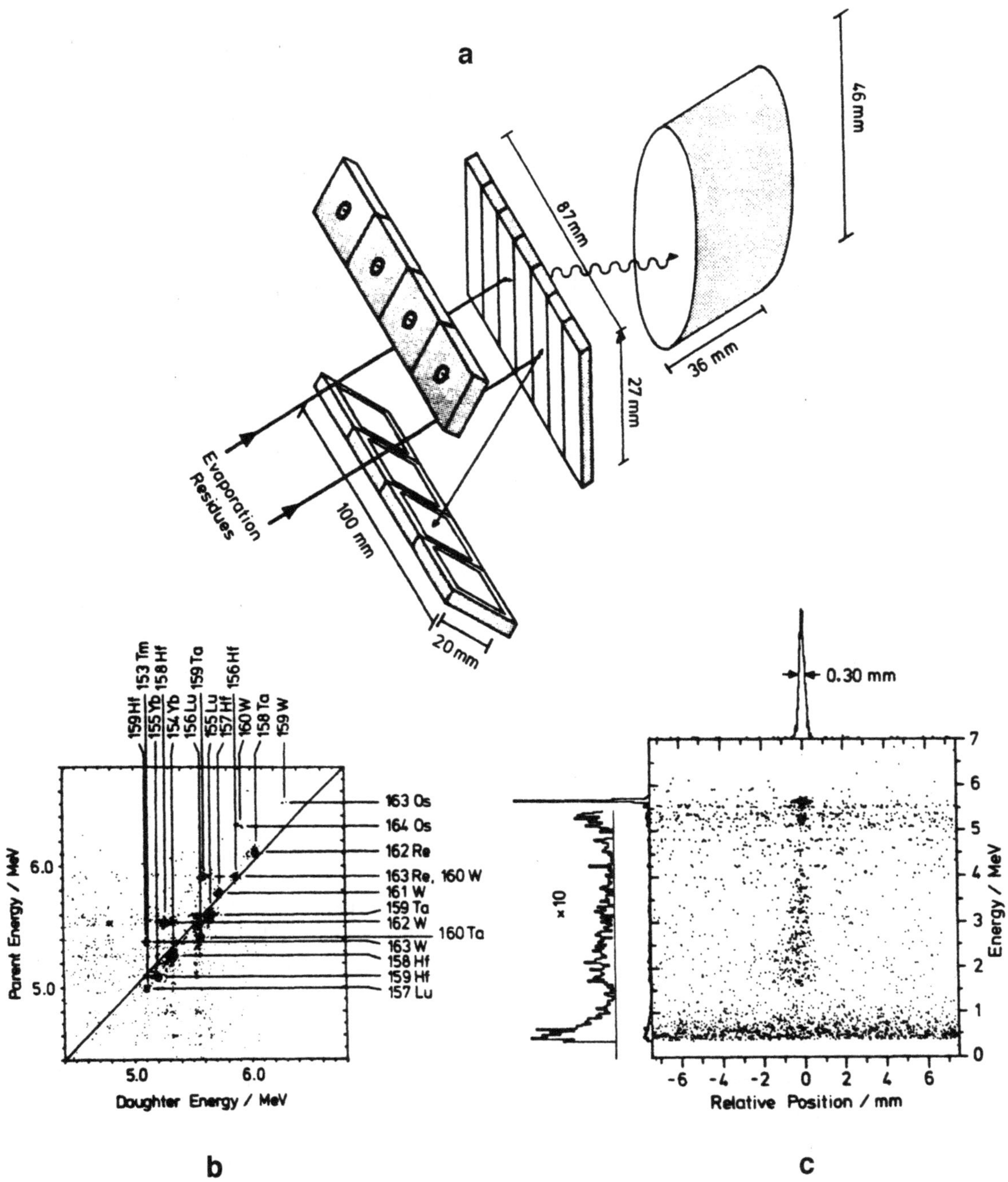

Fig. 3. Example of an α-decay measurement performed at the GSI velocity filter SHIP [65]. ^{110}Cd(^{58}Ni,xpyn) reactions were used for the production of neutron-deficient ytterbium-to-osmium isotopes [66, 67]. After velocity separation from the primary beam, the evaporation residues are implanted into an array of seven position sensitive detectors (a). Energy-energy (b) and energy-position (c) data from α-decay are shown as accumulated in these detectors during a 2.2 hours measurement with an average ^{58}Ni beam intensity of one particle-nA. The energy-energy coincidences were obtained with a time window of 0.2 s and a position window of 0.45 mm, showing clear evidence for 13 parent-daughter relations. For the energy-position plot, a window was set on the 5.60 MeV α-line of the (parent) ^{159}Ta decay. With a time window of 0.18 s, the correlated events are dominated by the 5.66 MeV line from α-decay of the direct α-daughter, 70 ms ^{155}Lu. The 5.21 MeV events are from α-decay of 1.65-s ^{155}Yb fed in β-decay of ^{155}Lu, the tail down to 1.5 MeV is due to ^{155}Lu α-particles emitted in backward direction. This technique is further improved by studying α-α time correlations beyond an integral time-condition applied to obtain the data (b) and (c) by additionally detecting γ/X-rays as well as backward α-emission and by mass analyzing the incoming beam via energy/time-of-flight measurements.

tion for detection of the recoil atom, agree very well with data from Ref. [89], while the α-particle energies from Refs. [65, 90] appear to systematically lie somewhat too high.

For the case of external-source geometry, high precision α-energy measurements using silicon semiconductor detectors require good knowledge of the energy lost due to finite source thickness as well as entrance window and dead layer regions. An additional correction has to be made for nonionizing processes in the sensitive volume itself which do not contribute to electron-hole pair production. While normally a linear relationship between the deposited energy (after applying the above corrections) and the pulse height is as-

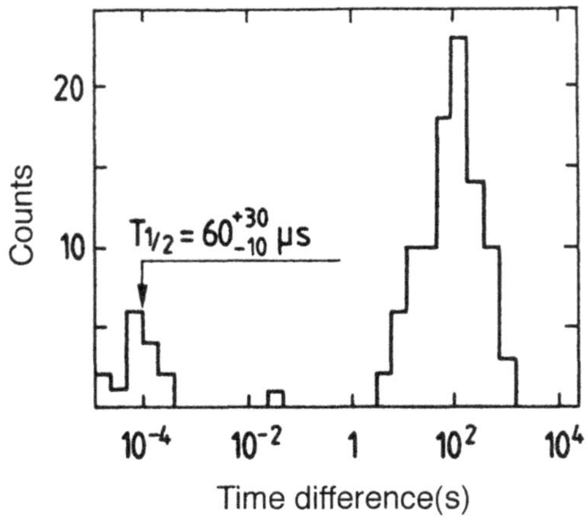

Fig. 4. Distribution of time differences between α-decay events of ^{110}Xe (3737 keV) and ^{106}Te (4160 keV) [48]. The peak at long time differences is due to random correlations, whereas the one at short time differences represents the $^{110}Xe \rightarrow {}^{106}Te$ correlation characterized by a 60-μs half-life of the daughter nucleus ^{106}Te. Note the logarithmic time scale.

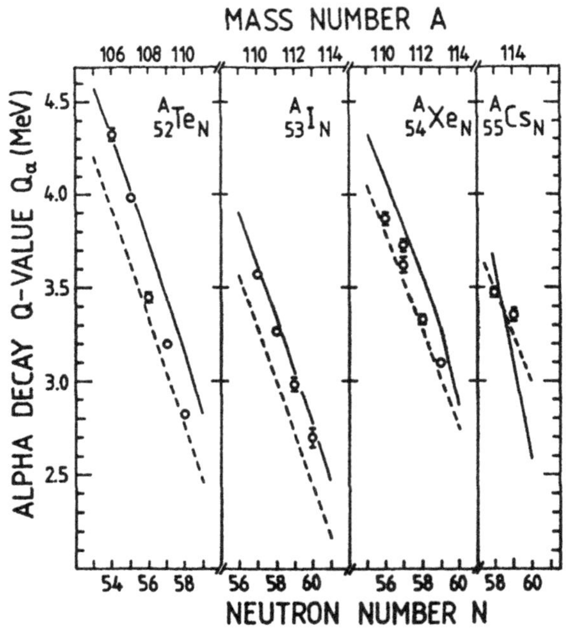

Fig. 5. Comparison of measured α-decay energies Q_α (circles) with predictions according to the droplet model of Myers [100] (solid line) and Hilf *et al.* [101] (dashed line). The experimental Q_α values are obtained [48] from measured α-particle energies by including a recoil correction. For ^{113}Cs also measurements of direct proton emission [102] were used. The recently observed α-decay of ^{108}I [97] and the remeasured Q_α values of some of these isotopes [97, 98] have not been taken into account.

sumed, Lennard *et al.* [91] recently proposed a model which implies a nonlinear dependence. On the basis of this model they were able to explain the unexpected effect of an increasing pulse height for increasing Z (for a given energy), which they observed in a systematic study using ^{1}H, ^{4}He, ^{7}Li and ^{16}O ions at low energies.

3.4 Coincidence measurements: Energy, time, and position correlations

Subsequent α-decays can be experimentally linked together by establishing an energy-coincidence relationship as demonstrated in Figure 3b. In order to apply this technique, the half-life of the daughter decay has to be short in order to keep the random-coincidence background low. Such *energy-energy correlations* can, of course, also be measured for any other type of radiation emitted "promptly" after an initial α-decay. Measurements of α-decay fine-structure, and of $\alpha-\gamma$ coincidences in particular, yield information about the nuclear structure of the states involved. Multipolarities of electromagnetic transitions between such states can be determined by measuring coincidences between α-particles and conversion electrons, and thus α X-ray coincidence data are of interest for making an unambiguous Z assignment.

Instead of setting integral time conditions, as it was done for obtaining the coincidence data shown in Figures 3b and 3c, one can determine time correlations in a differential way. By measuring time differences between subsequent α-decays of parent and daughter nuclei, the lifetime of the latter can be determined. The result of applying this *time-correlation* technique [92, 93] the $^{110}Xe \rightarrow {}^{106}Te$ α-decay is shown in Figure 4. Time differences were analyzed between pairs of ^{110}Xe and ^{106}Te α-decay events, detected either in the annular detector or in the ΔE detector in anticoincidence with the E detector (see Figure 4). The result of this correlation analysis (see Figure 5) clearly proves the parent-daughter relationship between the two α-decays, yielding a half-life of 60^{+30}_{-10} μs for ^{106}Te (the half-life of ^{110}Xe remains unknown).

The energy- and time-correlation technique becomes even more powerful if several subsequent α-decays, as well as competing decay channels such as β-decay or fission, are included. Moreover, the assignment of (rare) α-decay events can be facilitated by using position-sensitive detectors (see Figure 3). The measurement of *energy-time-position correlations* has successfully been performed for the heaviest elements [68], but also for the above-mentioned experiments at recoil separators and in particular for determining α-proton decay links [94, 95].

4. Example of recent alpha-decay studies

4.1 The island of alpha emitters above tin

Insight into the mass-energy surface and the underlying nuclear structure above ^{100}Sn has been gained by using ^{58}Ni beams from the UNILAC and the GSI on-line mass separator to conduct a systematic investigation of α-decay within the "island of α-emission" [48] above tin. Fourteen α-emitters were identified with $52 \leq Z \leq 55$, $54 \leq N \leq 60$, $2160\ \text{keV} \leq E_\alpha \leq 4160\ \text{keV}$ and $60\ \mu\text{s} \leq T_{1/2} \leq 18.6$ s, with 0.57 s ^{114}Cs marking the high-Z end of this island [95]. For the latter nuclide a b_α value of $1.8(6) \times 10^{-4}$ was measured. The corresponding partial production cross-section (for the α-decay of ^{114}Cs, produced through ^{58}Ni(^{58}Ni,pn) reactions) has been estimated to be 90 nb. For ^{114}Ba, an upper limit of $b_\alpha \leq 1.1 \times 10^{-3}$ and a ^{58}Ni(^{58}Ni,2n) cross-section of $0.20^{+0.13}_{-0.9}$ μb were measured [96]. Therefore, a value of 0.2 nb can be taken as an approximate measure of the lower cross-section limit of this method. (It is interesting to note that two orders-of-magnitude smaller production cross-sections have been reached in the SHIP experiments on the heaviest elements [68].) Recently, Page *et al.* [97] presented some experimental evidence for α-decay of ^{108}I with α-particle energies of 3730(25) and 3885(25) keV. As β-delayed proton emission can not be ruled out as an alternative assignment for these lines, this result is omitted in the following discussion. Page *et al.* [97] and Heine *et al.* [98] remeasured also some of the tellurium, iodine and xenon α-emitters and found minor inconsistencies with respect to the α energies determined at the GSI on-line mass separator.

The α-decay ^{105}Te→^{102}Sn (see Figure 2) represents the closest approach of charged-particle spectroscopy to the doubly-magic nucleus ^{100}Sn. For this decay, the reduced α-decay width of 3.7±1.6 (normalized to the ground state α-decay of ^{212}Po) is as high as expected for an even-even emitter close to a double shell closure and fits very well into the regular trend of the systematics of the widths (see Figure 1). So far there is hence no indication of a "superallowed" α-decay for this region, as was speculated [99] by assuming that the protons and neutrons involved in the decay originate from the same single-particle shells.

The experimental Q_α-values are in good overall agreement with those predicted by the droplet model (see Figure 5), a fact which has to be considered as a strong indication for the double-shell closure at $Z=N=50$: As a decomposition of the droplet-model Q_α values [100] shows, the main part of the large Q_α values of very neutron-deficient tellurium isotopes arises from the shell correction term rather than from the droplet term [103].

By linking measured Q_α, $Q_{EC}-S_p$ and Q_{EC} values to known masses of nuclei closer to the stability line, new mass values or mass differences, such as proton separation energies S_p very close to the proton drip line, have been derived [104, 105]. By using such Q-value links, even the masses of the proton-unbound nuclei ^{109}I and ^{113}Cs have been determined with an accuracy of 140 keV. On the one hand, only limited conclusions can be drawn from such a comparison between experimental results and predictions from mass formulae: The data are valid for a restricted area of the chart of nuclei only and tend to yield contradictory results dependent upon consideration of masses or mass differences. On the other hand, one can use extrapolations of empirical Q_α systematics such as the one shown in Figure 5 (or the results from model calculations after their verification against measurements, see also Figure 5) in order to predict properties of yet unknown, very proton-rich nuclides such as ^{105}Te or ^{109}Xe.

In concluding this chapter, I would like to emphasize that all discussions of Q_α values are based on the simplest, but experimentally unproven assumption that the observed α-decays represent ground state to ground state transitions; this assumption is tentatively supported by the nonobservation of fine-structure effects in this region expected for ^{111}Xe [48].

4.2 Alpha emitters with $N \geq 84$: The ^{146}Gd region and alpha-decay links towards lead isotopes

When moving away from the island of α-decay, above tin, towards higher-Z elements of the neutron-deficient rare earths with $N<82$, the influence of the $Z=50$ and $N=50$ closed shells becomes smaller, and correspondingly, the α-decay energies and decay widths are no longer enhanced.

For $N<82$ isotopes in the region of the elements Ba ($Z=56$) to W ($Z=74$), α-decay may become observable only for extremely neutron-deficient isotopes of the even-Z elements, while for the odd-Z elements the proton drip line will probably be reached before α-decay starts to contribute measurable decay branches. Under present experimental conditions, however, the production of these nuclei in sufficient amount seems to be difficult.

Under the strong influence of the $N=82$ shell closure, the (classical) onset of α-radioactivity occurs at $N=84$. Attempts to observe α-decay of $N=83$ isotones (in the region ytterbium to tungsten) across the closed shell have remained negative so far (see, e.g., Refs. [66, 106]).

An interesting feature of the very long-lived α-emitters of the elements Nd, Sm, and Gd is that β-decay is energetically not possible for these nuclei. Precise knowledge of the half-life and the α-decay energy then suffices to deduce the reduced α-widths. However, the experimental error on the small α-decay energies of these nuclei enters quite sensitively into the barrier penetration P (Eq. (6)) due to the strong exponential dependence.

Improved decay data have recently been reported for ^{146}Sm (Q_α = 2524(3) keV, $T_{1/2}$ = 1.031(45) × 10^8 y [20]) and ^{148}Gd ($T_{1/2}$ = 74.6(30) y [107]). These data are relevant to the possible occurrence of a $Z=64$ subshell closure at $Z=64$, separating the $d_{5/2}$, $g_{7/2}$ from the $h_{11/2}$, $s_{1/2}$, $d_{3/2}$ orbitals, which was first noticed as a small discontinuity in the Q_α systematic of $N=84$ isotones [108]. Theoretical calculations by Macfarlane *et al.* [109] suggested that the $Z=64$ gap should also lead to a small effect in the reduced-widths systematics. Although earlier experimental data [66, 110] on the even $N=84$ α-emitters gave a shallow minimum in the reduced α-widths at $Z=64$ (^{148}Gd), a definite conclusion could not be drawn due to the large uncertainties (in particular for the ^{146}Sm reduced width). The revised

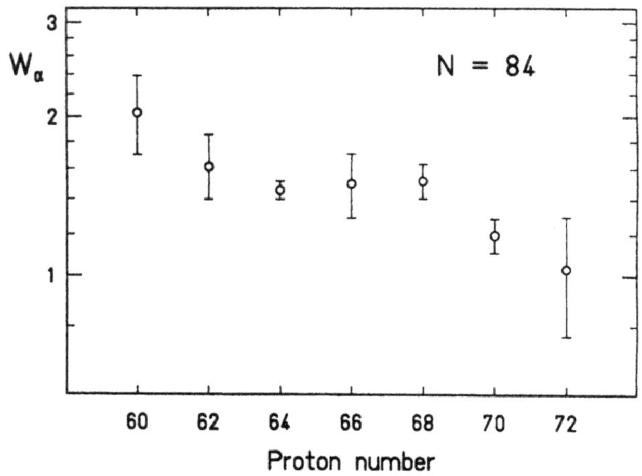

Fig. 6. Reduced width W_α for s-wave decay of even-even $N=84$ isotopes. The experimental data are taken from Ref. [19] except for ^{144}Nd (Q_α from Ref. [111]) and ^{146}Sm (Q_α and $T_{1/2}$ from Ref. [20]).

W_α values for the even $N=84$ emitters between $Z=60$ and 72, including the above-mentioned new results on ^{146}Sm and ^{148}Gd, have stayed almost constant within the experimental uncertainties (see Figure 6). In view of the rather small irregularity observed at $Z=64$ in the systematics of two-proton separation energies [111−113], the absence of a clear effect in the reduced α-widths, as deduced from the measured quantities combined with approximate barrier penetrability calculations, is certainly not surprising.

Many of the rare earth α-emitters above Gd were studied in the 1960's by heavy ion-induced fusion-evaporation reactions combined with helium-jet techniques. By using more advanced techniques (i.e. recoil spectrometers or on-line mass separators in combination with high-resolution silicon detectors), new spectroscopic information becomes available from more refined studies of these decays. This progress may be illustrated by (i) the detection of time and position correlations and α (or proton) radioactive nuclei in the lutetium-to-rhenium region [73, 74, 93], (ii) the determination of the isomeric state energy in ^{157}Lu by means of high-statistics (singles) α spectroscopy [63], and (iii) the detailed investigation of the ^{151,152}Ho α-decays [52, 53]. The latter α-γ-coincidence experiment allowed one to

– deduce the relative energies of the $\pi s_{1/2}$ and $\pi h_{11/2}$ isomers of both ^{151}Ho and ^{147}Tb with a precision of better than 1 keV, which is important for determining the proton single-particle level spacing above $Z=64$, in particular, for the one-proton nucleus ^{147}Tb;

– establish two previously unknown states above the ^{148}Tb 9^+ isomer, which represent the 7^+ and 8^+ members of the $\pi h_{11/2}\,\pi f_{7/2}$ multiplet; the experimental location of these lower-spin members, which is in good agreement with the prediction of shell model analyses, is important for determining the residual interactions needed for calculating prominent yrast states in the nuclei close to $N=82$ above $Z=64$.

Recent studies of α- and β-decays of neutron-deficient isotopes between gadolinium and lead (see, e.g., Refs. [114−119] and the more recent literature compiled in Section 3.1) have allowed one to experimentally establish long α-decay chains which include nuclei very far off stability. Such data yield rich systematics of α-decay energies, which can be extended to comprise masses and mass differences such as S_p, S_{2p} or the proton pairing energy Δ_p. A detailed discussion of this "mass-link" procedure and in particular the clarification of inconsistencies among the "base masses" around ^{146}Gd, are outside the scope of this paper. However, some highlights of these investigations can be summarized as follows:

– The precise determination of masses around ^{146}Gd yields an important input for shell-model analyses [112] of high-spin states of $N=82$ and 83 isotopes in this region. Refs. [120, 121] include a detailed comparison of the newly derived masses with results from β^+-endpoint and β^+/EC-ratio measurements.

– Based on the two long chains of even-even α-emitters terminating in ^{146}Gd and ^{148}Dy, the S_{2p} systematics can now be followed up to ^{182}Pb and ^{176}Hg. In order to obtain S_p values, the α-decay chains built on odd-Z $N=82$, 83 isotones have to be included and care has to be taken in correctly assigning α-decays involving isomeric states. The results obtained by using the mass-link procedure deviate systematically from model predictions. It has been argued [38] that this discrepancy might be due to problems with the mass measurement of ^{147}Tb. However, a recent evaluation of masses near ^{146}Gd [121] indicates that there might instead be a consistent failure of model predictions in this region.

– The mass-link procedure shows also that some of the lightest known Ir, Au and Tl isotopes are proton-unstable ($S_p<0$). Moreover, these data reveal [114, 116, 122] an unexpected increase of the proton pairing energy with increasing distance from the valley of stability (see Figure 7). By including a symmetry-dependent term in mass calculations, an improved prescription of the pairing energy has been obtained [123, 124].

4.3 Alpha decay in the lead region

Results obtained on the decays of Bi, Po, At, Fr and Ac isotopes are presented in the following as examples of recent α-spectroscopic studies in the lead region.

The neutron-deficient odd-A Tl isotopes are characterized by a $\pi s_{1/2}^{-1}$ ground state and a low-lying isomeric state which is known as the $\pi h_{9/2}$ intruder state [125]. Analogously, the odd-A Bi isotopes have a $\pi h_{9/2}$ ground state and a low lying $\pi s_{1/2}^{-1}$ (isomeric) intruder state. It was pointed out by Coenen *et al.* [126] that α-decay studies of light Bi isotopes across the $Z=82$ closed proton shell into Tl final states are the *only* way of determining the intruder-state excitation energies in these nuclei. The reason for this is

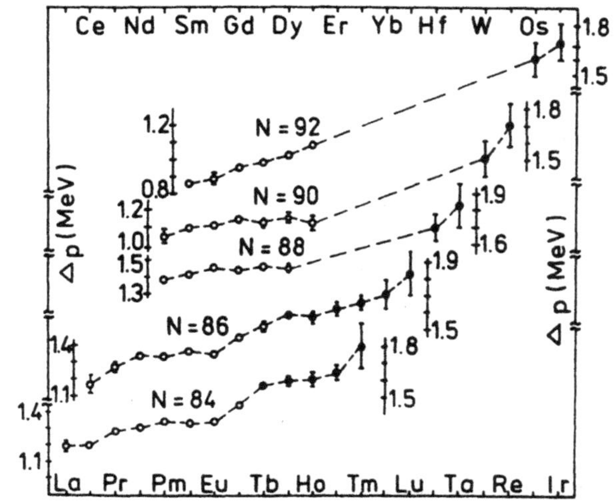

Fig. 7. Proton pairing energy Δ_p for neutron-rich gadolinium-to-rhenium isotopes as a function of the neutron number of the emitter [116]. The Δ_p values represent differences of experimental masses deduced from α- and β-decay Q-values.

that the $1/2^+ \to 9/2^-$ M4 transitions are the most strongly hindered M4 transitions known in any odd-mass nucleus [127]. As discussed by Vakthel *et al.* [128] the α-decays of the odd-A Bi isotopes typically proceed via three decay branches: (i) the unhindered $l=0$ transitions connecting the bismuth $9/2^-$ ground state with the $9/2^-$ isomer in the Tl daughter, (ii) the unhindered $l=0$ transition between the bismuth $1/2^+$ isomer with the Tl ground state, and (iii) the strongly hindered transition between the parent and daughter ground states. By combining these three energies, the intruder-state excitation energies can be determined in odd-A Bi and Tl isotopes.

The α-decays of neutron-deficient odd-A bismuth isotopes were investigated at the Leuven isotope separator LISOL. These measurements confirmed the basic picture given by Vakthel *et al.* [128]. The resulting systematics of $1/2^+$ and $9/2^-$ intruder states in Bi and Tl are shown in Figure 8. By including GSI data on ^{189m}Bi α-decay [118], the known intruder states can now be followed down to $N=104$ (midshell) for thallium and to $N=106$ for Bi isotopes.

An interesting example for studying intruder states via α-decay has recently been reported [34–36, 49, 50, 55]. By using the LISOL, the ISOLDE and the GSI on-line mass separators, they measured α singles as well as α-γ and α conversion-electron coincidences for the decays of the even-even nuclei ^{180,182,184}Hg, ^{186,188}Pb, ^{194,196,198}Po and ^{200,202}Rn. The results, which are compiled in Figure 9, allow one to draw conclusions on the nuclear structure of the states involved by using the reduced widths of groundstate-to-groundstate α-decay and the hindrance factors for α-decay to excited states (the hindrance factor is defined as the ratio of the δ^2 values for α emission into the 0^+ groundstate and into the 0_1^+ state, respectively; see Section 2). The first excited 0^+ state in ^{190,192,194}Pb, interpreted as a $\pi(2p-2h)$ configuration, lies at rather high excitation energies of 658, 769 and 931 keV, respectively. Therefore, only a small amount of mixing occurs between this level and the groundstate. The same is true for the intruder state and the groundstate of ^{198}Po. For ^{196}Po, however, there are indications from in-beam spectroscopy that the admixture of the $\pi(4p-2h)$ component to the normal $\pi(p)$ groundstate configuration may explain the reduced α-widths of ^{194,196}Po (see Section 2.2). Particle-hole excitations can also be used to explain the large hindrance factors observed for α-decay to excited states below the $Z=82$ shell gap (see Figure 9). The α-decay of ^{188}Pb to the ^{184}Hg groundstate demands the removal of a proton pair of the $Z=82$ shell, while α-decay to the first excited $0^+\pi(2p-4p)$ state of ^{184}Hg demands the removal of a proton pair of the $Z=82$ shell *and* excitation of a proton pair through the $Z=82$ shell gap. The latter process is of course strongly hindered, the measured hindrance factor being 21 (see Figure 9). The change of the hindrance factors when going from ^{182}Hg to ^{180}Hg can be understood in terms of the expected changes in shape and configuration between ^{178}Pt and ^{176}Pt, while the decrease of the hindrance factors when going from ^{198}Po to ^{194}Po can be explained by the above-mentioned mixing of the intruder $\pi(4p-2h)$ configuration and the $\pi(2p)$ groundstate in the Po isotopes [125]. This qualitative interpretation of the α hindrance factors measured for the decay of Hg, Pb and Po isotopes has been confirmed by microscopic calculations which include two-particle two-hole proton correlations and proton-neutron pairing vibrations [129]. The calculations reproduce also the trend towards large hindrance of the $0^+ \to 0_1^+$ decay of ^{200}Rn.

4.4 Alpha emission from oriented nuclei

Considerable progress has recently been made in studying α-emission from oriented nuclei. For the first time on-line experiments have been performed successfully on low-temperature nuclear orientation and on perturbed angular correlations, as well be discussed in the following.

By using the on-line mass separator at LISOL, Wouters *et al.* [130, 131] and Schuurmans *et al.* [132–134] implanted and oriented short-lived isotopes in a magnetized iron host foil at low temperature (down to 11 mK). While the few previous orientation experiments on long-lived α-emitters have hardly lead to reproducible results, these authors have succeeded in measuring, obviously in a reliable way, α-emission anisotropies for ^{211}Bi, $^{197-203}$Po, $^{198-211}$At, $^{205-209,219}$Rn, $^{205-209}$Fr and ^{223}Ra. ^{205}Fr with a half-life of 3.9 s represents the shortest-lived isotope which has ever been oriented on-line after direct implantation. One should note that in these experiments good energy resolution was reached for high purity germanium (and silicon surface-barrier) detectors, operated close to the iron foil at temperatures down to 4 mK. The anisotropy, observed for the $\pi h_{9/2} \to \pi h_{9/2}$ α-decay of odd-A astatine isotopes, is characterized by a change in sign as

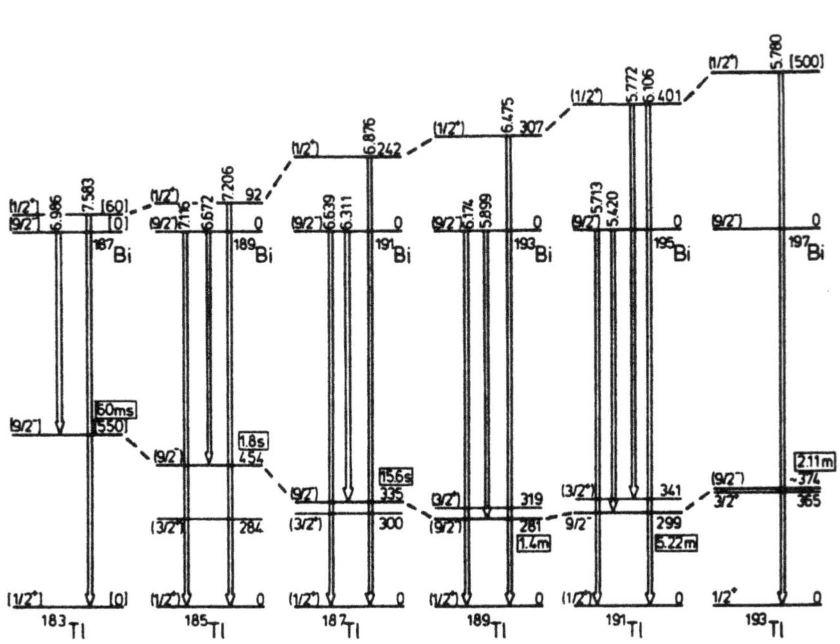

Fig. 8. Alpha decays of neutron-deficient odd-A bismuth isotopes [126]. The measured trend in excitation energies of the intruder states is shown as dashed lines for the $1/2^+$ level in bismuth ($N=106$ to 116) and for the $9/2^-$-level in thallium ($N=104$ to 114).

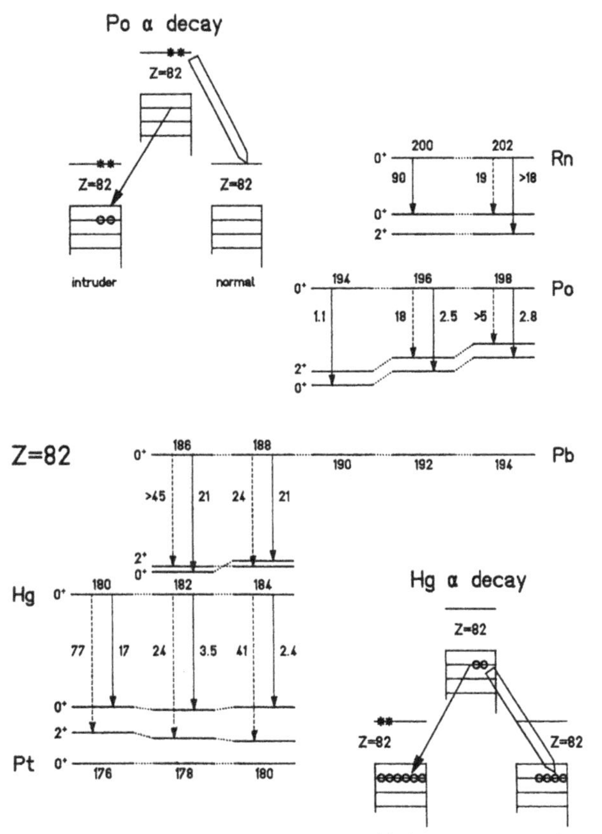

Fig. 9. Survey of alpha-decay fine structure observed in the lead region. Solid lines stand for $0^+ \rightarrow 0^+$, broken lines for $0^+ \rightarrow 2^+$ transitions. The $l=0$ transition to the excited 0^+ and the $l=2$ transition to the excited 2^+ state are indicated together with the respective hindrance factors. The latter quantities have been calculated by using the method of Rasmussen [17] and normalized to the groundstate alpha-widths. The data are from Ref. [49], but include more recent results for $^{180-184}$Hg, ^{186}Pb [50] and ^{200}Rn [55]. A schematic representation for the alpha decay above and below the $Z=82$ shell gap is shown in the insets.

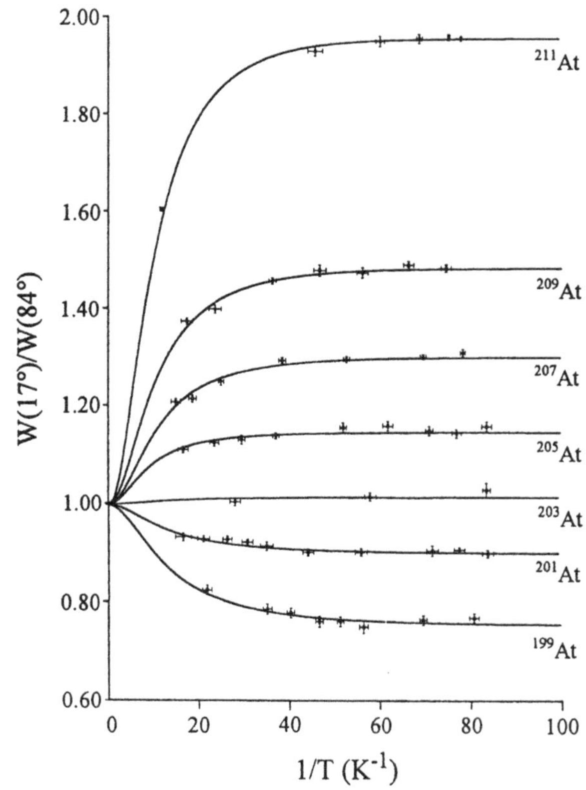

Fig. 10. Anisotropies as a function of $1/T$ for the α-particle emission in the decay of ^{199,201,203,205}At.

well as strong increase with increasing neutron number (Figure 10). This effect, which corresponds to a change from preferentially perpendicular α-emission with respect to the nuclear-spin direction, occurs when one goes further away from the $N=126$ closed shell. Qualitatively, one would tend to ascribe this experimental finding to the strong dependence of the barrier penetrability upon the nuclear shape. Quantitatively,

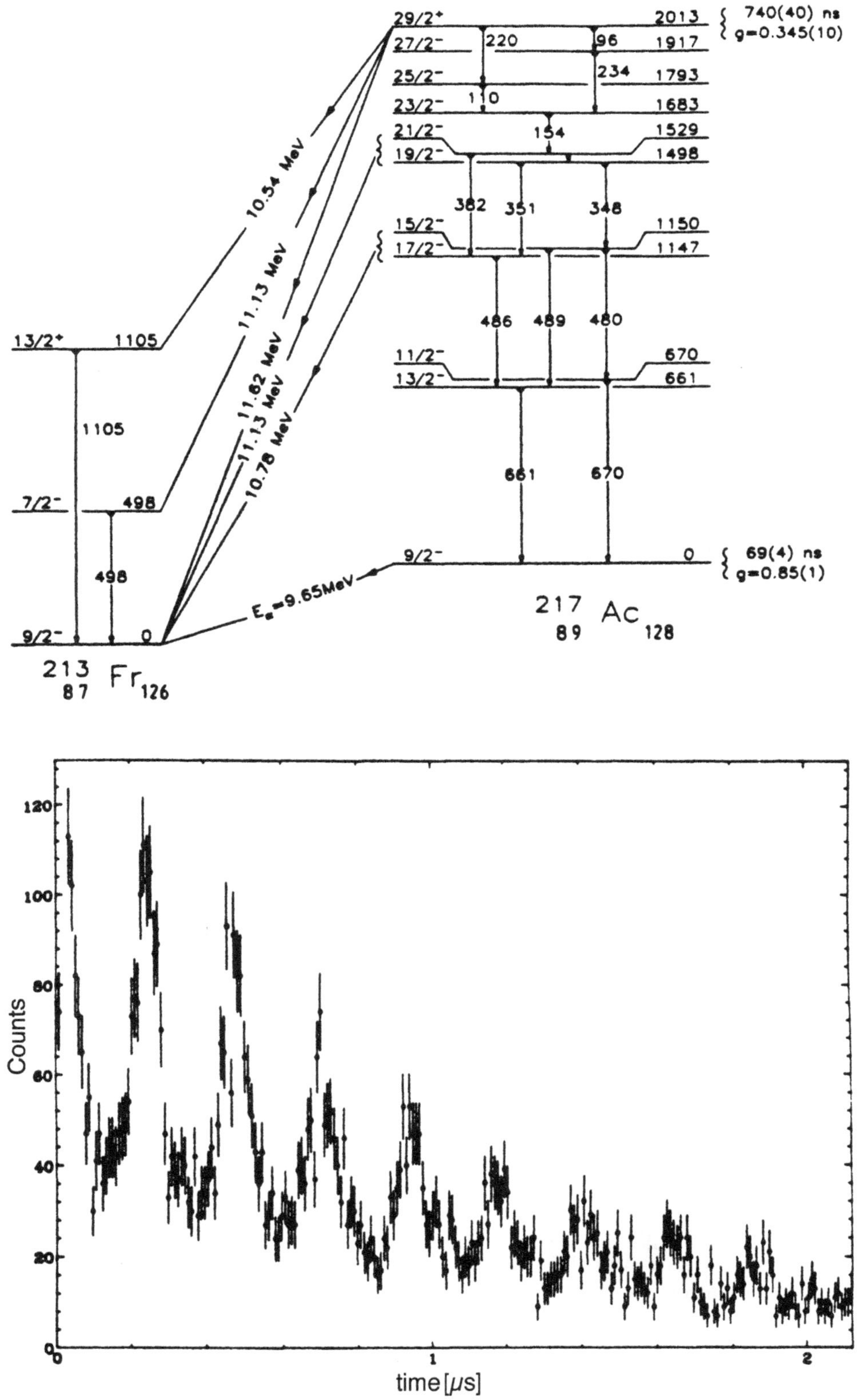

Fig. 11. The decay scheme of ^{217m}Ac, and the measured intensity distribution of the 10.54-MeV α-line as a function of time after the beam pulse in an external magnetic field. The measured characteristics of electron-capture (ε), alpha (α) and fission (f) decay are given in abbreviated form.

however, the interpretation in terms of changes in shape does not seem to be straight-forward, since it involves deformation parameter of parent and daughter nuclides and requires knowledge of sign and magnitude of mixing amplitudes [133, 134].

It is well known that the proton-neutron interaction can lead to drastic effects in nuclei in the immediate vicinity of closed shells. For example, the level scheme of $^{217}_{89}Ac_{128}$ (see Figure 11) does not resemble the one of $^{215}_{89}Ac_{126}$ with high-spin isomers typical of a spherical odd-even nuclide. The $9/2^-$, $13/2^-$, $11/2^-$, $17/2^-$, $15/2^-$ etc. levels of $^{217}_{89}Ac_{128}$ correspond rather to the 0^+, 2^+, 4^+ etc. sequences known in the neighbouring even nuclide $^{216}_{88}Ra_{128}$. The doublet structures of excited ^{217}Ac levels are thus interpreted as configurations of the type such as $\pi h_{9/2} \times \nu(g_{9/2})^2_{I_c}$, I_c being 2^+, 4^+ etc. This interpretation has recently been tested in a very elegant in-beam experiment performed by Decman *et al.* [135], who separated ^{217}Ac evaporation residues from the beam of heavy-ion projectiles (and from fission-products) by means of a novel electrostatic mass filter. During the beam pulses ^{217}Ac atoms were implanted into a catcher foil located in an external magnetic field. Between beam pulses, singles and coincidence measurements of α-particles, γ-rays, and conversion electrons were carried out as well as perturbed angular-correlation studies of α-particles (Figure 11) and γ-rays. Half-lives of 69 ns and 740 ns were determined for the ground state and the isomeric $29/2^+$ level, respectively, the former being the shortest-lived α-emitter known to date. Out of the detailed spectroscopic information obtained for the level scheme of ^{217}Ac (Figure 11), the measurement of g-factors from the perturbed α-particle angular distribution is particularly impressive. The above-mentioned interpretation of ^{217}Ac levels is confirmed by the measured g-factors. It is also in accordance with the observed hindrance of M1 transitions, while E2 transitions are enhanced compared to this model, indicating the onset of a small deformation for a nuclide with only two neutrons outside the closed shell $N = 126$.

5. Concluding remarks

It is obvious that α-decay studies yield valuable, systematic, and partly unexpected nuclear-structure information, which cannot be gained otherwise. This holds for ground state properties such as experimental masses and shell corrections near the doubly magic nuclide ^{100}Sn all the way to superheavy nuclei. But it also holds for isomers and other excited nuclear states, whose properties can be clarified by α-decay studies, especially if combined with other experimental techniques. The recent development of α spectroscopy is impressive indeed if one takes as examples the correlation techniques, the extension towards sub-μs half-lives, the measurements on oriented nuclei, and the successful search for new elements. I would like to emphasize in particular that the α-decay data form the most important basis for interpreting proton radioactivity [102] as well as the phenomenon of emission of clusters heavier than α particles from nuclear ground states. The latter decay mode is discussed in detail elsewhere in this volume [136]. Furthermore, I would like to recall that this report focused on α-decay studies of nuclear ground-states or long-lived isomers and hence excluded very little information on α emission from short-lived excited states (an up-to-date review on β-delayed particle emission has recently been written by Jonson and Nyman [137]).

On the other hand, α-decay is also of interest outside of nuclear physics. Solid-state information, such as the magnetic hyperfine field of 70(5)T for francium in iron, can be obtained from α-anisotropy measurements [131]. The natural thorium and uranium decay-chains are considered for dating purposes by measuring, e.g., α-recoil tracks or the growth of ^4He [138]. Alpha particles emitted from radioactive sources are used as ionizing particles in smoke detectors and, by means of Rutherford scattering, for chemical analyses of the lunar [139] and Martian [140] surfaces. Last but not least, α-spectroscopy is also relevant in detecting trace amounts of radiotoxic material, i.e. in particular plutonium in the biosphere. Most of the plutonium in the biosphere (5 to 10 t) is "man-made", i.e. produced by fall-out from atomic-bomb tests performed up to 1962. A second source of man-made origin is the peaceful use of atomic energy, and there are also contributions of natural origin [141]. In order to minimize risks in operating nuclear power plants and reprocessing factories, the plutonium distribution in their vicinity has to be controlled, including its migration into the environment. While other methods are being actively pursued, α-spectroscopy has reached a lower detection limit of 4.1×10^8 atoms for ^{239}Pu [142]. Even though the alternative technique of resonance ionization mass spectrometry [143, 144] is more sensitive by two orders of magnitude, α-counting is still competitive due to its technical simplicity.

In summary, I think it is fair to say that, almost exactly 100 years after its discovery, α-decay still represents a subject of considerable current interest mainly to basic research but also serves applied purposes.

References

1. Roeckl, E.: Alpha decay, in *Nuclear Decay Modes*, Poenaru, D. N. (Ed.), Institute of Physics, Bristol, in print.
2. Perlmann, I., Rasmussen, J. O.: Alpha-radioactivity, *Handbuch der Physik* 42, Springer-Verlag, Berlin 1957.
3. Mang, H. J.: Alpha decay, Annu. Rev. Nucl. Sci. **14**, 1 (1964).
4. Siegbahn, K. (Ed.), *Alpha-, Beta- and Gamma-Ray Spectroscopy*, Vol. 1, North-Holland, Amsterdam 1979.
5. Blendowske, R., *et al.*: Many body approach to alpha and cluster radioactivity, in *Nuclear Decay Modes*, Poenaru, D. N. (Ed.): Institute of Physics, Bristol, in print.
6. Poenaru, D. N., *et al.*: Traditional and fission approach to alpha-decay, in *Nuclear Decay Modes*, Poenaru, D. N. (Ed.), Institute of Physics, Bristol, in print.
7. Delion, D. S., *et al.*: Microscopic theory of cluster radioactivity, to be published (private communication by R. G. Lovas, 1994).
8. Gamov, G.: Zur Quantentheorie des Atomkerns, Z. Phys. **51**, 204 (1928).
9. Mang, H. J.: Zur Theorie des α-Zerfalls (Insbesondere der Kerne in der Umgebung von ^{208}Pb), Z. Phys. **148**, 582 (1957).
10. Mang, H. J.: Calculation of α-transition probabilities, Phys. Rev. **119**, 1063 (1960).
11. Fliessbach, T., Mang, H. J.: On absolute values of α-decay rates, Nucl. Phys. **A 263**, 75 (1976).
12. Tonozuka, I., Arima, A.: Surface α-clustering and α-decays of ^{212}Po, Nucl. Phys. **A 323**, 45 (1979).

13. Watt, A., *et al.*: Shell-model α-spectroscopic factors for ^{212}Po, J. Phys. **G 6**, 31 (1980).
14. Fliessbach, T., Okabe, S.: Surface alpha-clustering in the lead region, Z. Phys. **A 320**, 289 (1985).
15. Kadmenski, S. G.: On absolute values of α-widths for heavy spherical nuclei, Z. Phys. **A 312**, 113 (1983).
16. Dodig-Crnkovic, G., *et al.*: Absolute α-decay rates in Po isotopes, Nucl. Phys. **A 444**, 419 (1985).
17. Rasmussen, J. O.: Alpha-decay barrier penetrabilities with an exponential nuclear potential: even-even nuclei, Phys. Rev. **113**, 1593 (1959).
18. Hornshøj, P., *et al.*: Widths for s- and d-wave α decay of neutron-deficient isotopes with $Z \leq 82$, Nucl. Phys. **A 230**, 365 (1974).
19. Westmeier, W., Merklin, A.: Catalog of alpha particles from radioactive decay, Physics Data, Fachinformationszentrum Karlsruhe, Nr. 29-1, 1985.
20. Meissner, F., *et al.*: Half-life and α-ray energy of ^{146}Sm, Z. Phys. **A 327**, 171 (1987).
21. Hild, T., *et al.*: Spectroscopic investigations of ^{164,168}Ta, ^{166}W, and ^{164,165}Hf, Nucl. Phys. **A 492**, 237 (1988).
22. Meissner, F., *et al.*: Observation of a new α-ray emitter: ^{168}W, Z. Phys. **A 339**, 315 (1991).
23. Toth, K. S., *et al.*: ^{218}Ra α reduced width and its consequences for α clustering in the heavy elements, Phys. Rev. Lett. **56**, 2360 (1986).
24. Harmatz, B.: Nuclear data sheets for A = 154, Nucl. Data Sheets **26**, 281 (1979).
25. Greenwood, L. R.: Nuclear data sheets for A = 180, Nucl. Data Sheets **15**, 559 (1975).
26. Igo, C.: Optical model potential at the nuclear surface for the elastic scattering of alpha particles, Phys. Rev. Lett. **72**, 1 (1958).
27. Poenaru, D. N., Greiner, W.: Theories of cluster radioactivities, in *Nuclear Decay Modes,* Poenaru, D. N. (Ed.), Institute of Physics, Bristol, in print.
28. Poenaru, D. N., Ivascu, M.: Estimation of the alpha-decay half-lives, J. Physique **44**, 791 (1983) and references therein.
29. Hatsukawa, Y., *et al.*: Systematics of alpha decay half-lives, Phys. Rev. **C 42**, 674 (1990).
30. Buck, B., *et al.*: New look at α decay of heavy nuclei, Phys. Rev. Lett. **65**, 2975 (1990); Ground state to ground state alpha decays of heavy even-even nuclei, J. Phys. **G 17**, 1223 (1991).
31. Brown, B. A.: A simple relation for alpha decay half lives, Phys. Rev. **C 46**, 811 (1992).
32. Geiger, H., Nuttall, J. M.: The ranges of the α particles from various radioactive substances and a relation between range and period of transformation, Phil. Mag. **22**, 613 (1911); Geiger, H.: Reichweitenmessung an α-Strahlen, Z. Phys. **8**, 45 (1921).
33. Toth, K., *et al.*: Evidence from α decay that $Z = 82$ is not magic for light Pb isotopes, Phys. Rev. Lett. **53**, 1623 (1984).
34. Wauters, J., *et al.*: The alpha-branching ratios of the ^{188,190,192}Pb isotopes, Z. Phys. **A 342**, 277 (1992).
35. Wauters, J., *et al.*: Alpha-decay properties of neutron-deficient polonium and radon nuclei, Phys. Rev. **C 47**, 1447 (1993).
36. Wauters, J., *et al.*: The influence of shape chances on the alpha decay in the ^{191}Bi$^{m,g} \rightarrow {}^{187}$Tl $\rightarrow {}^{183}$Au chain, Z. Phys. **A 339**, 533 (1991).
37. Meissner, F., *et al.*: Decay of the new isotope ^{167}W, Z. Phys. **A 332**, 153 (1989).
38. Hagberg, E., *et al.*: Decays and masses of ^{162,163}Ta and some neighbouring nuclides, Phys. Rev. **C 45**, 1609 (1992).
39. Kratz, J. V., *et al.*: The new nuclide ^{263}Ha, Phys. Rev. **C 45**, 1064 (1992).
40. Liang, C. F., *et al.*: The alpha decay study of ^{223}Th ($T_{1/2}$ = 0.6 s), Z. Phys. **A 341**, 401 (1992).
41. Meissner, F., *et al.*: Revision of the decay data of $^{166-170}$Re, including new isomers ^{167m,169m}Re, Z. Phys. **A 343**, 283 (1992).
42. Schmitt-Ott, W.-D., *et al.*: Isomerism in neutron-deficient iridium isotopes: Alpha and beta-decay studies of $^{171-175}$Ir, Nucl. Phys. **A 545**, 646 (1992).
43. Liang, C. F., *et al.*: Alpha decay of ^{231}U to levels in ^{227}Th, Phys. Rev. **C 49**, 2230 (1994).
44. Hild, T., *et al.*: α and *beta* decays of $^{169-173}$Os and the nuclear structure of the daughter isotopes, Phys. Rev. **C 51**, 1736 (1995).
45. Schädel, M.: Chemistry of the Transactinide Elements, Radiochim. Acta **70/71**, 207–223 (1995).
46. Trautmann, N.: Fast Radiochemical Separations, Radiochim. Acta **70/71**, 237–245 (1995).
47. Bruske, C., *et al.*: Status report on the GSI on-line mass separator facility. Nucl. Instrum. Methods **186**, 61 (1981).
48. Schardt, D., *et al.*: Alpha decay of neutron-deficient isotopes with $52 \leq Z \leq 55$, including the new isotopes ^{106}Te ($T_{1/2}$ = 60 µs) and ^{110}Xe, Nucl. Phys. **A 368**, 153 (1981).
49. Wauters, J., *et al.*: Fine structure in the alpha decay of even-even nuclei as an experimental proof for the stability of the $Z = 82$ magic shell at the very neutron-deficient side, Phys. Rev. Lett. **72**, 1329 (1994).
50. Wauters, J., *et al.*: Alpha decay of ^{186}Pb and ^{184}Hg: The influence of mixing of 0^+ states on α-decay transition probabilities, Phys. Rev. **C 50**, 2768 (1994).
51. Guglielmetti, A., *et al.*: Identification of the new isotope ^{114}Ba and search for its alpha and cluster radioactivity, Phys. Rev. **C 52**, 740 (1995).
52. Bolshakov, V. A., *et al.*: Identification of the new Tl-nuclides at the proton-drip line, in Proc. 6th Int. Conf. on Nuclei far from Stability and 9th Int. Conf. on Atomic Masses and Fundamental Constants, Bernkastel-Kues, Neugart, R., Wöhr, A. (Eds.), Institute of Physics Conf. Ser. **132**, 743 (1993).
53. Liang, C. F., *et al.*: Relative energies of the $\pi h_{11/2}$ and $\pi s_{1/2}$ β-decay isomers of ^{147}Tb$_{82}$ and ^{151}Ho$_{84}$, Phys. Lett. **B 191**, 245 (1987).
54. Stycen, J., *et al.*: Proton-neutron interaction and the relative isomer masses in ^{148}Tb, in Proc. 7th Int. Conf. Nuclei far from Stability, Rosseau-Lake, Towner, I. S. (Ed.), AIP Conf. Proc. **164**, 489 (1988).
55. Sauvage, J., *et al.*: Decay of ^{181}Hg ($T_{1/2}$ = 3.6 s) and ^{181}Au ($T_{1/2}$ = 11,4 s), and low-spin states of ^{181}Pt and ^{177,181}Ir, Nucl. Phys. **A 540**, 83 (1992).
56. Van Duppen, P.: private communication, 1994.
57. Toth, K. S., *et al.*: Investigation of the ^{155}Tm $s_{1/2}$ isomer and of the ^{155}Yb β-decay branch, Phys. Rev. **C 44**, 1868 (1991).
58. Toth, K. S., *et al.*: α-decaying low-spin levels in ^{155}Lu and ^{157}Lu, Z. Phys. **A 340**, 343 (1991).
59. Akovali, Y. A., *et al.*: New mass differences and $^{181-186}$Au and $^{181-185}$Pt, Phys. Rev. **C 42**, 1130 (1990).
60. Bingham, C. R., *et al.*: New mass differences and $^{181-186}$Au and $^{181-185}$Pt, Phys. Rev. **C 44**, 1208 (1991).
61. Bingham, C. R., *et al.*: New mass differences and α-decay rates for Au and Pt isotopes, in Proc. 6th Int. Conf. on Nuclei far from Stability and 9th Int. Conf. on Atomic Masses and Fundamental Constants, Bernkastel-Kues, Neugart, R., Wöhr, A. (Eds.), Institute of Physics Conf. Ser. **132**, 735 (1993).
62. Toth, K. S., *et al.*: Investigation of proton-rich platinum isotopes with the fragment mass analyzer at ATLAS and the isotope separator at UNISOR, in Proc. 6th Int. Conf. on Nuclei far from Stability and 9th Int. Conf. on Atomic Masses and Fundamental Constants, Bernkastel-Kues, Neugart, R., Wöhr, A. (Eds.), Institute of Physics Conf. Ser. **132**, 589 (1993).
63. Bingham, C. R., *et al.*: New mass differences and α-decay rates for $^{181-186}$Au and $^{181-185}$Pt, Phys. Rev. **C 51**, 125 (1995).
64. Lewandowski, M., *et al.*: New ^{157}Lu isomer, Z. Phys. **A 340**, 107 (1991).

65. Münzenberg, G., *et al.*: The velocity filter SHIP, a separator of unslowed heavy ion fusion products, Nucl. Instrum. Methods **161**, 65 (1979).
66. Hofmann, S., *et al.*: Alpha decay studies of very neutron-deficient isotopes of Hf, Ta, W, and Re, Z. Phys. **A 291**, 53 (1979).
67. Hofmann, S., *et al.*: Detector system for investigation of proton radioactivity and new elements at SHIP, Nucl. Instrum. Methods **223**, 312 (1984).
68. Münzenberg, G.: Discovery, Synthesis and Nuclear Properties of the Heaviest Elements, Radiochim. Acta **70/71**, 193–206 (1995).
69. Leddy, M. J., *et al.*: α-decay of a new isotope, ^{204}Ra, Phys. Rev. **C 51**, R1047 (1995).
70. Leino, M.: Thesis, University of Helsinki, Report Series in Physics HU-P-D37, 1983, unpublished.
71. Yashita, S.: Thesis, Report LBL-15562, 1984, unpublished.
72. Ghiorso, A., *et al.*: Nucl. Instr. in Phys. Res. **A 269**, 192 (1988).
73. Wood, P. J., *et al.*: The discovery of new proton emitters, in Proc. 6th Int. Conf. on Nuclei far from Stability and 9th Int. Conf. on Atomic Masses and Fundamental Constants, Bernkastel-Kues, Neugart, R., Wöhr, A. (Eds.), Inst. Phys. Conf. Ser. **132**, 323 (1993).
74. Sellin, P. J., *et al.*: Alpha decay of the new isotope ^{172}Au, Z. Phys. **A 346**, 323 (1993).
75. Ninov, V., *et al.*: Identification of the new neutron-deficient isotopes ^{213,214}Pa, Z. Phys. **A 351**, 125 (1995).
76. Andreyev, A. N., *et al.*: Alpha decay of new U, Np and Pu isotopes and alpha spectroscopy for nuclei in Fr-Pa region, in Proc. 6th Int. Conf. on Nuclei far from Stability and 9th Int. Conf. on Atomic Masses and Fundamental Constants, Bernkastel-Kues, Neugart, R., Wöhr, A. (Eds.), Institut of Physics Conf. Ser. **132**, 759 (1993).
77. Andreyev, A. N., *et al.*: New nuclides ^{228,229}Pu, Z. Phys. **A 347**, 225 (1994).
78. Leino, M., *et al.*: Alpha decay of the new isotopes ^{207,208}Ac, Z. Phys. **A 348**, 151 (1994).
79. Uusitalo, J., *et al.*: Alpha decay of the new isotopes ^{210}Th and ^{211}Th, Phys. Rev. **C 52**, 113 (1995).
80. Leino, M., *et al.*: Alpha decay properties of very neutron-deficient isotopes with $85 \leq Z \leq 90$, in Proc. Int. Conf. on Exotic Nuclei and Atomic Masses ENAM 95, Arles 1995, Edition Frontières, Gif sur Yvette, in print.
81. Morita, K., *et al.*: New α-decaying neutron deficient isotopes ^{197}Rn and ^{200}Fr, Z. Phys. **A 352**, 7 (1995).
82. Nomura, T.: Heavy-element research at RIKEN, in Proc. Int. Conf. on Exotic Nuclei and Atomic Masses ENAM 95, Arles 1995, Edition Frontières, Gif sur Yvette, in print.
83. Kemmer, J.: Fabrication of low noise silicon radiation detectors by the planar process, Nucl. Instrum. Methods **169**, 499 (1980).
84. Coron, N., *et al.*: A composite bolometer as a charged-particle spectrometer, Nature (London) **314**, 75 (1985).
85. Rytz, A.: Recommended energy and intensity values of alpha particles from radioactive decay, At. Data Nucl. Data Tables **47**, 205 (1991).
86. Nichols, A. L.: Tables of alpha-particle emitters: energies and emission probabilities, in *Handbook of Nuclear Properties*, Poenaru, D. N., Greiner, W. (Eds.), Oxford University Press Oxford, in print.
87. Hofmann, S., *et al.*: Precision alpha energy calibration measurements in the decay of implanted isotopes, in GSI Sci. Rep. 1980, GSI-81-1, 204 (1981).
88. Hofmann, S., *et al.*: Stopping of alpha-recoil atoms in silicon, in GSI Sci. Rep. 1981, GSI-82-1, 241 (1982) and personal communication, 1981.
89. Bowman, J. D., *et al.*: Alpha spectroscopy of nuclides produced in the interaction of 5 GeV protons with heavy element targets, Phys. Rev. **C 25**, 941 (1982).
90. Della Negra, S., *et al.*: Alpha decay characteristics of neutron deficient isotopes of Pt isotopes produced in ^{63}Cu induced reactions on ^{112}Sn and ^{113}In targets, Z. Phys. **A 300**, 251 (1981).
91. Lennard, W. N., *et al.*: Nonlinear response of Si detectors for low-Z ions, Nucl. Instrum. Methods A, **248**, 454 (1986).
92. Schmidt, K.-H., *et al.*: Alpha decay properties of new protactinium isotopes, Nucl. Phys. **A 318**, 253 (1979).
93. Schmidt, K.-H., *et al.*: Some remarks on the error analysis in the case of poor statistics, Z. Phys. **A 316**, 19 (1984).
94. Livingston, K., *et al.*: Isomeric proton emission from the drip-line nucleus ^{156}Ta, Phys. Rev. **C 48**, R2151 (1993).
95. Davids, C. N.: New heavy proton radioactivities, in Proc. Int. Conf. on Exotic Nuclei and Atomic Masses ENAM 95, Arles 1995, Edition Frontières, Gif sur Yvette, in print.
96. Roeckl, E., *et al.*: The decay of 0.57 s ^{114}Cs, Z. Phys. **A 294**, 221 (1980).
97. Page, R. D., *et al.*: Evidence for the alpha decay of ^{108}I, Z. Phys. **A 338**, 295 (1991).
98. Heine, F., *et al.*: Proton and alpha radioactivity in the vicinity of ^{100}Sn, in Proc. 6th Int. Conf. on Nuclei far from Stability and 9th Int. Conf. on Atomic Masses and Fundamental Constants, Bernkastel-Kues, Neugart, R., Wöhr, A. (Eds.), Institute of Physics, Conf. Ser. **132**, 331 (1993).
99. Macfarlane, R. D., Siivola, A.: New region of alpha radioactivity, Phys. Rev. Lett. **14**, 114 (1965).
100. Myers, W. D.: *Droplet Model of Atomic Nuclei,* Plenum Press, New York 1977.
101. Hilf, E. R., *et al.*: Gross theory of nuclear masses and radii, in Proc. 3rd Int. Conf. on Nuclei Far From Stability, Cargèse, Report CERN 76-13, 142, 1976.
102. Hofmann, S.: Proton Radioactivity, Radiochim. Acta **70/71**, 93–105 (1995).
103. Schardt, D., *et al.*: Alpha decay studies of tellurium, iodine, xenon and cesium isotopes, Nucl. Phys. **A 326**, 65 (1979).
104. Plochocki, A., *et al.*: Masses of very neutron-deficient isotopes in the cadmium-to-cesium region, Nucl. Phys. **A 388**, 93 (1982).
105. Keller, H., *et al.*: β^+-endpoint measurements near ^{100}Sn and ^{146}Gd, Z. Phys. **A 340**, 363 (1991).
106. Hagberg, E., *et al.*: Alpha decay of neutron-deficient ytterbium isotopes and their daughters, Nucl. Phys. **A 293**, 1 (1977).
107. Prestwood, R. J., *et al.*: Half-life of ^{148}Gd, Phys. Rev. **C 24**, 1346 (1981).
108. Rasmussen, J. O., *et al.*: α-radioactivity in 82-neutron region, Phys. Rev. **89**, 33 (1953).
109. Macfarlane, R. D.: Theoretical calculations of properties of rare-earth alpha emitters, Phys. Rev. **134**, 1196 (1964).
110. Schmidt-Ott, W.-D., Toth, K. S.: Decay rates for doubly even $N=84$ α emitters and the subshell closure at $Z=64$, Phys. Rev. **C 13**, 2574 (1976).
111. Audi, G., Wapstra, A. H.: Nucl. Phys. **A 565**, 66 (1993).
112. Blomqvist, J., *et al.*: Atomic masses above ^{146}Gd derived from a shell model analysis, Z. Phys. **A 312**, 27 (1983).
113. Habenicht, W., *et al.*: Q_{EC} values of very proton-rich nuclei with $N \sim 82$ and favoured β transitions $\pi h_{22/2} < \rightarrow \nu h_{9/2}$, in Proc. 7th Int. Conf. Atomic Masses and Fundamental Constants, Darmstadt-Seeheim, Klepper, O. (Ed.), Technische Hochschule Darmstadt, Schriftenreihe Wissenschaft und Technik **26**, 244 (1984).
114. Hofmann, S., *et al.*: New neutron-deficient isotopes in the range of elements Tm to Pt, Z. Phys. **A 299**, 21 (1981).
115. Schrewe, U. J., *et al.*: Decay studies of the new isotopes ^{162,163}Hf, Phys. Rev. **C 25**, 3091 (1982).
116. Alkhazov, G. D., *et al.*: Identification of the proton drip line, Z. Phys. **A 311**, 245 (1983).
117. Schneider, J. R. H., *et al.*: Alpha decay of new neutron deficient gold, mercury and thallium isotopes, Z. Phys. **A 312**, 21 (1983).
118. Schneider, J. R. H.: Kernspektroskopische Untersuchungen extrem neutronenarmer Isotope in the Nähe der abgeschlossenen Protonenschale $Z=82$, Thesis, Technische Hochschule Darmstadt, GSI Report GSI-84-3, 1984, unpublished.

119. Runte, E., *et al.*: Completion of α-decay chains: The α-decays of ^{162}Ta and ^{163}Ta, Z. Phys. **A 324**, 119 (1986).
120. Rubio, B., *et al.*: The ground state mass of ^{147}Gd from single-neutron-transfer reactions and its impact on derived mass values, Z. Phys. **A 324**, 27 (1986).
121. Keller, H., *et al.*: Atomic masses near ^{146}Gd, in Proc. 6th Int. Conf. on Nuclei far from Stability and 9th Int. Conf. on Atomic Masses and Fundamental Constants, Bernkastel-Kues, Neugart, R., Wöhr, A. (Eds.), Institute of Physics Conf. Ser. **132**, 37 (1993); Beta-decay energy and atomic mass of ^{148}Tb, Z. Phys. **A 352**, 1 (1995).
122. Hofmann, S., *et al.*: Decay studies of new neutron deficient isotopes in the range of elements between gadolinium and lead, in Proc. 4th Int. Conf. on Nuclei far from Stability, Helsingør, CERN 81-09, 190 (1981).
123. Jensen, A. S., *et al.*: New mass relations and two- and four-nucleon correlations, Nucl. Phys. **A 431**, 393 (1984).
124. Vogel, P., *et al.*: Is there neutron-proton pairing in medium heavy nuclei?, Phys. Lett. **B 139**, 227 (1984).
125. Heyde, K., *et al.*: Coexistence in odd-mass nuclei, Phys. Rep. **102**, 293 (1983).
126. Coenen, E., *et al.*: α decay of neutron-deficient odd Bi nuclei: Shell model intruder states in Tl and Bi isotopes, Phys. Rev. Lett. **54**, 1783 (1985).
127. Braga, R. A., *et al.*: Very slow M4 transitions and shell-model intruder states in ^{199,201}Bi, Nucl. Phys. **A 349**, 61 (1980).
128. Vakthel, V. M., *et al.*: Alpha decay of neutron-deficient isotopes of Bi, Yad. Fiz. **28**, 1241 (1978); as cited in Sov. J. Nucl. Phys. **28**, 639 (1978).
129. Delion, D. S., *et al.*: Microscopic description of alpha decay to intruder 0_2^+ states in Pb, Po, Hg and Pt isotopes, Phys. Rev. Lett. **74**, 3939 (1995).
130. Wouters, J., *et al.*: Anisotropic alpha emission from on-line separated isotopes, Phys. Rev. Lett. **56**, 1901 (1986).
131. Wouters, J., *et al.*: Nuclear orientation techniques applied to on-line separated isotopes, Nucl. Instrum. Methods in Phys. Rev. **B 26**, 463 (1987).
132. Schuurmans, P., *et al.*: Anisotropic alpha-emission of strongly deformed nuclei, in Proc. 6th Int. Conf. on Nuclei far from Stability and 9th Int. Conf. on Atomic Masses and Fundamental Constants, Bernkastel-Kues, Neugart, R., Wöhr, A. (Eds.), Institute of Physics Conf. Ser. **132**, 601 (1993).
133. Schuurmans, P., *et al.*: Alpha anisotropy measurements on on-line separated At and Rn isotopes, in Proc. of the Intern. Conf. on Nuclear Shapes and Nuclear Structure, Antibes, Vergnes, M., Goutte, D., Heenen, P. H., Sauvage, J. (Eds.), Edition Frontières, Gif-sur-Yvette, 323, 1994.
134. Schuurmans, P., *et al.*: to be published.
135. Decman, D. J., *et al.*: Decay of the ground state and the $29/2^+$ isomer in ^{217}Ac and g-factor measurements from pertubed α-particle angular distributions, Nucl. Phys. **A 436**, 311 (1985).
136. Ardisson, G., Hussonnois, M.: Radiochemical Investigations of Cluster Radioactivities, Radiochim. Acta **70/71**, 123–133 (1995).
137. Jonson, B., Nyman, G.: Beta-delayed particle emission, in *Nuclear Decay Modes,* Poenaru, D. N. (Ed.), Institute of Physics, Bristol, in print.
138. Bishop, W. W., Miller, J. A. (Eds.): *Calibration of Hominoid Evolution,* Scottish Academic Press, Edinburgh 1972.
139. Turkevich, A. L.: Chemical composition of the lunar surface: alpha scattering, Acc. Chem. Res. **6**, 81 (1973).
140. Economou, T., *et al.*: Alpha particle experiments for chemical analysis of the Martian surface and atmosphere, J. Geophys. Res. **78**, 781 (1973).
141. Keller, C., de Alleluia, I. B.: Plutonium und andere schwere Radioelemente in der Natur, Chem. Ztg. **103**, 139 (1979).
142. Peuser, P., *et al.*: Detection methods for trace amounts of plutonium, in Methods of Low-Level Counting and Spectrometry, Vienna, Rep. IAEA-SM-252/40, 257 (1981).
143. Ruster, W., *et al.*: A resonance ionization mass spectrometer as an analytical instrument for trace analysis, Nucl. Instrum. Methods in Phys. Res. **A 281,** 547 (1989).
144. Erdmann, N., *et al.*: Resonance ionization mass spectrometry for trace analysis of plutonium, in Resonance Ionization Spectroscopy 1994, AIP Conf. Ser. **329**, 507 (1995).

Radiochimica Acta *70/71*, 123–133 (1995)

Radiochemical Investigations of Cluster Radioactivities

By G. Ardisson

Laboratoire de Radiochimie, Université de Nice Sophia Antipolis, 06108 Nice Cedex 02, France

and M. Hussonnois

Institut de Physique Nucléaire, B.P. 1, 91406 Orsay Cedex, France

(Received July 31, 1995; revised August 18, 1995)

Radioactive decays / Emission of heavy ions / Clusters

Abstract

The very rare spontaneous decays by emission of clusters with mass between that of α particles and of the light fission fragments are reviewed. The experimental tools, the detection techniques and the specific radiochemical procedures are described.

Introduction

Since the discovery of heavy-ion radioactivities by Rose and Jones in 1984 [1] several review papers have been published which describe the experimental techniques involved, the details of the new modes of radioactive decay as well as theoretical approaches [2–8]. In the present review, the pre-eminent role of radiochemistry in the discovery of heavy-ion radioactivities, similar to that in the discovery of the natural radioelements about 100 years ago, will be emphasized.

1. New spontaneous radioactive decay modes

Spontaneous decay modes belong to two main types:

– the leptonic decay modes resulting from the weak interaction, such as β decay, in which electrons with negative and positive charges are emitted simultaneously with antineutrinos and neutrinos, and the electron capture (e.g. K capture) of inner electrons by the atomic nucleus. These modes concern about 1500 nuclei in the chart of nuclides.

– the hadronic decay modes in which the emitted particles are bound by the strong interaction, among which the most frequent are α decays. The latter, predominant among the isotopes of heavy natural elements, were denominated by E. Rutherford in 1899 [9] and identified by the same author as helium ions in 1908 [10]. A second, also frequently occurring hadronic decay mode, was discovered about 40 years later by Petrjak and Flerov [11]. These authors found that the nucleus of ^{238}U could decay by spontaneous fission in two fragments of nearly equal masses.

The decay by emission of proton was discovered in 1970 by Jackson *et al.* on the nuclide ^{53m}Co [12]. This is a much rare decay mode. Up to now about 15 to 20 cases have been evidenced.

Between 1979 and 1985, Poenaru and Ivascu [13] proposed a new interpretation of α decay and treated this mode as an extremely asymmetric spontaneous fission process. On the basis of these considerations, Sandulescu, Poeanaru and Greiner [14, 15] were led to predict the possibility of new hadronic decay modes for heavy nuclei with emission of clusters of nucleons with masses in the range from 5 to about 70, intermediate between those of α particles and the light spontaneous-fission fragments. These spontaneous decay modes were first called *exotic* radioactivities, later heavy-ion or cluster emission or *cluster radioactivities.*

Finally, Poenaru *et al.* [16, 17] suggested that all nuclei in the universe beyond $Z = 40$, i.e. the so-called *trans-zirconium nuclides*, previously assumed to be stable, were actually metastable with respect to their partition in two asymmetric fragments. This process can be written as

$${}_{Z}A \rightarrow {}_{Z_1}A_1 + {}_{Z_2}A_2$$

in which the hadron numbers are conserved,

$$Z = Z_1 + Z_2 \qquad A = A_1 + A_2.$$

It is energetically allowed provided the energy balance

$$Q = [M(A, Z) - M(A_2, Z_2) - M(A_1, Z_1)]\, c^2 \qquad (1)$$

is positive. The *cluster* is the lighter fragment (A_2, Z_2). It is always a neutron-rich heavy ion and possesses in general an even mass number with $4 < A_2 \leq 70$. The heavy fragment (A_1, Z_1) is the *daughter nucleus.* Owing to shell effects, the probability of cluster emission is the highest when Z_1 and $N_1 = (A_1 - Z_1)$ are magic numbers or are close to magic numbers, i.e. $Z_1 = 82$ and $N_1 = 126$ (the "golden rule" of cluster decay). Hence, because the daughter nucleus is an isotope of lead, the parent emitter is a *trans-lead* nuclide.

With the Analytical Superasymmetric Fission Model (ASAFM), Poenaru and Ivascu [17] calculated that among 2272 nuclides, 501 would decay with a half-life less than 10^{30} s, 941 nuclides with a half-life less than 10^{50} s and 1330 nuclides with a half-life less than 10^{100} s. However, these relative unstabilities must be compared to the proton half-life, estimated to be 10^{38} to 10^{39} s, or to the age of the known universe which is (only) 10^{17} s. The test of these predictions is critically dependent on the ability to detect such extremely rare events.

2. Cluster radioactivities of trans-lead nuclides

a. The discovery of radioactive decay by emission of ^{14}C clusters

In a famous Note published in *Nature* with the title "A new kind of natural radioactivity" [1], H. J. Rose and G. A. Jones at Oxford University announced the discovery of a new rare *by-pass* process in the decay of ^{223}Ra. This nuclide is a well-known α emitter with a half-life of 11.435 d. The authors found that part of the parent nuclides decayed *directly* to ^{209}Pb by emission of a ^{14}C ion. This result was based on data collected in an experiment which lasted over one year and a half. The events were detected with a telescope counter $\Delta E - E$. In this device, charged particles cross a thin (8.2 μm) silicon layer losing a part ΔE of their energy, and reach a thick Si detector in which they are fully stopped. Rose and Jones had used this set up in the search of high energetic α particles ($E_\alpha \approx 12-15$ MeV) which might have been emitted by superheavy elements [18]. In a first experiment, a sample of the long lived ($T_{1/2} = 22$ y) ^{227}Ac, in equilibrium with its daughters and a total activity of 122 kBq was exposed in direct view of the telescope counter. The ΔE and E signals were simultaneously recorded on magnetic tapes over a period of 6 months and checked every 12 hours. Eight events were recorded in this first trial. Owing to the high fluence of α particles, the detector broke down at the end of the experiment. A new trial was performed over 6 months, and again 11 events were observed in the $\Delta E - E$ plane, grouped at a total energy of 30 MeV. A 6 months complementary experiment without source aimed at measuring the background (cosmic and electronic) of the set up. The authors concluded that among the various nuclides in the ^{227}Ac series, i.e. ^{227}Th, ^{223}Ra, ^{219}Rn etc. the only possible candidate for the observed particles could be ^{223}Ra. The energy and the energy loss of the particles were characteristic of ^{14}C ions. Accordingly the decay process could be

$$^{223}Ra \rightarrow {}^{14}C + {}^{209}Pb + 31.87\ \text{MeV}.$$

Other odd decays, such as $^{227}Ac \rightarrow {}^{15}N + {}^{212}Pb$, or $^{227}Th \rightarrow {}^{18}O + {}^{209}Pb$ were also considered and their decay probabilities estimated on the basis of the ratio of the Gamow factor for a given cluster emission to that of the corresponding α decay. Eventually, Rose and Jones concluded that the most probable decay mode in the series was the emission of ^{14}C ions.

The branching ratio, i.e. the competition of the $^{14}C/\alpha$ emission expressed by the ratio of the number of emitted particles $N(^{14}C)/N(\alpha)$ was estimated of the order of $(8.5 \pm 2.5) \times 10^{-10}$. This low value may well explain why this spontaneous decay mode was discovered only 88 years after Becquerel's initial observation of the emission of uranic rays.

b. Further experiments on the ^{14}C emission of radium isotopes

Although spectacular, the first experiments had several limitations, such as the high α counting rate, energy shifts owing to the long lasting counting, uncertainty in the energy calibration which was based on the extrapolation of the α particle peaks from an ^{241}Am source, etc. Thus several groups of investigators designed new experiments in order to confirm the ^{14}C emission of ^{223}Ra.

The same year 1984 of the parution of the publication by Rose *et al.* [1], a group of nuclear physicists and radiochemists led by S. Gales at the Orsay Institute of Nuclear Physics, used the large solid angle ($\Omega = 110$ msr) superconducting solenoidal coil spectrometer SOLENO to select and focus on a $\Delta E - E$ telescope the rare ^{14}C ions among an intense flux of α particles. The experiment was performed with a 9.14 MBq ^{227}Ac source, 300 times stronger than the source used by Rose and Jones. The source was mounted on an Al foil and covered with a 20 μg cm^{-2} carbon foil in order to prevent the escape of radon and other daughter products. The detector was calibrated with ^{12}C, ^{14}C and ^{16}O ions from the MP tandem accelerator. Eleven ^{14}C ions with a total energy of 29.4 MeV were counted in only 5 days. A branching ratio $b = (5.5 \pm 2.5) \times 10^{-10}$ was deduced [19], in good agreement with the original value of Rose and Jones [1].

Alexandrov *et al.* [20] also confirmed the ^{14}C emission of ^{223}Ra after a 30-day counting of a 3.14 MBq ^{227}Ac source in direct view of a telescope. Seven events were observed among a large number of triple and quadruple pile-up pulses. The branching ratio was close to the previous determinations.

The mass and the energy of the carbon clusters emitted by ^{223}Ra were measured at the Argonne National Laboratory by Kutschera *et al.* with an Enge split-pole magnetic spectrometer [21]. Owing to the limited solid angle of the device ($\Omega = 5.5$ msr) a very intense source was needed. To avoid long-term contamination by ^{227}Ac and in order to minimize the source thickness, the 18.7 d daughter ^{227}Th, parent of ^{223}Ra, was separated from 2 mg ^{227}Ac. An improved radiochemical separation and purification procedure was developed for handling the 26 GBq activity. However, inopportunely, the final source was prepared by direct evaporation, a method which is known to yield sources of poor quality. Moreover, the source was covered with a 100 μg cm^{-2} Al foil, which introduced important tailings of the α and ^{14}C peaks. The detector was a gas ionization counter which measured the energy loss ΔE and the total energy E. In a 6.09 day experiment, 24 events were recorded, but the fine structure (section 3) resulting from the feeding of excited states of ^{209}Pb could not be observed.

In 1984, P. B. Price and coworkers at Berkeley introduced the measurement of heavy-ion emission using solid state track detectors. The latter, consisting of polycarbonate foils, are selective for heavy ions with

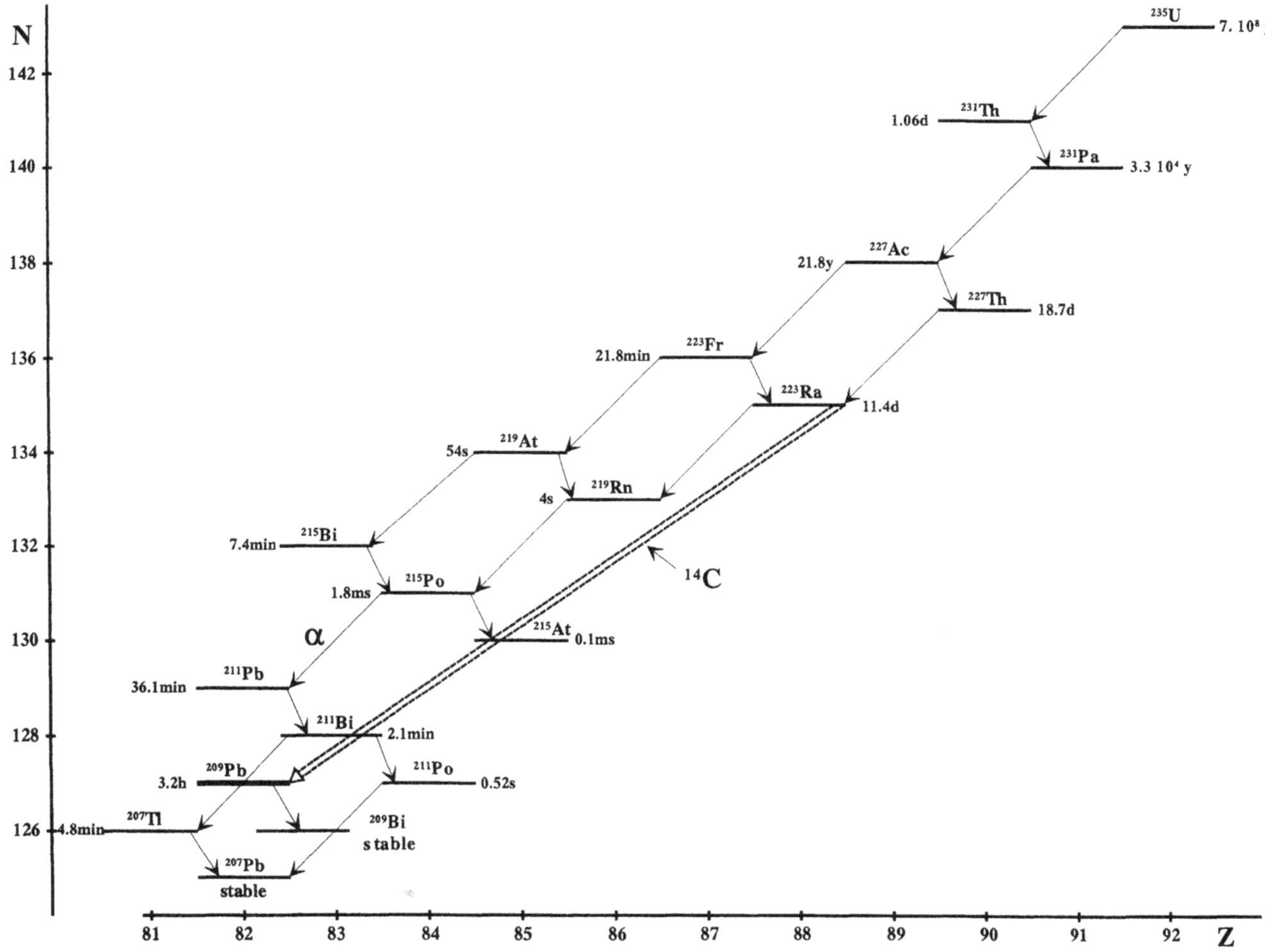

Fig. 1. ^{235}U natural series. Rose and Jones used a ^{227}Ac source in equilibrium with daughters. The ^{14}C rare by pass process (probability $< 10^{-9}$) which led to ^{209}Pb is also shown.

$Z > 2$, and likely to reach branching ratios much lower than that of ^{223}Ra owing to the larger solid angle (2 sr) viewed by the radioactive source in contact with the sheet. With the 600 MeV proton beam of the CERN Synchrocyclotron, the authors obtained odd and even Ra isotopes by a spallation reaction on a 55 g cm^{-2} ThC_2 target heated at 2100 °C. The isotopes of mass numbers 221 to 225 were separated with the ISOLDE electromagnetic separator and collected in different cups fitted with plastic detectors [22]. A total of 56 ^{14}C events were detected for ^{223}Ra; moreover, in the same experiment, emissions of ^{14}C clusters from the even ^{222}Ra ($T_{1/2}$ = 37.8 s) and ^{224}Ra ($T_{1/2}$ = 3.62 d) isotopes were discovered. The branching ratios were $(3.7 \pm 0.6) \times 10^{-10}$ and $(4.3 \pm 1.2) \times 10^{-11}$, respectively, i.e. one order of magnitude lower than that for ^{223}Ra.

The Orsay group pursued the experiments on ^{14}C radioactivity with the search for other emitters. First, the ^{14}C emission of the short-lived ^{222}Ra was confirmed by magnetic spectrometry with SOLENO using a source of the long-lived grandfather ^{230}U ($T_{1/2}$ = 20.8 d). The radiochemical preparation of the source is treated in section 3. Actually, the main objective of the project was a challenge: to search the emission of ^{14}C from ^{226}Ra ($T_{1/2}$ = 1600 y) discovered nearly hundred years ago by Pierre and Marie Curie. The 2.5 mg source was taken from the radium stored at the Institut du Radium. In a 9 day counting run, four ^{14}C events were recorded with SOLENO and the branching ratio $(3.2 \pm 1.6) \times 10^{-11}$ was determined [23]. This value was confirmed with improved statistics when the same source was investigated with polycarbonate foils in a Berkeley-Orsay collaboration [24].

The partial half-lives of ^{14}C decay of the three even Ra isotopes, compared to that of ^{223}Ra, revealed a marked odd-even effect for the probability of ^{14}C emission: the global transition emission rate of the odd radium isotope is hindered by a factor of about 100 with respect to the even mass isotopes [24].

c. Other ^{14}C emitters

Since the probability of cluster emission is the greatest when the daughter nucleus configuration is close to that of magic shells, attempts have been made to observe the phenomenon with parent nuclei near $Z = 88$, i.e. isotopes of the elements francium and actinium, whose daughters would be isotopes of Bi and Tl respectively, e.g. $^{221}Fr \rightarrow {}^{207}Tl_{126}$ and $^{225}Ac \rightarrow {}^{211}Bi_{128}$. Price *et al.* reported negative results for the two parent nuclides produced at CERN [22]. More recently, Bonetti *et al.* using a phosphate glass detector BP1 were successful and reported branching ratios of $(1.0\pm0.2)\times10^{-12}$ and $(6.0\pm1.3)\times10^{-12}$ for ^{221}Fr and ^{225}Ac respectively.

d. The single case of oxygen cluster emission

Following the "golden rule", isotopes of thorium, $Z=90$, should preferentially emit oxygen clusters, leaving daughter nuclei close to the doubly magic ^{208}Pb nucleus. The most convenient Th isotope to check the emission of oxygen is the long-lived ^{228}Th ($T_{1/2}$ = 1.91 y) for which the theory predicts a branching of about 10^{-14}. Bonetti *et al.* measured a 66.6 MBq ^{228}Th source supplied by the Chemistry Division at Harwell with a phosphate glass detector and reported a branching ratio of $(1.13\pm0.22)\times10^{-13}$ in good agreement with the theoretical value [26].

In the search of a fine structure in the energy spectrum of ^{222}Ra (section 3) carried out with SOLENO, Hussonnois *et al.* used a ^{230}U source in radioactive equilibrium. The chain comprises the daughter ^{226}Th ($T_{1/2}$ = 30.6 min). From a careful analysis of the energy spectrum of ^{18}O events [27], the upper limit of the branching ratio $b < 9.3\times10^{-13}$ could be determined for the decay

$$^{226}\text{Th} \rightarrow {}^{18}\text{O} + {}^{208}\text{Pb}.$$

e. $^{24-26}$Ne emissions of thorium, protactinium and uranium isotopes

From 1984, Price *et al.* at Berkeley and the Dubna group (Tretyakova, Sandulescu and coworkers) improved the solid state track detectors using polycarbonate foils and other materials such as phosphate glasses. The first emission of ^{24}Ne clusters by ^{232}U was observed at Berkeley in 1985 [28]. A hemisphere of polyethylene terephthalate (cronar) inside a vacuum chamber was exposed during one month to a commercial 18.5 MBq ^{232}U source. This type of detector is particularly sensitive to ions with $Z>6$ and is resistant to α ray fluences up to 2×10^{11} cm^{-2}. From the 31 significative events detected, the authors deduced a branching ratio $\lambda_{24Ne}/\lambda_\alpha$ of $(2.0\pm0.5)\times10^{-12}$.

In 1989, the Milano group investigated again the decay of ^{232}U in an experiment aimed at searching other decay modes of the nuclide, such as spontaneous fission and ^{28}Mg emission [29]. With a 23.8 MBq source from Harwell and a phosphate glass detector, the authors were able to detect 138 emissions of ^{24}Ne ions in a 61 day exposure time. The new branching ratio thus obtained was $(8.7\pm0.9)\times10^{-12}$ nearly four times higher than the previous value.

In 1991, the same group undertook a third determination of the branching [30]. The detector was again a PSK-50 phosphate glass; the exposure to a 15 ± 0.7 MBq source of ^{232}U, also from Harwell, during 38 days yielded the value $b = (9.16\pm1.1)\times10^{-12}$ consistent with the previous determination, and about 5 times higher than the Berkeley value. As the most likely explanation for the discrepancy, the authors proposed that "the activity of the first source was assumed to be that specified by the manufacturer rather than independently checked". With the revised branching ratio of ^{232}U, the Ne emissions of ^{234}U and ^{236}U were reevaluated.

In 1984–85, the Russian group discovered the ^{24}Ne decays of ^{233}U, ^{231}Pa and ^{230}Th with respective branchings of 7.5×10^{-13}, 6×10^{-12} and 5.6×10^{-13} [31–33].

The cluster radioactivity of ^{233}U was reinvestigated by Price *et al.* in 1991 [34]. Four ^{233}U sources, totalizing 799 mg uranium, were prepared by a displacement reaction from a hot acidic solution of uranyl nitrate onto aluminum plates according to

$$3\,UO_2^{2+} + 2\,Al + 2\,H_2O \rightarrow 2\,Al^{3+} + U_3O_8 + 2\,H_2.$$

The detector assembly consisted of 36 plates of PSK 50 phosphate glass detectors and the exposure lasted 582 days. About 1000 Ne clusters were recorded in the experiment. The derived branching ratio $(7.2\pm0.5)\times10^{-13}$ was consistent with the earlier value $(7.5\pm2.5)\times10^{-13}$. Moreover, the experiment which was designed for the search of competing emission of ^{28}Mg clusters, led to an upper limit of the branching ratio b (Mg/Ne) $< 1.8\times10^{-3}$.

In the case of ^{231}Pa, the most probable cluster emission, as indicated, is ^{24}Ne; the less probable emission of the odd mass cluster ^{23}F ($T_{1/2}$ = 2.2 s) should lead to the doubly magic nucleus $^{208}\text{Pb}_{126}$. With an improved counting statistics, the Berkeley group [35] found that among a thousand ^{24}Ne tracks, one event could be attributed to ^{23}F, with a branching ratio of 9.9×10^{-15}. This value disagrees with all theoretical expectations. The discrepancy might be due to a hindrance of the emission of ^{23}F, related to a weak overlap of the wave functions of the strongly deformed ^{231}Pa nucleus and the odd ^{23}F cluster, similar to that observed for the first time in the ^{14}C decay of ^{223}Ra [36, 37].

f. Multiple hadron emission: ^{234}U and ^{238}Pu

According to the "golden rule", the emission of ^{34}Si clusters should be favoured for transuranic nuclides such as ^{241}Am [38] or ^{240}Pu [39]. Up to now, the search of ^{34}Si radioactivity was negative. In the most recent trial, two types of phosphate glasses, LG760 and LG750, with an area of 68 cm^2 were exposed for 181 days to a 8 mg ^{241}Am source. The upper limit of the ^{34}Si branching ratio of ^{241}Am was estimated 7.4 $\times10^{-16}$, corresponding to a partial half-life larger than 5.8×10^{17} y [38].

According to the authors, the limit may be lowered with a ten-fold increase of the source activity and of

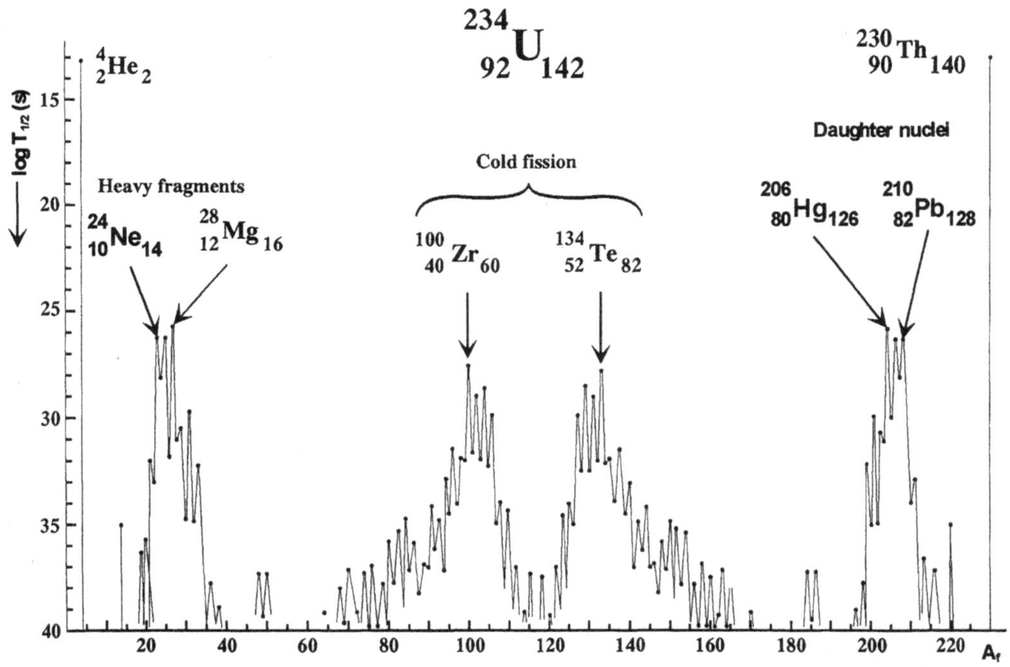

Fig. 2. ^{234}U partial half-lives for different fragmentation processes predicted by the Analytical Superasymmetric Fission Model.

the detector area. The main problem will be to sort out the rare ^{34}Si among more than 5×10^5 fission tracks. For ^{237}Np, the branching would be below 4×10^{-14} [40].

One may wonder if the detection limit of high Z clusters has now been reached. To clarify this point, two very sophisticated experiments performed by the Berkeley group are instructive. First the authors searched for Mg and Si cluster emissions of an enriched (99.84%) $^{234}U_3O_8$ source, highly purified by exchange chromatography. Three sources of 36 to 44 cm^2 area were measured with PSK-50 phosphate glass detectors during periods extending from 68 to 111 days. In addition to the α and spontaneous fission decays, two additional hadronic decay modes were discovered [41]. ^{28}Mg and $^{24-26}Ne$ clusters are emitted simultaneously, leading to the daughter nuclei $^{206}Hg_{126}$ and $^{210-208}Pb_{128-126}$. The branching ratios for these concomitant processes are respectively 1.4×10^{-13} and 4.4×10^{-13}, corresponding to partial half-lives of 1.1×10^{18} and 3.7×10^{17} y, to be compared to the spontaneous fission half-life of 1.9×10^{16} y. Such decay modes had been predicted since 1980 by the ASAFM model (Fig. 2).

The second experiment was performed with a 10.5 mg ^{238}Pu source and a hemispherical LG 750 phosphate glass detector [42]. The source was covered with an Al foil, thick enough to absorb the fission fragments, but traversed by 80 MeV, $Z = 14$ ions. Besides the α decay and spontaneous fission ($T_{1/2}$ = 5×10^{10} y), two more hadronic decays were observed: the emission of ^{28}Mg and ^{32}Si, the latter being the heaviest cluster observed up to now with a 1.4×10^{-16} branching and a partial half-life of 6×10^{17} y (2×10^{25} s). The partial half-life of the ^{28}Mg emission is 1.6×10^{18} y. It may be noted that these half-lives are considerably shorter than those calculated for metastable decays or suggested for the proton decay.

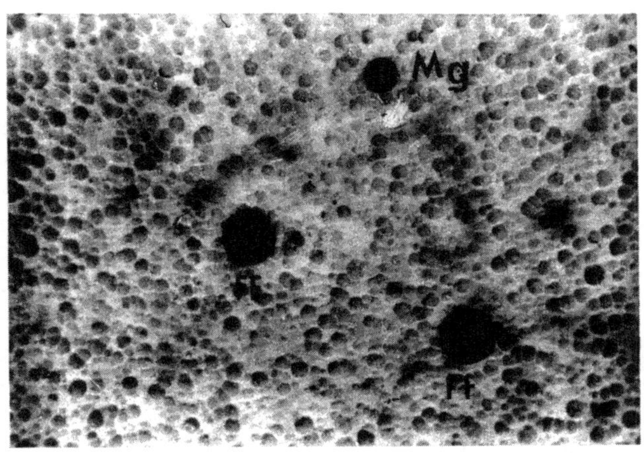

Fig. 3. Photomicrograph of one etch pit due to a ^{28}Mg ion emitted by ^{236}Pu, together with two spontaneous fission tracks.

A 690 day experiment was performed by Oglobin *et al.* with a 0.51 μg ^{236}Pu source [43] using a polyethylene terephtalate (lavsan) detector in vacuum. The observation of two ^{28}Mg clusters was consistent with a branching of 2×10^{-14} and a partial half-life of 1.5×10^{14} y.

This decay mode has been reinvestigated very recently by a collaboration of the Orsay, Dubna and Milano groups [44]. Over a period of 18 months, a series of 10 targets of natural uranium were irradiated with 34 MeV protons to produce an intense ^{236}Pu source. In each target, Pu and Np were separated from the bulk of uranium and from the fission products. In a next step, the Pu was purified from Np by anion exchange. The various Pu preparations were assembled. Owing to the ingrowth of daughter products, the source was purified immediately before electrodeposition on a Pt foil. The remarkably high radiochemical purity of the (74 ± 8) MBq ^{236}Pu source was demonstrated by γ spectroscopy of a freshly purified sample which revealed new γ-lines with relative intensities less than $5 \times 10^{-7}\%$ [45].

The source was placed in vacuum at the center of a hemispherical phosphate glass detector. After exposure during 216 days, fifteen ^{28}Mg tracks were found, corresponding to the branching $(2.7 \pm 0.7) \times 10^{-14}$, in good agreement with the previous determination.

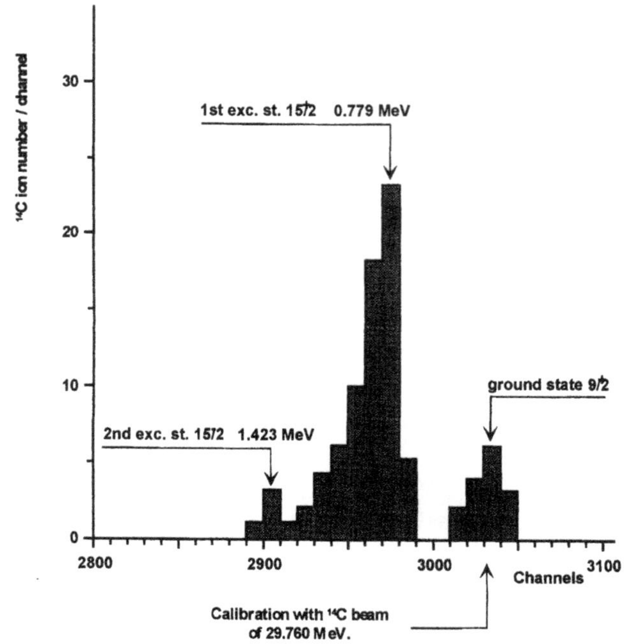

Fig. 4. The ^{14}C spectrum following the decay of a $^{227}Th + ^{223}Ra$ source.

3. The fine structure of the cluster emission of ^{223}Ra

The odd-even effect revealed by the partial half-lives of cluster decay of Ra isotopes was considered as a global effect, but the theorists wondered if the daughter nuclei could be formed in excited states, like in α decays. Scheid and Greiner [46] computed ^{14}C emission probabilities and concluded that in such case the total probability could increase five times relatively to decays which would leave the daughter nuclei in the ground state.

In 1989, the Orsay group undertook the measurement of the energy spectrum of ^{14}C clusters in order to detect a possible feeding of excited states of the daughter nuclei [47]. A relative probability of a few percent seemed desirable for observing the phenomenon. The most favorable case was the ^{14}C emission of ^{223}Ra. It was necessary to improve the energy resolution of the spectrometer because the separation between the ground and the first excited state of ^{209}Pb is only 779 keV. A strong ^{223}Ra source was also required to improve the counting statistics and record several hundred ^{14}C ions within a reasonable time.

About 70 µg (185 MBq) of ^{227}Ac were separated from a 150 mg sample of ^{231}Pa, which had not been treated chemically since 25 years. The actinium itself is not suitable for the preparation of a thin source. Therefore 150 MBq of the daughter ^{227}Th were separated from ^{227}Ac after its ingrowth and electroplated. Two weeks later, and during the experiment the ^{223}Ra activity remained fairly constant at (66 ± 2) MBq.

With the SOLENO spectrometer the ^{14}C ions could be well focalized and were detected with a large (450 mm^2) Si surface barrier detector with a FWHM resolution of 110 keV for heavy ions. In a 11-day run, 410 $^{14}C^{6+}$ ions were recorded. The histogram of the number of events versus the channel number showed three relatively well resolved peaks (Fig. 4). Thus the cluster decay has characteristics similar to those discovered about 80 years ago by Rosenblum for α decay.

An important feature of the spectrum is the high proportion (81%) of the ^{14}C decays which leave ^{209}Pb in the first excited state. Only 15% feed the ground level, while the remaining group of 4% was interpreted as decays to the second 1.423 MeV excited state (Fig. 5).

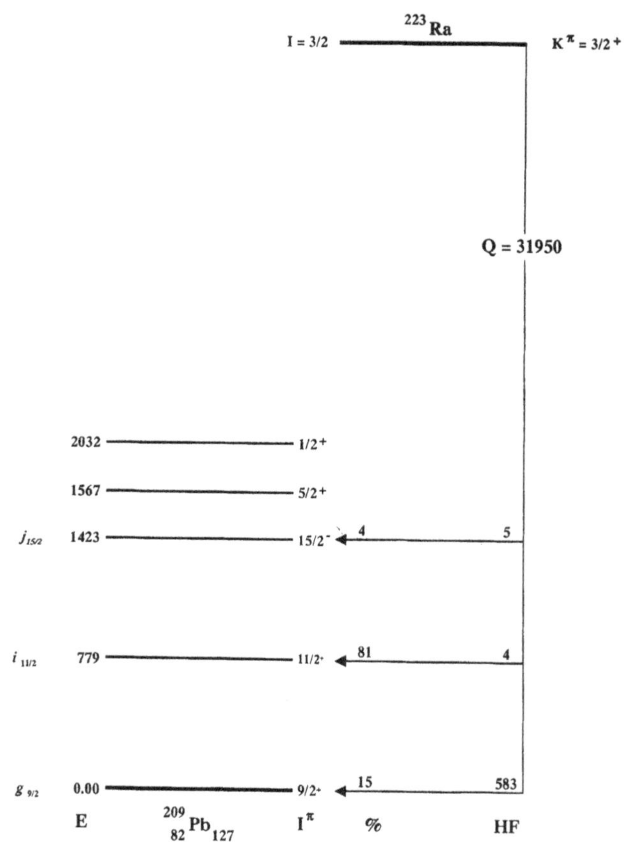

Fig. 5. The ^{223}Ra ^{14}C-decay scheme. Hindrance factors (HF) are calculated using the reduced widths γ^2 (see section 3 and Fig. 6).

It results that the odd-even effect observed for ^{223}Ra [24] was not a global, but rather a *structural* effect due to the differences of the odd nucleon configurations in the ground and excited states. In a log T-log $E^{-1/2}$ Geiger-Nutall plot, the points for even Ra

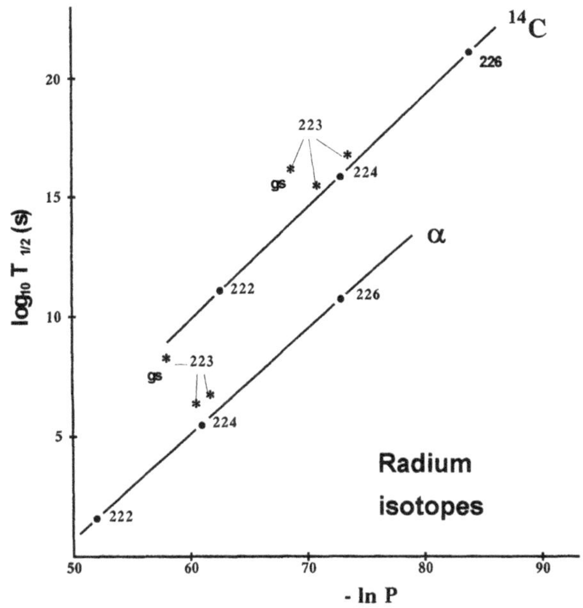

Fig. 6. Plot of the decimal logarithm of partial half-life (in seconds), versus the natural logarithm of penetrability for alpha and ^{14}C emissions. gs means ground-state. Hindrance factors are measured by the distance between the representative points of transitions to the ^{223}Ra odd nucleus and the even radiums straight line.

isotopes lie on a straight line, whereas the odd Ra point is located above the line. In a treatment similar to that applied for α decay, a cluster hindrance factor (HF) is defined as the ratio of the reduced width γ_i^2 for a given ^{14}C group emitted by the odd nucleus, to the reduced width γ_0^2 for the cluster decay to the ground states of even-even neighbouring isotopes [49]:

$$\gamma_i^2 = \hbar \ln 2/2\, P_{(i)}\, T_{1/2(i)}$$

in which $T_{1/2(i)}$ is in seconds and the barriers penetrabilities $P(i)$ expressed as

$$P_{(i)} = \exp(-2\, G_{(i)}).$$

The Gamow factor $G_{(i)}$ is

$$G_{(i)} = 1/\hbar \int (2\, \mu (E_{(i)} - Q_{(i)})^{1/2} dr)$$

where $\mu = A_1 A_2/(A_1 + A_2)$ is the reduced mass of the system. Q is given by Equation (1) and $E_{(i)}$ is the potential energy of the system (for details see [49]).

The calculated HF 583, 3.9 and 4.6 relevant to the respective transitions to the ground, first and second excited states of ^{209}Pb [49–50] fit well with the log (partial $T_{1/2}$)$-$ln P diagram (Fig. 6). The HF of the ground state transition has been explained in terms of the configuration of the $I^\pi = 3/2^+$ ground state of the parent ^{223}Ra [36, 37, 49, 50]. With the Asymmetric Rotor Model, the wave function of the odd neutron can be described as a mixed parity state arising from $1\, i_{11/2}$ and $1\, j_{15/2}$ shell-model orbits. In fact, nuclei with A between 220 and 230 are assumed to have an octupolar shape [51]. Thus the high HF to the ground state is attributed to the 127 th neutron in the $2\, g_{9/2}$ configuration (in the spherical shell model) of the semi-magic ^{209}Pb which has no counterpart in the ^{223}Ra wave function. Accordingly, the transition $\Pi\, (i_{11/2} \oplus j_{15/2}) \rightarrow \nu g_{9/2}$ hindered, while the transitions to the first ($\nu i_{11/2}$) and to the second ($\nu j_{15/2}$) excited states are allowed.

These considerations were based on the measured partial half-lives of even Ra isotopes, which so far were assumed to feed only ground states of the daughter. Hence it was of interest to measure the ^{14}C energy spectrum of even Ra isotopes, and in particular of ^{222}Ra which, as already indicated, is the granddaughter of ^{230}U in the ^{230}U (20.8 d) $\rightarrow$ ^{226}Th (31 min) $\rightarrow$ ^{222}Ra (37.5 s) chain. The preparation of a strong ^{230}U source is a challenging task which requires considerable skill. Radiochemically pure ^{230}U can be obtained from the 8% β^- branch of ^{230}Pa, a nuclide synthesized in the proton irradiation of thick Th targets via the $^{232}Th(p,3n)$ reaction. However, the competing (p,n) reaction generates ^{232}Pa ($T_{1/2}$ = 1.3 d) decaying to ^{232}U ($T_{1/2}$ = 70 y).

Therefore, to prepare ^{230}U free of interfering activities, the Pa fraction was separated after a cooling time of 15 days and purified on anion exchangers with HCl eluants of various concentrations. After ingrowth of ^{230}U in Pa, the two nuclides were again separated. ^{230}U was purified by liquid-liquid extraction and electrodeposited on a Pt disk. A strong source of 85 MBq was obtained from two irradiations performed on a time scale such to obtain the maximum activity.

The energy spectrum of the ^{14}C clusters was measured with a Si detector calibrated (for the first time) with the three ^{14}C peaks from the ^{223}Ra decay [27]. The spectrum of 210 events obtained in a 16 day run exhibited one single peak with a small energy tailing similar to that observed in an α spectrum (Fig. 7). The upper limit of the ^{14}C branching to the first excited state of ^{208}Pb at 2.614 MeV is below 2×10^{-12}, i.e. $T_{1/2} > 2 \times 10^{13}$ s.

4. About the future

a. Planned experiments

The detection of heavy ion emission from long-lived α emitters such as ^{232}Th ($T_{1/2} = 1.4 \times 10^{10}$ y) or ^{238}U ($T_{1/2} = 4.47 \times 10^9$ y) by direct measurement of tracks is rather limited because of the low specific activities. Recently (1995) the detection of $^{24-26}Ne$ clusters from ^{232}Th has been attempted by the Milano group [52]. The authors scanned a 1782 cm^2 area of Melinex detectors foils, exposed to Th sheets with a total area of 1000 cm^2. No event was detected during a 665 day experiment, from which a partial half-life $T_{1/2} > 10^{21}$ y is estimated for Ne cluster emission. On the other hand, the experiment performed in the Gran Sasso Underground Laboratory led to the discovery of spontaneous fission of ^{232}Th with a partial half-life $T_{1/2}$ = 1.22×10^{21} y.

A different approach is examined in the authors' laboratories. It is based on the radiochemical separation and measurement of the activity of ^{24}Na ($T_{1/2}$ =

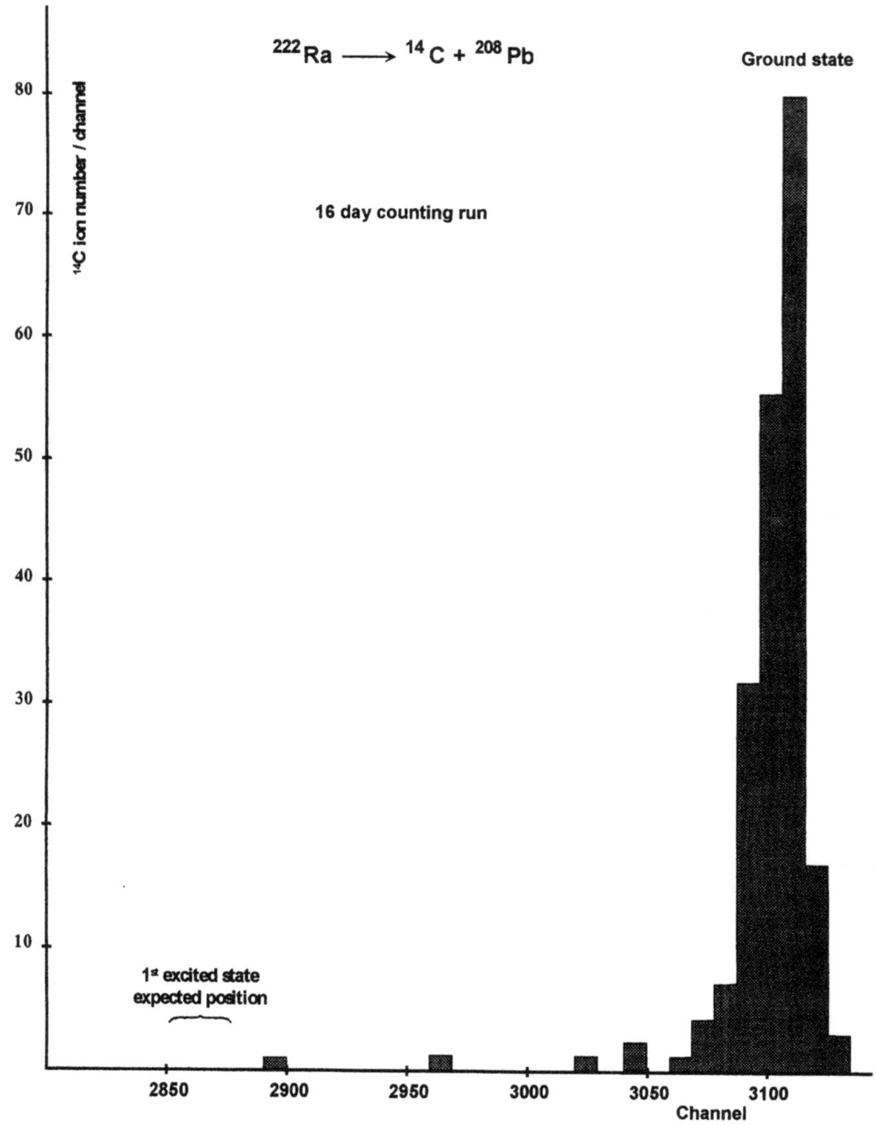

Fig. 7. ^{14}C spectrum following the ^{222}Ra decay.

15.0 h), daughter of ^{24}Ne ($T_{1/2}$ = 3.38 min). The extraction of a few ten ^{24}Na atoms, free of radioactive contaminants, from about 10 kg ^{232}Th is a big challenge for radiochemists.

Similarly, the search of ^{34}Si clusters emitted by ^{242}Cm ($T_{1/2}$ = 163 d) implies considerable radiochemical work. The required activity of 3 GBq (3 μg) must be produced by intense neutron irradiation of several mg of ^{241}Am, followed by the separation of Am and Cm at macroscale, and purification from fission products.

It is planned to search for a fine structure in the ^{14}C spectrum of odd-Z actinides, such as ^{225}Ac and ^{221}Fr. The purpose of the experiment is to check the cluster preformation theory suggested for the ^{223}Ra decay [49] and to investigate the structure of these nuclei [53]. Proton irradiation of a Th target with the 1 GeV beam available at CERN would not be sufficient to produce the required activity of ^{225}Ac (≈2 GBq). The radioelement could be extracted from several kg of ^{233}U kept for 20 years. In the decay chain $^{233}U \rightarrow {}^{229}Th \rightarrow {}^{225}Ra \rightarrow {}^{225}Ac$, about 0.7 GBq of ^{225}Ac is in equilibrium with one kg of ^{233}U. A major problem for radiochemists will arise from the very intense activity of ^{208}Tl generated in the ^{232}U series which is always present with ^{233}U.

Several unsuccessful experiments should be resumed, such as the search for $^{28-29}Mg$ decays of ^{235}U or the ^{34}Si emission of ^{238}U:

$$^{238}U \rightarrow {}^{34}Si + {}^{204}Pt_{126}$$

for which the predicted branching is $\approx 10^{-11}$. This value, however, may change within several orders of magnitude owing to the large uncertainty (hundreds of keV) of the mass of the daughter nucleus.

The emission of the heaviest clusters such as ^{48}Ca or ^{46}Ar is predicted for the ^{252}Cf and ^{254}Fm decays with $b \approx 10^{-14}$ and $T_{1/2} \approx 10^{18}$ s. The main problem, so far unresolved, is the discrimination of the heavy clusters from the fragments formed in ternary fission events.

Table 1. Summary of measured cluster emissions of trans-lead nuclei

Cluster emitted	Parent nucleus	Cluster energy (MeV)	log ($T_{1/2}$) (seconds)	Detection method	Year	Reference
	^{221}Fr	29.28	14.46	track detector	1993	Bonetti *et al.* [25]
	^{221}Ra	30.34	13.41	track detector	1993	Bonetti *et al.* [25]
			15.06	telescope	1984	Rose and Jones [1]
		29.85	15.1	telescope	1984	Alexandrov *et al.* [20]
	^{223}Ra		15.2	telescope, super conducting spectrometer	1984	Gales *et al.* [19]
			15.3	telescope, Enge magnetic spectrometer	1985	Kutschera *et al.* [21]
		29.85	16.0	Si detector, super conducting spectrometer		
		29.12	15.3	fine structure	1989	Brillard *et al.* [47]
		28.52	16.6			
^{14}C	^{222}Ra	30.97	11.02	track detector	1985	Price *et al.* [22]
	^{224}Ra	28.63	15.9	track detector	1985	Price *et al.* [22]
	^{226}Ra	26.46	21.3	telescope, magnetic super-conducting spectrometer	1985	Hourani *et al.* [23]
	^{225}Ac	28.57	17.15	track detector	1993	Bonetti *et al.* [25]
	^{226}Th	28.77	>15.3	Si detector, super conducting magnetic spectrometer	1991	Hussonnois *et al.* [27]
^{18}O	^{226}Th	42.09	>15.3	Si detector, super conducting magnetic spectrometer	1991	Hussonnois *et al.* [27]
^{20}O	^{228}Th	40.80	20.87	track detector	1993	Bonetti *et al.* [26]
^{23}F	^{231}Pa	46.67	>24.6	track detector	1992	Price *et al.* [35]
	^{230}Th	51.75	24.64	track detector	1985	Tretyakova *et al.* [33]
	^{232}Th	55.37	>29.2	track detector	1995	Bonetti *et al.* [52]
	^{232}U	55.86	21.06	track detector	1985	Barwick *et al.* [28]
^{24,26}Ne	^{233}U	54.27	24.82	track detector	1985	Tretyakova *et al.* [31]
	^{234}U	52.81	25.07	track detector	1987	S. Wang *et al.* [41]
	^{231}Pa	54.14	23.23	track detector	1984	Sandulescu *et al.* [32]
	^{234}U	65.26	25.55	track detector	1987	S. Wang *et al.* [41]
^{28}Mg	^{236}Pu	70.22	21.67	track detector	1990	Oglobin *et al.* [43]
	^{238}Pu	67.32	25.70	track detector	1989	S. Wang *et al.* [42]
^{30}Mg	^{236}U	63.29	>26.28	track detector	1989	Tretyakova *et al.* [60]
	^{237}Np	65.52	>27.3	track detector	1985	Tretyakova *et al.* [40]
^{32}Si	^{238}Pu	78.95	25.28	track detector	1989	S. Wang *et al.* [42]
	^{240}Pu	78.07		track detector	1986	Barwick [39]
^{34}Si	^{241}Am	80.60	>25.24	track detector	1987	Moody *et al.* [38]
	^{242}Cm	83.74	23.24	track detector	1992	Sandulescu and Greiner [7]

b. Cluster decays in the trans-tin region

Magic numbers in the trans-tin region are $N=50$ and 82, and $Z=50$. Shell effects are less pronounced than for $N=126$ and $Z=82$. Moreover, ^{208}Pb lies on the stability line, while the doubly magic closed shell nuclei ^{132}Sn and ^{100}Sn are far from that line. Emissions of ^{12}C or ^{16}O clusters have been predicted by Gupta *et al.* [54] and Poenaru *et al.* [55]. In an island of proton-rich nuclei around the Ba-Sm region, daughter nuclei formed in cluster decays are close to the doubly magic $^{100}Sn_{50}$ nucleus. Typical expected decays are

$$^{114}Ba \rightarrow {}^{12}C + {}^{102}Sn$$

$$^{128}Sm \rightarrow {}^{100}Cd + {}^{28}Si.$$

A search for the cluster radioactivity of the new nuclide ^{114}Ba produced in the ^{58}Ni(^{58}Ni, 2n) reaction was carried out at Dubna [56] and recently (1995) at the GSI in Darmstadt [57]. Three events were observed with track detectors, which together with the measured half-life of the parent nuclide, $T_{1/2} = 0.43^{+0.30}_{-0.15}$ s led Guglielmetti *et al.* to determine a partial half-life $T_{1/2} > 1.1 \times 10^3$ s for the spontaneous emission of ^{12}C by ^{114}Ba [57].

c. Further exotic decay modes

Tin radioactivity

Sandulescu pointed out that cold fission in which the available energy Q is shared between the two fission fragments, can be considered as a particular case of heavy-ion radioactivity. This process which leaves fragments in the $Z=50$ Sn region or close to that region is called *tin radioactivity*. The best example of this decay mode is

$$^{264}Fm_{164} \rightarrow {}^{132}Sn_{82} + {}^{132}Sn_{82}$$

in which the neutron-rich fermium (a dinuclear like molecule) splits into two identical doubly magic tin fragments with a probability comparable to that of α decays.

^{5}He emission

Although ^{5}He is an unbound nucleus ($T_{1/2} = 7.6 \times 10^{-22}$ s), its emission is energetically favoured. It could be observed in correlation experiments between the decay products. The best candidate for ^{5}He emission if ^{213}Po ($T_{1/2}=4$ µs), with an expected half-life $T_{1/2}$ (^{5}He) $= 10^{21}$ s and a branching $b=10^{-27}$. Hence the emission could be detected from excited states of ^{213}Po, for example fed in the β^- decay of ^{213}Bi, rather than from the ground state.

5. Concluding remarks

Since the discovery of radioactive decay by emission of ^{14}C, eleven years ago, over 14 emitters of clusters from ^{14}C to ^{32}Si have been identified among nuclei of trans-lead elements (Table 1). In all cases, the most probable emission is that of an even, neutron-rich cluster leaving a daughter nucleus with a configuration close to that of doubly magic shells, $Z=82$ and $N=126$.

With the improvement of solid state track detectors, scanning techniques and radiochemical preparations of increasingly active sources, the detected branching ratios of these rare decay events have been extended by six orders of magnitude, from 10^{-9} for ^{223}Ra to 10^{-15} for the limiting case of the ^{34}Si decay of ^{241}Am.

The fine structure of the ^{14}C emission of ^{223}Ra shows that the cluster decay is closer to α decay than to a fission process. However, the establishment of a unified model describing both cluster emission and neutronless cold fission need more experimental work, in particular on heavy clusters decay, such as ^{46}Ar or $^{48-50}$Ca. For this purpose, still more improvements will be required in the detection techniques and in the production of intense trans-plutonium sources.

The discovery of further trans-tin cluster emitters may confirm if the phenomenon can be considered a "*magic radioactivity*" as suggested by Sandulescu [6].

Price *et al.* [28] noticed that heavy ion radioactive decay might have been discovered already in the 50's. The spontaneous fission half-lives of several α emitters compiled by Vandenbosch and Huizenga [58] are of the same order of magnitude than those found for the most probable cluster decay. The misinterpretation of the recorded events was due to inaccuracy in the measurements of pulse heights.

In 1975, a brazilian group [59] claimed the discovery of spontaneous emission of heavy ions by ^{238}U. Nuclear emulsions loaded with uranium were stored during periods up to 8.5 years. A peak, distinct from the spontaneous fission tracks, in the mass range 20 to 70, was attributed to the emission of large nucleon clusters with Z between 10 and 20. Inopportunely, the partial half-live value $(6 \pm 3) \times 10^{22}$ s was considerably shorter than the upper limit measured for ^{232}Th [52].

Acknowledgment

G. A. expresses his gratitude to J. P. Adloff and C. Marsol for a critical review of the manuscript.

References

1. Rose, H. J., Jones, G. A.: Nature **307**, 245 (1984).
2. Greiner, W., Ivascu, M., Poenaru, D. N., Sandulescu, A.: *Treatise on Heavy-ion Science*, vol. **8**, edited by D. A. Bromley, Plenum Press, New York, p. 641 (1989).
3. Hourani, E., Hussonnois, M., Poenaru, D. N.: Ann. Phys. Paris **14**, 311 (1989).
4. Price, P. B.: Nucl. Phys. **A 502**, 41c (1989).
5. Price, P. B.: Annu. Rev. Nucl. Part. Sci. **39**, 19 (1989).
6. Sandulescu, A.: J. Phys. G: Nucl. Part. Phys. **15**, 529 (1989).

7. Sandulescu, A., Greiner, W.: Rep. Prog. Phys. **55**, 1423 (1992).
8. Gupta, R. K., Greiner, W.: Int. J. Mod. Phys. **E 3**, 335 (1994).
9. Rutherford, E.: Phil. Mag. ser. 5, **47**, 109 (1899).
10. Rutherford, E., Royds, T.: Phil. Mag. ser. 6, **17**, 281 (1909).
11. Petrjak, K. A., Flerov, G. N.: J.E.T.P. **10**, 1013 (1940).
12. Jackson, K. P., Cardinal, C. U., Evans, H. C., Jelley, N. A., Cerny, J.: Phys. Lett. **B 33**, 281 (1970).
13. Poenaru, D. N., Ivascu, M.: Report NP-17-80, Central Institute of Physics, Bucharest (1980).
14. Sandulescu, A., Poenaru, D. N., Greiner, W.: Sov. J. Part. Nucl. **11**, 528, (1980).
15. Poenaru, D. N., Ivascu, M.: Rev. Roum. Phys. **29**, 263 (1984).
16. Poenaru, D. N., Greiner, W., Ivascu, M., Sandulescu, A.: Phys. Rev. C **32**, 2198 (1985).
17. Poenaru, D. N., Ivascu, M.: *Particle emission from nuclei,* vol. **2**, 127 (1989) CRC Press, Boca Raton, Florida.
18. Jones, G. A.: Proc. *Interactions and Structures in Nuclei,* The University of Sussex, 7–9th Sept. 1987, edited by R. J. Blin-Stoyle and W. D. Hamilton, 1988, IOP Publishing Ltd.
19. Gales, S., Hourani, E., Hussonnois, M., Shapira, J. P., Stab, L., Vergnes, M.: Phys. Rev. Lett. **53**, 759 (1984).
20. Alexandrov, D. V., Beliatsky, A. F., Glukhov, Yu. A., Nikol'sky, E. Yu., Novatsky, B. V., Oglobin, A. A., Stepanov, D. N.: Pis'ma Zh. Eksp. Teor. Fiz. **40**, 152 (1984) [JETP Lett. **40**, 909 (1984)].
21. Kutschera, W., Ahmad, I., Armato, S. G. III, Friedman, A. M., Gindler, J. E., Henning, W., Paul, M., Rehm, K. E.: Phys. Rev. **C 32**, 2036 (1985).
22. Price, P. B., Stevenson, J. D., Barwick, S. W., Ravn, H. L.: Phys. Rev. Lett. **54**, 297 (1985).
23. Hourani, E., Hussonnois, M., Stab, L., Brillard, L., Gales, S., Shapira, J. P.: Phys. Letters **160 B**, 375 (1985).
24. Barwick, S. W., Price, P. B., Ravn, H. L., Hourani, E., Hussonnois, M.: Phys. Rev. **C 34**, 362 (1985).
25. Bonetti, R., Chiesa, C., Guglielmetti, A., Matheoud, R., Migliorino, C., Pasinetti, A. L., Ravn, H. L.: Nucl. Phys. **A 562**, 32 (1993).
26. Bonetti, R., Chiesa, C., Guglielmetti, A., Migliorino, C., Cesana, A., Terrani, M.: Nucl. Phys. **A 556**, 115 (1993).
27. Hussonnois, M., Ledu, J. F., Brillard, L., Dalmasso, J., Ardisson, G.: Phys. Rev. **C 43**, 2599 (1991).
28. Barwick, S. W., Price, P. B., Stevenson, J. D.: Phys. Rev. **C 31**, 1984 (1985).
29. Bonetti, R., Fioretto, E., Migliorino, C., Pasinetti, A., Barranco, F., Vigezzi, E., Broglia, R. A.: Phys. Lett. **B 241**, 179 (1990).
30. Bonetti, R., Chiesa, C., Guglielmetti, A., Migliorino, C., Cesana, A., Terrani, M., Price, P. B.: Phys. Rev. **C 44**, 888 (1991).
31. Tretiakova, S. P., Sandulescu, A., Zamyatnin, Yu S., Kovotchin, Yu. S., Micheev, V. L.: J.I.N.R. Rapid Commun. **7**, 23 (1985).
32. Sandulescu, A., Zamyatnin, Yu. S., Lebedev, J. A., Myasoedov, B. F., Tretiakova, S. P., Hasegan, D.: J.I.N.R. Rapid Commun. **5**, 5 (1984).
33. Tretiakova, S. P., Sandulescu, A., Micheev, V. L., Hasegan, D., Lebedev, J. A., Zamyatnin, Yu. S., Korotkin, Yu. S., Myasoedov, B. F.: J.I.N.R. Rapid Commun. **13**, 34 (1985).
34. Price, P. B., Moody, K. J., Hulet, E. K., Bonetti, R., Migliorino, C.: Phys. Rev. **C 43**, 1781 (1991).
35. Price, P. B., Bonetti, R., Guglielmetti, A., Chiesa, C., Matheoud, R., Migliorino, C., Moody, K. J.: Phys. Rev. **C 46**, 1939 (1992).
36. Ardisson, G., Hussonnois, M., Ledu, J. F., Brillard, L.: *Proceedings of the International School Seminar on Heavy Ion Physics*, Dubna, 336 (1989).
37. Ardisson, G., Hussonnois, M.: Compt. Rend. Acad. Sci. (Paris), **310**, 367 (1990).
38. Moody, K. J., Wang Shicheng, Hulet, E. K., Price, P. B., Barwick, S. W.: Phys. Rev. **C 36**, R2710 (1987).
39. Barwick, S. W.: Ph. D. Thesis, University of California, Berkeley (1986).
40. Tretiakova, S. P., Sandulescu, A., Micheev, V. L.: J.I.N.R. Rapid Commun. **13**, 34 (1985).
41. Wang Shicheng, Price, P. B., Barwick, S. W., Moody, K. J., Hulet, E. K.: Phys. Rev. **C 36**, 2717 (1987).
42. Wang Shicheng, Snowden-Ifft, D., Price, P. B., Moody, K. J., Hulet, E. K.: Phys. Rev. **C 39**, R1647 (1989).
43. Oglobin, A. A., Venikov, N. I., Lisin, S. K., Pirozhkov, S. V., Pchelin, V. A., Rodionov, Yu. F., Semochkin, V. M., Shabrov, V. A., Shvetsov, I. K., Shubko, V. M., Tretiakova, S. P., Mikheev, V. L.: Phys. Lett. **B 235**, 35 (1990).
44. Hussonnois, M., Bonetti, R., Guglielmetti, A., Tretiakova, S. P., Micheev, V. L., Golovchenko, A. N., Ponomarenko, V. A.: J.E.T.P. Letters (1995) in press.
45. Ardisson, G., Hussonnois, M., LeDu, J. F., Trubert, D., Lederer, M.: Phys. Rev. **C 49**, 2963 (1994).
46. Greiner, M., Scheid, W.: J. Phys. G: Nucl. Phys. **12**, L229 (1986).
47. Brillard, L., Elayi, A. G., Hourani, E., Hussonnois, M., Ledu, J. F., Rosier, L. H., Stab, L.: Compt. Rend. Acad. Sci. (Paris) **309**, 1105 (1989).
48. Rosenblum, S.: C. R. Acad. Sci. (Paris) **188**, 1401 (1929).
49. Hussonnois, M., Ledu, J. F., Brillard, L., Ardisson, G.: Phys. Rev. **42C**, R 495 (1990).
50. Hussonnois, M., Ledu, J. F., Brillard, L., Ardisson, G.: J. Phys.: Nucl. Phys. **G 16**, L77 (1990).
51. Chasman, R. R.: Phys. Lett. **B 96**, 7 (1980).
52. Bonetti, R., Chiesa, C., Guglielmetti, A., Matheoud, R., Poli, G., Mikheev, V. L., Tretiakova, S. P.: Phys. Rev. **C 51**, 2530 (1995).
53. Hussonnois, M., Ardisson, G.: Z. Phys. **A 349**, 311 (1994).
54. Gupta, R. K., Singh, S., Puri, R. K., Scheid, W.: Phys. Rev. **C 47**, 561 (1993).
55. Poenaru, D. N., Greiner, W., Gherghescu, R.: Phys. Rev. **C 47**, 2030 (1993).
56. Oganessian, Yu. Ts., Lazrev, Yu. A., Mikheev, V. L., Muzychka, Yu. A., Shirokovsky, I. V., Tretyakova, S. P., Utyonkov, V. K.: Z. Phys. **A 349**, 341 (1994).
57. Guglielmetti, A., Bonetti, R., Poli, G., Price, P. B., Westphal, A. H., Janas, Z., Keller, H., Kirchner, R., Klepper, O., Piechaczek, A., Roeckl, E., Schmidt, K., Plochocki, A., Szerypo, J., Blank, B.: Phys. Rev. Abstr. **26**, 54 (1995), Phys. Rev. **C 52**, 740 (1995).
58. Vandenbosch, R., Huizenga, J. R.: *Nuclear Fission,* Academic Press, New York, 46 (1974).
59. de Carvalho, H. G., Martins, J. B., de Souza, I. O., Tavares, O. A. P.: An. Acad. Bras. Cienc. **48**, 205 (1975).
60. Tretiakova, S. P., Zamyatnin, Yu. S., Kovantsev, N. N., Korotkin, Yu. S., Micheev, V. L., Timofeev, G. A.: Z. Phys. **A 333**, 349 (1989).

Radiochimica Acta 70/71, 135–145 (1995)

Spontaneous Fission

By Darleane C. Hoffman and Michael R. Lane

Chemistry Department, University of California, Berkeley and Nuclear Science Division, MS-70A/3307, Lawrence Berkeley Laboratory, Berkeley, CA 94720 USA

(Received July 3, 1995)

Spontaneous fission (SF) / SF properties: half-life systematics, mass division, kinetic energies, total kinetic energy, neutron and photon emission

Abstract

Recent experimental results for spontaneous fission half-lives and fission fragment mass and kinetic-energy distributions and other properties of the fragments are reviewed and compared with recent theoretical models. The experimental data lend support to the existence of the predicted deformed shells near $Z=108$ and $N=162$. Prospects for extending detailed studies of spontaneous fission properties to elements beyond hahnium (element 105) are considered.

I. Introduction

It has been 100 years since Henri Becquerel [1] found evidence for the natural decay chains in uranium by showing that uranium and its salts emitted radiation which blackened photographic plates even in complete darkness. This discovery led to extensive studies by Marie and Pierre Curie [2] who in 1898 proposed the name "radioactivity" for this new phenomenon and reported separation and discovery of the first radioactive elements radium and polonium from pitchblende. In addition to the alpha and beta decay exhibited by the ^{235}U and ^{238}U natural decay chains, it is interesting to note that spontaneous fission (SF), a much less common mode of natural radioactive decay, is also present. However, because of its small probability, SF was not discovered until 1940 when Petrzhak and Flerov [3] detected SF in ^{238}U with a partial half-life of some 10^{16} years, or only about one SF in 2×10^6 alpha decays. So far, SF has not been detected in any elements lighter than uranium (except for ^{8}Be which decays into two alpha particles), and upper limits for the probability of SF of 3.8×10^{-14} for ^{240}Th ($Z=90$) and 1.6×10^{-13} for ^{231}Pa ($Z=91$) have been set. Detailed studies of SF properties were not performed until much later when isotopes of higher Z elements with much shorter half-lives were synthesized. In the 1960's, sources of ^{252}Cf became available and measurements of the mass, charge, kinetic energies, neutron and photon emission of its fragments contributed much to our understanding of SF. Numerous reviews [4–11] of SF and low energy fission have been published and should be consulted for more detailed information.

II. Half-lives

In a recent review [11], we tabulated all of the SF half-lives reported as of mid-1992. The additional SF half-lives reported since then up until mid-1995 are given in Table 1 [Refs. 12–23]. The half-lives for e-e (even Z-even N) nuclides are plotted in Fig. 1, and those for e-o (even Z-odd N), o-e (odd Z-even N) and o-o (odd Z-odd N) nuclides are plotted in Fig. 2. In the case of the o-o nuclides, essentially no data except lower limits for SF half-lives have been reported and these lower limits are not included in the plot. It is extremely difficult to obtain measurements for the o-o isotopes because often they decay by electron-capture to e-e nuclides which decay via SF with very short half-lives; unless SFs in coincidence with the characteristic X-rays of the parent are measured, it is nearly impossible to tell which isotope is spontaneously fissioning. In order to assess the hindrances associated with the odd nucleons, the logarithm of the experimentally determined SF hindrance factor (HF) is often plotted as a function of the odd-neutron and odd-proton numbers as shown in Fig. 3. The HF is calculated relative to the geometric mean of the SF half-lives of the two adjacent e-e neighbors [11]; if the half-life of only one e-e neighbor is known, it is used in the calculation. It can be seen that the HFs are about 10^5 for either an odd proton or an odd neutron where actual measurements and not just limits are known. The high-spin, 9/2+[615], 157th neutron seems to be especially effective in lending extra stability to elements 100, 102, and 104.

As can be seen in Fig. 1, it appears that the strong effect of the deformed $N=152$ neutron subshell on SF half-lives has disappeared by $Z=104$. However, recent reports [21, 24] of longer than expected alpha and SF half-lives in the region of elements greater than or equal to 104 and near $N=162$ have been interpreted as a result of the predicted deformed shells [25–31] in the region of $N=162$ and $Z=108$. For example, the e-e isotope, ^{266}Sg ($Z=106$), was reported [21, 24] to decay primarily by alpha emission with an estimated half-life of 10–30 s, rather than by SF with a few millisecond half-life as might otherwise have been expected. The known SF half-lives for elements 104 and higher are plotted in Fig. 4 together with the recent predictions [25, 26] for these isotopes. As seen in Fig. 5, the predicted SF half-lives increase as a function of neutron number up to the deformed subshell at

Table 1. Recent measurements of half-lives, SF partial half-lives, and total kinetic energies.

Nuclide[a]	Half-life	$T^{SF}_{1/2}$ or %SF	TKE (MeV)[b]	Reference
^{238}Pu	87.7 ± 0.01 yrs	$(4.75\pm0.09)\times10^{9}$ yrs	177.0 ± 0.5	11
			177.0 ± 0.3	12
^{240}Pu	$(6.56\pm0.01)\times10^{3}$ yrs	$(1.14\pm0.01)\times10^{11}$ yrs	179.4 ± 0.5	11
			179.4 ± 0.1	12
^{242}Pu	$(3.75\pm0.02)\times10^{5}$ yrs	$(6.77\pm0.07)\times10^{10}$yrs	180.7 ± 0.5	11
			180.7 ± 0.1	12
^{248}Cm	$(3.48\pm0.06)\times10^{5}$ yrs	$(4.15\pm0.03)\times10^{6}$ yrs	182.2 ± 0.9	11
			182.0	13
^{237}Cf	2.1 ± 0.3 sec	~10%		14
^{238}Cf	If $T_{1/2}\approx1$ sec, then	>4 sec		11
	21 msec	~100%		14
^{240}Cf	0.9 ± 0.2 min	~2%		14
^{242}Cf	3.4 ± 0.2 min	≤0.014%		14
^{258}Md^{m}	57.0 ± 0.9 min	≤30%		15
^{252}No	$2.25^{+0.18}_{-0.16}$ sec	26.9%	202.4, 194.3*	11
	2.44 ± 0.12 sec		194.3*	16
		(21.6 ± 4.2)%		17
^{254}No	55 ± 5 sec	0.17%, 0.25%		11
	53 ± 20 sec	(0.17 ± 0.02)%	189.2*	16
^{256}Rf	6.7 ± 0.2 msec	$6.9^{+0.6}_{-0.2}$ msec	207 ± 13	11
	6.6 ± 1.1 msec		197.6 ± 1.1*	16
^{258}Rf	13 ± 2 msec	13 ± 2 msec $\leq T^{SF}_{1/2}\leq 15\pm2$ msec	220 ± 15	11
	14 ± 2 msec		198.9 ± 4.4*	16
^{262}Rf	47 ± 5 msec	47 ± 5 msec		11
	2.1 ± 0.2 sec	2.1 ± 0.2 sec	$215\pm2^{\#}$	18
^{263}Sg	0.9 ± 0.2 sec	1.3 sec??		11
	$1.1^{+2.8}_{-0.6}$ sec			19
		<30%		20
^{265}Sg	2−30 sec[c]	≤50%		21
^{266}Sg	10−30 sec[c]	≤50%		21
^{264}Ns	440^{+600}_{-160} msec	ND[d]		22
^{267}Hs	19^{+29}_{-10} msec	≤20%		20
^{268}Mt	70^{+100}_{-30} msec	ND[d]		22
^{269}110	270^{+1300}_{-120} μsec	ND[d]		23
^{272}111	$1.5^{+2.0}_{-0.5}$ msec	ND[d]		22

[a] This table lists half-lives, spontaneous fission branches, and TKE's measured since the review by Hoffman, Hamilton, and Lane [11]. For comparison, previous values (if available) from that review are listed along with the newly determined values.
[b] Average values of the pre-neutron emission TKE's except for those denoted by (*), which are average values of the TKE's based on a provisional mass analysis, and by (#), which is the most probable pre-neutron emission TKE.
[c] Calculated half-life based on observed alpha-decay energy and systematics.
[d] No spontaneous fission events were observed.

N = 162 followed by a decrease to about $N=170$ and then another increase in half-life. Fig. 6 shows the log of the SF half-lives together with the alpha half-lives for the e-e isotopes of elements 104 through 114. For element 104 the longest predicted SF half-life of a few seconds occurs at $N=162$ which is still several orders of magnitude shorter than the predicted alpha half-life. At heavier neutron numbers, the SF half-life becomes still smaller, and thus SF will dominate and determine the half-lives of these nuclides. At element 106, SF will not dominate until $N>164$, and at element 108, SF will not dominate until $N>166$. For $Z=110$, 112, and 114, the alpha half-lives are microseconds to milliseconds until about $N=172$ when they reach about a second and are increasing. Since the SF and alpha half-lives are comparable it seems likely that these isotopes will have appreciable SF branches and half-lives long enough to permit study if they can be produced in reasonable yields. However, SF half-lives of seconds or more may be expected for many odd isotopes of these elements due to HF's of 10^3 to 10^5 (Fig. 3) and should allow study of still heavier isotopes.

The possibility of SF in isomeric states of actinide nuclei has been discussed by Baran and Lojewski [32]. They calculated SF half-lives of K-isomeric states on the basis of the microscopic-macroscopic method for an isomeric state assumed to be a 2-quasiparticle excited state with high angular momentum. They performed calculations for e-e nuclei with $96<Z<110$ and $144<N<158$ and found that the SF half-life may be comparable to that of the SF half-life of the ground state. They found for $Z\geq104$ that the SF half-lives of the isomers and ground states may be comparable, provided the spin of the K-isomer is sufficient to prevent it from decaying by other processes. The recent assignment [33, 34] of a 2.1-s SF activity to ^{262}Rf which already has an ≈50 ms SF activity assigned to it [5] could be a case in point. Reasonable neutron and proton single-particle assignments for this nuclide could give a two-quasiparticle K-isomeric state with

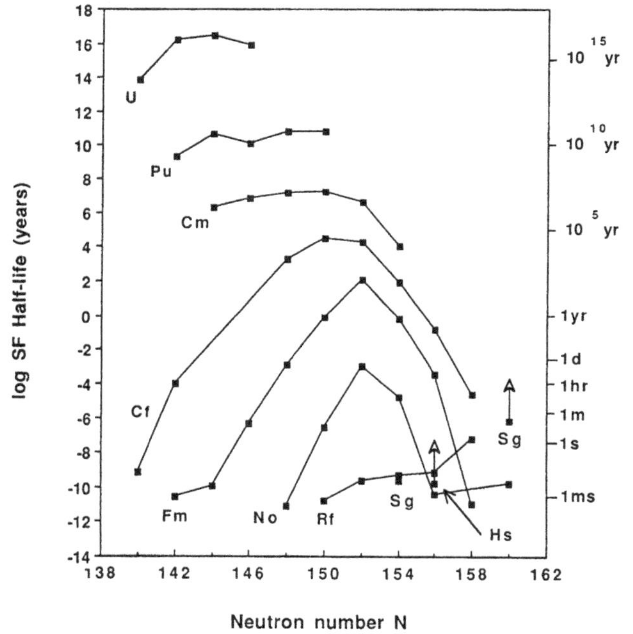

Fig. 1. Logarithms of SF half-lives of e-e nuclei plotted *vs.* neutron number. Arrows are used to indicate lower limits. [Data from Ref. [11] and Table 1.]

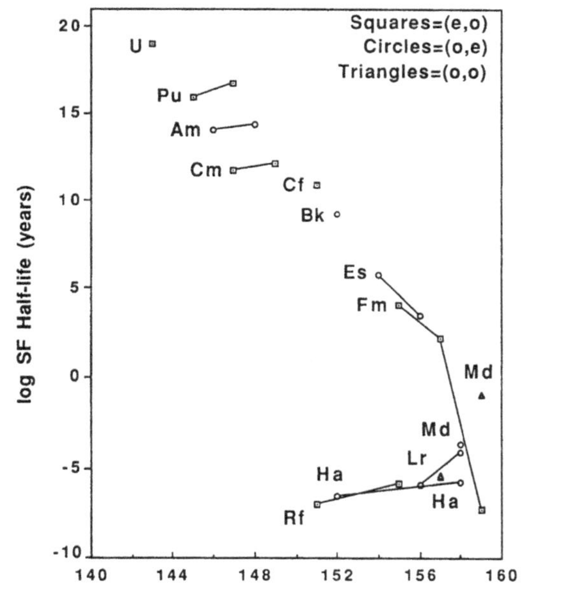

Fig. 2. Logarithms of SF half-lives of e-o, o-e, and o-o nuclei plotted *vs.* neutron number. (Lower limit values are not included.) [Data from Ref. [11].]

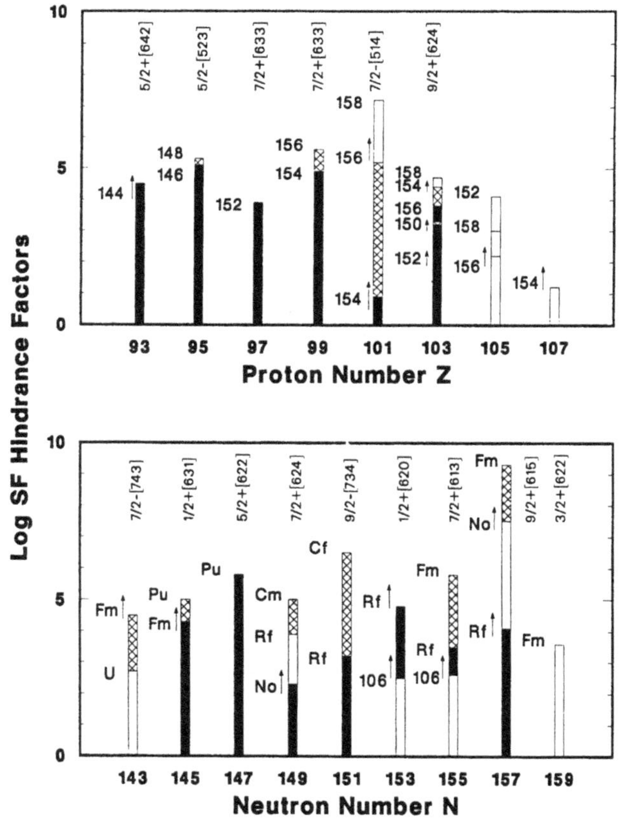

Fig. 3. Logarithms of SF hindrance factors (HF) for odd-neutron and odd-proton nuclides. Lower limit values are indicated by arrows. An open bar indicates that the HF was calculated relative to only one e-e neighbor. A filled or hacked bar indicates that the HF was calculated relative to two e-e neighbors. [From Ref. [11].]

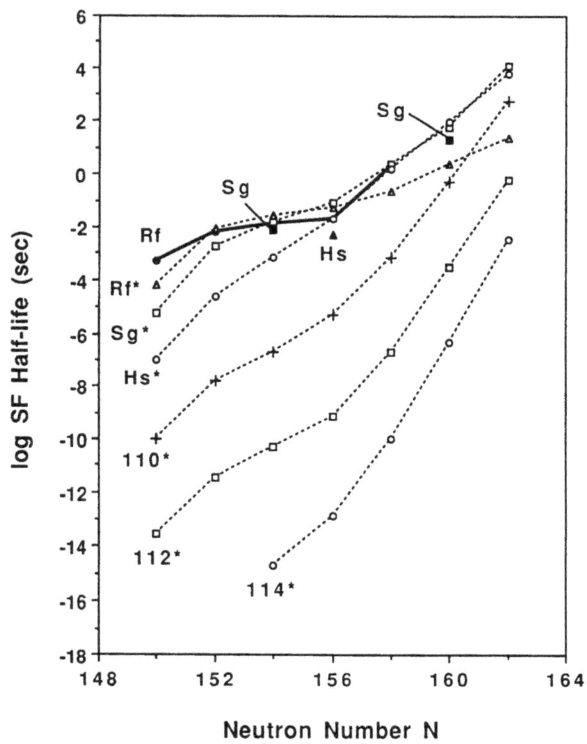

Fig. 4. Logarithms of experimental and theoretical SF half-lives for even-Z elements 104 through 114 as a function of even neutron number 150 through 162. [Data from Refs. [11] and [26] and Table 1.]

9+ or 10−. It is important to investigate whether or not this phenomenon can occur as there is the possibility that measured ground-state half-lives could actually be mixtures of the two states and could affect the interpretation of shell effects in these heavy nuclides [34].

III. Properties of the fission fragments

Spontaneous fission is especially sensitive to shell effects in both the fissioning system and the fragments

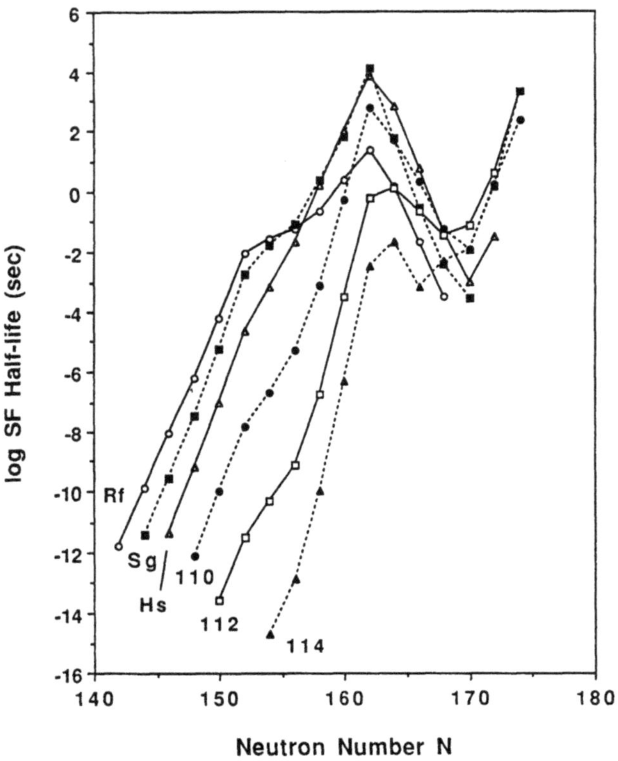

Fig. 5. Logarithms of the theoretical SF half-lives for even-Z elements 104 through 114 as a function of even neutron number 142 through 174. [Data from Ref. [26].]

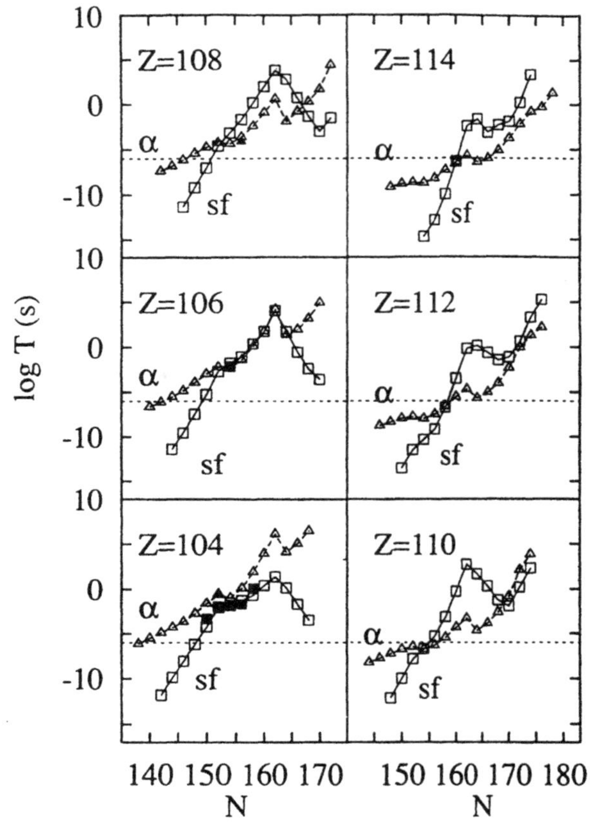

Fig. 6. Dependence of logarithm of the calculated SF half-lives, given in seconds, on the neutron number N, for elements 104–114. The α-decay half-lives are also shown, for comparison. Experimental values are given by full points. The horizontal dashed line indicates about the lowest half-life (1 μs) of a nucleus, which can be detected in a present-day set-up, after its synthesis. [From Ref. [26].]

because no additional energy is put into the nucleus before it fissions, as is the case even for thermal neutron-induced fission where the excitation energy due to the neutron capture is typically 5 or 6 MeV. This is particularly noteworthy in the SF of the heavy fermium isotopes where a change of only one or two neutrons results in a dramatic change from asymmetric mass division with "normal" total kinetic energy (TKE) to narrowly symmetric mass division with very high TKE approaching the Q-value for fission. Detailed studies of the mass, charge, and kinetic-energy distributions of the fragments at scission, as well as prompt neutron and photon emission from the excited fragments, have been performed in order to help understand the SF process and aid in developing and testing predictive theoretical models.

Fragment mass, kinetic-energy, and charge distributions

Early measurements of fragment mass-yield distributions from both SF and neutron-induced fission obtained by radiochemical and mass spectrometric methods were summarized by von Gunten [4] in 1969. Although such measurements have perfect Z and A resolution, they suffer from the fact that the distributions of the fragments after neutron emission and subsequent de-excitation by beta decay are measured and the distributions at scission must be inferred after correction for these processes. In addition, time-of-flight and kinetic-energy measurements have been performed for SF of ^{252}Cf and some lighter nuclides, but for the shorter-lived trans-Cf isotopes, most of the information about kinetic-energy and mass-yield distributions has been derived from measurements of coincident fission fragments with solid-state (SS) detectors. Again, distributions after prompt neutron emission are obtained, the mass of the fissioning nucleus must be assumed, and corrections must be applied for prompt neutron emission unless neutrons in coincidence with the fragments can be measured. Furthermore, the mass resolution of the SS detectors is not as good as with the other methods, but determination of major features and the most probable values of the mass and kinetic-energy distributions can be made. On-line systems using rotating wheels to collect short-lived SF nuclides on thin foils and move them between pairs of SS detectors for measurement have been devised [35–37] and have furnished much information on mass and kinetic-energy distributions. Other methods such as gas-filled detection systems [38] have been proposed, but are extremely difficult to implement for short-lived nuclides with low production yields which must be studied in nearly "on-line" experiments at the accelerators where they are produced.

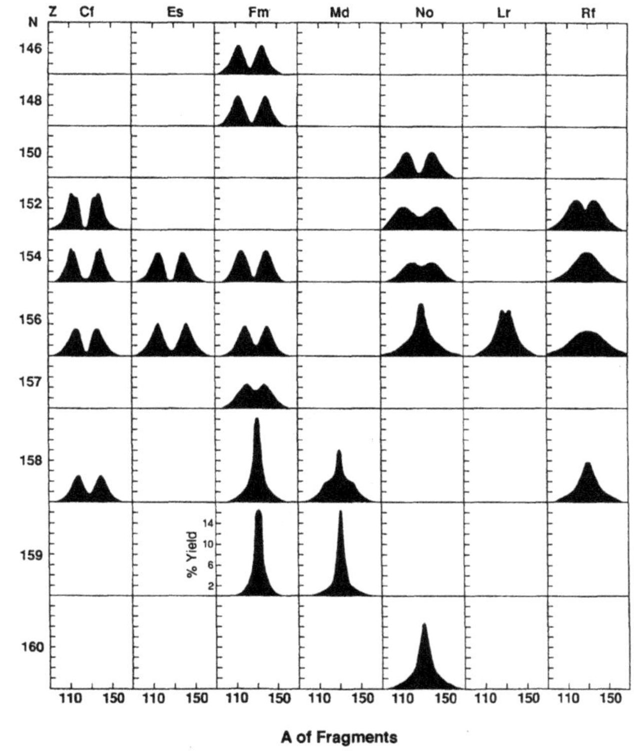

Fig. 7. Schematic representation of all known mass-yield distributions (normalized to 200% fission fragment yield) for SF of trans-Bk isotopes. [Data from Ref. [11] and Table 1.]

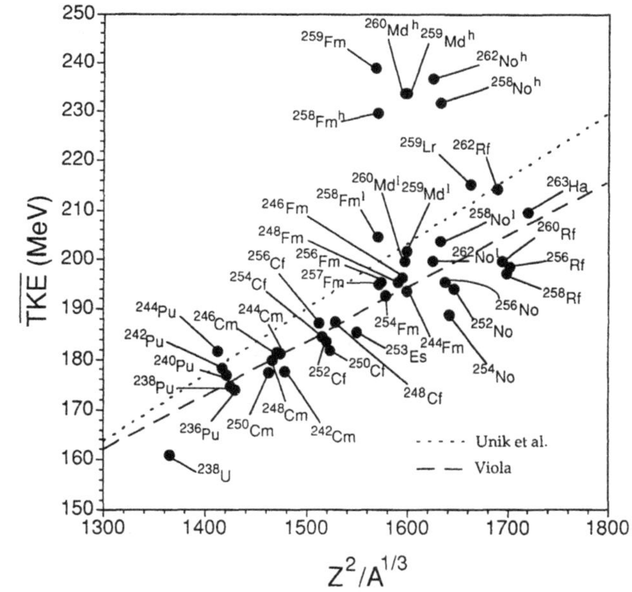

Fig. 8. Average or most probable TKE *vs.* $Z^2/A^{1/3}$. The solid line is the linear fit of Viola *et al.* [44] and the dashed line is the linear fit of Unik *et al.* [45]. Data are from Ref. [11] and Table 1 and have been corrected to the new Weissenberger parameters [46] as discussed in Ref. [11].

All of the mass distributions measured for low-energy and SF fission were found to be highly asymmetric until 1971, when much enhanced yields of near symmetric fission with high TKE were reported [39] from SF of ^{257}Fm. It was subsequently found [40, 41] that ^{258}Fm and ^{259}Fm exhibited narrowly symmetric mass distributions and very high TKEs. Since then measurements of many more trans-Bk isotopes have been made and there are a number of recent reviews [8–10] which concentrate on the SF properties of these heavy isotopes. Schematic representations of all of the measured mass distributions (normalized to 200% fragment yield) for SF of the trans-Bk isotopes are shown in Fig. 7. Recent results reported for ^{252,254}No, ^{256,258}Rf [42, 43] and for ^{262}Rf [33, 34] are included. It is interesting to compare the progression from asymmetric to symmetric mass distribution with increasing neutron number for Fm ($Z = 100$), No ($Z = 102$) and Rf ($Z = 104$) isotopes. The change from asymmetric to symmetric in No between $N = 154$ and 156 is not as abrupt as for Fm between $N = 157$ and 158, nor does the mass distribution become as narrowly symmetric even at $N = 160$ as it is for Fm at $N = 159$. Perhaps this is because the symmetric fragments for No cannot both have the $Z = 50$ closed shell configuration. It would be extremely interesting to measure the distributions for ^{260}No and ^{261}No, the $N = 158$ and 159 isotopes. The data for Rf are not as complete, but Hulet [42] postulated on the basis of the data for the $N = 152$, 154, and 156 isotopes of Rf that due to disappearance of the second barrier to fission these isotopes exhibit "liquid-drop model" (LDM) type fission with broadly symmetric mass distributions and TKE distributions with only one component, close to empirical TKE fits based on this model. (An updated plot of the most probable or average TKEs for SF is shown in Fig. 8 together with the empirical linear fits of Viola *et al.* [44] and Unik *et al.* [45].) However, the recent measurements of ^{262}Rf ($N = 158$) show a rather narrow, symmetric mass peak with "wings", not unlike the mass distribution observed for the transition nucleus ^{259}Md which also has 158 neutrons. Both of these nuclides have most probable TKEs which are rather high compared to the fit of Viola *et al.* shown in Fig. 8. In order to ascertain whether or not the properties of the Rf isotopes are being determined by disappearance of the second fission barrier, it is especially important to make measurements for ^{263}Rf to see if its mass distribution becomes still more narrowly symmetric or exhibits the broad distribution characteristic of LDM type fission and whether its TKE becomes still higher or is consistent with LDM fission.

Very little information is available for the odd-proton nuclides and additional measurements for Lr ($Z = 103$) would be most helpful. Although ^{261}Lr (39 m) and ^{262}Lr (216 m) are known [47, 48] and can be produced by transfer reactions between ^{254}Es and heavy ions such as ^{18}O and ^{22}Ne, their fission properties have not been measured because ^{262}Lr decays primarily via electron-capture (EC) to 5-ms ^{262}No which spontaneously fissions and masks the SF properties of both ^{261}Lr and ^{262}Lr. An upper limit of 10% was estimated [47] for the SF branch of ^{262}Lr.

The measured TKE distributions for some trans-Es isotopes are shown schematically on a semilog plot

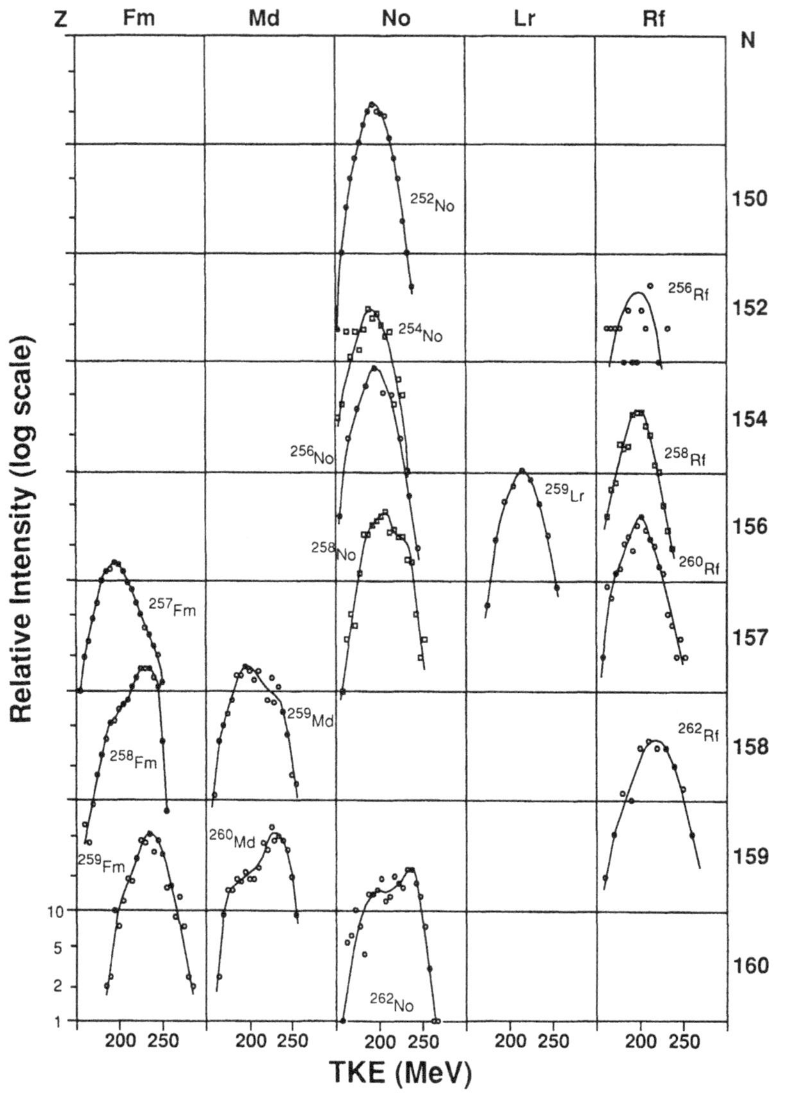

Fig. 9. TKE distributions for SF of some trans-Es isotopes. [Data from Ref. [11] and Table 1.]

in Fig. 9 so their shapes can be compared directly. It shows that many of the distributions, e.g., those for ^{257,258}Fm, ^{259,260}Md, and ^{258,262}No cannot be easily fit with a single Gaussian. Hulet and others [49, 50] have fit the distributions for ^{258}Fm, ^{259,260}Md, and ^{258,262}No with two Gaussians, centered around 235 MeV and the other around 200 MeV. They have called this "bimodal symmetric" fission, the higher TKE symmetric mode being associated with near spherical, symmetric fragments and the lower TKE mode with LDM type fission in which the second fission barrier has disappeared, resulting in broadly symmetric mass distributions and lower TKEs. However, several "transition" nuclei such as ^{256}No, ^{259}Lr, as well as ^{262}Rf show more or less symmetric TKE distributions with no evidence for more than one component and their mass distributions range from symmetric and asymmetric in ^{256}No to broadly symmetric in ^{259}Lr to narrowly symmetric with asymmetric "wings" in ^{262}Rf. The most probable TKEs and FWHMs are 196 MeV and 42 MeV, 215 MeV and 40 MeV, and 215 MeV and 43 MeV, respectively for ^{256}No, ^{259}Lr, and ^{262}Rf. These features can be seen more clearly in the contour plots shown in Fig. 10. The fragments from these transition nuclides may show not just "bimodal" fission consisting of two compact (spherical) or two deformed fragments, but combinations [10] consisting not only of two compact or two deformed fragments, but combinations including one compact fragment and one deformed fragment (with different deformations) as discussed in the scission-point model of Wilkins *et al.* [51] and Lee *et al.* [52]. Thus, because of the extremely large variances of the TKE values for symmetric mass division, we should perhaps speak of "multimodal" [7], rather than bimodal fission and a variety of combinations of different shapes, depending on the shell structure in the fissioning systems, could be involved.

In his recent review [25] of theoretical studies of ground-state properties of the heaviest nuclei, Sobiczewski has shown contour plots of the potential energy of ^{258}Fm as a function of β_2 and β_4 deformations. In the case where the energy is minimized in β_3, β_5, and β_6 degrees of freedom and the fission trajectory is cal-

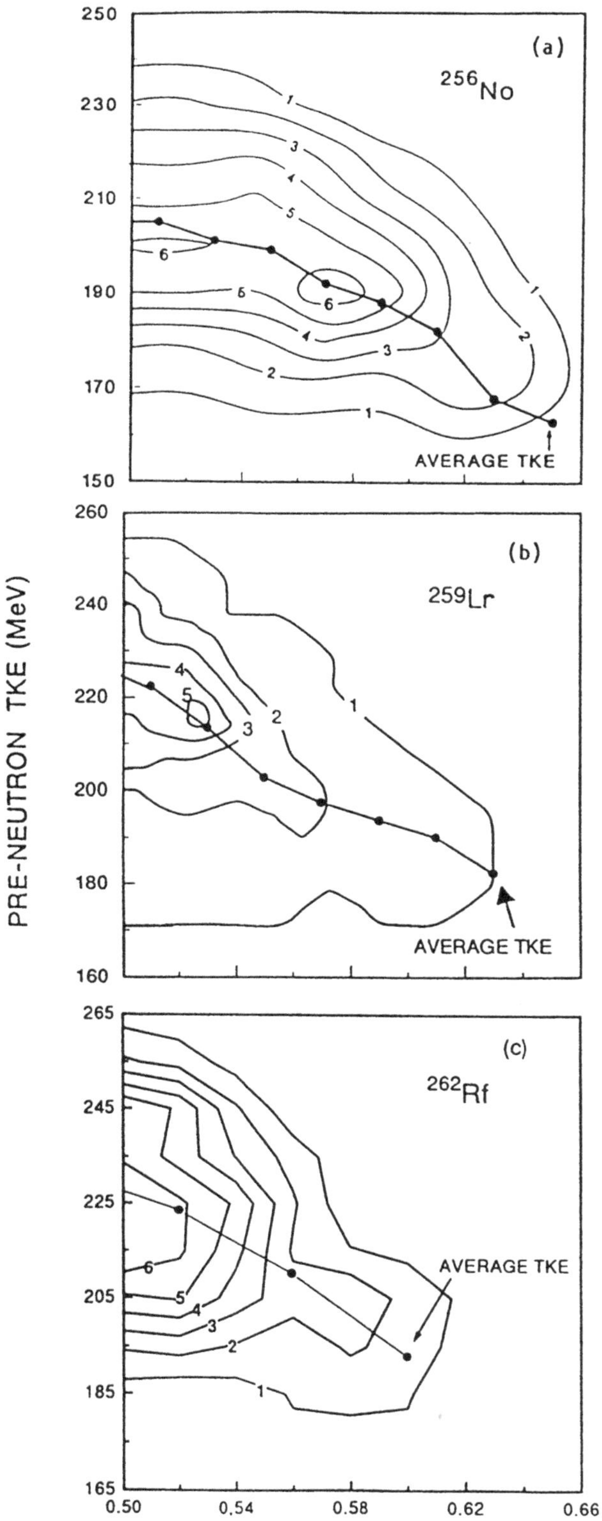

Fig. 10. Contour plots of pre-neutron-emission TKE *vs.* mass fraction. The connected points represent average TKE as a function of mass fraction. a) ^{256}No (346 SFs). The contours indicate equal numbers of events based on data groupings 20 MeV × 0.04 units of mass fraction. Contours labeled 1 through 6 represent 10 through 60 events, respectively. b) ^{259}Lr (442 SFs). The contours indicate equal numbers of events based on data groupings of 10 MeV × 0.02 units of mass fraction. Contours labeled 1 through 5 represent 10 through 50 events, respectively. c) ^{262}Rf (200 SFs). The contours indicate equal numbers of events based on data groupings of 10 MeV × 0.02 units of mass fraction. Contours labeled 1 through 5 represent 4 through 24 events, respectively. [From Refs. [11] and [18].]

culated statically, two paths of the same half-life result, one giving rise to compact, reflection symmetric shapes while the other leads to much more elongated shapes which are not necessarily reflection symmetric. This is consistent with the observation [49] of both high and low TKE fission in ^{258}Fm with nearly equal intensities. If only reflection symmetric shapes are included in the calculation, then the fission proceeds only to the compact shape valley. It would appear that a similar calculation, which takes into account reflection asymmetric shapes and also includes dynamical effects in the fission process following the barrier, might give paths resulting in the variety of asymmetric and symmetric fragments with different deformations indicated by the experimental results for the transition nuclei ^{256}No, ^{259}Lr, and ^{262}Rf.

The effect of neutron shells and fission channels in the SF of the even-even Pu isotopes 236, 238, 240, 242, and 244 have been carefully investigated by Wagemans *et al.* [53, 54]. They have found rapidly varying fission fragment mass and kinetic-energy distributions with the change of only a few neutrons. These were initially interpreted in the frame of the static scission-point model [51] as due to the changing relative importance of the $N = 82$ spherical fragment and the $N = 87$ deformed fragment shells and their combination with the $Z = 50$ spherical shell. More recently, these results have been interpreted in terms of the fission channel model of Brosa and coworkers [55, 56] and the relative fragment yields have been correlated in detail with this multimodal, random neck rupture model. In this model the potential-energy surface as a function of the deformation is calculated for the fissioning nucleus from ground state to scission. The pre-scission configuration is allowed to rupture in a random manner according to Rayleigh instabilities. The potential energy of the nucleus is calculated as a function of its nuclear deformation which is parameterized in terms of the half-length of the pre-scission shape, its neck radius and the position at which the neck ruptures. This gives rise to a number of different fission barriers which correspond to certain pathways within the calculated potential energy surface. These different pathways or channels give rise to different fragment mass and energy distributions, etc., and six different fission channels (three asymmetric channels called Standard I, II, and III, superasymmetric, superlong, and supershort) have been predicted for ^{252}Cf.

Essentially all of the information concerning charge division in SF is based on measurements of ^{252}Cf. Wahl [57] has comprehensively evaluated the data for SF of ^{252}Cf and thermal neutron-induced fission of ^{233}U, ^{235}U, and ^{239}Pu, and derived parameters for empirical models which describe charge dispersion for constant mass number and mass number dispersion for constant atomic number. He also notes the preference for formation of fragments with $Z = 50$ due to the effect of the 50-proton shell in the fragments, but also points out that the maximum fragment yields occur at

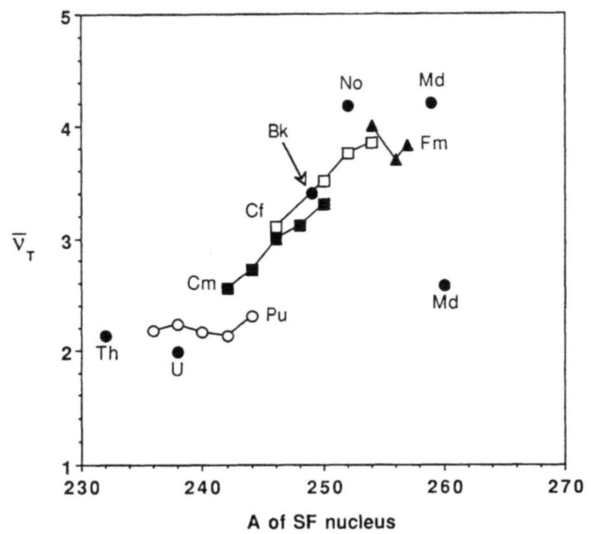

Fig. 11. Average total neutron emission per fission, $\overline{\nu}_T$, as a function of A of the spontaneously fissioning nucleus. [From Ref. [11], except for the addition of the value for ^{259}Md [58].]

higher Z-values of 52, 54, 56, and at the 82-neutron shell and above.

Prompt neutron and gamma-ray emission

In order to determine the pre-neutron emission fragment yields and kinetic energies, information concerning prompt neutron emission as a function of fragment mass is required. Unfortunately, for most spontaneously fissioning nuclides only the average number of prompt neutrons per fission, $\overline{\nu}_T$, and in many cases, not even that has been measured. The values measured for $\overline{\nu}_T$ for SF are plotted in Fig. 11 as a function of the mass number of the fissioning nuclide. It can be seen that, in general, the number of neutrons emitted increases with Z and A, and for trans-Pu nuclides, it increases with mass for a given Z. This trend is reversed in the Fm isotopes where the average neutron emission is actually lower for ^{256}Fm and ^{257}Fm than for ^{254}Fm. This is due to the increased yield of symmetric mass division with high TKE which means there is less energy left for prompt neutron and photon emission from the fragments. Early measurements [59] of prompt neutrons in coincidence with fission fragments were performed using a large Gd-loaded liquid scintillation detector and solid-state detectors to study neutron emission, multiplicities, and variances as a function of fragment mass ratio and TKE for several Cf and Fm isotopes. These studies showed that the neutron multiplicities vary greatly for these isotopes as a function of mass split and TKE. For example, for ^{252}Cf the average neutron emission for the most symmetric mass splits was found to range from 2.1 for the highest TKE (210−220 MeV) events to 5.9 for the lowest TKE (150−170) events; for the most asymmetric events, the range was from 1.7 for TKE from 190 to 210 MeV to 4.8 for TKE from 150 to 170 MeV, although the overall $\overline{\nu}_T$ was measured to be 3.735 with a variance of 1.55. For the most symmetric events from ^{257}Fm, $\overline{\nu}_T$ ranged from only 1.1 for TKE > 240 MeV (close to the Q-value for fission) to 4.9 for TKE from 160 to 180 while the average neutron emission over all mass splits and TKEs was 3.8 with a variance of 2.5. These data could probably now be interpreted on the basis of Brosa's multimodal model.

In the case of ^{259}Fm and ^{260}Md, whose most probable TKEs approach the Q-values for fission, it would be expected that prompt neutron emission from the symmetric, near-spherical fragments must also be very low. This was experimentally verified for ^{260}Md, which has a much more abundant high energy component than does ^{257}Fm, by Wild *et al.* [60] who measured a value of 2.58 for the average neutron emission from the fragments.

Recently, van Aarle *et al.* [61] performed similar measurements for ^{252}Cf of the neutrons emitted from the fragments as a function of fragment mass and kinetic energy. From an investigation of the correlations between the neutron multiplicity and the TKE of the fission event, they derived SF parameters in order to search for the six different fission channels predicted by Brosa *et al.* [56] and found evidence for all channels, although the intensity of the superasymmetric mode was only 0.3%. They also derived schematic pre-scission configurations for these various modes, including the number of neutrons to be emitted from the light and heavy fragments.

Because information about neutron emission as a function of fragment mass is not generally available for SF, various methods for use in correcting radiochemical data and SS measurements to pre-neutron emission values have been devised and fuller discussions of neutron emission and energy spectra are given in Refs. [6] and [10].

The excitation energy of the fragments can also be dissipated by gamma-ray emission. This phenomenon has been even less well investigated than neutron emission although it can give information concerning the deformation of the fragments and the configuration at scission. Most of the studies have been of ^{252}Cf, but there have been some studies of ^{238}U, ^{240}Pu, and ^{244}Cm; these are discussed in recent reviews. Sokol *et al.* [62] have reported one of the few studies of the trans-Cf isotopes and measured energies, intensities, total gamma-ray energy and average number of gamma-rays per fission. They found the number of photons per fission for ^{254}Cf and ^{259}Md to be about 5.3, somewhat smaller than the 6.5 to 7 measured for ^{248}Cm and ^{252,254}Cf. These data are exceedingly difficult to obtain but the information helps in inferring details about the shape of the system at scission and in deducing the deformation and excitation of the fragments and subsequent neutron emission.

IV. Future

Due to the stabilizing effect of the $Z=108$ and $N=162$ deformed shells, the SF and α half-lives of the

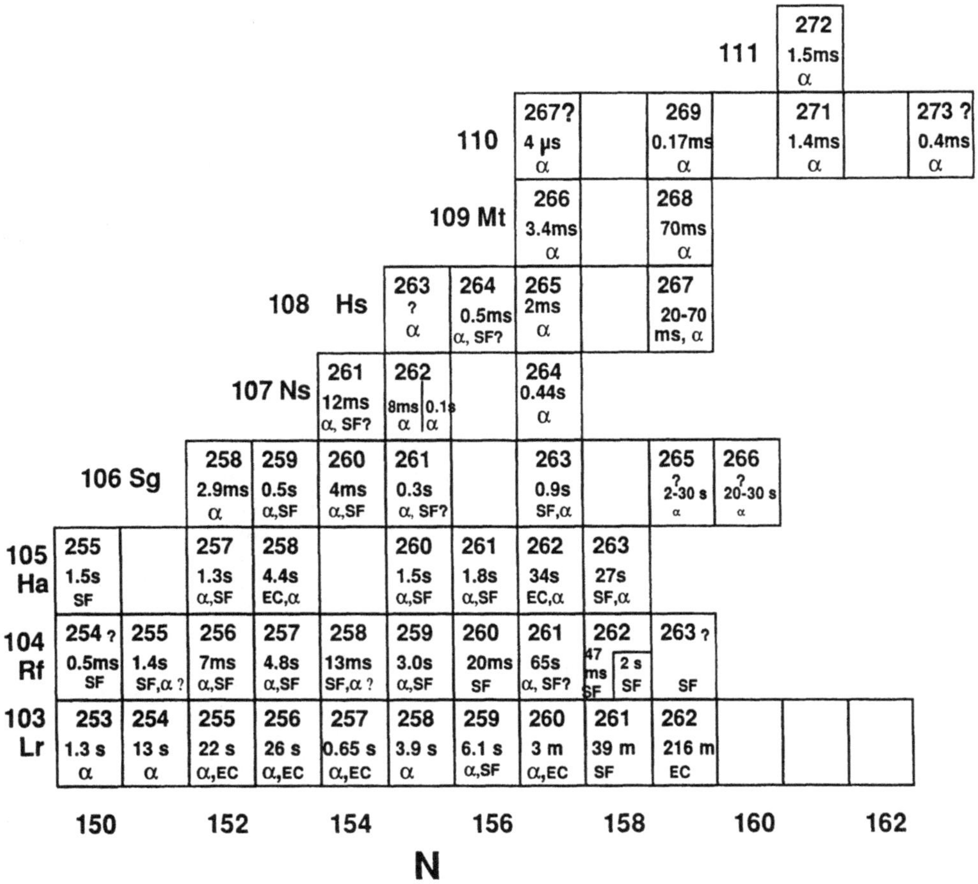

Fig. 12. Chart of the trans-nobelium isotopes.

e-e isotopes with $Z \geq 104$ are not decreasing as rapidly as previously anticipated (see Figs. 4–6). The recent discoveries of longer than expected half-lives for the e-e isotopes ^{262}Rf (2.1 s SF) and ^{266}Sg (20–110 s α) and isotopes of elements 107 through 111 [63] (see Fig. 12) which decay predominantly by alpha-emission support the results of these calculations [26]. It appears that SF should compete with alpha-decay in the e-e isotopes of $Z=108$ with N around 166 to 168, of $Z=110$ and 112 with $N \geq 170$ with half-lives in the range of 0.1 to 10 seconds. For $Z=114$, alpha decay appears to dominate for isotopes in this half-life range. The o-o isotopes would be expected to have still longer SF half-lives because of the odd particle hindrances (Fig. 3) discussed earlier; therefore, alpha-decay should predominate in these isotopes as well so the SF branches may be too small for study. One question to be answered will obviously be whether these odd-particle hindrances will be as large as previously observed for the lighter isotopes. Another question to be investigated is the possible existence of SF isomers and how to verify this experimentally.

Much progress has now been made in developing theoretical models with capabilities for predicting half-lives and scission configurations as well as properties of the fission fragments. Now, many challenges remain for the experimentalists seeking to obtain more information about SF in these new, still higher Z isotopes. The first will be to produce a sufficient number of atoms for study. Multiple-target systems utilizing the most neutron-rich actinides which can be made available, such as ^{244}Pu, ^{250}Cm, ^{249}Bk and ^{254}Es, together with appropriate high-intensity, neutron-rich heavy ion beams should greatly help in gaining access to isotopes with half-lives of seconds or longer. Another challenge will be to devise efficient, high-resolution techniques for positively assigning the atomic number and mass of nuclides which cannot be linked by alpha-decay genetics to known daughter or even grand-daughter isotopes. In some cases, such as for Sg (106), Ns (107), and Mt (109), chemical separations may be feasible [64] and experiments to perform chemical separations of Sg are being conducted by an international collaboration. For isotopes with half-lives much less than a second and production cross sections of picobarns or less, on-line or near on-line instrumentation techniques which can provide Z and A resolution sufficient for positive identification, as well as information about mass division, kinetic-energy distributions, and neutron and photon emission from the fragments need to be developed. Furthermore, these techniques need to be efficient enough to obtain statistically significant results in a finite length of time. However, the exciting possibility of extending our knowledge of the fission process, the influence of nuclear shells, and the limits to nuclear stability at the heaviest end of the chart of the nuclides appears especially promising.

Acknowledgments

We wish to thank Diana M. Lee for help with the graphics and calculations. This work was supported in part by the Director, Office of Energy Research, US Department of Energy under Contract No. DE-AC03-76SF00098.

References

1. Becquerel, H.: Compt. Rend. **122**, 501, 559, 689, 762, 1086 (1896).
2. Curie, P., Sklodowska Curie, M.: "Sur une Substance Nouvelle Radio-active, Contenue dans la Pechblende," Compt. Rend. **127**, 175 (1898).
3. Petrzhak, K. A., Flerov, G. N.: C. R. (Dokl.) Akad. Sci. URSS **28**, 500; J. Phys. USSR **3**, 275 (1940).
4. von Gunten, H. R.: Actinides Rev. **1**, 275 (1969).
5. Hoffman, D. C., Hoffman, M. M.: Annu. Rev. Nucl. Sci. **24**, 151 (1974).
6. Hoffman, D. C., Somerville, L. P.: Spontaneous Fission, in *Charged Particle Emission from Nuclei, Vol. III,* Poenaru, D. N., and Ivascu, M. S., Eds., CRC Press, Inc., Boca Raton, Florida, chap. 1 (1989).
7. Hoffman, D. C.: *50 Years with Nuclear Fission, Vol. I,* The American Nuclear Society, LaGrange Park, IL 83 (1989).
8. Hoffman, D. C.: *Proc. Int. Conf. on Fifty Years Research in Nucl. Fission, Berlin,* Nucl. Phys. **A502**, 21c (1989).
9. Hulet, E. K.: J. Radioanal. Nucl. Chem. Articles **142**, 79 (1990).
10. Wagemans, C.: Spontaneous Fission in *The Nuclear Fission Process*, CRC Press, Inc., Boca Raton, Florida, chap. 3, p. 35 (1991).
11. Hoffman, D. C., Hamilton, T. M., Lane, M. R.: Lawrence Berkeley Laboratory Report LBL-3300 (1992); *Nuclear Decay Modes,* Inst. of Phys. Publ., Bristol, England (In press, 1995).
12. Schillebeeckx, P., Wagemans, C., Deruytter, A. J., Barthelemy, R.: Nucl. Phys. **A545**, 623 (1992).
13. Benoufella, A., Barreau, G., Asghar, M., Audouard, P., Brisard, F., Doan, T. P., Hussonnois, M., Leroux, B., Trochon, J., Moore, M. S.: Nucl. Phys. **A565**, 563 (1993).
14. Lazarev, Yu. A., Shirokovsky, I. V., Utyonkov, V. K., Tretyakova, S. P., Kutner, V. B.: to be published in Nucl. Phys. A.
15. Moody, K. J., Lougheed, R. W., Wild, J. F., Dougan, R. J., Hulet, E. K., Hoff, R. W., Henderson, C. M., Dupzyk, R. J., Hahn, R. L., Summerer, K., O'Kelley, G. D., Bethune, G. R.: Nucl. Phys. **A563**, 21 (1993).
16. Wild, J. F., Hulet, E. K., Lougheed, R. W., Moody, K. J., Bandong, B. B., Dougan, R. J., Veeck, A. C.: J. Alloys Comp. **213/214**, 86 (1994).
17. Andreyev, A. N., Bogdanov, D. D., Chepigin, V. I., Kabachenko, A. P., Malyshev, O. N., Sagaidak, R. N., Salamatin, L. I., Ter-Akop'yan, G. M., Yeremin, A. V.: Z. Phys. **A345**, 389 (1993).
18. Lane, M. R., Gregorich, K. E., Lee, D. M., Mohar, M. F., Kacher, C. D., Sylwester, E. R., Hsu, M., Hoffman, D. C.: to be published.
19. Gregorich, K. E., Lane, M. R., Mohar, M. F., Lee, D. M., Kacher, C. D., Sylwester, E. R., Hoffman, D. C.: Phys. Rev. Lett. **72**, 1423 (1994).
20. Lazarev, Yu. A., Lobanov, Yu. V., Oganessian, Yu. Ts., Tsyganov, Yu. S., Utyonkov, V. K., Abdullin, F. Sh., Iliev, S., Polyakov, A. N., Rigol, J., Shirokovsky, I. V., Subbotin, V. G., Sukhov, A. M., Buklanov, G. V., Gikal, B. N., Kutner, V. B., Mezentsev, A. N., Sedykh, I. M., Vakatov, D. V., Lougheed, R. W., Wild, J. F., Moody, K. J., Hulet, E. K.: JINR Report E7-95-100, March 1995.
21. Lazarev, Yu. A., Lobanov, Yu. V., Oganessian, Yu. Ts., Utyonkov, V. K., Abdullin, F. Sh., Buklanov, G. V., Gikal, B. N., Iliev, S., Mezentsev, A. N., Polyakov, A. N., Sedykh, I. M., Shirokovsky, I. V., Subbotin, V. G., Sukhov, A. M., Tsyganov, Yu. S., Zhuchko, V. E., Lougheed, R. W., Moody, K. J., Wild, J. F., Hulet, E. K., McQuaid, J. H.: Phys. Rev. Lett. **73**, 624 (1994).
22. Hofmann, S., Ninov, V., Heßberger, F. P., Armbruster, P., Folger, H., Münzenberg, G., Schött, H. J., Popeko, A. G., Yeremin, A. V., Andreyev, A. N., Saro, S., Janik, R., Leino, M.: Z. Phys. **A350**, 281 (1995).
23. Hofmann, S., Ninov, V., Heßberger, F. P., Armbruster, P., Folger, H., Münzenberg, G., Schött, H. J., Popeko, A. G., Yeremin, A. V., Andreyev, A. N., Saro, S., Janik, R., Leino, M.: Z. Phys. **A350**, 277 (1995).
24. Lougheed, R. W., Moody, K. J., Wild, J. F., Hulet, E. K., McQuaid, J. H., Lazarev, Yu. A., Lobanov, Yu. V., Oganessian, Yu. Ts., Utyonkov, V. K., Abdullin, F. Sh., Buklanov, G. V., Gikal, B. N., Iliev, S., Mezentsev, A. N., Polyakov, A. N., Sedykh, I. M., Shirokovsky, I. V., Subbotin, V. G., Sukhov, A. M., Tsyganov, Yu. S., Zhuchko, V. E.: *Proc. Conf. Actinides-93, Santa Fe, NM;* J. Alloys and Comp. **213/214**, 61 (1994).
25. Sobiczewski, A.: Phys. Particles and Nuclei **25**, 295 (1994).
26. Smolanczuk, R., Skalski, J., Sobiczewski, A.: "Spontaneous-Fission Half-lives of Deformed Superheavy Nuclei", GSI Preprint 94-77, November 1994.
27. Cwiok, S., Sobiczewski, A.: Z. Phys. **A342**, 203 (1992).
28. Cwiok, S., Rozmej, P., Sobiczewski, A., Patyk, Z.: Nucl. Phys. **A491**, 281 (1989).
29. Patyk, Z., Sobiczewski, A.: Phys. Lett. **B256**, 307 (1991).
30. Patyk, Z., Sobiczewski, A.: Nucl. Phys. **A533**, 132 (1991).
31. Sobiczewski, A., Smolanczuk, R., Skalski, J.: J. Alloys and Comp. **213/214**, 38 (1994).
32. Baran, A., Lojewski, Z.: Nucl. Phys. **A475**, 327 (1987).
33. Hoffman, D. C., Lee, D. M., Gregorich, K. E., Mohar, M. F., Lane, M. R., Hannink, N. J., Kacher, C. D., Neu, M. P., Sylwester, E. R., Hsu, M., Yang, J. C.: "Production of New 2.1-sec SF Activity of ^{262}Rf", Nucl. Sci. Div. 1993 Annual Report, Lawrence Berkeley Laboratory Report LBL-35768, 74 (June 1994).
34. Gregorich, K. E., Hoffman, D. C., Lee, D. M., Mohar, M. F., Lane, M. R., Hannink, N. J., Kacher, C. D., Neu, M. P., Sylwester, E. R., Hsu, M., Yang, J. C.: "The $N=162$ Shell in ^{262}Rf", Nucl. Sci. Div. 1993 Annual Report, Lawrence Berkeley Laboratory Report LBL-35768, 75 (June 1994).
35. Hoffman, D. C., Lee, D., Ghiorso, A., Nurmia, M., Aleklett, K.: Phys. Rev. **C22**, 95 (1985).
36. Hulet, E. K., Wild, J. F., Dougan, R. J., *et al.*: Phys. Rev. Lett. **56**, 313 (1986).
37. Dougan, R. J., Lougheed, R. W., Hulet, E. K., Bethune, G. R.: Nucl. Chem. Div. FY-84 Annual Report, Lawrence Livermore National Laboratory Report UCAR 10062-84/1, 2-15 (1984) (unpublished).
38. Boissevain, J. G., Fowler, M. M., Gavron, A. I., Hoffman, D. C., Lysaght, P., Wilhelmy, J. B.: Isotope and Nuclear Chemistry Division Annual Report FY 1984, Los Alamos National Laboratory Progress Report LA-10366-PR, p. 147.
39. Balagna, J. P., Ford, G. P., Hoffman, D. C., Knight, J. D.: Phys. Rev. Lett. **26**, 145 (1971).
40. Hoffman, D. C., Wilhelmy, J. B., Weber, J., Daniels, W. R., Hulet, E. K., Lougheed, R. W., Landrum, J. H., Wild, J. F., Dupzyk, R. J.: Phys. Rev. **C21**, 972 (1980).
41. Hulet, E. K., Lougheed, R. W., Landrum, J. H., Wild, J. F., Hoffman, D. C., Weber, J., Wilhelmy, J. B.: Phys. Rev. **C21**, 966 (1980).
42. Hulet, E. K.: Phys. Atomic Nuclei **57**, 1099 (1994).
43. Wild, J. F., Hulet, E. K., Lougheed, R. W., Moody, K. J., Bandong, B. B., Dougan, R. J., Veeck, A.: J. Alloys and Comp. **213/214**, 86 (1994).

44. Viola, V. E., Kwiatowski, K., Walker, M.: Phys. Rev. **C31**, 1550 (1985).
45. Unik, J. P., Gindler, J. E., Glendenin, L. E., Flynn, K. F., Gorski, A., Sjoblom, R. K.: in *Proc. Third Internatl. IAEA Symp. on Phys. and Chem. Fission, Rochester, New York, 1973,* **II**, 19 (IAEA, Vienna 1974).
46. Weissenberger, E., Geltenbort, P., Oed, A., Gonnenwein, F.: Radiat. Eff. **96**, 47 (1986).
47. Lougheed, R. W., Moody, K. J., Dougan, R. J., Wild, J. F., Hulet, E. K., Dupzyk, R. J., Henderson, C. M., Gannett, C. M., Henderson, R. A., Hoffman, D. C., Lee, D. M., Sümmerer, K., Hahn, R. L.: 'Spontaneous fission and decay properties of ^{261,262}Lr and the new isotope, ^{262}No", Nuclear Chemistry Division FY87 Annual Report, UCAR 10062/87, Lawrence Livermore National Laboratory, 4-2 (1987).
48. Lougheed, R. W., Hulet, E. K., Wild, J. F., Moody, K. J., Dougan, R. J., Gannett, C. M., Henderson, R. A., Hoffman, D. C., Lee, D. M.: *50 Years with Nuclear Fission, Vol. II,* The American Nuclear Society, LaGrange Park, IL, 694 (1989).
49. Hulet, E. K., Wild, J. F., Dougan, R. J., Lougheed, R. W., Landrum, J. H., Dougan, A. D., Baisden, P. A., Henderson, C. M., Dupzyk, R. J., Hahn, R. L., Schädel, M., Sümmerer, K., Bethune, G. R.: Phys. Rev. **C40**, 770 (1989).
50. Hulet, E. K., Wild, J. F., Dougan, R. J., Lougheed, R. W., Landrum, J. H., Dougan, A. D., Schädel, M., Hahn, R. L., Baisden, P. A., Henderson, C. M., Dupzyk, R. J., Sümmerer, K., Bethune, G. R.: Phys. Rev. Lett. **56**, 313 (1986).
51. Wilkins, B. D., Steinberg, E. P., Chasman, R. R.: Phys. Rev. **C14**, 1832 (1976).
52. Lee, D. M., Wilkins, B. D., Hoffman, D. C.: Actinides 93 Conf., Santa Fe, New Mexico, Sept. 1993.
53. Wagemans, C., Schillebeeckx, P., Deruytter, A.: Nucl. Phys. **A502**, 287c (1989).
54. Demattè, L., Wagemans, C., D'hondt, P., Deruytter, A.: in *Seminar on Fission "Pont d'Oye III",* May 1995.
55. Brosa, U., Grossmann, S.: Z. Phys. **A310**, 177 (1983).
56. Brosa, U., Grossmann, S., Müller, A.: Phys. Reports **C197**, 167 (1990).
57. Wahl, A. C.: At. Data Nucl. Data Tables **39**, 1 (1988).
58. Ter-Akop'yan, G. M., Buklanov, G. V., Zeinalov, Sh. S., *et al.*, in: *Int. Workshop on the Physics of Heavy Ions* [in Russian], Dubna; published as JINR Report No. D7-87-68, 212 (1987).
59. Hoffman, D. C., Ford, G. P., Balagna, J. P., Veeser, L. R.: Phys. Rev. **C21**, 637 (1980).
60. Wild, J. F., van Aarle, J., Westmeier, W., Lougheed, R. W., Hulet, E. K., Moody, K. J., Dougan, R. J., Koop, E. A., Glaser, R. E., Brandt, R., Patzelt, P.: Phys. Rev. **C41**, 640 (1990).
61. van Aarle, J., Westmeier, W., Esterlund, R. A., Patzelt, P.: Nucl. Phys. **A578**, 77 (1994).
62. Sokol, E. A., Ter-Akop'yan, G. M., Krupman, A. I., Katkov, V. P., Nikonova, L. F., Eremin, N. V.: Sov. J. Atom. Energy **71**, 906 (1992) (translation of Sov. Atomnaya Energiya **71**, 422 (1991)).
63. Hofmann, S.: "Production and Stability of New Elements", GSI-95-25 Preprint, April 1995.
64. Wierczinski, B., Hoffman, D. C.: "Instrumentation for Atom-at-a-Time Chemistry of the Heavy Elements", Lawrence Berkeley Laboratory Preprint LBL-37463, June 1995 (to be published in IANCAS Frontiers in Nucl. Chem. Commemorating the Centennial of Becquerel's Discovery of Radioactivity).

Radiochimica Acta *70/71*, 147–161 (1995)

Radiochemical Studies of Complex Nuclear Reactions

By J. V. Kratz
Institut für Kernchemie, Universität Mainz, D-55099 Mainz, Germany

(Received July 23, 1995)

Radiochemical studies / Nuclear reactions / Fusion / Damped binary reactions, quasi-elastic transfer / Coulomb excitation at relativistic energies

Abstract

Low-energy nuclear fission, heavy-ion reactions close to the Coulomb barrier, proton-nucleus and nucleus-nucleus collisions at relativistic energies have traditionally been investigated by applying radiochemical techniques, i.e. by detecting heavy fragments produced in these reactions via their radioactive decay. This article presents a number of selected examples of studies in which various nuclear reaction mechanisms were investigated with different activation techniques. These do not always include chemical separations of the reaction product mixture even though the achievement of ultra-high sensitivity for the detection of rare reaction channels does require those. High-resolution spectroscopy of γ-rays, X-rays, α-particles and spontaneous-fission decays are being used to identify individual isotopes. Cross sections are determined with respect to proton number Z and neutron number N with respect to bombarding energy, with respect to emission angle and recoil range. The examples given refer to heavy-ion fusion, the reseparation channels in case of dynamically hindered fusion, damped binary reactions and the N/Z degree of freedom, the tilting mode and the excitation energy sharing in quasi-fission reactions, quasi-elastic transfer, and, at relativistic energies, to Coulomb excitation of the 2-phonon giant dipole resonance in heavy nuclei. The experimental techniques are presented and the pertinent results are discussed.

1. Introduction

Complex nuclear reactions are studied in a variety of ways. Historically, after the discovery of nuclear fission by Hahn and Strassmann [1], radiochemical methods became extremely important for the measurement of fission yields. While the cumulative fission product distribution after the decay of short-lived precursors was interesting for practical (military and economic) purposes [2] and could be determined by classical manual radiochemical separation techniques, independent yields [3] were of importance for the fundamental understanding of the fission process. For their measurement, methods of rapid chemical separations [4, 5] were being developed. They are still playing a role in areas where more modern physical techniques cannot be applied. The same change from radiochemical studies to instrumental counter techniques with complex detection systems of high granularity and large solid angle coverage can be reported for the investigation of proton-nucleus or nucleus-nucleus collisions from the barrier up to relativistic energies. Nevertheless, radiochemical studies have retained their importance in limited areas on the market, and these are defined by the unique Z and N resolution and extreme sensitivity of the radiochemical methods. In this article, a number of selected examples of studies of complex nuclear reactions using radiochemical techniques is given. They elucidate the types of experimental approaches and the information we usually wish to obtain such as cross sections for individual product nuclei, their variation with incident energy, and the energy spectra and angular distributions of the reaction products. For reasons of space limitations, radiochemical studies of low-energy fission are not covered in this article, and the reader is referred to extensive review articles [3, 6, 7]. The same holds for radiochemical studies of proton-induced reactions where the techniques are reviewed in [8] and the pertinent results in [9]. Thus, we concentrate on heavy-ion reactions close to the Coulomb barrier and include one example for a reaction study at relativistic energies.

2. Heavy-ion induced reactions at low energies

In a recent topical review [10] of heavy-ion reactions close to the Coulomb barrier it is outlined that the phenomena observable if two nuclei are brought together in a very gentle way are strongly influenced by subtle energy and nuclear structure dependences that disappear rapidly as the bombarding energy is raised above the barrier. The review covers fusion, complex reactions, and quasi-elastic reactions and emphasizes that these structure-dependent phenomena link together the possible reaction channels from elastic scattering all the way to fusion. Sub-barrier fusion is influenced by two important degrees of freedom in a dinuclear system: low-lying collective excitations of the partners and the exchange of nucleons between them. These 'elementary' processes do not seem to work independently, they rather cooperate in a correlated way to a more general degree of freedom that perhaps deserves the macroscopic name of 'neck formation'. For massive systems ($Z_pZ_t > 1000$) fusion is dynamically hindered and is competing increasingly with what is termed 'complex' reaction channels [10] which represent reseparation channels after a massive exchange of charge, mass, energy, and angular momentum. These reactions are seen as the missing link between the 'simple' reactions (inelastic scattering, one- and two-nucleon transfers), and fusion. The complex trans-

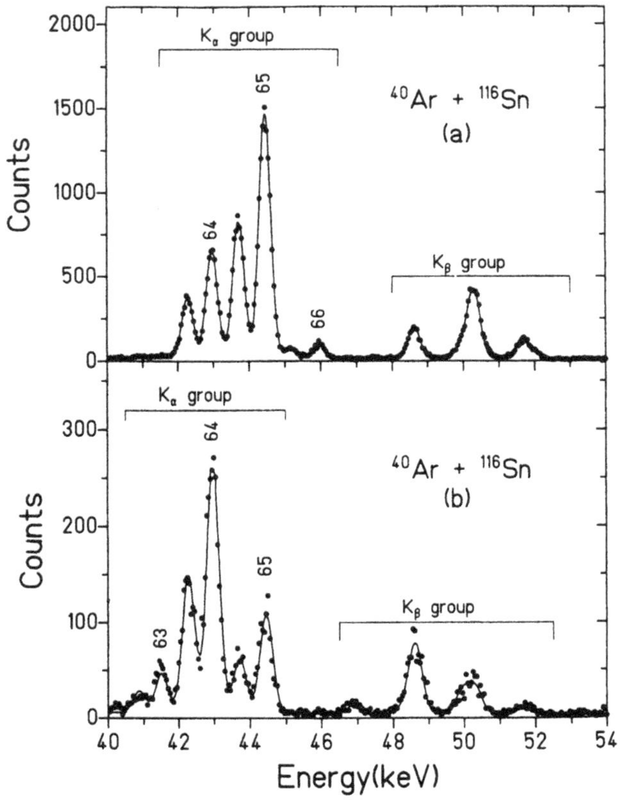

Fig. 1. K X-ray spectra [11] from an 80 min irradiation of ^{116}Sn with ^{40}Ar of 3.8 MeV/u. The spectra (a) and (b) were recorded about 70 min and 1.5 days, respectively, after the end of the irradiation. β-decay, mainly by electron capture, is clearly visible by a shift to lower Z in spectrum (b) as compared to spectrum (a). The Z-values are indicated at the positions of the respective $K_{\alpha 1}$ lines.

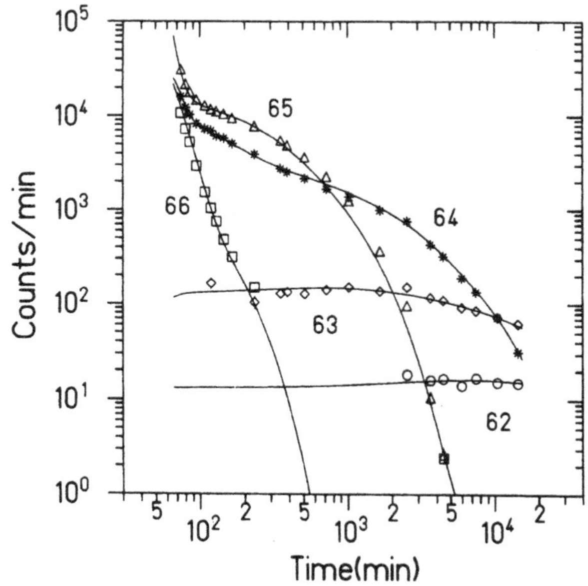

Fig. 2. Measured K X-ray decay curves [11] for the system $^{40}Ar + ^{116}Sn$ at 121.8 MeV (c.m.) sorted according to the indicated atomic numbers. The solid lines were obtained from a simultaneous least-squares fit to the data using known half-lives, parentages, and K X-ray multiplicities.

fer of mass and charge, evidence for neck formation even for the most massive systems, the influence of neck rupture on excitation energy sharing, cross section balances, quasi-fission time scales, and the apparent complexity of 'simple' (quasi-elastic) reactions at the barrier are discussed [10]. Much of the experimental evidence referred to in this review [10] results from radiochemical studies. This shows that one of the domains in which radiochemical reaction studies make outstanding contributions is in the field of heavy-ion reactions close to the Coulomb barrier.

2.1 Heavy-ion fusion

In a collaborative study [11] of sub-barrier fusion for relatively light systems with $Z_pZ_t < 1000$, the velocity filter SHIP at GSI and radiochemical activation techniques were combined in order to measure absolute fusion cross sections over 6 orders of magnitude. The role of the radiochemical measurements was essentially to 'calibrate' the transmission of the fusion-evaporation residues through SHIP and to provide an independent set of absolute cross sections. The activation method was as follows: During irradiations of 100 µm/cm² thick targets of isotopically enriched $^{112,116,122}Sn$ and $^{144,148,154}Sm$ on carbon backings of 40 µg/cm² by ^{40}Ar beams with a typical duration of one hour, evaporation residues were predegraded and stopped in water-cooled Al catcher foils of 2 and 3 mg/cm², respectively. Beam intensity normalization was done both by integrating the electrical charge of the ions in a Faraday cup, and by monitoring in two surface barrier detectors placed at ±45° behind the target the intensity of elastically scattered ^{40}Ar projectiles. Starting about 5 min after the end of irradiation, γ-ray as well as K X-ray spectra were recorded intermittently over a period of several days. The intrinsic Ge X-ray detectors had efficiencies of (6−15)% for K X-rays emitted from nuclei with charge $Z=40$ to 80 and a resolution of about 350 eV (FWHM). The Ge(Li) diodes for measuring the γ-ray spectra had efficiencies of (15−40)% relative to NaI and about 2 keV resolution (FWHM) at 1332 keV. The photoelectric peak efficiencies of the various detectors were known to within 4%.

Total evaporation-residue cross sections could be obtained both from the γ-ray and the K X-ray intensities. The K X-ray method has been described in detail in [12]. Typical spectra from a ^{116}Sn irradiation recorded at different times after the end of the irradiation are shown in Fig. 1. Also shown in the figure is a least-squares fit (solid curves) to the data to extract the yields as a function of the atomic number using the known energies and relative intensities of the characteristic K X-ray groups. Decay curves sorted according to atomic number obtained from many similar fits are shown on a log-log scale in Fig. 2. The production cross sections of the various β-active isotopes and in particular their sum, are the parameters which are then deduced from a simultaneous fit (also shown in Fig. 2) to all the decay curves using known half-lives and K

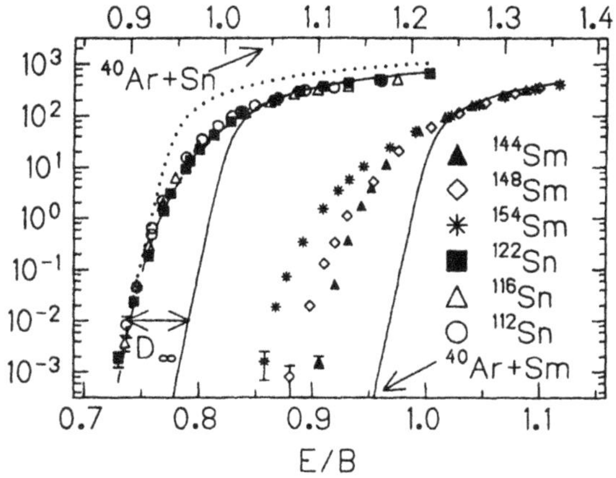

Fig. 3. Excitation functions [11] for ^{40}Ar-induced fusion on various indicated Sn (upper abscissa) and Sm (lower abscissa) isotopes. The abscissa are scaled in terms of the Bass-model barriers B. The full and the dotted curves are results of barrier penetration calculations using, respectively, the Bass potential and a modified potential.

X-ray multiplicities (typically 0.5 to 1.5). The latter are obtained from a careful inspection [12, 13] of their β- and γ-decay properties as tabulated in Nuclear Data Tables. These properties also enter into the evaluation of the γ-spectra. In addition to the activation measurements, the velocity filter SHIP determined cross sections for evaporation residues by 'ion counting' (implantation of the evaporation residues in a Si detector) and by adjusting the calculated transmission to bring the activation data and the velocity-filter data into agreement. Further, fusion-fission cross section were measured with a ΔE-E gas ionization chamber and the fission cross sections were added to the evaporation residue cross sections to yield complete fusion cross sections. Fig. 3 shows the resulting excitation functions [11] for the fusion of ^{40}Ar with the various Sn and Sm isotopes. In this figure, trivial size effects have been removed by dividing the energy E by the barrier value B calculated from a semi-empirical potential. Two conspicuous phenomena emerge when this type of scaling is used: (i) the tin data merge at all energies, while the samarium data merge only in the 'classical' region ($E/B > 1.04$); (ii) single-barrier penetration calculations on the basis of a one-dimensional potential (full curves) reproduce the data only in the classical region. If one readjusts the strength of the potential to better reproduce the low-energy part of the data one obtains the dotted curve (shown for the Sn isotopes only) which overestimates the high-energy data by a wide margin. The two potentials describing, alternatively, the 'classical' region or the 'pure tunneling' region differ in this case by 6.3±0.3 MeV in the barrier value. This is called the 'asymptotic' barrier shift, D_∞, as shown in the figure. Phenomenologically, the authors [11] introduced a barrier 'fluctuation' or 'distribution' $D(B)$ and calculated the fusion cross section from

$$\sigma_{\text{fus}}(E) = \int D(B)\sigma(B,E)\,\mathrm{d}B$$

where $\sigma(B,E)$ is calculated as before using a potential with barrier B. As an example, the Ar + Sn data are well described using a distribution of potentials generated from a Gaussian distribution of the radius parameter with a standard deviation of 3.7%. The remarkable fact emerging from such fits to the data is the 1:1 correlation of the standard deviations with nuclear structure (see also [14]). The different behaviour of the Sm isotopes as compared to the Sn isotopes is correlated with nuclear spectroscopy: the Sm isotopes span neutron numbers where one moves from spherical to deformed nuclei. Note that the asymptotic shifts for the deformed ^{154}Sm is larger (14.2 MeV) than that for the transitional nucleus ^{148}Sm (10.0 MeV), and the asymptotic shift for the latter is larger than that for the spherical ^{144}Sm (6.5 MeV). The analysis [11, 14] suggested that excitation of low-lying collective states in one or both nuclei play a decisive role in sub-barrier fusion. Also the exchange of nucleons between them boosts the fusion cross sections [10]. Whether this deserves the macroscopic name 'neck formation' is a question that a quantitative theory has to answer. Considerable progress has been made in the framework of coupled-reaction channels theory, where one tries to describe in a unified way all the observed couplings to collective excited states and transfer reactions [10]. This theory justifies the 'distribution of barriers' approach to parametrize phenomenologically experimental fusion excitation functions.

2.2 Competition between fusion and binary reseparation channels

For massive systems ($Z_p Z_t > 1000$) an increasing difference in energy between the standard-potential barrier and the experimental barriers is observed [10, 15]. This apparent fusion hindrance has been predicted by theory. Macroscopic models [16, 17] predict that an extra kinetic energy is needed if these massive systems are to fuse. According to [17] this energy in excess of the standard-potential barrier, E_{xx}, the 'extra-extra-push' energy, scales smoothly with a mean fissility. Quint *et al.* [15] have recently confronted this predicted scaling of E_{xx} with available fusion data with the following results: (i) there is evidence for finite E_{xx} values already below the theoretical threshold; (ii) it is true that a steep rise of E_{xx} occurs for systems with a large mean fissility, as predicted, however, the fissility parameter suggested by the model [17] is not a good sorting parameter; (iii) there is a conspicuous isotopic trend in the data opposite to the tendency predicted by the model [17]. More complex influences on the phenomenon must exist that are ignored so far by the macroscopic approach which describes the nuclear motion, after deep contact of the colliding nuclei is established. The model follows the evolution on a multidimensional potential energy surface along classical trajectories under the influence of friction. It is important to realize that the decision for fusion, in this model, occurs *after* the potential barrier has been passed [16, 17].

A very different suggestion [18] is that the system behaves diabatically in the approach phase, leading to a fast dynamical rise of the fusion barrier. The latter is predicted to be strongly nuclear structure dependent. This should give rise to a quasi-elastic reflection of the fragments from the diabatically raised barrier. It is remarkable that the diabatic theory [18] makes barrier shift predictions that are well correlated with the isotopic trends mentioned above.

A shift of the fusion barrier inevitably leads to a loss of fusion cross section. In the diabatic model [18] the missing fusion cross section should reappear in quasi-elastic, few-nucleon transfer channels. This is quite in contrast to the 'extrapush' models [16, 17]. In these models, the system's classical trajectory must, after passing the potential barrier, overcome two other 'milestones' [16] before fusion is ensured: the conditional (at fixed mass asymmetry) and the unconditional saddle-points. If the system does not pass behind the conditional saddle-point, one observes 'deep inelastic' collisions with nuclear mass and charge splits centered around the entrance channel configuration. 'Quasifission', i.e. a fission-like fragment distribution but with possibly larger mass width and different angular distributions, will be observed if the conditional, but not the unconditional saddle-point has been overcome.

In order to shed light onto the relevance of the scenarios discussed above, we used radiochemical techniques to obtain a general overview of all the binary reaction channels competing with fusion. We have determined mass- and charge distributions for binary reaction channels close to the standard-potential barrier in the systems ^{86}Kr + ^{76}Ge, ^{86}Kr + ^{104}Ru, and ^{86}Kr + ^{130}Te [19]. The systems span the region from apparently unhindered fusion via a slight onset of the 'extrapush' phenomenon, to a massive dynamical hindrance of fusion.

Isotopically enriched targets of ^{76}Ge, ^{104}Ru, and ^{130}Te on thin carbon backings were exposed to ^{86}Kr beams from the UNILAC, GSI, at incident energies of 3.22 MeV/u (Ge target), 3.64 and 3.84 MeV/u (Ru target), and 3.96 MeV/u (Te target). Due to the low melting point of Te, the Te targets were mounted on a rotating wheel synchronized with the pulsed beam. Typical beam intensities were 50 particle nA. The beam intensity as a function of time and the total charge collected were recorded by a current integrator. The actual target thicknesses were determined from the beam integral and from the intensities of elastically scattered projectiles in surface barrier detectors placed at $\pm 40°$. The bombardments lasted between 2 h and 8 h. The targets were positioned in the middle of a cylindrical (4 × 4 cm) catcher box lined in almost 4π geometry with 18 mg/cm^2 Cu foil. After bombardment, the catcher foil was removed and one half of the total area was subjected to chemical separations. The other half was counted without chemical separations.

The Cu catcher was dissolved in nitric acid and further processed in a chemical separation scheme that led ultimately to up to 17 chemical fractions containing the elements (Br, I), (Nb, Ag), (Rb, Cs), (Te), (Ru), (Ge), (Sb, Ga), (In, Sn), (As), (Se, Te, Pd), (Zr), (La-Pm), (Sm-Tb), (Dy-Ho, Y), (Er-Lu), (Sr, Ba), and (Zn, Cd). The separation scheme was modified and the number of isolated chemical fractions changed according to the specific requirements of the ^{86}Kr + ^{76}Ge, ^{86}Kr + ^{104}Ru, and ^{96}Kr + ^{130}Te reactions. The fractions were mounted for γ-counting as solid precipitates on membrane filters and covered with mylar foil. The complete chemical procedure lasted 3.5 h. After the end of the subsequent γ-ray measurements, the mylar foils were removed, and the chemical yields were determined by X-ray fluorescence analysis of the amount of inactive carrier of each element recovered in the final precipitates. Typically, 100 μg of carrier per element had been added to the radioactivities at the very beginning of the separation scheme. For Br and I the amount of carrier was 5 mg and the chemical yield was determined by neutron activation analysis. Chemical yields were between 60% and 90% with a few exceptions, i.e. Se, Te (30–35%), Pd, Sr, Ba (45–50%). The determination of chemical yields involved also, where possible, a comparison of the chemically isolated activities with those contained in the unseparated part of the Cu catcher foil. Systematic uncertainties in the determination of the chemical yield did in no case exceed 10%. Gamma-ray spectra of the target, the Cu-catcher, and the chemically separated fractions were counted periodically by 12 absolutely calibrated Ge detectors over a total period of one month. The counting times were increased from 15 min to 4 d. The spectral data were analyzed with an interactive version of the code GAMANAL. The data were further reduced with a set of computer programs using graphical display facilities and decay characteristics of known nuclides contained in a γ-ray library broken into subsets consisting of elements possibly present in a given sample. The decay rates were corrected for the variation of the beam intensity during the bombardment. Iterative corrections for precursor decay during the irradiation and between the end of irradiation and chemical separation were also applied. Cross sections were calculated from the corrected activities, the chemical yields, the target thickness, and the beam intensity. The procedures described above have yielded cross sections for a large number of isotopes: for the three different systems information on 18 (^{76}Ge), 29 (^{104}Ru) and 64 (^{130}Te) nuclides could be obtained. In particular, for the lightest system (^{86}Kr + ^{76}Ge), the identified nuclides represent 14 different partitions, illustrating that rather complex mass exchange reactions going well beyond the much studied one-nucleon transfer reactions can occur. This is remarkable since we are using an incident energy slightly below the fusion barrier for a system with a comparatively low Z_pZ_t product (1152). All these complex reaction channels are potentially competing with complete fusion, and one may therefore wonder whether the current practice to limit coupled-reaction channel calculations of sub-barrier fusion to just very few exit channels (usually inelastic and one-nucleon transfer) are realistic.

However, an even larger number of isotopes formed in the reactions remained undetected in this work because of unfavourable decay characteristics. Since we were primarily interested in obtaining global cross section balances, we have tried to reconstruct the complete nuclide distributions using as starting point the information gained from our experiment. Since the nuclei produced in such binary collisions are obviously centered around a narrow band in the (A vs. Z)-plane, one can without a great loss in generality parameterize the yields in terms of

$$Y(Z,A) = Y(Z) \cdot G_Z(A) \quad (1)$$

where $Y(Z)$ is the (mass integrated) nuclear charge distribution and $G_Z(A)$ is to a good approximation a (normalized) Gaussian distribution around a centroid mass A_Z with a variance σ_A^2, both of which are allowed to depend on Z. In applying this sort of parameterization to our present data we found that the Z dependence of σ_A, if any, is sufficiently weak for a given reaction system to be ignored within our experimental accuracy. Specifically, we were lead to choose full widths at half maximum (FWHM = 2.36 σ_A) equal to 3.0, 3.1 and 3.4 mass units, respectively, for the systems ^{86}Kr + ^{76}Ge, ^{104}Ru and ^{130}Te. The approximate constancy of the widths of the isotope distributions at fixed Z is in line with our earlier studies [20, 21] of reactions in the vicinity of the interaction barrier.

A second, safe assumption is that

$$Y(Z) = Y(Z_{cn} - Z) \quad (2)$$

where $Z_{cn} = Z_P + Z_t$ is the fused system's nuclear charge. This assumption implying that charged particle evaporation can be neglected, is confirmed by evaporation calculations: they show that the charged-particle branching ratios are at the 1% level for the relatively low excitation energies and spins involved here.

Another trivial, but sometimes useful restriction, is that one must have

$$\bar{A}_Z + \bar{A}_{Zcn-Z} \leq A_{cn} \quad (3)$$

The inequality, due to the 'missing mass' can be used to extract information on the average number of neutrons emitted by the fragments for a given charge split. We will come back to this below. Under the above assumptions, the number of parameters to be determined by the data is three per nuclear charge split. Although the number of data points we were able to determine is not quite as large, this does not cause any major difficulty in practice, since we can restrict the freedom of choice by requiring smooth evolutions (with Z) of the yields Y and the centroids $\bar{A}_Z$. In particular, we found that if one excludes one or two nucleon transfer products then $Y(Z)$ is well described by a superposition of two Gaussians centered close to, but not exactly at, the projectile, resp. target charge. Because we neglect charged-particle evaporation these two Gaussians are complementary and share the height and width parameters. For the system ^{86}Kr + ^{130}Te the more symmetric charge splits represent a third, rather broad gaussian component centered at symmetry that we have added. The deduced smooth parameterizations of the nuclide yields are plotted as solid curves in Fig. 4 for the ^{86}Kr + ^{130}Te reaction as an example. A close look at these plots for the three reactions reveals that there are conspicuous deviations from the smooth trends. As an example, for ^{86}Kr + ^{130}Te this concerns the yields of the isotopes ^{128}Sn, ^{129}Sb, ^{129,131,132}Te, and ^{131}I as well as their light complements i.e. one- and two-nucleon transfer reactions. This is illustrated in detail in Fig. 5 where enlarged plots of the yields for Sb, Te and I are shown again for better inspection. For the sake of definition we shall henceforth term these channels 'quasielastic transfer channels', whereas all the other channels will be called 'complex (multistep) reaction channels'.

Before going into the more detailed separate discussion of these two components, we point to Fig. 6 which contains the integrated mass distributions obtained by summing up the measured and interpolated yields. This figure illustrates well the evolution towards increasing complexity as the product Z_pZ_t is raised.

As mentioned in connection with the inequality (3), the average number of emitted neutrons associated with a given charge split can be deduced from the missing mass of complementary-charge fragments. For the quasi-elastic transfer channels, the missing mass is zero. For the complex reaction channels, the missing masses increase with increasing net transfer and reach values corresponding to kinetic energies known from low energy fission systematics. The combination of strongly subbarrier exit channel energies with a rather large exchange of nucleons between the two nuclei is a compelling reason for setting the dividing line between quasielastic and complex reactions as being *the passage of the potential barrier* [19]. If this interpretation is correct, then one should apply the barrier tunneling theory not to extract just the fusion cross section, but rather the sum of fusion plus complex reactions. The quantitative results using this idea [19] were indeed such that the sum of the fusion and complex-reaction cross sections in the three systems is convincingly compatible with the result of barrier passing calculations with parameters typical for unhindered fusion. Thus, coming back to the theories mentioned above, these results [19] support qualitatively the long-standing general idea [22] that, indeed, the lack of fusion cross section relative to the barrier-passing expectations is due to the dynamical evolution *following* the crossing of the potential barrier, rather than to a lack of penetration due to a diabatic rise of the barrier [18].

While the Gaussian in the center part of Fig. 6 is interpreted as true fusion-fission component, that in the bottom part of Fig. 6 is attributed [19] to quasifission (14 mb out of 80 mb for the total cross section

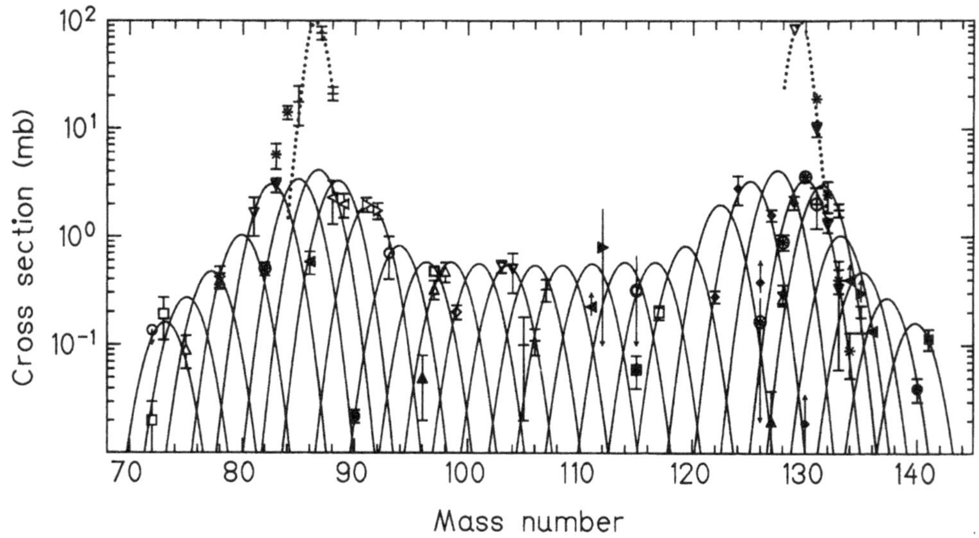

Fig. 4. Cross sections for the $^{86}Kr + ^{130}Te$ reaction at $E_{cm} = 203.0$ MeV. Closed and encircled symbols stand for measured independent yields, open symbols for measured cumulative yields corrected for precursor decay. The solid curves are Gaussians of FWHM = 3.4 u used to interpolate and integrate complex-reaction yields (see text). The dotted curves join the proton and neutron transfer channels and are just guides to the eye [19].

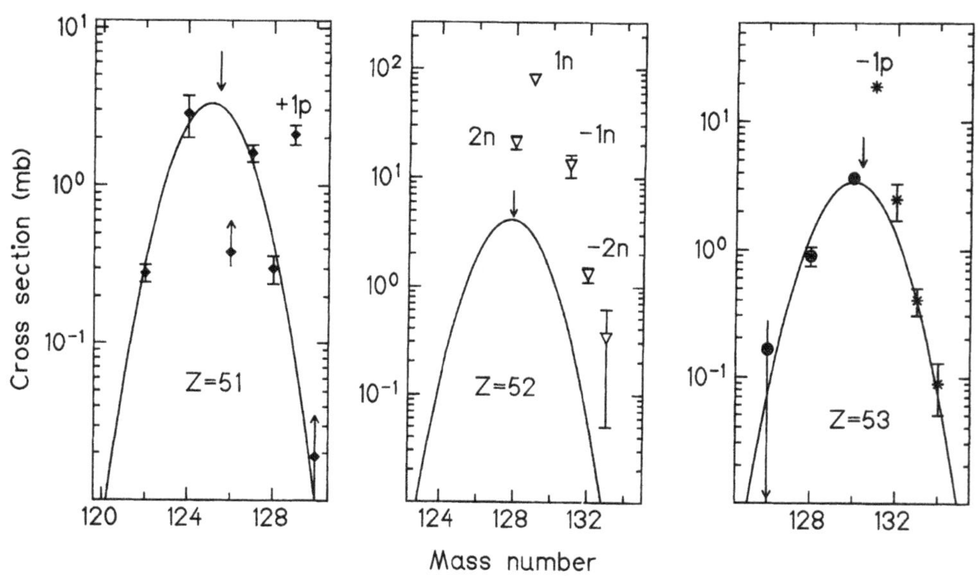

Fig. 5. Enlarged plots [19] of the yields for Sb, Te/Kr, and I ejectiles in the reaction $^{86}Kr + ^{130}Te$ at $E_{cm} = 203.0$ MeV. In the Te/Kr case averages of the complementary yields are shown. The Gaussian curves are explained in the text. The arrows close to the top of the curves indicate the most favoured mass at fixed charge from potential energy considerations. The predominantly 'quasielastic' channels are indicated ($1n$, $-1n$, $2n$, $-2n$, $+1p$, $-1p$).

for complex reactions). The small relative share of quasifission and the proximity of the complex reaction product distributions to the entrance channel partitions indicate furthermore, that the major part of the fusion deficit is produced by reseparation *in the early phase after passing the barrier* before the conditional saddle-point is overcome that would be initiating massive mass drifts. If the decision is made that early, shell structures are still partially intact playing a subtle role in setting the initial conditions for the complex further evolution. This is in line with the observations of Quint *et al.* [15] that the large scatter of extrapush energies evidenced when plotted against the mean fissility [17] as scaling parameter is strongly correlated with shell effects. To conclude, the radiochemical studies [19] have revealed for the first time that complex reactions compete with fusion and that these complex reactions are governed by a common mechanism, i.e. tunneling through the nucleus-nucleus potential barrier. More inferences from these data [19] concerning N/Z equilibration, excitation energy sharing and quasielastic transfer cross sections are discussed in Sections 2.3, 2.5 and 2.7.

2.3 N/Z equilibration in damped collisions

The mass-to-charge ratio of the fragments is an old theme in nuclear fission. For damped ('complex') heavy-

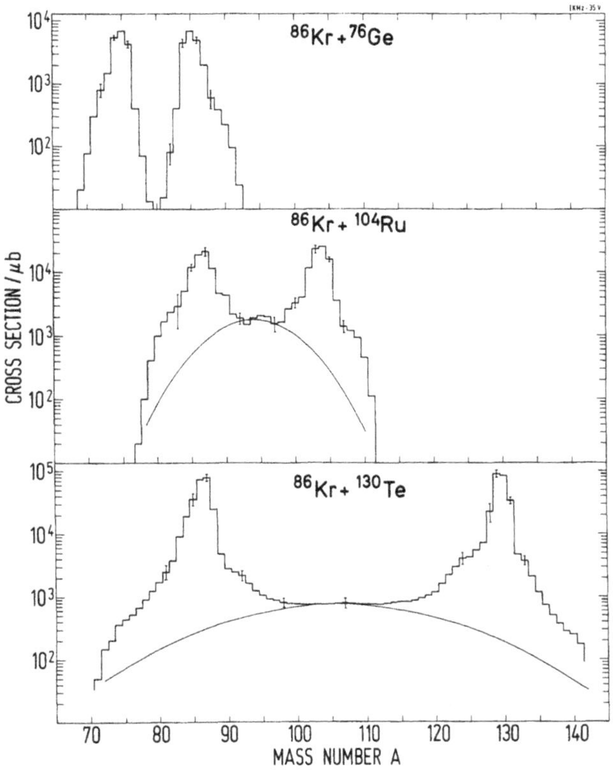

Fig. 6. Mass distributions [19] for $^{86}Kr + ^{76}Ge$ (top), ^{104}Ru (center), and ^{130}Te (bottom) at E_{cm} 129.1, 178.6, and 203.0 MeV, respectively. Error bars are given for representative mass numbers only. The Gaussians in the center and bottom part of the figure are used to estimate the cross section for fission or fission-like processes.

ion reactions, the present understanding of N/Z equilibration has been reviewed by Freiesleben and Kratz [23]. In the present article, we briefly refer to some contributions from radiochemical studies of damped collisions of various Xe isotopes with ^{197}Au. The techniques used were similar to those in [19], i.e. consisted of chemical separations and γ-ray spectroscopy to arrive at angle and kinetic-energy integrated cross sections [20]. As in the preceding Section, a smooth systematics of Gaussian distributions was found to be compatible with the data, except again for the quasi-elastic channels. The position of the centroids of these Gaussians and the resulting N/Z ratios as a function of Z were compared with model predictions, and it was found that potential energy considerations were useful to predict the most favoured N/Z ratios [20]. Referring back to Fig. 5, we like to point to the arrows close to the top of the Gaussian distributions representing the optimum N/Z ratios calculated from the consideration of the potential energy surface of two nuclei in close contact. One realizes that the peaks of the isotopic distributions are quite close to the predictions and that the rapid evolution towards N/Z equilibration evident for the complex reactions implies that a massive nucleon exchange, larger than the observed *net* exchange, must have taken place between the two nuclei. Clearly, the measured cross section for ^{124}Sb (Fig. 5) could not be explained by combining transfer probabilities for one-proton and one-neutron 'elementary' transfers, since it is larger than the cross section for ^{128}Sb which involves a smaller net transfer.

The apparently rapid N/Z equilibration might be connected to a collective E1 mode associated with an isovector type of hydrodynamical flow of neutrons and protons back and forth through the neck of the dinuclear system [24]. In such a hydrodynamical description, fluctuations in the division of charge are to be expected from the quantum-mechanical zero-point motion even when there is no excitation energy. If this harmonic oscillator is weakly damped through coupling to the intrinsic degrees of freedom of the system, which constitute a heat bath of temperature T, the eigenvalues of the mode are known to be

$$\varepsilon(T,\omega) = \frac{1}{2}\hbar\omega + \frac{\hbar\omega}{\exp(\hbar\omega/T)-1} \qquad (4)$$

where the collective frequency, ω, and the stiffness, c, are related through the inertial parameter B, $c = B\omega^2$, and B is assumed to be a constant. Using the harmonic oscillator eigenfunctions, one determines the charge variance at fixed mass asymmetry to be

$$\sigma_Z^2|_A = \frac{1}{c}\left(\frac{1}{2}\hbar\omega + \frac{\hbar\omega}{\exp(\hbar\omega/T)-1}\right). \qquad (5)$$

Two extreme predictions can be derived from (5) depending on the relative magnitude of the phonon energy $\hbar\omega$ compared to the temperature T:
If $\hbar\omega \gg T$, the zero-point motion dominates and

$$\sigma_Z^2|_A = \frac{\hbar\omega}{2c}. \qquad (6)$$

In this case the charge fluctuations are purely quantal. In the other extreme $\hbar\omega \ll T$,

$$\sigma_Z^2|_A = \frac{T}{c}. \qquad (7)$$

One is then dealing with classical statistical fluctuations. The remarkable difference between (6) and (7) is that $\sigma_Z^2|_A$ does not depend on the temperature (excitation energy) in the quantal case, while it should increase with the square root of the excitation energy in the classical case. This difference was addressed experimentally in the reaction of 900 MeV ^{132}Xe with ^{197}Au by two complementary radiochemical experiments by Poitou *et al.* [25]. The first experiment served to measure integral nuclide distributions: The reaction products were stopped in 4π-geometry in a thick aluminum catcher. After bombardment, the catcher was chemically separated into 14 chemical fractions containing the Au-like fragments which were further identified by their X-ray and γ-ray activities. From the cross sections $d^2\sigma/dZ\,dA$ a surface of independent yields in the $Z-A$ plane was constructed. For given atomic numbers Z the variances $\sigma_A^2|_Z$ were found to vary between 3.9 and 5.6 with the largest values being observed for the elements far removed from $Z=79$. Due to the physical integration over all excitation

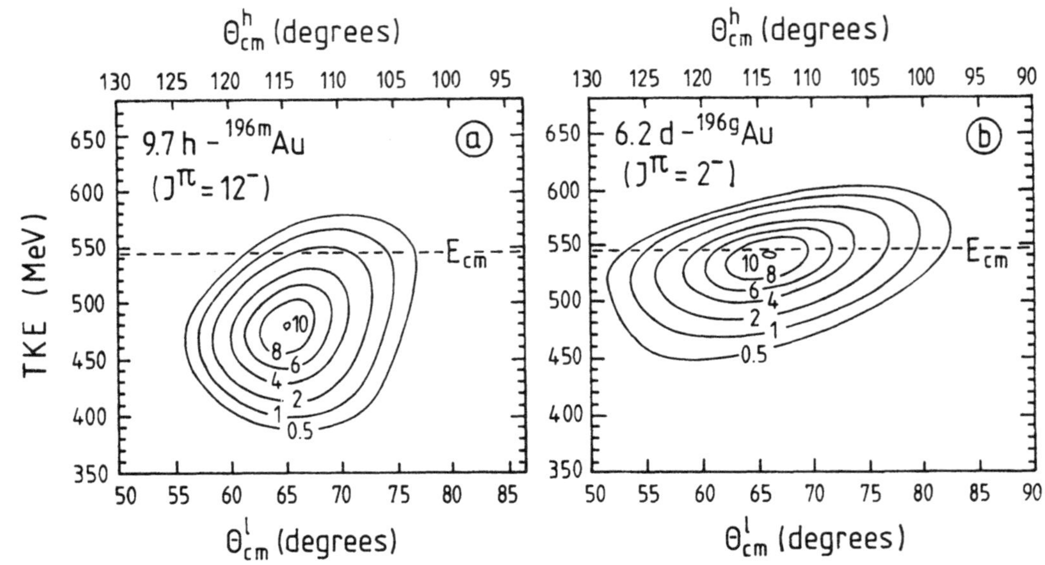

Fig. 7. Wilczynski diagrams for (a) $^{196m}Au(12^-)$ and (b) $^{196g}Au(2^-)$ in the $^{132}Xe + ^{197}Au$ reaction. The measured distributions are broader than the true distributions due to the limited resolution of the stacked-foil technique. The total kinetic energies (TKE) are not yet corrected for neutron evaporation. (From Ref. [26].)

energies these widths are broadened as compared to the pre-neutron emission widths. Information on this broadening was obtained from a second experiment measuring differential recoil ranges at given laboratory angles. The Au-like products were stopped in stacks of 6 μm thick mylar foils between $5° \leq \Theta_{lab} \leq 75°$. After bombardment, the foils were cut into slices each one covering 10° of lab angle. The products were identified by KX-ray and γ-ray singles and coincidence measurements, resulting in

$$\int_{R_1}^{R_2}\int_{\theta_1}^{\theta_2} \frac{d^4\sigma}{dA\, dZ\, d\theta dR}\, d\theta\, dR$$

cross sections in the laboratory system. The ranges R were converted into laboratory velocities. For each isotope these integrals were fitted both in the velocity and angle coordinate by a 5th-order polynomial and transformed into the center-of-mass system using two-body kinematics.

Technical details are given in [26]. Fig. 7b shows as an example the resulting Wilczynski plot for the low-spin (2^-) ground state of ^{196}Au resulting from direct (quasi-elastic) neutron stripping from ^{197}Au. Fig. 7a shows the Wilczynski plot for the "yrast trap" ^{196m}Au (12^-) which collects the flux of events associated with higher energy damping and higher angular momentum transfer ($> 12\,\hbar$) that ends at ^{196}Au after neutron evaporation. These distributions are broader than the true distributions due to the limited energy resolution of the stacked-foil technique. It was shown by Monte Carlo simulations that the instrumental resolution in TKE is about 55 MeV [26]. However, the positions of the maxima are quite accurately determined within uncertainties < 6 MeV and < 3°. Given the centroid of the distribution for each isotope (or isomer), iterative corrections of the average total kinetic energy, ⟨TKE⟩, for neutron evaporation were performed [26]. The energy carried away by each evaporated neutron was determined by statistical evaporation calculations and was on the order of 11 MeV. The average ⟨TKE⟩ associated with the population of the 2^- and the 12^- state of ^{196}Au (Fig. 7) are very different and result in average excitation energies of 3 MeV and 30 MeV, respectively. If the instrumental resolution is folded out, the width of the excitation energy distribution for $^{196g}Au(2^-)$ is <10 MeV. Thus, the majority of the observed yield of ^{196g}Au is consistent with its direct formation involving no neutron evaporation. Similar results are obtained for other quasi-elastic products, e.g. ^{198g}Au and ^{197}Pt. For the damped collision products including $^{198m}Au(12^-)$ and $^{196m}Au(12^-)$ continuously increasing excitation energies $\langle E \rangle$ are observed as their mass numbers decrease. Actually, for each element Z, one observes a linear increase of $\langle E \rangle$ with decreasing post-neutron-emission A, see Fig. 8. Also, the excitation energy widths of these products

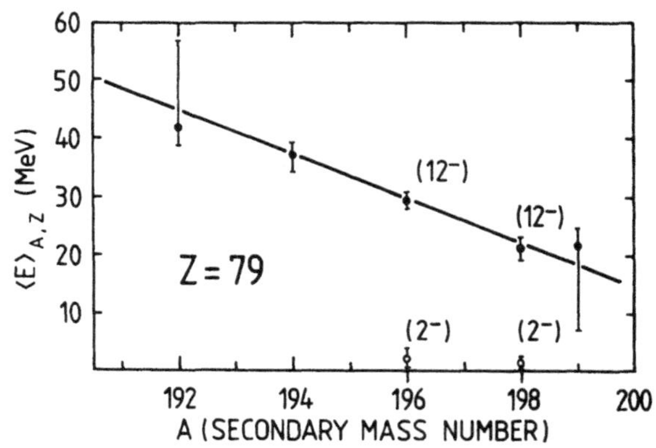

Fig. 8. Most probable heavy fragment excitation energies for $Z = 79$ associated with the formation of given post-neutron-emission isotopes. Open points are for quasi-elastic products, closed points for products from the damped collision process. (From Ref. [25].)

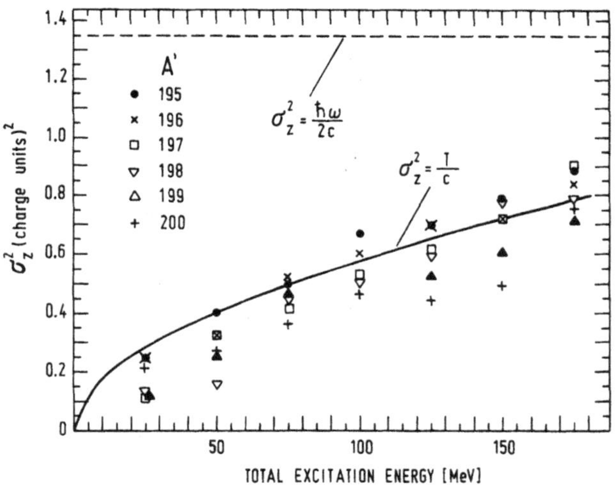

Fig. 9. Charge variances $\sigma_Z^2|_{A'}$ at fixed mass-asymmetry deduced from the reconstructed primary yield surface $\sigma(Z, A', E)$ for the $^{132}Xe + ^{197}Au$ reaction. The scatter of the points is representative for the uncertainties in $\sigma_Z^2|_{A'}$. The dashed curve indicates the expected variance due to quantal fluctuations. The solid curve is calculated for classical statistical fluctuations. (From Ref. [25].)

(after folding out the instrumental resolution) is larger, typically 20−30 MeV. This indicates [26] that these damped collision products are formed through neutron evaporation from several primary masses, e.g. ^{194}Au is formed from ^{195}Au at $\langle E\rangle \cong$ 15 MeV, from ^{196}Au at ~25 MeV, from ^{197}Au at ~35 MeV and so forth. This systematic correlation between the measured average excitation energies $\langle E\rangle$ and the secondary mass numbers A can be used to unfold the convolution of the primary distribution $\sigma(Z,A',E)$ by the excitation-energy-dependent evaporation process and to generate a mathematical cross section surface $\sigma(Z,A',E)$. The unfolding procedure for such radiochemical data was developed by Poitou *et al.* [26]. The resulting set of three equations reproduces the experimental observation, see Fig. 8, of a linear relation between the average excitation energies $\langle E\rangle|_{A,Z}$ at fixed Z and the post-neutron emission mass numbers A. Least-squares analyses of the measured quantities $P(Z)$ (the centroids of the post-neutron-emission isotope distributions), $\sigma_A^2|_Z$ (the associated variances), and several values of $\langle E\rangle|_{A,Z}$ (the average excitation energies associated with the formation of given isotopes) for each element Z using these equations gives a set of five Z-dependent parameters which defines the complete surface $\sigma(Z,A',E)$. It is possible with this surface to make any cut through the distribution and to calculate any zeroth, first, or second moment of one of the now primary variables conditioned by any of the remaining variables. Among other moments, this primary yield surface defines the conditional variances $\sigma_Z^2|_{A'}$, as a function of the total excitation energy, which are shown in Fig. 9 for a number of mass chains near A' = 197.

The solid line in Fig. 9 is the prediction of the dependence of $\sigma_Z^2|_{A'}$ vs E for full statistical equilibrium, Eq. (7), using a stiffness of 2.8 MeV and the temperature $T = \sqrt{8E/A_{tot}}$. The dashed line corresponds to the equilibrium variance, Eq. (6), for quantal fluctuations. The phonon energy for the $m=0$ component of the giant E1 resonance was estimated as $\hbar\omega$ = 78 MeV/$(A_1^{1/3} + A_2^{1/3})$ = 7.2 MeV. It was pointed out by Poitou *et al.* [25] that the disagreement of the predicted values $\sigma_{Z|A'}^2 \approx$ 1.35 with the experimental data can be reconciled if a considerably increased inertia associated with the E1 mode is assumed for the dinuclear complex (made up from a Xe-like fragment touching a Au-like fragment) as compared to a normal spherical nucleus. They referred to the work by Updegraff and Onley [27] where the $m=0$ component of the dipole vibration energy of ^{238}U in photofission was shown to decrease for increasing deformation and to be at least four times reduced for the very constricted shape of two touching fragments at the instant of scission. This would invalidate the condition $\hbar\omega \gg T$ for zero-point motion and explain the classical statistical behaviour of the charge fluctuations [25].

The conditional charge variances $\sigma_Z^2|_{A'}$ for $^{132}Xe + ^{197}Au$ are, of course, only one aspect of the whole yield surface $\sigma(Z,A',E)$ and correlations to other moments of this surface were also discussed [25]. This revealed [25] that the slow rise of the charge variances σ_Z^2 of the integral charge distribution with increasing total excitation energy is correlated with a hindrance of proton exchanges for energy losses up to about 80 MeV. In turn, the initial rearrangement of neutrons and protons in the $^{132}Xe + ^{197}Au$ system is achieved essentially by *neutron exchanges* and not by proton exchanges [25]. Thus, "charge equilibration" cannot be seen as an isolated, one-dimensional phenomenon in the $(N-Z)$-coordinate. It is one important aspect of the whole evolution of the $N-Z$-distribution in 'complex' reactions and its dependence on dynamical properties such as the distance of closest approach and the neck opening.

2.4 Neck formation: Information from the most massive systems

Complex reactions have also been observed [28] in very heavy systems such as $^{238}U + ^{238}U$ or $^{238}U + ^{197}Au$ at and below the Coulomb barrier. For such systems, standard potentials no *longer have a pocket*. Nevertheless rather 'exotic' transfer reactions leading to the formation of a number of Ra, Ac, Th, and Pa isotopes [28] were observed. These were stopped in Au-catcher foils that were dissolved after the end of bombardment. After removal of the Au by extraction into t-butylmethylether, an ion-exchange chromatography separation on four columns was performed [28]. Counting samples for α-particle spectroscopy were produced by electrochemical depositions on titanium discs.

The observation of such 'complex' reactions can be understood by assuming that a neck has formed between the two nuclei, possibly taking advantage of

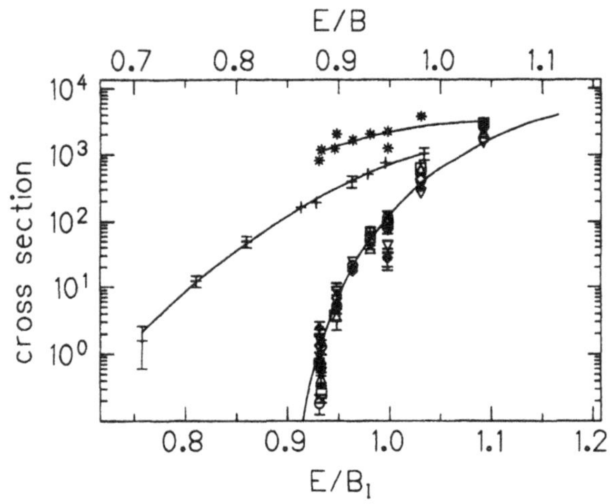

Fig. 10. Comparison of excitation functions for one-neutron transfer, massive transfer, and fusion. Crosses, one-neutron transfer (in mb) for $^{238}U + ^{238}U$; asterisks, one-neutron transfer (to ^{237}U, in mb) for $^{238}U + ^{197}Au$. All other symbols stand for complex transfer channels (in μb) ($^{223-225}Ra$, $^{225-226}Ac$, ^{227}Th) in the system $^{238}U + ^{197}Au$. While the smooth lines going through the neutron transfer data are just to guide the eye, the curve going through the complex transfer data represents the measured fusion cross section for $^{40}Ar + ^{154}Sm$ scaled down by a factor 100 and shifted towards 5% higher E/B_I values (see upper abscissa). The transfer data are from Funke *et al.* [28]. B_I is the interaction barrier, B is the adjusted 'neck formation' barrier. (From Ref. [10].)

a favorable tip-to-tip orientation. The analogy of the excitation function for such a 'neck formation' with fusion excitation functions is shown in Fig. 10 where a comparison is made with the fusion data for $^{40}Ar + ^{154}Sm$ [11], which, like $^{238}U + ^{197}Au$, is a combination of a medium soft, spherical nucleus with a well deformed nucleus. One finds that the 'barrier for neck formation' is at about 95% of the interaction barrier. From the analogy [10] of the excitation functions for 'neck formation' with fusion excitation functions we learn that subbarrier fusion physics continues all the way up to U + U, except that now, one can study more directly the coupling of the 'simple' to the 'complex' channels that lead to fusion in lighter systems. The life-time of the 'necked configuration' seems to be short ($<10^{-21}$ s): the angular distribution of the complex reaction products is sharply peaked backward [28].

2.5 Neck rupture: excitation energy sharing

Neck formation at barrier energies is also evidenced by the way the total excitation energy is shared at separation. *Well above the barrier*, the excitation energy division is discussed in terms of one-body dissipation models [29] where the processes are dominated by the heat generating deceleration, and a thermal equilibrium is gradually approached with increasing kinetic energy loss. The excitation energy is then shared in proportion to the fragment mass number. On the other hand, as is known from low-energy fission, the separation of two fragments which were linked by a neck can lead to a rather peculiar sharing of excitation energy: one fragment appears to be cold, while the other is hot, giving rise to the famous 'sawtooth' for the average number of emitted neutrons versus mass in asymmetric fission: It is decisive how the neck is ruptured, i.e. how the neck nucleons are shared at separation.

Evidence for a 'sawtooth'-like excitation energy division was obtained *at barrier energies* in a number of radiochemical studies by using the 'missing mass' information and potential energy considerations for the prediction of the primary mass numbers at a given Z: $^{48}Ca + ^{248}Cm$ [30], ^{50}Ti, ^{54}Cr, $^{58}Fe + ^{208}Pb$ [31], $^{40}Ar + ^{233.235.238}U$ [32], $^{86}Kr + ^{130}Te$ [19], $^{238}U + ^{197}Au$ [28]. The missing masses determined in these studies indicated that the acceptor of the neck nucleons obtained nearly the full excitation energy and the donor nuclei stayed cold.

Recent, more complete radiochemical studies by Klein *et al.* [33] confirm these observations for the system $^{51}V + ^{197}Au$ at the Bass-model barrier. These experiments involved the most extensive chemical separation scheme applied in nuclear reaction studies [34]. 22 chemical fractions were isolated from a 4π catcher foil (Cu) covering all elements from K through Bi. 270 cross sections were determined by low-level γ-ray spectroscopy with 20 absolutely calibrated Ge detectors with lead shielding. A section of the resulting yield systematics is shown in Fig. 11. The Gaussian isotopic distributions (FWHM = 4.5 u) were determined for 54 elements. Their centroids were compared with the primary masses calculated by the minimum potential energy approach and the resulting missing masses for the 54 values of Z were 'translated' into excitation energies $\langle E^* \rangle_Z$ by performing evaporation calculations. The total excitation energies E^*_{tot} resulting from energy conservation are used to define fractional excitation energies $\langle E^* \rangle_Z / E^*_{tot}$ *that are plotted in Fig. 12 as a function of* Z. The trends observed earlier [30, 31, 32, 19, 28] when mass is transferred in the direction of mass symmetry are confirmed [33]. The acceptor nuclei above ^{51}V are hot and the donor nuclei below ^{197}Au are cold. If ^{51}V changes its role to become the donor and if ^{197}Au becomes the acceptor the fractional excitation energies $\langle E^* \rangle_Z / E^*_{tot}$ *react consequently: The donor nuclei below* ^{51}V and the acceptor nuclei above ^{197}Au quickly attain low or high excitation energies, respectively. This is strong evidence for a collective mechanism in which the excitation energy sharing *at the barrier* depends on the way the neck nucleons are shared at separation. The random neck-rupture model by Brosa *et al.* [35] is in qualitative agreement with these observations. Already at a slightly higher bombarding energy (B + 25 MeV) the excitation energy division due to shape fluctuations at scission is apparently covered up by the dissipation of extra relative kinetic energy via nucleon exchange giving way to more equal energy sharing, see Fig. 13,

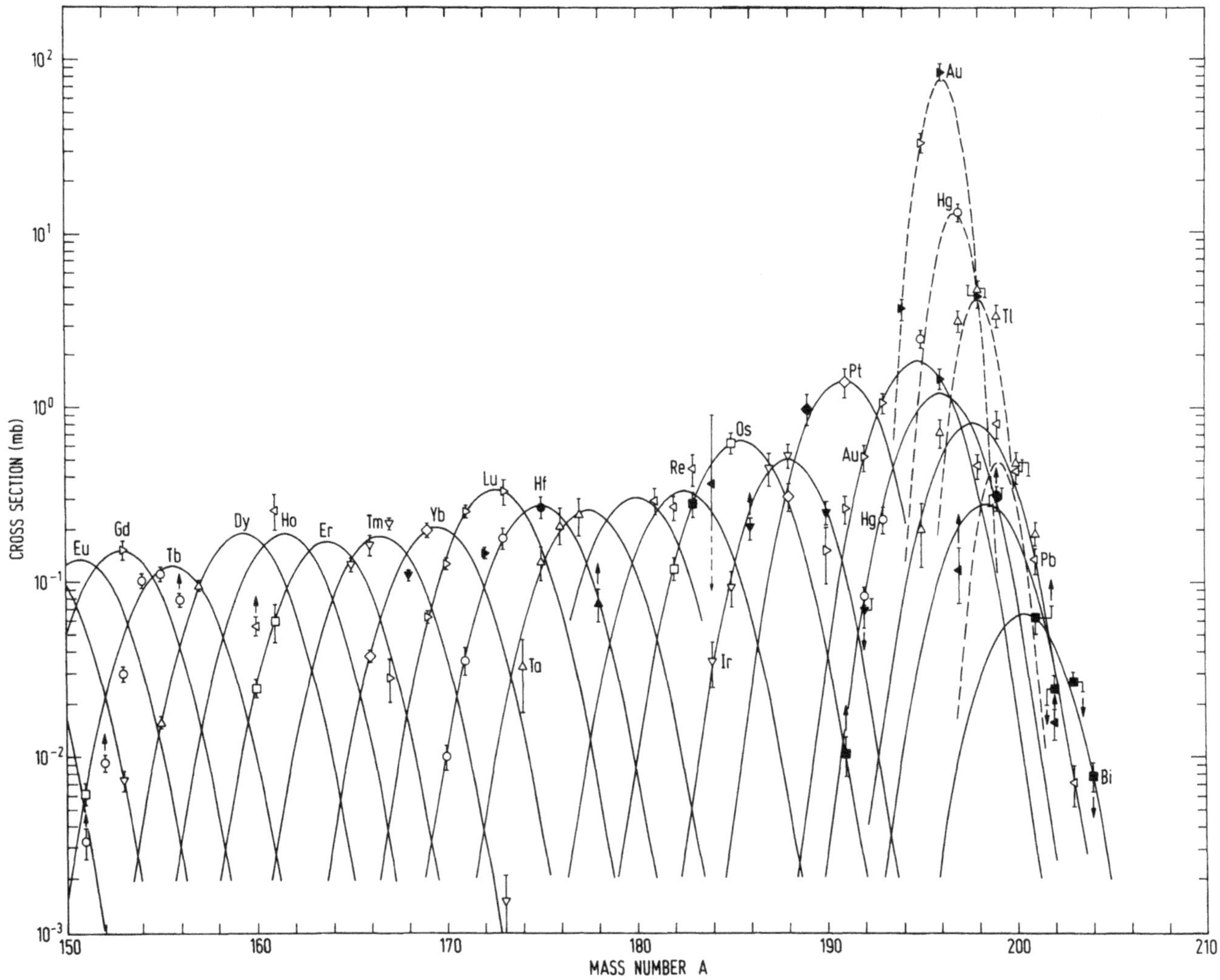

Fig. 11. A small section of the isotope distributions in the reaction $^{51}V + ^{197}Au$ at $E = B$. The dashed lines connect the cross sections for a separate 'quasielastic' component. The solid lines are Gaussians of FWHM = 4.5 u used to determine the most probable post-neutron emission mass numbers $\langle A \rangle_Z$. Closed symbols stand for measured independent yields, open symbols for measured cumulative yields corrected for precursor decay. (From Ref. [34].)

in accordance with the observations [29] at higher energy for modest total kinetic energy losses. These radiochemical experiments are being continued by stepwise increasing the bombarding energy further in order to see how quickly a thermal equilibrium is approached.

2.6 The tilting mode in quasi-fission reactions

Quasi-fission, a binary reaction mechanism in which the fission fragments have total kinetic energies consistent with the Viola systematics for fission, shows effects of memory of the mass and direction of the incident particles in the final distributions. This feature allowed Shen *et al.* [36] to derive a time scale for mass transfer in quasi-fission reactions from turning angles of the intermediate complex. It was found that the mass drift toward symmetry showed the characteristics of an overdamped motion with a rather long and excitation-energy independent time constant of $5 \cdot 10^{-21}$ s.

As was pointed out by Lützenkirchen *et al.* [37], these pre-equilibrium fission reactions might be responsible for the often-reported unusually large anisotropies in fission-fragment angular distributions in "fusion-fission" reactions in which the projectile was typically heavier than $A \cong 20$. These unusually large anisotropies had prompted several authors to put into question the validity of the saddle-point transition-state theory (see [37] for references) for very heavy systems and to suggest modifications thereof. In contrast, Lützenkirchen *et al.* [37] suggested that the unusually large anisotropies were associated with the inability of these heavy systems to form genuine compound systems. By applying the afore-mentioned K X-ray activation technique to the products of the reactions ^{50}Ti, ^{54}Cr, ^{58}Fe + ^{208}Pb implanted into a cylindrical arrangement of catcher foils that were cut into 5° slices of laboratory angles after bombardment, they made detailed measurements of angular distributions of quasi-fission products with fixed nuclear charge. They found that the distributions $d^2\sigma/d\Theta dZ$ exhibit

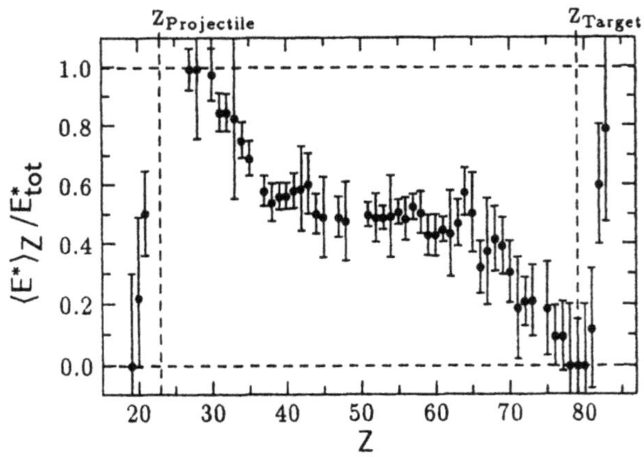

Fig. 12. Division of the total excitation energy at E_{lab} = 4.86 MeV/u (E_{BASS}) in the $^{51}V + ^{197}Au$ reaction. (From Refs. [33, 34].)

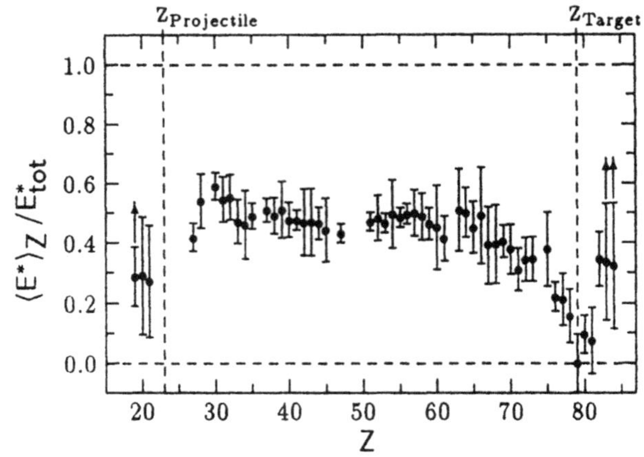

Fig. 13. Division of the total excitation energy at E_{lab} = 5.48 MeV/u (E_{BASS} + 25 MeV) in the $^{51}V + ^{197}Au$ reaction. (From Refs. [33, 34]).

forward-backward asymmetries that are strongly *Z*-dependent, even close to symmetry, see Fig. 14. The data were parameterized (full curves in the figure) by applying a generalization of the theory for fission angular distributions to non-equilibrated reactions such as quasi-fission [38]. The angular distribution close to the beam direction (0° and 180°) is sensitive to the 'tilting' mode which corresponds to a common rotation of both spheres of a dinucleus around the symmetry axis leading to a tilt of the most favourable rotation axis. The variance K_0^2 of the tilting angular momentum is treated as a dynamically accumulated quantity which is *Z*-dependent, see Fig. 15. From the rotational angles involved a rather long time-scale ($\sim 10^{-20}$ s), comparable to the rotational periods of the system, for the evolution of K_0^2 towards values comparable to typical equilibrium values known from fission was deduced [37]. One can conclude that the tilting mode in such quasi-fission reactions is not fully 'relaxed' resulting in much higher (*Z*- and time-dependent) anisotropies than expected from fusion-fission. Similar observations were reported even for ^{40}Ar-induced reactions on ^{197}Au and ^{208}Pb [39, 40].

2.7 'Simple' (quasi-elastic) reactions at the barrier

Quasi-elastic transfer reactions are known to be sensitive to the degree of 'orbit matching' achieved in the reaction. With some simplifications [10], this leads to the well known matching condition

$$Q_{eff} = Q + V_i(r_{cl}) - V_f(r_{cl}) = 0. \quad (8)$$

For neutrons, $V_i - V_f$ at the point of closest approach (cl) is usually ignored, for protons only the Coulomb part of $V_i - V_f$ is considered most of the time. Recent compilations [41, 42] of angle and energy integrated one-neutron transfer cross sections studied some 20–30% above the barrier energy fit very well into such a simple *Q*-value systematics if care is taken to remove the different neutron binding energies. It was proposed [41] that the latter could be done by plotting 'reduced' cross sections $\sigma_t \times (B_i B_f)^{1.1}$ vs. Q_{gg}, where $B_{i,f}$ are the neutron separation energies. This leads to a smooth and narrow band of data points following essentially an error function. The success of this smooth systematics indicated that aside from the *Q*-value and separation energy dependence, nuclear structure is essentially averaged out when integrated cross sections well above the barrier are considered.

Radiochemical yield measurements have played a key role in demonstrating that quasi-elastic transfer cross section measured *at the barrier* are in striking contrast to the smooth systematics at higher energies. As for fusion reactions, one is witnessing here also for the one-nucleon transfer reactions an increasing complexity that was suggested [19] to be due to coupling to complex reactions. We can illustrate this by looking again at Fig. 5 in Section 2.2. The 'reduced' cross sections in the $^{86}Kr + ^{130}Te$ reaction (in arbitrary units) are for the neutron pick-up σ'_{+1n} = 12700, and σ'_{-1n} = 4500 for the neutron stripping reaction. For the proton transfers one finds σ'_{+1p} = 1600 and σ'_{1p} = 13700. A look at Fig. 5 shows how this could be associated with coupling to the complex reactions represented by the smooth curves. The one-nucleon transfer channels ($\pm 1n$, $\pm 1p$) are seen to be large when the corresponding isotope is close to the maximum of the complex reaction channels and small when the contrary is the case. It suggests that the coupling to the more macroscopic trend followed by the complex reactions creates a polarization field that favours matching even for relatively negative *Q*-values. Note that the two-neutron transfer is also affected by this: σ_{+2n} is larger than σ_{-1n}! It is suggested [10] that in the matching Equation (8) one cannot neglect changes in the nuclear part of the potential which are due to polarization. The maximum possible 'macroscopic polarization' at fixed mass asymmetry is just the difference between the potential barrier and the conditional saddle-point on the potential energy surface. For symmetric systems with $Z_p Z_t$ between 1000 and 2000 this is on the order of 20 MeV!

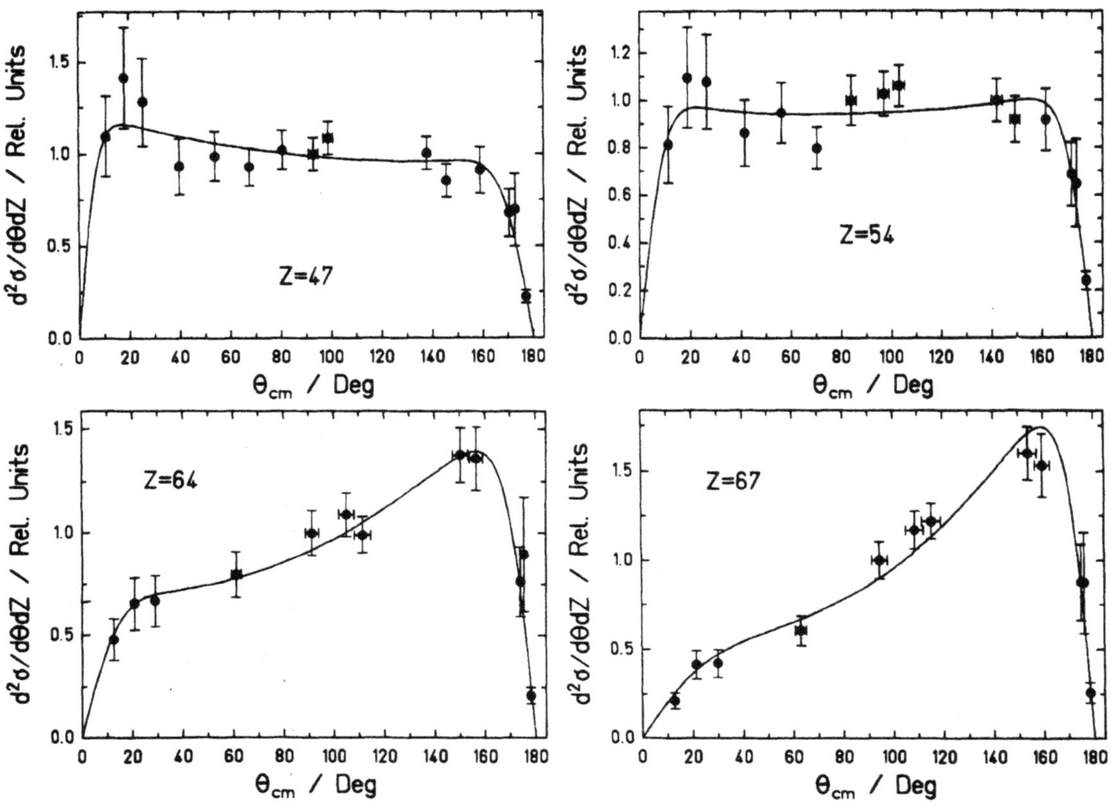

Fig. 14. Angular distributions in the centre of mass system for quasi-fission products in the reaction of ^{50}Ti with ^{208}Pb at 5.5 MeV/u. The distributions are asymmetric around 90° except for $Z \approx Z_{sym} = 52$. The full curves are theoretical fits (see text). (From Ref. [37].)

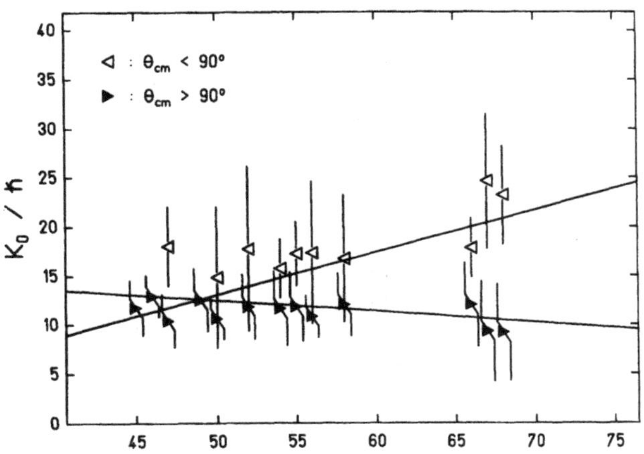

Fig. 15. K_0-values in the forward (◁) and backward (▶) hemispheres for fragment charges Z for the 5.22 MeV/u $^{40}Ar + ^{208}Pb$ reaction. The solid lines are linear fits intersecting at $Z_{sym} = 50$. (From Ref. [39].)

The break-down of semiclassical approximations for quasi-elastic transfer data was also revealed in very heavy systems by a series of detailed excitation function and angular distribution measurements using radiochemical techniques [43, 44]. For mismatched transfer reactions with negative Q-values, the back-angle cross sections were found to be depleted. It is tempting here to invoke again the coupling to the complex reactions discussed in Section 2.4 that we had interpreted as evidence for neck formation.

To conclude this section, we point out that even the 'simple' quasi-elastic transfer reactions are not at all simple at the barrier but are coupled to many complex channels producing a complexity resembling that of sub-barrier fusion reactions.

3. Electromagnetic dissociation of heavy nuclei at relativistic energies

The electromagnetic excitation of giant resonances by relativistic heavy-ion projectiles and their subsequent decay by neutron evaporation (electromagnetic dissociation, ED) has attracted much attention recently as it has been predicted that multiphonon giant dipole resonance states can be excited with large probabilities. The usual procedure to calculate cross sections for the ED process is the Weizsäcker-Williams (WW) method of virtual quanta. In this approach, the cross section is given by integrating over the product of the equivalent photon number $N_\gamma(E_\gamma)$ and the photonuclear cross section $\sigma_\gamma(E_\gamma)$:

$$\sigma_{\mathrm{ED}} = \int dE_\gamma\, N_\gamma(E_\gamma)\sigma_\gamma(E_\gamma).$$

As equivalent E1 and E2 quanta contribute at ~1 GeV/u, it is necessary to break up $\sigma_\gamma(E_\gamma)$ into different components corresponding to different multipolarities. In order to include multiphonon states, one adopts the method of Llope and Braun-Munzinger [45] of calculating multiple excitations in the harmonic approximation.

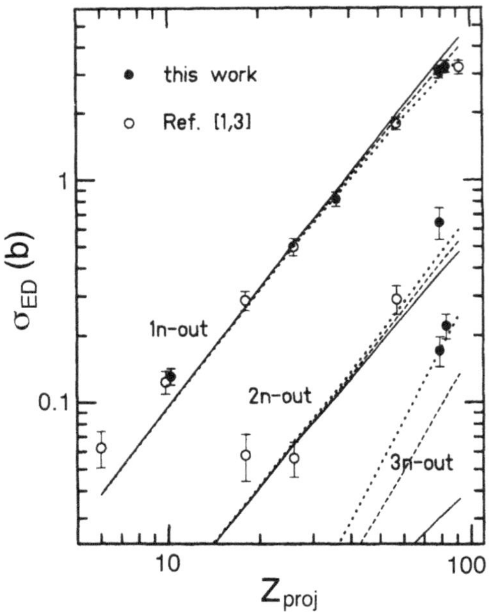

Fig. 16. ED cross sections for $1n$, $2n$, and $3n$ removal from a ^{197}Au target as a function of projectile charge. All experimental data points are scaled to the same bombarding energy of 1 GeV/nucleon. The full curve represents a simple WW calculation taking single-phonon GDR and GQR excitations into account. The dashed curve has been obtained by allowing for multiphonon GDR excitations in the harmonic approach. The dotted curve represents a calculation where the two-phonon GDR excitation probability is arbitrarily increased by a factor of 2.2. The full dots are from [46], the open dots are from previous work cited in [46].

The radiochemical procedure to investigate the electromagnetic excitation of giant resonances consisted of irradiating ^{197}Au targets with relativistic heavy ions between ^{20}Ne and ^{209}Bi and of a determination of the cross sections for the formation of $^{194-196}Au$ by observing the γ-ray activity following β decay of these reaction products. As no parallel momentum transfer is involved in these nucleus-nucleus encounters it is safe to simply measure the activities in the targets. The calculation of cross sections for the $1n$, $2n$, and $3n$ channels involved a correction for secondary reactions in the targets. The resulting cross sections as a function of Z_{proj} are shown in Fig. 16. They are compared with WW calculations of various complexity [46]. The solid lines represent a simple WW calculation taking single-phonon giant dipole resonance (GDR) and giant quadrupole (GQR) excitations into account. It is seen that this calculation overpredicts the $1n$-out cross section, underpredicts the $2n$-out cross section, and underpredicts the $3n$-out cross sections by a large factor of 5.7 and 6.8 for ^{197}Au and ^{209}Bi projectiles, respectively. As the latter is most sensitive to the 2-phonon GDR with excitation energies around 27 MeV, multiphonon excitations were allowed in the calculations yielding the dashed curves. As this calculation still underpredicts the $3n$-out cross sections for ^{197}Au and ^{209}Bi projectiles, the 2-phonon excitation probability was increased by a factor of 2.2 yielding the dotted curves in Fig. 16. Thus, the inclusive radiochemical measurements were the first to point to the excitation of the 2-phonon state of the GDR and to the fact that this excitation was more effective by a factor of 2 then in the harmonic approximation. It is interesting to note that, meanwhile, exclusive measurements of ^{136}Xe projectile dissociation with the LAND detector at GSI arrive at the same results [47].

4. Outlook

In this review, a subjective selection of radiochemical experiments and the pertinent techniques were presented that have been used to study mainly the physics of nucleus-nucleus collisions at the barrier. Generally, one determines cross sections for individual isotopes that are stopped during bombardment in a 4π catcher-foil arrangement. Depending on the goal of the experiment, integral cross sections or differential ones with respect to bombarding energy, scattering angle and recoil range in stacked foils are determined. These experiments have revealed that nuclear physics at the barrier is subtle and shows a richness of phenomena and influences that is only precedented by low-energy nuclear fission.

On the basis of the present knowledge of these reactions, a number of questions should be addressed by further studies:

Fusion cross sections should be followed further down to the nanobarn level in order to study fusion in the pure tunneling region. Interesting questions could be studied in this extreme regime that connect to spontaneous fission: there is the problem of inertial masses and of odd-particle hindrance that deserves attention. How do shell effects and deformations influence nuclear fusion and more generally large-scale dinuclear evolutions? It is necessary to study, in more detail, the competition between binary complex reactions and fusion below the barrier in medium heavy systems where complete fusion is still possible. What is the role of shell effects and isospin? The peculiar excitation energy sharing should be studied further by varying the nuclear structure of the partners. As shown in Section 2.4 massive systems can be used to study multidimensional tunneling processes involving dozens of nucleons under well controlled conditions such as varying the nuclear structure. The role of macroscopic polarization due to complex reactions in influencing the probabilities of single-nucleon transfer deserves further study.

Even though there is competition by 'counter techniques' in this field it is often observed that their abilities as to sensitivity, completeness, Z and A resolution, are limited. Therefore, in order to eventually answer the questions raised above, radiochemical studies of complex nuclear reactions at and below the Coulomb barrier will continue to be of importance. As shown by the example described in Section 3, there have also been other applications to detailed questions concerned with complex nuclear reactions, where radiochemical techniques are being used to advantage.

Acknowledgements

The author is indepted to his "fellow radiochemists" at GSI, W. Brüchle, M. Schädel, W. Reisdorf, K. Sümmerer, and G. Wirth, for many years of enjoyable collaboration, to his former graduate students T. Aumann, R. Bellwied, Th. Blaich, F. Funke, M. Gober, P. Klein, K. Lützenkirchen, and P. Zimmermann, for their decisive contributions to the work reviewed in this article, and to the Federal Minister for Research and Technology for continuous financial support, at this time under Contract No. 06 MZ 476.

References

1. Hahn, O., Strassmann, F.: Naturwissenschaften **27**, 11 (1939).
2. Coryell, C. D., Sugarman, N. (Eds.): National Nuclear Energy Series; Manhattan Project Technical Section, Radiochemical Studies: The Fission Products, Books 1–3, McGraw-Hill, New York 1951.
3. Denschlag, H. O.: Nucl. Sci. Eng. **94**, 337 (1986).
4. Herrmann, G., Trautmann, N.: Ann. Rev. Nucl. Part. Sci. **32**, 117 (1982).
5. Rengan, K., Meyer, R. A.: Ultrafrast Chemical Separations; Nuclear Science Series, NAS-NS-3118 (Radiochemical Techniques), National Academy Press, Washington, D. C., 1993.
6. Manohar, S. B., Goswami, A., Tomar, B. S.: J. Radioanal. Nucl. Chem., in press.
7. Denschlag, H. O.: Fission Fragment Mass, Charge and Energy Distributions, in Poenaru, D. N., Greiner, W. (Eds.) *Experimental Techniques in Nuclear Physics,* Walter de Gruyter, Berlin 1995.
8. Alexander, J. M.: Studies of Nuclear Reactions by Recoil Techniques in Yaffe, L. (Ed.), *Nuclear Chemistry, Vol. I,* Academic, New York 1968.
9. Hüfner, J.: Phys. Rep. **125**, 129 (1985).
10. Reisdorf, W.: J. Phys. G: Nucl. Part. Phys. **20**, 1297 (1994).
11. Reisdorf, W., Heßberger, F. P., Hildenbrand, K. D., Hofmann, S., Münzenberg, G., Schmidt, K. H., Schneider, J. H. R., Schneider, W. F. W., Sümmerer, K., Wirth, G., Kratz, J. V., Schlitt, K.: Phys. Rev. Lett. **49**, 1811 (1982); Nucl. Phys. **A438**, 212 (1985).
12. Stokstad, R. G., Reisdorf, W., Hildenbrand, K. D., Kratz, J. V., Wirth, G., Lucas, R., Poitou, J.: Z. Phys. **295**, 269 (1980).
13. Schlitt, K.: Diploma thesis, University Mainz (1984), unpublished.
14. Reisdorf, W., Hessberger, F. P., Hildenbrand, K. D., Hofmann, S., Münzenberg, G., Schmidt, K. H., Schneider, W. F. W., Sümmerer, K., Wirth, G., Kratz, J. V., Schlitt, K., Sahm, C. C.: Nucl. Phys. **A444**, 154 (1985).
15. Quint, A. B., Reisdorf, W., Schmidt, K.-H., Armbruster, P., Heßberger, F. P., Hofmann, S., Keller, J., Münzenberg, G., Stelzer, H., Clerc, H.-G., Morawek, W., Sahm, C.-C.: Z. Phys. **A346**, 119 (1993).
16. Swiatecki, W. J.: Nucl. Phys. **A376**, 275 (1982). Bjørnholm, S., Swiatecki, W. J.: Nucl. Phys. **A391**, 471 (1982).
17. Blocki, J. P., Feldmeier, H., Swiatecki, W. J.: Nucl. Phys. **A459**, 145 (1986).
18. Berdichevsky, D., Lukasiak, A., Nörenberg, W., Rozmej, P.: Nucl. Phys. **A499**, 609 (1989).
19. Reisdorf, W., Kratz, J. V., Bellwied, R., Brüchle, W., Keller, H., Lützenkirchen, K., Schädel, M., Sümmerer, K., Wirth, G.: Z. Phys. **A342**, 411 (1992).
20. Kratz, J. V., Brüchle, W., Franz, G., Schädel, M., Warnecke, I., Wirth, G., Weis, M.: Nucl. Phys. **A332**, 477 (1979).
21. Lützenkirchen, K., Kratz, J. V., Brüchle, W., Gäggeler, H., Sümmerer, K., Wirth, G.: Z. Phys. **A317**, 55 (1984).
22. Swiatecki, W. J.: J. Phys. C (Suppl.) **33**, 5 (1972); Swiatecki, W. J., Bjørnholm, S.: Phys. Rep. **C4**, 325 (1972).
23. Freiesleben, H., Kratz, J. V.: Phys. Rep. **106**, 1–120 (1984).
24. Brosa, U., Krappe, H. J.: Z. Phys. **A284**, 65 (1978).
25. Poitou, J., Lucas, R., Kratz, J. V., Brüchle, W., Gäggeler, H., Schädel, M., Wirth, G.: Phys. Lett. **88B**, 69 (1979); Kratz, J. V., Poitou, J., Brüchle, W., Gäggeler, H., Schädel, M., Wirth, G., Lucas, R.: Nucl. Phys. **A357**, 437 (1981).
26. Poitou, J., Kratz, J. V., Wirth, G., Lucas, R.: Nucl. Instrum. Methods **180**, 221 (1981).
27. Updegraff, W. E., Onley, D. S.: Nucl. Phys. **A161**, 191 (1971).
28. Funke, F., Kratz, J. V., Trautmann, N., Wiehl, N., Wirth, G., Brüchle, W., Fan Wo, Sümmerer, K.: Z. Phys. **340**, 303 (1991).
29. Töke, J., Schröder, W. U.: Ann. Rev. Nucl. Part Sci. **42**, 401 (1992).
30. Gäggeler, H., Brüchle, W., Brügger, M., Schädel, M., Sümmerer, K., Wirth, G., Kratz, J. V., Lerch, M., Blaich, Th., Herrmann, G., Hildebrand, N., Trautmann, N., Lee, D., Moody, K. J., Gregorich, K. E., Welch, R. B., Seaborg, G. T., Hoffman, D. C., Daniels, W. R., Fowler, M. M., von Gunten, H. R.: Phys. Rev. **C33**, 1983 (1986).
31. Keller, H., Bellwied, R., Lützenkirchen, K., Kratz, J. V., Brüchle, W., Gäggeler, H., Moody, K. J., Schädel, M., Wirth, G.: Z. Phys. **A328**, 255 (1987).
32. Scherer, U. W., Brüchle, W., Frink, C., Gäggeler, H., Herrmann, G., Kratz, J. V., Moody, K. J., Schädel, M., Sümmerer, K., Trautmann, N., Wirth, G.: Z. Phys. **A335**, 421 (1990).
33. Klein, P., Brüchle, W., Gober, M. K., Kratz, J. V., Schädel, M., Zimmermann, H. P.: Annual Report 1992, Institut für Kernchemie, Universität Mainz, IKMz 93-1, p. 27.
34. Klein, P.: Doctoral dissertation, University Mainz, GSI-93-37 Report (1993).
35. Brosa, U., Grossmann, S., Müller, A., Becker, E.: Nucl. Phys. **A502**, 423c (1989).
36. Shen, W. Q., Albinski, J., Gobbi, A., Gralla, S., Hildenbrand, K. D., Herrmann, N., Kuzminski, J., Müller, W. F. J., Stelzer, H., Töke, J., Back, B. B., Bjørnholm, S., Sørensen, S. P.: Phys. Rev. **36**, 115 (1987).
37. Lützenkirchen, K., Kratz, J. V., Wirth, G., Brüchle, W., Sümmerer, K., Lucas, R., Poitou, J., Gregoire, C.: Nucl. Phys. **A452**, 351 (1986).
38. Døssing, T., Randrup, J.: Nucl. Phys. **A433**, 215, 280 (1985), and Phys. Lett. **155B**, 333 (1985).
39. Keller, H., Lützenkirchen, K., Kratz, J. V., Wirth, G., Brüchle, W., Sümmerer, K.: Z. Phys. **A326**, 313 (1987).
40. Schulte, H., Jäckel, B., Esterlund, R. A., Knaack, M., Westmeier, W., Rox, A., Patzelt, P.: Phys. Lett. **232B**, 37 (1989).
41. van den Berg, A. M., Rehm, K. E., Kovar, D. G., Kutschera, W., Stephans, G. S. F.: Phys. Lett. **194B**, 334 (1987).
42. Rehm, K. E., Beck, C., van den Berg, A. M., Kovar, D. G., Lee, L. L., Ma, W. C., Videbaek, F., Wang, T. F.: Phys. Rev. **C42**, 2497 (1990).
43. Wirth, G., Funke, F., Brüchle, W., Wo Fan, Kratz, J. V., Sümmerer, K., Trautmann, N.: Phys. Lett. **253B**, 28 (1991).
44. Funke, F.: Doctoral thesis, University Mainz, GSI-90-10 Report (1990).
45. Llope, W. J., Braun-Munzinger, P.: Phys. Rev. **C41**, 2644 (1990), and Phys. Rev. **C45**, 799 (1992).
46. Aumann, T., Kratz, J. V., Stiel, E., Sümmerer, K., Brüchle, W., Schädel, M., Wirth, G., Fauerbach, M., Hill, J. C.: Phys. Rev. **C47**, 1728 (1993).
47. Schmidt, R., Blaich, Th., Elze, Th. W., Emling, H., Freiesleben, H., Grimm, K., Henning, W., Holzmann, R., Keller, J. G., Klingler, H., Kratz, J. V., Kulessa, R., Lambrecht, D., Lange, S., Leifels, Y., Lubkiewicz, E., Moore, E. F., Wajda, E., Prokopowicz, W., Schütter, Ch., Spies, H., Stelzer, K., Stroth, J., Walus, W., Wollersheim, H. J.: Phys. Rev. Lett. **71**, 963 (1993).

Radiochimica Acta 70/71, 163–175 (1995)

Radiochemical Studies of Complex Particle Emission in Low and Intermediate Energy Reactions

By S. M. Qaim

Institut für Nuklearchemie, Forschungszentrum Jülich GmbH, D-52425 Jülich, Germany

(Received August 21, 1995)

Complex particle emission / Energy and angular distributions / Radiochemical techniques / Integral cross section data / Nuclear model calculations / Reaction mechanisms

Summary

The probability of complex particle emission in low and intermediate energy nuclear reactions ($E \leq 70$ MeV) is generally small. The data are of significance from the fundamental point of view as well as for applications. Radiochemical and other interdisciplinary techniques like mass spectrometry and accelerator mass spectrometry play an important role in the study of these low-yield processes. To interpret the experimentally obtained integral data, comparisons are done with results of nuclear model calculations, based mainly on the statistical model. This paper briefly reviews the experimental and theoretical information available on each process. Several systematic correlations were developed which contribute to some mechanistic understanding and also allow a prediction of unknown data. Most extensive studies have been performed on ^{4}He-emission. In general, the compound and precompound models together adequately describe the ^{4}He-emission cross section; occasinally direct interactions need to be included. The ^{2}H-emission, on the other hand, appears to entail a strong contribution from direct processes. Radiochemical methods have contributed appreciably to the study of trinucleon emission reactions. Whereas in the case of ^{3}H-emission both compound and direct processes play an important role, the ^{3}He-emission appears to be dominated by direct interactions. The emission of heavier particles like ^{7}Be has been relatively little investigated. With the exception of ^{2}H and ^{4}He emissions, it appears that the higher the charge of the emitted complex particle, the lower is its emission probability.

Introduction

The interactions of neutrons and accelerated light ions with nuclei generally result in elastic and inelastic scattering, radiative capture and nucleon emission. Furthermore, emission of some complex particles like ^{2}H, ^{3}H, ^{3}He, ^{4}He, ^{6}Li, ^{7}Be etc., when energetically possible, is also observed, though with a much lower probability. Due to increasing Coulomb barrier effects this probability generally decreases with the increasing charge of the target nucleus. On the other hand, with the increasing projectile energy, the probability of complex particle emission also increases.

The emission of complex particles ^{3}H and ^{4}He was observed in neutron induced fission of heavy nuclei, and of ^{7}Be in high-energy proton induced spallation of several nuclei, already in the 1950s. However, in low and intermediate energy regions, where normal interactions dominate, studies on complex particle emission were primarily limited to the light mass nuclei. In the medium and heavy mass regions, due to high matrix activity and low cross sections, the purely physical methods could not be applied. Radiochemistry has traditionally played an important role in the study of such reactions, and interdisciplinary techniques have been of considerable significance. Of all the reactions involving complex particle emission, the probabilities of ^{2}H and ^{4}He emission are relatively high and on-line spectral studies in various target mass and projectile energy regions have been successfully performed. For the trinucleons ^{3}H and ^{3}He, and the heavier particles like ^{7,10}Be, on the other hand, radiochemical techniques have been advantageously applied.

The study of complex particle emission is of considerable significance, both for enhancing our understanding of nuclear theory and for practical applications. Reactions like (d,α), (^{3}He,t), (^{3}He,α) etc., with relatively high cross sections, are known to populate higher lying levels of the product nucleus. On-line spectroscopic studies of such reactions thus often furnish some useful information on the high-spin levels of the product. The best information on the reaction mechanism is generally obtained through investigations of the energy and angular distributions of the emitted particles. For low yield products, however, radiochemically obtained integral cross sections lead to some useful mechanistic information via a comparison of the experimental data with the results of nuclear model calculations. For practical applications, like estimation of radiation damage, nuclear heating, tritium breeding, activation of structural materials etc., the radiochemical data are generally of greater importance.

The work on trinucleon emission in neutron induced reactions was reviewed several times [cf. Refs. 1–3]. This paper presents a more general survey of the field of complex particle emission reactions. The emphasis is, however, on radiochemical methods of investigation, and particular attention is paid to the low and intermediate energy reactions ($E \leq 70$ MeV). The emission of complex particles in fission and spallation processes is not considered.

Experimental techniques

On-line methods

On-line spectral techniques involve measurement of the energy and angular distributions of the emitted complex particles. A detailed description of those techniques is beyond the scope of the present review. A few salient features relevant to an understanding of the subject are, however, given below.

The ranges of emitted complex particles are rather short, the target samples have therefore to be thin. Due to the low intensities of the available fast neutron sources, and due to the low charged particle fluxes which can be applied without damaging the target sample, the detection of emitted complex particle is difficult. This is all the more so the smaller the cross section of the nuclear process or the higher the background.

For detection of emitted complex particles nuclear emulsions, thin CsI(Tl) crystals, small semiconductor detectors and, in particular, several dE/dx-E type counter telescopes have been used. The dE/dx-thin detector gives the specific energy loss in the detector and thus leads to an identification of the complex particle, and the E-thick detector defines the total energy of the registered complex particle. Information on the angular distribution is obtained by placing an array of such detector systems at several angles relevant to the direction of the projectile.

In work with neutrons, the technique has been applied at 14 MeV to study the emission of ^{2}H, ^{3}H and ^{4}He, mainly from some light mass target nuclei [cf. Refs. 4–6]. The emission of ^{2}H and ^{4}He, on the other hand, has been studied also from heavier nuclei at several other energies [cf. Refs. 7–10]. With very fast neutrons ($E \geq 25$ MeV) and with high energy charged particles as projectiles ($E \geq 50$ MeV) the emission of several complex particles (like ^{3}H, ^{3}He, ^{4}He, ^{6}Li, ^{7}Be etc.) has also been studied [cf. Refs. 11–15].

Many of the counter telescopes used are incapable of detecting low-energy particles. A quadrupole spectrometer system [cf. Refs. 16], on the other hand, transfers the emitted charged particle to a position of low background where it is detected advantageously via a dE/dx-E detector system. However, such an arrangement has intensity problems and to date only ^{2}H and ^{4}He could be investigated.

Off-line methods

The off-line methods of investigation involve accumulation and subsequent identification of the emitted particle or the residual nuclide, using radiometric analysis, mass spectrometry (MS) or accelerator mass spectrometry (AMS). The techniques of sample preparation, irradiation and projectile flux determination, as well as the radiochemical separations involved, are the same, no matter which detection method for the reaction product is used. The major part of the work, however, has been carried out using the activation technique, i.e. via assay of the induced radioactivity.

Sample preparation

Sample preparation is a very critical step in the study of complex particle emission reactions. Due to the very low cross sections of such processes a careful check of the impurities in the sample is mandatory. It is not uncommon that a strong reaction channel on an impurity leads to the same product as expected from the weak nuclear process under investigation. A neglect of the impurity would thus often impair the whole result. In work with monoenergetic neutrons, due to relatively low available fluxes, generally thick samples are used. This calls upon a consideration of the role of secondary particles. Finally, in work with charged particles, thin uniform samples are needed. These are generally produced using several physico-chemical techniques like electrodeposition, sublimation, sedimentation etc.

Irradiations and flux determination

Irradiations and projectile flux determination demand considerable care since quantitative measurements are involved. A considerable amount of work has been performed with neutrons, involving both monoenergetic and spectral sources. Monoenergetic neutrons in the energy range of 3 to 12 MeV are produced via the $^{2}H(d,n)^{3}He$ reaction and those in the energy range of 13 to 20 MeV via the $^{3}H(d,n)^{4}He$ reaction. The high energy deuteron beam is furnished by a Van de Graaff machine or a cyclotron. In the case of the former reaction deuterium gas targets have been commonly used. The high energy spectral neutron sources are generally d/Be based, i.e. neutrons are produced via breakup of high energy deuterons on Be. For example, a 53 MeV deuteron beam from a cyclotron incident on a 1-cm thick Be disc provides a neutron spectrum extending from about 10 MeV to 53 MeV, with a maximum at about 23 MeV. Irradiations with charged particles are done using beams extracted from cyclotrons or other accelerators. Generally thin samples placed in a row are irradiated together ("stacked-foil technique"). Due to energy degradation in the intervening matter the effective projectile energy in each sample is different. It is thus possible to obtain cross sections at several energies in just one experiment. In radiochemical studies involving both neutrons and charged particles the projectile flux is measured using suitable monitor reactions.

Radiochemical separations

Specific radiochemical separations facilitate isolation of transmutation products. A variety of radiochemical methods like precipitation, coprecipitation, solvent extraction, ion-exchange chromatography, thermochromatography, distillation etc. have been employed [cf. Refs. 17, 18]. In particular the high performance ion-exchange chromatography and thermochromatography have been finding increasing applications, for example in the study of (n,α) reaction on lanthanoids [19] and the (p,^{7}Be) process on ^{209}Bi [20]. In addition to the removal of the matrix activity, the radiochemical separations also facilitate the preparation of thin sources for β^- counting or soft X-ray spectrometry. The excitation function of the ^{58}Ni(n,α)^{55}Fe reaction, for example, was determined via measurement of the 5.9 keV X-ray of radiochemically separated ^{55}Fe [21].

In studies on triton emission, on the one hand the residual nuclide, if radioactive, was radiochemically separated and assayed and, on the other, the accumulated tritium was isolated and counted. When the cross section was high, the assay of tritium could be done via liquid scintillation counting. In irradiated samples with very low count rates of tritium, however, a vacuum extraction method was developed [cf. Ref. 22]. The idea for this technique came from the experience available in handling tritium in hot atom chemistry work [cf. Ref. 23]. Counting was done in the gas phase using an anticoincidence circuit.

Measurement of radioactivity

In investigations on complex particle emission reactions the level of radioactivity encountered was extremely low. Even after chemical separations its measurement often posed great problems. Very low background detectors were thus mandatory. Besides well-shielded NaI(Tl), Ge(Li) and HPGe detectors, $\gamma\gamma$-coincidence and triple coincidence methods were applied. In the case of low energy γ-rays small-sized Ge detectors (with Be-windows) were used, and for X-rays Si(Li) detectors. The pure β^- emitters were assayed using anticoincidence counters and the very soft β^- emitters (like tritium, ^{37}Ar) via gas counting (see above).

Other methods of product assay

As mentioned above, some of the reaction products could be identified also by MS and AMS. These techniques are especially suitable when the transmutation product is stable or very long-lived. In nuclear reactions studies, mass spectrometry has been occasionally applied to the determination of inert gases and several other elements. However, regarding complex particle emission, it has found application so far exclusively to the estimation of ^{3}He and ^{4}He. For detecting ^{3}He-particles a quadrupole spectrometer in a dynamic mode was used [cf. Ref. 24] and for ^{4}He-particles a magnetic spectrometer in a static mode [cf. Ref. 25]. The technique constitutes a sensitive method for the detection of light mass stable gaseous products and about 10^8 atoms can be detected. The application of the AMS technique to cross section work is more recent [cf. Ref. 26]. It involves mass separation after acceleration of ions to several MeV energies. This is a very sensitive technique and has been successfully applied in recent years to the determination of cosmogenic nuclides like ^{26}Al, ^{36}Cl, etc. As far as complex particle emission reactions are concerned, to date the AMS technique has found application in the detection of ^{10}Be [cf. Ref. 27].

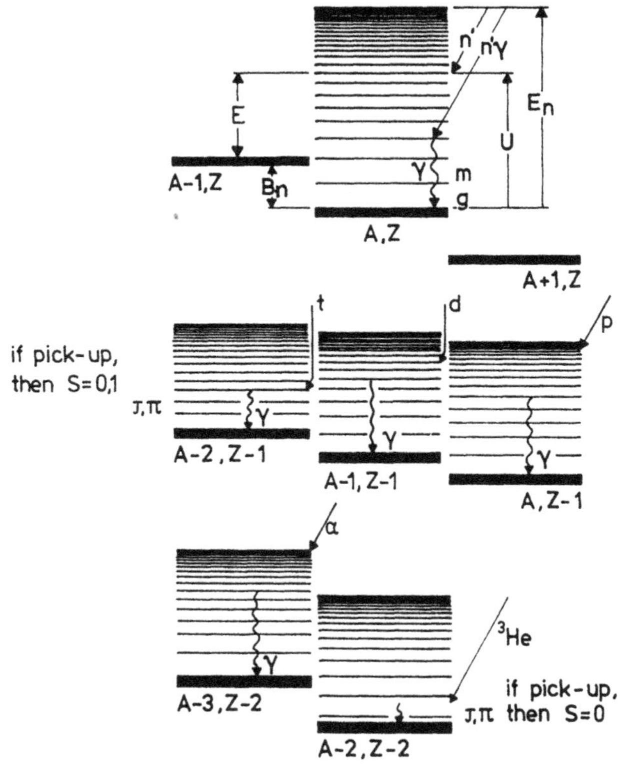

Fig. 1. Schematic representation of the energy scheme of neutron induced nucleon and complex particle emission reactions.

Nuclear model calculations

The energy spectra and the angular distributions of the emitted ^{4}He-particles have been generally understood in the light of compound and precompound models. In the case of other emitted complex particles, whenever such information is available, the results have been interpreted in terms of direct reaction theories. In particular the DWBA (Distorted Wave Born Approximation) calculation has been successfully applied [cf. Refs. 4, 5]. The total cross section for the emission of a particular complex particle, i.e. the cross section integrated over all angles and energies, is more difficult to reproduce via model calculations. Since radiochemical techniques lead to integral cross sections, we consider the calculational attempts to interpret those data in more detail.

At excitation energies above 30 MeV the precompound model has been very successful in describing radiochemical data on nucleon emission. However, it does not give satisfactory results on complex particle emission. One of the major difficulties lies in the assumption on the preformation factor of a complex particle, the ^{4}He-particle being an exception. Furthermore, nuclear structure effects are expected to play a stronger role than in the case of nucleon emission.

In the energy region up to about 30 MeV the statistical model has been partially successful in describing the emission of complex particles. A simplified energy scheme of various nuclear processes is given in Fig. 1. The interaction of a neutron with the target nucleus (A, Z) leads to the formation of a compound nucleus $(A+1, Z)$ in an excited state. In the subsequent deexcitation process both nucleons and complex particles (d,t,^{3}He,α) may be emitted. Due to the relatively high thresholds of complex particle emission reactions, the transitions occur not only to the continuum but also to discrete levels of the product nucleus. The Hauser-Feshbach calculational method, based on the statistical model, takes into account the level structures of the products concerned. In this method the total cross section for the excitation of a level with an excitation energy of E_B^* and spin I_B is given by the equation

$$\sigma(E_B^*, I_B) = \sum_{J}^{J_{max}} \sigma_{form}(J) G(E_B^*, I_B, J)/g)J),$$

where $\sigma_{form}(J)$ gives the formation cross section of the compound nucleus with spin J.

The total width $g(J)$ for the decay of the compound nucleus in all the open channels is divided into two parts

$$g(J) = \sum_{b'} \sum_{l'_2} \sum_{S'_2 = |l'_2 - J|}^{l'_2 + J} \sum_{I_{B'} = |S'_2 - I_{b'}|}^{S'_2 + I_{b'}} \left[\sum_{E_{B'}^* = 0}^{E_c} T_{l'_2} + \int_{E_{B'}^* = E_c}^{E_B^{*'}\max} T_{l'_2} \rho(E_{B'}^*, I_{B'})\, dE_{B'}^* \right],$$

i.e., the summation of transmission coefficients T_l of discrete levels is replaced by an integration over level density ρ at the beginning E_c of the "continuum region".

The two major sources of uncertainty in these calculations are the optical model parameters used for determining transmission coefficients, and the form of level density applied. The uncertainties in the former, especially at low excitation energies, are the limiting factors in the calculation of complex particle emission cross sections.

Integrated ^{2}H and ^{4}He emission cross sections have been analysed in terms of multistage Hauser-Feshbach calculations with reasonable success [cf. Refs. 28, 29]. In the case of triton emission cross sections, besides purely Hauser-Feshbach calculations [cf. Refs. 30, 31], limited attempts were also made to introduce precompound effects [cf. Refs. 32–35]. Regarding the activation data on ^{4}He emission, an incorporation of the direct interaction contribution has also been attempted [cf. Refs. 36, 37]. All those results are discussed below.

Status of data

Spectral data

As mentioned above, the best mechanistic information on the emission of complex particles is obtained through studies on energy spectra and angular distributions. Such studies have generally been performed in the following cases:

a) (n,d), (n,t) and (n,α) reactions on light mass target nuclei at E_n = 14 MeV. Due to the relatively high cross sections and high neutron fluxes the studies have been quite successful and revealed that complex particle emission (except α-emission) involves predominantly direct interactions (one or two nucleon pick-up, knock-out, etc.)

b) charged particle induced reactions on light and medium mass target nuclei at projectile energies of $\geq$20 MeV. Here a large number of reactions with appreciable cross sections are possible, e.g. (p,d), (p.t), (p,α), (p,^{7}Be), (d,t), (d,α), (^{3}He,t), (^{3}He,α), (α,d), (α,t) etc. Some of them have been investigated only qualitatively while in other cases more quantitative work has been done. The general conclusion from all those studies is that, except for ^{4}He, the complex particle emission reactions entail strong direct components.

Despite the useful mechanistic information provided by the spectral studies, there remains an inherent deficit in those investigations. Most of the measurements are done in the forward direction, covering an angle of up to about 70°, and hence the information available in the backward direction is relatively weak. The forward-peaked emission spectra depict the role of direct interactions, the lack of information in the backward direction, however, implies that the compound nucleus contribution cannot be estimated. Furthermore, an integration of the cross section over all angles, based on measurements done in the forward direction, is rather uncertain. Such integrated data are needed for developing systematics and for some practical applications.

Radiochemical data

The radiochemical data are primarily integral data. In the present review they will be considered in more detail. In general, radiochemistry has contributed appreciably to this field of study over the last 20 years.

A large number of systematic radiochemical studies have been performed using 14 MeV neutrons, quasi-monoenergetic neutrons in the energy ranges of 5–10 MeV and 15–20 MeV, and d(Be) breakup spectral neutrons. Several investigations have also been done using light ions (p,d,^{3}He,α) as projectiles. We consider the available information on each emitted complex particle individually.

^{2}H-emission

The deuteron emission is difficult to characterize via radiochemical methods. Since the binding energy of a deuteron is rather low, the Q-values for ^{2}H-emission and pn-emission do not differ appreciably. The determination of the residual nucleus thus gives a sum of (d+pn)-emissions. Extensive radiochemical studies performed at Jülich on 26 target nuclei established the systematics of the [(n,d) + (n,pn) + (n,n'p)] cross sections [cf. Ref. 38] at 14 MeV, and showed that in

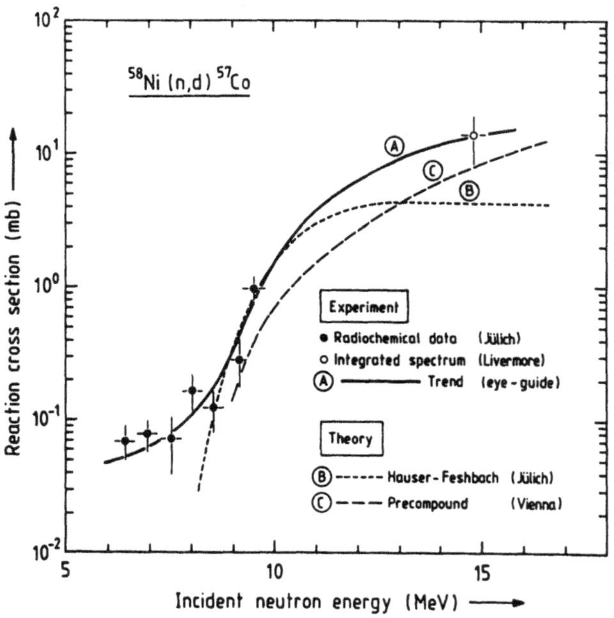

Fig. 2. Excitation function of the ^{58}Ni(n,d)^{57}Co reaction. Points and the solid line describe the experimental data and trend. The dashed lines give results of Hauser-Feshbach and precompound calculations (taken from [Ref. 39]).

the case of the lightest mass isotope of an element there is a strong competition between the (n,2n) and (n,n′p) processes, but did not shed much light on the emission of a deuteron. A mass spectrometric characterization of the emitted deuteron would be ideal. However, due to strong background effects the technique has hitherto not been developed.

A large number of integrated ^{2}H-emission cross sections at 14 MeV were deduced from spectral measurements done at Livermore [cf. Refs. 28, 29]. Those data are comparable to the radiochemical data. For elements with $Z = 13$ to 42 the cross section was found to be almost constant. Hauser-Feshbach calculations [38] suggested that ^{2}H-emission contains large contributions from direct processes.

As far as the energy dependence of deuteron emission in neutron induced reactions is concerned, the available information is sparse. Radiochemical measurements in the neutron energy range of 14 to 20 MeV are ambiguous since the (n,pn) and (n,n′p) contributions become much stronger than the (n,d) contribution. Only below 10 MeV can some information be deduced. The only excitation function reported so far [39] is reproduced in Fig. 2. Hauser-Feshbach calculations show that beyond 11 MeV the relative contribution of the statistical processes decreases and at about 15 MeV it is <30% of the total (n,d) cross section. Calculations involving precompound effects seem to agree with the experimental trend within a factor of 2.

Investigations with charged particles involving reactions like (p,d), (α,d) etc. have concentrated primarily on spectral studies. In a detailed investigation at Oak Ridge, integrated ^{2}H-emission cross sections for 62 MeV protons on several target nuclei were deduced [cf. Ref. 14]. An attempt was made to understand the results in the frame of the precompound model. Some radiochemical data are also available. However, as discussed above, an unambiguous information on ^{2}H-emission cannot be deduced from those data.

^{3}H-emission

Triton emission could be investigated either by measuring the β^- activity of the accumulated tritium or by assaying the radioactivity of the residual product (if not stable). In contrast to the deuteron, the binding energy of the triton is high. The Q-value of the (n,t) process thus substantially differs from those of the (n,p2n) and (n,dn) processes, all of which lead to the same product nucleus. At an incident neutron energy of about 14 MeV, the (n,t) process could thus be conveniently studied via estimation of the residual product.

Early spectral studies on triton emission, performed using 14 MeV neutrons, had shown that the (n,t) cross section for light mass target nuclei is exceptionally high and the deuteron pick-up process dominates. Investigations on the medium and heavy mass target nuclei, on the other hand, could only be done via radiochemical methods. Systematic studies performed in the early 1970s at Jülich [40, 41] showed that most of the earlier data were wrong, mainly because the disturbances from the interfering impurities were not properly considered. As a result of careful measurements on 18 target nuclei a systematic trend was developed which is given in Fig. 3. Some complementary studies performed at Debrecen [30, 42] showed that the data for odd-mass target nuclei follow a somewhat different trend (cf. Fig. 3). Presumably the large differences in Q-values for the (n,t) reactions on odd- and even-mass target nuclei are responsible for this effect.

Using the radiochemical technique it was also possible to determine the isomeric cross-section ratio ($\sigma_m/\sigma_m + \sigma_g$) for the population of several isomeric pairs in (n,t) reactions at 14 MeV [43]. A plot of the data against the spin of the isomeric state (J_m) showed the typical dumb bell structure. The most favoured spin in the (n,t) process, however, appeared to be different from that observed in reactions involving nucleon emission.

The energy dependence of triton emission was investigated in three ways: (a) integral measurements with d(Be) breakup spectral neutrons, (b) excitation function measurements using monoenergetic neutrons, (c) excitation function measurements using protons, deuterons and α-particles as projectiles. We discuss these in some detail below.

Integral measurements of (n,t) cross sections were done using 53 MeV d(Be) and 30 MeV d(Be) breakup neutrons [cf. Refs. 22, 44, 45]. In particular, with 53 MeV d(Be) neutrons about 40 target elements between Li and Bi were investigated, both via tritium counting and measurement of the radioactive product.

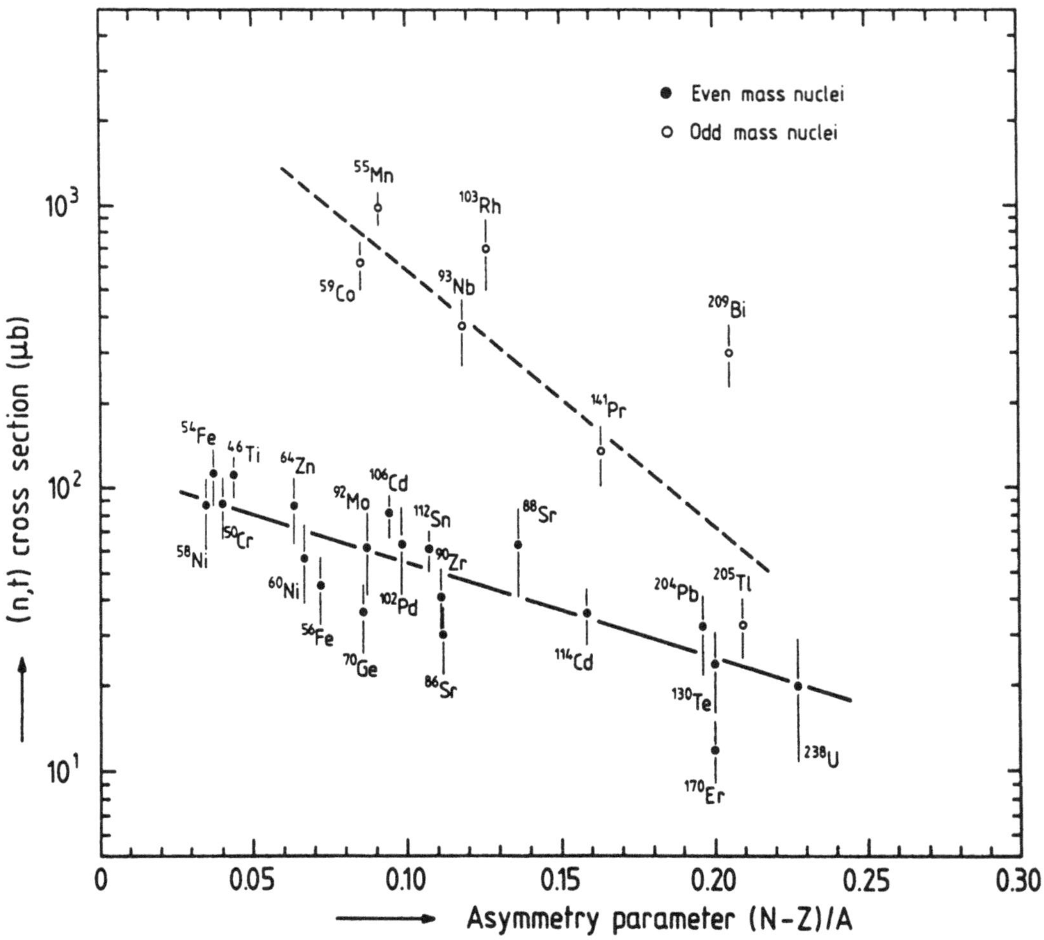

Fig. 3. Updated version of the systematics of (n,t) reaction cross sections at 14.6±0.4 MeV for nuclei with $Z \geq 22$ first reported by Qaim and Stöcklin [41]. The trend for even-mass nuclei was suggested by the Jülich group [cf. Refs. 40, 41] and that for odd-mass nuclei by the Debrecen group [cf. Refs. 30, 42]. The diagram is taken from Ref. [43].

The results are shown in Fig. 4. Obviously the (n,t) cross section for the lightest elements is exceptionally large. For elements with $Z > 20$, however, the emission of a triton is relatively independent of the target element. A comparison of the tritium counting data with those obtained via residual activity measurement shows that for nuclei with $A > 40$ the emission of three particles (p2n) is much more favoured than the emission of a bound trinucleon (^{3}H).

Excitation functions of (n,t) reactions were measured using quasi-monoenergetic neutrons and tritium counting [cf. Refs. 46–52]. Attempts have also been made to determine the excitation function using breakup neutrons. The integral count rates of tritium, determined experimentally after irradiations of the target material in several well-known neutron fields, were used to calculate the (n,t) excitation function via spectrum unfolding codes [cf. Refs. 53, 54]. Information is now available on (n,t) excitation functions for 13 target nuclei in various mass regions. The data for a few selected medium and heavy mass nuclei are reproduced in Fig. 5. As expected, the (n,t) cross section increases with the increasing energy of the incident neutron. At a given energy, the experimental cross section decreases over the mass range 27 to 59, thereafter it is almost constant. The results of Hauser-Feshbach calculations are also given in Fig. 5. Evidently, the statistical model reproduces the (n,t) excitation function of Al very well. With the increasing mass of the target nucleus, however, the statistical contribution appears to decrease drastically. In those cases possibly direct interactions play an important role.

Excitation functions of (p,t), (d,t) and (α,t) reactions on Al, V, Nb and Au were measured using the "stacked-foil" technique and tritium counting [34, 35]. The trends for these processes as well as the (n,t) reaction are shown in Fig. 6. Apparently the cross sections of (p,t), (α,t) and (n,t) reactions near the thresholds are almost equal. The difficulty in studies on (n,t) reactions originates from the very low fluxes of monoenergetic neutrons (by a factor of about 10^5 lower than for charged particles). At higher energies, as depicted by studies with d(Be) breakup neutrons [cf. Refs. 22, 44], the (n,t) cross section is much lower than charged particle induced triton emission cross sections. The cross section is the highest for the (d,t) reaction, which possibly involves a neutron pick-up process. Similar to the (n,t) reaction, the charged particle induced triton emission processes could also not be satisfactorily described by compound and precompound models.

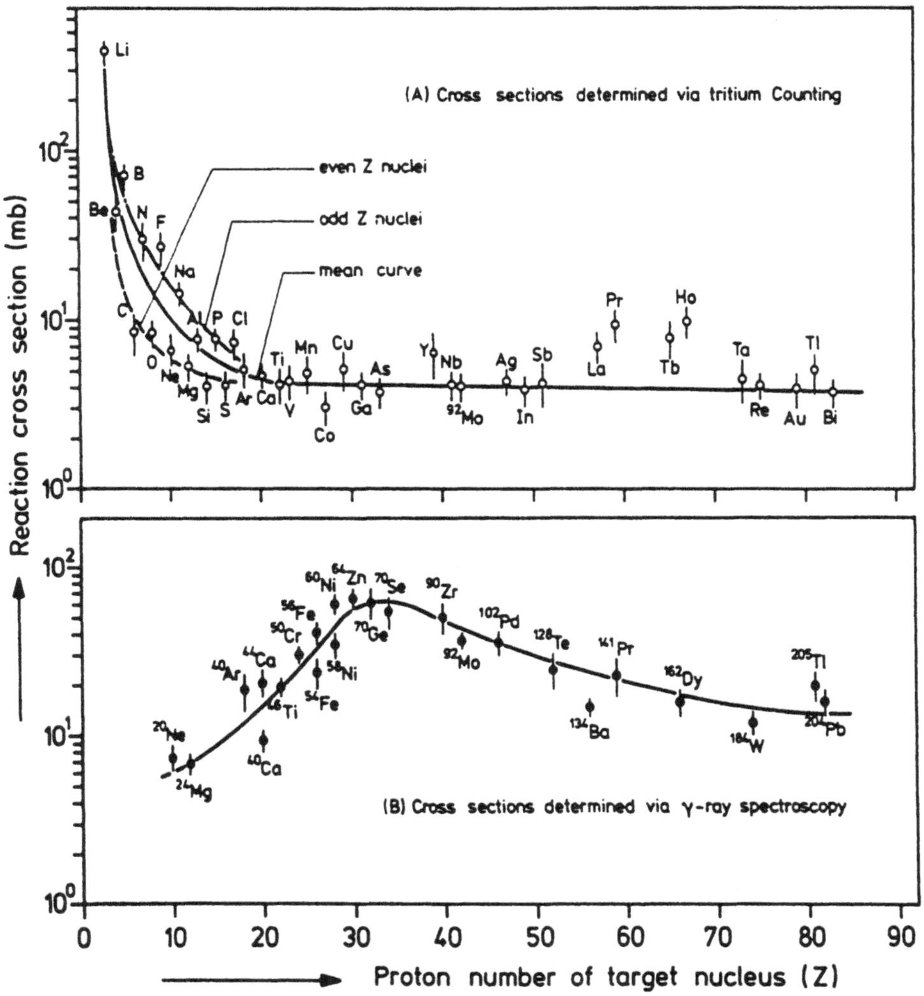

Fig. 4. Cross sections of (n,t) reactions induced by 53 MeV d(Be) breakup neutrons plotted as a function of Z of the target element. Data were obtained, (A) via tritium counting, (B) via measurement of the residual product (taken from [Ref. 44]).

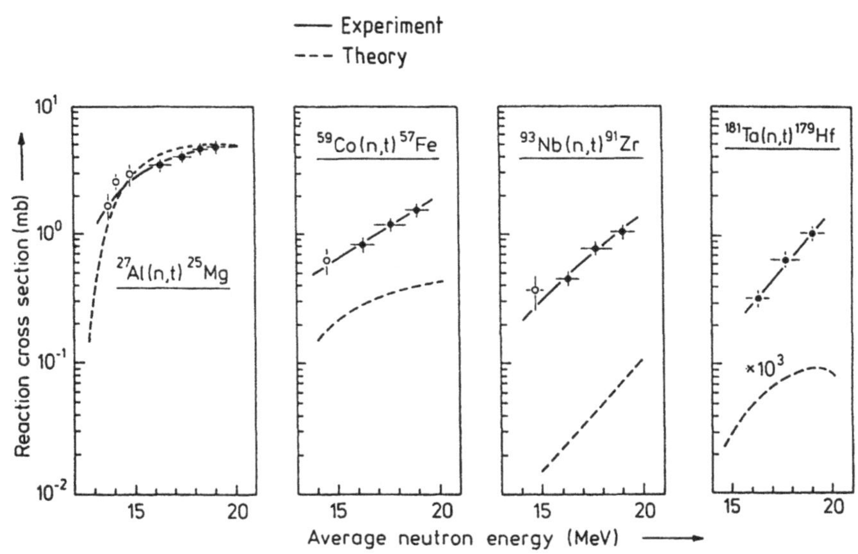

Fig. 5. Excitation functions of (n,t) reactions on ^{27}Al, ^{59}Co, ^{93}Nb and ^{181}Ta. The solid and open points give the experimental data. The solid line is an eye-guide, and the dashed curve denotes the results of Hauser-Feshbach calculations (taken from Refs. [30, 51, 52]).

The (^{3}He,t) reaction is expected to have a rather high cross section since it possibly entails a charge transfer process. A few angular distribution studies appear to support this hypothesis [cf. Ref. 55]. However, as far as we know, no radiochemical studies have been performed on this reaction.

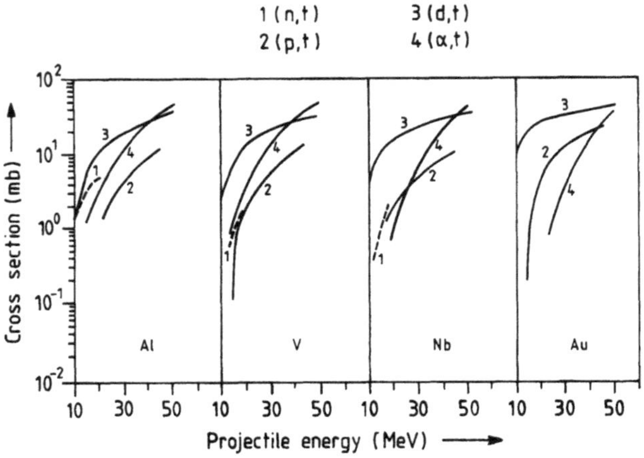

Fig. 6. Trends in excitation functions of (p,t), (d,t), (α,t) and (n,t) reactions on Al, V, Nb and Au (taken from Refs. [30, 33, 51, 52]).

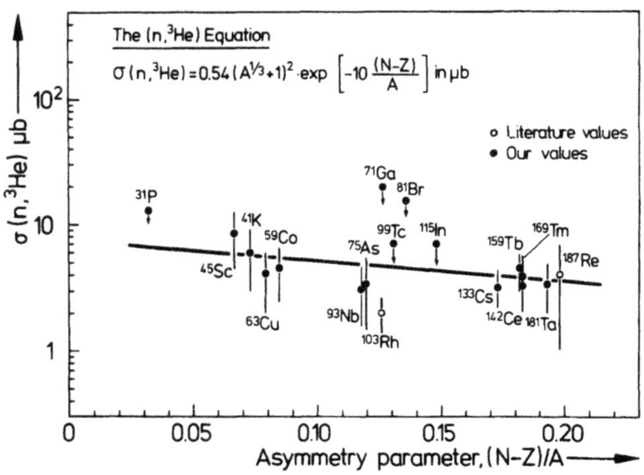

Fig. 7. Systematics of (n,^{3}He) reaction cross sections at 14 MeV for medium and heavy mass nuclei (taken from [Ref. 58]).

^{3}He-emission

The emission of ^{3}He-particles in neutron induced reactions is extremely difficult to study. At 14 MeV this process constitutes almost the weakest reaction channel. Except for a single angular distribution study [56], which suggests that direct interactions are involved, almost all the information to date has been obtained radiochemically. Similar to the (n,t) reaction, at 14 MeV the (n,^{3}He) process can be investigated via measurement of the residual nucleus, since the Q-values of the possible contributing processes (n,^{3}He), (n,dp) and (n,2pn) differ appreciably. The pitfalls encountered in the study of (n,t) processes, however, were even more drastic in the case of (n,^{3}He) reaction. Extensive and careful measurements done at Jülich [cf. Refs. 57, 58] in the 1970s showed that, except for some isolated reliable values [cf. Refs. 59, 60], most of the literature data were wrong. The experimental results, and the systematics developed therefrom are shown in Fig. 7. The trend is somewhat similar to that for (n,t) cross sections on even mass nuclei (cf. Fig. 3); in terms of absolute magnitude, however, the

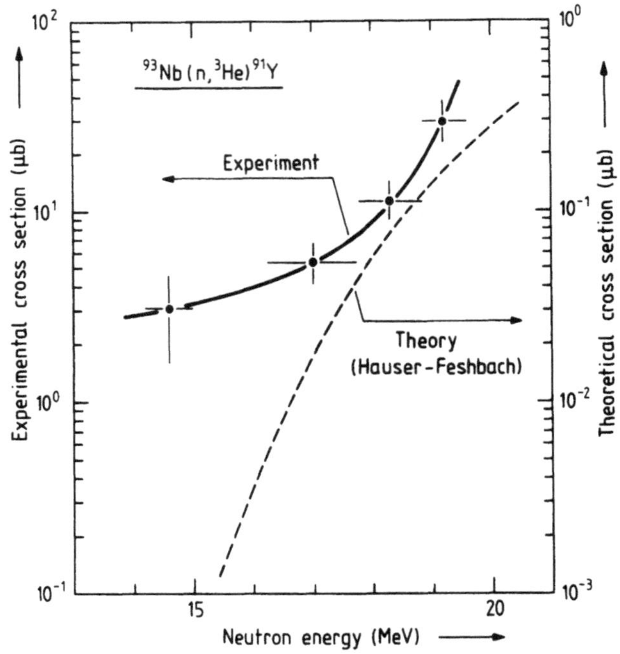

Fig. 8. Excitation function of ^{93}Nb(n,^{3}He)^{91}Y process. Points and the solid line describe the experimental data and trend; the dashed line gives results of Hauser-Feshbach calculations (taken from Ref. [61]).

(n,^{3}He) cross section is by an order of magnitude smaller than the (n,t) cross section.

The energy dependence of ^{3}He-emission in neutron induced reactions could be studied radiochemically up to 20 MeV only for the target nuclei ^{93}Nb and ^{139}La [cf. Refs. 61, 62]. Although some contributions (<20%) from (n,2pn) and (n,dp) processes could not be ruled out, the major portion of the measured cross section was attributed to the (n,^{3}He) process. The results for ^{93}Nb are shown in Fig. 8. The cross section increases at first slowly, later, however, rapidly with the increasing neutron energy. Also shown in the figure are the results of Hauser-Feshbach calculations on the first chance emission of a ^{3}He-particle. The experimental and theoretical curves differ by a factor of >10^2 over the entire investigated energy range; apparently the contribution of statistical processes is small and direct interactions play an important role.

Radiochemical studies with neutrons or charged particles of energies above 20 MeV demand that the accumulated ^{3}He gas be determined mass spectrometrically. This was done in investigations using 53 MeV d(Be) neutrons [24, 63]. In general the ^{3}He/^{4}He emission ratio was determined both mass spectrometrically and via assay of the activation product. The results are reproduced in Fig. 9. Evidently the ratio depends on mass and charge of the target nucleus; interesting, however, is the observation that the ^{3}He-emission may amount up to 40% of the ^{4}He-emission. Also worth pointing out is the fact that the activation and mass spectrometric data agree, implying that the activation product is formed primarily via ^{3}He-emission and not via (dp) or (2pn) emission. Furthermore, a comparison of the (n,t) and (n,^{3}He) data shows that,

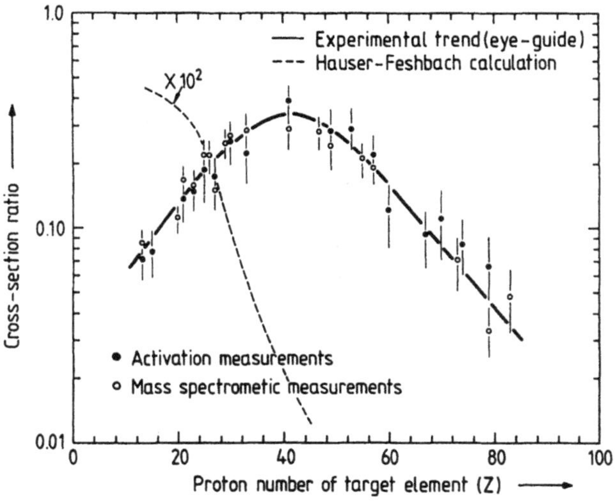

Fig. 9. ^{3}He/^{4}He emission cross-section ratio as a function of Z of the target element. The solid line describes the experimental trend; the dashed curve gives the result of Hauser-Feshbach calculation (taken from Ref. [63]).

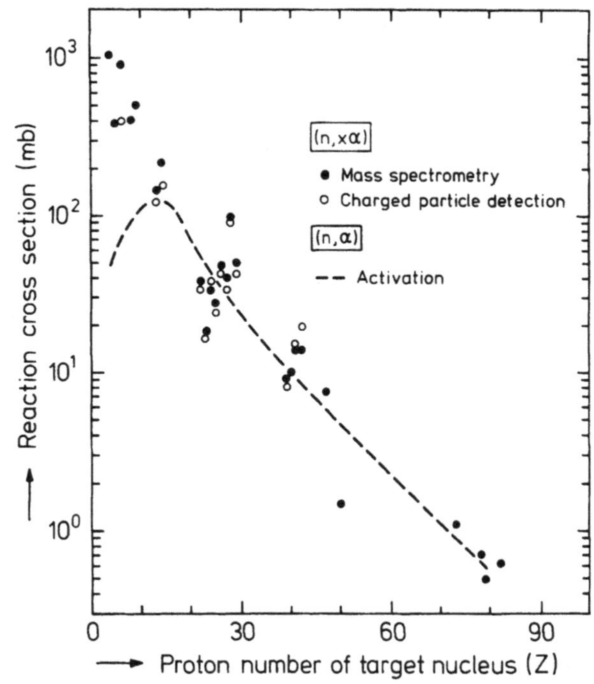

Fig. 10. ^{4}He-emission cross sections at 14 MeV as a function of Z of the target nucleus. The open circles describe (n,xα) data obtained via integration of the emitted ^{4}He-spectra (cf. Refs. [28, 29]) and the closed circles those via mass spectrometry (cf. Ref. [64]). The curve describing the activation data relates to the pure (n,α) contribution and is based on more than 100 data points.

of the two trinucleons, the emission of ^{3}He is intrinsically weaker.

In Fig. 9 the results of Hauser-Feshbach calculations on the ^{3}He/^{4}He emission ratios, averaged for the whole neutron spectrum, are also given. The calculated ratio is much smaller than the experimental ratio, suggesting that the contribution of statistical processes to ^{3}He-emission is much smaller than in ^{4}He-emission. Furthermore, the calculated ratio decreases rapidly with increasing Z, indicating that, with increasing Z of the target nucleus, the contributions of statistical processes decrease much more sharply in the case of ^{3}He-emission than in ^{4}He-emission.

The emission of a ^{3}He-particle in proton, deuteron and α-particle induced reactions is possible, especially at energies above 20 MeV. Some spectral studies on the (p,^{3}He) reaction at E_p = 62 MeV were reported and integrated cross sections deduced [cf. Ref. 14]. There appear to be more direct contributions than in ^{3}H-emission. Cross sections via mass spectrometric identification of the emitted ^{3}He-particles in charged particle induced reactions have hitherto not been reported.

^{4}He-emission

Due to its exceptionally strong binding energy, the ^{4}He-emission is comparable to the emission of nucleons, especially when light mass target nuclei are involved. For heavier nuclei, due to the increasing Coulomb barrier the probability of ^{4}He-emission decreases rapidly.

Considerable effort has been devoted to the study of ^{4}He-emission in reactions induced by neutrons as well as charged particles. Both spectral and radiochemical measurements were performed and the data base is fairly strong. In addition, extensive mass spectrometric measurements on ^{4}He, emitted in the interactions of 14 MeV neutrons with a large number of elements, were carried out at Atomics International [cf. Ref. 64]. A summary of all the available 14 MeV data for elements of natural isotopic composition is presented in Fig. 10. The (n,xα) data obtained via integration of the emitted ^{4}He-spectra [cf. Refs. 28, 29] and via mass spectrometry [64] are shown as individual points. The trend for the pure (n,α) reaction based on activation measurements is given as a curve. Obviously for very light nuclei ($Z<10$) the pure (n,α) contribution is small, the (n,2α) and multiparticle breakup processes leading to the emission of several α-particles being dominant. For nuclei with $Z>10$, however, a greater part of the measured (n,xα) cross section is furnished by the (n,α) reaction.

The energy and angular distribution studies on (n,α) and (p,α) reactions suggest that the major part of the cross section is described by compound and precompound models, direct processes making only small contributions. In the case of (d,α) and (^{3}He,α) reactions, on the other hand, the contributions of direct processes appear to be stronger. The radiochemically obtained data, especially the excitation functions of (n,α) and (p,α) reactions, have been often compared with results of nuclear model calculations. In general, the compound and precompound models adequately reproduce the experimental data. The ^{4}He-emission process, however, is sensitive to Q-value effects, and the changing contributions of neutron and proton emission may often affect the mechanism of ^{4}He-emission. In some recent studies [cf. Refs. 36, 37] an attempt

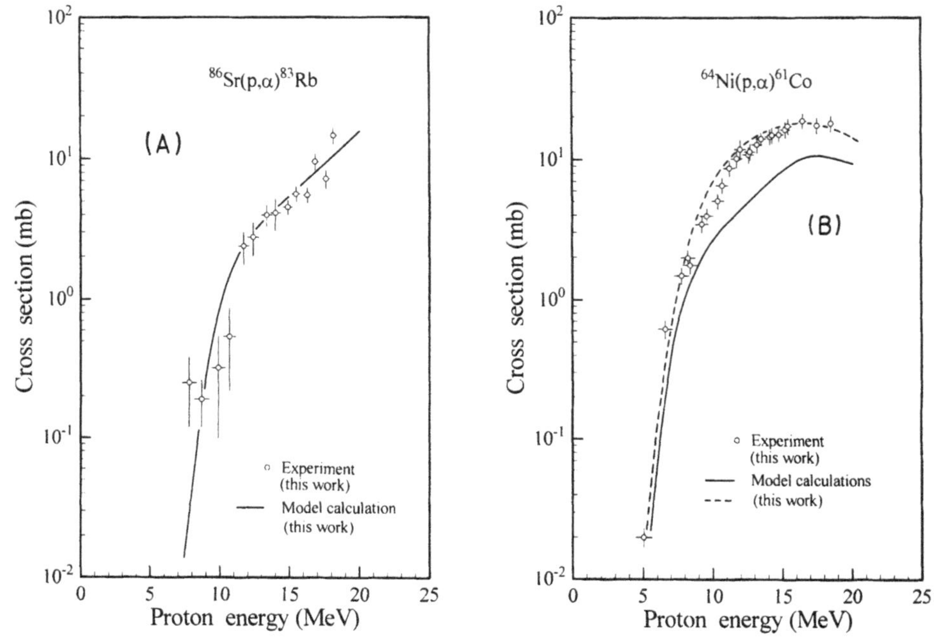

Fig. 11. Excitation function of (p,α) reaction on, (A) ^{86}Sr, (B) ^{64}Ni. The open circles give experimental data. The solid curve gives the sum of the compound nucleus evaporation and precompound emission. The dashed curve includes an additional direct pickup component (taken from Ref. [37]).

has been made for the first time to include direct interaction contribution to the measured activation cross section. The results for the (p,α) reactions on ^{86}Sr and ^{64}Ni are shown in Fig. 11. Whereas the conventional compound and precompound calculations reproduce the excitation function of the $^{86}Sr(p,\alpha)^{83}Rb$ reaction fairly well, inclusion of a direct component in the $^{64}Ni(p,\alpha)^{61}Co$ process appears mandatory. Presumably, due to the very low threshold of 2.46 MeV for the ^{64}Ni(p,n) process, the statistical ^{64}Ni(p,α) contribution is considerably diminished.

^{7}Be-emission

The ^{7}Be-emission in neutron induced reactions is extremely difficult to study since the Q-value of the reaction is highly negative and the available monoenergetic neutrons in the energy region above 15 MeV have low fluxes. Radiochemical studies were performed so far only on eight elements between Al and Au using 53 MeV d(Be) breakup neutrons [cf. Ref. 18]. In each case the chemically separated ^{7}Be activity was determined. The cross section was found to be in the range of 0.5 to 1.0 μb (see below).

Studies with charged particles have been more extensive. As mentioned above, ^{7}Be was detected in the spallation process already in the 1950s. At energies below 200 MeV ^{7}Be-emission spectra were investigated in reactions induced by protons, ^{3}He- and α-particles. The forward peaked nature of those spectra suggests the existence of direct processes. Radiochemical studies were carried out mostly on light mass target elements like C, N, O, F, Al etc. using protons, deuterons, ^{3}He- and α-particles as projectiles. In the case of medium and heavy mass target nuclei, on the other hand, the use of protons has been more common. Measurements at two energies (65 and 85 MeV) were done already in the 1960s and some systematic trends were reported [cf. Ref. 65]. More recently detailed excitation functions of (p,^{7}Be) reactions were measured up to about 100 MeV [cf. Ref. 66]. Some typical results are shown in Fig. 12. In the light mass region, ^{7}Be is expected to be formed as a product nucleus. For medium and heavy mass target nuclei, however, the energetics of various competing processes suggest that

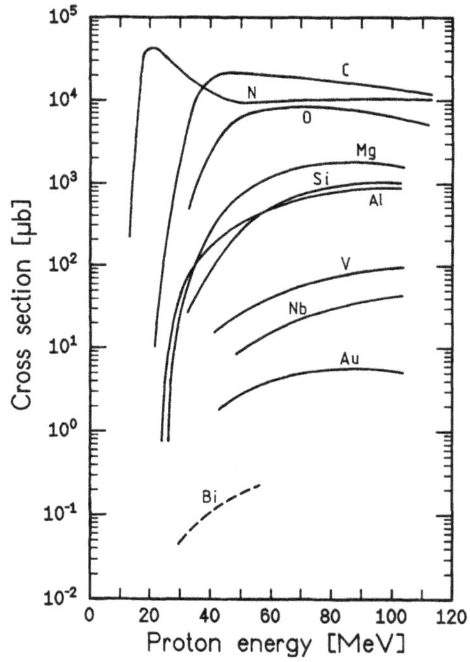

Fig. 12. Systematics of (p,^{7}Be) excitation functions on target nuclides of elements C to Bi (taken from Ref. [66]).

the emission of the complex particle (^{7}Be) is dominant. These systematics of the excitation functions reveal that for proton energies above 40 MeV the emission of ^{7}Be decreases with the increasing mass of the target nucleus. The trend is valid also for heavy elements (Au and Bi) which undergo fission with high energy protons.

Similar to the trinucleon emission reactions discussed above, the investigation of ^{7}Be-emission demands a very carful check of the impurities, since some of the light elements, possibly present as impurities, have a high ^{7}Be-emission cross section. In the case of ^{209}Bi, for example, the (p,^{7}Be) cross section was found to be ambiguously high. A check of the oxygen impurity level in ^{209}Bi and a careful investigation of the product nucleus ^{203}Hg revealed [20] that the real ^{209}Bi(p,^{7}Be) cross section is much smaller and fits well into the systematics described above.

To date no attempt has been made to reproduce the radiochemically determined ^{7}Be-emission cross section via a nuclear model calculation. The inadequacy of the compound nucleus model, the problem in applying the precompound model, and the inability to treat the direct component are the main reasons.

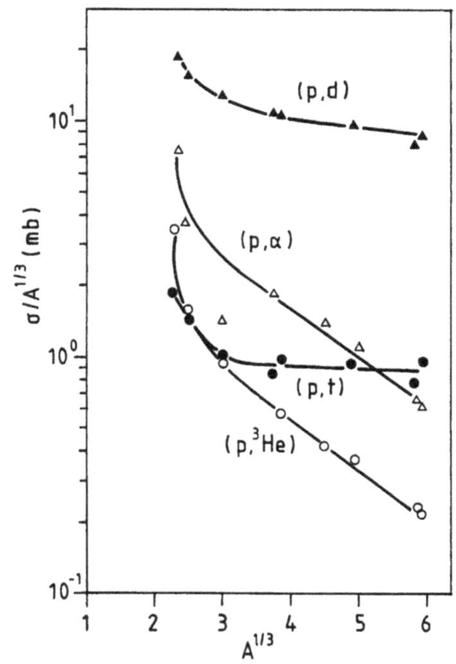

Fig. 13. Systematic trends in normalized (p,d), (p,t), (p,^{3}He) and (p,α) reaction cross sections induced by 62 MeV protons (taken from Ref. [14]).

Emission of other complex particles

Limited spectral studies on the emission of several other complex particles like ^{6,7}Li, ^{9,10}Be, ^{10,11}B etc. in charged particle induced reactions have been performed. However, radiochemical techniques could not be applied since all of those emitted particles are stable. In recent years a few investigations on ^{10}Be-emission could be successfully performed via the AMS technique [cf. Refs. 27, 67]. Still heavier particles like ^{12,14}C, ^{22}Na etc. have also been observed but mostly at proton energies above 300 MeV, where the fragmentation process starts making a significant contribution.

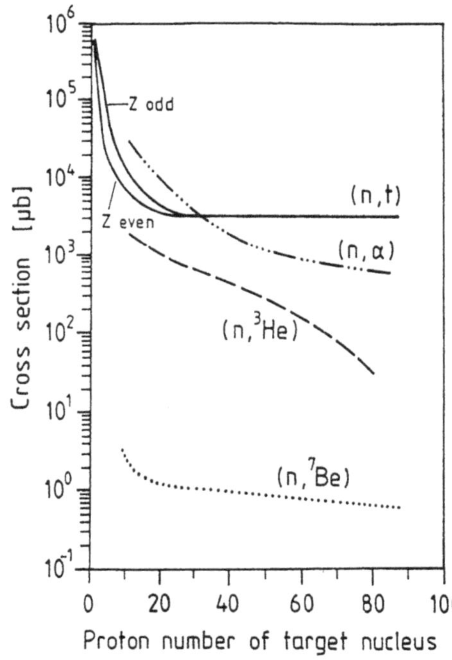

Fig. 14. Systematic trends in (n,t), (n,^{3}He), (n,α) and (n,^{7}Be) reaction cross sections induced by 53 MeV d(Be) breakup neutrons (taken from Ref. [18]).

Comparison of some complex particle emission reaction cross sections

For a few target elements several complex particle emission reactions were investigated and the integral data obtained. To date most extensive studies were performed with 62 MeV protons and 53 MeV d(Be) breakup neutrons. The systematic trends in proton induced reaction cross sections are given in Fig. 13 [cf. Ref. 14]. The values were obtained by an integration of the spectral data and were normalized to $A^{1/3}$. The trends for the (p,d) and (p,t) reactions are somewhat similar. The same appears to be true for the (p,α) and (p,^{3}He) reactions. A plot of the (p,^{7}Be) data for 62 MeV protons taken from Fig. 12 gave a sharply decreasing trend (not shown in Fig. 13), possibly due to stronger Coulomb barrier effects than in ^{3}He- or ^{4}He-emission.

The results for the neutron induced reactions are reproduced in Fig. 14 [cf. Ref. 18]. The data for (n,t) reactions were obtained by tritium counting, for (n,^{3}He) reactions by mass spectrometry and for (n,^{7}Be) reactions by ^{7}Be measurement. Only in the case of (n,α) reactions activated products were studied. Since the (n,nα) processes are sufficiently strong, the activation product nucleus assay does not describe the total α-emission. The real (n,xα) cross section may therefore be somewhat higher.

The experimental trends in Figs. 13 and 14 suggest that ^{2}H and ^{4}He emissions constitute rather exceptional cases. As discussed above, the ^{2}H-emission generally involves a single nucleon transfer process and has therefore a higher probability than ^{3}H, ^{3}He and ^{7}Be emissions, which would entail many particle transfer reactions. The ^{4}He-emission is an exception due to its extremely high binding energy (see above), its probability of emission being comparable to those of nucleons. The trends for ^{3}H, ^{3}He and ^{7}Be emissions, on the other hand, follow a definite pattern; i.e. the higher the charge of the emitted complex particle, the lower is its emission probability.

Conclusions

Investigations of complex particle emission reactions constitute an interesting field of study. Most of the experimental and theoretical work to date has been performed on ^{4}He-emission, which is a relatively strong process in the light and medium mass regions. In general, the compound and precompound models together can adequately describe the ^{4}He-emission cross section. Only in some special cases direct interactions need to be considered. Very little radiochemical work has been done on ^{2}H-emission. On the other hand, radiochemical methods have contributed appreciably to the study of trinucleon emission reactions. Whereas in the case of ^{3}H-emission both compound and direct processes play an important role, the ^{3}He-emission appears to be dominated by direct interactions. The emission of heavier particles like ^{7}Be has been relatively little investigated. With the exception of ^{2}H and ^{4}He emissions, it appears that the higher the charge of the emitted complex particle, the lower is its emission probability.

Acknowledgement

The author thanks Professor (Emeritus) Dr. G. Stöcklin warmly for his active support of this research over a period of two decades and for stimulating discussions. The cooperation and assistance of several coworkers over the years is gratefully acknowledged.

References

1. Qaim, S. M.: "A survey of fast neutron induced reaction cross-section data", in Proc. Conf. on Nuclear Cross Sections and Technology, Washington, D. C., March 1975 (R. A. Schrack and C. D. Bowman, Editors), NBS Gaithersburg, Special Publication 425, 1975, p. 664.
2. Qaim, S. M.: "Recent advances in the study of some neutron threshold reactions", in Proc. Int. Conf. on Neutron Physics and Nuclear Data for Reactors and other Applied Purposes, Harwell, England, September 1978, OECD-NEA Paris, 1979, p. 1088.
3. Qaim, S. M.: "Complex particle emission in fast neutron induced reactions", in Proc. Europhysics Topical Conf. on Neutron Induced Reactions, Smolenice, Czechoslovakia, June 1982 (P. Oblozinsky, Editor), Institute of Physics EPRC, Bratislava 1982, p. 291.
4. Miljanic, D., Valkovic, V., Rendic, D., Furic, M.: Nucl. Phys. **A156**, 193 (1970).
5. Miljanic, D., Valkovic, V.: Nucl. Phys. **A176**, 110 (1971).
6. Perroud, J. P., Sellem, Ch.: Nucl. Phys. **A227**, 330 (1974).
7. Fischer, R., Traxler, G., Uhl, M., Vonach, H.: Phys. Rev. **C30**, 72 (1984).
8. Paulsen, A., Widera, R., Arnotte, F., Liskien, H.: Nucl. Sci. Eng. **72**, 113 (1979).
9. Graham, S. L., Ahmad, M., Grimes, S. M., Satyanarayana, H., Saraf, S. K.: Nucl. Sci. Eng. **95**, 60 (1987).
10. Bharuth-Ram, K., Bawa, A. C., McMurray, W. R.: Phys. Rev. **C36**, 1749 (1987).
11. Kropp, L., Forti, P.: Nucl. Instrum. Methods **104**, 381 (1972).
12. Subramanian, T. S., Romero, J. L., Brady, F. P., Fitzgerald, D. H., Garrett, R., Needham, G. A., Ullmann, J. L., Watson, J. W., Zanelli, C. I., Brenner, D. G., Parel, R. E.: Phys. Rev. **C34**, 1580 (1986).
13. Goeckner, F., Grimes, S. M., Brient, C. E., Lee, T. M., Sterbenz, S. M., Bateman, F. B., Haight, R. C., Young, P. G., Chadwick, M. B., Wasson, O. A., Vonach, H.: "The ^{59}Co(n,α) reaction from threshold to 30 MeV", in Proc. Int. Conf. on Nuclear Data for Science and Technology, Gatlinburg, USA, May 1994 (J. K. Dickens, Editor), American Nuclear Society, La Grange Park, Illinois 1994, p. 318.
14. Bertrand, F. E., Peelle, R. W.: Phys. Rev. **C8**, 1045 (1973).
15. Padgett, M. L., Sobotka, L. G., Wozniak, G. J., Guarino, G., Pacheco, A. J., Moretto, L. G., Chan, Y., Stockstad, R. G., Tserruya, I., Ward, S.: Nuclear Science Division Annual Report, Lawrence Berkeley Laboratory, California, USA, August 1984, LBL-16870, p. 107.
16. Grimes, S. M., Haight, R. C., Anderson, J. D.: Nucl. Sci. Eng. **62**, 187 (1977).
17. Qaim, S. M., Wölfle, R., Stöcklin, G.: J. Radioanalyt. Chem. **30**, 35 (1976).
18. Scholten, B., Qaim, S. M., Stöcklin, G.: Radiochim. Acta **62**, 107 (1993).
19. Qaim, S. M., Ollig, H., Blessing, G.: Radiochim. Acta **26**, 59 (1979).
20. Neumaier, B., Rösch, F., Qaim, S. M., Stöcklin, G.: Radiochim. Acta **65**, 1 (1994).
21. Qaim, S. M., Wölfle, R., Rahman, M. M., Ollig, H.: Nucl. Sci. Eng. **88**, 143 (1984).
22. Qaim, S. M., Wölfle, R., Stöcklin, G.: J. Inorg. Nucl. Chem. **36**, 3639 (1974).
23. Stöcklin, G.: *Chemie heißer Atome, Chemische Folgereaktionen von Kernprozessen,* Verlag Chemie, Weinheim, 1969.
24. Wu, C. H., Wölfle, R., Qaim, S. M.: Nucl. Phys. **A329**, 63 (1979).
25. Farrar IV, H., McElroy, W. N., Lippincott, E. P.: Nucl. Techn. **25**, 305 (1975).
26. Suter, M.: "Accelerator mass spectrometry and its impact on nuclear data studies", in Proc. Int. Conf. on Nuclear Data for Science and Technology, Jülich, Germany, May 1991 (S. M. Qaim, Editor), Springer Verlag, Berlin 1992, p. 381.
27. Bodemann, R., Lange, H.-J., Leya, I., Michel, R., Schiekel, T., Rösel, R., Herpers, U., Hofmann, H. J., Dittrich, B., Suter, M., Wölfli, W., Holmqvist, B., Condé, H., Malmborg, M.: Nucl. Instrum. Methods **B82**, 9 (1993).
28. Grimes, S. M., Haight, R. C., Alvar, K. R., Barschall, H. H., Borchers, R. R.: Phys. Rev. **C19**, 2127 (1979).
29. Haight, R. C., Grimes, S. M., Johnson, R. G., Barschall, H. H.: Phys. Rev. **C23**, 700 (1981).
30. Sudár, S., Csikai, J.: Nucl. Phys. **A319**, 157 (1979).
31. Qaim, S. M., Klapdor, H. V., Reiss, H.: Phys. Rev. **C22**, 1371 (1980).
32. Uhl, M.: IRK Progress Report 1986 (E. Wild and G. Dirniger, Editors), Vienna, 1986, p. 26.
33. Ignatyuk, A. V., Mihaly, K., Grudzevich, O. T., Bödy, Z. T.: Report INDC(HUN)-025/L (1988).
34. Merkel, M., Münzel, H.: Nucl. Phys. **A333**, 173 (1980).

35. Merkel, M., Münzel, H.: Radiochim. Acta **29**, 173 (1981).
36. Qaim, S. M., Uhl, M., Molla, N. I., Liskien, H.: Phys. Rev. **C46**, 1398 (1992).
37. Qaim, S. M., Uhl, M., Rösch, F., Szelecsényi, F.: Phys. Rev. **C52**, 733 (1995).
38. Qaim, S. M.: Nucl. Phys. **A382**, 255 (1982).
39. Qaim, S. M., Wölfle, R.: Phys. Rev. **C32**, 305 (1985).
40. Qaim, S. M., Stöcklin, G.: J. Inorg. Nucl. Chem. **35**, 19 (1973).
41. Qaim, S. M., Stöcklin, G.: Nucl. Phys. **A257**, 233 (1976).
42. Biró, T., Sudár, S., Miligy, Z., Dezsö, Z., Csikai, J.: J. Inorg. Nucl. Chem. **37**, 1583 (1975).
43. Qaim, S. M.: Nucl. Phys. **A438**, 384 (1985).
44. Qaim, S. M., Wölfle, R.: Nucl. Phys. **A295**, 150 (1978).
45. Wölfle, R., Khatun, S., Qaim, S. M.: Nucl. Phys. **A423**, 130 (1984).
46. *Neutron Cross Sections, Vol. 2, Neutron Cross Section Curves,* compiled by V. McLane, C. L. Dunford and P. F. Rose, Academic Press, Inc., New York (1988).
47. Qaim, S. M., Wölfle, R.: Nucl. Sci. Eng. **96**, 52 (1987).
48. Liskien, H., Widera, R., Wölfle, R., Qaim, S. M.: Nucl. Sci. Eng. **98**, 266 (1988).
49. Suhaimi, A., Wölfle, R., Qaim, S. M., Stöcklin, G.: Radiochim. Acta **40**, 113 (1986).
50. Suhaimi, A., Wölfle, R., Qaim, S. M., Warwick, P., Stöcklin, G.: Radiochim. Acta **43**, 133 (1988).
51. Qaim, S. M., Wölfle, R., Liskien, H.: Phys. Rev. **C25**, 203 (1982).
52. Wölfle, R., Qaim, S. M., Liskien, H., Widera, R.: Radiochim. Acta **50**, 5 (1990).
53. Wölfle, R., Sudár, S., Qaim, S. M.: Nucl. Sci. Eng. **91**, 162 (1985).
54. Wölfle, R., Suhaimi, A., Qaim, S. M.: Nucl. Sci. Eng. **115**, 71 (1993).
55. Schulte, A.: "Studium der (^{3}He,t)-Reaktion und der (p,n)-Reaktion bei intermediären Energien", KFA Jülich, Report Jül-Spez-371 (1986).
56. Miljanic, D., Zadro, M., Rendic, D.: Nucl. Phys. **A419**, 351 (1984).
57. Qaim, S. M.: J. Inorg. Nucl. Chem. **36**, 239 (1974); Erratum **36**, 3886 (1974).
58. Qaim, S. M.: Radiochim. Acta **25**, 13 (1978).
59. Csikai, J., Szalay, A.: Nucl. Phys. **68**, 546 (1965).
60. Husain, L., Bari, A., Kuroda, P. K.: J. Inorg. Nucl. Chem. **30**, 3145 (1968).
61. Qaim, S. M., Wölfle, R., Liskien, H.: Phys. Rev. **C34**, 489 (1986).
62. Qaim, S. M., Wölfle, R.: Acta Phys. Hungarica **69**, 137 (1991).
63. Qaim, S. M., Wu, C. H., Wölfle, R.: Nucl. Phys. **A410**, 421 (1983).
64. Kneff, D. W., Oliver, B. M., Farrar IV, H., Greenwood, L. R.: Nucl. Sci. Eng. **92**, 491 (1986).
65. Lafleur, M. S., Porile, N. T., Yaffe, L.: Can. J. Chem. **44**, 2749 (1966).
66. Scholten, B., Qaim, S. M., Stöcklin, G.: Radiochim. Acta **65**, 81 (1994).
67. Lange, H.-J., Hahn, T., Michel, R., Schiekel, T., Rösel, R., Herpers, U., Hofmann, H.-J., Dittrich-Hannen, B., Suter, M., Wölfli, W., Kubik, P. W.: Appl. Radiat. Isot. **46**, 93 (1995).

Radiochimica Acta 70/71, 177–189 (1995)

The Physics and (Radio)chemistry of Solar Neutrino Experiments

By Richard L. Hahn

Chemistry Department, Brookhaven National Laboratory*, Upton, New York 11973-5000, U.S.A.

(Received September 26, 1995)

Solar neutrino problem / Neutrino oscillations / Neutrino detectors / Solar models / Low-level radioactive impurities

Abstract

The situation in solar neutrino science has changed drastically in the past several years, with results now available from four neutrino experiments that use different methods to look at different regions of the solar-neutrino energy-spectrum. While the goal of all of these experiments is physics, they all rely heavily on chemistry and radiochemistry. Three of these experiments are radiochemical; the ^{37}Cl detector and the two different forms of ^{71}Ga detectors used in GALLEX and SAGE are based on the chemical isolation and counting of the radioactive products of neutrino interactions. The fourth, Kamiokande, detects neutrinos in real time; however, it also depends on radiochemistry in that (as in all the solar neutrino detectors) radioactive contaminants must be controlled at very low levels. It is noteworthy that all of these experiments report a deficit of observed neutrinos relative to the predictions of standard solar models, the so-called "solar neutrino problem". This paper reviews the basic principles of operation of these neutrino detectors (as well as some new detectors currently under construction), reports their recent results, and discusses some of the theoretical interpretations that are now in vogue.

1. Introduction

It is fascinating that neutrinos, which were "invented" in the 1930's to account for the then-anomalous observation of the continuous energy spectrum observed in radioactive beta-decay, are used today to probe the mechanisms of energy production in stars. In fact, solar neutrinos are being actively studied in several current or planned experiments because of the possibility that they may reveal "New Physics", i.e., new properties of these fundamental particles, such as non-zero mass.

Compared to the situation in solar neutrino physics several years ago, when there was only one operating solar neutrino experiment, namely the radiochemical ^{37}Cl detector operated by R. Davis and colleagues at the Homestake Mine [1], today the field is flourishing with three additional experiments providing data on solar neutrinos: Kamiokande [2], which is an on-line detector, and GALLEX [3, 4] and SAGE [5], which are both radiochemical detectors. That all four neutrino experiments have detected solar neutrinos is a significant accomplishment that is sometimes taken for granted. In addition, at least three more on-line detectors, Super-Kamiokande [6], SNO [7] and Borexino [8], and one radiochemical detector [9], ^{127}I, are under construction.

The great continuing interest in neutrino physics of course is due to the fact that, even with the new data from these solar neutrino experiments, the "solar neutrino problem", first identified in the ^{37}Cl detector, still persists: the different experiments, which are sensitive to different regions of the solar-neutrino energy-spectrum, all report deficits compared to the predictions of standard solar models (SSM). The most recent data seem to indicate that the neutrino deficit (a) cannot be explained in terms of possible modifications to the SSM, and (b) may be indicative of the existence of new neutrino properties (called "New Physics").

In this paper, I review the basic principles of operation of these neutrino detectors (as well as some new detectors currently under construction), report on their latest results, and discuss some of the theoretical interpretations that are now in vogue.

2. Basic operation of solar neutrino detectors

Because the cross-sections for neutrino interactions in matter are extremely small, $\leq 10^{-40}$ cm^2, all solar-neutrino detectors have certain features in common: They are very large, low-level experiments containing many tons of target material. They are situated in low-background environments, such as deep underground laboratories, so as to minimize cosmic-ray backgrounds. And strict controls of radioactive contaminants, especially of naturally occurring elements such as U, Th, Ra and Rn, are applied in selecting both the site of the laboratory and the materials that make up the detector. Two broad categories of detectors are used in solar neutrino experiments: radiochemical detectors and on-line detectors. The characteristics of these two different detector types are discussed below.

2.1 Radiochemical detectors

The operation of radiochemical detectors is governed by the principle that, as a result of an interaction with

* Research sponsored by the U. S. Department of Energy, Office of High Energy and Nuclear Physics, under contract DEAC0276CH00016 with Brookhaven National Laboratory.

a neutrino, the target nucleus is transformed into a different nuclear species that can be chemically separated and removed from the bulk of target atoms with high specificity and efficiency. This method is employed to study the charged-current weak interaction (CC), such as neutrino capture (also called inverse beta-decay), $^{A}Z + \nu_e \rightarrow {}^{A}(Z+1) + \beta^-$, where the product nucleus, because of its change in atomic number, has different chemistry from the target. Often in neutrino capture, the product nucleus is radioactive, so that after it has been removed from the target and purified, its radioactive decay can be observed. The combination of the measured energy spectrum and half-life of the known product nuclide then serves as a characteristic signature for neutrino capture in the target (note that the other neutrino types, ν_μ and ν_τ, do not induce inverse beta-decay).

It has been suggested at times that other means of identification of the chemically separated product nucleus be used, such as (accelerator-based) mass-spectrometry or resonance-ionization mass-spectrometry, particularly when the product is not radioactive. However, these alternate methods have not as yet been applied to an operating solar-neutrino detector.

Two types of thresholds determine the characteristics of a radiochemical neutrino detection system. The primary one is the Q-value of the chosen neutrino-capture reaction (equivalent to that of the inverse beta-decay process); often a very low threshold, well below 1 MeV, can be achieved. The second threshold is that of the counting system used to detect the radioactive decay of the product; this instrumental threshold can usually be set comfortably below the relevant decay energies, e.g., ~0.5 keV for gas-filled proportional counters.

During a "solar exposure" or "run" with a radiochemical detector, the target passively absorbs neutrinos, collecting the radioactive products during a period chosen to be one or more half-lives. At the end of the exposure, the product is chemically extracted from the target, and a new exposure period begins. Because the radioactive product is detected off line, the target is a "black box", yielding no real-time information about the details of the neutrino interactions. Instead, the result of the experiment is a single number, namely the rate of formation of the product nucleus, the "production rate", given in Solar Neutrino Units, SNU (1 SNU = 1 neutrino capture per second per 10^{36} target atoms).

Offsetting this disadvantage are two very important factors: The first has already been mentioned, the very low neutrino-capture thresholds that are attainable. The second advantage is the extremely high selectivity and sensitivity of the method; typically, from a target that contains $\sim 10^{30}$ atoms, on the order of 10 product atoms can be quantitatively removed and identified.

Thus, it is no accident that three of the four solar neutrino detectors that have obtained results to date are radiochemical, namely ^{37}Cl and ^{71}Ga (two detectors), with respective neutrino-capture Q-values of 0.814 and 0.233 MeV. It is also worth noting here that another radiochemical detector had been put into operation with the aim of searching for Tc produced over geological time by neutrino capture in deep Mo ores. Unfortunately, the initial results were unreasonably high, and were attributed to contamination problems [10].

2.2 On-line detectors

As their name states, on-line detectors observe neutrino events as they occur in real time in the target. Many schemes for such experiments depend upon placing radiation detectors (counters, phototubes, etc.) either around or in the target material (usually a transparently clear liquid). Observations are then made of interactions in the target, e.g., (a) of ν_e capture, by detecting the resulting β particle and/or the radiation emitted from the excited product nucleus, or (b) of neutrino elastic-scattering, by detecting the resulting scattered electron.

The major advantage of measuring on-line events is that detailed information about the interaction, such as the energy, direction and arrival time of the neutrino, can be obtained. However, there is an associated serious disadvantage of these detectors, namely that their operating thresholds are usually quite high, several MeV, so that only a small fraction of the full energy-range of solar neutrinos is amenable to measurement. In addition, the ratio of the desired neutrino signal to other events in these detectors is usually very small. Typically, only about 1 in 10^4 original events survives the data cuts and is associated with an incoming neutrino.

Characteristic of the data that can in principle be obtained with on-line detectors are a value of the neutrino flux integrated over energy (reminder: the unit of flux is neutrinos $cm^{-2}\ s^{-1}$, not SNU), a neutrino (or electron) energy-spectrum, and a neutrino angular distribution that points in the direction of the neutrino source (in this case, the Sun).

3. Predictions of standard solar models

The literature of nuclear fusion reactions in the sun, and of the concomitant production of neutrinos, is quite extensive. As an example, Figure 1 shows the energy spectra of the neutrino fluxes at the Earth from the solar "hydrogen burning" reactions that produce *pp*, *pep*, ^{7}Be, and ^{8}B neutrinos, as predicted by the standard solar model of Bahcall and coworkers [11–13]. Not shown are the spectra expected from the solar CNO cycle.

The measured solar radiation flux at the Earth, 0.139 J $cm^{-2}\ s^{-1}$, is equivalent to a value of the solar energy output or luminosity of 3.86×10^{33} erg s^{-1} (Earth-Sun distance = 1.5×10^{13} cm). Since the over-

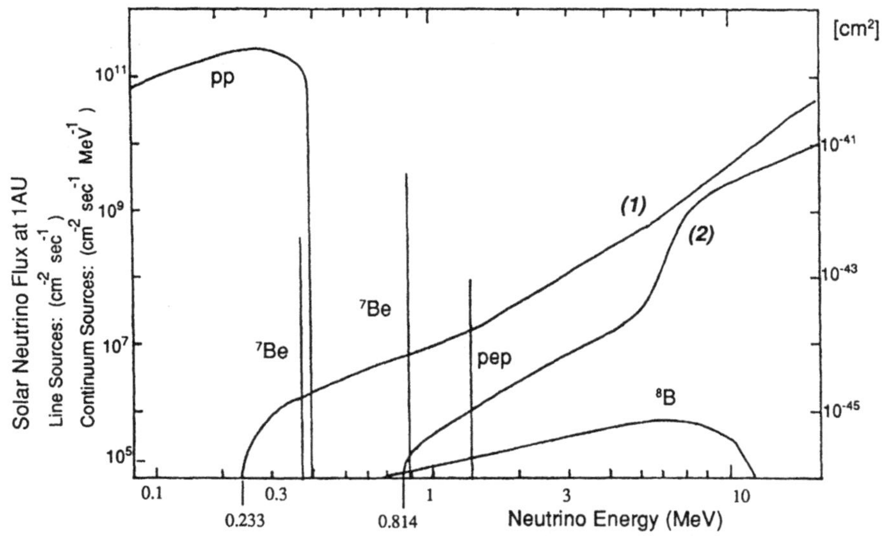

Fig. 1. Neutrino energy spectra from the standard solar model for solar hydrogen burning [11]. Cross-sections for ν_e capture are shown for ^{71}Ga (curve 1) and ^{37}Cl (curve 2). Flux units are given on the left ordinate scale; cross-sections, on the right. Thresholds of different neutrino detectors are shown.

all reaction involved in the hydrogen burning sequence, $4p \rightarrow {}^4He + 2\beta^+ + 2\nu_e$, produces 2 neutrinos and some 26 MeV in energy, we can use these values to obtain a useful estimate of the flux of solar neutrinos at the Earth's surface, 6.5×10^{10} cm^{-2} s^{-1}.

Figure 1 also contains the excitation functions (cross-sections vs. energy) for the ^{37}Cl and ^{71}Ga detectors, labeled (2) and (1) respectively, with cross-sections $\leq 10^{-40}$ cm^2. These cross-sections are not directly measured for the neutrino interactions, but are obtained from β-decay data for ^{37}Ar and ^{71}Ge, (p,n) reactions on ^{37}Cl and ^{71}Ga, etc. [11]. The importance of the *Q*-values of the selected reactions is apparent in the figure. The ^{37}Cl detector, with its threshold of 0.814 MeV, is sensitive to the *pep*, ^{7}Be and ^{8}B, but not to the *pp*, neutrinos. On the other hand, the ^{71}Ga detector, because of its threshold of 0.233 MeV, can see a major portion of the *pp* branch, as well as the higher-energy neutrinos. Detecting the *pp* (and *pep*) neutrinos is especially important because they are produced in the first step in solar energy generation and their intensity is known to within a few percent from the measured solar luminosity, almost independently of solar model assumptions. Note also that the lowest instrumental threshold attained by the Kamiokande on-line detector is 7.0 MeV, so that it is sensitive only to the upper portion of the ^{8}B spectrum.

The production rate of a radiochemical detector is equal to the integral over energy of the product of the energy-dependent neutrino fluxes and cross-sections, taken from the *Q*-value to the appropriate maximum. For the ^{37}Cl and ^{71}Ga detectors, Table 1 lists the predicted rates in SNU from two different SSM's [12, 14], for the individual neutrino branches, including the CNO cycle (^{13}N and ^{15}O). The predicted neutrino fluxes are also listed, for comparison with data from on-line detectors.

4. Experimental details and results

This section discusses significant experimental details of each of the solar neutrino detectors, and summarizes recent results.

4.1 ^{37}Cl

The radiochemical ^{37}Cl detector, which was developed by Davis and colleagues at Brookhaven National Laboratory [1], was the first experiment built to observe solar neutrinos. The detector, which has been operating in the Homestake Mine (USA) since 1971, consists of 615 tons of the organic liquid, C_2Cl_4, perchloroethylene; the natural abundance of ^{37}Cl is 24.23%. The neutrino-capture reaction is $^{37}Cl + \nu_e \rightarrow {}^{37}Ar + \beta^-$, with a *Q*-value as previously noted of 0.814 MeV.

Because Ar is a noble gas, its chemical extraction can be effectuated in high yield by periodically bubbling helium gas through the liquid to sweep out the Ar atoms. After purification, the ^{37}Ar is added to P-10 counting gas, 90% Ar and 10% CH_4; this gas mixture is placed in a miniaturized proportional counter so as to optimize the counting efficiency. The signature of ^{37}Ar is the observation of the peak due to *K* X-ray and Auger-electron emission in orbital electron-capture, decaying with the characteristic 35.2-day half-life.

Crucial to this experiment is the demonstration that the radioactive ^{37}Ar gas is quantitatively removed from the liquid target. Many tests were done with Ar carrier and tracers to prove that this goal could be achieved. A key experiment was done with C_2Cl_4 that contained radioactive ^{36}Cl, which decays to ^{36}Ar. The counting data for the extracted Ar showed that no losses occurred of the ^{36}Ar hot atoms that were formed by the decay process (and by extension, by neutrino capture).

Table 1. Predictions of standard solar models.

ν_e branch	ν_e Flux[a] (cm^{-2} s^{-1})	^{37}Cl Rate[a] (SNU)	^{71}Ga Rate[a] (SNU)	ν_e Flux[b] (cm^{-2} s^{-1})	^{37}Cl Rate[b] (SNU)	^{71}Ga Rate[b] (SNU)
pp	6.0×10^{10}	0.0	70.8	6.0×10^{10}	0.0	71.1
pep	1.4×10^{8}	0.2	3.1	1.3×10^{8}	0.2	3.0
^{7}Be	4.9×10^{9}	1.2	35.8	4.4×10^{9}	1.1	30.9
^{8}B	5.7×10^{6}	6.2	13.8	4.5×10^{6}	4.6	10.8
^{13}N	4.9×10^{8}	0.1	3.0	3.5×10^{8}	0.1	2.4
^{15}O	4.3×10^{8}	0.3	4.9	2.9×10^{8}	0.2	3.6
Sum	$\mathbf{6.60 \times 10^{10}}$	**8.0**	**132**	$\mathbf{6.55 \times 10^{10}}$	**6.3**	**122**

[a] Values from BP [12].
[b] Values from TCL [14].

Ever since the first ^{37}Cl results were announced, they have caused quite a stir. Recent cumulative results, from 107 solar exposures during the time period 1970–1993, yield a mean production rate of 0.481 ± 0.048 atoms per day, or 2.55 ± 0.25 SNU [15]. In Table 1, the predicted values from two different solar models are listed for ^{37}Cl, 8.0 ± 1.0 SNU by Bahcall and Pinsonneault (BP) [12], and 6.3 ± 1.0 SNU by Turck-Chieze and Lopes (TCL) [14]; both are in serious disagreement with the experimental value. This well-known discrepancy has come to be known as the "solar neutrino problem". It has raised many intriguing questions, stimulated a great amount of theoretical research, and led to the development of several other solar neutrino experiments.

An interesting separate issue is the fact that Davis has pointed out that the data taken with the ^{37}Cl detector over two decades seem to show an anticorrelation with sunspot activity [16]. However, as reported in Ref. [15], averages of the ^{37}Cl runs, taken for 7-year periods, show little variation with time.

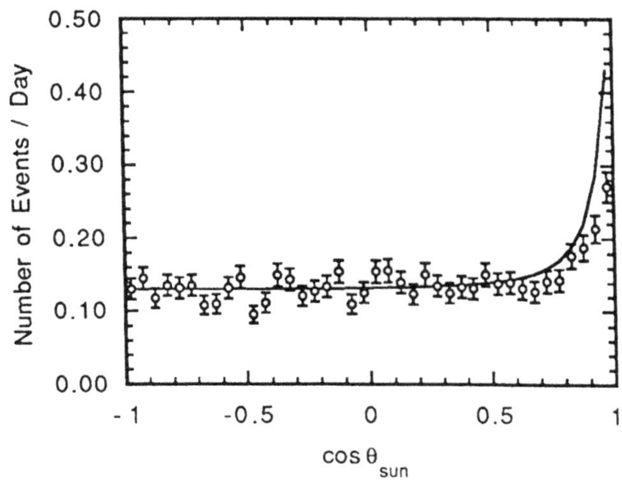

Fig. 2. Angular distribution of neutrino events relative to the sun's direction recorded by Kamiokande-2 and -3 [18]. The curve is the SSM prediction.

4.2 Kamiokande

For many years, the ^{37}Cl detector was the only one looking at solar neutrinos. Then in 1987, a second detector, Kamiokande, began operating as an on-line solar neutrino detector in the Kamioka mine (Japan) [2]. It was originally constructed to observe the postulated decay of the proton, but was then modified to be a "solar neutrino observatory" after it detected energetic neutrinos from the supernova, SN 1987A [17]. The target is purified light water (H_2O), with a total mass of 3000 tons; of this, 680 tons provide the fiducial volume of the target for solar neutrino detection, with the remaining mass of water serving as shielding around the target.

Neutrinos are detected in Kamiokande as a result of their elastic scattering from electrons in the target medium; the forward-scattered electrons emit Cerenkov radiation that is recorded in the multiple phototubes that view the target. The trajectories of these electrons, and therefore of the incoming neutrinos, are then reconstructed from these signals. Kamiokande has gone through several modifications to improve its operation and to reduce backgrounds, such as the installation of chemical purification equipment to remove naturally occurring radioactive contaminants, such as Rn, from the water. As these improvements have occurred, the threshold for neutrino detection has dropped. In Kamiokande-2, the threshold was 9.3 MeV; in the improved Kamiokande-3, it was lowered to 7.5 and then to 7.0 MeV. As noted earlier, in all of these stages, Kamiokande has been sensitive only to ^{8}B neutrinos.

Figure 2 shows the composite results from Kamiokande-2 and 3, obtained during some 1700 days of data-taking [18]. These data, which have survived the various cuts that remove non-neutrino-related events, are plotted vs. cosine Θ_{sun} (the known angle between the detector and the sun, since the position of the sun in the sky is known from the recorded time of each event). The solid curve in the figure is the prediction of the standard solar model of BP [12]. The data (and the theoretical prediction) are seen to peak at $\Theta_{sun} = 0$, whereas the rest of the measured angular distribution is essentially flat. Kamiokande has thus confirmed that the sun is the source of the observed neutrinos, something that was assumed but not proven in the ^{37}Cl experiment.

The value of the ^{8}B flux measured by Kamiokande [18] is 2.74×10^6 neutrinos cm^{-2} s^{-1}. The ratio of this measured flux to the prediction of BP [12] is 0.48 ± 0.07; if the comparison is instead made with TCL [14], the flux ratio is 0.62 ± 0.09. In either case, a substantial fraction of the predicted neutrino flux is missing. Qualitatively speaking, we conclude that these Kamiokande results confirm the solar neutrino deficit first recorded by the ^{37}Cl experiment (at this point, we neglect important details such as the different energy thresholds of these two neutrino detectors).

Note that an experimental test of neutrino-flux variations over time, and of the proposed anticorrelation [16] of the solar neutrino flux with sunspot activity, has also been an ongoing activity of the Kamiokande group. A recent analysis [18] of their data, taken during the period 1987–1993 when the sunspot activity was changing markedly, concluded that there was no strong variation over time of the solar neutrino flux.

4.3 ^{71}Ga: GALLEX and SAGE

There are two operating radiochemical detectors, SAGE and GALLEX, that use ^{71}Ga (natural abundance, 39.89%) as the target in the reaction, $^{71}Ga + \nu_e \rightarrow {}^{71}Ge + \beta^-$. SAGE is located in the Baksan Neutrino Observatory (Russia); GALLEX, in the Gran Sasso National Laboratory (Italy). This situation of two radiochemical detectors that use the same neutrino reaction is quite unusual because such neutrino-detection experiments are very expensive, and require many years of intensive effort before any reliable data are forthcoming. Yet, having two very similar, if not identical, detectors offers the unique possibility of independent verification of a solar neutrino measurement.

The basic operation of the gallium detector is analogous in many ways to that of the ^{37}Cl detector. This situation is understandable when it is realized that much of the early research, especially of the chemistry, of the ^{71}Ga detector was done at Brookhaven [19]. However, the product of the neutrino capture, ^{71}Ge, is not a noble gas as is ^{37}Ar, so some ingenuity had to be used to find successful chemical procedures to separate Ge from the target, and to prepare a Ge compound that is suitable for low-level counting.

The chemistry of Ge^{4+} in aqueous medium is complex, depending upon such factors as the acidity and the Cl concentration. Thus, in strong acid (HCl) with an excess of Cl, the volatile compound, $GeCl_4$, is formed, while in weak acid, this compound hydrolyzes, forming hydroxychloro complexes that are soluble in solution. So, the key to the removal of the radioactive ^{71}Ge product from the Ga target is to establish conditions to favor formation of the $GeCl_4$, which can be entrained in a stream of gas. The extraction of this compound is quantitative. The extracted Ge is then removed from the gas by dissolving it in H_2O so it can be concentrated and purified. Finally, it is converted to another gaseous compound, GeH_4, which is a suitable counter-filling gas; this compound is purified by gas chromatography, mixed with Xe and put into a (miniaturized) proportional counter. The Xe is added to increase the counting efficiency for the energetic X-rays emitted in the decay of ^{71}Ge. The signature of ^{71}Ge is the observation of the *K* and *L* peaks from electron capture, decaying with the characteristic 11.43-day half-life.

The chemical forms of the gallium targets used in GALLEX and SAGE are different, and so their initial chemical procedures differ. The GALLEX target contains 30 tons of gallium in a concentrated 101-ton aqueous solution of $GaCl_3$ (8 *M*) and HCl (2 *M*). Under these conditions, any Ge, such as that formed by ν_e capture, will automatically tend to form volatile $GeCl_4$ [20, 21]. This product is then extracted by bubbling an inert gas, N_2, through the liquid target. SAGE uses liquid Ga metal (melting point, 29.8 °C) as its target; the experiment began with 30 tons of metal, which was increased to 57 tons in 1991. To remove any Ge formed in the target, an aqueous mixture of HCl and the oxidant, H_2O_2, must be vigorously mixed with the molten Ga metal [5]. $GeCl_4$ is formed and migrates to the aqueous phase, from which it can then be removed in a stream of gas. From this stage on, essentially all of the ensuing steps used in the chemistry and the counting are very similar in SAGE and GALLEX.

However, one difference to be noted between SAGE and GALLEX is in the ^{71}Ge energy-spectra obtained with their low-level counting systems. Since its solar data-taking began, GALLEX has had the capability of detecting both the *L* (1.17 keV) and *K* (10.37 keV) peaks, while SAGE until recently [27] has been able to count only the *K* peak, because of excessive background and/or noise in the *L*-region of the spectrum.

It is of interest to note that, as in the case of the ^{37}Cl detector, extensive radiochemical tests have been done of the efficiency of the removal of Ge from the different forms of Ga targets, especially to investigate potential hot-atom effects that might trap individual atoms of Ge. For example, in GALLEX, radioactive Ge was formed *in situ* in small samples of the $GaCl_3$ target solution by (a) introducing purified ^{71}As that decayed by electron capture to ^{71}Ge, which was then extracted and monitored by X-ray proportional counting [22], and (b) irradiating the solution with low-energy protons, using the (p,n) reaction to form ^{69}Ge, whose extraction from the solution was monitored by γ-ray counting [23]. In SAGE, samples of Ga metal were irradiated with neutrons; the resulting radioactive nuclides, ^{70}Ga and ^{72}Ga, β decayed to stable ^{70}Ge and ^{72}Ge, which were then assayed by mass spectrometry [24]. All of these tests were consistent in showing at the few percent level that no atoms of Ge were anomalously held back in the Ga targets.

All of the above-noted tests attempted to mimic the conditions that occur during the formation of ^{71}Ge by neutrino capture in Ga. However, the ultimate test of the chemistry consists of irradiating the gallium target with an "artificial sun", an intense neutrino source with a known disintegration rate. These tests have been done recently, and are discussed below in Section 5.

In the start-up phases of both SAGE and GALLEX, even though great care was taken to avoid radioactive contaminants, similar problems were encountered. The first, which was initially rather serious, was due to the fact that radioactive germanium isotopes were produced by cosmic-ray bombardment while the gallium target material was above ground, before being brought into the underground laboratory. One isotope, ^{68}Ge, has a very long half-life, 271 days, and a decay mode that is very similar to that of ^{71}Ge. It was found that an extremely small ($\leqslant 1\%$) but non-negligible fraction of this cosmogenic ^{68}Ge unaccountably remained in the gallium target, being removed only slowly by the usual Ge extraction steps. The resulting number of extracted ^{68}Ge atoms in the initial solar runs was high enough to raise the counter backgrounds to a level where it was not possible to see the decay of the ~1 atom per day of ^{71}Ge expected. Special steps had to be taken, such as heating the 101 tons of GALLEX target solution or cleaning anew the SAGE target containers, to reduce the ^{68}Ge to acceptable levels; these steps were done successfully. However, the schedules of both SAGE and GALLEX were delayed by several months before their "^{68}Ge problems" were solved.

A second problem, common to both experiments, was due to the fact that ^{222}Rn is an environmental contaminant. For example, the radioactive non-volatile daughters of Rn can deposit on the outside of the counters and contribute to the background. In addition, GALLEX found that a few atoms of ^{222}Rn somehow manage to find their way into the counters, even though the gas chromatography does separate Rn from GeH_4 with high efficiency. Since some Rn decay events fall in the energy regions of interest for ^{71}Ge, it is of vital importance that Rn events not be mistaken for those from ^{71}Ge. GALLEX data analysis procedures can identify Rn counts definitively; typically, ~3 atoms of Rn are found in a counter, which on average produce 1 distinctive count that can be vetoed.

Of the two gallium detectors, SAGE was the first to begin taking solar data. Their first published results in 1991 [5], which came from five runs, contained no clear evidence for ^{71}Ge, and therefore, no evidence for solar neutrinos entering the target. The best fit to the SAGE data gave a production rate of 20 ± 17 (stat.) ±32 (syst.) SNU, while a more conservative analysis set an upper limit, <55 SNU at the 68% confidence level.

Then in 1992, GALLEX published its initial results from 14 solar exposures [3]. The average value, 83 ± 19 (stat.) ±8 (syst.) SNU, provided the first positive evidence of solar neutrino-capture in gallium. GALLEX also claimed that its data represented the first direct observation of solar *pp* neutrinos [25].

Since these initial results appeared, both experiments have updated their cumulative results periodically, and the apparent discrepancy between SAGE and GALLEX has disappeared. As we shall see below, their results now are in agreement, within the stated experimental errors.

Because I am a member of the GALLEX collaboration, it is simplest for me to illustrate data that are "typical" for all the radiochemical detectors by showing representative results published by GALLEX [4]. Figure 3 shows the cumulative energy spectrum measured in GALLEX solar runs, containing 2 peaks at energies that agree with those expected for the *L* and *K* peaks in the electron-capture decay of ^{71}Ge. The open bars are the data obtained during the first 33 days of counting (2 mean lives of ^{71}Ge); the closed bars, during the last 33 days of a counting period that is typically 6 months long. The *L* and *K* peaks, which clearly decay during the counting period, are readily distinguishable from the long-term background. Figure 4 presents the corresponding decay curve for the combined *L* and *K* peaks. The solid line is a maximum-likelihood (M-L) analysis, done with contributions from the 11.43-day ^{71}Ge, a small amount of the cosmogenic 271-day ^{68}Ge, and a small, constant background. The χ^2 fit to the data, taking the theoretical values from the M-L curve, is quite acceptable for 20 degrees of freedom (22 time bins).

Recent overall values from the gallium detectors (with statistical and systematic errors combined in quadrature) are: 77 ± 10 SNU from 39 completed GALLEX solar runs [26]; and 74 ± 15 SNU from 21 SAGE runs [27]. These values are considerably less than those from the standard solar models listed in Table 1: 132 SNU from BP [12] and 122 SNU from TCL [14].

5. Tests with an intense radioactive neutrino source

While the possibility of doing a full-scale test of a radiochemical neutrino detector with an intense, calibrated radioactive source has been discussed many times, the first such test to be done was recently accomplished by GALLEX; see Ref. [28] and the recent update, Ref. [29]. The aim of course was to show whether all of the ^{71}Ge atoms produced by neutrinos in the gallium are actually recovered.

Because the source was to emit neutrinos, not antineutrinos, it had to be a neutron-deficient isotope. The nuclide selected was 28-day ^{51}Cr, which decays by electron capture, emitting neutrinos with energies of 426, 431, 746 and 751 keV, all above the 233-keV threshold of the ^{71}Ga detector. [Note three relevant points: (1) SAGE is in the process of doing an analogous test with ^{51}Cr, but its results have not yet been

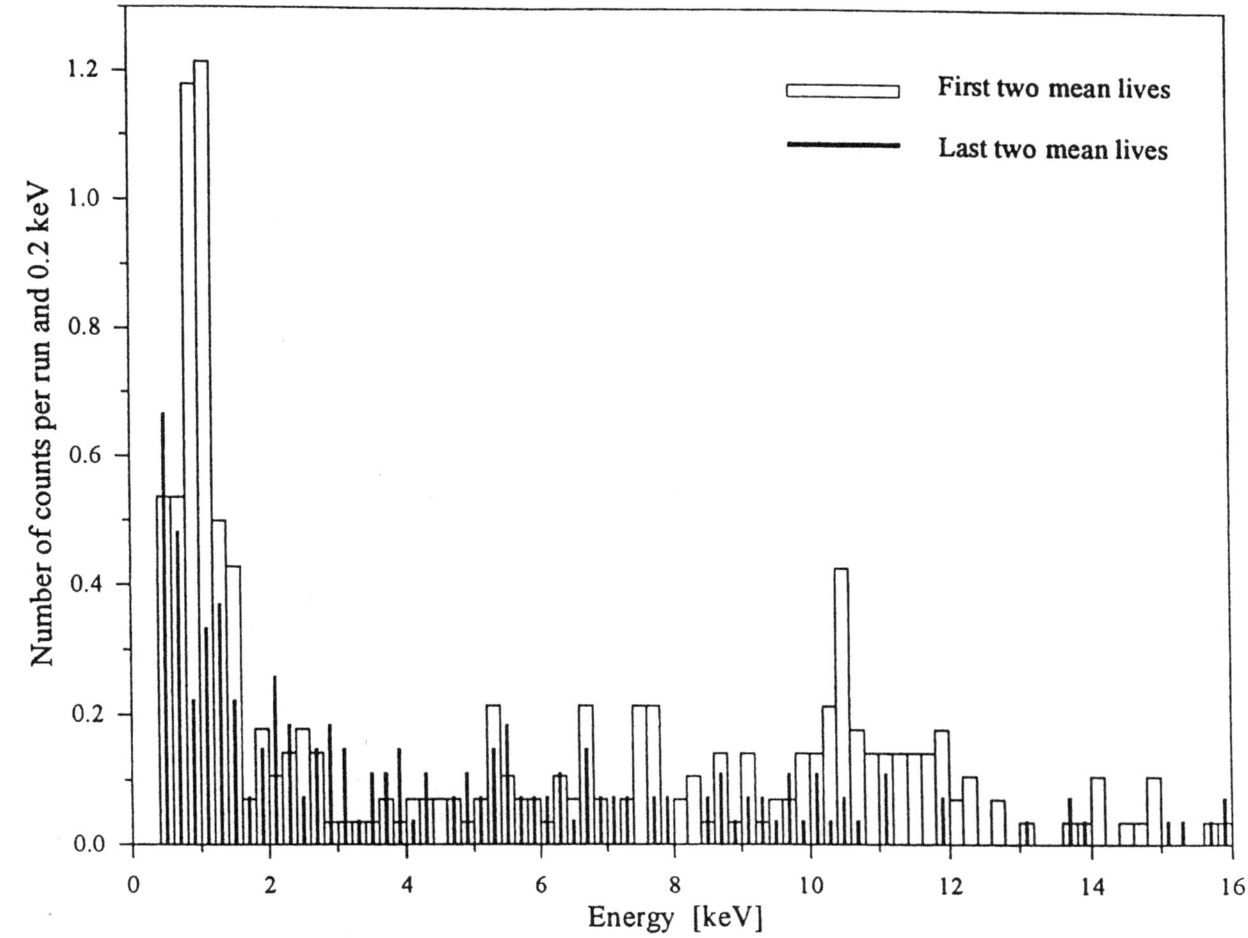

Fig. 3. Cumulative energy spectrum from GALLEX solar exposures [4], plotted for 2 different time periods of similar duration: open bars – during the first 33 days (2 mean lives of ^{71}Ge) after start of counting; closed bars – during the last 33 days of the counting period that is typically 6 months long.

published*; (2) to test the ^{37}Cl experiment, with its threshold of 814 keV, a different radioactive source that emits higher energy neutrinos, such as ^{65}Zn, would be needed; (3) Kamiokande, which detects scattered electrons, can instead be tested with energetic electrons or γ rays.]

Since the neutrino signal from the Sun cannot be "turned off", the GALLEX ^{51}Cr source intensity had to be ≥50 PBq, in order to produce a signal in the gallium detector about one order of magnitude greater than that from the Sun. Production of this source was a major endeavor in its own right. First, ≈36 kg of isotopically enriched ^{50}Cr (39%) had to be produced by gas centrifugation. This material was then converted to Cr metal and irradiated with a high flux of thermal neutrons for 24 days in the Siloe' reactor in France. The neutrino-emission rate of the source was

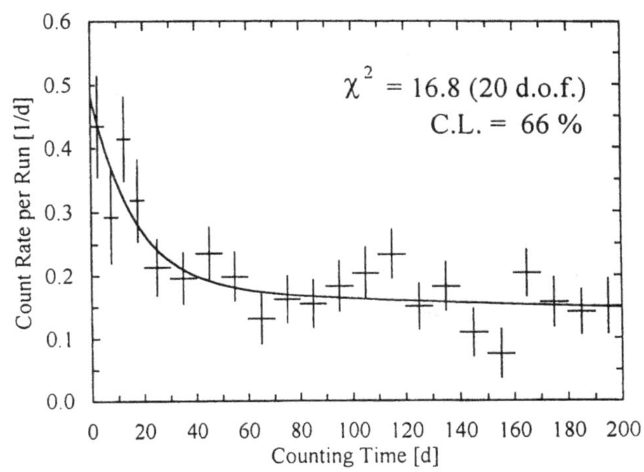

Fig. 4. Average count rate per run from GALLEX solar exposures plotted as a function of time after the start of counting [4]. The counting-time bins are shown as horizontal bars. The solid curve is the maximum-likelihood solution for ^{71}Ge plus 2 longer components, as described in the text.

obtained by directly measuring its disintegration rate by a variety of techniques, e.g., by calorimetry, by absolute counting of the 320-keV γ-ray of ^{51}Cr, and by determination of the concentration of the ^{51}V daughter produced by the ^{51}Cr decay. The average value of the

* *Note added in proof:* SAGE (S. R. Elliott *et al.*) announced its initial results from its ^{51}Cr test at TAUP (IV International Workshop on Theoretical and Phenomenological Aspects of Underground Physics) in Toledo, Spain, September 1995. The ^{51}Cr source, produced from 0.5 kg of 92% ^{50}Cr, had a strength of 509 kCi (measured by calorimetry) when it was inserted into the gallium target. The preliminary result was that 0.93 ± 0.16 of the ^{71}Ge atoms produced by the neutrino source were recovered.

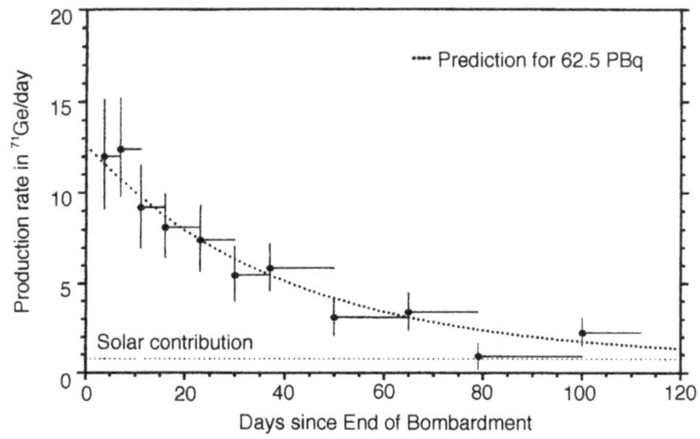

Fig. 5. Number of ^{71}Ge atoms produced per day during the course of the ^{51}Cr source experiment [29]. The points for each run are plotted at the beginning of each exposure, with the horizontal lines showing the length of the exposures. The predicted curve decreases with the known half-life of ^{51}Cr, and also includes the constant contribution due to solar neutrinos and side reactions.

directly measured source strength [29] was 62.5 ± 0.4 PBq (1.69 ± 0.01 MCi) at end-of-bombardment (EOB), when the source was removed from Siloe'.

This source was transported to the Gran Sasso Laboratory, and placed in the GALLEX target tank for 109 days (3.9 half-lives of ^{51}Cr). During this time, 11 extractions of ^{71}Ge were done, with the experimental conditions kept exactly as in the solar runs. The results are shown in Figure 5, where the production rate determined in each run is plotted vs. the time since EOB. The curve in the figure, which decreases with the known half-life of ^{51}Cr, is the prediction from the known source strength at EOB. It clearly follows the trend of the data. The best fit to these ^{71}Ge data gives a mean initial equivalent source strength [29] of 60.4 ± 6.6 PBq.

The ratio [29] of the source strength extracted from the ^{71}Ge data relative to the directly measured strength is 0.97 ± 0.11. Clearly, to within the 11% uncertainty, all of the ^{71}Ge atoms produced by the neutrino source were recovered. The large, ~40% deficit in the GALLEX solar neutrino signal cannot be attributed to experimental artifacts. Moreover, this source experiment directly demonstrates the validity of the basic principles of the radiochemical methods used in solar-neutrino detection: that on the order of 10 radioactive atoms produced by neutrino capture can be removed from tons of target material, purified and counted with no significant losses.

6. Interpretation of the data

Discussions have been presented many times of what has been learned from consideration of the solar neutrino problem; see e.g. Reference [30]. Although the scope of this article is mainly experimental, some theoretical ideas are included here to put the experimental results in perspective, and to illustrate why this field has generated such wide interest.

6.1 Calculations with solar models

All four solar neutrino experiments have reported clear deficits compared to solar model predictions. The ratios of the experimental results to the theoretical values of Table 1 (with their statistical errors neglected) are, for BP [12] [and TCL [14]]: 0.32 ± 0.03 [0.40 ± 0.04] for ^{37}Cl; 0.48 ± 0.07 [0.62 ± 0.09] for Kaminokande; 0.59 ± 0.07 [0.63 ± 0.08] for GALLEX; and 0.56 ± 0.11 [0.60 ± 0.12] for SAGE. We see that by now the experimental uncertainties have been reduced to the point where these ratios are several standard deviations away from unity. It is becoming increasingly difficult to accommodate these results in the framework of the solar models.

These ideas have been amplified in recent analyses of the GALLEX results [26, 31]. Simply stated, if we take the SSM prediction for the (*pp* + *pep*)-branch in Table 1 as being correctly fixed by the luminosity constraint and convert the Kamiokande experimental result for ^{8}B into SNU's, the sum of (*pp* + *pep*) + ^{8}B = 86 SNU, for either BP [12] or TCL [14]. This value is so very close to the measured GALLEX rate of 77 ± 10 SNU that little room is left for contributions from ^{7}Be or CNO neutrinos, a result that poses a serious problem for the SSM's.

A quantitative upper limit can be extracted from the GALLEX result for the allowable fraction of ^{7}Be neutrinos [26], f_{Be}; of course, the SSM requires that $f_{Be} = 1$. As an example of this approach, the calculated gallium signal in terms of the BP [12] values is $[73.9 + 5.8(1-f_{Be})] + [35.8 f_{Be}] + [6.7 \pm 1.0] + [7.9]$ SNU, where the terms in brackets, taken from Table 1, are respectively: the (*pp* + *pep*) values plus a correction that keeps the luminosity constant as ^{7}Be decreases; the SSM value for ^{7}Be modified by f_{Be}; the Kamiokande result for ^{8}B, equal to the SSM value modified by the measured reduction factor, f_B; and the SSM value for CNO.

Setting this sum equal to the present GALLEX value, with its total 2σ-error of ±20 SNU, produces a value of $f_{Be} \leq +0.09$ at the 95% confidence level. This result poses two serious problems for the astrophysical theories: (1) $f_{Be} \ll 1$, and (2) $f_{Be} < f_B$. Not only is the derived fraction of ^{7}Be much less than the SSM value, but it is less than the measured ^{8}B fraction. This second conclusion states a paradox, since ^{7}Be precedes ^{8}B in the SSM fusion cycle.

Figure 6 illustrates these conclusions [26]. Part (a) compares TCL [14] and BP [12] predictions (columns at left) with the published cumulative solar results [3, 4, 26] of GALLEX. Also shown is the anticipated result from 60 runs after 4 years of data taking, with an expected 2σ-error of ±16 SNU (assuming no change in the present mean value). Note that, while the measured mean SNU values have changed only slightly over time, their 2σ-errors have decreased substantially. The "heuristic model" value (column at right), which is the 86 SNU's cited above from only (*pp* + *pep*) + ^{8}B, is close to the 2σ experimental limit.

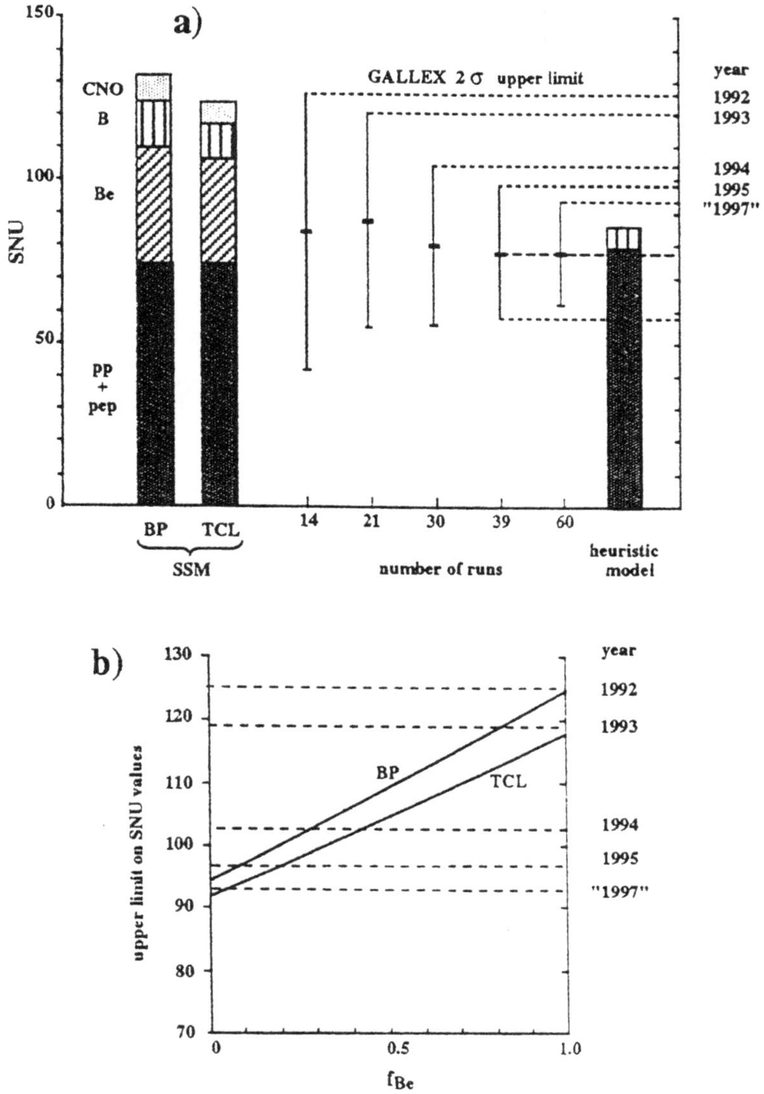

Fig. 6. (a) GALLEX results published after 14, 21, 30 and 39 runs (Refs. [3, 4, and 26]), and the expected result after 60 runs (in 1997), with the corresponding $\pm 2\sigma$-error ranges. The 2 left columns indicate SSM predictions [12, 14] for gallium, with contributions from the different solar neutrino branches. The column on the right, the "heuristic model", contains only the predicted contribution from the ($pp + pep$) neutrino branches plus a reduced contribution from ^8B as measured by Kamiokande. The sum of these, 86 SNU, is very close to the current GALLEX result, leaving little room within the 2σ error bars for any contributions from ^7Be and CNO neutrinos. (b) The maximum calculated SNU values for gallium, based on the GALLEX and Kamiokande results, plotted vs. the values of f_{Be} for BP [12] and TCL [14] (see text). Also shown are the 2σ upper limits from (a). It is clear that as the GALLEX errors have decreased, the evidence has become stronger for a small value of f_{Be}.

In part (b) of the figure, the maximum calculated SNU values for gallium are plotted vs. the values of f_{Be}, for TCL [14] as well as BP [12] (refer to the above example). The decreasing GALLEX 2σ-error limits have drastically reduced the allowable values of f_{Be}. In fact, for the expected future case of $2\sigma = \pm 16$ SNU, the extracted values of f_{Be} will be ≈ 0 or even negative. GALLEX is now at a point where a further 20% improvement in statistics (from ± 20 to ± 16 SNU) will make a drastic difference in setting an upper limit on the ^7Be flux and defining the scope of what is being called the "^{7}Be solar neutrino problem".

These calculations have been extended [31] to include all 3 types of neutrino detectors, by considering both the ^8B and ^7Be fluxes to be variable parameters. Figure 7 shows an example of the results for the case of the calculation based on TCL [14], plotted as functions of the parameters f_{Be} and f_B [the fraction of the SSM fluxes, labeled here as $\Phi(i)/\Phi_0(i)$]. The lines labeled $\langle Cl \rangle + \ldots$ are for the mean and $\pm 2\sigma$ values from the ^{37}Cl detector; $\langle K \rangle + \ldots$, the corresponding values from Kamiokande; and $\langle Ga \rangle + 2\sigma$, for the GALLEX cases where $2\sigma = 20$ and 16(*) SNU. The mean GALLEX line is located off scale, where the parameters are negative. Just as in Figure 6, the values of f_{Be} and f_B in Figure 7 are generally <1 for all of the detectors, and there are regions where $f_{Be} < f_B$.

The maximum values of the parameters that are allowed by the combination of any two of the experiments can be found in Figure 7 at the intersection of the appropriate lines. For example, the intersection of the Ga + 2σ line (20 SNU) with the $\langle Cl \rangle$ line gives

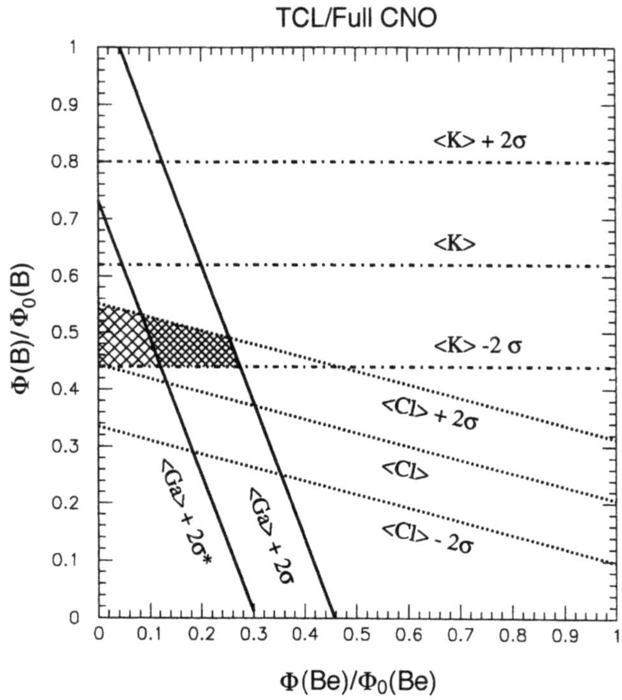

Fig. 7. Values of f_B and f_{Be} [the fraction of the SSM fluxes, labeled here as $\Phi(i)/\Phi_0(i)$] from the SSM fit (TCL [14] with full CNO contribution) to the data from the 3 solar neutrino experiments, Kamiokande (K), ^{37}Cl(Cl), and GALLEX (Ga) [31]. See text for details.

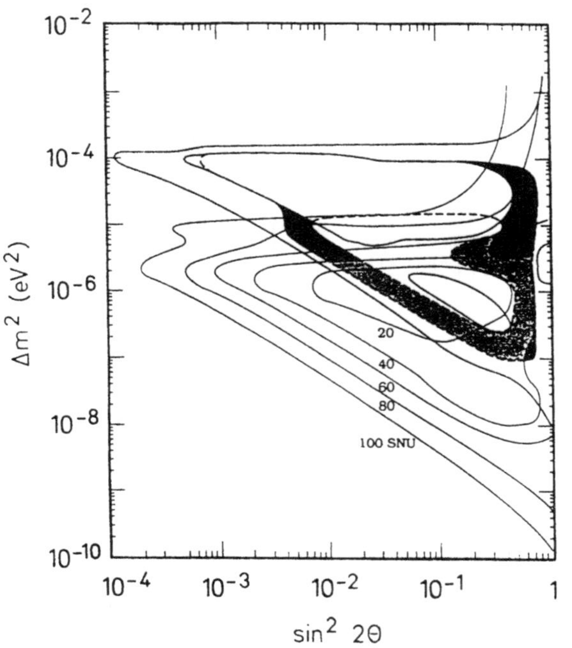

Fig. 8. MSW plot as presented by GALLEX in Ref. [25]. The contour plots were calculated for the ^{71}Ga detector. The gray regions are consistent with the results from ^{37}Cl and Kamiokande, while the smaller black regions also include the GALLEX results.

$f_{Be} \leq 0.30$, whereas the combination of the same GALLEX line with the $\langle K \rangle$ line gives $f_{Be} \leq 0.20$. The combination of GALLEX with Kamiokande gives more stringent limits for f_{Be} than does GALLEX with ^{37}Cl.

Also note that the lines for $\langle Cl \rangle$ and $\langle K \rangle$ do not intersect, except at their opposite 2σ-limits. This situation reflects the fact that, although the ^{37}Cl detector samples part of the ^{7}Be neutrino energy-spectrum as well as all of the ^{8}B spectrum while Kaminokande detects only the upper end of the ^{8}B spectrum, the $\langle Cl \rangle$ value is only about half of the $\langle K \rangle$ value. These results are unexplainable in terms of the SSM's.

Of course, the tightest constraints from this analysis are obtained from the common intersection region of all 3 experiments (cross-hatched), which is delineated by the $\langle K \rangle - 2\sigma$, and the respective $\langle Cl \rangle + 2\sigma$ and $\langle Ga \rangle + 2\sigma$, lines. Reduction of the GALLEX error to 16(*) SNU will constrain this limited region further, to values of f_B centered around 0.5 and $f_{Be} \leq 0.12$ in the particular analysis shown in the figure.

6.2 Possibility of neutrino oscillations

Alternative explanations have been put forth to explain the observed solar neutrino deficits; e.g., see Ref. [11]. One especially attractive idea is based on the fact that the solar-neutrino detectors respond almost exclusively to ν_e. Thus, if a ν_e from the Sun can be transformed ("oscillate") into another neutrino type ("flavor"), ν_μ or ν_τ, before it arrives at the detector on Earth, it will not be observed. The neutrino will still exist, but there will be a measured deficit of ν_e. Present-day particle theory assumes that all of the neutrino flavors have zero mass. For these oscillations to occur, at least one of the neutrino types must take on an important new characteristic, a mass >0. It is the allure of this new property of fundamental particles that has excited the scientific community.

A particular variation of neutrino-oscillation theory, that of Mikheyev, Smirnov and Wolfenstein, (MSW) [32] predicts that the probability of a neutrino oscillation is enhanced, going through a resonance, when it occurs in the presence of a high density of electrons, e.g., in the Sun ("matter oscillations" as opposed to "vacuum oscillations"). In their theory, this resonance depends, among other factors, on a proper combination of the electron density and the neutrino energy. Because the density in the Sun varies over a great range, from an estimated 148 g/cm^3 at the core to ~1 g/cm^3 at the surface, there is an appropriate range of densities available to match up with the wide range of neutrino energies produced by solar fusion. Thus, the Sun is an optimal site for neutrino oscillations.

Two other parameters in the theory are $\Delta m^2 = m_1^2 - m_2^2$, the difference of the squares of the masses of the 2 neutrino-states involved in the oscillation (where m_i is the mass of the ith neutrino mass-eigenstate), and the mixing angle, Θ, which, with its cosine varying between 0 and 1, tells us the relative fractions of the 2 eigenstates. As an illustration, Figure 8 presents the results of MSW theory, taken from Ref. [25], with calculated contours drawn for ^{71}Ga as functions of Δm^2 and $\sin^2(2\Theta)$. Unlike the problems that are encountered by using only the SSM's to analyze the data, the MSW calculations can readily account for the results of all of the experiments. The gray regions are consistent with the results from ^{37}Cl and Kamiokande, while the black regions, which also include the early GALLEX results, narrow the allowable MSW solutions further. We see that the results from the three different detectors fix the available combinations of Δm^2 and Θ to two to three limited regions in the MSW space. The Δm^2 values fall in the range, $10^{-5}-10^{-6}$ eV^2. This result has been interpreted as indicating approximate masses for the 3 neutrino flavors: $\nu_e \sim 0$, $\nu_\mu \sim 10^{-3}$ eV, and $\nu_\tau \sim 25$ eV. This last neutrino mass is large enough to be of cosmological interest, since it is well known that the visible mass in the universe is too small, only ~10% of that needed to account for the gravitational pull of galaxies that is inferred from their observed motions. Can neutrinos provide part of the missing mass?

The probability that a ν_e emitted in the Sun will still be a ν_e when it arrives on Earth, i.e., the "survivability", is energy dependent in the MSW theory. This characteristic means that different portions of the neutrino energy-spectrum will be reduced by different amounts, resulting not only in reduced ν_e fluxes but also in distortions of the observed spectral shape from what we expect from our knowledge of nuclear physics. This energy-dependent reduction of flux can account for the noted discrepancy between the ^{37}Cl and Kamiokande results. Such spectral distortions should not occur if the cause of the solar neutrino deficit were the solar models, which include the "normal" nuclear physics of the relevant fusion processes in the Sun. Observation of a distorted neutrino energy spectrum can thus serve as a signature of neutrino oscillations.

7. New detectors

All of the solar neutrino experiments have noted the disappearance of ν_e to the point where there is a serious discrepancy with the SSM's. However, none has provided direct evidence for oscillations, the so-called "smoking gun", such as (a) the distortion of the ν_e energy-spectrum noted above, or (b) the observation of the other neutrino flavors, ν_μ or ν_τ, that should appear to make up for the neutrino deficit. Such data can only be provided by the new generation of on-line solar neutrino detectors now under construction.

In this section, I will briefly discuss three on-line detectors that are expected to become operational in the near future: Super-Kamiokande, SNO and Borexino. Two goals of all of these detectors are (a) to achieve much higher data collection rates than the neutrino detectors that are operating today, and (b) to lower their operating thresholds by reducing the levels of radioactive contaminants, e.g., ^{238}U, ^{232}Th, ^{226}Ra and ^{222}Rn. In principle, all of these detectors will be able to measure the energy spectrum of the solar neutrinos.

7.1 Super-Kamiokande

A major expansion at the Kamioka site (Japan), Super-Kamiokande, is being undertaken at present [33]. Its principles are identical to those of Kamiokande. However, the new detector will contain 50000 tons of purified H_2O; the fiducial volume containing 22000 tons of H_2O will be viewed by 11200 large photomultiplier tubes to detect neutrino-electron elastic scattering. Improved electronics and purification processes to remove radon and other naturally occurring radioactive impurities will be installed.

Super-Kamiokande's goal is to achieve significant improvements relative to Kamiokande: to reduce the detection threshold from 7 to 5 MeV and to accumulate about 100× as much data. The high data rate and low threshold should allow it to recognize distortions in the electron-energy spectrum that should result from neutrino oscillations.

7.2 SNO

SNO is an acronym for the Sudbury Neutrino Observatory [34], currently being constructed in the Sudbury mine (Canada) at the 6800-feet level (2073 m). The target in SNO is 1000 tons of heavy water, D_2O, placed in a thin-walled transparent vessel that will float in an outer container filled with H_2O. Some 9500 large photomultiplier tubes will surround the D_2O to observe the Cerenkov radiation resulting from neutrino-induced reactions in this medium.

Great care is being taken to control the sources of radioactivity and other backgrounds in SNO; for example, an industrial clean room has been constructed in the mine at the detector site, as has an elaborate water purification system that employs reverse osmosis, ion exchange and hyperfiltration. Nevertheless, its detection threshold will still be on the order of 5 MeV, so it too will be sensitive only to ^8B neutrinos.

Those reactions that occur between neutrinos and the deuterium in the target give SNO its unique characteristics. The ν_e-capture can be detected in SNO on-line, via the Cerenkov light from the electrons emitted in the reaction, $^2D + \nu_e \rightarrow p + p + \beta^-$. This reaction should provide an accurate measure of the β-energy spectrum. This β-decay process, the charged-current (CC) interaction, occurs only for ν_e, not for the other neutrino flavors.

On the other hand, the neutral-current (NC) interaction, $^2D + \nu_x \rightarrow n + p + \nu_x$, can be caused with equal probability by all three neutrino flavors, ν_x. The Q-value, 2.23 MeV, is simply the binding energy of the deuteron. This NC reaction is probably the most important one in SNO since the appearance of other neutrino flavors besides ν_e will be definitive proof of neutrino oscillations. No other planned detector has this capability. However, observation of the NC signal depends upon the detection of the resulting neutrons, not an easy feat. Three potential methods of neutron detection in D_2O (which acts as a neutron moderator) are being investigated in SNO: (1) Neutron capture by the deuteron will produce ^3H + a prompt ~6-MeV γ-ray, which in turn will produce Cerenkov light in the detector. (2) Addition of a chloride salt, such as $MgCl_2$, to the D_2O will enhance the neutron signal from the NC since neutron capture in ^{35}Cl produces a prompt ~8-MeV γ-ray that also gives rise to Cerenkov radiation. (3) Another method, which does not depend on production of Cerenkov light, is the direct counting of the neutrons with arrays of ^3He-filled proportional-counters suspended in the D_2O. However, all of these methods are susceptible to the problem that any radioactive contamination that emits γ rays with >2.23 MeV, e.g., ^{208}Tl from the ^{232}Th decay series, will photodissociate the deuteron, producing background neutrons.

If the ratio of the signals, CC/NC <1, then SNO will not only verify the disappearance of ν_e but will record for the first time the appearance of the other neutrino flavors, ν_x. If the CC/NC ratio = 1, then only ν_e will be present and the possibility of neutrino oscillations will be ruled out.

7.3 Borexino

The proposal to build Borexino [8, 35] at the Gran Sasso Laboratory (Italy) is based on using an organic liquid scintillator with high photon yield to do neutrino spectroscopy down to extremely low energies, ≥0.4 MeV (in the original plan, the liquid would have contained boron, thus the name Borexino). This threshold is 10× smaller than the 5-MeV goal of both Super-Kamiokande and SNO. Borexino will be unique if it proves successful at making measurements at such low energies, and will be mainly sensitive to the ^7Be solar neutrinos.

The so-called "5-MeV (threshold) wall" that affects both Super-Kamiokande and SNO is primarily the result of decays of radioactive impurities, especially in the naturally occurring ^{238}U and ^{232}Th series, where energetic α particles and γ rays are emitted. In order to breach this wall and achieve the extremely low threshold that is its *sine qua non*, Borexino will have to attain impurity levels that are orders of magnitude lower than those of any other neutrino detector, e.g., 10^{-16} g of uranium or thorium per g of detector and 10^{-14} g of potassium per g (^{40}K can cause problems). This is a formidable challenge, especially since analytical chemical methods cannot at present achieve the required sensitivity levels for some of these elements.

In Borexino, the energy of the recoil electrons resulting from neutrino elastic-scattering will be measured by the liquid scintillator. The current plan is to have ~300 tons of liquid scintillator in a transparent spherical vessel, placed in an outer container filled with high purity H_2O. Phototubes mounted in the H_2O will view the scintillator. The fiducial region will consist of ~100 tons of scintillator; the remaining scintillator and the outer H_2O will act as shielding.

Because of the newness of its approach and the severe technical requirements of ultra-low impurity levels, Borexino has built and tested an intermediate stage at Gran Sasso, known as the Counting Test Facility (CTF), to serve as a proof-of-principle facility. The CTF contained ~4 tons of liquid scintillator, surrounded by H_2O. On-line chemical systems periodically purified the organic scintillator and the water, to prevent the buildup of radioactive contaminants in these liquids caused by leaching from the vessel walls, phototubes and other construction materials, or by deposition from the air.

The initial tests at the CTF of the on-line purification systems, done to control and measure the concentrations of U, Th, K and ^{14}C at ultra-low levels, were very promising [36]. Attention is now being given to plans for the full-scale detector.

8. Summary

As the experimental uncertainties on the results from the present generation of 4 solar neutrino experiments have continued to improve, our focus has sharpened on the nature of the solar neutrino problem. In terms of the solar model predictions, there seem to be deficits, which cannot be accommodated in the theory, in the fluxes of ^8B and ^7Be, with the indication that the latter is especially low. Conversely, the mechanism of neutrino oscillations, especially as embodied in the MSW theory, provides a natural explanation of the solar neutrino deficits.

However, the MSW theory requires non-zero neutrino mass, a fundamentally new property of these elementary particles. As such, it is very important to obtain definitive data to prove or disprove this assumption. The new generation of on-line neutrino detectors is designed to provide such data. It is conceivable that, in a few years, we will have a solution to the solar neutrino problem.

References

1. Rowley, J. K., Cleveland, B. T., Davis, R.: AIP Conf. Proc. No. 126, *Solar Neutrinos and Astronomy*, ed. M. L. Cherry, K. Lande, W. A. Fowler, p. 1, AIP, New York (1985).
2. Hirata, K. S. *et al.*: Phys. Rev. Lett. **65**, 1297 (1990).
3. GALLEX Collaboration: Anselmann, P. *et al.*: Phys. Lett. **B285**, 376 (1992).
4. GALLEX Collaboration: Anselmann, P. *et al.*: Phys. Lett. **B314**, 445 (1993) and Phys. Lett. **B327**, 377 (1994).
5. Abazov, A. I. *et al.*: Phys. Rev. Lett. **67**, 3332 (1991).
6. Hirata, K. S. *et al.*: Phys. Rev. Lett. **66**, 9 (1991).
7. Ewan, G. T. *et al.*: *Sudbury Neutrino Observatory Proposal*, Report **SNO-87-12** (1987); Aardsma, G. *et al.*: Phys. Lett. **B194**, 321 (1987).
8. Raghavan, R. S. *et al.*: in *Proc. 25th Intl. Conf. High Energy Phys.*, ed. K. K. Phua and Y. Yamaguchi, p. 482, South Asia Phys. Soc. (1991).
9. Cleveland, B. T. *et al.*: *Proceedings of the 23rd International Cosmic Ray Conference*, Calgary, Alberta, Canada (July, 1994).
10. Wolfsberg, K., Schroeder, N.: private communications.

11. Bahcall, J. N.: *Neutrino Astrophysics*, Cambridge University Press, Cambridge (1989).
12. Bahcall, J. N., Pinsonneault, M. H.: Rev. Mod. Phys. **64**, 885 (1992).
13. Bahcall, J. N., Ulrich, R.: Rev. Mod. Phys. **60**, 297 (1988).
14. Turck-Chieze, S., Lopes, I.: Astrophys. J. **408**, 347 (1993).
15. Cleveland, B. T. *et al.*: *Proceedings of Neutrino 94*, Eilat, Israel (May 1994); Nucl. Phys. **B (Proc. Suppl.) 38**, 47 (1995).
16. See, e.g., Davis, R.: in *Frontiers of Neutrino Astrophysics, Proc. Intl. Symp.*, ed. Y. Suzuki and K. Nakamura, p. 47, Universal Acad. Press, Tokyo (1993).
17. See, e.g., Bahcall, J. N. *et al.*: Nature **327**, 682 (1987).
18. Hirata, K. S. *et al.*: Phys. Rev. **D44**, 2241 (1991); Suzuki, Y.: Nucl. Phys. **B (Proc. Suppl.) 35**, 407 (1994).
19. See, e.g., Dostrovsky, I.: *Proceedings of an Informal Conference on Status and Future of Solar Neutrino Research*, ed. G. Friedlander, p. 231, Brookhaven National Laboratory Report **BNL-50879** (1978).
20. Rowley, J. K., Stoenner, R. W.: Brookhaven National Laboratory Report **BNL-40818** (1987).
21. Henrich, E., Ebert, K. H.: Angew. Chem., Intl. Ed. **31**, 1283 (1992).
22. Hahn, R. L., Stoenner, R. W.: to be published.
23. Anselmann, P.: Thesis, University of Heidelberg (1992).
24. Gavrin, V. N. *et al.*: Los Alamos National Laboratory Report **LA-UR 94-25** (1994).
25. GALLEX Collaboration: Anselmann, P. *et al.*: Phys. Lett. **B285**, 390 (1992).
26. GALLEX Collaboration: Anselmann, P. *et al.*: Phys. Lett. **B357**, 237 (1995).
27. Abdurashitov, J. N. *et al.*: *Proceedings of Neutrino 94*, Eilat, Israel (May 1994); Nucl. Phys. **B (Proc. Suppl.) 38**, 60 (1995); Abdurashitov, J. N. *et al.*: Phys. Lett. **B328**, 234 (1994).
28. GALLEX Collaboration: Anselmann, P. *et al.*: Phys. Lett. **B342**, 440 (1995).
29. GALLEX Collaboration *Internal Note* **GX 79** (July 1995).
30. Hata, N., Langacker, P.: Phys. Rev. **D52**, 420 (1995).
31. Hahn, R. L., Rowley, J. K., Weneser, J.: Brookhaven National Laboratory Report **BNL-62208** (1995).
32. Mikheyev, S. P., Smirnov, A. Yu.: Sov. J. Nucl. Phys. **42**, 1441 (1985); Wolfenstein, L.: Phys. Rev. **D20**, 2634 (1979).
33. See, e.g., Suzuki, Y.: *Proceedings of Neutrino 94*, Eilat, Israel (May 1994); Nucl. Phys. **B (Proc. Suppl.) 38**, 54 (1995).
34. McDonald, A. B., Sinclair, D.: private communications.
35. "*Borexino at Gran Sasso—a proposal*", ed. G. Bellini *et al.* (1991).
36. Deutsch, M., Hartmann, F. X.: private communications.

C. New Synthetic Elements

Radiochimica Acta *70/71*, 193–206 (1995)

Discovery, Synthesis and Nuclear Properties of the Heaviest Elements

By Gottfried Münzenberg

Gesellschaft für Schwerionenforschung mbH, Planckstr. 1, D-64220 Darmstadt

(Received July 17, 1995)

Heavy elements / Fission / Alpha decay / In-flight separation

Abstract

The most significant progress in heavy-element research, the discovery of a new region of shell stabilized nuclei in the transactinoides and the extension of the Periodic Table to 111 elements, are the subject of this paper. The advance of experimental development for the investigation of the heavy elements allowing physics with single atoms to obtain the production cross-sections and gross nuclear properties will be presented and discussed in the frame of theoretical interpretations.

1. Introduction

The driving force behind heavy-element research is the question for the upper end of the Periodic Table. A first relevant answer was given by Meitner and Frisch [1] already when they explained nuclear fission immediately after its discovery by Hahn and Straßmann. Regarding the nucleus as a charged droplet, the stability of the heaviest elements decreases due to the increasing number of protons as the adhesive nuclear forces cannot any more be counterbalanced by the repulsive Coulomb forces. The nucleus promptly disintegrates by fission. Spontaneous fission is the decay mode specific to the elements at the top of the nuclear table, uranium and beyond. According to the first rough estimate by Meitner and Frisch the Periodic Table would be terminated close to 100 protons. Detailed investigations on the fissioning nuclear droplet by Bohr and Wheeler [2] and Feenberg [3] located the upper end of the periodic table close to 110 protons.

The particular importance of shell effects for nuclear stability was realized by Goeppert-Mayer [4] and Haxel, Jensen and Suess [5]. To investigate theoretically whether shell effects could extend the nuclear table was possible with the refinement of nuclear models and the possibility of using computers. Calculations by Meldner [6] showed that the next double shell closure above the doubly magic lead isotope ^{208}Pb is located at 114 protons and 184 neutrons, far in the liquid-drop unstable region. Whether this shell would be strong enough to stabilize nuclei against fission in this region of macroscopic instability was discussed in terms of a semiempiric approach by Myers and Swiatecki [7], and on the basis of improved theoretical models after the development of methods to calculate nuclear levels in deformed nuclei by Nilsson [8] and the possibility to handle shell corrections for deformed nuclei as well as the introduction of the macroscopic-microscopic description superimposing shell effects to the nuclear droplet by Strutinski [9]. The theoretical models predicted an island of superheavy nuclei at the double shell closure $Z=114$ and $N=184$. This island was predicted to be separated from the upper end of the periodic table by a "sea of instability", where the nuclei were unstable in terms of the charged nuclear drop, the shell effects however not yet strong enough to guarantee their existence. A number of calculations on the stability of the superheavy nuclei was performed by Mosel and Greiner [10], Fizet and Nix [11], and Randrup [12] to give some important examples. The predicted halflifes ranged up to 10^{15} y, long enough for superheavy nuclei to exist in nature.

The first experimental results on the transactinoides were likely to support at least the macroscopic part of this picture insofar as the shore of the sea of instability seemed to be approached for the heaviest elements. For all even-even isotopes of element 104 spontaneous fission was observed with halflifes of the order of some tens of milliseconds. A number of fission activities with halflifes of the order of milliseconds was observed and assigned to the elements 106 and 107 [13], apart from the most neutron rich isotope of element 106 with mass 263, which was identified as an α-emitter [14]. This picture started to change, when in later experiments with the recoil-separator SHIP isotopes of elements 107 and 109 were identified as α-emitters [15].

In 1984 Demin and collaborators reinvestigated the fission activities assigned to element 106 and found, that they primarily originate from isotopes of element 104, the α-daughters of the corresponding isotopes of element 106 [16]. This result was confirmed by the direct observation of the α-decay of the corresponding isotopes of element 106 [17]. For all known isotopes of the heaviest elements from 105 to 111 α-decay was observed [18–21]. These exciting experimental results indicated an enhancement of stability against fission in the transactinoide region [22]. An enhanced stability, due to a hexadecapole deformation had been predicted by Cwiok *et al.* [23], and was corroborated by Leander *et al.* [24] and Sobiczewski *et al.* [25].

The transactinoide elements are produced by complete fusion of heavy ions [18–20]. Two types of reactions have been successful so far: The use of actinoide targets and corresponding light beams from carbon to sulphur starting from the heaviest possible target close to the final nucleus, and the fusion of targets from the doubly magic lead region, lead or bismuth, with neutron rich metal isotopes from titanium to nickel leading to weakly excited compound systems.

The discovery of the new region of deformed nuclei with enhanced stability is the subject of this paper. It will be discussed in the frame of theoretical models. The experimental techniques will be described briefly. The possibilities to further extend the nuclear table by fusion of heavy ions, and the new possibilities with radioactive nuclear beams will be outlined.

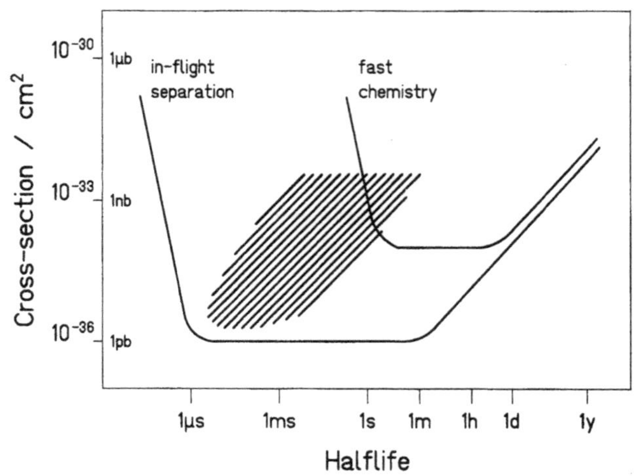

Fig. 1. Sensitivity-ranges for the techniques applied to heavy-element research.

2. Experimental techniques for heavy-element research

2.1 Accelerator development

The prerequisite for heavy-element research is a powerful heavy-ion accelerator. As the neutron excess increases for heavy nuclei, heavy ion fusion in general leads to neutron deficient species. This can be compensated partly by using neutron rich targets and projectiles. The most neutron rich projectiles appear in nature with poor enrichments. We find for example ^{18}O with an abundance of 0.2%, ^{58}Fe with 0.3%, and ^{64}Ni with 0.9%. Only the recently developed Electron Cyclotron Resonance (ECR) ion sources with material consumptions of less than 4.0 mg/h for average beam intensities of $5 \cdot 10^{12}$ particles per second permit experiments of several weeks duration, necessary to run excitation functions even for nuclei produced on the level of some picobarns [26]. Another important necessity for the cold heavy-ion fusion is the continuous energy variation of the accelerated beam.

2.2 Separation techniques

A number of techniques has been applied to the discovery and investigation of heavy elements. The classical chemical separation to identify a new element was used only up to element 104. Methods not involving any separation were the helium-jet, a gasflow, to transport the heavy nuclei from the target area to silicon surface barrier detectors which has been successful up to element 106 in combination with the technique of parent-daughter correlation, and the rotating drum to detect spontaneous fission by means of nuclear track detectors [19, 20]. Beyond element 105 sensitive and rapid detection techniques are needed which are selective to the heavy elements. Cross-sections drop below the nanobarn region. The corresponding production rates are less than one atom per hour and nuclear halflifes fall below one second down to milliseconds.

Present heavy-element research is physics with single atoms. This requires a clean separation of the desired species and a safe identification atom by atom. Such a technique developed in the last ten years being the only method applied today in transactinoide research – apart from fast on-line chemistry – is the recoil separation in-flight combined with the $\alpha-\alpha$ parent–daughter correlation. Fig. 1 shows the sensitivity of this method in cross-section and halflife. It presently ranges down to one picobarn and covers halflifes from one millisecond to about one hour. The sensitivity-range for on-line chemistry is also indicated.

The separation in-flight and implantation into surface-barrier detectors, where the decay schemes of the embedded nuclei can be studied in situ, was already discussed in the first heavy-ion research projects end of the sixties [27]. In-flight separators have been firstly applied to fission fragments [28, 29]. For transactinoide research kinematic separators such as velocity- or energy filters and gasfilled separators are in use [30]. Mass-resolving modified Mattauch-Herzog type recoil mass separators, also applicable in this field, are presently only used for the investigation of lighter fusion products (Table 1) [30].

Kinematic separation is most efficient as it separates all fusion products irrespective of their specific mass or ionic chargestate from the background of other energetic particles such as scattered projectiles or nuclei from other nuclear reactions than fusion. The key to the kinematic separation in-flight is that only the nuclei formed by complete fusion recoil from the target with the full momentum transfer from the projectile. The average recoil velocity is:

$$\langle v \rangle = v_P m_P / m_C \tag{2.1}$$

with $v_P m_P$ and m_C denoting the velocity of the projectile and the masses of the projectile and the compound nucleus, respectively. In-flight separation with a velocity filter has been developed at SHIP (Separator for Heavy Ion reaction Products) [31, 32] at GSI Darmstadt to a sensitivity in the range of picobarns. SHIP (see Fig. 4 in S. Hofmann "Proton Radioactivity", this volume page 97) is a mirror-symmetric combination of two velocity filters with separated electrostatic and

Table 1. Ion optical in-flight separators with a programme for heavy element investigation. Separators used for the separation of trans-actinoide research are marked with an asterisk

Filters:		
SHIP*	Velocity filter	GSI Darmstadt (Germany)
VASILISSA*	Energy filter	JINR Dubna (Russia)
Separators:		
FMS		ANL Argonne (USA)
Gasfilled Separators:		
GNS*		JINR Dubna (Russia)
GARIS*		RIKEN Tokio (Japan)
RITU		JYFL Jyväskylä (Finland)
HECK*		GSI Darmstadt (Germany)

magnetic deflection, and quadrupole triplets to focus the recoils onto the detector system. A small deflection magnet at the exit improves the separation quality [21].

The recoils are indentified by parent−daughter $\alpha-\alpha$ correlation. This method, which has been developed and used by Ghiorso *et al.* successfully up to element 106 in connection with the helium-jet gas transport system [14], relies on the identification of an unknown heavy nucleus on the basis of the known daughter activities. The new element as the parent at the top of that chain can be identified unambiguously. The nuclear table can be extended step by step, producing new elements with isotopes located along specific α-decay chains.

The parent−daughter correlation method is now used with even higher reliability for the energetic recoils implanted into a position-sensitive silicon-detector [33]. The energy signal of the nucleus impinging the detector and a time-of-flight measurement allow a rough mass determination of discriminate the heavy-mass fusion products from the background of energetic ions [34]. All decays from this very nucleus will be observed at this position of implantation, and can be followed over all generations until the chain is terminated by long-lived nuclei, spontaneous fission, or β-decay not observable in the detector. The history of each single nucleus sticking at a well defined position in the silicon detector can be studied in situ over hours or even days. This new technique opened up physics with single atoms which is the experimental basis of present heavy-element research, and recently also applied to other exotic nuclei at the limits of stability [35].

2.3 Recent experimental results

In this section experimental results of importance for our present picture of the heavy elements including groundstate properties and production of the heaviest elements are summarized. More details are given in Refs. [18−21, 36]. An evaluation of the criteria for the discovery of a new element and the priority of the discovery of the transfermium elements up to element 109 has been prepared by the "Transfermium Working Group", headed by D. H. Wilkinson. The group had been installed by the International Union of Pure and Applied Physics (IUPAP) and the International Union of Pure and Applied Chemistry (IUPAC) [37]. This paper also contains a complete biography.

Element 106 is the key to the present understanding of the stability of the heaviest elements. It was discovered already in 1974 in the reaction $^{249}Cf(^{18}O,4n)$ by the identification of the isotope $^{263}106$ with the $\alpha-\alpha$ parent−daughter correlation method as an α-emitter with 0.9 s halflife [14]. Spontaneous fission observed in first irradiations of lead targets with ^{54}Cr assigned to the odd-mass isotope $^{259}106$ [38] was found not to originate from the decay of an isotope of element 106, but from daugher isotopes of element 104 proceeding from α-decay [16]. This result, based on the measurement of spontaneous fission with nuclear track detectors in cross bombardements, was confirmed by the direct observation of the isotopes $^{259,260,261}106$ in irradiations of ^{208}Pb and ^{207}Pb, respectively, with ^{54}Cr [17]. All of them were identified as α-emitters including the doubly even isotope with mass 260 for which a fission branch of 50% was determined. The experimental results indicated an enhancement of nuclear stability against spontaneous fission as compared to the existing predictions which gave fission halflifes of less than 0.1 ms [39] for this isotope. This finding was confirmed in all later experiments which showed that all isotopes of the heavier elements including elements 110 and 111 are α-emitters. In irradiations of ^{248}Cm with ^{22}Ne the most neutron rich isotopes of element 106 known at present, $^{266}106$ and $^{265}106$, were observed. The halflifes, estimated from the measured decay energy on the basis of half-life systematics, an advanced Geiger-Nuttal formula, are 10−20 s [40]. Two examples of single-atom decay chains from this experiment are displayed in Fig. 2 [49]. The chain starting from the even-even nucleus $^{266}106$ ends with a fission of the daughter. The chain from $^{265}106$ can be followed over three generations of α-decays.

For element 107 presently three isotopes are known. The Isotopes $^{261}107$ and $^{262}107$ were produced in irradiations of ^{209}Bi with ^{54}Cr [43]. The most neutron rich isotope $^{264}107$ was discovered as granddaughter in the decay chain starting from $^{272}111$ [44]. The halflifes of these α-emitting nuclei for which no fission was observed range between 8 ms and 440 ms. The odd-odd nucleus $^{262}107$ has an isomeric transition. Early results reporting a fission activity with 1−2 ms halflife for the isotope $^{261}107$ [45] could not be confirmed.

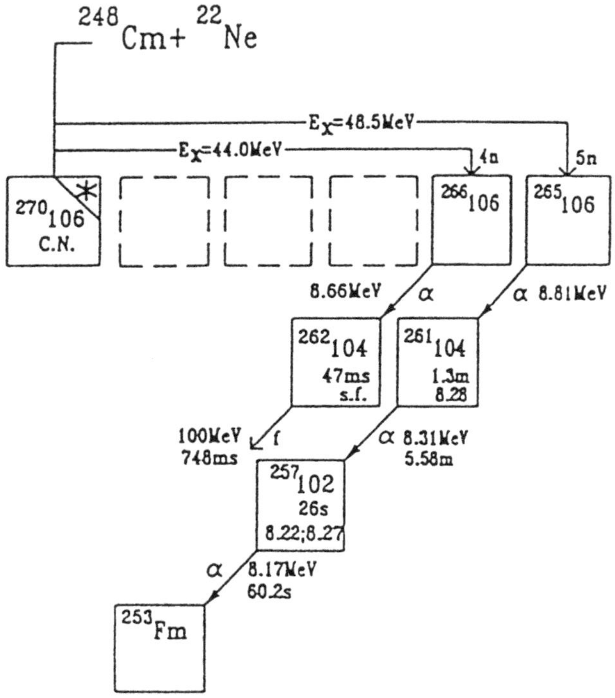

Fig. 2. Two out of ten single-atom decay chains from ^{265}106 and ^{266}106. The data for the daughter decays known from literature are displayed in the boxes, the measured time intervals between the α-decays and the corresponding energies are listed next to the arrows. The signal from the implantation of the mother nucleus was not registered, therefore no time intervals are given for the first generation [49].

Element 108 was first identified in the reaction ^{208}Pb(^{58}Fe,n)^{265}108 on the basis of three atoms [34]. This result was confirmed with 75 α-decays of this isotope produced in the same reaction [26]. For the doubly even isotope ^{264}108 α-decay [34] was found. The heaviest isotope ^{267}108 was produced in irradiations of ^{238}U with ^{34}S [46] and confirmed when it appeared as daughter in the decay chain of ^{271}110 [26]. The halflifes range from 0.45 ms for the doubly even isotope with mass 264 to 60 ms for the most neutron rich one.

Element 109 was discovered in irradiations of ^{209}Bi with ^{58}Fe by measuring the decay chain originating from one single atom which was assigned to the isotope with mass 266 on the basis of a careful statistical analysis of the decay sequence [15], and confirmed in a later experiment [47]. The isotope ^{268}109 was found in the decay chain from element 111 [44]. The halflifes of the two α-emitting isotopes are 3.4 ms and 70 ms, respectively.

Elements 110 and 111, the two most recent discovered elements, form the present top of the nuclear table. Production cross-sections fall below 10^{-35} cm^2. Element 110 was first identified in the reaction ^{108}Pb(^{62}Ni,n) by the observation of the isotope ^{269}110 [26]. The isotope ^{271}110 was synthesized with a ^{64}Ni beam. With this more neutron rich isotope, the production cross section for the synthesis of element 110 was enhanced from 3.5 pb to 15 pb [44]. The halflifes of the two isotopes are 170 µs and 1.4 ms, respectively.

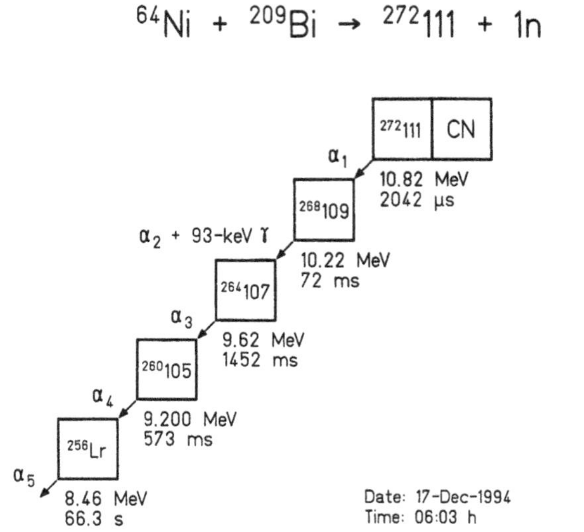

Fig. 3. Decay chain of the third atom of element 111, observed on Dec. 17, 1994 at 6:03 h [44].

There is strong evidence for a second transition in ^{271}110 with about 60 ms halflife [21]. Two more isotopes are reported, ^{267}110, produced in the irradiation of ^{209}Bi with ^{59}Co [48] which needs confirmation, as the first daughter generations of the decay chain are not observed, and ^{273}110 produced by irradiation of ^{244}Pu with ^{34}S and the evaporation of five neutrons. This isotope is of special interest as it is the most neutron rich one [49].

Element 111 was identified in the reaction ^{209}Bi (^{64}Ni,n) on the basis of three α-chains. The observed isotope ^{272}111 has a halflife of 1.5 ms [44, 21]. Fig. 3 shows the decay chain of the third atom of element 111. It starts with the implantation of the mother nucleus, 2.04 ms later the α-decay is observed at an energy of 10.82 MeV. After another 72 ms the daughter nucleus decays with an energy of 10.22 MeV, accompanied by a γ of 93 keV. It is followed by the granddaughter decay of 9.62 MeV after 1.45 s. Then, 573 ms later, an α-decay of 9.2 MeV is measured. The α-sequence ends after 66 s with an α-decay of 8.46 MeV. This decay sequence does not only confirm the discovery of a new element but also that of two new isotopes, ^{268}109 and ^{264}107. The assignment of the chain is fixed by the last two decays assigned to the known isotopes ^{260}105 and ^{256}Lr.

The transactinoide region is displayed in Fig. 4. Decay modes, halflifes and α-decay energies are given for each isotope. The chemical symbols and names proposed by the discoverers[1] are: Rutherfordium (Rf) and Kurchatovium (Ku) for element 104, respectively (no common name has been found yet); Hahnium (Ha) for element 105; Seaborgium (Sg) for element 106;

[1] The interdivisional committee of the "Union of Pure and Applied Chemistry" did not accept the name proposals yet and refused the right of the discoveres to name an element. It proposed new names [51]. The severe protest of the discoverers, namely from the american and german groups caused the IUPAP to reconsider the proposal of the subcommission and to postpone the decision on the acceptance of the names to 1997.

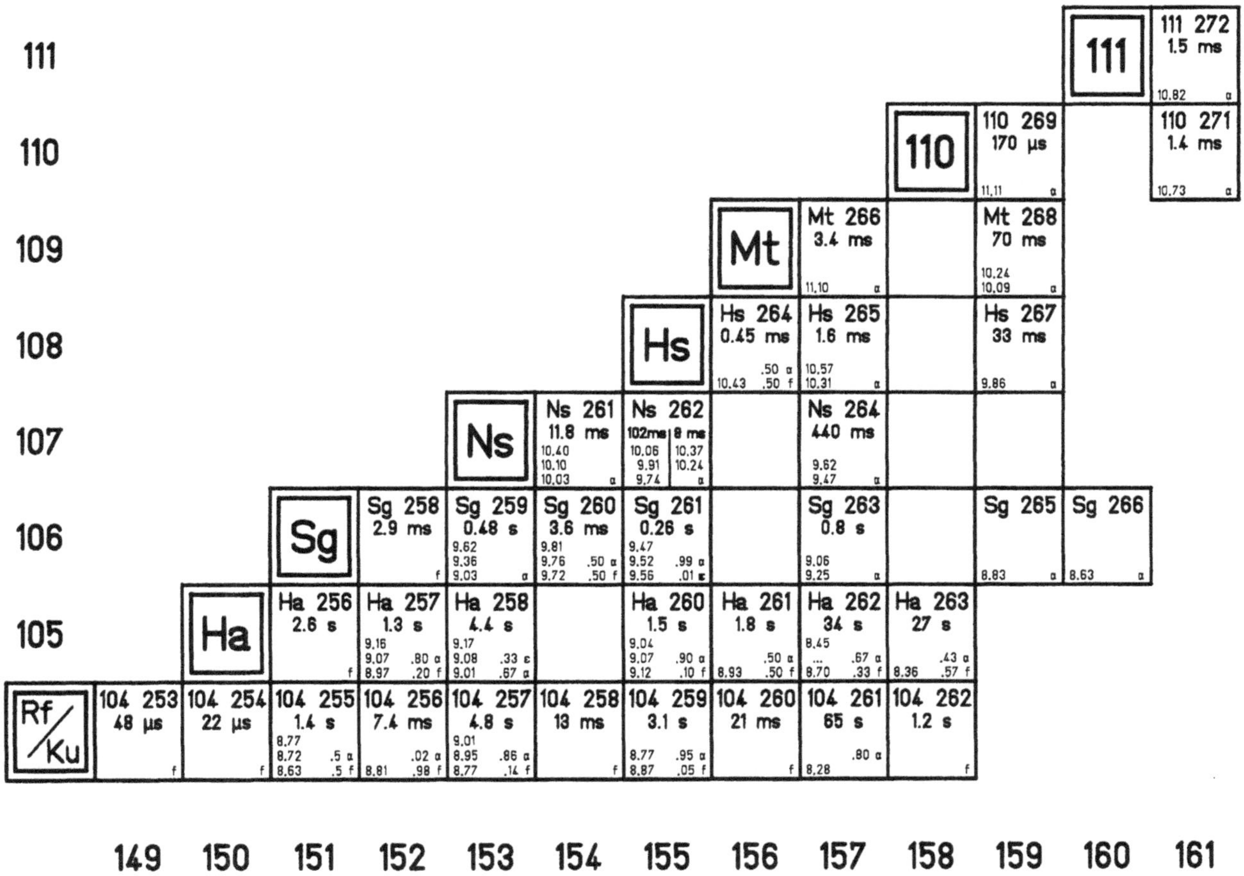

Fig. 4. The upper end of the nuclear table.

Nielsbohrium (Ns) for element 107; Hassium (Hs) for element 108 and Meitnerium (Mt) for element 109. For the elements 110 and 111 no names are proposed up to now.

3. Groundstate properties of the heaviest elements

3.1 Physics with single atoms – gross properties of the heavy elements

The experimental data obtained for the heavy elements rely on the observation of single-atom decays. Nevertheless gross properties such as dominant decay modes, halflife, decay energy and groundstate mass, and even groundstate deformations are obtained. With the help of the macroscopic-microscopic model describing nuclear stability in terms of the charged nuclear drop for the smooth part with a shell correction to account for the nuclear structure, shell correction energies and fission barriers can be extracted from the data. It is noteworthy to mention that the data evaluation from few events needs careful statistical treatment [15, 19, 50].

To summarize the most striking results, already pointed out (Fig. 4):

– spontaneous fission is most important in the region near element 104, but disappears for the heaviest elements.

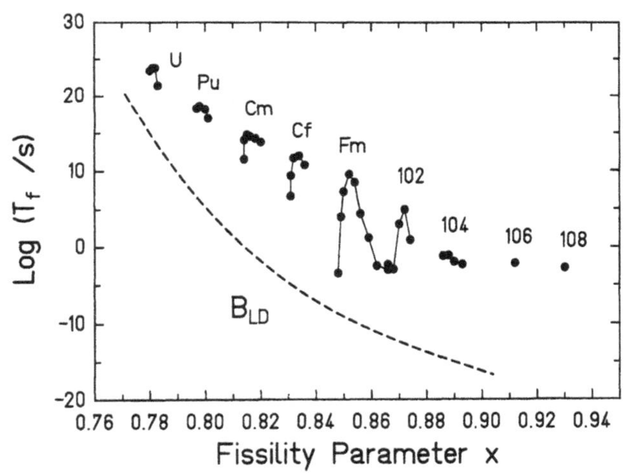

Fig. 5. The partial halflifes for spontaneous fission for the doubly even isotopes of the transuranium elements. The dashed curves show the macroscopic fission halflife.

– the halflifes of the isotopes in the transactinoide region grow significantly towards increasing neutron numbers.

These observations suggest an unexpectedly high stability at the upper end of the periodic table and towards 160 neutrons, which is corroborated by the partial fission halflifes of the doubly even uranium and transuranium isotopes displayed in Fig. 5. The fission

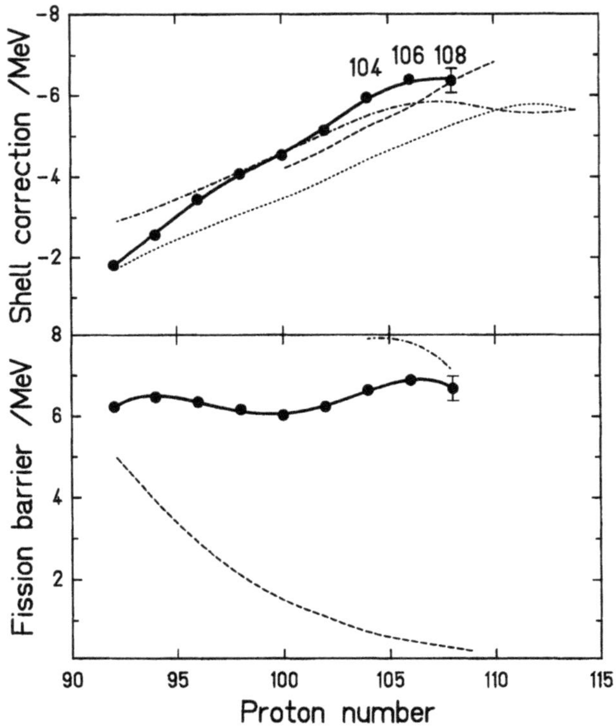

Fig. 6. Upper panel: The experimental shell correction energies for the even-even isotopes along the $N-Z=48$ α-chain down to uranium (dots) compared to current theoretical predictions. Cwiok *et al.* [23] (dashed-dotted line), Möller *et al.* [52] (dotted line), and Patyk *et al.* [54] (dashed line). Lower panel: Experimental (dots) and calculated [56] (dashed-dotted line) fission barriers for the same isotopes. The dashed line represents the liquid-drop part of the fission barrier.

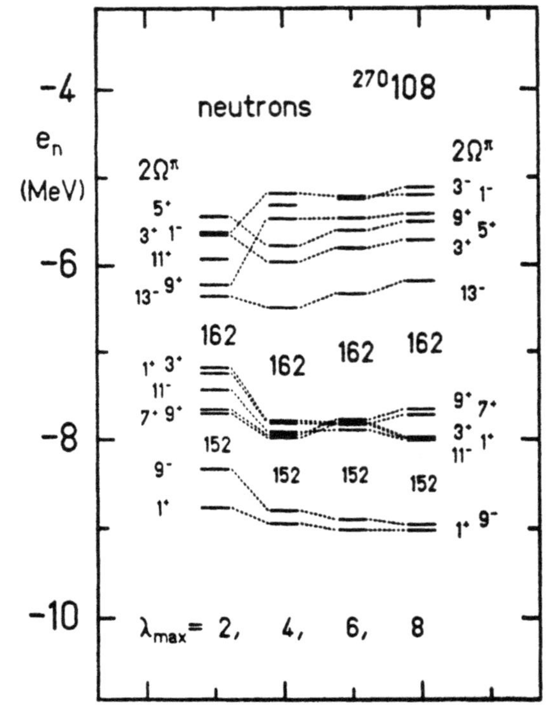

Fig. 7. The dependence of the shell correction on the degrees of freedom in nuclear deformation [54].

halflifes are plotted in a logarithmic time scale versus the fissility parameter. From uranium to nobelium halflifes decrease with increasing fissility parameter over more than twenty orders of magnitude: from the age of the universe down to seconds, as expected from the nuclear liquid drop model. The structure observed in the isotopic chains of the heavy actinoides is due to a small shell effect at 152 neutrons, the influence of which increases towards the heavier elements. A complete change of the systematics appears in the transactinoide elements. Fission halflifes become nearly constant with a slight enhancement towards increasing neutron number. They do not depend on the fissility parameter, which is a macroscopic scaling parameter. Nuclei in this region exist only by shell effects. The comparison of the experimental data to the calculated liquid-drop halflifes, shows an enhancement of the fission halflifes in the transactinoide region by 10 to 20 orders of magnitude caused by the shell stabilization.

From the α-decay energies, Q-values and masses can be evaluated, with sufficient precision however only for the doubly even isotopes where the α-decays are expected to be groundstate transitions. In the macroscopic-microscopic approach, the shell effect is:

$$E_{SH} = m - m_{LD}, \quad (3.1)$$

where m is the experimental mass. The liquid-drop mass m_{LD} must be taken from a theoretical model. Fig. 6 shows the experimental shell effects from uranium to element 108 along the $N-Z=48$ α-chain taken from the measured masses with the finite-range droplet model [52, 53]. They increase from 2 MeV for ^{232}U to more than 6 MeV for ^{264}108. This is about half the shell strength of the most strongly bound doubly magic nucleus ^{208}Pb. The microscopic calculations from Cwiok *et al.* [23] and from Patyk *et al.* obtained with a Woods-Saxon potential [54], as well as from Möller and Nix with the folded Yukawa potential plus exponential term [52, 53] show the same trend.

The microscopic theory reveals that the shell effect originates from a negative hexadecapole deformation in the groundstate, e.g. a barrel-like shape of the nuclei. This is nicely demonstrated in Fig. 7 where the gaps in the neutron levels in the nucleus ^{270}108 are plotted for increasing degrees of freedom in deformation space [54]. The shell gap grows when the hexadecapole degree of freedom ($\lambda=4$) is included in the model. Hexadecapole deformation in this region is not unexpected. It is well known that in regions where quadrupole deformation starts or ends, positive and negative hexadecapole deformations, respectively, are observed. It must be noted here that already the early calculations include hexadecapole deformation. The result of the recent calculations is the unexpectedly strong influence of the hexadecapole deformation on the shell correction energy.

The calculated groundstate microscopic corrections [55] for the heavy elements between the two double magic shell closures at ^{208}Pb and ^{284}114 are displayed in Fig. 8. The hexadecapole stabilized region centered at $Z=106$ and $N=164$ extends from the upper end of the periodic table to the double shell closure

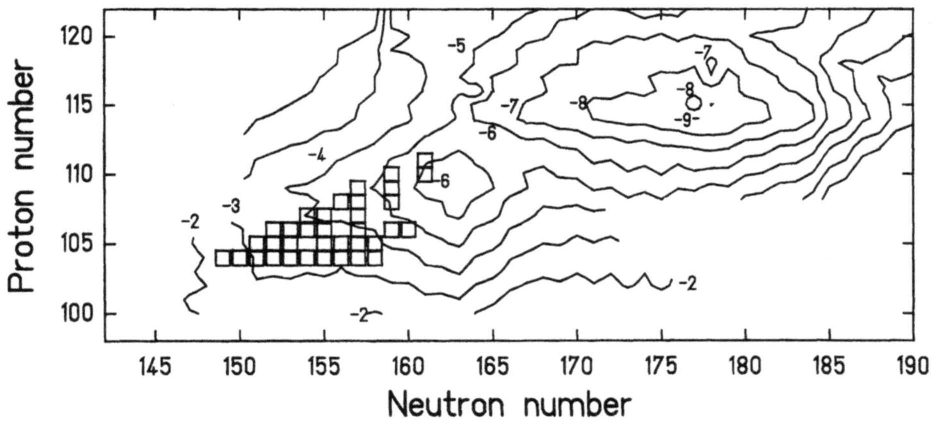

Fig. 8. Microscopic corrections for the heavy elements [55].

at $Z=114$ and $N=184$. The shell correction is as high as 7−8 MeV. The maximum stability at the spherical shell closure is downshifted to $Z=114$ and $N=178$. The shell effect of 9 MeV is not much higher than that at the maximum of the hexadecapole stabilization.

3.2 Fission barriers and halflife

The height of the fission barrier B_F can be obtained from the shell correction energy and the liquid drop barrier to good approximation:

$$B_F = B_{LD} - E_{SH}. \tag{3.2}$$

Fig. 6 shows the experimental fission barriers, again for the doubly even isotopes along the $N-Z=48$ chain. They are almost constant from uranium to element 108 as the liquid drop part of the barrier decreases in the same way as the shell correction energies increase and rise to 6−7 MeV. Above element 104 the macroscopic part of the fission barrier falls below 0.5 MeV and is practically no more effective. The nuclei in this region are shell stabilized. The calculations from Sobiczewski [56] nicely reproduce the experimental data.

The explanation for the decrease of fission halflife despite a constant barrier hight is the barrier width [39]. The light actinoides are known to have broad, double humped barriers, a superimposition of a thick liquid drop barrier and the microscopic correction (see inserts in Fig. 9). The shell correction oscillates with increasing deformation of the nucleus due to periodic crossings of the single-particle levels which cause variations in the level density. For the transactinoides the liquid drop barrier is practically absent. The outer hump of the barrer sitting on top of the liquid drop barrier falls below the groundstate, only the inner hump of the barriers is left and remains as a thin barrier. In the parabolic approach of the barrier the halflife is:

$$T_{1/2} = \left[1 + \exp\left(\frac{2\pi B_F}{\hbar\omega_F}\right)\right]\frac{\ln 2}{n} \tag{3.3}$$

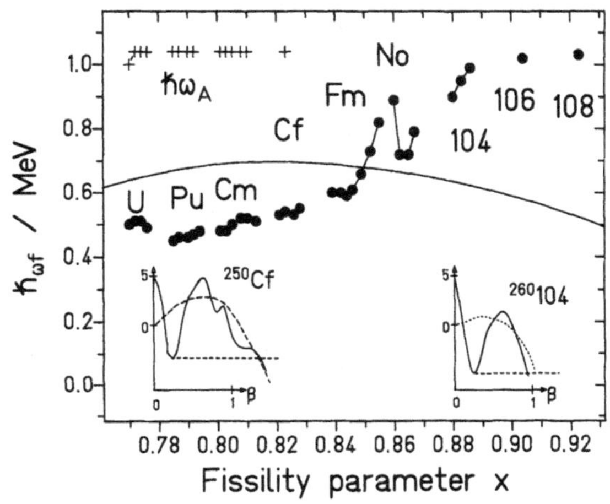

Fig. 9. The fission-barrier curvature parameter, dots: experimental data for the heaviest elements, crosses: curvature of the inner barrier obtained from induced fission, line: calculated liquid-drop barrier. The barrier shapes for a typical actinoide and transactinoide nucleus are indicated schematically.

$\hbar\omega_F$ is the barrier curvature parameter and n the number of barrier assaults, $n = 2.5 \cdot 10^{20}$/s. Fig. 9 shows the experimental barrier curvatures plotted versus the fissility parameter [47]. In the uranium region we find weakly curved, thick liquid-drop dominated barries. Their curvature is close to that of the liquid drop barrier [58]. The curvature parameter of the transactinoide barriers is much higher, about 1 MeV and compatible to the curvature of the inner barrier of the light actinoides [59]. This strongly supports our picture that for the typical shell barriers only the inner hump of the barrier is effective.

While there is no direct experimental proof of the hexadecapole deformation, the quadrupole deformation can be derived from α-transitions to rotational bands. Fig. 10 shows the nuclear moments of inertia from the spherical lead region to the deformed actinoides. The nuclear moment of inertia for $^{256}104$ fits well to that of the actinoides, suggesting at similar quadrupole deformation [57].

A halflife prediction for the doubly even isotopes near the $N=162$ shell [56] is given in Fig. 11. The

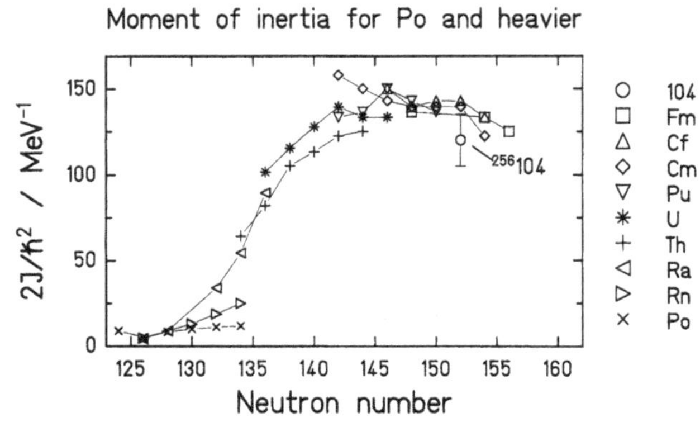

Fig. 10. Moments of inertia for the nuclei above the doubly magic ^{208}Pb.

halflifes for odd and odd-odd nuclei will be considerably longer. For the known isotopes the predictions agree well with the experimental data. The nuclei in this region are predicted to be α-emitters. The longest-lived isotope ^{268}106 not far away from the known ^{266}106 has a halflife of about 2 h. The limit of 10^{-6} s for recoil separators is marked as thick dashed line. The hexadecapole stabilized region forms a bridge between the upper end of the periodic table and the spherical double shell closure at $Z=114$ and $N=184$. These nuclei, existing only due to microscopic stabilization can be considered as "deformed superheavy nuclei".

4. Production of the heaviest elements

4.1 Hot and cold fusion

The only method known presently to synthesize transactinoide elements is the complete fusion of heavy ions, the amalagamation of target and projectile. The problems related to the fusion of heavy nuclei are discussed in a number of reviews [60, 61, 62].

For target and projectile to get into close contact that the nuclear forces become effective and drive the dinuclear into a mononuclear system, the projectile needs sufficient energy to overcome the Coulomb repulsion. This energy in our cases exceeds the Q-value for the formation of the compound nucleus. The energy excess remains as excitation energy. It is dissipated by particle-evaporation or fission, and γ-emission. In the reactions discussed here, charged-particles: α or proton evaporation, are strongly suppressed because of the high Coulomb barrier of the compound nucleus and the small excitation energy. Neutron evaporation is the channel leading to the heavy nuclei. For the heavy elements the deexcitation process is dominated by fission competing with neutron emission.

The fusion-evaporation reaction is a two-step process determined by the fusion cross section σ_{fus} and the survival probability w_x of the compound system after passing through a cascade of evaporation of x neutrons. The production cross-section of the heavy nucleus in its groundstate is:

$$\sigma = \sigma_{\text{fus}}(E_P - E_B, l)\, w_x(E^*, l). \qquad (4.1)$$

The fusion cross-section depends on the projectile energy with respect to the barrier. The survival probability is determined by the excitation energy of the compound nucleus. The angular momentum l of the fusing system cuts-off the fusion cross-section and limits the survival probability as well. The angular momentum limited fusion cross-section is:

$$\sigma_{\text{fus}} = \pi \lambda^2 \sum_{0}^{l_{\text{lim}}} (2l+1) T_l(E_P - E_B) \qquad (4.2)$$

where λ is the de Broglie wavelength of the dinuclear system and T_l is the transmission of the l-th partial wave through the fusion barrier. The limiting angular momentum l_{lim} of the compound system is obtained from the rotating liquid drop model [63]. For the highly fissionable systems discussed here it is below 15 $\hbar$.

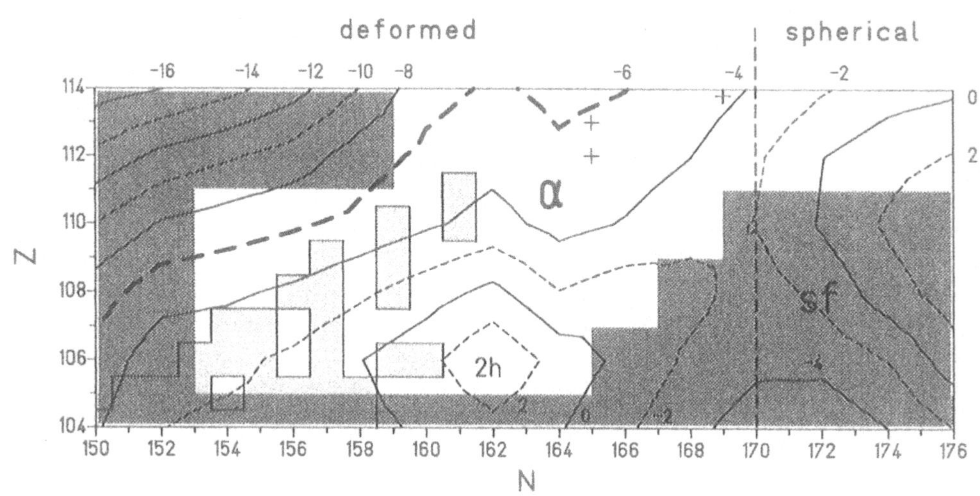

Fig. 11. Predicted halflifes of the even-even isotopes of the heavy and superheavy elements. The thick dashed line indicates the halflife of 10^{-6} s. The known isotopes of the heavy elements are marked by squares [56].

The nuclei at the upper end of the periodic table are produced in central collisions.

Two classes of fusion reactions have been successful for the synthesis of heavy elements. The historically first one was the use of actinoide targets such as ^{248}Cm, ^{249}Bk, and ^{249}Cf and beams of neutron rich light ions as ^{15}N or ^{18}O. The second type started from targets of the doubly magic lead region, ^{208}Pb and ^{209}Bi and neutron rich beams of e.g. ^{54}Cr, ^{58}Fe, and ^{62,62}Ni.

The idea behind the use of the actinoide targets was, to start from a target as close to the final nucleus as possible: This was partly due to the limited capabilities of the accelerators to provide energetic beams of heavy ions, at least in the early experiments. There was however also an important physics motivation to use the lightest possible beams – and consequently the heaviest available targets – for heavy-element synthesis. The necessary condition for the amalgamation of target and projectile is that the dinuclear system has to pass inside the fission configuration of the compound nucleus. For highly fissionable compound systems the fission saddle is located at small deformations. The distance from the point of nuclear contact, the Coulomb barrier, for these very asymmetric systems is much closer to that point as compared to symmetric systems. The extra energy needed to drive the system on its way to fusion inside the fission saddle shifts the barrier to higher energies, a phenomenon called "Extra Push" [64].

The use of lead targets and the corresponding metal-ion beams was motivated by the idea to produce the compound nucleus as cold as possible. The symmetry of the system and the strongly bound doubly magic lead nucleus allows to form to cold compound systems with excitation energies of 10 to 20 MeV at the barrier. They have to survive only one or two steps of evaporation resulting in a high survival probability, even for the most heavy and fissile nuclei at the upper end of the periodic table [65]. The actinoide systems are formed with excitation energies near 40 to 50 MeV and hence have to survive the fission competition over four to five evaporation steps which dramatically reduces their chance of survival. The first and striking success of the cold heavy ion fusion was the high cross section for the production of ^{254}No by the fusion of the two doubly magic nuclei ^{48}Ca and ^{208}Pb after the evaporation of only two neutrons. Nevertheless the extra-push idea persisted as objection against the cold heavy-ion fusion of massive nuclei [66]. The first convincing evidence for the production of a transactinoide nucleus by cold fusion was the identification of the well known ^{257}104 in the irradiation of ^{208}Pb with ^{50}Ti. The evaporation of only one neutron was an unambiguous proof the cold formation of the compound nucleus [67].

4.2 Fusion of massive systems: The Extra Push

The measured production cross-sections for the heavy elements starting from nobelium are displayed in

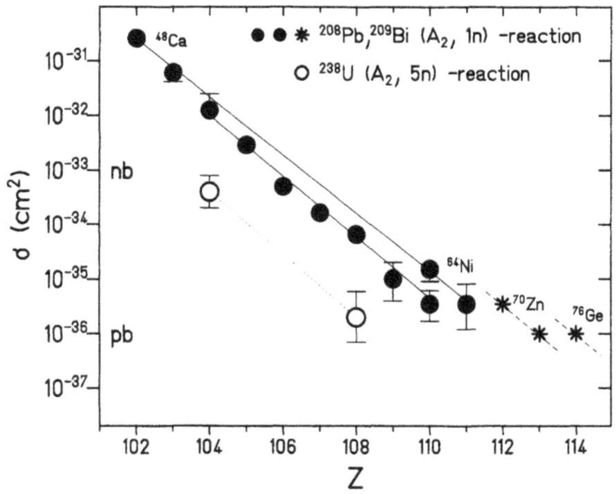

Fig. 12. Experimental (dots and circles) and extrapolated (asterisks) production cross-sections for the heavy elements. The full line connects the points for cold fusion by one-neutron evaporation with lead or bismuth targets, the dotted line corresponds to actinoide targets and evaporation of five neutrons [21].

Fig. 12. The cross sections decrease from about 1 μb for the production of nobelium to below 10 pb for the transactinoide region down to a fraction of less than 10^{-6} of the fusion cross-section for zero angular momentum (Eq. 4.2). Attempts to reproduce the cross-sections with fusion-evaporation calculations failed so far [67]. One approach to explain the experimental data in the heavy-element region was to shift the fusion barrier to higher energies which reduces the cross section at the conventional barrier. The shift above the Coulomb barrier was explained on terms of a dynamical fusion-hindrance, the extra-push [64]. A number of experiments on lighter systems, including the investigation of fusion-fission [68] and the direct observation of the heavy fusion residues was carried out [60]. Because of the fast decrease of the production cross section and the difficulty to correct for the fission competition, the experiments observing the surviving heavy residues were made only for compound systems in the region from mercury to uranium [60]. For those systems however the onset of extra-push can be accessed with symmetric systems only.

If the complete heavy-ion fusion to a first approximation is described in terms of inversed fission, the fusability can be expressed by a macroscopic parameter in analogy to the fissility, including asymmetric target-projectile combinations:

$$x_E \propto \frac{Z_P \cdot Z_T}{A_P^{1/3} \cdot A_T^{1/3} \cdot (A_P^{1/3} + A_T^{1/3})} \tag{4.3}$$

where x_E is the "entrance fissility".

A number of, predominantly macroscopic, models has been developed [69] to describe the fusion dynamics. The extra-push for the nuclei on their way to the mono-nucleus is described in terms of deformable bodies connected by a neck through which nucleons

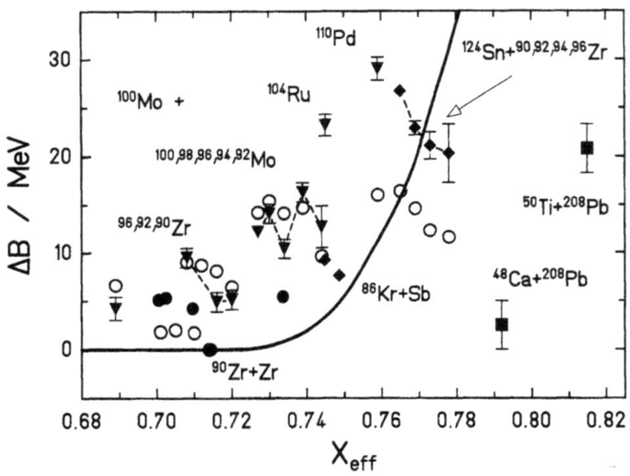

Fig. 13. The experimental extra-push data (dots) compared to the macroscopic model [71] (solid line), and the microscopic calculations [73] (circles). The squares refer to heavy-elements.

are exchanged. The dissipative losses are explained by interaction of the nucleons with the moving nuclear walls and the transport of energy through the neck by nucleon flow.

To extract the extra-push[2] experimentally from the study of evaporation-residues, fusion-evaporation calculations are needed to reproduce the experimental cross sections with appropriate barrier distributions. The experimental barrier shifts obtained with this method [60] are plotted in Fig. 13 versus the entrance fissility parameter [71]. This "effective" fusability parameter x_{eff} is an average of the entrance fissility x_E and the compound nucleus fissility x_C.

$$x_{\text{eff}} = x_E/3 + 2x_C/3. \tag{4.4}$$

The reference barriers for the energy shifts in Fig. 13 have been taken from the Bass model [70]. The general trend of increasing extra push beyond the critical value of x_{eff} seen in the experimental data is nicely explained by the theory [71]. The isotopic chains obtained by variation of the irradiated target isotope show a structure, especially for the irradiation of the isotopes ^{90,92,94,96}Zr with ^{124}Sn opposite to the general extra-push trend. The barrier shift for the reaction involving the spherical ^{90}Zr or ^{92}Mo is below the general trend, that for the soft deformed ^{96}Zr or ^{100}Mo above. These structure effects were interpreted in terms of microscopic two-center model calculations [73]. They reproduce the observation that stiff shell nuclei show less dynamical heating than the deformable mid-shell nuclei [72]. The explanation is that in the phase of approach the degeneracy of levels in the approaching nuclei is removed and the nucleons staying on their orbits can be raised to higher levels, which is not possible for shell nuclei where all levels are filled. The attempt to extract a barrier shift for the heavy elements, indicated as squares in Fig. 13, fails completely. It is noteworthy to mention here that the data for symmetric systems are not a stringent test of the model, as they are not sensitive to the scaling parameter describing the asymmetry of the amalgamating system.

Certainly the application of a macroscopic approach such as the extra-push model to explain the fusion of the purely shell stabilized transactinoide nuclei is at least questionable [19]. A macroscopic-microscopic description including structure effects certainly would be more appropriate. There is no doubt, that the small production cross-sections for the heavy elements can only be explained by a small fusion probability and not by the fission competition in only one neutron-evaporation step [72] (Eq. 4.1). The competition between neutron evaporation and fission is determined by the time scale for both processes. A decreasing survival probability requires that the time for fission must decrease in the same way. Fission, as a collective process, however, cannot become infinitely short. In contrast, recent experiments and calculations indicate that fission is a comparatively slow process.

The question, important for the production of the heaviest elements by cold heavy ion fusion, is whether there is a shift of the excitation functions for the production of increasingly heavier elements to higher energies. In this case cold fusion would be terminated near elements 110 and 111 due to the strongly increasing fission competition towards the heavier elements at high excitation energies. A clear answer to this question came in a recent series of experiments, where excitation functions for the production of elements 108, and 110 were measured [21]. The data, displayed in Fig. 14 show a downshift of the production cross-section maxima to smaller excitation energies for the heavier elements. They follow the unshifted fusion barrier, as the arrows marking the Bass-barrier show. The small halfwidth of about 5 MeV [44] indicates the strong competition between barrier penetration and fission. The low-energy side of the excitation function is cut by the barrier penetrability, the high energy side by the fission competition with the neutron evaporation. The enhancement of the fission competition is manifested by the decrease and finally the disappearance of the two- and three neutron evaporation channels above element 104, and 108, respectively. For the production of the shell stabilized actinoides it is important that the excitation functions are located below the critical excitation energy of about 20 MeV where shell effects are damped out [74].

Up to now there exists no theoretical approach to explain the fusion of the heavy elements in terms of a macroscopic theory. First attempts were calculations in the frame of the fragmentation theory, which predict the most favourable target-projectile combinations, and cold heavy ion fusion as the most suited reaction for the production of the heaviest elements [75]. For

[2] Correctly the extra-push necessary for compound formation is the "Extra-Extra push" [71].

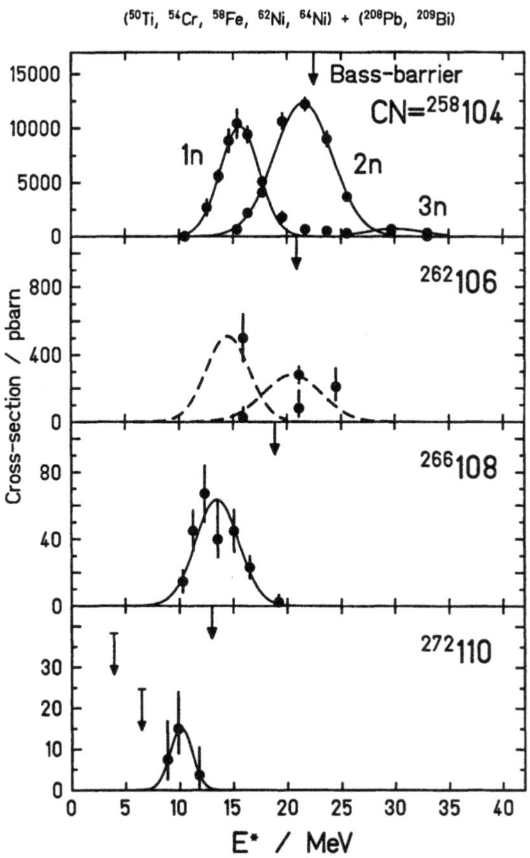

Fig. 14. Excitation functions for the production of the transactinoide elements 104, 106, 108, and 110 by irradiations of ^{208}Pb with ^{50}Ti, ^{54}Ni, ^{58}Fe and ^{64}Ni, respectively [21].

the shell stabilized elements beyond 106 the fusion has to rely on shell effects because of the absence of a macroscopic pocket in the nucleus-nucleus potential of target and projectile. The fusion probability for the capture of the target nucleus is zero – as expected from the extra-push model. The flat potential-energy surface of the fusing system where the projectile nucleus at the Coulomb barrier practically comes to rest in the touching configuration, gives the chance also to very slow and normally improbable processes. In close analogy to the atomic reactions, the filling of nuclear levels of the interacting nuclei leads to the combined system. The influence of structure effects on nuclear transfer has been investigated experimentally as well as theoretically in terms of coupled channels calculations for deformed and spherical nuclei as the doorway to fusion [76]. The question is however how the system will further develop on its way to fusion. Cold multinucleon transfer initiated by transfer of superfluid protons [77] to fill the free nuclear orbitals above the closed shells of the magic ^{208}Pb was proposed as possible fusion mechanism [21]. Of course it is difficult to give an interpretation of a reaction channel observed with a fraction of less than 10^{-10} of the geometric cross section, nevertheless it is most likely that a process of such kind will occur.

4.3 Prospects for new elements in the closed-shell region $Z = 114$ and $N = 184$

Fig. 11 shows the predicted halflifes for the heaviest elements around the deformed shell at $Z = 106$ and $N = 164$ from the upper end of the periodic table to the spherical shell closure. It predicts that the whole region between the present upper end of the nuclear table and the spherical region is accessible by in-flight separation with a lower experimental time limit of 10^{-6} s, even fast on-line chemistry with separation times of seconds should be possible. The production cross sections are on the picobarn level. For the projectiles above nickel the most neutron rich isotopes such as ^{64}Ni, ^{70}Zn, and ^{76}Ge allow to add four neutrons to the beam while stepping up by two protons. For element 110 the isospin change in the projectile from ^{62}Ni to ^{64}Ni yielded an increase of the cross-section by a factor of about four. This enhancement may be due to the smaller excitation energy reduced by about two MeV. If a similar enhancement can be expected for the heavier beams, the expected cross-sections will stay on the level shown in Fig. 12 for the elements beyond 111 [21].

There is however a severe drawback of this extrapolation: Beyond element 112 the excitation energy of the compound nucleus at the fusion barrier falls below the neutron separation energy of about 7 MeV. If the fusion relies on the one-neutron channel, we will not profit from the increasing cooling effect, as we have to put sufficient excitation energy into the compound system to evaporate the neutron, the formation cross-section could be reduced. On the other hand, with decreasing excitation energy the radiative fusion [78] becomes more probable. Moreover the excitation energy of the compound system will fall below the fission barrier. Certainly the system needs to pass inside the fission saddle. The question is, which way it would however take in the multidimensional deformation space. The production of the elements beyond element 111 will give more insights into the formation of heavy elements, for which, as the recent experiments showed, experiments in other nuclear regions will not allow to draw final conclusions.

One idea might be to go directly to the spherical region and to produce the element 116 in the irradiation of ^{208}Pb with ^{82}Se to profit from the predicted enhanced groundstate stability of this nucleus [21]. This reaction would lead for the first time to the spherical superheavies. A problem will be its identification by parent–daughter correlation as practically all daughter nuclei are unknown and the α-chain rides through the longlived isotopes near the maximum hexadecapole stability.

The hot fusion, the production of heavy elements from actinoide targets and after evaporation of five neutrons evidently is less favourable in view of the present data. Fig. 12 shows that the production cross sections are by more than one order of magnitude below those for the hot fusion. The advantage that this type of reaction gives access to the neutron rich species decreases for the heavier elements due to the availability of increasingly neutron rich projectiles. New schemes are also tested, such as the production of heavy elements by cooling down the compound nucleus via α-evaporation. First results on element 106 are reported [79]. The still open question is, how to reach the magic neutron number $N=184$. The reaction $^{82}Se(^{208}Pb,n)$ leads to the isotope $^{289}116$ with 173 neutrons which is already close to the 178 neutrons, slightly below the spherical shell closure, where the maximum shell correction is predicted by Möller *et al.* (Fig. 8), but still more than 10 neutrons from the spherical shell. The reaction $^{48}Ca(^{248}Cm,4n)$ leads to the isotope $^{292}116$ with three more neutrons. There are two possibilities for complete fusion reactions to come closer to the shell:

Overshooting the proton number, automatically would lead to more neutrons, due to the increasing neutron excess for the heavier projectile-isotopes. The magic region would then be crossed by the α-decay chain. A nice example is already the chain starting from $^{292}116$ crossing the center of the hexadecapole stabilized region. The second possibility, now discussed in context with the new radioactive beam facilities providing exotic nuclear beams with energies above the Coulomb barrier, is the use of neutron rich radioactive beams [35]. Existing and planned laboratories such as GSI (Germany), GANIL and the SPIRAL project (France) [80], RIKEN (Japan) and its upgrade [81], MSU and its upgrade (Michigan State, USA), and the Argonne Project (USA) [82] deliver already or will provide beams of instable neutron rich nuclei with intensities up to 10^7/s for the running facilities, and up to 10^{10}/s for the projected secondary-beam facilities.

By simply downscaling the present sensitivity of below 10 pb accomplished with stable beams of $5 \cdot 10^{12}$/s, the limit for the use of secondary beams for complete fusion reactions would be 0.1 μb to 1 nb. The transfermium region up to element 104 will be accessible with secondary-beam induced heavy ion fusion (Fig. 12). Scientific goals are the investigation of neutron rich species and the influence of nuclear structure on the production of heavy elements.

5. Conclusion

In-flight separation is now the established method for heavy-element research, together with the new developed detection techniques which permit to identify new elements and to extract gross nuclear properties from the decay of single atoms. The new generation of experiments led to the discovery of deformed, shell stabilized nuclei creating a region of enhanced stability extending from the upper end of the nuclear table to the magic proton number 114. The closed proton shell has already been closely approached with the discovery of element 111. Improved theories predict half-lifes of much less than the age of the earth. According to our present knowledge it is not expected to find superheavy nuclei in nature, plutonium is the heaviest primordial element.

Concerning the man-made elements, the recent experimental results on the production of heavy elements by cold heavy ion fusion are promising. The nuclei in the spherical superheavy region around element 114 seem within reach.

Acknowledgement

I gratefully acknowledge fruitful discussions with P. Armbruster, S. Hofmann, F. P. Heßberger, P. Möller, and A. Sobiczewski. I am indebted to F. Nickel for reading the manuscript, V. Ninov for providing recent material, and E. Pfeng for arranging the manuscript in final form.

Bibliography

1. Meitner, L., Frisch, O.: Nature **143**, 239–240 (1939).
2. Bohr, N., Wheeler, J. A.: Phys. Rev. **56**, 426–450 (1939).
3. Feenberg, E.: Phys. Rev. **55**, 504–505 (1939).
4. Goeppert-Mayer, M.: Phys. Rev. **74**, 235–239 (1948).
5. Haxel, O., Jensen, J. H. D., Suess, H. D.: Phys. Rev. **75**, 1769 (1949).
6. Meldner, H.: Ark. Fysik **36**, 593–597 (1967).
7. Myers, W. D., Swiatecki, W. J.: Nucl. Phys. **81**, 1–60 (1966).
8. Nilsson, S. G., Nix, J. R., Sobiczewski, A., Szymanski, Z., Wizech, S., Gustafson, C., Möller, P.: Nucl. Phys. **A115**, 545–562 (1968).
9. Strutinski, V. M.: Nucl. Phys. **A95**, 420–442 (1967).
10. Mosel, U., Greiner, W.: Z. Phys. **222**, 261–282 (1969).
11. Fizet, E. O., Nix, J. R.: Nucl. Phys. **A193**, 647–671 (1972).
12. Randrup, J., Larsson, S. E., Möller, P., Sobiczewski, A., Lukasiak, A.: Phys. Scr. **10A**, 60–64 (1974).
13. Flerov, G. N., Ter-Akopian, G. M.: Rep. Prog. Phys. **7**, 817–875 (1983).
14. Ghiorso, A., Nitschke, J. M., Alonso, I. R., Alonso, C. T., Nurmia, M., Seaborg, G. T., Hulet, E. K., Lougheed, R. W.: Phys. Rev. Lett. **33**, 1490–1493 (1974).
15. Münzenberg, G., Reisdorf, W., Hofmann, S., Agarwal, Y. K., Heßberger, F. P., Poppensieker, K., Schneider, J. R. H., Schneider, W. F. W., Schmidt, K. H., Schött, H. J., Armbruster, P., Sahm, C. C., Vermeulen, D.: Z. Phys. **A315**, 145–158 (1984).
16. Demin, A. G., Tretyakova, S. P., Utyonkov, V. K., Shirokovsky, I. V.: Z. Phys. **A315**, 197–200 (1984).
17. Münzenberg, G., Hofmann, S., Folger, H., Heßberger, F. P., Keller, J. G., Poppensieker, K., Quint, B., Reisdorf, W., Schmidt, K. H., Schött, H. J., Armbruster, P., Leino, M. E., Hingmann, R.: Z. Phys. **A322**, 227–235 (1985).
18. Armbruster, P.: Ann. Rev. Nucl. Sci. **25**, 135–194 (1985).
19. Münzenberg, G.: Rep. Prog. Phys. **51**, 57–104 (1988).
20. Seaborg, G. T., Loveland, W. D.: *The Elements beyond Uranium*, John Wiley and Sons, Inc., New York (1990).
21. Hofmann, S.: Proc. XV divisional EPS conference "Low Energy Dynamics", Apr. 18–22, 1995, St. Petersburg, Russia, World Scientific Publishing Company, Singapore, in press.
22. Armbruster, P.: From Nuclei to Stars, proc. 91st Corso Soc. Italiana di Fisica, Bologna, Italy, North-Holland, Amsterdam, 222–240 (1985).

23. Cwiok, S., Pashkevich, V. V., Dudek, J., Nazarewicz, W.: Nucl. Phys. **A410**, 254–270 (1983).
24. Leander, G. A., Möller, P., Nix, J. R., Howard, W. M.: Proc. 7th int. Conf. on Atomic Masses and Fundamental Constants, Darmstadt, Germany, Int. Rep. Technische Hochschule Darmstadt, 466–471 (1984).
25. Sobiczewski, A., Patyk, Z., Cwiok, S.: Phys. Lett. **B186**, 6–8 (1987).
26. Hofmann, S., Ninov, V., Heßberger, F. P., Armbruster, P., Folger, H., Münzenberg, G., Schött, H. J., Popeko, A. G., Yeremin, A. V., Andreev, A. N., Saro, S., Janik, E., Leino, M. E.: Z. Phys. **A350**, 277–280 (1995).
27. HILAB, A Proposal to establish a Heavy-Ion Laboratory at Burlington, Massachusetts, Chairman: Grodzins, A.: Int. Rep. Hig Voltage Corporation (1968).
28. Fulmer, C. B., Cohen, B. L.: Nucl. Phys. **6**, 547–564 (1958).
29. Armbruster, P.: Nukleonik **3**, 188–194 (1961).
30. Münzenberg, G.: Greiner, W., Poenaru, D. N., Eds., De Gruyter, Berlin (1995).
31. Münzenberg, G.: Int. J. Mass Spectr. Ion Phys. **14**, 363–378 (1974).
32. Münzenberg, G., Faust, W., Hofmann, S., Armbruster, P., Güttner, K., Ewald, H.: Nucl. Instrum. Methods **161**, 65–82 (1979).
33. Hofmann, S., Münzenberg, G., Heßberger, F. P., Schött, H. J.: Nucl. Instrum. Methods **223**, 312–318 (1984).
34. Münzenberg, G., Armbruster, P., Berthes, G., Folger, H., Heßberger, F. P., Hofmann, S., Keller, J., Poppensieker, K., Quint, A. B., Reisdorf, W., Schmidt, K.-H., Schött, H. J., Sümmerer, K., Zychor, I., Leino, M. E., Hingmann, R., Gollterthan, U., Hanelt, E.: Z. Phys. **A328**, 49–59 (1987).
35. Geissel, H., Münzenberg, G., Riisager, K.: Ann. Rev. Nucl. Part. Sci., in press.
36. Flerov, G. N., Ter-Akopian, G. M.: Prog. Part. Nucl. Phys. **19**, 197–239 (1988).
37. Barber, R. C., Greenwood, N. N., Hrynkiewicz, A. Z., Jeannin, Y. P., Lefort, M., Sakai, M., Ulehla, I., Wapstra, A. A., Wilkinson, D. H.: Prog. Part. Nucl. Phys. **29**, 453–529 (1992).
38. Oganessian, Yu. Ts., Tretyakov, Yu. P., Iljinov, A. S., Demin, A. G., Pleve, A. A., Tretyakova, S. P., Plotko, V. M., Ivanow, M. P., Danilov, N. A., Korotkin, Yu. Ts., Flerov, G. N.: JETP. Lett. **20**, 265–266 (1974).
39. Randrup, J., Larsson, S. E., Möller, P., Nilsson, S. G., Pomorski, K., Sobiczewski, A.: Phys. Rev. **C13**, 229–239 (1976).
40. Lazarev, Yu. A., Lobanov, Yu. V., Oganessian, Yu. Ts., Utyonkov, V. K., Abdullin, F. Sh., *et al.*: Phys. Rev. Lett. **73**, 624–627 (1994).
41. Oganessian, Yu. Ts.: Nucl. Phys. **A583**, 823–836 (1995).
42. Münzenberg, G., Armbruster, P., Heßberger, F. P., Hofmann, S., Poppensieker, K., Reisdorf, W., Schneider, J. R. H., Schneider, W. F. W., Schmidt, K.-H.: Z. Phys. **A309**, 89–90 (1982).
43. Münzenberg, G., Armbruster, P., Hofmann, S., Heßberger, F. P., Folger, H., Keller, J. G., Ninov, V., Poppensieker, K., Quint, A. B., Reisdorf, W., Schmidt, K.-H., Schött, H. J., Sümmerer, K., Zychor, I., Leino, M. E., Ackermann, D., Gollerthan, U., Hanelt, E., Morawek, W., Vermeulen, D., Fujita, Y., Schwab, T.: Z. Phys. **A333**, 163–175 (1989).
44. Hofmann, S., Ninov, V., Heßberger, F. P., Armbruster, P., Folger, H., Münzenberg, G., Schött, H. J., Popeko, A. G., Yeremin, A. V., Andreev, A. N., Saro, S., Janik, E., Leino, M. E.: Z. Phys. **A350**, 281–282 (1995).
45. Oganessian, Yu. Ts., Demin, A. G., Danilov, N. A., Flerov, G. N., Ivanov, M. P., Iljinov, A. S., Kolesnikov, N. N., Markov, B. N., Plotko, V. M., Tretyakova, S. P.: Nucl. Phys. **A273**, 505–522 (1976).
46. Lazarev, Yu. A., Lobanov, Yu. V., Oganessian, Yu. Ts., Tsyganov, Yu. S., Utyonkov, V. K., Abdullin, F. Sh., Jliev, S., Polyakov, A. N., Rigol, J., Shirokovski, I. V., Subottin, V. G., Sukhov, A. M., Buklanov, G. V., Gikal, B. N., Kutner, V. B., Mezentzev, A. N., Sedykh, I. M., Vakatov, D. V.: Phys. Rev. Lett., in press.
47. Münzenberg, G., Hofmann, S., Heßberger, F. P., Folger, H., Ninov, V., Poppensieker, K., Quint, A. B., Reisdorf, W., Schött, H. J., Sümmerer, K., Armbruster, P., Leino, M. E., Ackermann, D., Gollerthan, U., Hanelt, E., Morawek, W., Fujita, Y., Schwab, T., Türler, A.: Z. Phys. **A330**, 435–436 (1988).
48. Ghiorso, A., Lee, D., Sommerville, L. P., Loveland, W., Nitschke, J. M., Ghiorso, W., Seaborg, G. T., Wilmarth, P., Leres, R., Wydler, A., Nurmia, M., Gregorich, K., Galord, R., Hamilton, T., Hannik, N. J., Hoffmann, D. C., Jarzynski, C., Kacher, D., Kadkhodayan, B., Kreek, S., Lane, M., Lyon, A., McMahan, A., Neu, M., Sikkeland, T., Swiatecki, W., Türler, A., Walton, J. T., Yasgita, S.: Nucl. Phys. **A583**, 861–866 (1995).
49. Oganessian, Yu. Ts.: Proc. XV divisional EPS conference "Low Energy Dynamics", Apr. 18–22, St. Peterburg, Russia, World Scientific Publishing Company, Singapore, in press.
50. Schmidt, K. H., Sahm, C. C., Pielenz, K., Clerc, H. G.: Z. Phys. **A316**, 19–26 (1984).
51. "Names and Symbols of the Transfermium Elements", Pure Appl. Chem. **66**, 2419–2421 (1994).
52. Möller, P., Nix, J. R.: At. Data Nucl. Data Tables **39**, 213–233 (1988).
53. Möller, P., Nix, J. R.: J. Phys. **G20**, 1547–1681 (1994).
54. Patyk, Z., Sobiczewski, A.: Nucl. Phys. **A533**, 132–152 (1991).
55. Möller, O., Nix, J. R.: At. Data Nucl. Data Tables, 60pp (1995).
56. Sobiczewski, A., priv. Comm., 1995 and Proc. XV divisional EPS conference "Low Energy Nuclear Dynamics", Apr. 18–22, St. Petersburg, Russia, World Scienific Publishing Company, Singapore, in press.
57. Münzenberg, G., Armbruster, P.: *Exotic Nuclear Spectroscopy,* W. C. McHarris, Ed., Plenum Press, New York, USA, 181–203 (1990).
58. Nix, J. R.: Ann. Phys. **41**, 52 (1967).
59. Bjornholm, B., Lynn, J. E.: Rev. Mod. Phys. **52**, 725 (1980).
60. Schmidt, K.-H., Morawek, W.: Rep. Prog. Phys. **54**, 949–1003 (1991).
61. Vandenbosch, R.: Ann. Rev. Nucl. Part. Sci. **42**, 477–481 (1992).
62. Reisdorf, W.: J. Phys. **G20**, 1297–1353 (1994).
63. Cohen, S., Plasil, F., Swiatecki, W. J.: Ann. Phys. **82**, 557–596 (1974).
64. Swiatecki, W. J.: Nucl. Phys. **A376**, 275–291 (1982).
65. Oganessian, Y. Ts.: Lect. Notes, Phys. **33**, 221–252 (1974).
66. Viola, V. E. Jr., Mignerey, A. C., Breuer, H., Wolf, K. L., Glagola, B. G., Wilcke, W. W., Shröder, W. U., Huizenga, J. R., Hilscher, D., Birkelund, J. R.: Phys. Rev. **C22**, 122–127 (1980).
67. Heßberger, F. P., Münzenberg, G., Hofmann, S., Reisdorf, W., Schmidt, K. H., Schött, H. J., Armbruster, P., Hingmann, R., Thuma, B., Vermeulen, D.: Z. Phys. **A321**, 317–327 (1985).
68. Shen, W. Q., Albinski, J., Gobbi, A., Gralla, S., Hildenbrand, K. D., Herrmann, N., Kuzminski, J., Müller, W. F. J., Stelzer, H., Töke, J., Back, B. F., Bjørnholm, S., Sørensen, S. P.: Phys. Rev. **C36**, 115–141 (1985).
69. Feldmeier, H.: Prog. Phys. **50**, 915 (1987).
70. Bass, R.: Lect. Notes Phys. **117**, 420 (1980).
71. Blocki, J. P., Feldmeier, H., Swiatecki, W. I.: Nucl. Phys. **A459**, 145–172 (1986).
72. Armbruster, P.: Proc. Int. Nucl. Phys. Conf., Sao Paolo, Brasil 1989, World Scientific, New Jersey (1989) 313–335.
73. Berdichewski, D., Lukasiak, A., Nörenberg, W., Rozmej, P.: Nucl. Phys. **A499**, 609–636 (1989).
74. Iljinov, A. S., Cherepanov, E. A.: Int. Ber. Dubna, JINR P 7-84-68 Dubna (1984).

75. Sandulescu, A., Greiner, W.: J. Phys. **G3**, 189–193 (1977).
76. Corradi, L., Ackermann, D., Beghini, S., Montagnoli, G., Mueller, L., Napoli, D. R., Petrache, C., Pollarrolo, G., Rowley, N., Scarlassara, F., Segato, G. F., Signorini, C., Spolaore, P., Soramel, F., Stefanini, A. M.: *Frontier topics in Nuclear Physics,* Eds.: Scheid, W., Sandulescu, A., Plenum, New York 145–146 (1994).
77. Von Oertzen, W.: Z. Phys. **A342**, 177–128 (1992).
78. Keller, J. G., Schmidt, K.-H., Heßberger, F. P., Münzenberg, G., Reisdorf, W., Clerc, H. G., Sahm, C. C.: Nucl. Phys. **A452**, 173–204 (1986).
79. Nomura, T.: 2nd Riken – INFN – Symposium '95, RIKEN, Tokyo, May 22–26, World Scientific Publishing Company, to be published.
80. The SPIRAL project, Int. Rep., GANIL, Caen, France (1994).
81. RI-Beam Factory, RIKEN – Accelerator research facility, Int. Rep., RIKEN, Tokio, Japan (1994).
82. The Argonne Radioactive Beam Project, Int. Rep., Argonne Natl. Lab., Argonne, USA (1995).

Radiochimica Acta 70/71, 207–223 (1995)

Chemistry of the Transactinide Elements

By Matthias Schädel
Gesellschaft für Schwerionenforschung, D-64291 Darmstadt, Germany

(Received July 12, 1995)

Transactinides / Element 104 / Element 105 / Element 106 / Periodic Table / Atom-at-a-time chemistry / Heavy element synthesis / Molecular calculations

Abstract

The experimentally known chemical properties of the transactinide elements 104 and 105, and the experimental techniques used to study these properties on an atom-at-a-time base, are reviewed. The production of transactinides in heavy ion reactions and the specific aspects of chemical separations with single atoms is briefly discussed. The chemical properties of the first two transactinide elements are compared with the lighter element homologs in group 4 and 5 of the Periodic Table and with the expected behaviour obtained from most recent atomic and molecular calculations which take the increasingly strong relativistic effect into account. Elements 104 and 105 behave as transactinide elements and, in general, exhibit properties characteristic of their position in group 4 and 5 of the Periodic Table. However, surprising deviations of the chemical properties from expectations based on simple extrapolations have been observed. It is shown that the chemical properties of the heaviest elements cannot reliably be predicted by simple extrapolations. Prospects to extend our chemical knowledge at the top end of the Periodic Table are discussed.

1. Introduction

It was Wilhelm Conrad Röntgen's discovery of the X-rays in November 1895 which gave Henri Becquerel's research a new scope. He was motivated by his ambition to clarify the character of the X-rays and to find a link to his own scientific field, the study of phosphorescent bodies. Only a few months later, in 1896, Henri Becquerel discovered a new radiation [1]. Becquerel's discovery pushed the door wide open for a new science. Already two years later, in 1898, Pierre and Marie Sklodowska Curie interpreted Becquerel's observation correctly as an atomic phenomenon, coined the term "Radio-active" [2], and found the two new chemical elements polonium and radium.

In the centenary since Becquerel's discovery and half a century after Glenn T. Seaborg introduced the actinide concept [3] the Periodic Table of the Elements was completed and extended, see Fig. 1. Since the recent discovery of elements 110 and 111 [4, 5] we know 19 man-made transuranium elements. The first 11 artificial elements beyond uranium belong to the actinides. Their chemistry is documented in a large number of books and articles, including the ones appearing in this volume. Additional overviews can be found in [6–8]. All experimental findings about the chemistry of element 103, lawrencium (Lr), show that, as expected, the actinide series ends at Lr and that a new 6d transition series begins with element 104, rutherfordium (Rf)[1]. Hence, the currently known eight transactinide elements have been placed in the Periodic Table under their lighter homologs of the 5d series Hf, Ta, Re, Os, Ir, Pt, and Au.

Investigating the chemical properties of the transactinide elements, both experimentally and theoretically, is the most important and challenging goal to determine the architecture of the Periodic Table at its upper end. Increasing deviations from the periodicity of chemical properties based on extrapolations from lighter homologs in the Periodic Table have been predicted for some time, see e.g. [9], as a consequence of increasingly strong relativistic effects. They originate from the increasingly strong Coulomb field of the highly-charged atomic nucleus. Thus, electron configurations different from those usually responsible for the chemical behaviour of the lighter homologs may become decisive in the transactinide elements. By a detailed comparison of their chemical behaviour with that of their lighter homologs, and by comparison with results of relativistic molecular-orbital (MO) calculations, it is hoped to explore the influence of "relativistic effects" in the chemistry of the heaviest elements.

The pioneering studies of some basic chemical properties of element 104 showed that it behaves like a transactinide element. In general, element 104 exhibits chemical properties similar to those of zirconium and hafnium, as expected for an element located in group 4 of the Periodic Table, see Fig. 1. These early experiments exploited the chloride volatility in gas-phase experiments [10–13], and studied in aqueous solution the complexing behaviour of α-hydroxyisobutyrate (α-HIB) [14] and of chloride anions [15]. Although, all

[1] Until unanimously accepted final names for the heaviest elements are recommended by the International Union of Pure and Applied Chemistry, the names endorsed by the Nomenclature Committee of the American Chemical Society for use in the U.S. are employed in this article without any prejudice about the priority in the discovery. It shall be noted that the names kurchatovium (Ku) and nielsbohrium (Ns) were used in some of the original publications of the Dubna group for elements 104 and 105. Since 1992 nielsbohrium (Ns) is used for element 107 leaving hahnium (Ha) for element 105.

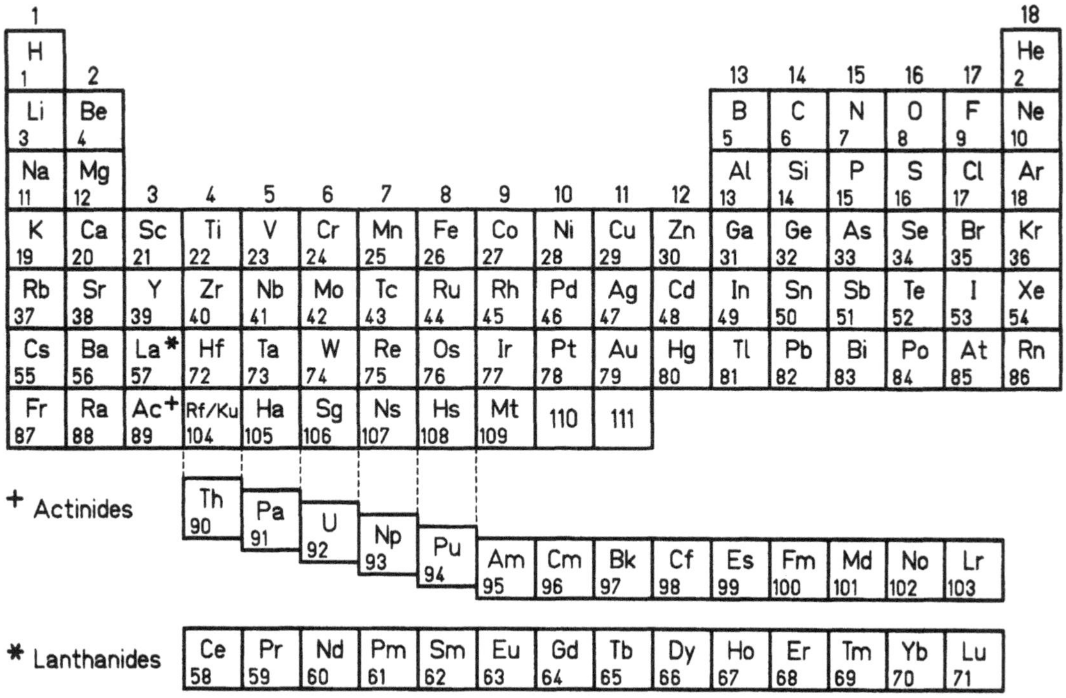

Fig. 1. Periodic Table of the Elements. While in group 3 the chemistry of all lanthanides follows quite uniformely that of their prototype element lanthanum such a behaviour is found only for the second half of the actinides. The arrangement of the actinides reflects that the first actinide elements still resemble, to a decreasing extent, the chemistry of the other groups: Th the fourth group below Hf, Pa the fifth group below Ta, U the sixth group below W, Np the seventh group below Re, and Pu the eighth group below Os. The known transactinide elements 104 through 111 shall take the positions below Hf in group 4 and Au in group 11.

these were key experiments to demonstrate that the transactinides begin with element 104, none of these experiments enabled a detailed study of the rutherfordium chemistry. First generation gas-phase experiments with chlorides [16] and bromides [17, 18] of element 105, hahnium (Ha), again indicated for the Ha halides a much greater volatility than that of the respective actinide compounds. These much higher volatilities of the Ha compounds are characteristic for a transactinide element. But these experiments also unveiled the lower volatility of the Ha compounds as compared with the ones from Nb, a lighter homolog in group 5. This gave the first hint that transactinide elements exhibit chemical properties which can not always be directly inferred from those of the lighter homologs. The first generation gas-phase separations with element 105, as earlier experiments on element 104, were plagued with model-dependent corrections of the observed deposition temperatures to deduce comparable volatilities. This was needed, because the deposition temperature depends strongly on the isotope half-life and on experimental parameters when using the thermochromatographic techniques [19, 20]. Therefore, it was not possible to make detailed comparisons of the halide volatilities of the group 5 elements Nb, Ta, and Ha. The results from the first generation experiments on the transactinide elements were summarized in a number of review articles [21–24].

Since the late 1980s a renewed interest in studying in more detail the chemical properties of the transactinide elements has rejuvenated the entire field of research, both experimentally and theoretically, see [25–29] for recent reviews. The main thrust for series of new, and much more detailed experiments over the last decade mainly came from new technical developments of the experimental apparates. Computer-controlled automated systems have greatly enhanced the ability to perform rapidly and reproducibly large numbers of chromatographic separations on miniaturized columns in the liquid phase. A new quality of gas-phase experiments began with the development of transport and detection systems to identify unambiguously heavy-element nuclides through their characteristic α-decay on-line immediately after they have passed a gas-chromatographic column. This allows a much more direct comparison with the behaviour of the lighter element homologs. Simultaneously with the manifestation of interesting and sometimes surprising new experimental results also new developments of relativistic quantum-chemical codes at the Dirac-Fock level took place, and these are now becoming capable of modelling complex molecules.

In this article, a brief discussion of the synthesis and decay of the transactinide elements and a short section about specific aspects of the single-atom chemistry will precede an outline of recent developments of the experimental techniques. This will be followed by a review on the results of both aqueous and gas-phase studies of the heaviest elements investigated, elements 104 and 105. Current attempts and prospects for studies of still heavier elements are briefly considered.

2. Synthesis and decay of transactinide elements

The synthesis of transactinide elements depends on nuclear fusion reactions with heavy ion beams. Details of the production and the decay of the heaviest-element nuclides are discussed in this volume [30], and were summarized in a number of review articles [31–37]. Product yields are rapidly decreasing from about one atom per minute for element 104 to about one atom per day for the heaviest man-made element with atomic number 111 [5]. Half-lives decreasing from about one minute to one millisecond for the longest-lived known isotopes of these elements pose an additional challenge, especially for chemical investigations. All transactinides have both short half-lives and such low production rates that, on the average, each synthesized atom has decayed before a new one is made. The term 'one-atom-at-a-time' was coined for this situation. Its consequences for chemical investigations will briefly be discussed in the next section.

The heaviest elements were synthesized with highest cross-sections in reactions with ^{54}Cr, ^{58}Fe, ^{62}Ni, and ^{64}Ni projectiles together with targets of ^{208}Pb and ^{209}Bi. Except for 4-s ^{257}Rf and 3-s ^{258}Ha, only nuclides with half-lives of less than one second are produced in these reactions. This is too short for existing chemical techniques of transactinide research. Their speed and efficiencies allow sensitive chemical studies with radionuclides of a few seconds half-life, and some techniques require a half life of at least ten seconds; see below for a detailed discussion. More neutron-rich and, hence, longer-lived nuclides are accessible in heavy-ion reactions with actinide targets. Typical experimental conditions are heavy-ion beam currents of 3×10^{12} particles per second and a maximum useful target thickness of about 700 μg cm^{-2}. Recent chemical studies of element 104 use the longest-lived known isotopes 65-s ^{261}Rf produced in the ^{248}Cm(^{18}O,5n) reaction with a cross section of 5 nb. Abundant use of 34-s ^{262}Ha, and partially of 27-s ^{263}Ha [38, 39], is made to shed light on the chemistry of element 105. These isotopes are synthesized in 5n- and 4n-reactions of ^{18}O beams on ^{249}Bk targets with a cross section of 6 nb and 2 nb, respectively. These cross sections yield production rates of the order of one atom per minute. Thus, all chemical procedures with transactinide elements are truly performed with single atoms, or "one-atom-at-a-time" [23, 26].

Recent successes to synthesize the more neutron-rich isotopes ^{265}Sg and ^{266}Sg [40] of element 106, seaborgium (Sg), with deduced half-lives of the order of ten seconds, and cross sections of up to a few hundred picobarns, have paved the way to extend our knowledge about the chemistry of the heaviest man-made elements in future experiments. Nuclides with half lives sufficiently long to explore the chemistry of elements beyond $Z=106$ can, presumably, be found around the neutron number 162 for elements up to $Z=107$ and, possibly, in the $N\leq174$ region for elements 108 through 114 [41]. Provided that nuclear reactions are at hand to produce these nuclides with cross sections of at least several tens or hundreds of picobarns, chemists can hope to get insights into the chemistry of transactinide elements beyond seaborgium. See [37] for more details of the nuclear aspects of transactinide chemistry.

3. Single-atom chemistry

The one-atom-at-a-time situation of the transactinides poses some unique problems for the chemistry at the top end of the Periodic Table. The classical law of mass action is no longer valid [42, 43] because a single atom cannot exist in different chemical forms taking part in the chemical equilibrium at the same time. For single atoms the concept of chemical equilibrium needs to be substituted by an equivalent expression in which concentrations or activities are replaced by probabilities to find the atom in one state or the other. The atom can sample these states with a high frequency of hundreds or millions of exchange reactions per second if the chemical system was selected such that the free enthalpy of activation is below 17 kcal [44]. The time one atom spends in one state or the other, the measure for its probability to be in either phase, can be determined in dynamical partition experiments. These are characterized by the flow of a mobile phase relative to a stationary phase while a single atom is frequently changing between these two phases. This is realized in many chromatographic separations, e.g. in the exchange between a gaseous and a solid phase (wall adsorption) in gas chromatography or between a mobile aqueous phase and a stationary ion exchange resin in liquid chromatography. Here the retention or elution times provide the information about what fractional average time one atom has spent in either phase. Such a characterization of the behaviour of a single atom will already yield information close to the equilibrium constant which would be obtained from macroscopic amounts of this element.

4. Experimental techniques

A review on fast chemical separation procedures can be found in an accompanying article in this volume [45]. Here, only these techniques will be briefly outlined which are directly related to the chemical investigations of transactinide elements.

4.1 Production and transport

The target- and recoil-chamber arrangement is schematically shown in Fig. 2, and is described in more detail in [46, 47]. Heavy-ion beams passed through a vacuum isolation window, a volume of nitrogen cooling-gas, and a target-backing before interacting with ^{248}Cm or ^{249}Bk as a target material. Reaction products recoiling out of the target were thermalized in a vol-

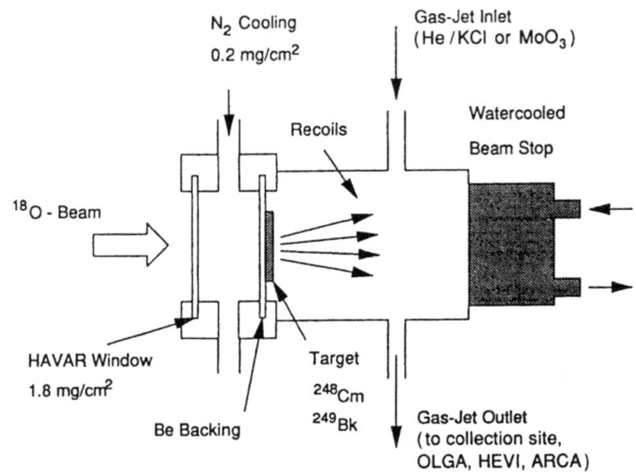

Fig. 2. Schematic representation of the target and recoil-chamber arrangement in transactinide chemistry experiments. Fig. from [28, 60].

ume of helium loaded with 50 to 200 nm size aerosols to collect all products. As the first step common to all experiments, a continuous He flow ($\leq 2\,l\,min^{-1}$) through capillary tubes (≈ 1.5 mm inner diameter (i.d.)) provided a fast transport of the transactinides, and other reaction products, from the site of the heavy-ion reaction into the chemistry set-up. He (aerosol)-jets allowed for a transportation over distances of up to several tens of meters with yields of about 50% [47–50]. Overall transport times were of the order of two to five seconds.

So far, none of the experiments has exploited the potential of chemical reactions with the aerosols. Materials were selected to minimize their influence on the chemical procedures. Separations carried out in the aqueous phase used KCl as aerosols while MoO_3 or carbon was often preferred in gas-phase separations [51].

To circumvent problems with chemically interfering aerosol materials in some earlier frontal gas-chromatographic experiments the separation columns were directly coupled to the recoil chamber [52, 53]. In such a configuration a still high enough transport yield can be achieved without using any aerosols. Transportation times as short as 0.1 second were measured [54], and the recoil chamber can simultaneously be used as a reaction chamber to form chemical compounds by adding a reactive gas to the inert carrier gas [55]. This technique, which was not used in recent gas-chromatographic studies, was applied in thermochromatographic separations to investigate the volatility of element 105 bromides [17] and in an attempt to determine the volatility of element 104 in its atomic state [56].

4.2 Gas-phase chemistry

Early gas-phase separations of elements 104 and 105 halides used the principle of thermochromatography, see [20] and references therein, in which a negative temperature gradient was formed along the chromatographic tube. Compounds with different volatilities separated from each other while moving downstream the column in the gas flow and deposited at a particular position inside the tube. These deposition zones at a given temperature are related to the adsorption enthalpy. The final region of deposition within the column was determined radiochemically for long-lived nuclides, e.g. for radioactive tracers of the lighter-element homologs. For short-lived isotopes of elements 104 and 105 fission-track detectors were used to determine the region of tracks from the spontaneous fission (SF) decay. Real-time detection of the nuclear decay of transactinides was not possible and the method was limited to fission-track counting which provided no unique signal about the nuclide or the element. Model dependent corrections were necessary to relate observed deposition temperatures of species with different half-lives to comparable volatilities.

More recent experiments used the isothermal gas-phase chromatography to study the retention times of Rf and Ha chlorides and bromides in real-time together with their lighter-element homologs [51, 57]. The On-Line Gas Chemistry Apparatus (OLGA II) [49], and the Heavy Element Volatility Instrument (HEVI) [58], were the applied apparates, see Fig. 3 for a schematic layout of both systems. Reaction products continuously transported by the He(aerosol)-jet to a 6 mm i.d. quartz column were stopped in a quartz-wool plug located in the first, high-temperature section of the column. At temperatures of up to about 1000 °C the aerosol particles were destroyed and the heavy element formed a volatile compound with a reactive gas like HBr, BBr_3, HCl or Cl_2 added at this point with a flow rate of about 100–200 ml min^{-1}.

In earlier gas-phase separations, KCl was used as an aerosol material with the disadvantage that the KCl recondensed downstream in the colder, isothermal part of the column and the chromatography was, at least partially, carried out on a KCl surface [59]. In later experiments, MoO_3 or graphite aerosols were used [60]. This had the advantage that volatile species were formed which did not condense in the chromatographic column when chlorinating agents or oxygen were added as a reactive gas.

The second and much longer part of the quartz tube represented a lower and adjustable, isothermal temperature section for the gas-chromatographic separation. All compounds are transported through the column by means of the flowing He. The specific velocity of each compound depends on its volatility and on the temperature of the isothermal part of the column. The transit- or retention time of the volatile species was determined using the half lives of the radionuclides as a "clock". The measure for the volatility was the specific temperature $T_{50\%}$ at which the corresponding compound exited the chromatographic column with 50% yield. At this temperature, the retention time, as defined in classical gas chromatography, is equal to the half life. To measure the separated species, they were, at the outlet of the chromatographic column,

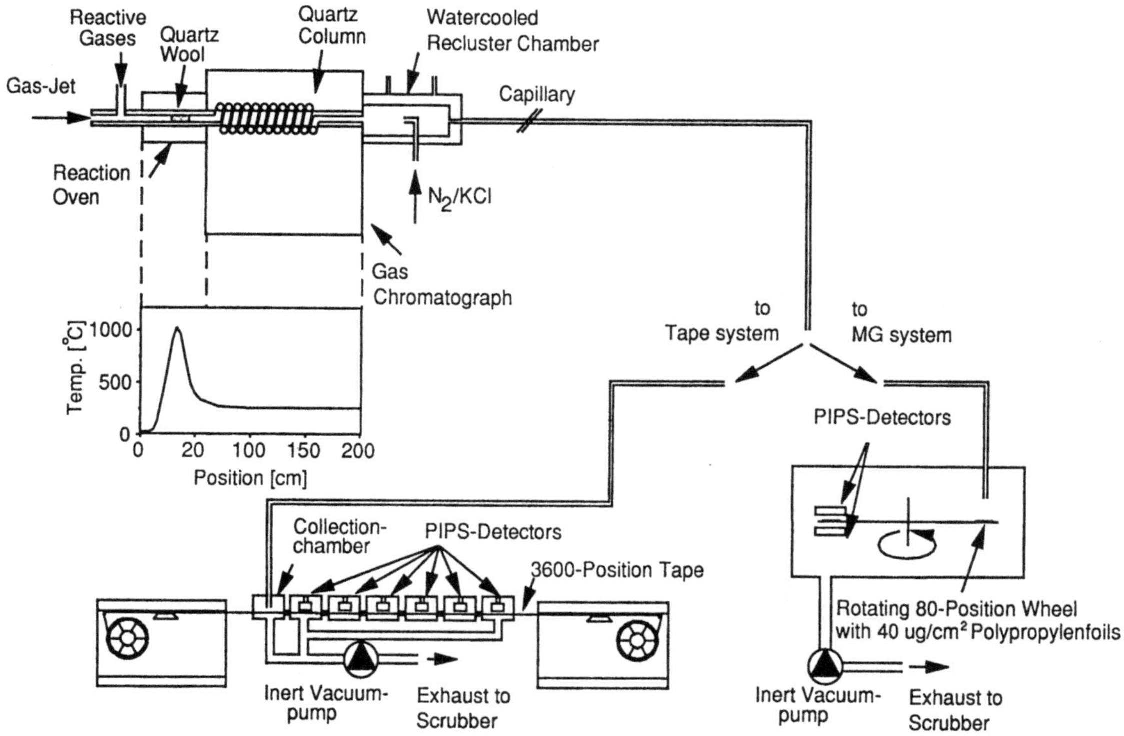

Fig. 3. Schematic of the on-line isothermal gas chemistry set-ups OLGA II and HEVI coupled to two alternative detection devices for α-spectroscopy and SF-fragment kinetic-energy measurements, a tape system and a rotating wheel. Fig. from [57].

ejected through an orifice into a recluster chamber. Here they reattached to new aerosols which were added for the transport of the separated products through a capillary to a detection set-up. Recent developments to miniaturize and cool the recluster chamber in OLGA III [61] have significantly reduced the time necessary for reattachment. Radionuclides with half lives between 1 and 4 s [62] became accessible with OLGA III.

The two different detection systems used in the gaschromatographic experiments were equipped with a rotating catcher wheel [58, 63] or a tape system [49, 64]. Both were operated stepwise to transport collected products in front of PIPS (passivated implanted planar silicon) detectors to measure characteristic α-decay energies and the kinetic energy of SF fragments. Thus, single atoms of ^{261}Rf and ^{262}Ha were detectable. Only the rotating wheel was equipped with catcher foils thin enough to measure the nuclear decay on both sides of the foil. This gave the advantage of a high detection efficiency and allowed to measure the total kinetic energy of coincident fission fragments. The tape system had the advantage that, because of its length and the collection on unused spots, no long-lived background was building up which could have obscured the detection of single atoms of transactinide elements.

The energies of all observed events were stored together with a time and detector information for off-line processing of these data. This allowed to search for time-correlated parent-daughter pairs of α-events to assure an unambiguous identification of single events, see e.g. [38, 39, 64].

4.3 Aqueous-phase chemistry

While all separations in the gas-phase were carried out as a continuous process, so far mainly discontinuous, batch-wise chemical separations were performed in the aqueous phase. Recent discontinuous studies of the transactinides were either performed manually [48, 65, 66] or with the Automated Rapid Chemistry Apparatus (ARCA) [39, 67–70].

While one of the manual separations made use of the characteristic adsorption of the group 5 elements Nb, Ta and Ha from nitric acid on glass surfaces [48] all other separations were manually performed liquid-liquid extractions. These procedures usually contained the following steps: reaction products were collected on a platinum or Teflon™ disc from the He(KCl)-jet for about 60 to 90 s, picked up with 10 µl of the aqueous phase (mostly HCl or HNO_3 of various concentrations), transferred to a 1 ml centrifuge cone which contained 20 µl of the organic phase, mixed ultrasonically for 5 s, and centrifuged for 10 s to separate the phases. The organic phase with one of the extractants methylisobutyl ketone (MIBK), triisooctylamine (TIOA), tributylphosphate (TBP), or 2-thenoyltrifluoroacetone (TTA) was transferred to a glass cover slip, dried on a hot plate, and placed over a PIPS detector. This procedure took about 1 min and was mainly applied to study the aqueous chemistry of element 104 with the 65-s ^{261}Rf [65, 66].

Automated separations with ARCA II have focused on the investigation of the aqueous-phase chemistry of element 105 with the 34-s ^{262}Ha. ARCA is a

Fig. 4. Photograph of the computer-controlled Automated Rapid Chemistry Apparatus, ARCA II [67]. See text for description.

micro-computer controlled apparatus for fast, repetitive, and reproducible chromatographic separations in a miniaturized and chemically inert high-performance liquid-chromatography system [67]. A photograph of ARCA II is shown in Fig. 4. ARCA II consists of a central part to collect the aerosols from the jet and to perform chromatographic separations of the reaction products in 8 mm long micro-columns of 1.6 mm i.d. Two movable magazines, in Fig. 4 partially protruding from the central part, contain 20 of these columns each. Depending on the chemistry, these columns were filled with a cation exchange resin (17.5±2 μm particle size Aminex A6) or with TIOA or diisobutylcarbinol (DIBC) as an organic extractant on an inert support material. Incorporated into the central part of ARCA II are three sliders. The one on top shuttles the reaction products after their collection alternating on top of one of the microcolumns, whereby the reaction products are dissolved, complexed, and loaded onto the column. Two sliders downstream of the columns allow to collect a Ha fraction on a Ta disc. The sample is dryed by intense IR-light and a stream of hot He-gas. Peripheral components are three chemically inert chromatography pumps (not shown) and pneumatically driven 4-way slider valves (cylinders on top). Each pump pumps one pre-selected eluent through Teflon tubings to the 4-way valves. These allow a fast and reproducible stepwise changing of the eluents on the columns. The low dead volume of about 35 μl including the free column volume gives a fast access to desired fractions. Within 3 s to 4 s, element 105 was separated from trivalent actinides and group 4 elements in a 35 μl fraction of 0.05 M α-HIB, eluted from an Aminex A6 column [39]. In the most time consuming separation, Ha has was eluted after 25 s as the third eluent fraction from a TIOA column [68]. Sample preparation and mounting on PIPS detectors required about 20 s. Experiments with ARCA were performed in 1 min cycles. After each separation the magazine was moved by one step, thus introducing a new column in the elution position. By this means, time-consuming reconditioning of used columns and cross-contaminations from previous separations are avoided.

First successful attempts were made to directly detect events from the α- and SF-decay from a thin film of solution in a large area PIPS-detector [39]. Start of counting was 0.8 s after end of collecting the reaction products on the top slider SL1, i.e. 0.8 s after the start of chemistry. This can be a useful technique as long as no high resolution α-spectroscopy is needed for a clear identification. To guarantee this, the chemical separation has to provide a clean fraction which does not contain any radionuclides showing similar or higher α-decay energies.

So far, only one continuous isolation procedure, first developed for nuclear decay studies of Hf, Ta, and W isotopes [71], was applied to an element 104 separation [72] in combination with a discontinuous sample preparation step. Reaction products transported by an aerosol jet were continuously dissolved in an aqueous phase which was pumped through three consecutive ion exchange columns. Trivalent actinides were retained on the first cation exchange column followed by an anion exchange column to adsorb anionic complexes of element 104, and a third column filled with a cation exchange resin to collect all the trivalent daughter products from the α-decay of 65-s ^{261}Rf. At the end of an experiment, all element 104 daughter products were stripped from this third column to prepare a sample for the α-spectroscopy and to find out how strongly element 104 had been retained on the anion exchange column.

The centrifuge system SISAK III [73] allows a continuous separation of nuclides with half-lives down to $\approx$1 s. The separation is based on fast multistage solvent extractions using specially designed mini-centrifuges for phase separation. SISAK has successfully been applied to a large number of γ-spectroscopic studies of lighter elements. Most recently, a completely new detection system, based on liquid scintillation spectroscopy, was developed for on-line α-spectroscopy and SF-detection in a flowing liquid behind SISAK [74]. An energy resolution of 200 to 500 keV and a high β- and γ-suppression was achieved in preparatory experiments [75], and the entire set-up was put to a test in two experiments aimed to separate 2-s ^{262}Rf and 1.8-s ^{261}Ha [76]. Preparations are under way to apply this system for an on-line separation and detection of ^{265,266}Sg with SISAK [77].

5. The theoretical situation in general

At present each comparison of experimental results and theoretical calculations will always involve empirical systematics because no theoretical apparatus is at hand which can directly compute the chemical behaviour. The present situation of theoretical models and their application to the transactinide chemistry is reviewed in [78, 79].

On the atomic level, early relativistic calculations for atoms and ions [80–83] have already shown that there is a strong relativistic stabilization of $s_{1/2}$- and $p_{1/2}$-orbitals as a function of atomic number, and, as a secondary effect, a destabilization of *d*-orbitals and their spatial expansion. On the current level of multiconfiguration Dirac-Fock (MCDF) calculations important information about electronic configurations, ionization potentials, promotion energies, and atomic and ionic radii were obtained [84–88]. One of the important roles of atomic data from the MCDF calculations are their contribution as key input data to predict chemical properties, like oxidation states or volatilities, based upon empirical models which mainly exploit thermodynamic correlations, see e.g. [23]. However, it will not be possible to find any straightforward correlation between computable atomic properties and the chemical behaviour of a heavy element. This is, because the interplay of *s*-, *p*-, *d*-, and *f*-orbitals can change molecular properties like the relative amount of ionic versus covalent bonding, the bond length in the molecule, and the effective charge of the central atom. These for the chemistry decisive quantities are not directly extractable from the atomic calculation.

With progress in molecular calculations, especially from the Dirac-Slater (DS) calculations using the relativistic Discrete Variational Method with self-consistent charge (SCC) approximation [89, 90] much more refined approaches begin to appear for a better theoretical understanding of the chemistry of the transactinide elements. For example, a lower effective charge on the central atom and a higher covalency in the bonding, as obtained in such calculations, strongly indicated that chlorides and bromides of transactinides are more volatile than their lighter homologs in group 4 and 5 of the halides [85, 90, 91]. However, theory predicts already for element 105 a preferential formation of oxyhalides over the formation of pure halides [92], a tendency known for elements of group 6 and onward. Such a trend can lead to the formation of different compounds with different volatilities for the respective transactinide element as compared with its lighter homologs when halides and oxygen are simultaneously present in gas-chromatographic experiments. Unfortunately, a complete model to calculate volatilities from basic atomic and molecular structure data has, up to now, not been developed.

Molecular DS DVM calculations indicated [93] that the first three transactinide elements will be tetra-, penta-, and hexavalent as their lighter homologs. The stability of the maximum oxidation state, however, was predicted to decrease from element 104 to 106 [93].

Empirical correlations which completely rely on the known physicochemical properties of elements and their compounds in the groups and periods of the Periodic Table, e.g. [18, 20, 94, 95], allow classical extrapolations of chemical properties of the transactinide elements. These extrapolations provide information on what would be expected if no increasingly strong relativistic effect would be present in the transactinide elements because only the weaker relativistic effect already present in the lighter elements is inherent in this procedure. Such data are important to judge how strong relativistic effects contribute to the observed chemistry of transactinide elements.

6. Element 104

6.1 Theoretical predictions

According to the normal continuation of the Periodic Table, element 104 is expected to be a member of group 4 below Ti, Zr, and Hf with an electronic ground-state configuration of [Rn] $5f^{14}\,6d^2\,7s^2$. This configuration was also indicated by early relativistic calculations [81, 82] in a Dirac-Fock computation with a single configuration approximation [81].

After multiconfiguration Dirac-Fock (MCDF) calculations had indicated that because of relativistic stabilization of the $p_{1/2}$-orbital, the electronic structure of element 103 is [Rn] $5f^{14}\,7s^2\,7p_{1/2}$ rather than [Rn] $5f^{14}\,6d\,7s^2$ [96, 97] it was extrapolated that element 104 should have a $7s^2\,7p^2$ rather than a $6d^2\,7s^2$ configuration [23]. This raised the question whether element 104 might be a "*p*-element" with properties similar to Pb that has a ground state configuration of [Xe] $5f^{14}\,5d^{10}\,6s^2\,p^2$. It is important to note two crucial differences between the Pb and Rf configuration: contrary to element 104, Pb has a closed 5*d*-shell and a great energy gap of more than 4 eV between its ground- and next-higher state.

A more recent multiconfiguration Dirac-Fock (MCDF) calculation, using 468 *jj*-configurations, gave a (80%) $6d\,7s^2\,7p$ ground-state configuration [84] with a level only 0.5±0.2 eV higher consisting of the $6d^2\,7s^2$ configuration (95%). This calculation obtained the $7s^2\,7p^2$ state 2.9 eV above the ground state. The $6d\,7s^2\,7p$ ground-state configuration was confirmed in a similar MCDF calculation [86] with an energy gap of 0.24 eV between the ground-state and next higher state.

Most recent atomic calculations with the relativistic coupled cluster method based on the Dirac-Coulomb-Breit Hamiltonian (CCSD) which include dynamical correlations [88] reverted to the $6d^2\,7s^2$ configuration as the ground state of element 104. Here, the $6d\,7s^2\,7p$ state is 0.274 eV above the ground state. Contrary to earlier calculations [85, 86] it is now assumed [88] that a 6*d* electron is first ionized instead of the 7*p* electron.

From MCDF calculations ionization potentials (IP), atomic radii, and ionic radii were deduced [85, 86]. It was found that radii, orbital energies, and ground levels of ionized states of element 104 are similar to Hf but different from Pb and its homologs [85]. From this it was concluded that there is no basis for expecting any distinct "*p*-character" in the chemical properties of element 104 and that its chemistry will resemble the Zr and Hf chemistry. The authors of [85] calculated that the first IP of element 104 (6 eV) is lower than for Zr and Hf and should result in a relatively greater stability of the lower oxidation state and/ or a lower volatility of binary compounds for monovalent Rf [85]. From a relativistic SCF-Xα scattering wave Dirac-Fock computation for the tetrachloride molecule of element 104 a lower effective charge of the central Rf atom and a higher covalency in the bonding was obtained [85]. This could explain a higher volatility of $RfCl_4$ in comparison with $ZrCl_4$ or $HfCl_4$.

The DS DVM was applied to study the electronic structure of (i) the tetrachlorides of Pb and group 4 elements [98] and (ii) the highest chlorides of group 4, 5, and 6 elements [93]. These calculations agree that the electronic structures of $RfCl_4$ and $HfCl_4$ are similar and that the bonding is typical of the transition *d*-element compounds. Both calculations [93, 98] show the same trend in the ionicity within the group: a decrease in the values of the effective charges from Zr to Hf and a small increase from Hf to Rf. These for very heavy elements most advanced molecular calculations [93, 98] show the $RfCl_4$ to be a rather covalent compound, with the covalency being greater in $RfCl_4$ than in $HfCl_4$. This leads to a stronger metal-ligand bond in $RfCl_4$ compared with $HfCl_4$. $RfCl_4$ should be thermally more stable than $ZrCl_4$ or $HfCl_4$ [93]. In these compounds it is expected that the tetravalent state of element 104 will be predominant.

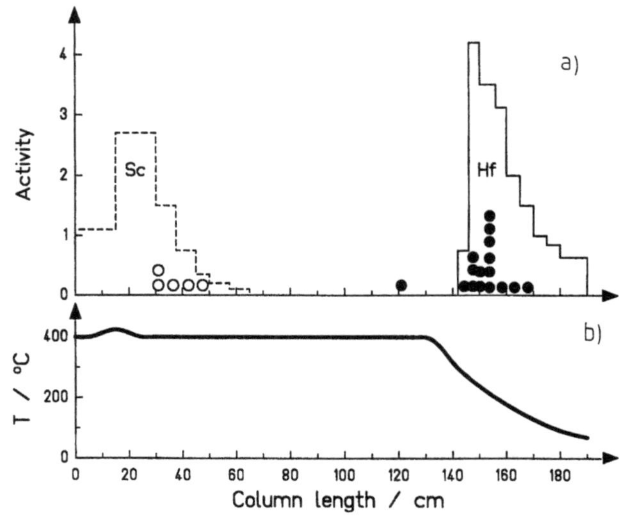

Fig. 5. a) Thermochromatographically determined deposition zone of element 104 chlorides (black dots) and Hf (solid line) compared with Sc (dashed line) and heavy actinides (open circles). b) Temperature gradient in the column. From [13].

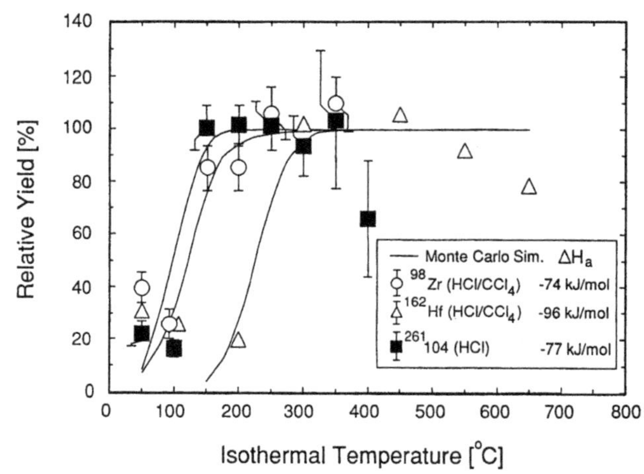

Fig. 6. Gas-chromatographically determined relative chemical yield for Zr (open circles), Hf (triangles) and Rf (solid squares) chlorides as a function of the temperature in the isothermal part of the column together with Monte Carlo simulations (solid line). Inserted in the figure are the chlorinating gases used and the deduced adsorption enthalpies [99]. Figure from [57].

6.2 Experimental results

The pioneering chemical experiments on element 104 in the gas phase [11, 12] showed that element 104 forms volatile chlorides, and that the Rf chloride is much more volatile than the actinide chlorides and similar to $ZrCl_4$ and or $HfCl_4$ [13, 53]. This experimental series gave the first indication that element 104 behaves like Hf, and that it is a group 4 element. The strongly enhanced retention after having introduced KCl into the gas-solid column at 300 °C was taken as an indication that the tetrachloride was formed as the most stable compound [53]. After applying half-life corrections to the experimentally observed distributions for $RfCl_4$ a volatility was obtained similar or even slightly higher than the one for $HfCl_4$ [13, 53]. Small differences in volatility were not observable in these experiments because of a steep temperature gradient, see Fig. 5.

Modern gas chromatographic experiments are apt to determine the relative volatilities between various elements more precisely and unambiguously through time- and energy-correlated events from the α-decay of mother and daughter nuclei [51]. Fig. 6 shows the relative (maximum yield = 100%) chemical yield of volatilized Rf versus temperature in the isothermal section of the gas-chromatographic column in comparison with the behaviour of Zr and Hf chlorides [51, 57, 99]. MoO_3-aerosols were used in the transport jet. The chlorinating gases are listed in the figure together with the respective elements. The solid lines in Fig. 6 resulted from Monte Carlo simulations [57, 99] based on the microscopic model of gas-solid chromatography [19]. The simulation accommodated the influence of the carrier gas flow, the temperature distribution in the column, and the different half-lives of the investigated nuclei. Calculated was the expected yield of a large number of sample molecules of one chemical species with a given adsorption enthalpy (ΔH_a) at each isothermal temperature. A specific curve of yield vs.

temperature resulted for each value of ΔH_a chosen as an input parameter. The curve for the ΔH_a value which best fits the measured data was chosen by a least-squares method, and these ΔH_a values are given in the insert in Fig. 6. To obtain a consistent set of adsorption enthalpies from various experiments, data from the literature were reanalyzed using a calculated value for the adsorption entropy (ΔS_a) according to [100], and all these data were compiled together with the most recent experimental results [51, 57, 99].

The results for the volatilities of group 4 chlorides showed two groups: one with a higher volatility and ΔH_a of about -70 kJ mol^{-1} and a second one with a lower volatility and ΔH_a of -90 kJ mol^{-1} or lower. The Zr and Rf chlorides shown in Fig. 6 exhibit a similar and rather high volatility. Such a behaviour was thought to be characteristic for the pure tetrahalides $ZrCl_4$ and $RfCl_4$. The lower volatility for the Hf compound was suggested to indicate that the oxyhalide compound $HfOCl_2$ instead of the pure tetrachloride was studied in this experiment [51, 99]. The formation of oxyhalides as a function of oxygen content in the gas will be discussed in more detail for the group 5 elements, see below.

The compilation of reanalyzed adsorption enthalpies [51, 99] indicated that in at least one of the previous experiments [11] the formation of the pure tetrahalide of Zr and Hf together with the formation of the Rf oxyhalide was observed. The formation of a Zr oxyhalide in a different experiment can also be inferred [51, 99].

A first and preliminary result from a gas-chromatographic study of group 4 bromides indicated a lower volatility of the Rf bromide as compared with the chloride. The experimental result hinted to a higher volatility of the Rf bromide by about 100 °C as compared with the Hf compounds [60]. In this experiment it was observed that the yield of, presumably, the tetrabromide of element 104 remained stable at temperatures above 300 °C and that it steeply decreased between 200 °C and 100 °C. In the same experiment, the yield of Hf bromides dropped to about 60% in the temperature range from 600 °C to 550 °C, remained constant from 550 °C to 400 °C, and then dropped to a level of a few percent at 300 °C. The presence of two regions of sharply decreasing yield was taken as evidence for the formation of two different Hf compounds with two different volatilities. The $HfBr_4$ is a candidate for the compound with the higher volatility and $HfOBr_2$ for the one with the lower volatility. Zirconium was not included in this investigation. The results of this experiment are not conclusive and can only be taken as evidence (i) because of large error bars on the rutherfordium data, and, more important, (ii) because of the use of a He(KCl)-jet in this experiment, which may have covered the quartz surface of the chromatographic column with KCl leading to much higher retention temperatures [53, 59].

A comparison of the chloride and bromide volatilities of group 4 element compounds exhibited, in general, a higher volatility of the chlorides. The summary of all experimental data showed that, within the presently achieved experimental resolution, no trend in the adsorption enthalpies, and, hence, no differences in the volatilities of the group 4 halides of Zr, Hf, and Rf can be observed as long as the same compound is formed [51]. This indicates that the measured volatilities are in agreement with results from the relativistic DS DVM calculations [93] while empirical extrapolations in group 4 predict [95] a slightly less volatile $RfCl_4$ [57]. So far, the chemical composition of the investigated species has not been governed decisively enough by the experiments performed until now. Further experiments under absolutely oxygen-free conditions, including the material of the chromatographic column, are needed before any final conclusion on the influence of the relativistic effect on the volatility of element 104 halides can be drawn. It is also premature to conclude which group 4 element has the highest and which one has the lowest tendency to form a pure tetrahalide or an oxyhalide.

One experiment was carried out in the gas phase to probe a *p*-element like behaviour of element 104 in its atomic state [56]. It was speculated that a much higher volatility of element 104 as compared with the group 4 elements Ti, Zr, and Hf would indicate such a character. In analogy to a similar experiment on element 103 [101, 102] the retention of element 104 was measured along a quartz column at 1170 °C in a hydrogen containing Ar carrier-gas. All fission tracks from the 2.6-h ^{256}Fm, an unwanted by-product produced in the nuclear reaction, and from 4.5-s ^{259}Rf were observed in a typical *d*-element region. A lower limit for the sublimation enthalpy of $\Delta H_S^\circ(\text{Rf}) \geq 370$ kJmol^{-1} was determined. This result is in agreement with predictions [84, 85, 98] that despite of a possible $6d\,7s^2\,7p$ ground-state configuration the chemical properties of Rf will be similar to Hf.

The first aqueous phase separations of element 104 were carried out in a column liquid chromatography [14]. Rf was eluted together with Zr and Hf from a cation exchange resin with α-HIB. It behaved entirely different from No and the trivalent actinides which were retained on the column. To exclude hydrolysis reactions at low acidities the chloride complexation of Rf was investigated in an automated solvent extraction chromatography with the quarternary amine Aliquat 336 as a stationary phase and 12 M HCl as a mobile phase [15]. Under these conditions, Zr, Hf, and Rf extracted into the organic phase while trivalent actinides were not adsorbed on the column. The Rf, together with the Hf tracer, was then eluted with 6 M HCl. Although only 6 α-decays from ^{261}Rf were detected this experiment demonstrated that the chloride complexation of Rf is similar to Hf and much stronger than that of the trivalent actinides [15].

In an experiment which combined a fast, continuous separation of element 104 on an anion exchange column with the off-line detection of longer-lived daughter products it was shown that element 104, like

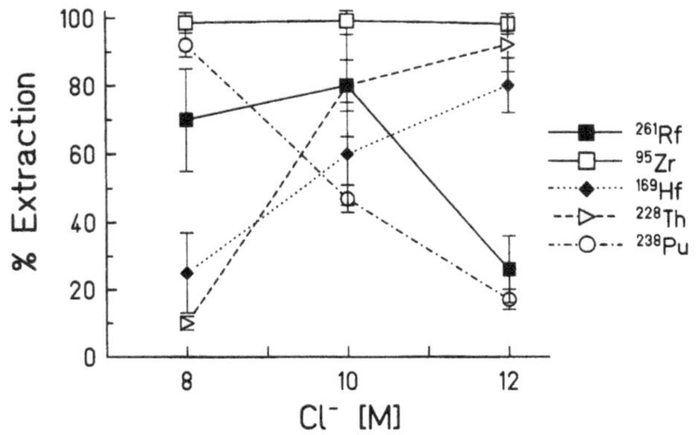

Fig. 7. Extraction of ^{261}Rf, ^{95}Zr, ^{169}Hf, ^{228}Th, and ^{238}Pu into 0.25 M TBP in benzene as a function of chloride concentration. The H^+ concentration was kept constant at 8 M. Adapted from [66].

Hf, forms stable anionic complexes in 0.2 M HF solution [72]. This result established additional evidence, that element 104 is a transactinide element which belongs into group 4.

It was a series of manually performed separations which recently enlightened new and interesting properties of Rf in aqueous solution [25, 65, 66]. Extraction studies with TIOA showed that anionic complexes of tetravalent Rf are nearly 100% extracted from 12 M HCl as is Zr but not Th [65]. Cationic species were investigated by extraction into thenoyltrifluoroacetone (TTA). A distribution coefficient for Rf between those of Th and Pu indicated that the hydrolysis of element 104 is less than that for Zr, Hf, and Pu [103].

The effect of chloride concentration and pH on the extraction of Rf into tributylphosphate (TBP) in benzene and into triisooctylamine (TIOA) in benzene was compared with that of group 4 homologs Zr and Hf, the pseudo-group 4 element Th, tetravalent Pu, Nb as a member of group 5, and trivalent actinides. In general, Rf was found to behave as a group 4 element and, unlike the trivalent actinides, extracts efficiently as a neutral complex into TBP from concentrated HCl. Between 8 M and 12 M HCl Rf extracts better into TBP than does Hf. Surprising results were obtained when the chloride concentration was varied at constant [H^+]. Fig. 7 shows the distribution of ^{261}Rf, ^{95}Zr, ^{169}Hf, ^{228}Th, and ^{238}Pu into a TBP phase from an aqueous phase being 8 M in [H^+] as a function of the Cl^- concentration [66]. In Cl^- concentrations above 10 M at constant [H^+] of 8 M, the extraction of tetravalent Rf decreases and behaves differently from Hf and Th, and most like Pu^{4+}. This indicates that the chloride complexation of Rf is stronger than that of Hf and Th and similar to that of tetravalent Pu. Consequently, Rf forms neutral salt complexes more readily than Zr and Hf, and in TBP inextractable anionic species at higher Cl^- concentrations. The compound $RfCl_6^{2-}$ was suggested as a candidate [25]. Extraction studies at a constant concentration of 12 M Cl^- showed that the Rf extraction is sharply increasing with increasing [H^+] between 8 M and 12 M [66]. Such a behaviour is not exhibited by Zr and Hf.

As it was found earlier for element 105, see next section for details, also these results on the aqueous chemistry of element 104 indicate unexpected trends in the chemical behaviour of the transactinide elements as compared with classical extrapolations. This demonstrates that specific properties of the lighter homologs can not be simply extrapolated to obtain the chemical properties of the transactinide elements in detail. As some of the Rf extraction experiments suffered from small but possibly not unimportant differences in the chemical procedures applied to the individual elements, e.g. different contact times and volumes used, it is important to confirm these very interesting findings in an experiment that establishes identical separation conditions for all of these elements under investigation.

7. Element 105

7.1 Theoretical predictions

A normal continuation in the Periodic Table puts element 105, hahnium (Ha), in group 5 below Ta. From this analogy to Ta, the electronic ground state configuration of Ha is expected to be [Rn]$5f^{14}\,6d^3\,7s^2$. This is in agreement with earlier relativistic calculations [82] and with the most recent MCDF calculation [87]. Contrary to elements 103 and 104, a dominating contribution of the $7p_{1/2}$ state is not expected for Ha. Atomic and ionic radii and the first through fifth ionization potentials were obtained in the MCDF calculations [87]. Differences in the chemical behaviour of element 105 as compared with the classically expected properties of eka-tantalum may possibly arise from (i) a greater atomic or ionic radius, (ii) differences in the covalency and the effective charges of the central ion, and/or (iii) differences in the radial extension of *d*-wave functions which are most important in molecular bondings [90–93, 104, 105].

Standard redox potentials were estimated for group 5 elements and the pseudo-group 5 element Pa on the basis of ionization potentials from MCDF calculations [87, 106]. It was shown [106] that the stability of the pentavalent state is increasing along the group from V to Ha, while that of the tetra- and trivalent state decreases in this direction. Although the stability of the maximum oxidation state is decreasing along the period from element 103 to 105, no extra stability of the trivalent state of Ha was found in the estimates [106]. It would be interesting to probe this prediction by reduction experiments. These could be carried out in aqueous solution similar to the experiments performed on element 103 [107].

Relativistic molecular DS DVM calculations allow to determine the degree of covalency in the bonding and the effective charge on the central metal atom in halide and oxyhalide compounds of the group 5 elements V, Nb, Ta, and Ha [90–92, 104]. From the results of these calculations it was inferred [91] that

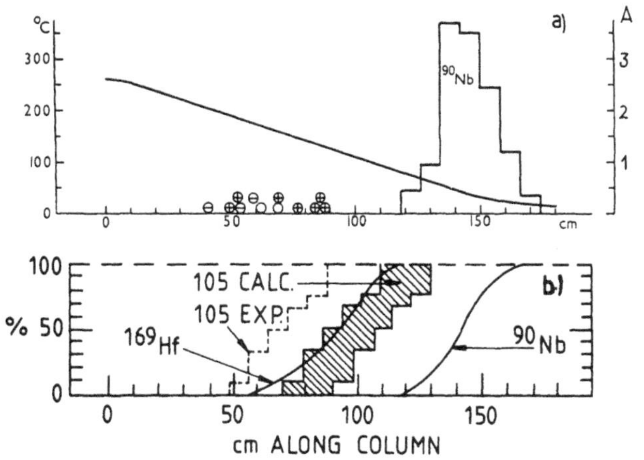

Fig. 8. a) (upper panel) Distribution of fission tracks (circles) attributed to element 105 bromides from three thermochromatographic experiments compared with the ^{90}Nb distribution, and the temperature gradient along the column. b) (lower panel) Measured integral distribution of 14.6-h ^{90}Nb and 3.3-min ^{169}Hf (smooth solid lines) are compared to the measured integral fission distribution (dashed step function), and a calculated region (dashed area) which compensates for the short half-life of 1.8-s ^{261}Ha. Adapted from [17].

$HaBr_5$ should have a higher volatility compared with its lighter homologs. Such a distinct higher volatility of $HaBr_5$ was not obtained in an extrapolation of physicochemical properties of elements and their compounds within the groups and periods of the Periodic Table [94]. Only the smaller relativistic effects which are already present in the lighter homologs are extrapolated in this approach which does not account for the increasingly strong relativistic effect in the heaviest elements. Generally, volatilities predicted from relativistic calculations are higher than those extrapolated from physicochemical properties [57].

The electronic structure of the group 5 element anionic complexes $[MCl_6]^-$, $[MOCl_4]^-$, $[M(OH)_2Cl_4]^-$, and $[MOCl_5]^{2-}$, were calculated with the relativistic DS DV method [108]. The charge density distribution analysis showed (i) that Ta has the highest tendency to form the pure halide complex $[TaCl_6]^-$, (ii) that Pa has a strong Pa=O bond in its oxyhalide complex, and (iii) that Nb and Ha show about equal trends to form the pure hexahalide and the oxyhalide complex. The preferential formation of $[HaOCl_5]^{2-}$ was expected at medium HCl concentrations. At the highest HCl concentrations $[HaCl_7]^{2-}$, like $[PaCl_7]^{2-}$, was expected to be the most stable species [108].

7.2 Experimental results

The volatilities of element 105 chlorides and bromides were first studied [16, 17] with 1.8-s ^{261}Ha in a similar thermochromatographic set-up as was used for element 104. The results from these early studies of the bromides are shown in Fig. 8. In these experiments, the SF-track distribution along the column was assumed to characterize the element 105 behaviour. The measured distribution of ^{261}Ha was corrected for the much shorter half-life before comparing it with the adsorption zones of longer-lived tracer activities of various elements including 3.3-min ^{169}Hf to represent a group 4 behaviour and 14.6-h ^{90}Nb as a group 5 element.

It was apparent from these experiments (i) that element 105 behaves like a typical transactinide element, (ii) that the chlorides and bromides of element 105 are less volatile than the Nb compounds, and (iii) that the element 105 compounds show a volatility which is rather similar to, or only slightly higher than the volatility observed for the Hf compounds (see above for a discussion of the Hf behaviour in group 4). The authors of the experiment concluded that element 105 is a homolog of the group 5 elements Nb and Ta [16, 17]. Their results were differently interpreted as showing that Ha behaves more like a group 4 element or, as an alternative interpretation, that these results much closer resemble the behaviour of the tetrachloride of V or the oxychlorides of V and Nb [21, 109].

First gas-chromatographic studies on the bromides of the group 5 elements Nb, Ta, and Ha [51, 59] were performed with the OLGA II set-up [49]. The results are shown in Fig. 9 where the chemical yield at a given temperature is a measure for the volatility. The curve for Nb was measured with pure HBr as a brominating agent, for Ta with HBr containing BBr_3 vapor, and for Ha with both, pure HBr and HBr with BBr_3 vapor. These data seem to establish that the bromides of Nb and Ta have a similar volatility while the Ha bromide is less volatile. Good agreement was found with reanalyzed earlier thermochromatographic data [99]. However, both sets of experimental results seem, at the first glance, to disagree with results from relativistic molecular calculations [91]. Theory predicted that $HaBr_5$ has the highest covalency and lowest effective charge among the group 5 bromides which should make $HaBr_5$ more volatile than its homologs. Also a "non-relativistic" thermodynamical model [94] estimated a band of possible $HaBr_5$ volatilities which are close to the $NbBr_5$ and $TaBr_5$ volatilities. This situation is depicted in Fig. 10. Here, the vapor pressures as a function of temperature [51, 91] are used as a measure of the relative bromide volatility.

The current interpretation of the observed low Ha-bromide volatility assumes the formation of an oxybromide compound like $HaOBr_3$. This is supported by theoretical calculations which showed for Ha a much stronger tendency to form oxyhalide compounds as for Nb and Ta [92]. While the generally lower volatility of the oxyhalide compounds can be inferred from the DS CVM calculations it remains open whether the $HaOBr_3$ volatility will be higher or lower than that of Nb and Ta oxyhalides [92]. Further experiments are needed for a comparison of all group 5 oxybromides. This would enable theory to determine which interaction is determining the volatility of Ha halides [92].

In a first series of experiments to investigate the chloride compounds of group 5 elements in the gas

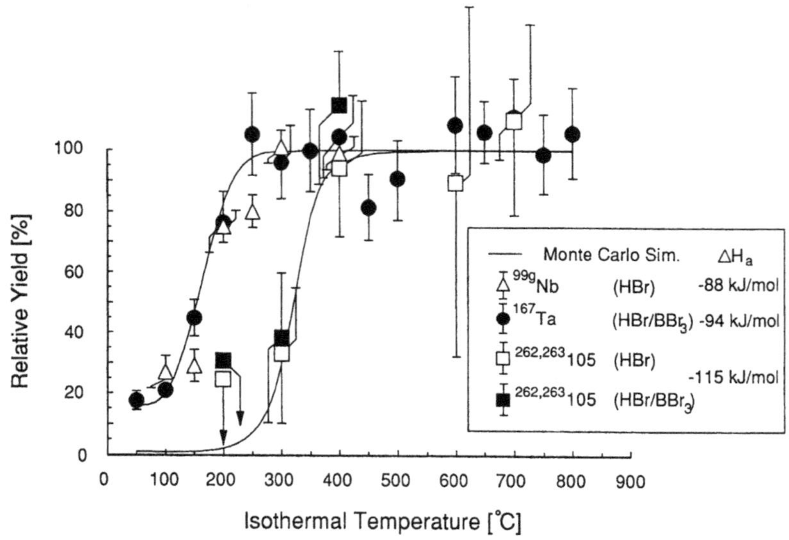

Fig. 9. Relative yield of Nb (triangles), Ta (solid dots) and Ha (squares) bromides as a function of the temperature in the isothermal part of the gas-chromatography column [51, 59]. Inserted in the figure are the brominating agents and the deduced adsorption enthalpies which were used for the Monte-Carlo simulations (solid lines). Figure from [51].

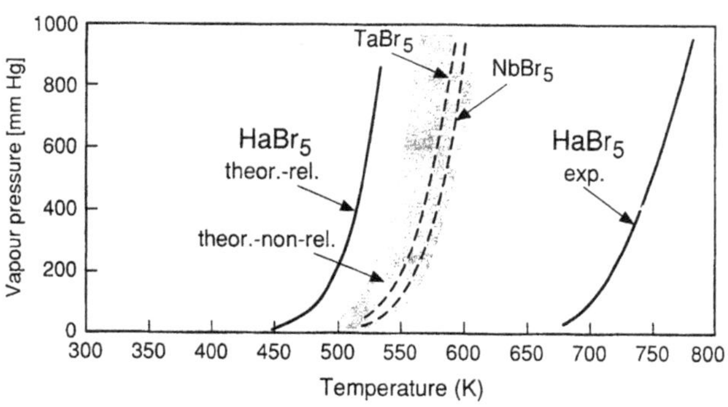

Fig. 10. Vapor pressure curves for Nb, Ta, and Ha bromides. The curve $HaBr_5$ exp. was deduced from the experimental data shown in Fig. 9. The curve theor.-rel. is from [91] and the area denoted theor.-non-rel. from [94]. Figure from [51].

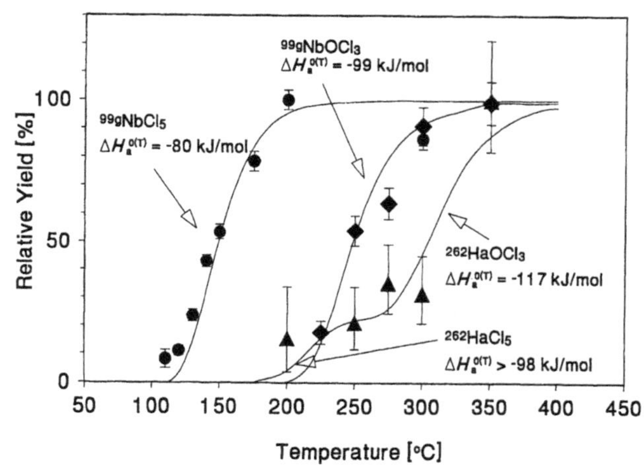

Fig. 11. Relative yield of $NbCl_5$ (dots) and $NbOCl_3$ (diamonds) in comparison with the relative yield of $HaCl_5$ and $HaOCl_3$ (triangles) as a function of the temperature in the isothermal part of the gas-chromatography column [57]. The solid lines are from Monte Carlo simulations with the adsorption enthalpies given in the figure. Figure from [57].

phase, it was observed that Ha chlorides are more volatile than their respective bromides, probably similar in volatility to $NbCl_5$ [57, 110]. This is in agreement with the finding that the group 4 and 5 chlorides are more volatile than the respective bromides. For Ta, only a species with a much lower volatility was observed, presumably $TaOCl_3$ [57].

In a most recent experiment [57] precautions were taken to reduce the partial pressure of free O_2 as far as possible. Carbon aerosols were used in the He-jet to transport the reaction products into OLGA III. Purified HCl was added as a chlorinating agent. The relative yield curves for Ha chlorides are shown in Fig. 11 together with yield curves measured for $NbCl_5$ and $NbOCl_3$ under identical conditions [57]. The Monte Carlo simulations (solid line in Fig. 11) reproduced the positions and shapes of the measured yields very well. The yield curve for Ha chlorides was analyzed assuming that two chemical species were present, namely $HaCl_5$ and $HaOCl_3$ [57]. Compared with $NbOCl_3$, $HaOCl_3$ became volatile at an approximately 50 °C higher temperature. This is in qualitative agreement with extrapolations within the Periodic Table. An investigation of the Ta-chloride volatilities is needed to complete the picture in group 5. Presently, only a lower limit can be given for the adsorption enthalpy of $HaCl_5$. Future experiments are needed to extend the yield measurements of Ha chlorides to lower temperatures for a comparison with theoretical predictions. Both methods, the DS DVM calculation and the empirical extrapolation predict the $HaCl_5$ as being equally or even more volatile than $NbCl_5$.

The first aqueous chemistry of element 105 was an experiment to study the adsorption onto glass surfaces after fuming twice with concentrated HCl and HNO_3 solutions [48]. In manual separations, it was shown that Ha adheres with high yield to glass surfaces from HNO_3 solution, a behaviour very characteristic for the group 5 elements Nb and Ta and for Pa as pseudo-group 5 element. The next experiment was an attempt to study the extraction of Ha fluoride complexes from a mixture of HNO_3 and HF into MIBK [48] under conditions in which Ta extracts but Nb does not. From an extrapolation in group 5 it was expected that Ha would behave more like Ta than Nb but, surprisingly, Ha did not extract.

To investigate this unexpected finding and more facets of the Ha chemistry in greater detail, on-line separations were conducted with the Automated Rapid Chemistry Apparatus ARCA II [67]. In the first experiments, liquid extraction chromatography separations were performed with the anion exchange liquid TIOA on an inert support [68]. Preparatory tests had shown that TIOA extracts all group 5 elements and Pa, irrespective of the formation of mono- or polynegative anions, from HCl solutions of 10 M and above. At lower concentrations, the selective back extraction allowed to distinguish between the chemical behaviour of Nb, Ta, Pa, and Ha. A diagram of the results from the chromatographic separations of Ha and its lighter homologs is given Fig. 12. Element 105 was shown [68] to extract into the TIOA on the chromatographic columns in ARCA from (i) either 12 M HCl/0.02 M HF or (ii) 10 M HCl, as do Nb, Ta, and Pa. This result clearly indicated the formation of anionic Ha complexes. To obtain an element 105 fraction in this experiment, the amine, along with the extracted species, was stripped from the column with acetone/0.02 M HF, see upper panel in Fig. 12, and from this fraction a sample was prepared for the α-spectroscopic identification of ^{262}Ha. Elution positions of element 105 relative to those of Nb, Ta, and Pa were determined [68] in 10 M HCl/0.025 M HF (lower panel in Fig. 12), in 4 M HCl/0.02 M HF (middle panel), and in 0.5 M HCl/0.01 M HF [70] (not shown). In some experiments, 6 M HNO_3/0.015 M HF was used to complete the back extraction of Ha into the mobile aqueous phase.

At HCl concentrations of 4 M and above, Ha was found to behave similarly to Nb and Pa while at the lower concentration of 0.5 M HCl Ha follows the pseudo-group 5 element Pa in its behaviour. The results of these chromatographic studies of Ha are plotted (bars) in Fig. 13 as the percentage extraction together with results from batch extractions performed with tracers of the lighter homologs (curves).

At all HCl concentrations below 12 M, Ha was found to behave differently from its closest homolog Ta. Apparently, in contrast to straightforward extrapolations, the trend in the chemical properties of the group 5 elements from Nb to Ta does not continue but is reversed in going from Ta to Ha. This is mainly because of the exceptional behaviour of Ta within group 5. Only for Ta, strong largely covalent complexes of the pure halide type such as $[TaX_6]^-$ (X≡Cl, F) are predominant over a wide range of HCl concentrations. From the non-Ta-like behaviour of element 105 and its similarity to Nb and Pa, it was concluded [68, 70] that the complex structure of Ha must be of the Nb and Pa type, presumably $[HaOX_4]^-$,

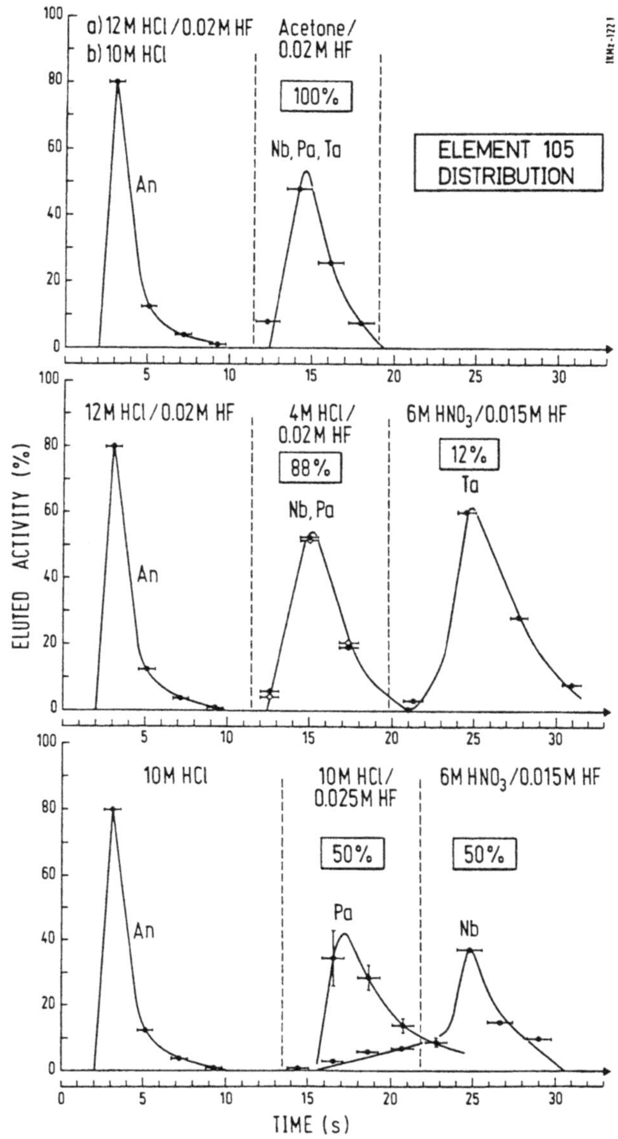

Fig. 12. Diagram of elution curves for trivalent actinides (An), Nb, Ta, and Pa from TIOA columns (1.6 mm × 8 mm), under the same conditions as in the Ha experiments. The inserts show the relative amount of element 105 found in each specific eluent. Figure from [27] adapted from [68].

in hydrolysis equilibrium with $[Ha(OH)_2X_4]^-$, or $[HaOX_5]^{2-}$. This preferential formation of oxyhalide complexes of element 105 is in agreement with results from relativistic DS DVM calculations [108]. The oxygen-containing complexes have higher effective charges and carry dipole moments, both contributing to a lower extractability of these ions into the TIOA as compared with the largely covalent and therefore more hydrophobic $[TaX_6]^-$ [28, 108]. The formation of $[HaX_7]^{2-}$ in concentrated HCl, as expected from theory [108], was not excluded in these experiments. The use of mixed HCl/HF solutions in these experiments prohibits a clear assignment to one specific complex, and it clearly calls for experiments with only one possible halide ligand. The non-Ta-like behaviour of Ha observed in the extraction experiments with MIBK [48] can be understood in the frame of these

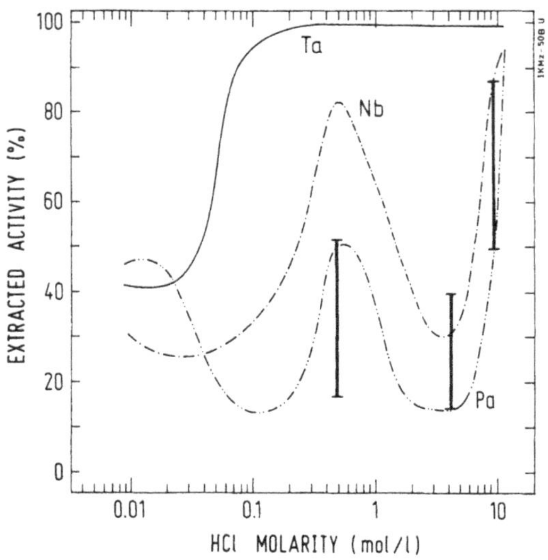

Fig. 13. Fractional extracted activity of Ta, Nb, and Pa tracers (curves) as a function of HCl concentration in the system TIOA-HCl/0.03 M HF. The bold bars encompass the upper and lower limits deduced for the Ha extraction from the elution positions in the chromatography experiments, see text. The bar for the complete extraction of Ha from 12 M HCl/0.02 M HF is not included in the figure for clarity. Figure from [28] adapted from [70].

results as the procedure was specific for Ta and did not include a Nb-like behaviour.

Extractions of Nb, Pa, and Ha into DIBC from concentrated HBr solutions showed [69] that the tendency to form polynegative anions is increasing in the sequence Pa < Nb < Ha. Elutions from cation exchange columns with unbuffered 0.05 M α-HIB solution showed that Ha, together with Nb and Ta, elutes promptly [39] while tri- and tetravalent ions are strongly retained under these conditions. This provided an unambiguous proof that pentavalent Ha is the most stable valence state in aqueous solution [39], and it supported the position of element 105 as a member of group 5 of the Periodic Table.

The chemistry of element 105, to a greater extent than the chemistry of element 104, provided many surprises. Its non-Ta-like behaviour and its similarity to Nb and/or Pa, depending on its chemical environment, demonstrated that the chemical properties of this second transactinide element cannot be reliably extrapolated from the trends observed in its lighter homologs.

8. The future – element 106 and beyond

Elements 106 through 112 are expected to complete the 6*d* transition series. As such they should be placed in groups 6 through 12 of the Periodic Table as homologs of the elements W through Hg. However, based on our experience with the transactinide elements 104 and 105, it appears that the electronic configuration of these elements will be difficult to predict from a simple extrapolation. From the relativistic stabilization of the *s*-orbitals it is expected that all transactinides will have a $7s^2$-configuration. This was confirmed in a most recent atomic calculation [111] for element 111 which, in contrast to its lighter homologs in group 11, has a $6d^9\,7s^2$ ground state configuration. However, the knowledge of the electronic ground state configuration provides not enough information to predict the chemical properties of the heavier transactinides. In addition, these elements may possibly exhibit chemical properties which are in their details significantly different from simple extrapolations within their groups. It was pointed out [7, 25] that the electronic orbitals in these elements are close enough in energy and extension that a great variety of chemical behaviours may result, in its diversity similar to the beginning of the actinide series. Some of the surprising facets of elements 104 and 105 chemistry may have given us the first glimpse of such diverse properties. Presently, it would be premature to expect that the general ordering principle of the Periodic Table will not hold anymore but more and more unexpected properties may be observable for elements 106 and beyond. A comparison with up-to-date atomic and molecular calculations, may reveal the contribution of the increasingly important influence of relativistic effects on the chemistry of the heaviest elements.

Existing separation technologies, like ARCA II [67], OLGA III [61, 62], and SISAK III [73] with its newly developed on-line detection system [74], provide means to extend our studies of the transactinides to element 106, seaborgium (Sg). The discovery of the longer-lived isotopes ^{265}Sg and ^{266}Sg [40] has eased the situation of operating the separation apparates at their limits. Several attempts are under way to begin these investigations [77] with all three of the above mentioned techniques. The research goals are to determine the hexavalency of Sg ions and to characterize the formation and behaviour of oxyhalides in aqueous solution and in the gas phase. Results from DS DVM calculations are at hand [93, 108, 112] to compare experimental findings with theoretical predictions. This is planned to be complemented by a Sg extraction from α-HIB solution into trioctylamine. Reports about first results from a thermochromatographic separation of 0.9-s ^{263}Sg [113] were not unanimously accepted because of an equivocal assignment of the observed fission tracks.

Although elements 107–111 have been discovered by physical methods, none of these elements has been identified after chemical separation because of the very short half-lives of the presently known nuclides. A number of experiments have been prepared and tested to separate longer-lived isotopes of element 107, nielsbohrium (Ns), see [114, 115] and references therein, and heavier elements. From our knowledge of heavy element nuclei it appears that longer lived species of the heaviest elements probably exist. The biggest problem now is to find the best method to synthesize these nuclei. One of the promising pathes are heavy ion reactions with ^{254}Es as a target [37]. It will be intellectually most interesting and experimentally most challenging to extend chemical studies to still heavier elements to get further insights into the importance of relativistic effects on chemical properties at the top end of the Periodic Table. Ultimately, perhaps nuclear and chemical research may even reach the elusive island of spherical superheavy element around $Z=114$ and $N=178$.

Acknowledgments

Its a great pleasure to thank, among many other colleagues quoted in this article, H. W. Gäggeler, J. V. Kratz, and A. Türler for many discussion and for providing me with several figures including the latest unpublished results. Enlightening discussions with V. Pershina about the theory of transactinide chemistry are gratefully acknowledged. I like to thank B. Schausten for preparation of many drawings and W. Brüchle for reading the manuscript.

References

1. Becquerel, H.: Compt. Rend. **122**, 420, 501, 559, 689, 762, 1086 (1896), and Compt. Rend. **123**, 855 (1896).
2. Curie, P., Sklodowska Curie, M.: Compt. Rend. **127**, 175 (1898).
3. Seaborg, G. T.: Chem. Eng. News **23**, 2190 (1945).
4. Hofmann, S., Ninov, V., Heßberger, F. P., Armbruster, P., Folger, H., Münzenberg, G., Schött, H. J., Popeko, A. G., Yeremin, A. V., Andreyev, A. N., Saro, S., Janik, R., Leino: Z. Phys. A **350**, 277 (1995).
5. Hofmann, S., Ninov, V., Heßberger, F. P., Armbruster, P., Folger, H., Münzenberg, G., Schött, H. J., Popeko, A. G., Yeremin, A. V., Andreyev, A. N., Saro, S., Janik, R., Leino: Z. Phys. A **350**, 281 (1995).
6. Seaborg, G. T.: Radiochim. Acta **61**, 115 (1993).
7. Katz, J. J., Seaborg, G. T., Morss, L. R.: *The Chemistry of the Actinide Elements, 2nd ed.*, Chapman and Hall, London, 1986.
8. Seaborg, G. T., Loveland, W. D.: *The Elements Beyond Uranium*, John Wiley and Sons, Inc., New York 1990.
9. Fricke, B., Greiner, W.: Phys. Lett. **30B**, 347 (1969).
10. Zvara, I., Chuburkov, Yu. T., Caletka, R., Zvarova, T. S., Shalaevsky, M. R., Shilov, B. V.: At. Energy **21**, 8 (1966); J. Nucl. Energy **21**, 60 (1966); Sov. J. At. Energy **21**, 709 (1966).
11. Zvara, I., Chuburkov, Yu. Ts., Caletka, R., Shalaevsky, M. R.: Sov. Radiochem. **11**, 161 (1969).
12. Chuburkov, Yu. T., Zvara, I., Shilov, B. V.: Sov. Radiochem. **11**, 171 (1969).
13. Zvara, I., Belov, V. Z., Chelnokov, L. P., Domanov, V. P., Hussonnois, M., Korotkin, Yu. S., Shegolev, V. A., Shalaevsky, M. R.: Inorg. Nucl. Chem. Lett. **7**, 1109 (1971).
14. Silva, R. J., Harris, J., Nurmia, M., Eskola, K., Ghiorso, A.: Inorg. Nucl. Chem. Lett. **6**, 871 (1970).
15. Hulet, E. K., Lougheed, R. W., Wild, J. F., Landrum, J. H., Nitschke, J. M., Ghiorso, A.: J. Inorg. Nucl. Chem. **42**, 79 (1980).
16. Zvara, I., Eichler, B., Belov, V. Z., Zvarova, T. S., Korotkin, Yu. S., Shalaevskii, M. R., Shegolev, V. A., Hussonnois, M.: Sov. Radiochem. **16**, 709 (1974).
17. Zvara, I., Belov, V. Z., Domanov, V. P., Shalaevskii, M. R.: Sov. Radiochem. **18**, 328 (1976).
18. Zvara, I., 1976, In: *Transplutonium Elements,* eds. W. Müller and R. Lindner, North-Holland, Amsterdam, pp. 11–20.
19. Zvara, I.: Radiochim. Acta **38**, 95 (1985).
20. Zvara, I.: Isotopenpraxis **26**, 251 (1990).
21. Keller, O. L. Jr., Seaborg, G. T.: Ann. Rev. Nucl. Sci. **27**, 139 (1977).
22. Hulet, E. K., In: *Actinides in Perspectives,* ed. N. M. Edelstein, Pergamon Press, Oxford (1981), pp. 453–490.
23. Keller, O. L., Jr.: Radiochim. Acta **37**, 169 (1984).
24. Silva, R. J., In: *The Chemistry of the Actinide Elements,* eds. Katz, J. J., Seaborg, G. T., Morss, L. R., 2nd ed., Vol. 2, Chapman and Hall, London (1986), pp. 1085–1117.
25. Hoffman, D. C.: Proc. *Fifty Years with Transuranium Elements,* The Robert A. Welch Foundation Conf. on Chemical Research XXXIV, Houston, TX, 22–23 October 1990, pp. 255–274, and Lawrence Berkeley Laboratory Report **LBL-29815** (1990).
26. Hoffman, D. C.: Radiochim. Acta **61**, 123 (1993).
27. Schädel, M., In: Nuclei Far From Stability/Atomic Masses and Fundamental Constants 1992, ed. R. Neugart and A. Wöhr, Institut of Physics Conf. Series 132, Institut of Physics Publ., Bristol (1993), pp. 413–422.
28. Kratz, J. V.: Proc. Int. Conf. *Actinides-93*, Santa Fe, NM, 19–24 September 1993, Elsevier, Amsterdam (1994), and J. Alloy Comp. **213/214**, 20 (1994).
29. Hoffman, D. C.: Chem. Eng. News **72**(18), 24 (1994).
30. Münzenberg, G.: Radiochim. Acta **70/71**, 193 (1995).
31. Armbruster, P.: Ann. Rev. Nucl. Part. Sci. **35**, 135 (1985).
32. Münzenberg, G.: Rep. Prog. Phys. **51**, 57 (1988).
33. Herrmann, G.: Angew. Chem. Int. Ed. Engl. **27**, 1417 (1988).
34. Reisdorf, W., Schädel, M.: Z. Phys. A – Hadrons and Nuclei **343**, 47 (1992).
35. Oganessian, Yu. Ts.: J. Alloy Comp. **213/214**, 50 (1994).
36. Münzenberg, G.: *The Prospects of Heavy Element Research.* In: Frontier Topics in Nuclear Physics (W. Scheid, A. Sandulescu Eds.) Plenum Publ. Comp., New York, 1995 (in press).
37. Schädel, M., Hofmann, S.: J. Radioanal. Nucl. Chem., in press (1995).
38. Kratz, J. V., Gober, M. K., Zimmermann, H. P., Schädel, M., Brüchle, W., Schimpf, E., Gregorich, K. E., Türler, A., Hannink, N. J., Czerwinski, K. R., Kadkhodayan, B., Lee, D. M., Nurmia, M. J., Hoffman, D. C., Gäggeler, H., Jost, D., Kovacs, J., Scherer, U. W., Weber, A.: Phys. Rev. C **45**, 1064 (1992).
39. Schädel, M., Brüchle, W., Schimpf, E., Zimmermann, H. P., Gober, M. K., Kratz, J. V., Trautmann, N., Gäggeler, H., Jost, D., Kovacs, J., Scherer, U. W., Weber, A., Gregorich, K. E., Türler, A., Czerwinski, K. R., Hannink, N. J., Kadkhodayan, B., Lee, D. M., Nurmia, M. J., and Hoffman, D. C.: Radiochim. Acta **57**, 85 (1992).
40. Lougheed, R. W., Moody, K. J., Wild, J. F., McQuaid, J. H., Lazarev, Yu. A., Lobanov, Yu. V., Oganessian, Yu. Ts., Utyonkov, V. K., Abdullin, F. Sh., Buklanov, G. V., Gikal, B. N., Iliev, S., Mezentsev, A. N., Polyakov, A. N., Sedykh, I. M., Shirikovsky, I. V., Subbotin, V. G., Sukov, A. M., Tsyganov, Yu. S., Zhuchko, V. E.: Proc. Int. Conf. *Actinides-93*, Santa Fe, NM, 19–24 September 1993, Elsevier, Amsterdam (1994), J. Alloy Comp. **213/214**, 61 (1994), and Lazarev, Yu. A., Lobanov, Yu. V., Oganessian, Yu. Ts., Utyonkov, V. K., Abdullin, F. Sh., Buklanov, G. V., Gikal, B. N., Iliev, S., Mezentsev, A. N., Polyakov, A. N., Sedykh, I. M., Shirikovsky, I. V., Subbotin, V. G., Sukov, A. M., Tsyganov, Yu. S., Zhuchko, V. E., Lougheed, R. W., Moody, K. J., Wild, J. F., McQuaid, J. H.: Phys. Rev. Letters **73**, 624 (1994).
41. Smolanczuk, R., Skalski, J., Sobiczewski, A.: Preprint GSI-94-77 (1994).
42. Guillaumont, R., Adloff, J. P., Peneloux, A.: Radiochim. Acta **46**, 169 (1989).
43. Guillaumont, R., Adloff, J. P., Peneloux, A., Delamoye, P.: Radiochim. Acta **54**, 1 (1991).
44. Borg, R. J., Dienes, G. J.: J. Inorg. Nucl. Chem. **43**, 1129 (1981).
45. Trautmann, N.: Radiochim. Acta **70/71**, 237 (1995).
46. Nitschke, J. M.: Nucl. Instrum. Methods **138**, 393 (1976).

47. Schädel, M., Brüchle, W., Haefner, B.: Nucl. Instrum. Methods in Phys. Res. **A264**, 308 (1988).
48. Gregorich, K. E., Henderson, R. A., Lee, D. M., Nurmia, M. J., Chasteler, R. M., Hall, H. L., Bennett, D. A., Gannett, C. M., Chadwick, R. B., Leyba, J. D., Hoffman, D. C., Herrmann, G.: Radiochim. Acta **43**, 223 (1988).
49. Gäggeler, H. W., Jost, D. T., Baltensperger, U., Weber, A., Kovacs, A., Vermeulen, D., Türler, A.: Nucl. Instrum. Methods in Phys. Res. **A309**, 201 (1991).
50. Günther, R., Becker, H. U., Zauner, S., Nähler, A., Kratz, J. V., Schädel, M., Brüchle, W., Jäger, E., Schimpf, E., Schausten, B.: Institut für Kernchemie, Univ. Mainz, Jahresbericht 1994, IKMz 95-1 (1995) p. 7.
51. Gäggeler, H. W.: J. Radioanal. Nucl. Chem. **183**, 261 (1994).
52. Chuburkov, Yu. T., Belov, V. Z., Caletka, M., Shalaevsky, M. R., Zvara, I.: Sov. Radiochem. **11**, 386 (1969), and Chuburkov, Yu. T., Zvara, I., Shilov, B. V.: Sov. Radiochem. **11**, 171 (1969).
53. Zvara, I., Chuburkov, Yu. T., Belov, V. Z., Buklanov, G. V., Zakhvataev, B. B., Zvarova, T. S., Maslov, O. D., Caletka, R., Shalaevsky, M. R.: J. Inorg. Nucl. Chem. **32**, 1885 (1970).
54. Belov, V. Z., Zvarova, T. S., Shalaevsky, M. R.: Sov. Radiochem. **17**, 87 (1975).
55. Zvara, I., Keller, O. L. Jr., Silva, R. J., Tarrant, J. R.: J. Chromatogr. **103**, 77 (1975).
56. Zhuikov, B. L., Chuburkov, Yu. T., Timokhin, S. N., Kim U Jin, Zvara, I.: Radiochim. Acta **46**, 113 (1989).
57. Türler, A.: Radiochim. Acta **72**, 7 (1996).
58. Kadkhodayan, B., Türler, A., Gregorich, K. E., Nurmia, M. J., Lee, D. M., Hoffman, D. C.: Nucl. Instrum. Methods in Phys. Res. **A317**, 254 (1992).
59. Gäggeler, H. W., Jost, D., Kovacs, J., Scherer, U. W., Weber, A., Vermeulen, D., Türler, A., Gregorich, K. E., Henderson, R. A., Czerwinski, K. R., Kadkhodayan, B., Lee, D. M., Nurmia, M. J., Hoffman, D. C., Kratz, J. V., Gober, M. K., Zimmermann, H. P., Schädel, M., Brüchle, W., Schimpf, E., Zvara, I.: Radiochim. Acta **57**, 93 (1992).
60. Türler, A., Gäggeler, H. W., Gregorich, K. E., Barth, H., Brüchle, W., Czerwinski, K. R., Gober, M. K., Hannink, N. J., Henderson, R. A., Hoffman, D. C., Jost, D. T., Kacher, C. D., Kadkhodayan, B., Kovacs, J., Kratz, J. V., Kreek, S. A., Lee, D. M., Leyba, J. D., Nurmia, M. J., Schädel, M., Scherer, U. W., Schimpf, E., Vermeulen, D., Weber, A., Zimmermann, H. P., Zvara, I.: J. Radioanal. Nucl. Chem. **160**, 327 (1992).
61. Eichler, B., Türler, A., Jost, D. T., Gäggeler, H. W.: Annual Report 1993, Labor für Radio- und Umweltchemie, Paul Scherrer Institut (1994), p. 42.
62. Türler, A., Jost, D. T., Piguet, D., Eichler, B., Mendel, M., Nähler, A., Eberhardt, K., Trautmann, N.: Annual Report 1994, Labor für Radio- und Umweltchemie, Paul Scherrer Institut (1995), p. 43.
63. Hoffman, D. C., Lee, D. M., Ghiorso, A., Nurmia, M. J., Aleklett, K., Leino, M.: Phys. Rev. C **24**, 495 (1981).
64. Jost, D. T., Vermeulen, D.: IEEE, Trans. Nucl. Sci. **39**, 186 (1992).
65. Czerwinski, K. R., Gregorich, K. E., Hannink, N. J., Kacher, C. D., Kadkhodayan, B., Kreek, S. A., Lee, D. M., Nurmia, M. J., Türler, A., Seaborg, G. T., Hoffman, D. C.: Radiochim. Acta **64**, 23 (1994).
66. Czerwinski, K. R., Kacher, C. D., Gregorich, K. E., Hamilton, T. M., Hannink, N. J., Kadkhodayan, B., Kreek, S. A., Lee, D. M., Nurmia, M. J., Türler, A., Seaborg, G. T., Hoffman, D. C.: Radiochim. Acta **64**, 29 (1994).
67. Schädel, M., Brüchle, W., Jäger, E., Schimpf, E., Kratz, J. V., Scherer, U. W., Zimmermann, H. P.: Radiochim. Acta **48**, 171 (1989).
68. Kratz, J. V., Zimmermann, H. P., Scherer, U. W., Schädel, M., Brüchle, W., Gregorich, K. E., Gannet, C. M., Hall, H. L., Henderson, R. A., Lee, D. M., Leyba, J. D., Nurmia, M. J., Hoffman, D. C., Gäggeler, H., Jost, D., Baltensperger, U., Ya Nai-Qi, Türler, A., Lienert, Ch.: Radiochim. Acta **48**, 121 (1989).
69. Gober, M. K., Kratz, J. V., Zimmermann, H. P., Schädel, M., Brüchle, W., Schimpf, E., Gregorich, K. E., Türler, Hannink, N. J., Czerwinski, K. R., Kadkhodayan, B., Lee, D. M., Nurmia, M. J., Hoffman, D. C., Gäggeler, H., Jost, D., Kovacs, J. Scherer, U. W., Weber, A.: Radiochim. Acta **57**, 77 (1992).
70. Zimmermann, H. P., Gober, M. K., Kratz, J. V., Schädel, M., Brüchle, W., Schimpf, E., Gregorich, K. E., Türler, Czerwinski, K. R., Hannink, N. J., Kadkhodayan, B., Lee, D. M., Nurmia, M. J., Hoffman, D. C., Gäggeler, H., Jost, D., Kovacs, J., Scherer, U. W., Weber, A.: Radiochim. Acta **60**, 11 (1993).
71. Bruchertseifer, H., Eichler, B., Estevez, J., Zvara, I.: Radiochim. Acta **47**, 41 (1989).
72. Szeglowski, Z., Bruchertseifer, H., Domanov, V. P., Gleisberg, B., Guseva, L. J., Hussonnois, M., Tikhomirova, G. S., Zvara, I., Oganessian, Yu. Ts.: Radiochim. Acta **51**, 71 (1990), and Radiokhimiya **33**, 90 (1991).
73. Persson, H., Skarnemark, G., Skålberg, M., Alstad, J., Liljenzin, J. O., Bauer, G., Haberberger, F., Kaffrell, N., Rogowski, J., Trautmann, N.: Radiochim. Acta **48**, 177 (1989), and Alstad, J., Skarnemark, G., Haberberger, F., Herrmann, G., Nähler, A., Pense-Maskow, M., Trautmann, N.: J. Radioanal. Nucl. Chem. **189**, 133 (1995).
74. Wierczinski, B., Eberhardt, K., Herrmann, G., Kratz, J. V., Mendel, M., Nähler, A., Rocker, F., Tharun, U., Trautmann, N., Weiner, K., Wiehl, N., Alstad, J., Skarnemark, G.: Nucl. Instrum. Methods A, submitted for publication (1995).
75. Eberhardt, K., Alstad, J., Eichler, B., Gäggeler, H. W., Herrmann, G., Jost, D. T., Kling, H. O., Mendel, M., Nähler, A., Piguet, D., Skarnemark, G., Tharun, U., Trautmann, N., Türler, A., Weiner, K., Wiehl, N.: Institut für Kernchemie, Univ. Mainz, Jahresbericht 1994, IKMz 95-1 (1995) p. 11.
76. Gregorich, K., Wierczinski, B., Mendel, M., Nähler, A., Alstad, J., Skarnemark, G.: private communication 1995.
77. Schädel, M., for a Darmstadt–Mainz–Göteborg–Oslo–Villigen–Bern–Dresden–Leipzig–Kassel–Berkeley-collaboration: UNILAC-proposal *Studies of Element 106 Properties,* GSI Darmstadt 1994.
78. Malli, G. L. (ed.): *Relativistic and Electron Correlation Effects in Molecules and Solids,* NATO ASI Series Vol. 318, Plenum Press, New York (1994).
79. Pershina, V.: Chem. Rev., accepted for publication (1995).
80. Fricke, B., Waber, J. T.: Actinides Rev. **1**, 433 (1971).
81. Fricke, B., In: *Structure and Bonding,* eds. Dunitz, J. D., *et al.*, Vol. 21, Springer Verlag, Berlin (1975), pp. 90–142.
82. Desclaux, J. P.: At. Data Nucl. Data Tables **12**, 311 (1973).
83. Desclaux, J. P., Freeman, A. J.: In: *Handbook on the Physics and Chemistry of the Actinides,* eds. Freeman A. J., Lander, G. H., North-Holland, Amsterdam (1984), pp. 1–77.
84. Glebov, V. A., Kasztura, L., Nefedov, V. S., Zhuikov, B. L.: Radiochim. Acta **46**, 117 (1989).
85. Zhuikov, B. L., Glebov, V. A., Nefedov, V. S., Zvara, I.: J. Radioanal. Nucl. Chem. **143**, 103 (1990).
86. Johnson, E., Fricke, B., Keller, O. L. Jr., Nestor, C. W., Tucker, T. C.: J. Chem. Phys. **93**, 8041 (1990).
87. Fricke, B., Johnson, E., Rivera, G. M.: Radiochim. Acta **62**, 17 (1993).
88. Eliav, E., Kaldor, U., Ishikawa, Y.: Phys. Rev. Letters **74**, 1079 (1995).
89. Rosen, A., Ellis, D. E.: J. Chem. Phys. **96**, 3039 (1975).
90. Pershina, V., Sepp, W. D., Fricke, B., Rosen, A.: J. Chem. Phys. **96**, 8367 (1992).
91. Pershina, V., Sepp, W. D., Fricke, B., Kolb, D., Schädel, M., Ionova, G. V.: J. Chem. Phys. **97**, 1116 (1992).
92. Pershina, V., Sepp, W. D., Bastug, T., Fricke, B., Ionova, G. V.: J. Chem. Phys. **97**, 1123 (1992).
93. Pershina, V., Fricke, B.: J. Phys. Chem. **98**, 6468 (1994).

94. Eichler, B.: Annual Report 1993, Labor für Radio- und Umweltchemie, Paul Scherrer Institut (1994), p. 38.
95. Eichler, B., Türler, A., Gäggeler, H. W.: Annual Report 1994, Labor für Radio- und Umweltchemie, Paul Scherrer Institut (1995), p. 34 and p. 35.
96. Desclaux, J. P., Fricke, B.: J. Phys. **41**, 943 (1980).
97. Brewer, L.: High Temp. Sci. **17**, 1 (1984).
98. Ryzhkov, M. V., Gubanov, V. A., Zvara, I.: Radiochim. Acta **57**, 11 (1992).
99. Türler, A., Gäggeler, H. W., Eichler, B., Jost, D. T., Kovacs, J., Scherer, U. W., Kadkhodayan, B., Gregorich, K. E., Hoffman, D. C., Kreek, S. A., Lee, D. M., Schädel, M., Brüchle, W., Schimpf, E., Kratz, J. V.: Proc. School Seminar on Heavy Ion Physics, Dubna, May 1993, and Paul Scherrer Institut Rep. PSI-PR-93-12, 1993.
100. Eichler, B., Zvara, I.: Radiochim. Acta **30**, 233 (1982).
101. Jost, D. T., Gäggeler, H. W., Vogel, Ch., Schädel, M., Jäger, E., Eichler, B., Gregorich, K. E., Hoffman, D. C.: Inorg. Chim. Acta **146**, 255 (1988).
102. Eichler, B., Hübener, S., Gäggeler, H. W., Jost, D. T.: Inorg. Chim. Acta **146**, 162 (1988).
103. Czerwinski, K. R.: *Studies of Fundamental Properties of Rutherfordium (Element 104) Using Organic Complexing Agents,* Ph. D. Thesis, Berkeley (1992); Rep. **LBL-32233** (1992).
104. Pershina, V., Fricke, B.: J. Chem. Phys. **99**, 9720 (1993).
105. Pyykkö, P.: Chem. Rev. **88**, 563 (1988).
106. Ionova, G. V., Pershina, V., Johnson, E., Fricke, B., Schädel, M.: J. Chem. Phys. **96**, 11096 (1992).
107. Scherer, U. W., Kratz, J. V., Schädel, M., Brüchle, W., Gregorich, K. E., Henderson, R. A., Lee, D. M., Nurmia, M. J., Hoffman, D. C.: Inorg. Chim. Acta **146**, 249 (1988).
108. Pershina, V., Fricke, B., Kratz, J. V., Ionova, G. V.: Radiochim. Acta **64**, 37 (1994).
109. Hyde, E. K., Hoffman, D. C., Keller, O. L. Jr.: Radiochim. Acta **42**, 57 (1987).
110. Kadkhodayan, B.: *On-Line Gas Chromatographic Studies of Rutherfordium (Element 104), Hahnium (Element 105), and Homologs,* Ph. D. Thesis, Berkeley (1993); Rep. **LBL-33961** (1993).
111. Eliav, E., Kaldor, U., Schwerdtfeger, P., Hess, B. A., Ishikawa, Y.: Phys. Rev. Letters **73**, 3203 (1994).
112. Pershina, V., Fricke, B.: J. Phys. Chem. **99**, 144 (1995).
113. Zvara, I.: Contribution to the Third Int. Conf. on Nuclear and Radiochemistry, Wien, September 7–12, 1992, and Timokhin, S. N., Zvara, I.: Private communications.
114. Schädel, M., Jäger, E., Schimpf, E., Brüchle, W.: Radiochim. Acta **68**, 1 (1995).
115. Schädel, M., Jäger, E., Brüchle, W., Sümmerer, K., Hulet, E. K., Wild, J. F., Lougheed, R. W., Moody, K. J.: Radiochim. Acta **68**, 7 (1995).

Radiochimica Acta *70/71*, 225–236 (1995)

Actinide Separation Science

By Gregory R. Choppin

Department of Chemistry, Florida State University, Tallahassee, FL 32306-3006, USA

and Kenneth L. Nash

Chemistry Division, Argonne National Laboratory, 9700 South Cass Avenue, Argonne, IL 60439-4831, USA

(Received August 17, 1995)

Actinides / Separations / Ion exchange / Solvent extraction / Plutonium

The beginnings

Separations involving actinide elements began with the discovery of radioactivity. W. Crookes and H. Becquerel found that upon addition of carbonate anion to a solution in which some uranium was dissolved, the uranium remained in the supernate as a soluble uranyl carbonate complex while a precipitate formed which contained the β, γ radioactivity. E. Rutherford and F. Soddy obtained a similar separation of thorium and the β, γ radioactivity. In both cases, following separation, the actinide portion of the sample showed an increase in radioactivity while the radioactivity in the precipitated fraction decreased with time. These separations and the associated observations led Rutherford and Soddy to propose that, in radioactive decay, atoms of the original elements are transformed into atoms of different elements. Marie and Pierre Curie began a program of separation of components of pitchblende to isolate the non-uranium radioactive elements and in 1898 announced the discovery of the new element "Polonium" whose radioactivity was 400 times greater than that of uranium. The separation methods used by these pioneers involved precipitation, which remained the dominant separation technique until the Manhattan Project of World War II.

Another important milestone in the early history of actinide element separations occurred in 1939. Between 1934 and 1939, approximately fifty research papers described the discovery and study of transuranium elements with $Z=93$, 94, 95 and 96. Then O. Hahn and F. Strassmann conducted very careful separations on irradiated uranium samples and were able to prove that these transuranium elements were, in fact, products of nuclear fission with atomic numbers below 60. This led to new experiments in 1940 in which neptunium ($Z=93$) and plutonium ($Z=94$) were synthesized and separated. These new transuranium elements were isolated and identified using an oxidation/reduction cycle with bromate as the oxidizing agent followed by precipitation in which the reduced cations were coprecipitated with crystalline LaF_3. Both the redox and fluoride precipitation separations remain standard procedures in actinide separation science.

In the Manhattan Project, a major task involved the production of plutonium from uranium by neutron irradiation. The separation of plutonium was accomplished by precipitation with bismuth phosphate. This process, pioneered by S. G. Thompson, involves coprecipitation of Pu(IV) by $BiPO_4$ followed by oxidation to Pu(VI) which is not coprecipitated by $BiPO_4$. This is a batch process and has the disadvantage of losing the uranium to the waste stream. Continuous solvent extraction processes were developed, one of which uses tributyl phosphate (TBP) as the extractant of U(VI) and Pu(IV) from nitric acid. In this PUREX process (which remains the principal method used internationally for reprocessing of spent reactor fuel), the Pu(IV) is reduced subsequently to Pu(III) and leaves the TBP phase, effecting the separation of the Pu and U [1].

In one research effort of the Manhattan Project, separation of pure fractions of the individual trivalent lanthanides was accomplished using elution from columns of cation exchange resin. The original citric acid eluant was soon replaced by aminopolycarboxylic acid eluants, including ethylenediaminetetraacetic acid (EDTA). The cation exchange resin-citrate elution system, however, survived as the separation method used in research on the discovery of the transplutonium elements from americium ($Z=95$) through fermium ($Z=100$). These elements exist normally in aqueous solution in the trivalent state and their elution sequence and relative separation factors mimic those of their lanthanide analogs. This allowed the researchers to predict with good accuracy the elution position to be expected for a new element. Appearance of new radioactivity at the predicted elution position has been the major initial chemical evidence for a new element. For the transuranium elements, greater separation factors were required and α-hydroxyisobutyric acid replaced citric acid as the eluant of choice. A comparison is shown in Fig. 1 of the Ln(III) and An(III) elution positions used to confirm the discovery of the element $_{101}$Md, in which a combined total of 17 atoms were observed in about a nine hour irradiation experiment [2].

In the following sections we describe techniques used at present in separations of actinide elements. Many of these are similar to, or are direct descendants of these early methods. The ease of change between

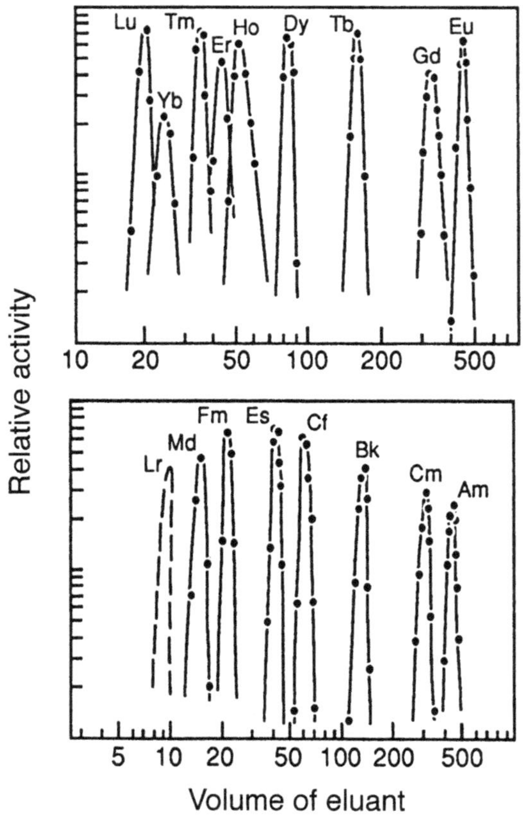

Fig. 1. Elution of Ln(III) and An(III) from a column of cation exchange resin by a solution of ammonium α-hydroxyisobutyrate at pH 4.0 [2].

the oxidation states found in the earlier members of the actinide elements and the stability of the trivalent state of the transplutonium members make it convenient to separate this review into two sections. The first section deals with separations utilizing differences in oxidation states and is concerned mainly with the elements of $Z=90$ to 95. The second section reviews separations of the trivalent ($Z=95-101$, 103) elements. Non-trivalent states can be utilized in separations of some of the transplutonium elements but this is a less common approach.

Redox separations

In this paper, separation of Pa is not reviewed as relatively little attention has been given to this element. For the elements beyond plutonium, except for nobelium ($Z=102$), the trivalent state is the most stable although Am(V) and Bk(IV) can be utilized in separations in basic systems. For thorium, IV is the only oxidation state of interest in aqueous separations. For U, Np and Pu, the redox chemistry is varied and the different oxidation states are of major use in separation schemes. The reduction potentials for various redox couples of these three elements are listed in Table 1 at pH 0 and pH 8. In general, acidic solutions favor lower oxidation states while basic media promote the stability of the higher states. For Np, the IV, V and VI states are rather easily achieved although Np(V) is the most

Table 1. Reduction potentials for U, Np and Pu

Element	pH	Redox couple ($E°$ in V)			
		III/IV	IV/V	V/VI	IV/VI
U	0	−0.61	0.61	0.063	0.34
	8	−1.95	0.08	−0.23	−0.07
Np	0	0.16	0.74	1.14	0.94
	8	−1.13	0.15	0.81	0.48
Pu	0	0.98	1.17	0.92	1.04
	8	−0.39	0.70	0.60	0.65

commonly encountered state in aqueous media. Plutonium is unusual as it is possible to have several of the common oxidation states (III, IV, V, VI) present simultaneously in a solution. For Pu, as expected, the higher oxidation states gain stability as the pH increases. The use of redox reactions is a major feature of plutonium solution chemistry.

Disproportionation reactions leading to several oxidation states in solution can play a significant role in the chemical behavior of U, Np and Pu. At the trace concentrations involved in many separations of environmental samples ($<10^{-6}$ M), disproportionation is an unlikely reaction as it is dependent on the square of the concentration of the disproportionating species and the fourth power of the H^+ concentration. The rates of the disproportionation reactions as well as the rates of many of the redox reactions of the actinides are influenced markedly by the chemical nature of the reactants and the products. For example, the change between An(IV) and An(VI) requires rearrangement of the primary coordination sphere of the actinide cations prior to electron transfer. This results in a large activation barrier, causing disproportionation reactions such as

$$3\,Pu^{4+} + 2\,H_2O \rightarrow 2\,Pu^{3+} + PuO_2^{2+} + 4\,H^+$$

to be slow. By contrast, the reaction

$$Pu^{3+} + PuO_2^{2+} \rightarrow Pu^{4+} + PuO_2^{+}$$

is faster, presumably because it involves only electron exchange with no change in the bonding in the two half reactions

$$Pu^{3+} \rightarrow Pu^{4+} \text{ and } PuO_2^{2+} \rightarrow PuO_2^{+}.$$

The different oxidation states of the actinides form complexes whose strengths vary (usually) in the sequence IV>VI>III≫V. It is possible to form complexes of the general formula MX_n where n may vary from 0 to 6 and is largest for An(IV) and smallest for An(V) species. This variation in n can be used with ion exchange chromatography and solvent extraction partition techniques to achieve separations of these actinide elements. An example of the use of these differences in complexation strengths involves separations using columns of anion exchange resin and solutions of hydrochloric acid [3]. Anionic chloro complexes

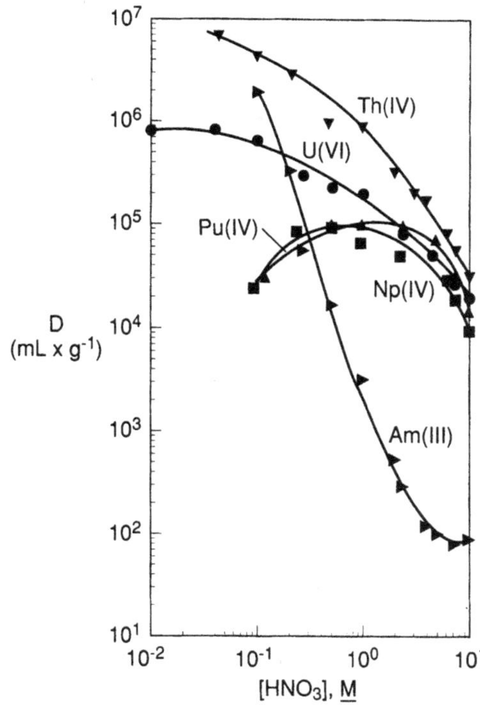

Fig. 2. Uptake of actinide ions by Diphonix resin as a function of $[HNO_3]$ [6].

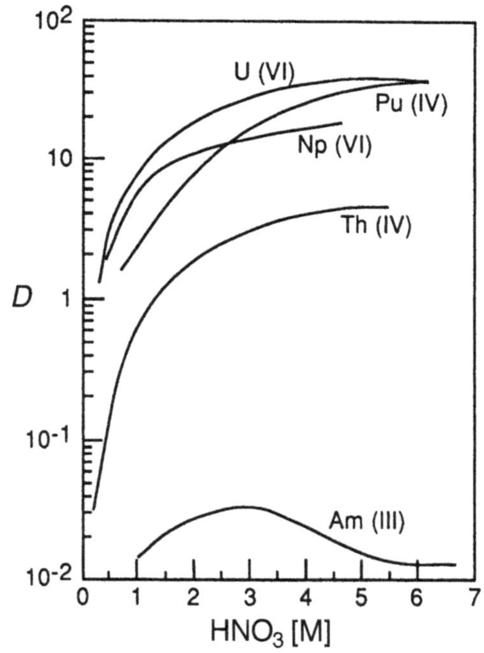

Fig. 3. Dependence of the extraction of several actinides with varying nitric acid concentrations and 30% (v/v) TBP in kerosene ($D = [An]_o/[An]_a$) [1].

such as $An^{VI}O_2Cl_3^-$ and $An^{IV}Cl_5^-$ sorb to the resin and are separated from non-complexed and cationic complexed species. The $An^{VI}O_2Cl_3^-$ species is sorbed strongly in 1 M HCl solution whereas the An(IV) species, requiring formation of at least $AnCl_5^-$, is sorbed only at concentrations of HCl above 2 M. The formation of chloro complexes by the An(III) and An(V) cations is too weak to allow appreciable sorption of anionic species even in 12 M HCl. The nitrate complexing is stronger, and both An(IV) and An(VI) species are sorbed strongly by anion exchange resin even in 1 M HNO_3 [4].

A new chelating ion exchange resin, Diphonix, combines a methylenediphosphonic acid chelating group with carboxylic and benzene sulfonic acid groups in a styrene-divinylbenzene matrix. This combination results in good metal ion kinetics and effective sorption of actinide metal ions in all oxidation states from moderate to strongly acid solutions [5]. Even from 10 M HNO_3, the hexavalent and tetravalent species are so strongly retained by the resin that they can only be removed by elution with a moderately concentrated solution of a structurally related diphosphonate chelating agent (1-hydroxyethane-1,1-diphosphonic acid, HEDPA). The distribution ratios for Am(III), U(VI), Pu(IV), Np(IV) and Th(IV) as a function of HNO_3 are shown in Fig. 2 [6]. The acid dependence for Am(III) uptake indicates normal cation exchange behavior while that for Th(IV) and U(VI) has been interpreted in terms of coordination of these cations via the phosphoryl oxygens of the fully protonated methylenediphosphonate groups.

In solvent extraction, complexation can be used to enhance or to retard extraction of cations. Trioctylphosphine oxide, TOPO, is a neutral extractant and forms extracted species with the general formula $AnX_n \cdot (TOPO)_i$ (where $i = 1-3$). In such a system, formation of the neutral species, AnX_n, in the aqueous phase can be considered as a prerequisite to extraction. The extraction of actinides in different oxidation states by 0.1 M TOPO in cyclohexane as a function of the concentration of HCl agrees with the elution patterns with 6 M HCl from anion resin for oxidation states, i.e., V < III < IV < VI.

The use of extraction with tributylphosphate in commercial reprocessing has been mentioned earlier. The variation of the extraction coefficient for actinides in the III, IV and VI oxidation states with nitric acid concentration is shown in Fig. 3 [1]. In 4 − 8 M HNO_3, uranium exists as U(VI), plutonium as Pu(IV) and americium as Am(III), with large differences in *D* between each of these elements. In the PUREX process, following dissolution of the fuel elements, the nitric acid solution (4 − 6 M HNO_3) is contacted with a solution of TBP in kerosene into which $UO_2(NO_3)_2(TBP)_2$ and $Pu(NO_3)_4(TBP)_2$ are extracted. The organic solution is contacted with a reducing (Fe^{2+} or U^{4+}) nitric acid solution which causes the plutonium to be stripped to the aqueous phase as Pu(III). Subsequently, the uranium is stripped (without redox) into a dilute HNO_3 aqueous solution [1]. Am(III) is extracted to a negligible extent from nitric acid.

To compensate for the lack of extraction of trivalent actinides in the PUREX process, a new extractant system was developed in the 1980's which has the capability of extracting actinides in the tri-, tetra-, and hexavalent oxidation states from nitric or hydrochloric acid media. The extractant of choice is octyl(phenyl)-*N,N*-diisobutylcarbamoylmethylphosphine oxide (CMPO)

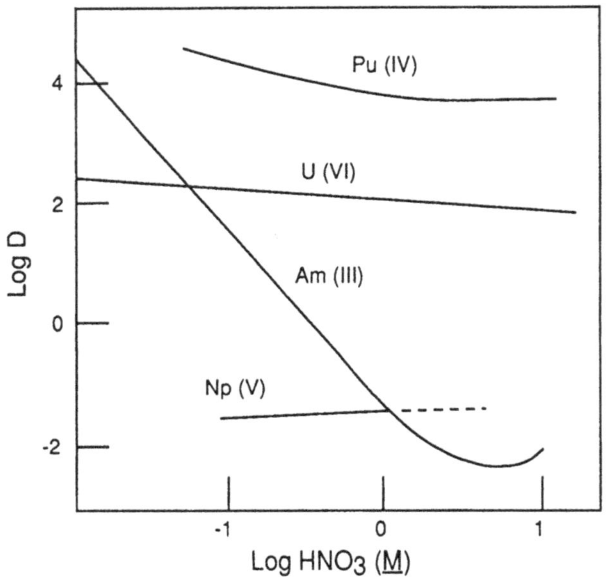

Fig. 4. Dependence of Pu(IV), U(VI), Am(III) and Np(V) extraction on nitric acid concentration using 0.5 M HDEHP in isooctane [9].

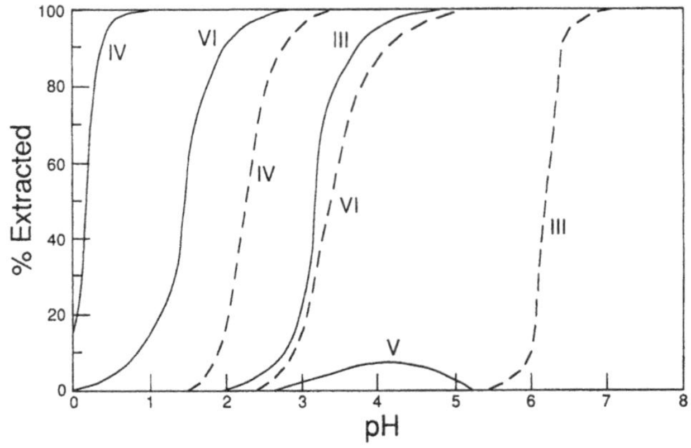

Fig. 5. The percent extraction by 0.5 M HTTA (solid curves) and 0.2 M DBM (dashed curves) in xylene or benzene, respectively, as a function of aqueous phase pH for An(III), An(IV), AnO_2^+ and AnO_2^{2+} cations [10–12].

and the process based on its use is the TRUEX process [7]. The TRUEX process solvent consists of 0.2 M CMPO and 1.2 M TBP in a normal paraffinic hydrocarbon diluent. Because this solvent can simultaneously extract the actinides in the important process oxidation states, it could have significant impact in the reduction of the volume of high-level waste requiring burial in a geological repository. An oxidation state selective separation of actinides based on the use of TRUEX extraction and a diphosphonic acid complexant has been proposed [8].

Solvent extraction systems utilizing acidic extractants also provide excellent separations. Fig. 4 shows the variation in extraction coefficient with nitric acid concentration for one such extractant, diethylhexylphosphoric acid or HDEHP, which is present in the diluent as a dimer [9]. The decrease in the value of D with increasing HNO_3 concentration is consistent with the reaction

$$An^{n+}_{(a)} + n(HDEHP)_{2(o)} \leftrightarrow An(DEHP)_n(HDEHP)_{n(o)} + nH^+_{(a)}.$$

Another class of cationic extractants are the β-diketones such as thenoyltrifluoroacetone (TTA). At pH 0, TTA solutions extract Pu(IV) but not Am(III), Np(V) or U(VI). However, at pH 3–4, the III, IV and VI species are extracted strongly. Other β-diketones which are useful extractants are dibenzoylmethane (DBM) and 1-phenyl-3-methylpyrazol-5-one (PMBP). DBM is useful at pH>4 as it is less soluble in water than TTA at neutral and basic pH's. By contrast, PMBP provides better extraction from strongly acidic solutions than does TTA. Fig. 5 compares the relation of percent extraction and pH for TTA and DBM solutions for different oxidation states [10–12].

The use of alkaline solutions to effect separations of the actinide elements has been reviewed by Russian scientists [13]. Alkylpyrocatechols such as (α,α-dioctylethyl)pyrocatechol, DOP, have been shown to be useful actinide extractants from aqueous solutions in which complexants suppress hydrolysis of the cations. Generally, for An(III)/Ln(III) systems, the extraction of the An(III) cations by DOP increases with increasing pH while that of the Ln(III) cations decreases. The largest separation factor reported for the Eu/Am pair in such a system was 70. Amines such as Aliquat 336 can also be used, but the separation factors between trivalent cations are usually less than four. Generally, in these amine systems the extraction order follows the sequence: An(III)>An(IV)>An(V)>An(VI). The separation factors between different oxidation states are quite useful (e.g., for Am(III)/U(VI), 100; for Am(III)/Pu(VI), 75). In contrast, the DOP extraction systems have the sequence: An(IV)>An(III)>An(V)>An(VI). The ability to oxidize some transplutonium actinides in alkaline solutions has provided excellent separations. For example, in a mixture of Am(III) and Cm(III), the Am(III) can be oxidized by chemical or electrolytic means to Am(VI) at pH 10. In a solution of 0.5 M $Na_4P_2O_7$, when contacted with an organic phase of PMBP in chloroform, the americium reduces to Am(V) and remains in the aqueous phase while the Cm(III) extracts. A separation factor of about 10^3 is reported [14]. Similar methods have been used to separate Bk(IV) from trivalent actinides. For Bk(IV), separations can be achieved in acid solutions using TOPO as well as TBP. Fig. 6 shows the variation of log D for actinides in several oxidation states as a function of NaOH concentration using Aliquat 336 in cyclohexane as the extractant [15].

Polyethylene glycols have been demonstrated to produce two-phase systems (called aqueous biphases) when equilibrated with concentrated salt solutions of water-structuring anions. The most unique aspect of these systems is that the immiscible phases are largely

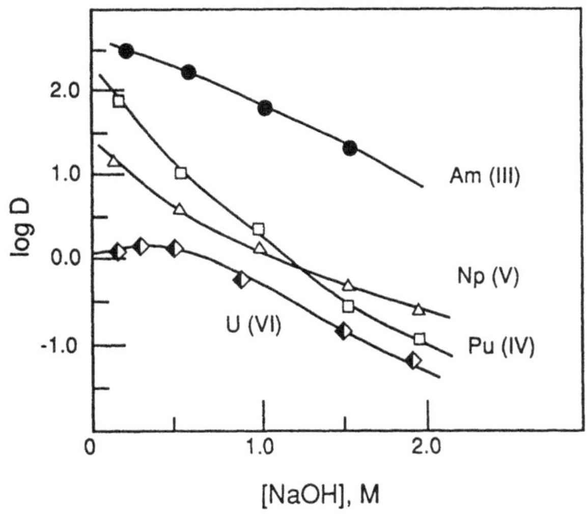

Fig. 6. The dependence of extraction of Am(III), Pu(IV), Np(V) and UO_2^{2+} on NaOH concentration with Aliquat 336 in cyclohexane; the aqueous phase also contained 0.002 M dihydroxydiaminebutyltetraacetic acid [15].

aqueous in nature. Actinide separations in these systems have been reviewed [16] with Myasoedov and coworkers reporting conditions for the separation of transplutonium elements from uranium, thorium and the lanthanides [17]. The polyethylene glycol phase itself does not extract actinides strongly, but Arsenazo III is a highly effective carrier for these metal ions. A method for preparation of an isotope generator for ^{239}Np from ^{243}Am using an ammonium sulfate solution containing potassium phosphotungstate has also been reported [18].

Oxidation state speciation

A major question of concern over the possible release of plutonium to the environment is the speciation of plutonium within its four possible oxidation states. Such speciation depends on the pH, the E_h, the presence of various complexing agents, the concentration of plutonium, biological action, etc. Several procedures have been reported for separating plutonium into its different oxidation states without, hopefully, shifting the redox equilibria during the separation. As noted earlier, redox is generally fast between the III and the IV states and the V and VI states but slower for (III, IV) ↔ (V, VI) reactions.

This difference in redox rates is utilized in methods in which Pu(III + IV) are removed from solution by precipitation with LaF_3, leaving Pu(V + VI) in the solution. However, there are some problems with this technique as some reduction can occur during the precipitation if it is carried out in an excess of fluoride. The limitations of this method can be overcome if Pu(IV + VI) are sorbed from basic solutions onto silica gel, leaving Pu(V) in solution [19, 20]. Alternatively, Pu(IV + V) can be sorbed on $CaCO_3$ from solutions of pH 9 (Pu(VI) remains in solution in this case) [20]. These sorption separations can be complemented by LaF_3 precipitation to obtain the oxidation state speciation distribution. Another method uses sorption of Pu(V) onto TiO_2 solid at pH 9 following removal of Pu(III + IV) from the solution by precipitation with $BiPO_4$ [21].

Oxidation state separations can be achieved simply and rather cleanly with solvent extraction. Np(IV), Np(V) and Np(VI) were separated by 40% (v/v) TBP in benzene. From 3 M HCl solution in 5–10 minutes, 90% extraction of Np(VI) occurred with less than 5% of Np(IV) or Np(V) coextracted [19]. TTA solutions have been proposed for Pu redox speciation [11] since a 0.5 M TTA solution in toluene extracts Pu(IV) to 100% from a pH 0.6 solution in which Pu(V) and Pu(VI) are retained quantitatively in the aqueous solution. Adjustment of the pH to 4 with acetate allows separation of the Pu(VI) completely from the Pu(V) (which remains totally in the aqueous phase). In order to minimize any redox change which might occur prior to separation by increased acidity, a separation scheme for neutral and basic aqueous phases using DBM solutions has been developed [12]. Again, quite complete separation of actinides in the III, IV, V and VI oxidation states is achieved.

Separations of the trivalent cations

The trivalent oxidation state is the thermodynamically stable state in solution for the transplutonium actinides. Like their lanthanide congeners, the radii of the trivalent actinide cations contract with increasing atomic number. Moreover, the radii of the trivalent actinide series overlap those of the lanthanides (e.g., $r_{Am} \approx r_{Nd}$). This similarity in radii and the strong similarity in chemical behavior between the two families pose moderately difficult problems in the separation of individual actinide ions, and in the separation of the transplutonium actinides from the lanthanides.

Separation of the individual actinide elements utilizes the small differences in their radii. By contrast, lanthanide/actinide separations more often depend upon the slightly stronger interaction of the trivalent actinides with ligands containing soft donor bases (S, Cl or N). This difference formed the basis of several separation techniques developed between 1945 and 1960, some of which still constitute the "state-of-the-art" in the field, and continues to be a principal focus in development of new separation systems.

Separations of trivalent actinides from lanthanides

Whether in reactors or accelerators, production of actinides is always accompanied by formation of trivalent lanthanide cations in at least comparable amounts. The separation of An(III) from Ln(III) was first achieved by cation exchange from concentrated chloride media. Street and Seaborg [22], Diamond *et al.* [23], and later, Choppin and Chetham-Strode [24] showed that the elution from cation exchange resin using HCl solutions

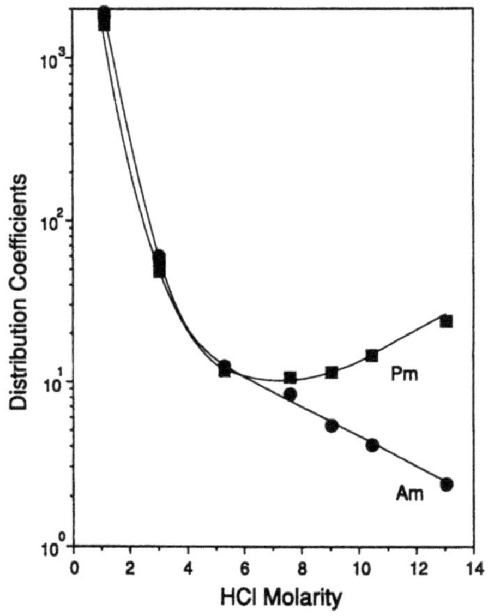

Fig. 7. Variation of the distribution coefficients from Dowex 50 cation exchange resin of Am and Pm as a function of [HCl] [23].

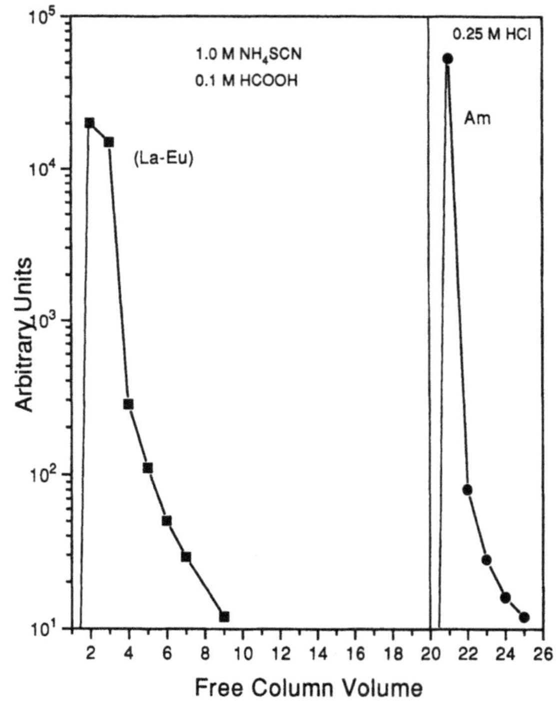

Fig. 8. Lanthanide/Americium separation using a quaternary amine extraction chromatographic resin and aqueous solutions of SCN^-/formate and HCl [35].

provided no separation of the two families below 6 M HCl, but increasing separation occurred as the HCl concentration increased from 6 to 12 M (Fig. 7) [23]. Even better separation efficiency was obtained using salt solutions (e.g., LiCl) or HCl dissolved in an alcohol-water solvent [25]. Choppin *et al.* [2] reported larger actinide/lanthanide separation factors for elution from anion resin with 10 M LiCl aqueous solutions. The increase in separation in these media coincides with formation of polychloro inner sphere complexes. Surls and Choppin [26] reported similar results with 2 M NH_4SCN as eluant. The LiCl anion exchange process is used currently for actinide/lanthanide separation in actinide production [27].

An efficient separation of trivalent actinides from all matrix elements (lanthanides and other fission products) with both cation and anion exchange resins was achieved using H_2O-C_2H_5OH solutions of sulfuric acid [28, 29]. Usuda and coworkers [30, 31] have proposed a separation scheme for trivalent actinides using a three step ion exchange partition from light actinides and fission products.

Quaternary ammonium extractants are analogous to tetraalkylammonium based anion exchange resins. The first application of quaternary amines to lanthanide/actinide group separations was by Moore [32], with the system Aliquat 336/xylene using relatively low concentrations of salts. Other applications of the method have been summarized by Weaver [33]. The Aliquat class of extractants has been immobilized on an inert support to create extraction chromatographic materials which combine the best features of the extractants with the increased efficiency of extraction chromatography [34]. With 1.0 M NH_4SCN/0.1 M HCO_2H solution, An(III) cations are sorbed quantitatively while Ln(III) ions are not retained. The An(III) can be eluted with 0.25 M HCl, Fig. 8 [35].

The extraction strength of the class of compounds collectively known as bifunctional extractants (CMP's and CMPO's) does not lend itself readily to selective separations. However, there has been at least one report of the use of such bifunctional extractants for separation of Ln(III) from An(III). Muscatello *et al.* [36] reported that dilute NH_4SCN with 0.244 M DHDECMP (dihexyldiethylcarbamoylmethylphosphonate) resulted in americium/europium separation factors as high as 10.8.

A more efficient, but considerably more complex, approach to actinide/lanthanide separation was developed by Weaver and Kappelmann [37]. For a 1 M carboxylic acid (pH 1.8) solution, HDEHP extraction of Am is depressed relative to that for the lanthanides. At higher pH's (≈ 3) the separation factors are increased, with the most consistent enhancement observed for lactic acid. Addition of 0.05 M diethylenetriamine-*N,N,N′,N″,N″*-pentaacetic acid (DTPA) to the solutions of carboxylic acids at pH 3 resulted in separation factors for Ln(III) from Am(III) and Cm(III) of greater than 100. While the extracted species may have one bound lactate, the DTPA complexes remain in the aqueous phase. This separation process has been termed TALSPEAK (Trivalent Actinide Lanthanide Separation by Phosphorus Reagent Extraction from Aqueous Komplexes).

A number of reports have discussed applications, process experience and modifications of TALSPEAK. Kosyakov and Yerin [38] report that, for Cm extraction with TALSPEAK-type aqueous solutions using HDEHP/decane solutions, both extractant (HDEHP) and acid dependencies are reduced to 2, indicating the

Table 2. Europium/Americium separation factors for a series of acidic extractants [41]

Extractant	Diluent	$S_{Eu/Am}$
dibutylphosphoric acid	$CHCl_3$	22.9
dioctylphosphoric acid	$CHCl_3$	14.1
1-phenyl-3-methyl-4-acetyl-pyrazolone	$CHCl_3$	3.47
thenoyltrifluoroacetone	$CHCl_3$	3.02
neocupferron	$CHCl_3$	1.74
N-benzoylphenylhydroxylamine	$CHCl_3$	1.66
N-2,4-dichlorobenzoyl-phenylhydroxylamine	$CHCl_3$	1.32
β-isopropyltropolone	$CHCl_3$	0.98
1-hydroxy-2-naphthoic acid	hexone	1.07
2-hydroxy-1-naphthoic acid	hexone	1.02
3-hydroxy-2-naphthoic acid	hexone	0.95
5,7-dichloroxine	$CHCl_3$	0.10

extraction of the $AnLac^{2+}$ complex in the aliphatic diluent. Similar extractant dependencies were reported for Bk, Eu and Ce. Bond and Leuze [39] describe a lanthanide/actinide removal and separation scheme using a 1 M lactic acid-0.05 M DTPA strip to separate trivalent actinides from the lanthanides which is adaptable to liquid membrane separations, as illustrated in a report by Novikov and Myasoedov [40]. They report the use of a "supported liquid membrane" impregnated with HDEHP for the separation of Am/Cm, Eu/Tb and Am/Eu using DTPA, citric acid and the potassium salt of a heteropolyacid as aqueous complexants. The optimum separation factors reported are $S_{Cm/Am} = 5.0$, $S_{Tb/Eu} = 10.8$ and $S_{Am/Eu} > 100$.

An alternative approach is to consecutively apply two unrelated separations processes. Sekine and Dyrssen reported Eu/Am separation factors for 12 acidic extractants, and Table 2 is adapted from their work [41]. Based on these separation factors, a counter-current separation procedure was proposed which initially involves extraction by dibutyl phosphoric acid (HDBP) to remove Am preferentially, followed by contact of the aqueous phase with 5,7-dichloroxine to extract Eu. When HDEHP is substituted for HDBP, Kasting *et al.* [42] found that such a separation procedure could result in actinide/lanthanide separation factors of several thousand.

French researchers have been active in the use of soft donor extractants and complexants for An/Ln group separations [43–45]. Musikas showed that good Am/Eu separation is obtained with hydrazoic acid as a result of enhanced retention of Am(III) in the aqueous phase due to complexation with the azide anion. Choppin and Barber [46] found that An(III)-azide stability constants are somewhat larger than those for Ln(III); however, the enthalpies for complex formation do not seem to support a model of significantly enhanced covalent bonding contribution in the actinide complexes. The question of the role of covalency effects with soft donors in An(III) vs. Ln(III) complexation may be, at least partially, a matter of definition and deserves further study.

Table 3. Distribution ratios and separation factors for americium and europium extraction by 0.0297 M 4-benzoyl-2,4-dihydro-5-methyl-2-phenyl-3H-pyrazol-3-thione in toluene from 1.0 M $NaClO_4$ as a function of DPPHEN (synergist) [51]

[DPPHEN]	D_{Am}	D_{Eu}	$S_{Am/Eu}$
0.00269	25.3	0.138	183
0.00215	21.9	0.112	196
0.00144	14.8	0.077	192
0.00108	10.3	0.059	174
0.000718	6.1	0.039	156
0.000359	2.7	0.021	129

A soft donor extractant system of *o*-phenanthroline + nonanoic acid [45] extracts Am by an order of magnitude more strongly than Eu from 0.1 M $NaNO_3$ solutions at pH 4.5–5.1 ($S_{Eu/Am} = 17.4(\pm 0.9)$). To accomplish the separation at higher acidity, attention has been directed to the complexant/extractant 2,4,6-tris(2-pyridyl)-1,3,5-triazine (TPTZ), used with carboxylate and sulfonate coextractants to offset the hydrophilicity of the $Am(NO_3)_3 \cdot TPTZ$ complex.

Extractant molecules which incorporate S as a donor atom are often plagued by poor stability when contacted with acidic (particularly nitric) solutions [45]. Reports of good actinide/lanthanide separation factors for solvent extraction by thio derivatives of HDEHP were dismissed by Freiser [47] as resulting from hydrolysis of the extractant to produce the oxygenated derivative. In general, *f* elements are poorly extracted by simple thio extractants. Recent results by Jarvinen *et al.* [48] indicate moderate separation factors for Am from Eu using very pure dithiophosphinic acid extractants, Cyanex 301, dicyclohexyldithiophosphinic acid and diphenyldithiophosphinic acids.

Other classes of sulfur donor extractants appear to be more resistant to hydrolysis, and have demonstrated some potentially useful selectivity for An(III) over Ln(III); e.g., thiopyrazolones. The mixture of 0.3 M BMPPT (4-benzoyl-2,4-dihydro-5-methyl-2-phenyl-3H-pyrazol-3-one)/0.01 M TOPO/benzene preferentially extracts Am(III) from 0.1 M $LiClO_4$ at pH 3 ($S_{Eu/Am} = 68$) [49]. The analogous system based on the oxygen donor analog and TOPO [50] gives stronger extraction but no significant separation of curium from europium. Substitution of the soft donor synergist 4,7-diphenyl-1,10-phenanthroline (DPPHEN) for TOPO [51] results in even greater selectivity for americium ($S_{Eu/Am} = 190$). The extracted species is $M(BMPPT)_3(DPPHEN)$ (Table 3). This is the only known example of a system which contains soft donor atoms in both the primary extractant and in the synergist. Choppin *et al.* [52] have reported on the separation of Am from Eu using various combinations of thiothenoyltrifluoroacetone, tributylphosphate, tributylphosphine sulfide and *N,N*-dimethyl-*N,N'*-dihexyl-3-oxapentanediamide as coextractant ligands in a synergistic extraction system. The soft donor ligands show little enhancement of Am/Eu separation factors.

Table 4. Elution of transplutonium elements from Dowex 50 cation exchange resin using carboxylic acids, 87 °C, pH = 3.0–4.5, 2 minutes/drop, 2 mm × 10–20 mm column [2, 54]

Carboxylic acid	Retention time (drop number)					
	Am	Cm	Bk	Cf	Es	Fm
0.25 M ammonium citrate	94.0	80.8	50.0	38.3	32.5	26.7
0.4 M ammonium lactate	58.5	49.0	33.0	22.0	18.0	13.6

Separations of individual trivalent actinides

The key parameter for the isolation of individual An(III) cations is the steady decrease in the radius with increasing atomic number. It is important to note that considerable variation is observed in the relative stability of trivalent lanthanide/actinide complexes as a function of cation radius. In general, linearity in correlations of stability constants and cation radii of either 4*f* or 5*f* elements are not observed [53]. The nonlinear trends suggest that variations in the dehydration of the cation in its complex makes a substantial contribution to the relative stability of aqueous complexes.

The earliest separation processes for individual trivalent ions were based on inorganic ion exchangers which demonstrated separation factors for adjacent ions of 1.01 to 1.05, barely acceptable for chromatographic separations using large columns. This situation was only slightly improved with the development of polymeric organic ion exchange materials. However, the introduction of water-soluble chelating agents led to the development of the first efficient separations methods for isolation of transplutonium elements [54]. Am, Cm, Bk, Cf, Es and Fm were eluted in reverse order from Dowex 50 cation exchange resin when the eluting solution was 0.25 M ammonium citrate at pH 3.0–3.5 or 0.4 M ammonium lactate at pH 4.0–4.5 (Table 4).

These observations led to the development of the eluant α-hydroxyisobutyric acid which provides average separation factors for adjacent lanthanides or trivalent actinides of about 1.3–1.5. The kinetics of the equilibration on the Dowex 50 column with α-hydroxyisobutyrate is superior to that of ligands such as the aminopolycarboxylates which had been employed previously. This "BUTT" elution method is still acknowledged as the most effective ion exchange separation method for *f* elements. A unique characteristic of this system is that separation is achieved almost solely as a result of the difference in stability of the aqueous complexes of the cations with α-hydroxyisobutyric acid since the α-hydroxyisobutyrate complexes of the trivalent *f* elements exhibit the most consistent trends across the *f* series of any known carboxylate complexants.

Early studies of the extraction of actinides (and lanthanides) by TTA demonstrated one of the first solvent extraction methods for the isolation of individual

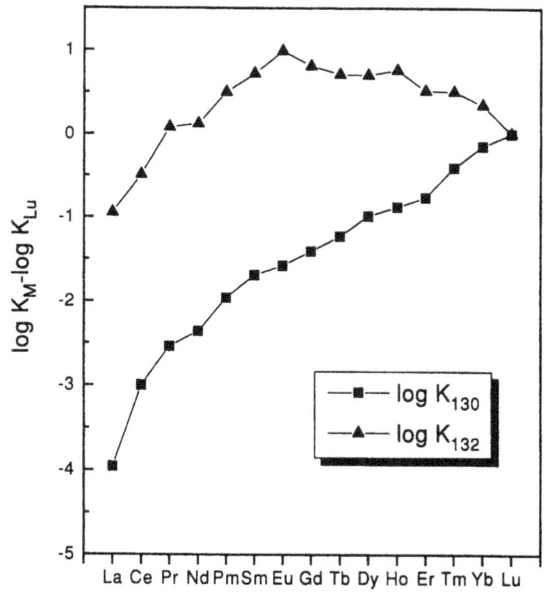

Fig. 9. Normalized extraction equilibrium constants for lanthanide extraction of $Ln(TTA)_{3n}$ (log K_{130}) and $Ln(TTA)_3(TBP)_2$ (log K_{132}) by HTTA (■) and TBP (▲) [57].

trivalent actinide ions [55]. The separation factors for adjacent ions are slightly greater than those observed in a BUTT column, but the separation factors are less consistent for adjacent ions. TTA has found frequent application in studies of synergistic solvent extraction of lanthanides and actinides because the stoichiometry is usually consistent and predictable and the synergistic adduct (e.g., TBP, TOPO) completes the dehydration of the actinide cation. The need to dehydrate, fully or partially, the metal ion before extraction into the organic phase is an important factor in extractant success.

The literature on trivalent lanthanide/actinide separations by synergistic systems has been reviewed by Mathur [56]. A common feature of synergistic extraction systems is increased extraction strength, generally at the expense of selectivity. This can be illustrated by considering the results of Farbu *et al.* [57] for the extraction of lanthanides from xylene solutions of TTA and TTA/TBP mixtures. The equilibrium constants for extraction of $Ln(TTA)_3(H_2O)_n$ (n = 2, 3) (normalized to Lu^{3+}) are presented in Fig. 9 and show an approximately linear correlation with lanthanide atomic number [57]. By contrast, for extraction of $Ln(TTA)_3 \cdot (TBP)_2$, the separation factors vary with the value of Z(Ln) for the light lanthanides but not for the heavier ones. With or without a synergist, the trivalent actinides show no consistent trend relating the separation factors to the value of Z(An) [58]. Larger synergistic effects are observed for stronger synergistic ligands (like phosphine oxides), but at the expense of reduced separation factors.

An interesting report of synergism in lanthanide separations using LIX 51 (a fluorinated β-diketone), LIX 54 (p-dodecylbenzoylacetone) and DBM as primary extractants, and TBP or TOPO as synergists indi-

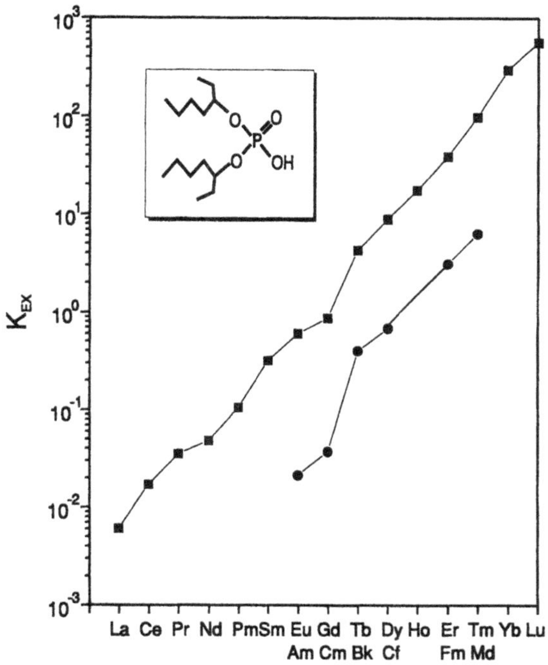

Fig. 10. Equilibrium coefficients for lanthanide/actinide extraction by HDEHP in toluene [55].

Table 5. Separation factors (relative to Am) for solvent extraction of trivalent lanthanides and actinides with 0.817 M DHDECMP in diisopropyl benzene/1.0 M HNO_3 [68]

Actinides	$S_{Am/R}$	$S_{R'/R}$	Lanthanides	$S_{Am/R}$	$S_{R'/R}$
Fm	0.575		Lu	0.048	
		1.04			1.41
Es	0.595		Yb	0.067	
		1.15			1.34
Cf	0.685		Tm	0.090	
		1.05			1.15
Bk	0.719		Y	0.103	
		0.90			1.29
Cm	0.649		Er	0.133	
		1.54			1.40
Am	1.00		Ho	0.186	
					1.43
			Dy	0.266	
					1.32
			Tb	0.351	
					1.15
			Gd	0.403	
					1.49
			Eu	0.602	
					1.22
			Sm	0.735	
					1.16
			Pm	0.855	
					1.12
			Nd	0.962	
					1.26
			Pr	1.21	
					1.12
			Ce	1.36	
					1.04
			La	1.41	

cates mean separation factors for adjacent lanthanides of 1.7 to 2.3 [59]. The use of this system for actinide separations has not been reported.

Extensive investigations by Chmutova and coworkers of PMBP show that the synergistic enhancement for TBP and TOPO adducts (in benzene) of the trivalent actinides (extracted from HNO_3) increase in the order Am<Cm<Bk<Cf [60]. On the other hand, Khopkar and Mathur [61] do not report any particular order in the extraction of these metal ions from hydrochloric acid solutions into the same extractant mixtures in xylene. Continued investigations by Mathur and Khopkar [62, 63] and by Dukov and Genov [64, 65] (Ln(IV) only) showed a general increase in extraction but few useful separation systems.

Extraction of metal ions by pyrazolones and a related extractant, *N*-benzoyl-*N*-phenylhydroxylamine, have been studied by Mathur and Khopkar [58, 63, 66]. The principal difference between the two extractants lies in the formation of 6-membered rings in the pyrazolones vs. 5-membered rings for the complexes with the hydroxylamine extractant. The distribution data for the pyrazolones indicates some potential for intragroup separations, but successful lanthanide/actinide separation using pyrazolones alone appears unlikely.

Some selectivity for light/heavy (i.e., light lanthanides from heavy lanthanides or light actinides from heavy actinides) intragroup separations may be derived from the tendency of the light members of the two series to form self-adducts. This self-adduct results from coordination of a fourth ligand without ionization to form an extracted complex $M(Ex)_3(HEx)$ with 1:4 metal to extractant stoichiometry. Presumably, the light/heavy fractionation results from steric restriction caused by the shrinking radii of the heavier members of the respective series. An extraction chromatographic method for the separation of Bk from the trivalent actinides, lanthanides and inactive impurities using PMBP has been suggested [62].

One of the principal commercial methods for isolation of individual lanthanides is solvent extraction with HDEHP, a reagent first reported in 1957 [67] (Fig. 10). Adjacent lanthanide separation factors are, on average, nearly twice as high for HDEHP separations than for α-hydroxyisobutyrate cation exchange. Trivalent actinides are similarly well separated by this reagent.

The neutral bifunctional extractants (CMP's, CMPO's and related compounds) have been applied to some cation size selective separations. Horwitz *et al.* [68] have studied the separation of the lanthanides and trivalent actinides from Am to Fm (Table 5) using dihexyl-*N,N*-diethylcarbamoylmethylphosphonate (DHDECMP) and aqueous nitrate solutions. Steadily decreasing distribution ratios are observed for the lanthanides, but nearly constant *D*'s are found for the trivalent actinides. These results suggest possible interlanthanide (but probably not interactinide) separations. There have been two additional reports on the application of CMP's to intragroup separation schemes

[69, 70]. Alkylene phosphine dioxides [71] have been shown to exhibit some ability to enhance the separation of californium from americium ($S_{Cf/Am} > 25$), and berkelium from curium ($S_{Bk/Cm} > 10$). The adaptation of CMPO's to extraction chromatography may result in development of additional selective separations methods.

Finally, ligands designed for size selective metal ion coordination, the crown ethers and related species, are being studied in a number of laboratories. There are a number of reports of the use of crown ether extractants alone for lanthanide separations and others on the use of acyclic polyethers in cooperation with coextractant molecules. Some of these discuss trivalent actinide extraction. Samy *et al.* [72] investigated the extraction of lanthanides from 1 M lithium trichloroacetate (pH 3–4.5) with 0.01 M 18-crown-6 in 1,2-dichloroethane, and reported that separation factors for the lanthanides from La to Tb exceed those reported for HDEHP extraction. Separation factors for the lanthanides heavier than Tb are much smaller, possibly indicating a poor fit of the lanthanide in the crown. Tang and Wai [73] report intragroup separations of the lanthanides lighter than Tb using sym-benzo-18-crown-5-oxyacetic acid in 80% chloroform-20% heptanol.

Despite the nominal good match between the cation radii and the ligand cavity, recent NMR data [74] indicate that only three of the six oxygens are bound strongly to lanthanum in aqueous solutions. Separations using the 18-crown-6 ligand are limited to its aqueous solubility. Alterations of the basic ligand structure have been made to improve the hydrophobicity of the derivatives and their complexes.

The work of Mathur and Khopkar [75] on the use of benzo-15-crown-5 and dicyclohexano-18-crown-6 as synergists for lanthanide and actinide extraction with 1-phenyl-3-methyl-4-trifluoroacetyl-pyrazol-5-one (PMPFT)/chloroform suggest some potential for such mixtures to accomplish certain specific group and inter-actinide separations. Aly *et al.* [76] found significant enhancement of the extraction, but little separation with the synergistic system of TTA and 15-crown-5. In both studies, 1:3:2 (M:TTA:CE) adducts are indicated for the 15-crown-5 synergists while Mathur and Khopkar report 1:3:1 adducts for the 18-crown-6 synergistic reaction.

There is evidence that non-cyclic polyethers can function as synergists in lanthanide/actinide separations. Ensor and Shah [77, 78] report that 1,13-bis[8-quinolyl]-1,4,7,10,13-pentaoxatridecane (Kryptofix-5, or K-5) enhances the extraction of Ce(III), Eu(III), Tm(III), Am(III), Cm(III) and Bk(III) into chloroform solutions of TTA. The extracted complex has the stoichiometry $M(TTA)_3 \cdot K\text{-}5$, and the synergistic enhancement is comparable to that of TBP under the conditions studied. Intragroup separation factors for lanthanides and actinides are approximately 2, while the usual synergistic leveling effect is observed for Am/Eu separations (i.e., the separation factors are poorer than for TTA alone).

Future directions

With the end of the cold war, the demand for large-scale separation of purified transuranic elements has largely disappeared. The legacy of 50+ years of actinide production for nuclear weapons is a large stockpile of fissile metals (which could be used in nuclear reactors to generate electricity) and an immense volume of radioactive chemical wastes confined in storage tanks. The result has been a fundamental shift in the motivation for continued development of separations methods, and in the types of separations which are needed.

New separations methods must address the additional demands brought on by the need to adapt to the widely varying materials requiring treatment. For example, waste storage tanks left over from PUREX processing contain a variety of salt solutions, partially decomposed chelating agents, and conditions ranging from moderately acidic to strongly alkaline. Most such tanks contain mixtures of transuranic elements, fission products and substantial concentrations of non-radioactive materials. The chemical composition in the waste tanks varies with each tank individually, making the treatment of each a unique separations challenge. The principal obstacle to minimization of the volume (and hence the cost) of wastes requiring disposal in a geological repository is segregation of the radioactive components from the non-radioactive bulk. Hence, methods which are highly specific for actinides (in all oxidation states) over iron, zirconium and aluminum (which are common metallic constituents of the wastes) are desirable. Because the contents of many of the storage tanks are alkaline, selective separations methods which exhibit acceptable kinetics and operate under alkaline conditions would be useful. Reliable separations schemes for identification of actinide ions in various oxidation states are needed for environmental monitoring and speciation of actinides and for remediation of contaminated material, buildings, grounds, etc. An important characteristic of any new reagent for rad-waste treatment is compatibility with the target waste form. A number of elements such as phosphorus and chromium are undesirable in wastes to be immobilized in glass for permanent disposal.

Separation of the trivalent ions (of lanthanides from actinides and of individual actinides) is presently important only for the production of pure transplutonium elements for research purposes. If "actinide burnup" strategies are pursued for destruction of transuranium elements in a nuclear reactor or an accelerator, large scale separations methods for lanthanides from actinides will be required. Since the separations of these ions are based on the regular change in the cation radii across the series and/or in the stronger interaction of actinides with soft donor ligands, enhancing size-selective complexation and developing ligands (or combinations of ligands) with the optimum mix of hard and soft donor groups seem to be the key factors required for more efficient separation processes. Developing such improved processes for the separation of ions with such similar properties demands in-depth understanding of the basic solution chemistry associated with complex formation of these ions.

Metal ion complexation reactions in both the extractant/resin phase and the aqueous phase of separation systems can be manipulated to achieve improved separations; hence, the thermodynamics and kinetics of complexation reactions in polar and nonpolar media are important. The relative unpredictability of solvation effects suggests that basic research into the energetics of solvation reactions (in aqueous, nonaqueous and polyelectrolyte media) could have significant impact in the development of improved separation processes. In particular, the relationship between total system entropy, solvation of metals, ligands and complexes and the number, character and spatial orientation of ligand donor atoms are important. Advances in the basic science of actinide solution chemistry will undoubtedly result in the developent of more efficient separation processes.

Acknowledgment

Preparation of this paper was assisted by grants from the USDOE-OBES Division of Chemical Sciences to Argonne National Laboratory and to Florida State University. The authors wish to thank Andrew H. Bond for assistance in preparing the final version of this manuscript.

References

1. Choppin, G. R., Rydberg, J.: *Nuclear Chemistry: Theory and Applications,* Pergamon Press, Oxford 1980, p. 257.
2. Choppin, G. R., Thompson, S. G., Harvey, B. G., Seaborg, G. T.: J. Am. Chem. Soc. **76**, 6229 (1954).
3. Marcus, Y.: Report # Pub/UP/R-20, Israel Atomic Energy Commission (1960).
4. Buchanan, R. F., Faris, J. P.: IAEA Conference on Use of Radioisotopes in the Physical Sciences and Industry, International Atomic Energy Agency (1960).
5. Chiarizia, R., Horwitz, E. P., Alexandratos, S. D.: Solvent Extr. Ion Exch. **12**, 211 (1994).
6. Horwitz, E. P., Chiarizia, R., Diamond, H., Gatrone, R. C., Alexandratos, S. D., Trochimczuk, A. Q., Crick, D. W.: Solvent Extr. Ion Exch. **11**, 943 (1993).
7. Schulz, W. W., Horwitz, E. P.: Sep. Sci. Technol. **23**, 1191 (1988).
8. Nash, K. L., Rickert, P. G.: Sep. Sci. Technol. **28**, 25 (1993).
9. Myasoedov, B. F., Guseva, L. I., Lebedev, I. A., Milyukova, M. S., Chmutova, M. K.: *Analytical Chemistry of Transplutonium Elements,* John Wiley and Sons, Inc., New York 1974, p. 256.
10. Foti, S. C., Freiling, E. C.: Talanta **11**, 385 (1964).
11. Bertrand, P. A., Choppin, G. R.: Radiochim. Acta **31**, 135 (1982).
12. Saito, A., Choppin, G. R.: Anal. Chem. **55**, 2454 (1983).
13. Karalova, Z. K., Myasoedov, B. F., Bukina, T. I., Lavrinovich, E. A.: Solvent Extr. Ion Exch. **6**, 1109 (1988).
14. Karalova, Z. I., Lavrinovich, E. A., Myasoedov, B. F.: J. Radioanal. Nucl. Chem. **159**, 259 (1992).
15. Peretroukhin, V. F., Shilov, V. P., Pikaev, A. K.: Alkaline Chemistry of Transuranium Elements and Technetium and the Treatment of Alkaline Radioactive Wastes, Institute of Physical Chemistry of Russian Academy of Sciences (1994).
16. Rogers, R. D., Bond, A. H., Bauer, C. B.: Sep. Sci. Technol. **28**, 1091 (1993).
17. Myasoedov, B. F., Chmutova, M. K.: New Methods of Transplutonium Elements Isolation, Purification and Separation from Rare Earth Elements and Selected Fission Products. In: *Separations of f Elements* (K. L. Nash, G. R. Choppin, Eds.). Plenum Press, New York 1995, p. 11.
18. Molochnikova, N. P., Frenkel, V. Y., Myasoedov, B. F.: J. Radioanal. Nucl. Chem. **121**, 409 (1988).
19. Inoue, Y., Tochiyama, O.: J. Inorg. Nucl. Chem. **39**, 1443 (1977).
20. Kobashi, A., Choppin, G. R., Morse, J. W.: Radiochim. Acta **43**, 211 (1988).
21. Bondietti, E. A., Trabalka, J. R.: J. Radiochem. Radioanal. Lett. **42**, 169 (1980).
22. Street, K., Jr., Seaborg, G. T.: J. Am. Chem. Soc. **72**, 2790 (1950).
23. Diamond, R. M., Street, K., Jr., Seaborg, G. T.: J. Am. Chem. Soc. **76**, 1461 (1954).
24. Choppin, G. R., Chetham-Strode, A.: J. Inorg. Nucl. Chem. **15**, 377 (1960).
25. Guseva, L. I., Tikhomirova, G. S.: Radiokhimiya **14**, 188 (1972).
26. Surls, J. P., Choppin, G. R.: J. Inorg. Nucl. Chem. **4**, 62 (1957).
27. King, L. J., Bigelow, J. E., Collins, E. D.: In: *Actinide Separations, ACS Symposium Series, Vol. 117* (J. D. Navratil, W. W. Schulz, Eds.). American Chemical Society, Washington, DC 1980, p. 133.
28. Guseva, L. I., Tikhomirova, G. S., Stepushkina, V. V.: Radiokhimiya **29**, 211 (1987).
29. Guseva, L. I., Tikhomirova, G. S., Stepushkina, V. V.: Radiokhimiya **29**, 629 (1987).
30. Usuda, S., Shinohara, S. N., Yoshikama, H.: J. Radioanal. Nucl. Chem. **109**, 353 (1987).
31. Usuda, S.: J. Radioanal. Nucl. Chem. **123**, 619 (1988).
32. Moore, F. L.: Anal. Chem. **36**, 2158 (1964).
33. Weaver, B.: Solvent Extraction in the Separation of Rare Earths and Trivalent Actinides. in: *Ion Exchange and Solvent Extraction, Vol. 6* (J. A. Marinsky, Y. Marcus, Eds.). Marcel Dekker, New York 1974, p. 189.
34. Horwitz, E. P., Suaro, L. J., Bloomquist, C. A. A.: J. Inorg. Nucl. Chem. **29**, 2033 (1967).
35. Horwitz, E. P., Dietz, M. L., Chiarizia, R., Diamond, H., Maxwell, S. C., III, Nelson, M. R.: Anal. Chim. Acta **310**, 63 (1995).
36. Muscatello, A. C., Horwitz, E. P., Kalina, D. G., Kaplan, L.: Sep. Sci. Technol. **17**, 859 (1982).
37. Weaver, B., Kappelman, F. A.: J. Inorg. Nucl. Chem. **30**, 263 (1968).
38. Kosyakov, V. N., Yerin, E. A.: J. Radioanal. Chem. **56**, 93 (1980).
39. Bond, W. D., Leuze, R. E.: Removal of Curium from High-Level Wastes. In: *Actinide Separations, ACS Symposium Series, Vol. 117* (J. D. Navratil, W. W. Schulz, Eds.). American Chemical Society, Washington, DC 1980, p. 441.
40. Novikov, A. P., Myasoedov, B. F.: Solvent Extr. Ion Exch. **5**, 111 (1987).
41. Sekine, T., Dyrssen, D.: Talanta **11**, 867 (1964).
42. Kasting, G. B., Hulet, E. K., Heppert, J. A., Wild, J. F.: J. Inorg. Nucl. Chem. **41**, 745 (1979).
43. Vitorge, P.: Report # 1984 CEA-R-5270 (1985).
44. LeMarois, G., Fitoussi, R., Cuillerdier, C.: Properties and Uses of Nitrogen and Sulfur Donor Ligands in Actinide Separations. In: *Actinide Separations, ACS Symposium Series, Vol. 117* (J. D. Navratil, W. W. Schulz, Eds.). American Chemical Society, Washington, DC 1980, p. 131.

45. Musikas, C.: Actinide-Lanthanide Group Separation Using Sulfur and Nitrogen Donor Extractants. In: *Actinide/Lanthanide Separations* (G. R. Choppin, J. D. Navratil, W. W. Schulz, Eds.). World Scientific Press, Singapore 1985, p. 19.
46. Choppin, G. R., Barber, D. W.: J. Less Common Met. **149**, 231 (1989).
47. Freiser, H.: Solvent Extr. Ion Exch. **6**, 1093 (1988).
48. Jarvinen, G. D., Barrans, R. E., Jr., Schroeder, N. C., Wade, K. L., Jones, M. M., Smith, B. F., Mills, J. L., Howard, G., Frieser, H., Muralidharan, S.: Selective Extraction of Trivalent Actinides from Lanthanides with Dithiophosphinic Acids and Tributylphosphate. In: *Separations of f Elements* (K. L. Nash, G. R. Choppin, Eds.). Plenum Press, New York 1995, p. 43.
49. Smith, B. F., Jarvinen, G. D., Miller, G. G., Ryan, R. R., Peterson, E. J.: Solvent Extr. Ion Exch. **5**, 895 (1987).
50. Chmutova, M. K., Kochetkova, N. E.: Zh. Anal. Khim. **25**, 710 (1970).
51. Ensor, D. D., Jarvinen, G. D., Smith, B. F.: Solvent Extr. Ion Exch. **6**, 439 (1988).
52. Choppin, G. R., Yao, J., Wharf, R. M.: Extraction of Eu(III) and Am(III) with Thio and Amide Extractants. In: *Separations of f Elements* (K. L. Nash, G. R. Choppin, Eds.). Plenum Press, New York 1995, p. 31.
53. Nash, K. L., Rickert, P. G., Lessman, E. P., Mendoza, M. D., Feil, J. F., Sullivan, J. C.: New Water-Soluble Phosphonate and Polycarboxylate Complexants for Enhanced f Elements Separations. In: *Separations of f Elements* (K. L. Nash, G. R. Choppin, Eds.). Plenum Press, New York 1995, p. 125.
54. Thompson, S. G., Cunningham, B. B., Seaborg, G. T.: J. Am. Chem. Soc. **72**, 2798 (1950).
55. Stary, J.: Talanta **13**, 421 (1966).
56. Mathur, J. N.: Solvent Extr. Ion Exch. **1**, 349 (1983).
57. Farbu, L., Alstad, J., Augustson, J. H.: J. Inorg. Nucl. Chem. **36**, 2091 (1974).
58. Mathur, J. N., Khopkar, P. K.: Solvent Extr. Ion Exch. **1**, 597 (1983).
59. Nakamura, S., Surakitbanbarn, Y., Akiba, K.: In: *Proceedings of the Symposium on Solvent Extraction (Japanese Association of Solvent Extraction)* (Hamamatsu, Japan) 1987.
60. Chmutova, M. K., Pribylova, G. A., Myasoedov, B. F.: Zh. Anal. Khim. **28**, 2340 (1973).
61. Khopkar, P. K., Mathur, J. N.: Sep. Sci. Technol. **17**, 985 (1982).
62. Mathur, J. N., Khopkar, P. K.: J. Radiochem. Radioanal. Lett. **57**, 259 (1983).
63. Mathur, J. N., Khopkar, P. K.: Polyhedron **6**, 2099 (1987).
64. Dukov, I., Genov, L.: Solvent Extr. Ion Exch. **4**, 999 (1986).
65. Dukov, I., Genov, L.: Solvent Extr. Ion Exch. **6**, 447 (1988).
66. Mathur, J. N., Khopkar, P. K.: Radiochim. Acta **39**, 77 (1986).
67. Peppard, D. F., Mason, G. W., Maier, J. L., Driscoll, W. J.: J. Inorg. Nucl. Chem. **4**, 334 (1957).
68. Horwitz, E. P., Muscatello, A. C., Kalina, D. G., Kaplan, L.: Sep. Sci. Technol. **16**, 417 (1981).
69. Aly, H. F., Latimer, R. M.: J. Inorg. Nucl. Chem. **32**, 3081 (1970).
70. Yan, J., Ye, W.: Zhongguo Xitu Wuebao **7**, 39 (1989).
71. Chmutova, M. K., Nesterova, N. P., Koiro, O. E., Myasoedov, B. F.: Zh. Anal. Khim. **30**, 1110 (1975).
72. Samy, T. M., Imura, H., Suzuki, N.: J. Radioanal. Nucl. Chem. **126**, 153 (1988).
73. Tang, J., Wai, C. M.: Anal. Chem. **58**, 3233 (1986).
74. Mathur, J. N., Choppin, G. R.: Solvent Extr. Ion Exch. **11**, 1 (1993).
75. Mathur, J. N., Khopkar, P. K.: Solvent Extr. Ion Exch. **6**, 111 (1988).
76. Aly, H. F., Khalifa, S. M., Navratil, J. D., Saba, M. T.: Solvent Extr. Ion Exch. **3**, 623 (1985).
77. Ensor, D. D., Shah, A. H.: Solvent Extr. Ion Exch. **1**, 241 (1983).
78. Ensor, D. D., Shah, A. H.: Solvent Extr. Ion Exch. **2**, 591 (1984).

Radiochimica Acta 70/71, 237–245 (1995)

Fast Radiochemical Separations for Heavy Elements

By N. Trautmann

Institut für Kernchemie, Universität Mainz, D-55099 Mainz, Germany

(Received August 31, 1995)

Rapid chemical methods / One-atom-at-a-time chemistry / Discontinuous and continuous separation procedures / Solution chemistry / Gas phase chromatography / Actinide and transactinide elements

Abstract

Rapid radiochemical separation procedures for the investigation of the chemical behaviour of the heavy elements on a one-atom-at-a-time scale are reviewed. Besides the microcomputer controlled apparatus ARCA II as an example for a discontinuous technique, the fast on-line solvent extraction system SISAK 3 connected with liquid scintillation counting and the on-line gas chemistry apparates OLGA II and HEVI are described in more detail. Some results on the chemical properties of the heaviest actinides and the first two transactinide elements obtained with these techniques are presented and prospects for chemical studies of still heavier elements are briefly discussed.

Introduction

Since the discovery of natural radioactivity by H. Becquerel [1] chemical separation methods have played an important role for the identification and study of radioactive isotopes of the elements. Radiochemical techniques provide an unambiguous way for unravelling mixtures of radioactive species and firmly assigning atomic numbers. Among others, the use of such techniques has led to the discovery of the first radioactive elements in nature, polonium and radium [2, 3] and of nuclear fission [4], the identification of technetium, the first synthetic element [5] and of the first transuranium element, neptunium [6]. In the early decades chemical separations requiring several hours were used routinely but already Hahn and Strassmann [7] developed fast procedures and discovered nuclides with half-lives down to less than one minute. In the last thirty years most attention has been paid to the study of nuclides with half-lives in the range of seconds to minutes and therefore strong efforts were made to speed up chemical procedures by automation.

There are two main approaches to fast radiochemical separations: the batchwise discontinuous and the continuous operation. In discontinuous procedures a nuclide is produced, chemically separated and measured sequentially whereas in continuous procedures the target is permanently irradiated and extraction of the species from the target, chemical separation, and counting are performed on-line. Several reviews of rapid radiochemical methods and their applications to nuclear chemistry have been published, see e.g. [8–13].

Chemical studies of the heaviest elements have recently received considerable interest because of expected relativistic effects [14, 15] which should have an influence on the structure of the electron shells, yielding deviations from the periodicity of chemical properties based on extrapolations from the lighter homologs in the Periodic Table. Among the difficulties in studying the chemistry of the heaviest elements are the short half-lives of the nuclides produced in heavy-ion reactions, the low production rates and the large quantities of undesired nuclides formed simultaneously. This requires fast and selective chemical procedures and unambiguous detection techniques. At the moment 111 elements are known, the last two were just discovered [16, 17] and are out of scope for chemical investigations due to their short half-lives; e.g. 1.5-ms ^{272}111. In designing rapid chemical separations for heavy elements it has to be taken into account that the elements heavier than fermium ($Z=100$) are produced in quantities of one-atom-at-a-time by bombarding suitable targets in heavy-ion accelerators with high-intensity beams.

This paper gives a review on fast chemical methods developed in the last decade for chemical studies of the heaviest actinide and the transactinide elements 104, 105 and 106. Some results are presented together with a brief outlook for studies of still heavier elements. More details on the chemistry of the transactinide elements can be found in another article of this volume [18].

Single atom chemistry

The one-atom-at-a-time production rate for the elements heavier than fermium arises the question whether the chemical behaviour of individual atoms is different from the behaviour of bulk amounts. In order to reproduce macroscopic chemistry at a level of a few atoms, the separation step has to be repeated very often until the individual differences are balanced out and hence, multistep procedures should be used. It was shown experimentally with 3.05-min ^{218}Po that at a level of 5000 to 40 atoms the electrochemical deposition of polonium on silver and its coprecipitation with elementary tellurium or arsenic sulphide give no deviation in the chemical behaviour compared to macroscopic amounts [19]. In a theoretical study on the validity of single atom chemistry Borg and Dienes [20] found that a surprisingly small number of atoms, about ten, should be sufficient to establish chemical identity under the usual conditions of experimentation. They considered a simple exchange reaction and could show that equilibrium is rapidly established if the free energies of activation are small, and a statistical analysis showed that the probability of incurring a serious error is very small.

Guillaumont *et al.* [21, 22] have pointed out that for a single atom the classical law of mass action expressed in terms of the concentration or the number of reacting species is inadequate for describing the equilibrium state. They suggested for single atom chemistry the introduction of a specific thermodynamic function, the single-particle free enthalpy. An expression equivalent to the law of mass action is derived in which concentrations are replaced by the probability of finding the atom in a given state and in a given phase. In this way partition methods can still be used, and the distribution coefficients are expressed in terms of probabilities. With a new model the equilibrium state of reactions proceeding in a closed system is described. The results show that for a particular reaction in which the stoichiometric coefficients of all species of a given element are equal to one, the law of mass action is still valid if expressed in terms of the average number of reacting species.

In most chemical separation procedures, the principal step is a distribution between two phases, gaseous, liquid, or solid, in such a way that the element of interest occurs in the one and the bulk of the interfering elements in the other phase. It is advantageous if a single atom changes frequently between the two phases as it is realized in chromatographic methods like gas or liquid chromatography thus giving the same results as in a macroscopic experiment.

Experimental

Target and transport system

In heavy element experiments a target arrangement as shown schematically in Fig. 1 is used very often. The heavy-ion beam passes through a vacuum isolation foil (e.g. 1.8 $mg \cdot cm^{-2}$ HAVAR), separating the target chamber from the vacuum of the beam transfer line, then through a volume of nitrogen cooling gas and a beryllium backing before interacting with the target material. The reaction products recoiling out of the target are stopped in helium or another gas loaded with aerosols of 50−200 nm diameter. The products attached to the clusters are transported through a capillary to the chemistry system yielding, as a rule, transport times of a few seconds. For separations in the aqueous phase KCl is mostly used as cluster material whereas MoO_3 or carbon are often applied in gas phase separations [23] in order to prevent chemical reactions with the aerosols.

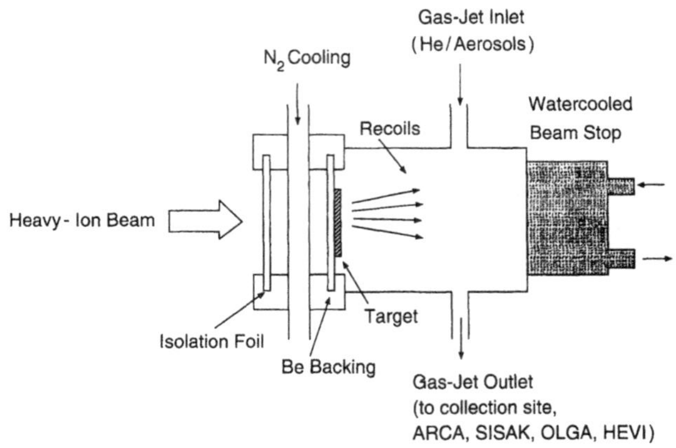

Fig. 1. Schematic diagram of a target arrangement as used in heavy-ion reactions.

In the isolated fractions the atoms of the heavy elements are detected by α-particle and spontaneous fission (SF) pulse height analyses using either chemically inert passivated ion implanted planar silicon (PIPS) detectors or liquid scintillation spectroscopy. With the latter method up to four parameters per event are generated and registered, namely one parameter for the energy spectra, two for β- and γ-background discrimination and one for time correlation. The data are transferred via a list mode buffer to a computer system and stored either on a magnetic tape or a disc enabling direct on-line as well as off-line analysis. This allows the detection of time-correlated parent-daughter pairs of α-α or α-SF-events for an unambiguous assignment of single events.

Solution chemistry

Most of the recent aqueous phase chemistry studies of the heavier actinide and the transactinide elements have been performed manually or with automated high pressure liquid chromatography (HPLC) systems in a discontinuous batch-wise mode.

Batch experiments

For manual liquid-liquid extraction studies of element 104 the recoiling products from the $^{248}Cm(^{18}O,5n)^{261}104$ reaction were stopped in helium gas and transported with a helium gas-jet transport system to the chemistry collection site. There, the activity was collected on a Teflon disc for 90 s and then removed with 10 µl of aqueous phase which were added to 20 µl of the organic phase in a centrifuge cone and mixed in an ultrasonic bath for 5 s. The solution was centrifuged for 10 s to separate the phases, the organic phase was removed and put on a hot disc for evaporation. The disc was placed over a PIPS detector and counted for α- and SF-events. This procedure took about 1 min and was applied for investigating the extraction behaviour of element 104 with triisooctyl-

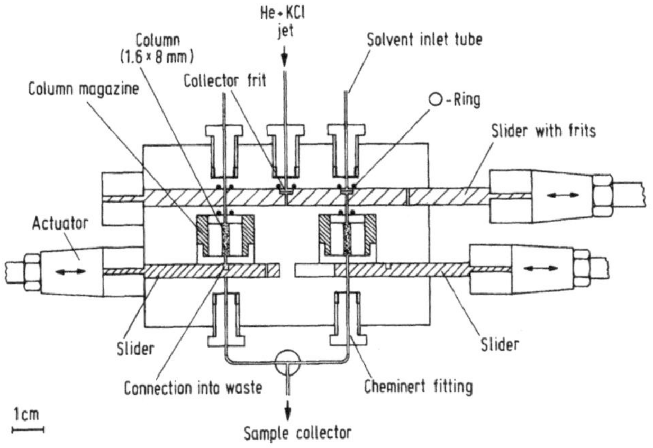

Fig. 2. View of the catcher/chemistry unit of the microprocessor controlled **A**utomated **R**apid **C**hemistry **A**pparatus ARCA II [29].

amine (TIOA) and tributylphosphate (TBP) [24, 25] with 65-s $^{261}104$.

Studies of the aqueous phase chemistry of element 105 were performed on 34-s $^{262}105$ produced by the $^{249}Bk(^{18}O,5n)$ reaction in a rather similar way using the characteristic property of the group 5 elements to adhere to glass surfaces upon fuming with concentrated nitric acid. In further experiments the extraction behaviour of element 105 from mixed nitric acid/hydrofluoric acid solutions into methyl isobutyl ketone (MIBK) was found to be different from that of its direct homolog, tantalum [26].

High performance liquid chromatography can be used for fast separations of actinide [27] and transactinide elements. The microprocessor controlled **A**utomated **R**apid **C**hemistry **A**pparatus, ARCA [28], applies this principle in combination with a gas jet transport system. For fast and repetitive separations to be carried out in a chemically inert environment ARCA II has been designed [29]. ARCA II consists of i) a catcher/chemistry unit that collects the activity from the gas-jet and carries out the chemical separation, ii) the peripheral components for high performance liquid chromatography (pumps, valves etc.), and iii) the electropneumatic part together with a control unit coupled to a microprocessor. The solvents contact only inert materials. The apparatus is built symmetrically and allows operations in an altering mode.

The catcher/chemistry part of ARCA II is shown in Fig. 2. The body of this unit consists of three teflon parts pressed against each other to obtain a leaktight seat between the stationary and the movable components. Three inlet tubes are connected to the upper part by fittings and the solvents are fed into the system through 0.3 mm (i.d.) capillaries connected to the fittings on the left and right side. The middle fitting connects the gas-jet capillary to one of the two polyethylene frits pressed in a Kel-F slider. Below the frit in the central position the bore has a connection to a ventilation system to pump off the carrier gas during collection of the clusters with the attached reaction products on the frit. Simultaneously, previously collected reaction products are washed from the other frit and a chemical separation takes place on the right side in Fig. 2. At the same time a magazine carrying 20 columns can be pushed forward to bring a new column into position (left side). After the separation is finished on the right column, the slider with the frits is moved to the left to start a new activity collection on the right frit, to bring a new column into position and to begin with a chemical separation on the left column. The columns are made by drilling 8 mm long holes with a diameter of 1.6 mm into Kel-F plugs pressed into holes in the stainless steel magazine. Pneumatically driven actuators push the magazine forward in a stepwise mode. Depending on the position, the effluent from the column can be directed either to the sample collector or into the waste. Each pump moves one eluent through Teflon tubings to the central catcher/chemistry unit allowing a reproducible and fast changing of the effluents on the columns.

The separation procedure determines whether the columns are filled with a cation exchanger or with an organic extractant adsorbed on an inert supporting material. The low dead volume of the columns of about 35 µl gives fast access to the isolated fractions. Already within a few seconds a separation can be finished. The sample preparation for α- and fission-fragment pulse height analysis occurs by fast evaporation of the effluent to dryness by means of intense infrared light and hot helium-gas followed by a flaming and cooling step. Sample preparation and mounting in front of the detector requires about 20 s, so that the measurement can be started ~30 s after the end of collection. This is much faster than with the first version of ARCA where columns of 500 to 40 mm in length and from 5 to 1.7 mm i.d. were used. ARCA I was applied to study the elution behaviour of ^{260}Lr, ^{254}Md and ^{255}Md from di-(2-ethylhexyl)-phosphate (HDEHP) columns with HCl and the reduction of Lr^{3+} with V^{2+} and Cr^{2+} [30].

ARCA II was applied for detailed studies of the chemical behaviour of element 105 with 34-s $^{262}105$. From sorption and elution experiments on TIOA columns it was found that element 105 behaves similarly to niobium and the pseudo-homolog protactinium but different to the closest homolog, tantalum [31, 32]. Elution experiments of element 105 from cation exchange resin columns with α-hydroxyisobutyric acid (α-HIB) resulted in the same behaviour of element 105 as niobium, tantalum and protactinium providing an unambiguous proof that the most stable oxidation state of element 105 in aqueous solutions is the pentavalent state [33].

For repeated high precision chemical separations of short-lived nuclides of the heavy elements the **A**utomated **C**hromatographic **C**hemical **E**lement **S**eparation **S**ystem, ACCESS, and an automated injection system have been constructed [34]. The system is modular and can be used either in an off-line or on-line mode with manual injection or with the automated injection in-

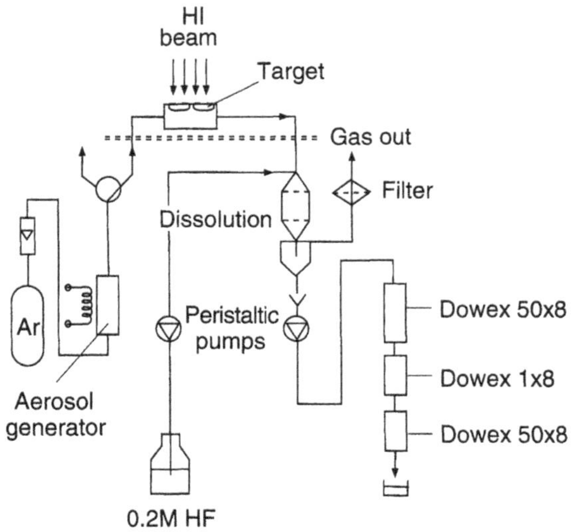

Fig. 3. Experimental set-up for the separation of the elements 104 and 106 by fast continuous ion exchange [36].

terface. The size of a typical column is 20 mm in length and 2 mm in diameter. The eluate can be collected directly on heated metal foils. Typically, with this unit it takes a few minutes from the production of the nuclides to the measurement of the sample.

Continuous techniques

The continuous separation of element 104 in combination with a discontinuous sample preparation has been used to study its solution chemistry in hydrofluoric acid [35]. The reaction products were transported from the target position to the separation apparatus with a gas-jet. The nuclides were dissolved in 0.2 M HF and the solution was passed through three columns filled with ion exchangers. In the first, cation exchange column the actinides were removed whereas in the next column filled with anion exchanger material, element 104 was sorbed, and in the third cationic column the decay products of element 104 were retained. A schematic diagram of the experimental set-up is presented in Fig. 3. The reaction products attached to KCl clusters and transported in an argon jet were dissolved in 0.2 M HF. The solution passed through three combined columns filled with Dowex 50x8, Dowex 1x8 and Dowex 50x8 at a flow rate of 1.5 to 2.0 ml/min using a peristaltic pump. The transportation time from the target to the separation apparatus was about 5 s and the hold-up time through the aerosol absorber and the ion exchange columns was less than 30 s. The identification of 65-s ^{261}104 was performed via its descendants ^{253}Fm$-^{253}$Es which were collected on the third column, later eluted by means of 6 M HNO_3 and purified. The counting sample was prepared by electrospraying an acetic acid solution of the Fm-Cf fraction onto the surface of a Si/Au detector. In this experiment it was shown that in HF solution element 104 forms stable anionic complexes which can be sorbed on an anion exchange resin, i.e., its chemical behaviour is similar to zirconium and hafnium, the homologs in the Periodic Table. With the same experimental set-up model experiments have been performed with short-lived tungsten isotopes in order to explore the applicability of this procedure for the isolation of element 106 [36].

The centrifuge system SISAK 3 [37, 38], "**S**hort-lived **I**sotopes **S**tudied by the **AK**UFVE Technique", allows the continuous separation of nuclides with half-lives down to ~1 s from complex nuclear reaction product mixtures [39]. The method is based on fast multistage solvent extractions using high-speed centrifuges for phase separation. The reaction products are continuously transported from the production site to the SISAK equipment with a gas-jet system. A degassing centrifuge is used to transfer the reaction products from the gas-jet to a liquid phase and to separate the carrier gas and other volatile products from the liquid. The solution is then fed into a centrifuge battery of up to 4 H-centrifuges. From the last solvent extraction step the solution containing the isolated element is pumped to the measuring unit. For the construction of the centrifuges two different materials were used, namely titanium passivated with palladium or polyetheretherketone (PEEK). The performance with respect to time behaviour is equal for both materials.

Fig. 4 shows a cut-away drawing of a H-0.3 centrifuge. The two liquid phases are mixed in a Y-shaped inlet channel and the "emulsion" enters centrally the centrifuge and is accelerated in the inlet chamber to a rotation speed of 15000−30000 rpm. After acceleration, the mixture is forced into the separation volume which contains four small chambers symmetrically arranged around the axis of rotation and isolated from each other. Here the mixture has to follow a zig-zag motion imposed by peripheral partition walls and interspersed baffle ridges. The separated phases are discharged from the light and heavy phase collection chambers by stationary pump wheels. The outlets are equipped with observation windows of transparent teflon and with valves to be able to maintain a proper pressure balance in the centrifuge and thus get well separated outgoing phases. In a HG-0.1 degassing unit the gas and the liquid leaving a static mixer are separated at ~20000 rpm. A stationary pump wheel pumps the liquid out of the centrifuge via a throttle valve. The gas phase is discharged at an outlet surrounding the central shaft and then fed into an exhaust system. Both types of centrifuges are driven by 3-phases asynchronous motors. The hold-up time for a one step procedure is about 2 s including the transport time of the reaction products in the jet system (1.0 s) and the time for the transportation of the separated fraction to the measuring unit (0.3 s).

Very recently the possibility of applying SISAK 3 to study the chemistry of elements 105 and 106 has been explored [40]. As a characteristic chemical property of these elements the formation of negatively charged complexes with α-HIB and lactic acid in weakly acid solutions and the solvent extraction of

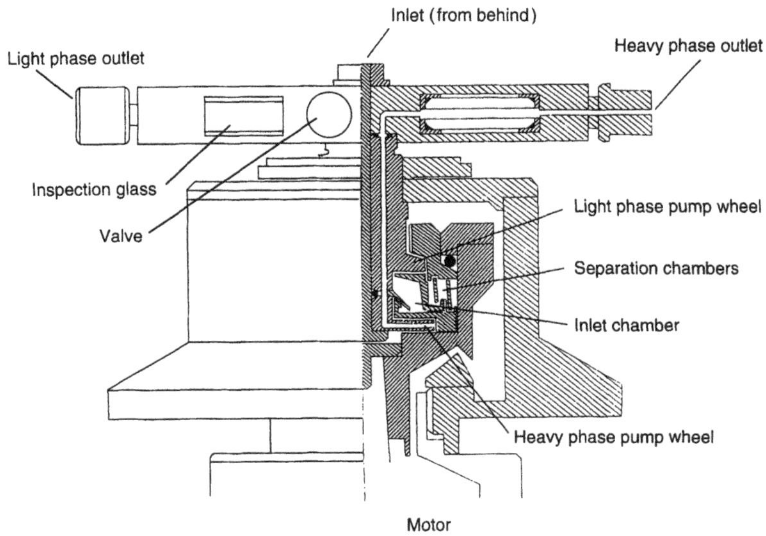

Fig. 4. Cut-away drawing of the H-03 centrifuge as used in the SISAK 3 system [37].

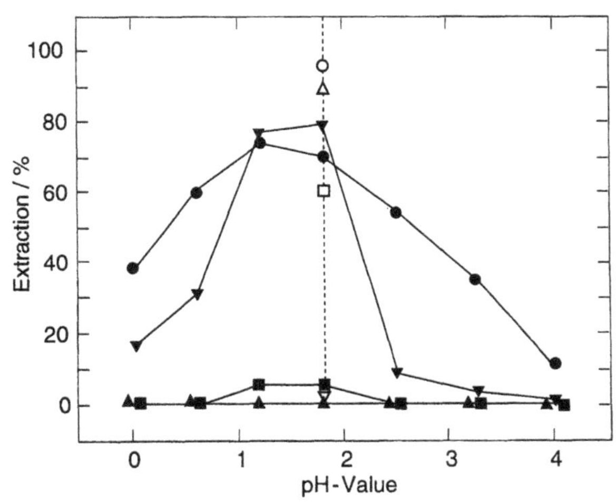

Fig. 5. Extraction yields of zirconium (■), niobium (●), molybdenum (▼), lanthanum (▲), hafnium (▽), tantalum (○), tungsten (□), and protactinium (△) from 1.0 M α-HIB at various pH-values with 0.046 M TOA/Shellsol T in SISAK on-line experiments [40].

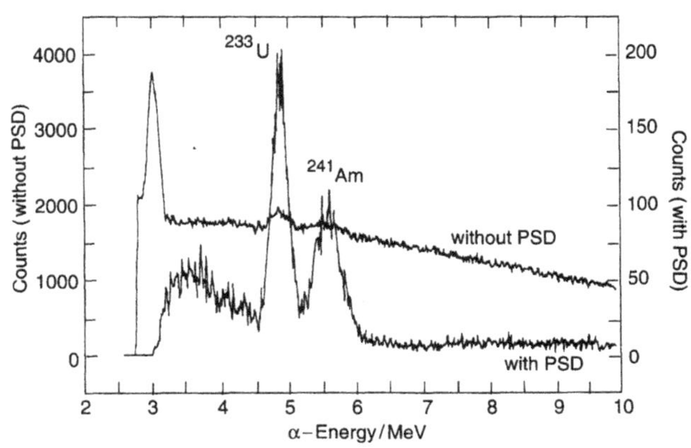

Fig. 6. Energy spectrum of ^{233}U and ^{241}Am α-particles measured with the extractive HDEHP scintillator in a flowing organic solution after separation with SISAK 3 in presence of strong β-γ-activities without and with pulse shape discrimination (PSD) [41].

these complexes by the liquid anion exchanger trioctylamine (TOA) was chosen. The on-line studies were carried out using Zr-Nb-Mo-La activities produced by fission and Hf-Ta-W activities from α-particle induced reactions. The highest extraction yields for the Group 5 homologs of element 105 and the Group 6 homologs of 106 have been obtained for an unbuffered 1.0 M α-HIB solution at pH 1.8 as can be seen in Fig. 5. Group 4 homologs of element 104 and the lanthanides remain in the aqueous phase.

Most of the nuclides of the heaviest elements decay by emission of α-particles or by spontaneous fission. Therefore, a method for fast measurement of α-particle energies and spontaneous fission events is required as an inherent part of the SISAK experiments. A liquid-scintillation counting system has been explored [41] for this purpose allowing the continuous internal measurement of α-particle energies and the detection of spontaneous fission events in flowing organic solutions. Dimethyl-POPOP and naphthalene dissolved in toluene together with the extracting agent act as an extractive scintillator. An average α-energy resolution of 4.8% FWHM was measured in a flowing organic solution. The detection system consists of three spherical quartz cells optically coupled to a photomultiplier and connected by loops for the flowing solution in order to increase the detection efficiency.

The interference of β- and γ-radiation in the detection of α-particles and fission fragments is minimized by electronic pulse shape discrimination and pile-up rejection. Correlations between successive α-α- or α-fission-events in parent-daughter pairs can be used for an unambiguous identification of single events. The effect of pulse shape discrimination is shown in Fig. 6 for the α-energy spectra of ^{233}U and ^{241}Am in the presence of β- and γ-ray fission product emitters. For the on-line extraction of ^{233}U and ^{241}Am with SISAK 3 from 0.01 M HCl, HDEHP was applied as extractant. Furthermore, an on-line model experiment for element 104 was carried out with SISAK 3 in combination with

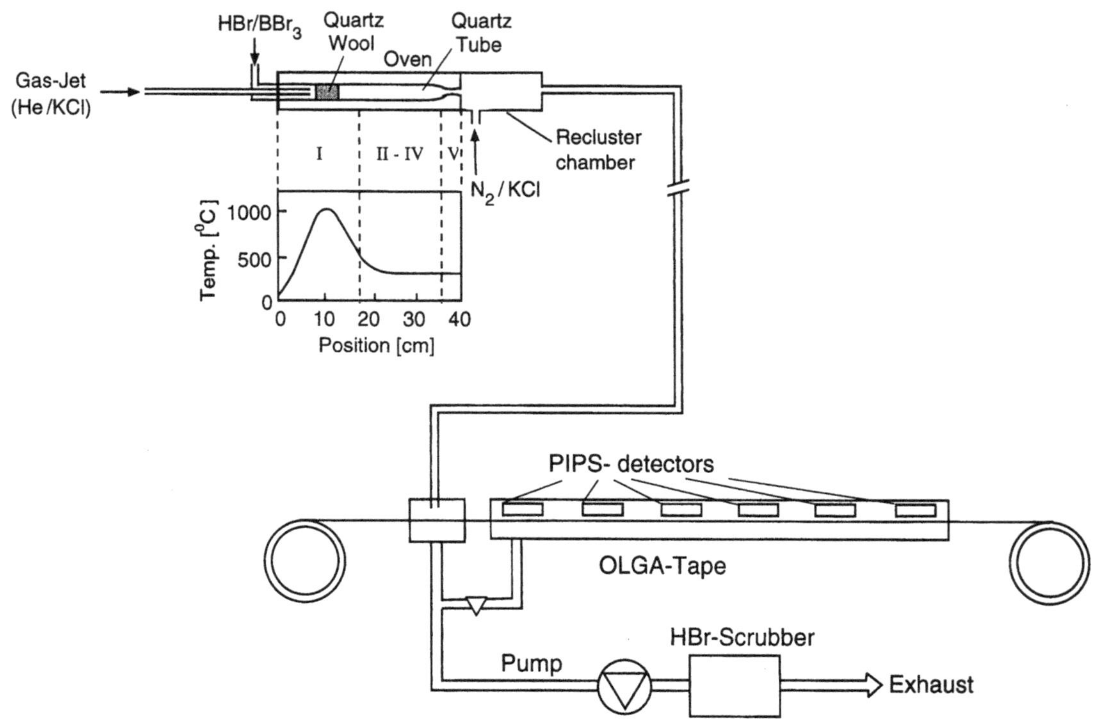

Fig. 7. Scheme of the on-line gas chemistry set up OLGA II in combination with a tape system and PIPS detectors [43].

liquid scintillation spectroscopy using the α-emitter 17-s ^{161}Hf produced in the bombardment of a samarium target with Ne-ions. Hafnium was transferred into an extractive scintillator with 0.25 M dibutylphosphate as extractive agent and the organic phase was pumped through three detector cells in series. In this way the α-particle peak of 17-s ^{161}Hf at 4.60 MeV could clearly be identified.

SISAK 3 in connection with the liquid scintillation spectroscopy system will be applied to study chemical and nuclear properties of the elements 104 and 105 and in a collaborative effort to investigate for the first time the chemistry of element 106.

Gas phase chemistry

Fast gas chemical procedures have proven to be powerful tools for the separation of the heavy elements and for determining adsorption enthalpies and entropies. The early gas phase separations of element 104 and 105 chlorides and bromides used the principle of thermochromatography [42] in which a negative temperature gradient was established along a chromatographic tube. The volatile species were deposited at a particular position inside the tube and from the deposition temperatures adsorption enthalpies could be determined. The disadvantage of this technique is that a real-time detection of the nuclear decay of the short-lived nuclides of the heavy elements is not possible and the used fission track detectors provide no unequivocal assignment. Correction were necessary to correlate the volatility with short nuclear half-lives.

In recent years the on-line isothermal gas chromatography in quartz columns has been applied to study the retention times of the 104 and 105 chlorides and bromides [23]. Hereby the reaction products attached to small aerosol particles are transported through a capillary and continuously injected into a quartz column. In the column the particles with the reaction products are collected on a quartz wool plug heated to about 900 to 1000 °C. Reactive gases are added to form volatile halide molecules. Depending on the temperature of the second, the isothermal part of the chromatography column the volatile species are transported through the tube or adsorbed. The species leaving the column are reattached to new aerosol particles and transported to the counting unit. Two devices are in operation, the **O**n-**L**ine **G**as Chemistry **A**pparatus, OLGA II [43] and the **H**eavy **E**lement **V**olatility **I**nstrument, HEVI [44].

A schematic view of OLGA II is shown in Fig. 7. The entire set-up includes three parts, a gas-jet system for the transport of the reaction products to OLGA II, a gas chromatograph as the main part and the detection system for the radioactivity. The apparatus OLGA II consists of a quartz column of 50 cm length and 6 mm i.d. with a narrow exit tip situated inside an oven with five individually controlled heating sections. The reaction products transported on clusters are stopped in a quartz wool plug of about 5 mm length inserted into the column. Section I of the oven at the position of the quartz wool plug is kept at a fixed temperature, normally 900 °C. Sections II–IV, which heat the isothermal chromatography section (length 25 cm) and the exit tip of the column, are kept at a certain temperature for a given experiment. Depending on the experiment this temperature is varied between 100 °C and 800 °C. Due to the high temperature in the first section of the column, the isothermal length decreases from about 20 cm at $T_{isothermal}$ = 700 °C to about 6 cm at 100 °C. At the position of the quartz wool plug reactive gases like HBr/BBr_3 or HCl can be added with a variable flow rate. Volatile species formed are then transported along the cooler chromatographic section of the column undergoing sorption and desorption steps with retention times indicative of their volatility at the temperature of the isothermal part of the column. Very volatile products leave the column through the tip whereas the less volatile products decay in the chromatography section.

The products leaving the column are thermalized in a recluster chamber (31 cm length and 7 cm i.d.) which is flushed with a N_2/KCl jet or another gas-jet at a rate of 2 l/min. The reattached products are then transported with a second capillary to a moving tape system. The activity is deposited on a regular computer tape and subsequently stepped in front of six large-area PIPS detectors, which are chemically inert. All α- and SF-events can be recorded and analyzed in an event-by-event mode with the OLGA II data acquisition system [45]. One tape allows the collection of about 3000 samples free from activities deposited from previous runs. The collection and counting chambers are evacuated with an inert vacuum pump. The exhaust of the pump is connected to a gas scrubbing system to remove the reactive gases.

The retention time of the volatile species is determined by using the half-lives of the radionuclides as a "clock". The measure for the volatility of a compound is the specific temperature $T_{50\%}$ at which the compound passes the chromatographic column with 50% yield. At this temperature the retention time is equal to the half-life.

For studies with very short-lived nuclides OLGA II has been improved in several aspects leading to OLGA III [46]. The main changes are the column length which has been increased to 1.9 m and the redesignment of the recluster unit. Now the two gas streams meet head on in a small cooled chamber of 15 cm^3 volume. In this way the residence time in the recluster unit has been reduced from about 20 s to ~2 s so that radionuclides with half-lives of 1 to 4 s can be investigated with OLGA III.

HEVI is similar to OLGA II but was designed with a larger quartz column to provide a longer region of isothermal temperature. The differences between OLGA II and HEVI in addition to the longer quartz columns are the heat sinks and the detection system. HEVI uses a horizontal rotating wheel system (Merry-Go-round) with 80 collection positions around its periphery [47]. The aerosols carrying the separated activities are deposited on thin polypropylene foils which are stopped at preset intervalls between six pairs of PIPS detectors placed above and below the wheel. The wheel system allows the registration of α-events from both sides of the deposited spot together with the detection of single and coincident SF events with an efficiency of ~60%. OLGA II has also been equipped with this wheel system in studies of the chlorides of element 104 [48].

On-line isothermal gas phase chromatography was applied to investigate the halides of element 104 and 105 via the nuclides ^{261}104 and ^{262}105 produced on a one-atom-at-a-time scale in the reactions ^{248}Cm(^{18}O,5n) and ^{249}Bk(^{18}O,5n), respectively [23, 48, 49]. Using HBr and HCl gas as halogenating agents, volatile bromides and chlorides of the elements 104 and 105 could be formed and their behaviour compared to the lighter homologs of the Group 4 and 5 elements. The data reveal that element 104 chloride has a volatility resembling very much those of the chlorides of hafnium and zirconium. Element 105 bromides were found to be less volatile than the bromides of the Group 5 elements niobium and tantalum. This is in contrast to relativistic molecular calculations [50] which predict a higher volatility of 105 bromide than its homologs. It was also found that 105 chlorides are more volatile than the bromides [51] which is also true for element 104.

Outlook

With the described fast chemical separation techniques it appears possible to extent our knowledge on the chemical behaviour of the transactinide elements to element 106 and, maybe, to element 107. Until recently the longest-lived known isotope of element 106 was 0.9-s ^{263}106. SISAK 3 seems to be the only system to be fast enough for performing chemical experiments with such a short-lived nuclide but it is just at the limit. The very recent discovery of the longer-lived isotopes ^{265}106 and ^{266}106 [52, 53] with half-lives in the range of 2–30 s and 10–30 s, respectively, has made the situation easier also for the other separation techniques. Gas chromatographic studies with OLGA III, liquid-liquid extractions using SISAK 3 in connection with liquid scintillation counting, and high pressure liquid chromatography with ARCA II will be applied to prove the hexavalency of element 106 and to characterize the behaviour of various complexes and oxyhalides in aqueous solutions and in the gas phase.

The prospects of going beyond element 106 depend on the access to longer-lived species of these elements. There is strong evidence for the enhanced stability predicted theoretically in the region of the deformed shells at N = 162 and Z = 108. This gives hope for producing isotopes of elements heavier than 106 with half-lives long enough for chemical experiments, despite of the expectation that the cross sections are getting smaller and smaller. Various experiments have been proposed and already tested [54, 55] for the separation of longer-lived isotopes of element 107 and still heavier elements. However, it cannot be excluded that new chemical procedures on a time scale so far inaccessible have to be developed to extend the studies on the chemical behaviour of the heaviest elements.

Acknowledgement

Financial support from the Bundesministerium für Bildung, Wissenschaft, Forschung und Technologie under Contract Number 03 HE3 MAI is gratefully acknowledged.

References

1. Becquerel, H.: Compt. Rend. **122**, 420 (1896).
2. Curie, P., Curie Sklodowska, M.: Compt. Rend. **127**, 175 (1898).
3. Curie, P., Curie Sklodowska, M., Bemont, G.: Compt. Rend. **128**, 1215 (1898).
4. Hahn, O., Strassmann, F.: Naturwissenschaften **27**, 89 (1939).
5. Perrier, C., Segrè, E.: J. Chem. Phys. **5**, 715 (1937).
6. McMillan, E., Abelson, P. H.: Phys. Rev. **57**, 1185 (1940).
7. Hahn, O., Strassmann, F.: Naturwissenschaften **28**, 54 (1940).
8. Herrmann, G., Trautmann, N.: Ann. Rev. Nucl. Part. Sci. **32**, 117 (1982).
9. Meyer, R. A.: J. Radioanalyt. Nucl. Chem. **142**, 135 (1990).
10. Skarnemark, G., Alstad, J., Kaffrell, N., Trautmann, N.: J. Radioanalyt. Nucl. Chem. **142**, 145 (1990).
11. Rengan, K.: J. Radioanalyt. Nucl. Chem. **142**, 173 (1990).
12. Hampton, C. V., McHarris, Wm., C.: J. Radioanalyt. Nucl. Chem. **142**, 183 (1990).
13. Rengan, K., Meyer, R. A.: Ultrafast Chemical Separations, Nuclear Science Series, NAS-NS-3118, National Academy Press, Washington (1993).
14. Pershina, V., Fricke, B., Kratz, J. V., Ionova, G. V.: Radiochim. Acta **64**, 37 (1994).
15. Kratz, J. V.: J. Alloys Compounds **213/214**, 20 (1994).
16. Hofmann, S., Ninov, V., Heßberger, F. P., Armbruster, P., Folger, H., Münzenberg, G., Schött, H. J., Popeko, A. G., Yeremin, A. V., Andreyev, A. N., Saro, S., Janik, R., Leino, M.: Z. Phys. **A350**, 277 (1995).
17. Hofmann, S., Ninov, V., Heßberger, F. P., Armbruster, P., Folger, H., Münzenberg, G., Schött, H. J., Popeko, A. G., Yeremin, A. V., Andreyev, A. N., Saro, S., Janik, R., Leino, M.: Z. Phys. **A350**, 281 (1995).
18. Schädel, M.: Radiochim. Acta **70/71**, 207–223 (1995).
19. Reischmann, F.-J., Rumler, B., Trautmann, N., Herrmann, G.: Radiochim. Acta **39**, 185 (1986).
20. Borg, R. J., Dienes, G. J.: J. Inorg. Nucl. Chem. **43**, 1129 (1981).
21. Guillaumont, R., Adloff, J. P., Peneloux, A.: Radiochim. Acta **46**, 169 (1989).
22. Guillaumont, R., Adloff, J. P., Peneloux, A., Delamoye, P.: Radiochim. Acta **54**, 1 (1991).
23. Gäggeler, H. W.: J. Radioanalyt. Nucl. Chem. **183**, 261 (1994).
24. Czerwinski, K. R., Gregorich, K. E., Hannink, N. J., Kacher, C. D., Kadkhodayan, B. A., Kreek, S. A., Lee, D. M., Nurmia, M. J., Türler, A., Seaborg, G. T., Hoffman, D. C.: Radiochim. Acta **64**, 23 (1994).
25. Czerwinski, K. R., Kacher, C. D., Gregorich, K. E., Hamilton, T. M., Hannink, N. J., Kadkhodayan, B. A., Kreek, S. A., Lee, D. M., Nurmia, M. J., Türler, A., Seaborg, G. T., Hoffman, D. C.: Radiochim. Acta **64**, 29 (1994).
26. Gregorich, K. E., Henderson, R. A., Lee, D. M., Nurmia, M. J., Chasteler, R. M., Hall, H. L., Bennett, D. A., Gannett, C. M., Chadwick, R. B., Leyba, J. D., Hoffman, D. C., Herrmann, G.: Radiochim. Acta **43**, 233 (1988).
27. Schädel, M., Brüchle, W., Haefner, B., Kratz, J. V., Schorstein, W., Trautmann, N., Herrmann, G.: Radiochim. Acta **25**, 111 (1978).
28. Schädel, M., Brüchle, W., Haefner, B.: Nucl. Instr. Meth. Phys. Res. **A264**, 308 (1988).
29. Schädel, M., Brüchle, W., Jäger, E., Schimpf, E., Kratz, J. V., Scherer, U. W., Zimmermann, H. P.: Radiochim. Acta **48**, 171 (1989).
30. Brüchle, W., Schädel, M., Scherer, U. W., Kratz, J. V., Gregorich, K. E., Lee, D., Nurmia, M., Chasteler, R. M., Hall, H. L., Henderson, R. A., Hoffman, D. C.: Inorg. Chim. Acta **146**, 167 (1988).
31. Kratz, J. V., Zimmermann, H. P., Scherer, U. W., Schädel, M., Brüchle, W., Gregorich, K. E., Gannett, C. M., Hall, H. L., Henderson, R. A., Lee, D. M., Leyba, J. D., Nurmia, M. J., Hoffman, D. C., Gäggeler, H., Jost, D., Baltensperger, U., Nai-Qi, Ya, Türler, A., Lienert, Ch.: Radiochim. Acta **48**, 121 (1989).
32. Zimmermann, H. P., Gober, M. K., Kratz, J. V., Schädel, M., Brüchle, W., Schimpf, E., Gregorich, K. E., Türler, A., Czerwinski, K. R., Hannink, N. J., Kadkhodayan, B., Lee, D. M., Nurmia, M. J., Hoffman, D. C., Gäggeler, H., Jost, D., Kovacs, J., Scherer, U. W., Weber, A.: Radiochim. Acta **60**, 11 (1993).
33. Schädel, M., Brüchle, W., Schimpf, E., Zimmermann, H. P., Gober, M. K., Kratz, J. V., Trautmann, N., Gäggeler, H., Jost, D., Kovacs, J., Scherer, U. W., Weber, A., Gregorich, K. E., Türler, A., Czerwinski, K. R., Hannink, N. J., Kadkhodayan, B., Lee, D. M., Nurmia, M. J., Hoffman, D. C.: Radiochim. Acta **57**, 85 (1992).
34. Kreek, S. A., Nurmia, M. J., Kadkhodayan, B., Gregorich, K. E., Lee, D. M., Hoffman, D. C., Hall, H. L., Palmer, C. E. A., Baisden, P. A., Henderson, R. A.: Nucl. Instr. Meth. Phys. Res. **A317**, 251 (1992).
35. Szeglowski, Z., Bruchertseifer, H., Domanov, V. P., Gleisberg, B., Guseva, L. J., Hussonnois, M., Tikhomirova, G. S., Zvara, I., Oganessian, Yu. Ts.: Radiochim. Acta **51**, 71 (1990).
36. Szeglowski, Z., Bruchertseifer, H., Brudanin, V. B., Buklanov, G. V., Constantinescu, O., Lein, Dinh Thi, Domanov, V. P., Guseva, L. I., Husssonnois, M., Tikhomirova, G. S., Zvara, I., Oganessian, Yu. Ts.: J. Radioanalyt. Nucl. Chem. Lett. **186**, 353 (1994).
37. Persson, H., Skarnemark, G., Skålberg, M., Alstad, J., Liljenzin, J. O., Bauer, G., Haberberger, F., Kaffrell, N., Rogowski, J., Trautmann, N.: Radiochim. Acta **48**, 177 (1989).
38. Alstad, J., Skarnemark, G., Haberberger, F., Herrmann, G., Nähler, A., Pense-Maskow, M., Trautmann, N.: J. Radioanalyt. Nucl. Chem. **189**, 133 (1995).
39. Altzitzouglou, T., Rogowski, J., Skålberg, M., Alstad, J., Herrmann, G., Kaffrell, N., Skarnemark, G., Talbert, W., Trautmann, N.: Radiochim. Acta **51**, 145 (1990).
40. Wierczinski, B., Alstad, J., Eberhardt, K., Eichler, B., Gäggeler, H., Herrmann, G., Jost, D., Nähler, A., Pense-Maskow, M., Reddy, A. V. R., Skarnemark, G., Trautmann, N., Türler, A.: Radiochim. Acta **69**, 77 (1995).
41. Wierczinski, B., Eberhardt, K., Herrmann, G., Kratz, J. V., Mendel, M., Nähler, A., Rocker, F., Tharun, U., Trautmann, N., Weiner, K., Wiehl, N., Alstad, J., Skarnemark, G.: Nucl. Instr. Meth. Phys. Res. A, in press.
42. Zvara, I.: Isotopenpraxis **26**, 251 (1990).
43. Gäggeler, H. W., Jost, D. T., Baltensperger, U., Weber, A., Kovacs, A., Vermeulen, D., Türler, A.: Nucl. Instr. Meth. Phys. Res. **A309**, 201 (1991).
44. Kadkhodayan, B., Türler, A., Gregorich, K. E., Nurmia, M. J., Lee, D. M., Hoffman, D. C.: Nucl. Instr. Meth. Phys. Res. **A317**, 254 (1992).
45. Jost, D. T., Vermeulen, D.: IEEE, Trans. Nucl. Sci. **39**, 186 (1992).
46. Eichler, B., Türler, A., Jost, D. T., Gäggeler, H. W.: Labor für Radio- und Umweltchemie der Universität Bern und des Paul Scherrer Instituts, Annual Report 1993, 42 (1994).
47. Hoffman, D. C., Lee, D., Ghiorso, A., Nurmia, M. J., Aleklett, K., Leino, M.: Phys. Rev. **C24**, 495 (1981).
48. Türler, A., Gäggeler, H. W., Gregorich, K. E., Barth, H., Brüchle, W., Czerwinski, K. R., Gober, M. K., Hannink N. J., Henderson, R. A., Hoffman, D. C., Jost, D. T., Kacher, C. D., Kadkhodayan, B., Kovacs, J., Kratz, J. V., Kreek, S.

A., Lee, D. M., Leyba, J. D., Nurmia, M. J., Schädel, M., Scherer, U. W., Schimpf, E., Vermeulen, D., Weber, A., Zimmermann, H. P., Zvara, I.: J. Radioanalyt. Nucl. Chem. **160**, 327 (1992).

49. Gäggeler, H. W., Jost, D. T., Kovacs, J., Scherer, U. W., Weber, A., Vermeulen., D., Türler, A., Gregorich, K. E., Henderson, R. A., Czerwinski, K. R., Kadkhodayan, B., Lee, D. M., Nurmia, M., Hoffman D. C., Kratz, J. V., Gober, M. K., Zimmermann, H. P., Schädel, M., Brüchle, W., Schimpf, E., Zvara, I.: Radiochim. Acta **57**, 93 (1992).
50. Pershina, V., Sepp, W.-D., Fricke, B., Kolb, D., Schädel, M., Ionova, G. V.: J. Chem. Phys. **97**, 1116 (1992).
51. Türler, A.: Radiochim. Acta **72**, 7 (1996).
52. Lougheed, R. W., Moody, K. J., Wild, J. F., Hulet, E. K., McQuaid, J. H., Lazarev, Yu. A., Lobanov, Yu. V., Oganessian, Yu. Ts., Utyonkov, V. K., Abdullin, F. Sh., Buklanov, G. V., Gikal, B. N., Iliev, S., Mezentsev, A. N., Polyakov, A. N., Sedykh, I. M., Shirokovsky, I. V., Subbotin, V. G., Sukhov, A. M., Tsyganov, Yu. S., Zhuchko, V. E.: J. Alloys Compounds **213/214**, 61 (1994).
53. Lazarev, Yu. A., Lobanov, Yu. V., Oganessian, Yu. Ts., Utyonkov, V. K., Abdullin, F. Sh., Buklanov, G. V., Gikal, B. N., Iliev, S., Mezentsev, A. N., Polyakov, A. N., Sedykh, I. M., Shirokovsky, I. V., Subbotin, V. G., Sukhov, A. M., Tsyganov, Yu. S., Zhuchko, V. E., Lougheed, R. W., Moody, K. J., Wild, J. F., Hulet, E. K., McQuaid, J. H.: Phys. Rev. Letters **73**, 624 (1994).
54. Schädel, M., Jäger, E., Schimpf, E., Brüchle, W.: Radiochim. Acta **68**, 1 (1995).
55. Schädel, M., Jäger, E., Brüchle, W., Sümmerer, K., Hulet, E. K., Wild, J. F., Lougheed, R. W., Dougan, R. J., Moody, K. J.: Radiochim. Acta **68**, 7 (1995).

D. Applications of Radioactivity

Radiochimica Acta 70/71, 249–272 (1995)

The Impact of Radioactivity on Medicine

By G. Stöcklin*, S. M. Qaim and F. Rösch
Institut für Nuklearchemie, Forschungszentrum Jülich GmbH, D-52425 Jülich, Germany

(Received December 18, 1995)

Dedicated to Alfred P. Wolf, the father of modern radiopharmaceutical chemistry

Radioactive tracers in medicine / Medical radionuclide production / Cyclotron and nuclear reactor / Positron emitter / Single photon emitter / Radiopharmaceutical / Therapeutic radionuclide / PET / SPET / Endoradiotherapy

Summary

Radioactivity has revolutionized life sciences during the last century, and it is still an indispensable tool. Nuclear medicine today is a well established branch of medicine. Radionuclides and radiopharmaceuticals play a key role both in diagnostic investigations and therapy. Both cyclotron and reactor produced radionuclides find application, the former more in diagnostic studies and the latter in therapy. With the advent of emission tomography (PET and SPET), new vistas for probing biochemistry in vivo have been opened. The radiochemist faces an ever-increasing challenge of designing new tracers for diagnostic and therapeutic applications. Rapid, efficient and automated methods of radionuclide and precursor production, labeling of biomolecules, and quality control need to be developed.

The beginning

The most important application of radioactivity has been in life sciences, especially in medicine, and it is still increasing. The reason is simple: in the ideal physiological experiment the subject or system should not be disturbed. Radioactive tracers due to their sensitivity and due to the fact that isotopic labeling does not significantly alter the chemical properties of the molecule, provide the perfect tool for this approach. Thus, it did not take long, after the discovery of radioactivity until it was used in life sciences. The reader, who is interested in a complete early history, is referred to an excellent chronology of nuclear medicine by Marcel Brucer [1], which contains many details and gossip of the early days, a real gold mine, particularly with respect to the period up to about 1970.

The start with natural radioactivity

It was George de Hevesy (Fig. 1), a born Hungarian and a physical chemist by training, who carried out the first biological experiments with natural radioactivity in 1920. During his stay in Niels Bohr's institute in Copenhagen, he collaborated with a botanist, who was growing Horse Bean plants in a culture solution [cf. 1]. He added a small amount of ThB(^{212}Pb) to the culture solution and, after a period of growth, the various parts of the plant were removed, ashed, and measured under the electroscope, thus showing the movement of lead atoms through the plant [2]. De Hevesy had discovered the concept of metabolic turnover. In 1924 he extracted RaE(^{210}Bi) from RaD(^{210}Pb) and studied the metabolism of Bi in rabbits [3]. Later the metabolism of lead was also studied [4]. Hevesy was critical of his own use of toxic heavy metals in biology [cf. 1], and soon got interested in radioisotopes of body constituents which were not available in the early twenties.

Fig. 1. George de Hevesy, Chemist 1885–1966, "Father of Nuclear Medicine", Nobel Prize 1943.

The first real nuclear medical experiment with a natural radioisotope was reported by Blumgart and

* *Present address:* Nuklearmedizinische Klinik der TU München, Klinikum rechts der Isar, Ismaninger Str. 22, D-81675 München, Germany.

Weiss in 1927. They used radon-226 for measuring the speed of blood flow [5]. Blumgart spent a year on cardiology services in England where he learned of a few attempts to measure flow, but they all tampered with blood vessels, physiologically forbidden. Upon his return to Harvard in 1926 he heard of William Duane's radon cow for cancer therapy [cf. 1]. Duane was Marie Curie's first assistant, who invented the radium-radon extraction cow, to cut the cost of radon which was extensively used in cancer therapy. Duane measured the radiation dose from outside by an electroscope. So did Blumgart, who didn't know about de Hevesy's studies [cf. 1].

Small neutron sources produce first useful amounts of artificial radioisotopes

After the discovery of the neutron by James Chadwick in 1932, Enrico Fermi made the first small neutron sources, using (α,n) reactions on light elements such as beryllium. With a radium-beryllium source he produced many short-lived radioisotopes [6]. De Hevesy also used this method, and with almost a week's exposure he produced phosphorous-32 [cf. 1]. Later he got access to a 600 mg radium-beryllium neutron source which Niels Bohr had received on his fiftieth birthday from the Union Haute Minière [cf. 1], and de Hevesy was able to produce microcuries of ^{32}P for his biological studies [7]. With the help of a Geiger-Müller tube an easy detection of β-particles of ^{32}P became possible. In 1935 Chievitz and de Hevesy reported their first use of ^{32}P for the study of phosphorus metabolism in rats [8].

Small neutron sources were also used by Evans, who made the first few microcuries of iodine-128 for biological studies [9]. Hertz, Roberts and Evans injected iodine-128 into rabbits and started investigations on the metabolism of the thyroid [10], an important background for present day thyroid diagnostics.

The first cyclotron era

With the invention of the cyclotron by E. O. Lawrence a wealth of radioisotopes became available in millicurie quantities, and their application in life sciences rapidly increased. E. O. Lawrence supplied de Hevesy with sufficient ^{32}P (via normal mail), made by Martin Kamen at Berkeley [cf. 1]. This led Gilbert Lewis, the Dean of Chemistry at Berkeley, to set up the first school of radiochemistry in conjunction with the Radiation Laboratory at Berkeley [cf. 1].

John Lawrence, the MD brother of E. O. Lawrence, was interested in the therapeutical application of ^{32}P, and first studies on the treatment of leukemia were initiated [11]. A couple of offshoots of J. Lawrence's discovery survived: Polycythemia vera is still treated with ^{32}P.

After the discovery of iodine-131 by Livingood and Seaborg [12], this radioisotope was soon used for thyroid cancer therapy. It was produced in large quantities at cyclotrons by bombarding tellurium with deuterons. Depending on the energy available, a mixture of iodines was obtained and pure iodine-131 was not available until it was separated from fission products in the late 1940s [cf. 1]. Already in 1939 Hamilton and Soley studied the metabolism of radioiodine from outside, using a Geiger-Müller tube. By the end of 1940 the MIT cyclotron at Boston produced iodine-130 and iodine-131 in quantities sufficient for therapy [cf. 1], which were successfully used by Hamilton and Soley [13].

Soon another therapeutic radioisotope, strontium-89, was produced at the cyclotron and already in 1940 Pecher applied it to cure prostate metastases in bone [14]. This third therapeutic agent, now routinely produced in a nuclear reactor, is still used in palliative therapy of bone metastases.

In 1934 Lauritsen and Crane produced the first 2 mCi of carbon-11 via the (p,n) reaction on boron [15]. The short-lived "organic" positron emitters used in positron emission tomography (PET) today were all discovered in the early thirties: nitrogen-13 by Crane and Lauritsen [16], oxygen-15 by Livingston and McMillan [17] and fluorine-18 by Snell [18]. In 1940 fluorine-18 was used to study the absorption of fluoride by enamel, dentine, bone and hydroxyapatite [19].

In 1945 Tobias *et al.* applied ^{11}C labeled carbon monoxide to humans by inhalation in order to see a possible metabolism to CO_2, which they did not observe [20]. It is interesting to note that 30 years later the fixation, retention and exhalation of ^{11}C labeled CO was studied with a whole body counter and radiogas chromatography [21], which among others confirmed the result of Tobias *et al.* that CO is not metabolized.

The reactor era

Tracer applications in general, and nuclear medicine in particular, received an impetus after the announcement in the June 14, 1946, issue of Science that radioactive isotopes were available for public distribution. On August 2, 1946, the first peacetime shipment of radioisotopes from the Oak Ridge nuclear reactor went to the Barnard Free Skin and Cancer Hospital in St. Louis [cf. 1] to Martin Kamen, who together with S. Ruben had discovered carbon-14 in 1940 at the Berkeley cyclotron [22]. Soon thereafter the long-lived β-emitters tritium, carbon-14, sulfur-35 and phosphorus-32 as well as iodine-125 became available in large quantities and in the form of labeled compounds. The introduction of scintillation counters and scanners together with the practice of labeling drugs and biomolecules with these long-lived "organic" radioisotopes, made the tracer technique an essential tool in life sciences. De Hevesy's radioindicator concept significantly contributed to the discoveries in cellular and molecular life processes. The availability of isotopically labeled compounds made it possible to demonstrate that macromolecules, such as proteins, nucleic acids, and complex lipids are synthesized from simple cellular metabolites. Nucleic acids were labeled with tritium and carbon-14, thus allowing to follow the behavior and fate of individual DNA and RNA molecules. Obviously, the tracer technique made important contributions to cancer research and genetics. A major application has been in the field of drug development, pharmacokinetics and metabolic studies which depend to a large extent on carbon-14 and tritium labeled compounds. In addition, autoradiography provided pictures of the regional distribution of drugs and metabolites in organs and cells and allowed the mapping of receptor density and distribution.

The availability of these radionuclides in the 1950s led Salomon Berson and Rosalyn Yalow [23] to the invention of the radioimmuno assay (RIA), based on the principle of competitive binding of natural and labeled antigen to an antibody. This approach has become the basis of an increasing number of assays in routine diagnosis of endogenous and exogenous body constituents. This extremely sensitive and simple method was and is still used to some extent to measure hormones, enzymes, hepatitis virus and certain serum proteins, drugs and many other substances in a few drops of blood serum. Rosalyn Yalow received the Nobel Prize in 1977. According to the IAEA, in 1981 a total of some 40 million tests were performed per annum in the USA alone [24]. Iodine-131 became available for the widely used thyroid cancer endotherapy and sealed ^{60}Co irradiation sources for implantation – or teletherapy. It was first used in cancer treatment in August 1948 at the Ohio State University. Since then thousands of teleradiocobalt-60 sources have been installed worldwide and tens of millions of treatments have been given for cancer [25]. Today modern accelerators have substituted these sources to a large extent, but endotherapy with radionuclides is still a growing field (see below).

A variety of diagnostic radiopharmaceuticals also became available. For example, Tubis introduced [^{131}I]Iodohippurate for measuring kidney functions [26]. The ^{99}Mo/^{99m}Tc- and ^{90}Sr/^{90}Y generators were developed by Tucker *et al.* [27] at the Brookhaven National Laboratory, and in 1960 the first ^{99m}Tc-generator was shipped to Harper at the University of Chicago [1]. The ^{99m}Tc-radiopharmaceuticals became the workhorse in nuclear medicine (see chapter by A. Jones). Xenon-133 was used for circulation studies and the measurement of perfusion in the brain.

In 1954 the Society of Nuclear Medicine was formed in the US. In the early 1960s the first Anger scintillation cameras became readily available commercially and imaging revolutionized diagnostic nuclear medicine.

The renaissance of the cyclotron

The cyclotron was still required for the production of a number of neutron deficient radionuclides, which could not be made in a reactor. However, in the 1960s a real renaissance of the cyclotron took place. Among others, short-lived positron emitters such as carbon-11, nitrogen-13, oxygen-15 and fluorine-18 were produced even years before the advent of positron emission tomography (PET). Of particular interest was the production of short-lived radioactive gases such as [^{15}O]O_2, [^{15}O]CO, [^{13}N]N_2, [^{11}C]CO. Due to the short half-life the cumulative radiation dose per unit of administered radioactivity is low. Thus relatively large amounts of radioactivity can be injected or inhaled into the body giving high count rates and allowing to follow rapid dynamic processes. One of the first uses of oxygen-15 ($T_{1/2}$ = 2 min) was to follow the removal rate of [^{15}O]CO_2 from specific regions of the lung [28]. In addition, tracers such as ^{11}CO or C^{15}O can be used to label red blood cells at high activities, thus allowing studies of blood flow in a variety of organs. Furthermore, the emitted positron allows to use $\gamma\gamma$-coincidence counting with a high inherent resolution. First medical "in hospital" cyclotrons were installed at the Hammersmith Hospital, London, and in the Washington University Medical Center, St. Louis [29]. Many other important medical radionuclides were made with cyclotrons (see below and chapter by Weiner and Thakur). In 1973 thallium-201 was introduced as an analog of potassium [30]. It is still one of the major radionuclides used in nuclear medicine to measure regional myocardial blood flow.

For the radiochemists an era of fast chemistry started which made use of the new analytical techniques such as gas chromatography (GC) and later high performance liquid chromatography (HPLC).

The cyclotron produced radionuclide iodine-123 ($T_{1/2}$ = 13.2 h), with its optimum and abundant (83%) γ-ray energy of 159 keV, was introduced into nuclear medicine among others through the initiative of William G. Meyers and Harold Anger [31], and was subsequently used for labeling many diagnostic radiopharmaceuticals (see below), because of its considerably lower dose constant. Only recently can it be produced in sufficient radionuclidic purity and yield to allow its commercial supply (see below, medical radionuclide production). New high-current dedicated small cyclotrons for medical radionuclide production were developed to be used as black boxes. Today there are more than 200 cyclotrons worldwide, many of them in hospitals, producing short-lived radionuclides for medical applications. We describe below some diagnostic and therapeutic applications of radionuclides. It is not intended to give an exhaustive review; instead, only some typical examples are given.

Medical radionuclide production

The development of a production process involves a study of several aspects such as nuclear data, high current targetry, chemical processing, automation and quality control. A very important criterion is the high

Table 1. Routine methods of production of some commonly used positron emitters [34]

Radioisotope	$T_{1/2}$	Mode of decay	Main γ-ray energy in keV (%)	Production data: Nuclear reaction	Energy range (MeV)	Thick target yield MBq(mCi)/μAh
^{11}C	20.4 min	β^+ (99.8) EC (0.2)	511 (199.6)	$^{14}N(p,\alpha)$	13→3	3820 (103)
^{13}N	10.0 min	β^+ (100)	511 (200)	$^{16}O(p,\alpha)$	16→7	1665 (45)
^{15}O	2.0 min	β^+ (99.9) EC (0.1)	511 (199.8)	$^{14}N(d,n)$ $^{15}N(p,n)$	8→0 10→0	2368 (64) 2220 (60)
^{18}F	109.6 min	β^+ (97) EC (3)	511 (194)	$^{18}O(p,n)$ $^{20}Ne(d,\alpha)$	16→3 14→0	2960 (80) 1110 (30)
^{68}Ge ↓ (generator)	271 d	EC (100)		RbBr(p,spall)	800, 500	0.15 (0.004)
^{68}Ga	68 min	β^+ (90) EC (10)	511 (180) 1077 (3)			
^{82}Sr ↓ (generator)	25 d	EC (100)		Mo(p,spall) $^{85}Rb(p,4n)$	800 60→40	3.7 (0.1) 14.8 (0.4)
^{82}Rb	1.3 min	β^+ (96) EC (4)	511 (192) 776 (13.4)			

specific activity of the final product. The whole process has therefore to be designed accordingly. The (n,γ) reaction, generally occurring in a nuclear reactor, has thus a disadvantage unless the activation product decays by β^- emission or EC to a medically important daughter nuclide which can be chemically separated. The other processes advantageously employed include (n,p), (n,α) and (n,fission). In contrast, the cyclotron production of radionuclides leads, at least in principle, to high specific activity products. However, the large number of competing nuclear reactions demand a detailed and accurate knowledge of nuclear data. This knowledge is essential for calculating thick target yields, estimating levels of radionuclidic impurities and designing suitable target systems. High current targets are needed to produce radionuclides in sufficient quantities. Due to the rapid degradation of charged particle energy in the target material, considerable heat is generated which needs to be effectively dissipated. Thus the major difference between the reactor and cyclotron production of radionuclides lies in the demands on nuclear data and sophisticated targetry techniques. A further difference lies in the time scale. Since most of the cyclotron products are short-lived, the chemical processing has to be fast. In general, however, the techniques of chemical processing and quality control are similar for both reactor and cyclotron production of radionuclides.

Positron emitters

The number of positron-emitting radionuclides is large. However, most of the PET studies to date are carried out with short-lived "organic" positron emitters, e.g. ^{11}C ($T_{1/2}$ = 20 min), ^{13}N ($T_{1/2}$ = 10 min), ^{15}O ($T_{1/2}$ = 2.0 min) and ^{18}F ($T_{1/2}$ = 110 min). There is considerable interest in ^{62}Cu ($T_{1/2}$ = 9.7 min), ^{68}Ga ($T_{1/2}$ = 68 min) and ^{82}Rb ($T_{1/2}$ = 1.3 min), which are obtained via longer-lived generator systems. Several other radionuclides, for example, ^{38}K ($T_{1/2}$ = 7.6 min), ^{73}Se ($T_{1/2}$ = 7.1 h), ^{75}Br ($T_{1/2}$ = 1.6 h), ^{94m}Tc ($T_{1/2}$ = 52 min) etc. also find some applications. The routine production methods of some most commonly used positron emitters are summarized in Table 1.

"Organic" β^+ emitters

The excitation functions for the formation of short-lived "organic" positron emitters have been measured in several laboratories. Typical data for the $^{16}O(p,\alpha)^{13}N$ [cf. 32] and $^{18}O(p,n)^{18}F$ [cf. 33] reactions are depicted in Fig. 2. The cross sections show strong fluctuations, probably due to the population of known discrete levels of the product nucleus. Evidently, for experimental studies on such reactions involving light mass nuclei, charged particle beams of well defined energies are required. It is also worth mentioning that nuclear model calculations cannot *a priori* describe such excitation functions. It is possible to fit the calculational results to known experimental data but, in general, theory is not capable of providing unknown data with the required accuracy in the light mass region.

The theoretically expected thick target yields of the four major "organic" positron emitters via the most important production routes are given in Fig. 3 [cf. 34]. Evidently, the nuclear structure effects observed in the excitation function are smoothed out in the integral yield calculation.

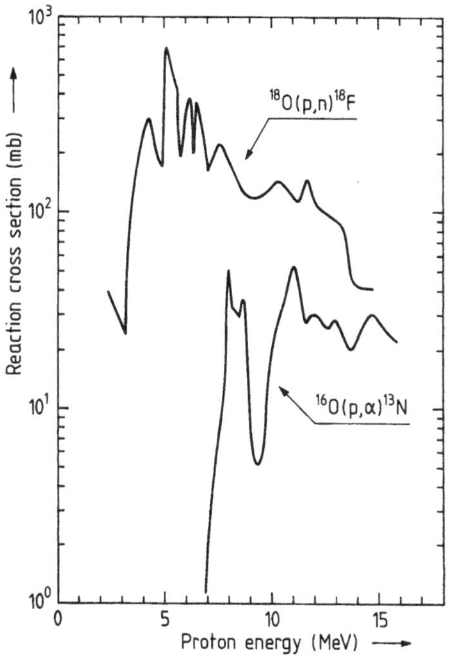

Fig. 2. Excitation functions of $^{16}O(p,\alpha)^{13}N$ and $^{18}O(p,n)^{18}F$ reactions [Refs. 32, 33].

Targetry constitutes a very important aspect of the production process. The choice of the target and structural materials is based on several considerations, such as their capability to withstand high intensity charged particle beams, the thick target yield and chemical form of the desired radioactive product, and above all its recovery and chemical reactivity. Both gas and liquid targets are commonly used. A typical gas target system is reproduced in Fig. 4 [cf. Ref. 34]. It consists of a conical cylinder made of aluminum-magnesium alloy containing 3% magnesium. The double window in front of the target consists of a 0.6 mm thick aluminum sheet (closing the target) and a 50 μm thick aluminum foil (separating the cyclotron vacuum). A helium stream (0.8 bar, −7 °C) flows through the two foils and cools the window. The target body is cooled by flowing water. The target filling and evacuation systems are also shown. Suitable target material and purity of the target gas are important prerequisites for obtaining high specific activity. The gas target shown in Fig. 4 is operated in the batch mode. It is filled with high purity nitrogen gas to a pressure of several bar. For ^{11}C-production a proton beam is used, the proton energy range within the gas being $E_p = 13 \rightarrow 3$ MeV. For ^{15}O-production, on the other hand, a deuteron beam is utilized, and the deuteron energy range within the gas corresponds to $E_d = 8 \rightarrow 0$ MeV. The formation of respective ^{11}C- and ^{15}O-precursors depends on the impurities and additives present in the nitrogen gas. The beam intensity and length of irradiation, i.e. the radiation dose, have also some effect on the chemi-

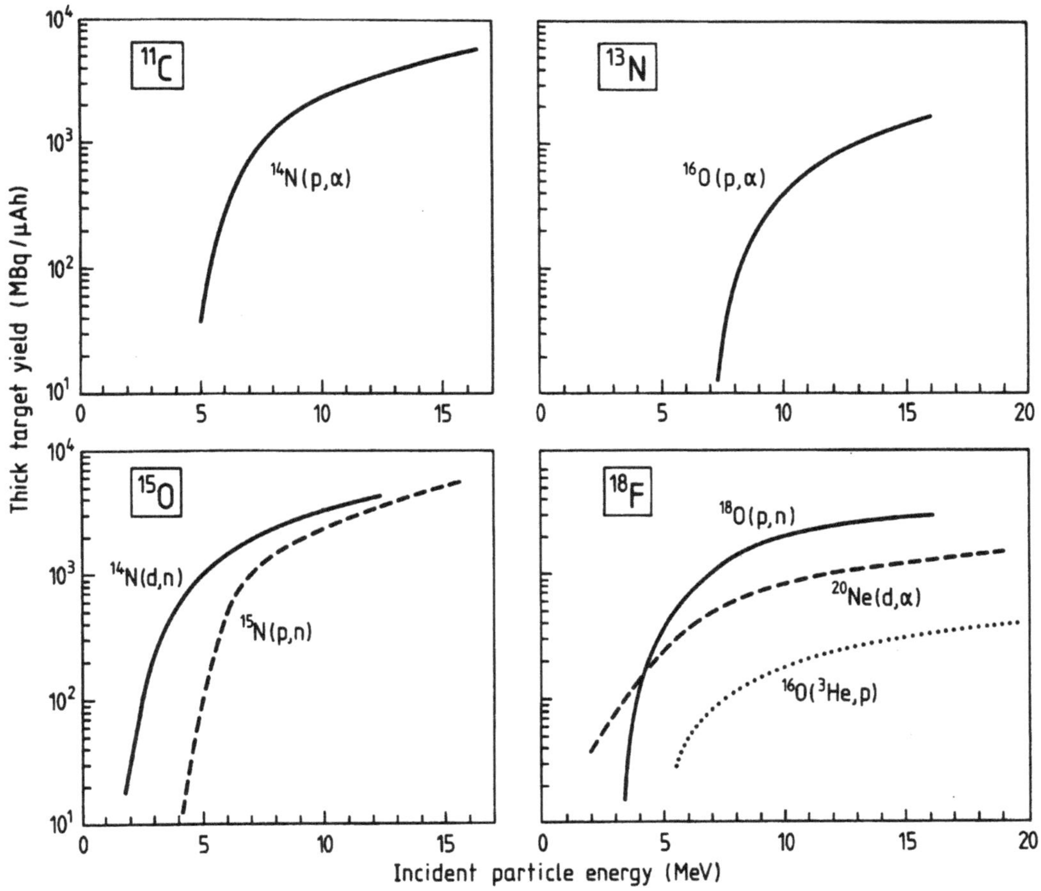

Fig. 3. Integrated thick target yields of the commonly used positron emitters (^{11}C, ^{13}N, ^{15}O, ^{18}F) expected from the most common production routes [Ref. 34].

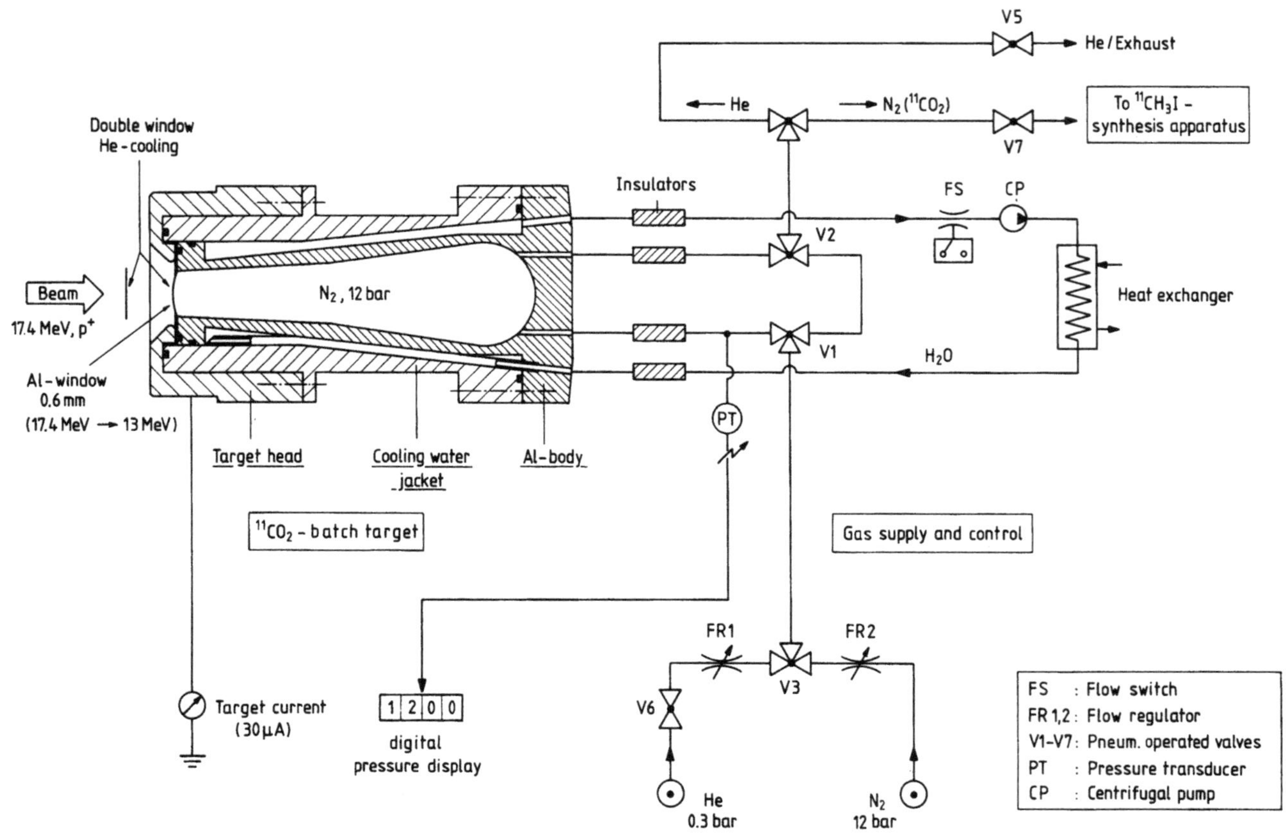

Fig. 4. Typical gas target system used in batch mode for production of ^{11}C and ^{15}O [Ref. 34].

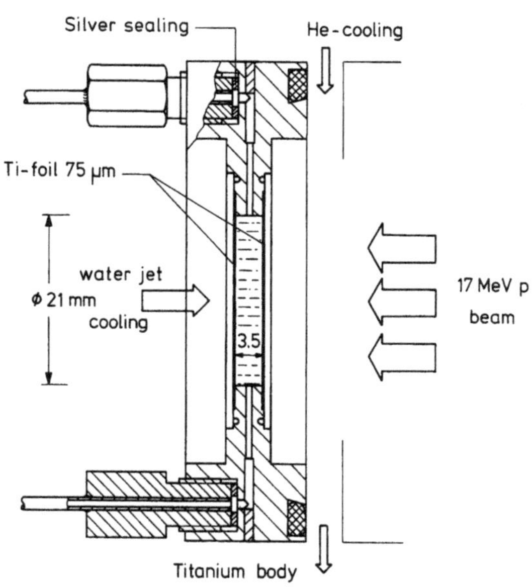

Fig. 5. Typical high pressure water target system for production of ^{13}N and ^{18}F [Ref. 34].

cal species formed in the gas target (for a review on in-target-chemistry see [80]). In general beam currents of 20 – 30 μA are used.

A typical water target is shown in Fig. 5 [cf. Ref. 34]. It consists of a titanium body, electron-beam welded to two titanium foils (75 μm thick), which act as front and back windows. The target takes 1.3 ml of water with no expansion space and the thickness of the water filling amounts to 3.5 mm. During irradiation, the back window is water-cooled, typically to 8 – 10 °C, and the front window is helium-cooled to −7 °C. The target can be remotely loaded with water. A polyethylene-polypropylene copolymer tube with an i.d. of 0.8 mm and a He drive pressure of 1.3 bar is considered a reliable transfer system. Transfer over 40 m through this tubing from the target to the nearest hot-cell takes only 2 min, and >90% of the product activity is recovered. The same target can be used for the production of both ^{13}N and ^{18}F. In the former case natural water ($H_2{}^{16}O$) is used and the proton energy range within the target corresponds to $E_p = 16 \rightarrow 7$ MeV. In the latter case, however, enriched water ($H_2{}^{18}O$) is employed and the effective projectile energy is $E_p = 16 \rightarrow 3$ MeV.

Using the two types of target systems described above it is possible to produce the four major "organic" positron emitters in GBq amounts at small cyclotrons accelerating two particles ($E_p \leq 18$ MeV; $E_d \leq 10$ MeV) with beam currents of 30 – 40 μA. Even smaller cyclotrons accelerating only protons ($E_p = 11 - 12$ MeV) deliver the required amounts of positron emitters, the only difficulty being with ^{15}O. For this radioisotope an alternative reaction $^{15}N(p,n)^{15}O$ [cf. Ref. 35] on highly enriched ^{15}N is used. Today fully automated cyclotrons and target systems are commercially available and provide the short-lived positron emitters very reliably. Presently attempts are underway to make use of mini-accelerators with $E_p = 3.7$ MeV or $E_d = 3.2$ MeV [cf. 36]. Whereas ^{15}O can be produced in sufficient quantities, high current (up to 1 mA) targets need to be developed to produce ^{18}F for subsequent labeling work.

Starting from the basic precursors (like $^{11}CO_2$, $^{13}NO_3^-$, ^{15}OO, $^{18}F^-$, etc.), the synthesis of radiopharmaceuticals follows using automated methods. Some aspects of those syntheses are described below. In the case of $^{18}F^-$, it is worth mentioning that presently not all syntheses can be performed via nucleophilic reactions. In some cases electrophilic reactions involving $[^{18}F]F_2$ are also used. This chemical form of ^{18}F is produced using a Ne gas target containing about 0.1% F_2 [cf. 37], or via an $^{18}O_2$ gas target with a two-step irradiation involving isotopic exchange between F^- and F_2 [cf. 38]. The chemical recovery and the subsequent reactivity of the product are, however, more difficult to control. The production of "organic" positron emitters thus involves some interesting aspects of gas phase and aqueous chemistry.

Generator produced β^+ emitters

The generator produced positron emitters find application mostly in PET studies at centers without a cyclotron. Of all the generator systems suggested only the $^{62}Zn/^{62}Cu$, $^{68}Ge/^{68}Ga$ and $^{82}Sr/^{82}Rb$ are widely used. Considerable effort has been devoted in recent years to the investigation of chemical aspects of generator preparation (for reviews see [39, 40]). A consideration of all those areas is beyond the scope of this article. Here a very brief account is given of the production of the three parent radionuclides concerned.

The β^+ emitter ^{62}Cu ($T_{1/2}$ = 9.7 min) forms a copper(II) pyruvaldehyde bis (N^4-methylthiosemicarbazone) complex (^{62}Cu-PTSM) which is used for regional blood flow measurement in brain and heart [cf. 41]. The parent ^{62}Zn is generally produced via the $^{63}Cu(p,2n)$ process at a medium-sized cyclotron [cf. Refs. 41, 42], the optimum energy range being E_p = 26 → 21 MeV. The irradiated copper is dissolved in HCl and the ^{62}Zn separated by ion-exchange chromatography. The product is then loaded onto an anion-exchange generator column for periodic removal of the daughter ^{62}Cu.

The β^+ emitter ^{68}Ga ($T_{1/2}$ = 68 min) is widely used for PET attenuation correction and also for labeling. The parent ^{68}Ge is rather difficult to produce. The $^{68}Ga(p,2n)^{68}Ge$ reaction has a high cross section but due to the long half-life of the product the yield is low [cf. 43, 44]. The method of choice is thus the spallation process [cf. 45]. To date it is therefore exclusively produced at large accelerators where long irradiations in parasitic positions are possible. The purification of the product involves several multi-step wet chemical processes.

The β^+ emitter ^{82}Rb ($T_{1/2}$ = 1.3 min) has found application in myocardial blood flow studies with PET mainly in hospitals without an on site cyclotron. The possible production methods for the parent ^{82}Sr (for a review see [46]) include the ^{3}He- and α-particle induced reactions on natural krypton [cf. 47], the (p,4n) reaction on ^{85}Rb [cf. 46, 48, 49] and, above all, the spallation of molybdenum with high energy protons [cf. 46]. In practice, however, the spallation process is the method of choice, and the major quantities of ^{82}Sr are presently produced at high energy accelerators where, similar to ^{68}Ge production, long parasitic irradiations are possible. The product is purified via multi-step wet chemical separations.

Less commonly used and potentially useful β^+ emitters

The number of these radionuclides is relatively large. However, they are regarded as research tools and their relative importance keeps on changing. Some of the areas where they find application include:

a) blood flow tracers (e.g. ^{38}K, ^{82m}Rb etc.)
b) labeling of biomolecules: analog tracers (e.g. $^{75,76}Br$ and ^{124}I labeled products, ^{73}Se as substitute for S, etc.)
c) metal complexes (e.g. ^{64}Cu, ^{45}Ti, ^{52}Fe, etc.)
d) quantitation of SPET radiopharmaceuticals with PET (e.g. ^{124}I for ^{123}I, ^{94m}Tc for ^{99m}Tc, ^{110g}In for ^{111}In, etc.)
e) quantitation of uptake kinetics of radiotherapeuticals with PET (e.g. ^{86}Y for ^{90}Y, ^{83}Sr for ^{89}Sr, etc.).

The production of many of these radionuclides demand larger cyclotrons. In most of the cases machines accelerating four particles, viz. p, d, ^{3}He and α-particles, have proven to be very versatile. Evidently, the development of a production process at such a machine involves considerable amount of fundamental nuclear chemistry work [cf. 50]. Besides nuclear data and targetry related studies, efficient and remotely controlled chemical processing methods need to be developed. Since in many cases highly enriched expensive target material is used, its recovery is an important feature of the whole process.

A variety of chemical techniques, involving both wet and dry methods have been applied in the separation of radioactive products. Some of the very commonly used methods consist of solvent extraction [cf. 51], ion-exchange chromatography [cf. 52, 53], distillation and thermochromatography. The latter techniques are simple and very often effective for the initial isolation of radioactive products from strong matrix activities. In recent years the thermochromatography has been very successfully applied in the production of ^{73}Se (cf. 54], ^{75}Br [cf. 55–57], ^{94m}Tc [cf. 58] etc. Thus the production of potentially important β^+ emitters requires a continuous research and development work.

Photon emitters

A large number of γ-ray emitting radionuclides have found application in diagnostic nuclear medicine, us-

Table 2. Routine methods of production of some commonly used photon emitters [50]

Radio-isotope	$T_{1/2}$	Mode of decay	Main γ-ray energy in keV (%)	Production data		
				Nuclear process	Energy range (MeV)	Thick target yield MBq(mCi)/μAh
^{67}Ga	3.26 d	EC (100)	93 (37) 185 (20)	$^{68}Zn(p,2n)$	26→18	185 (5)
^{99}Mo ↓ (generator)	2.75 d	β^- (100)	181 (6) 740 (12)	$^{235}U(n,f)$ $^{98}Mo(n,\gamma)$		a) b)
^{99m}Tc	6.0 h	EC (100)	141 (87)			
^{111}In	2.8 d	EC (100)	173 (91) 247 (94)	$^{112}Cd(p,2n)$	25→18	166 (4.5)
^{123}I	13.2 h	EC (100)	159 (83)	$^{123}Te(p,n)$ $^{124}Te(p,2n)$ $^{127}I(p,5n)^{123}Xe$ [c] $^{124}Xe(p,x)^{123}Xe$ [c]	14.5→10 26→23 65→45 29→23	137 (3.7) 392 (10.6) 777 (21) [d] 414 (11.2) [d]
^{201}Tl	3.06 d	EC (100)	69–82 (X-rays) 166 (10.2)	$^{201}Tl(p,3n)^{201}Pb$ [e]	28→20	18 (0.5) [f]

a) TBq (hundreds of Ci) amounts of this radioisotope are produced at centers with well-developed technology.
b) Method used in less-developed laboratories.
c) ^{123}Xe decays by EC (87%) and β^+ emission (13%) to ^{123}I.
d) This is ^{123}I yield expected from the decay of ^{123}Xe over an optimum time of about 7 h.
e) ^{201}Pb decays by EC (100%) to ^{201}Tl.
f) This is ^{201}Tl yield expected from the decay of ^{201}Pb over an optimum time of 32 h.

ing either γ-cameras or, in recent years, single photon emission tomography (SPET). The production methods of some important photon emitters are summarized in Table 2. The most commonly used diagnostic radioisotope ^{99m}Tc ($T_{1/2}$ = 6.0 h) is produced using a nuclear reactor. Its widespread use is mainly based on its convenient availability as a $^{99}Mo/^{99m}Tc$ generator. The ^{99m}Tc-radiopharmaceuticals are treated in more detail elsewhere (see chapter by A. Jones). Many of the accelerator produced γ-ray emitting radionuclides like ^{67}Ga, ^{111}In, ^{123}I and ^{201}Tl are also finding increasingly wide applications. All of those radioisotopes are produced using a medium-sized cyclotron. In recent years very high intensity medium-sized cyclotrons have been developed with the specific aim of medical radioisotope production. Extracted beams of about 0.5 mA are available; they fall slantingly on solid targets. Such completely automated cyclotrons are now commercially available and allow the production of several SPET-radioisotopes in large quantities.

The metallic radionuclides ^{67}Ga and ^{111}In form strong metal complexes and are of considerable medical interest. Both of them are produced via (p,2n) reactions, the former on ^{68}Zn and the latter on ^{112}Cd (for recent reviews see [59, 60]). Targetry and chemical processing problems have been well studied. The medical applications are well documented (see chapter by Weiner and Thakur).

Iodine-123 is a very useful analog label for biomolecules. Its nuclear properties are also almost ideal for SPET studies. The radionuclide ^{201}Tl is very broadly used for myocardial blood flow measurements (see chapter by Weiner and Thakur) and has a relatively long shelf-life. We discuss the production of these two important radioisotopes in rather more detail.

^{123}I ($T_{1/2}$ = 13.2 h)

For the production of ^{123}I about 25 nuclear processes have been investigated (for review see [61]). Among them only four are of significance. At low energy cyclotrons the $^{123}Te(p,n)$ reaction over E_p = 14.5 → 10 MeV is applied [cf. 62, 63], and at medium energy cyclotrons the $^{124}Te(p,2n)$ process over E_p = 26 → 23 MeV (for earlier references see the above mentioned review). In either case highly enriched target material is used. The recovery of ^{123}I is rather easy. The irradiated $^{123}TeO_2$ or $^{124}TeO_2$ target is heated at 755 °C (just above the m.p. of TeO_2) in a stream of oxygen whereby radioiodine is removed and taken up in a small amount of water. The target is regenerated and is ready for reuse. This is a well established process. The disadvantage of the (p,n) process is the low ^{123}I yield and that of the (p,2n) route the high level of ^{124}I impurity.

The third route of ^{123}I-production consists of the $^{127}I(p,5n)^{123}Xe \xrightarrow[2.1\,h]{EC,\beta+} {}^{123}I$ process (for references see the above mentioned review). The optimum energy range for the process is E_p = 65 → 45 MeV. Thus it needs a higher energy cyclotron. The removal of radioxenon from the irradiated NaI target is done either on-line, i.e. during the irradiation in a flow system, or off-line, i.e. after the end of the irradiation in a batch mode.

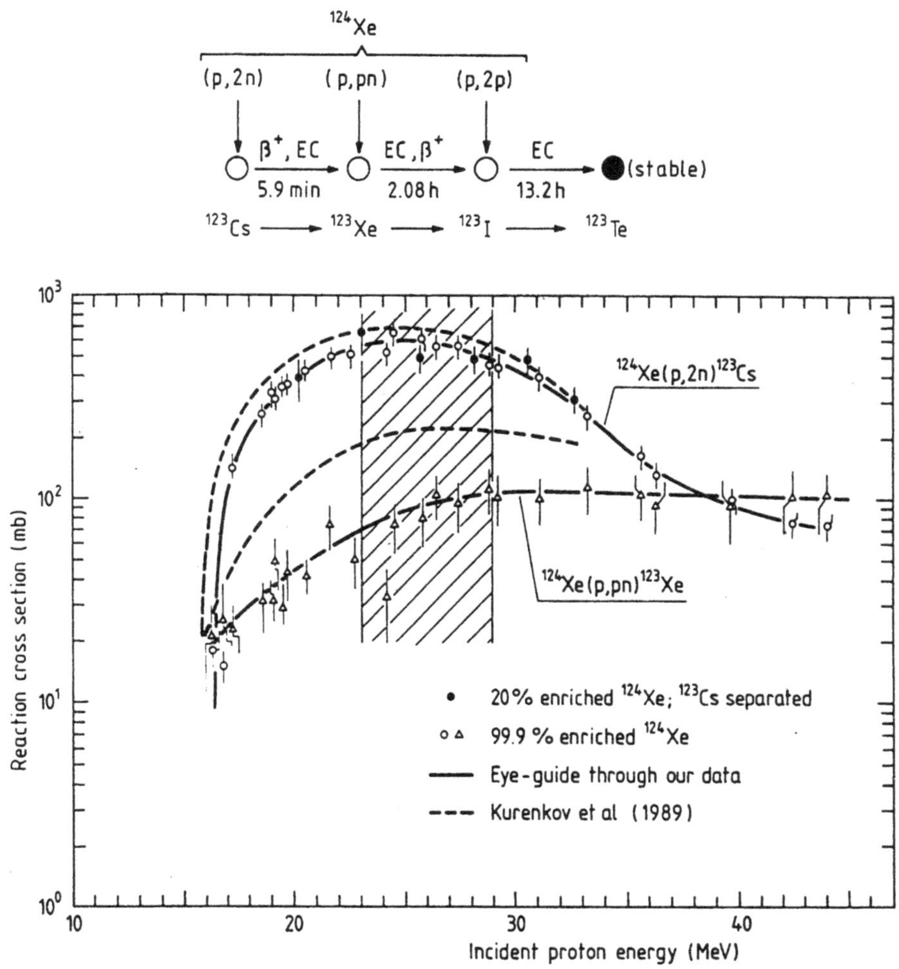

Fig. 6. Possible production routes of ^{123}I via protons on highly enriched ^{124}Xe. The lower part of the diagram gives the excitation functions of $^{124}Xe(p,2n)^{123}Cs$ and $^{124}Xe(p,pn)^{123}Xe$ processes [Ref. 65].

The collected radioxenon is allowed to decay for about 7 h during which period the maximum growth of ^{123}I occurs. Thereafter ^{123}I is removed from the vessel containing radioxenon. This is a relatively high yield process and the only impurity observed is ^{125}I (~0.25%).

The fourth route of ^{123}I-production consists of the $^{124}Xe(p,x)$-process. It involves highly enriched ^{124}Xe but requires only a medium-sized cyclotron. The development of this process clearly demonstrates the changing demands on the quality of the medically important radionuclides. In contrast to the above mentioned three routes, this process leads to ^{123}I of the highest purity. The natural abundance of ^{124}Xe is only 0.1%; consequently the highly enriched ^{124}Xe is very expensive. What has become a common production technique today, was not imaginable about 15 years ago.

The proton irradiation of ^{124}Xe leads to the formation of ^{123}I via three possible routes (cf. Fig. 6). Experimentally, the contribution of the (p,2p) process was found to be negligible, so the major production routes are (p,2n) and (p,pn). The cross section data for these two routes were measured both radiochemically and via direct counting of irradiated cells. The results for the major contributing reaction $^{124}Xe(p,2n)^{123}Cs$ reported by two groups [64, 65] are in good agreement; for the $^{124}Xe(p,pn)^{123}Xe$ reaction, however, there is some disagreement. The ^{123}I yields measured in a few laboratories [64–66] are, however, somewhat discrepant. Despite some of these discrepancies in the basic data, the method has been well developed. It involves highly sophisticated technology which is now commercially available.

^{201}Tl ($T_{1/2}$ = 73.5 h)

Several methods have been investigated for the production of this radionuclide. The most common route, however, consists of the $^{203}Tl(p,3n)^{201}Pb \xrightarrow[9.4\,h]{EC} {}^{201}Tl$ process. Extensive studies on this process have been done in several laboratories [cf. 30, 67–70]. The optimum energy range for production is $E_p = 28 \rightarrow 20$ MeV. Thus in this case also a medium-energy cyclotron is required.

The production of no-carrier added ^{201}Tl via irradiation of ^{203}Tl or ^{nat}Tl involves a two step process and elaborate wet chemical processing. At first ^{201}Pb is separated from the matrix activity and inactive Tl. Thereafter it is allowed to decay for an optimum time

Table 3. Methods for the production of some radioisotopes used in endotherapy[a]

Radio-isotope	$T_{1/2}$	$E_{\beta^-,max}$ in MeV	Main γ-emission in keV (%)	Production route
^{32}P	14.3 d	1.7		$^{31}P(n,\gamma)$, $^{32}S(n,p)$
^{67}Cu	2.6 d	0.6	185 (49)	$^{67}Zn(n,p)$,Ga(p,spall)
^{89}Sr	50.5 d	1.5		$^{88}Sr(n,\gamma)$
^{90}Y	2.7 d	2.3		$^{90}Sr/^{90}Y$ Generator
^{125}I	60.2 d	Auger electrons	35 (7)	$^{124}Xe(n,\gamma)^{125}Xe \xrightarrow{EC}$
^{131}I	8.0 d	0.6	364 (81)	$^{130}Te(n,\gamma)^{131}Te \xrightarrow{\beta^-}$ $^{235}U(n,f)$
^{124}I	4.2 d	2.0 (β^+)	603 (61)	$^{124}Te(p,n)$, $^{124}Te(d,2n)$
^{153}Sm	1.9 d	0.8	103 (30)	$^{152}Sm(n,\gamma)$
^{186}Re	3.7 d	1.1	137 (9)	$^{185}Re(n,\gamma)$, $^{186}W(p,n)$
^{188}Re	17 h	2.0	155 (15)	$^{186}W(n,\gamma)^{187}W(n,\gamma)^{188}W \xrightarrow{\beta^-}$

[a] For α-emitters see below (Table 7).

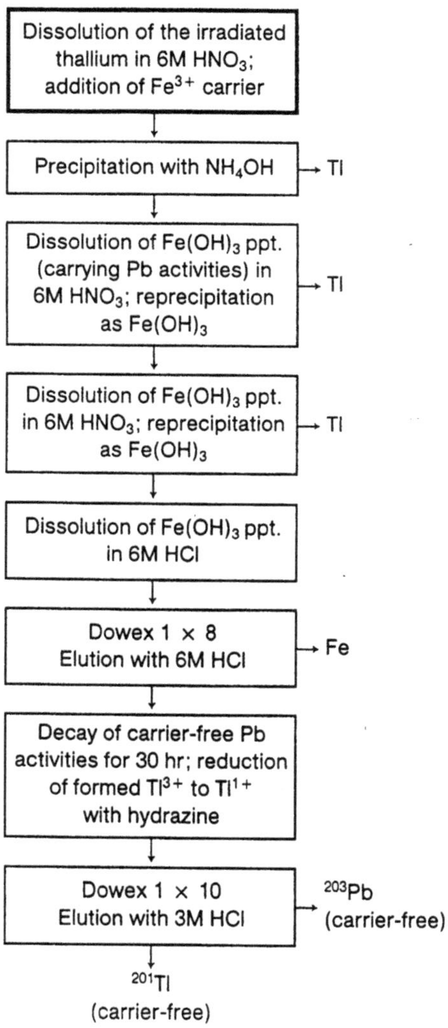

Fig. 7. Flow sheet of the separation procedure for the $^{203}Tl(p,3n)^{201}Pb \rightarrow {}^{201}Tl$ production process [Ref. 70].

of 32 h whereby ^{201}Tl grows in. The removal of the daughter is then effected via anion-exchange chromatography. A flow sheet of the separation procedure [cf. 70] is given in Fig. 7. In a slightly modified form it is used for commercial production. Thus even classical separation methods like coprecipitation and ion-exchange occasionally find widespread applications. It should be mentioned that Tl itself is a toxic heavy metal. The final no-carrier added ^{201}Tl should therefore not contain any appreciable quantities of the target material. Furthermore, for medical application only the product $^{201}Tl^+$ is of interest. It is also necessary that the level of the radionuclidic impurity ^{200}Tl should be as small as possible. The demands on the chemical, radiochemical and radionuclidic purity of the final product are thus rather stringent.

Therapy related radionuclides

Production routes of some therapy related radionuclides are listed in Table 3. Most of them are produced in a nuclear reactor via the (n,γ) process. In several cases the product decays by β^- emission or EC to the desired radionuclide, typical examples being the production of ^{131}I and ^{125}I. Occasionally the (n,γ) product captures a further neutron which leads to the parent of a suitable therapy nuclide (e.g. ^{188}W). This double neutron capture process is, however, significant only in a high flux nuclear reactor [cf. 71]. Some of the therapeutic radionuclides are produced via (n,p) and (n,α) reactions or via fission of ^{235}U. Similar to diagnostic radionuclides, a few therapy related radioisotopes are also obtained via a generator system [cf. 40]. The parent activity is produced via one of the above mentioned methods.

In recent years attempts have been underway to develop new therapeutic radionuclides which can be produced at cyclotrons. ^{211}At has been known for a long time and is produced via the $^{209}Bi(\alpha,2n)$ process at 28 MeV (see below, Table 7). Typical new examples are furnished by ^{124}I and ^{186}Re. Whereas in the former case both (d,2n) and (p,n) reactions have been applied [cf. 72–74], the latter makes use only of the (p,n) process [cf. 75]. Since the half-lives are relatively long and the yields rather low, high current targetry is man-

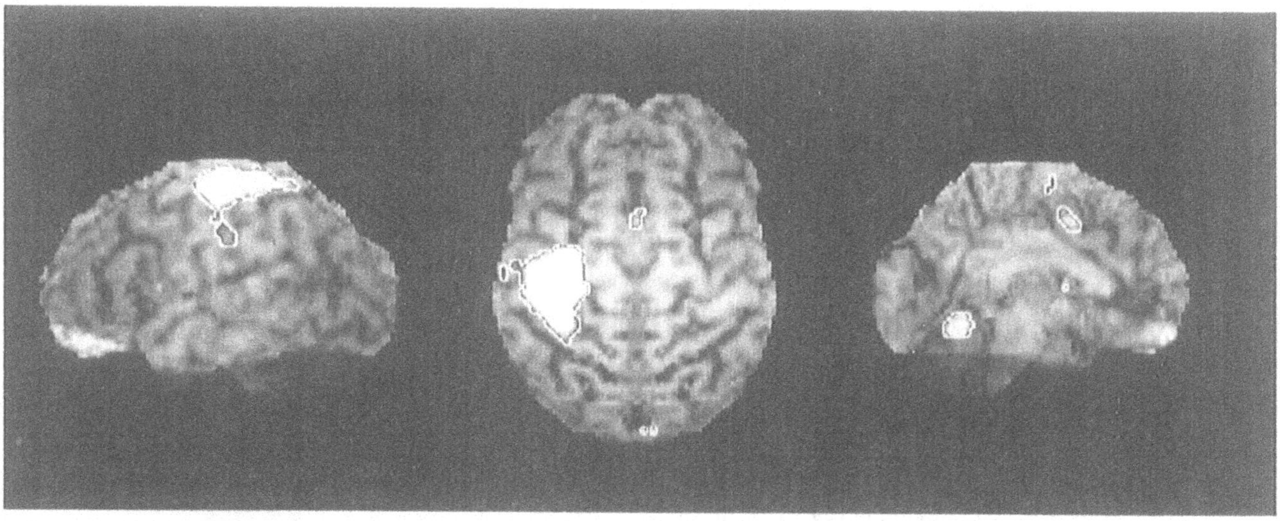

Fig. 8. Brain activation study with PET using $[^{15}O]H_2O$ as flow tracer: Finger opposition task of the right hand ($n = 9$) demonstrating incease of regional cerebral blood flow in left brain areas associated with motor function. (Courtesy, P. Bartenstein, Klinik für Nuklearmedizin, Technische Universität München).

datory. Irradiations in a parasitic mode, especially using slanting beams, are very advantageous.

Probing regional biochemistry in living man

With the advent of positron emission tomography (PET) in the mid seventies it became possible to quantitatively measure regional activities of a positron emitter labeled molecule with high sensitivity and a spatial resolution of a few millimeters. With the help of biomathematical models, turnover rates of metabolic substrates could be determined by means of positron emitter labeled biomolecules and PET. A new era of nuclear medicine was initiated. The most important example of this beginning is the introduction of 2-$[^{18}F]$Fluoro-2-deoxy-D-glucose (2-FDG). Based on the model developed by Sokoloff *et al.* for 2-$[^{14}C]$Deoxyglucose (DG) for the determination of the regional glucose metabolic turnover rate in rat by autoradiography [76], the Brookhaven National Laboratory prepared and evaluated the fluorine-18 analog 2-FDG [77, 78], which like DG is transported into the brain and phosphorylated but does not undergo further metabolic degradation. This metabolic trapping allows quantitation of the regional glucose metabolism with the help of a three-compartment model [79], for example in healthy and diseased brain areas, a technique which is widely used today for the noninvasive diagnosis of brain, heart and tumors (see below).

The short-lived "organic" β^+ emitters, in particular carbon-11 and fluorine-18, were subsequently used to study a wide range of biochemical processes in living man, and a new concept of tracer design started, which allowed the measurement of

- metabolic turnover of biological substrates and drugs
- enzyme and neurotransmitter concentration
- receptor density and -occupancy
- immuno reactions,

whenever it is possible to synthesize the labeled substrate or drug with sufficient total and specific activity; a new challenge for the radiochemists!

Fast PET chemistry at the cyclotron

Labeling with carrier-free, i.e. no-carrier-added (n.c.a.) short-lived radionuclides of high starting activity, having the penetrating 511 keV annihilation radiation, demanded a new type of chemistry. It is interesting to note that many of the pioneers came from hot atom chemistry. In fact, target chemistry and precursor production are often based on hot atom processes (for a review see [80]). A new technology of automated fast chemistry was developed during the past 20 years (for a recent monograph see [81]).

An extreme example is oxygen-15 with its 2 minute half-life. It is extensively used in simple chemical forms, such as $[^{15}O]O_2$ for the study of oxygen metabolism in the brain or $[^{15}O]H_2O$ for measuring blood flow. The sensitivity of the technique is demonstrated in Fig. 8 by a so called "activation study". The movement of fingers clearly shows a local increase of blood flow in brain areas associated with motor functions. Oxygen-15 labeled n-butanol, because of its optimum lipophilicity, is also of interest for measuring cerebral blood flow by PET. The synthesis of n-$[^{15}O]$butanol by reaction of molecular oxygen with tri-n-butylborane was chosen to prepare the oxygen-15 labeled molecule [82]. The synthesis was subsequently improved by loading the butylborane onto an alumina column [83, 84]. Finally a procedure was developed which allowed to synthesize the product in sufficient activity and purity for clinical use. The radioactive oxygen taken out of the cyclotron target is passed through an alumina cartridge, preloaded with tri-n-butylborane:

$$B(n\text{-}Bu)_3 \xrightarrow{^{15}OO} B(n\text{-}Bu)(^{15}O\text{-}n\text{-}Bu)(^{16}O\text{-}n\text{-}Bu).$$

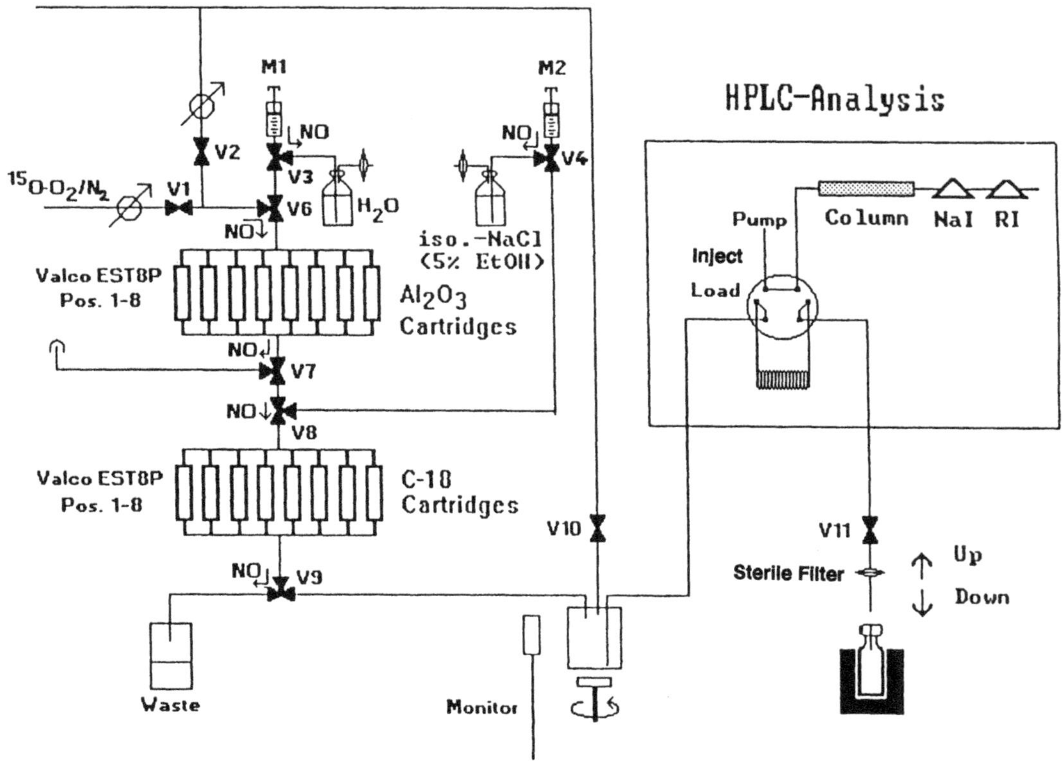

Fig. 9. Schematic set-up of the automated synthesis of n-[^{15}O]butanol [Ref. 85]. V1–V11 are diaphragm valves. M1 and M2 are pneumatic syringe pumps. Up to eight productions of n-[^{15}O]butanol in isotonic saline containing 5% ethanol can be performed within a turn-around time of 15 min.

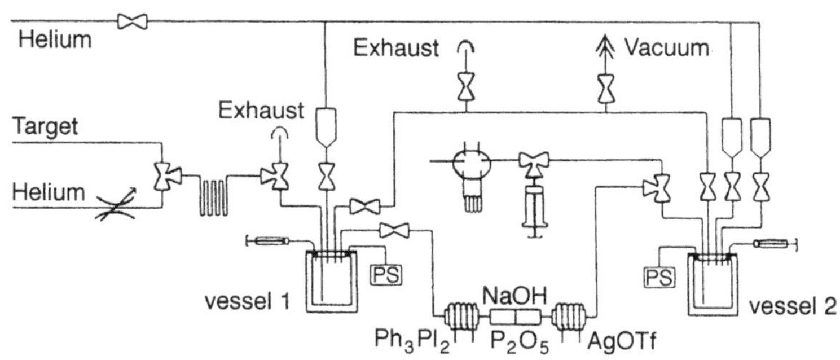

Fig. 10. Scheme of the automated on-line preparation of [^{11}C]methyl triflate [Ref. 81]. The $^{11}CO_2$ is transfered from the gas target by a flow of helium into reaction vessel 1 and reduced by lithium aluminium hydride. Subsequent addition of a mixture of diethyleneglycol monobutyl ether and water releases [^{11}C]methanol, which is then converted into [^{11}C]methyl iodide or into [^{11}C]methyl triflate by a column technique (see text).

The hydrolysis of the oxygen-15 labeled n-butylborane complex is effected with water within 56 seconds:

$$\mathrm{B(n\text{-}Bu)\text{-}^{15}O\text{-}(n\text{-}Bu)\text{-}O\text{-}(n\text{-}Bu)} \xrightarrow{H_2O} \mathrm{n\text{-}Bu^{15}OH} + \mathrm{n\text{-}BuOH} + \mathrm{B(n\text{-}Bu)}.$$

A schematic set-up of the automated synthesis with built-in quality control by HPLC is shown in Fig. 9 [85]. The synthesis is operated by a PC and is capable of producing up to eight batches, each in four minutes.

In the case of carbon-11, with a half-life of 20 min, elaborate n.c.a. syntheses of very complex compounds have been carried out [85a]. Not only the available time but also the high specific activity needed in many cases demand a sophisticated methodology. Labeled ligands for receptor mapping require specific activities greater than 18.5 TBq/mmol (500 Ci/mmol). Carbon-11 labeling, just as carbon-14 labeling, starts with precursors such as CO, CO_2, HCN, CH_3I etc.; however, ^{11}C-precursor production proceeds on-line or off-line in automated synthesizers. The major methylation agent is methyl iodide, since many radiopharmaceuticals are prepared via ^{11}C-methylation of NH or OH functional groups of the corresponding desmethyl compound. Methyl iodide is generally prepared from $^{11}CO_2$ via [^{11}C]methanol. A solid state flow system has been developed for on-line production of [^{11}C]methyl iodide or the more reactive methyl triflate [86, 87]. Triphenylphosphine diiodide adsorbed on γ-alumina is used as iodinating agent [88]. The [^{11}C]methyl iodide

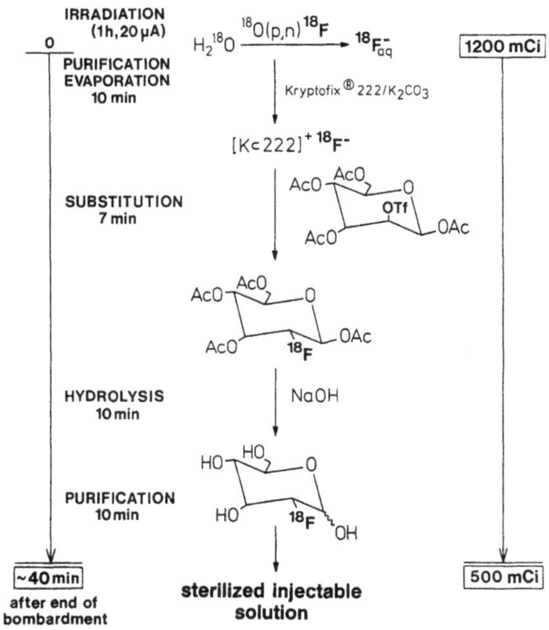

Fig. 11. Scheme of 2-[^{18}F]FDG synthesis [Ref. 91]. Aqueous n.c.a. [^{18}F]fluoride is used for the kryptate-mediated nucleophilic synthesis of 2-[^{18}F]fluoro-2-deoxy-D-glucose. The fluorination of the triflate precursor is followed by the base catalyzed hydrolysis of the tetraacetylated 2-[^{18}F]FDG according to a recent modification [91a].

can be passed through an alumina column charged with silver trifluoromethanesulfonate for conversion to [^{11}C]methyl triflate [87]. A scheme of the automated on-line preparation of these methylation agents is shown in Fig. 10. Recently, a new method of production of ^{11}C-labeled methyl iodide from [^{11}C]methane was reported [89]. The [^{11}C]CO_2 is reduced by passing it with hydrogen gas over a NiO catalyst. The [^{11}C]CH_4 then reacts at 725 °C with iodine:

$$I_2 \rightleftharpoons 2\,I^{\cdot}$$

$$^{11}CH_4 + I^{\cdot} \rightarrow {}^{11}CH_3^{\cdot} + HI$$

$$^{11}CH_3^{\cdot} + I_2 \rightarrow {}^{11}CH_3I + I^{\cdot}$$

Compared to carbon-11, fluorine-18, having a 110 minute half-life, allows more elaborate syntheses and applications. Fluorine-18, which can be produced in large quantities at a low energy cyclotron (see above, radionuclide production) also lends itself to shipment to places which can be reached within a few hours. For hospitals without a cyclotron this satellite concept is well advanced in the USA, Europe and Australia, since it is more cost effective than running a cyclotron locally. In addition, fluorine-18 tracers are increasingly used with scanners less expensive than modern PET, such as collimated SPET or simple γ,γ-coincidence imaging devices.

This is particularly true for the workhorse 2-[^{18}F]FDG. Only nucleophilic [^{18}F]fluoride can be produced in a no-carrier-added form, while electrophilic fluorine is inherently carrier-added [^{18}F]F_2. Therefore most ^{18}F-fluorination procedures (for a review see [90]) are based on nucleophilic substitution reactions. As an example the preparation of 2-[^{18}F]FDG is outlined in Fig. 11, even though n.c.a. preparation of this metabolic tracer is not needed, and the first synthesis at the Brookhaven National Laboratory was a carrier-added electrophilic one [77, 78]. Present day routine synthesis is based on phase transfer catalyst, e.g. the amino polyether Kryptofix 2.2.2. mediated nucleophilic fluoride-for-triflate exchange on the tetraacetylated mannose [91]. Fig. 11 shows that several hundred millicuries of the radiopharmaceutical can be produced within about forty minutes. Automated synthesizers are commercially available.

Table 4. Major PET radiodiagnostics

Radiopharmaceutical	Application	
	Neurology	
[^{15}O]O_2	Oxygen consumption	dementia, ischemia, stroke
[^{15}O]H_2O	Blood flow	
[^{15}O]Butanol	Blood flow	
2-[^{18}F]FDG	Glucose metabolism	
L-6-[^{18}F]FDOPA	Presynaptic dopaminergic function	parkinsonism
L-[S-methyl-^{11}C]Methionine	Amino acid metabolism and -transport	brain tumors
[^{11}C]Raclopride	D_2-receptor density or -occupancy	schizophrenia, therapy control
[^{18}F]N-Methylspiperone	D_2-receptor density or -occupancy	schizophrenia, therapy control
[^{11}C]Flumazenil	Benzodiazepine receptor mapping	epilepsy
	Cardiology	
[^{13}N]NH_3	Blood flow	ischemia viability
2-[^{18}F]FDG	Glucose metabolism	
[^{11}C]Acetate	Oxidative metabolism (Oxygen consumption)	
[^{11}C] or [^{18}F]Fatty acids	β-Oxidation	
[^{11}C]Hydroxyephedrine (HED)	Sympathetic nerve viability	heart transplant infarction
	Oncology	
[^{15}O]H_2O	Blood flow	
2-[^{18}F]FDG	Glucose metabolism	
[^{11}C]Methionine	Amino acid metabolism and -transport	
5-[^{18}F]Fluoruracil (FU)	Therapy control	
5-[^{18}F]Fluorodeoxyuridine (FUdR)	Cell proliferation	
[^{18}F]Fluoromisonidazol (FMISO)	Hypoxia	

Tracers for probing biochemistry in vivo

The applicability of cyclotron-produced radionuclides and their corresponding radiopharmaceuticals in both

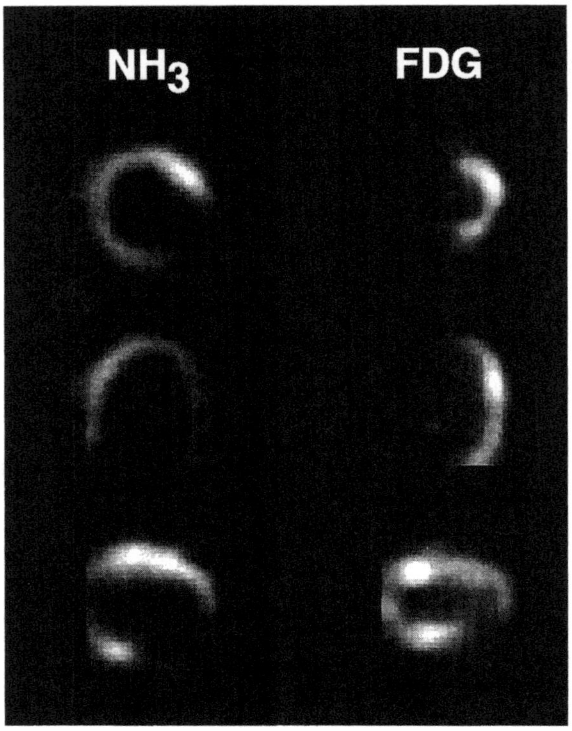

Fig. 12. PET study of blood flow with [^{13}N]ammonia and glucose metabolism with 2-[^{18}F]FDG in a patient with coronary artery disease. Flow regionally decreased, metabolism regionally preserved (flow-metabolism-mismatch). From upper row: Short axis views, horizontal and vertical long axis. (Courtesy M. Schwaiger, Klinik für Nuklearmedizin, Technische Universität München).

basic medical research and clinical practice has already been demonstrated. Several positron emitting radiopharmaceuticals are being used in diagnostic nuclear medicine (see Table 4). 2-[^{18}F]FDG remains as the workhorse. In the field of cardiology viability of ischemic tissue can be shown in cases where the regional blood flow has already decreased. The latter is measured with [^{13}N]ammonia. Fig. 12 shows a typical PET study of a flow-metabolism mismatch in a patient with coronary artery disease. Metabolic imaging is helpful in selecting patients for surgery or coronary angioplasty. [^{11}C]Acetate is used for measuring regional oxidative metabolism in the heart. [^{11}C]Hydroxyephedrine (HED), a catecholamine analog which is stored in cardiac presynaptic sympathetic nerve terminals, is an attractive research tool in neurocardiology (for review see [92]), and is used among others to evaluate reinnervation of heart transplant. Many brain disorders can be evaluated by measuring regional glucose or oxygen turnover. Fig. 13 shows a typical PET scan of a patient with Alzheimer's disease as obtained with 2-[^{18}F]FDG. Local hypometabolism (arrow) is observed. PET also allows in vivo mapping of receptor areas, the measurement of receptor densities and neurotransmitter activities with suitable labeled receptor ligands or neurotransmitters. Before the advent of PET this was only possible by ex vivo animal- or post mortem studies using autoradiography. Most of the PET studies have been carried out for the evaluation of the central dopaminergic system, which plays a role in Parkinson's disease and schizophrenia. ^{11}C- or ^{18}F-labeled antagonists, such as neuroleptics and other centrally active drugs, with high affinity for a specific subtype receptor, can be used for therapy control [93–95]. The receptor occupancy by the drug given to the patient can be "titrated" with the radioactive ligand and PET. This allows a better setting of the drug dose for the individual patient than from the measurement of the drug plasma level. Fig. 14 gives an example of this approach. Carbon-11 labeled raclopride, a salicyl amide derivative with a high affinity for the D_2 receptor has been used for this purpose [93]. The figure shows that blockade of the dopaminergic receptors is

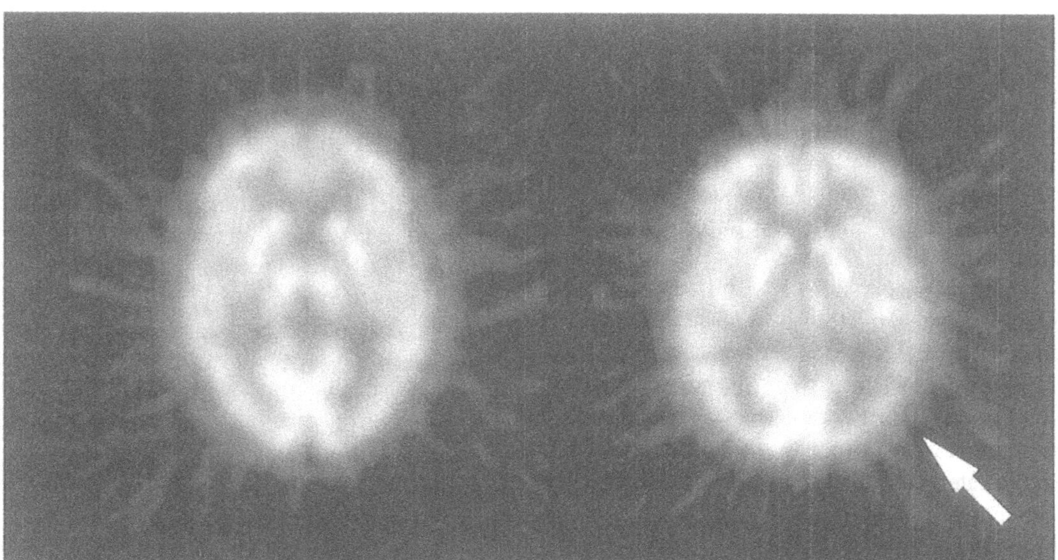

Fig. 13. Glucose metabolism study with PET using 2-[^{18}F]FDG in a normal control (left) and in a patient with Alzheimer's disease. The transversal slice shows the typical bilateral tempoparietal hypometabolism (arrow). (Courtesy P. Bartenstein, Klinik für Nuklearmedizin, Technische Universität München).

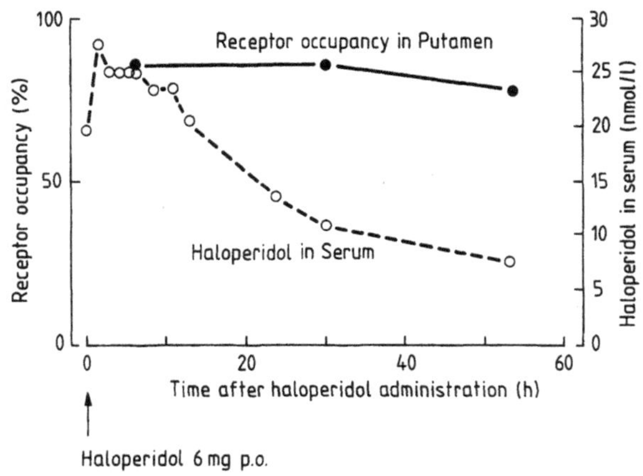

Fig. 14. Measurement of D_2 receptor blockade after application of 6 mg of the neuroleptic haloperidol to a schizophrenic, using [^{11}C]raclopride and PET. Haloperidol concentration in patient's plasma is also shown [93]. (Courtesy L. Farde, Karolinska Hospital, Stockholm).

already reached at very small plasma concentrations of the neuroleptic haldol and that the occupancy at the site of action, i.e. the Putamen in the brain, lasts much longer than indicated by the plasma concentration.

The neurotransmitter dopamine plays an important role in Parkinson's disease, where its concentration is decreased. Since dopamine does not cross the blood brain barrier, a pro-drug approach is applied and the amino acid L-DOPA is given for therapy. L-DOPA is transported through the blood brain barrier and decarboxylated by the DOPA decarboxylase (DD) to dopamine, thus increasing the dopamine pool (cf. equation given below). By using fluorine-18 labeled 6-[^{18}F]2-DOPA as tracer, the presynaptic dopaminergic function can be monitored with PET [96].

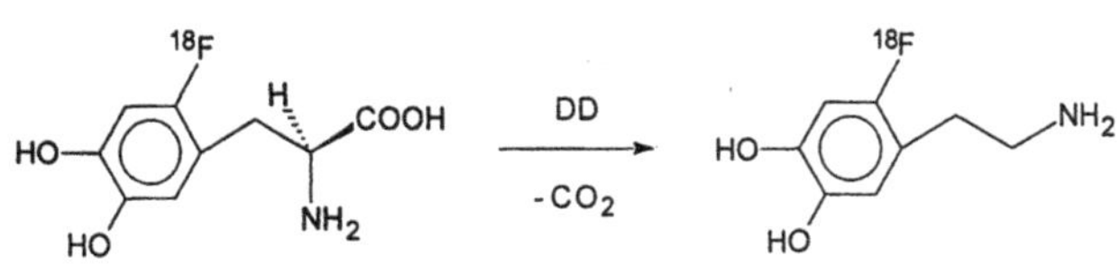

Since the scanner cannot distinguish between different chemical species, the metabolic fate of the label is another important information, which also has to be provided by the radiochemist. Another application of radiolabeled central receptor ligands was found to be in the diagnosis of epileptic foci using the ^{11}C labeled benzodiazepine receptor ligand flumazenil [97]. The epileptic focus can be localized by a receptor binding defect, an information which is of importance for surgery [98].

Oncology studies have become increasingly important in nuclear medicine. Radiopharmaceuticals have been used for the diagnosis of occult lesions, staging of disease, selection of patients for cancer therapy, and the determination of the pharmacokinetics of chemotherapeutic agents. The metabolic state of the tumor can be evaluated by 2-FDG or amino acids, the cell surface by suitable receptor ligands such as peptides or monoclonal antibodies, and the proliferation

Table 5. Major Iodine-123 radiodiagnostics for SPET

Radiopharmaceutical	Application
Oncology	
[^{123}I]meta-Iodobenzylguanidine (MIBG)	Pheochromocytoma, neuroblastoma
3-[^{123}I]Iodo-α-methyl-L-tyrosine (IMT)	Amino acid transport (brain tumors)
[^{123}I]Iodo-Tyr^3-octreotide	Somatostatin-receptor bearing tumors
[^{123}I]Monoclonal antibodies (mAb)	Immunoscintigraphy
[^{123}I]mAb fragments (e.g. $F(ab')_2$)	Immunoscintigraphy
Neurology	
[^{123}I]Benzamides (e.g. IBZM)	D_2-receptor mapping or -occupancy
[^{123}I]β-CIT	D_2-dopamine and serotonine transporter (neuro-degenerative and psychiatric disorders)
p-[^{123}I]Iododexetimide (IDEX)	Muscarinic-cholinergic receptor mapping (epilepsy)
[^{123}I]Iomazenil	Benzodiazepine receptor mapping (epilepsy)
MIBG	Sympathetic nerve function
Cardiology	
15-(p-[^{123}I]Iodophenyl pentadecanoic acid (IPPA)	β-Oxidation (viability)
15-(p-[^{123}I]Iodophenyl-3-methylpentadecanoic acid (BMIPP)	oxidative metabolism (viability)

by DNA precursors (for a review see [99]). In addition blood flow in the tumor can be evaluated with suitable tracers in order to predict the likelihood of drug access. Hypoxia, limiting the efficacy of radiation therapy can be studied with labeled nitroimidazole conjugates (for review see [100]).

Iodine-123 radiopharmaceuticals for SPET

With the advent of Single Photon Emission Tomography as a device allowing semiquantitative tomography with radioisotopes having a dominant γ-ray with energies in the range of about 70 to 200 keV, a wider and more economical way of functional imaging became available. The generator isotope technetium-99m and some longer-lived radionuclides, such as iodine-123, do not require an on-site cyclotron. While technetium-99m radiopharmaceuticals are almost exclusively limited to flow tracers (see chapter by A. Jones), and useful metabolic tracers are not yet available, iodine-123 as an "organic" isotope lends itself to the design of metabolic radiopharmaceuticals applying the analog approach as in the case of fluorine-18. Unfortunately,

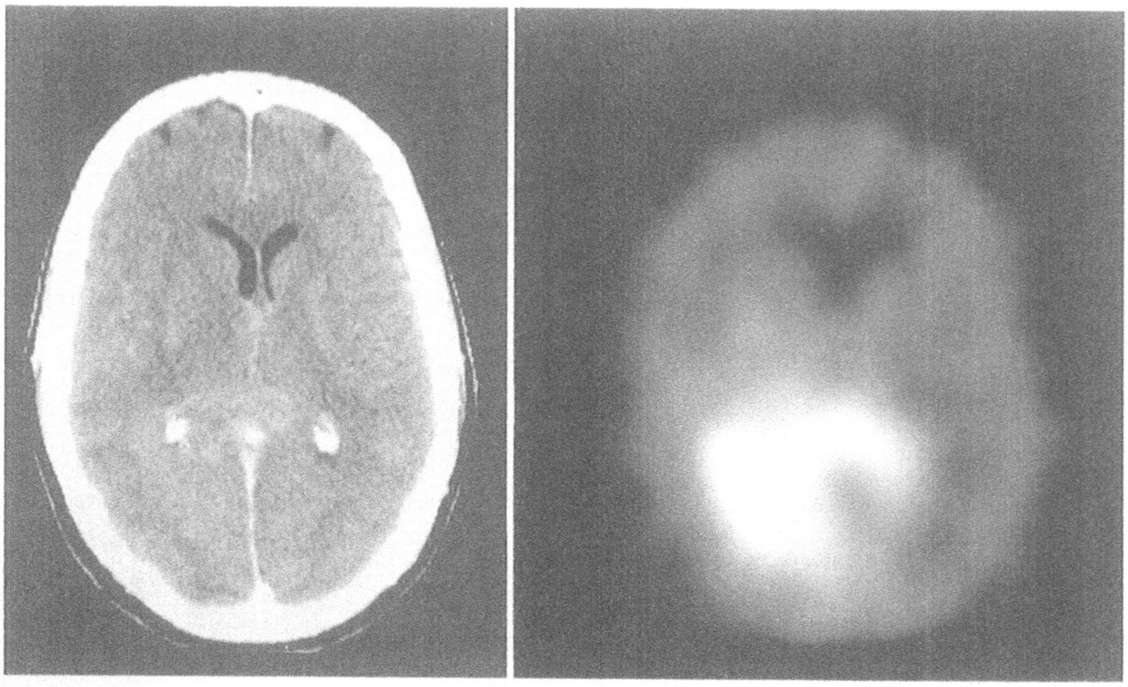

Fig. 15. Left: X-ray computed tomography (CT) after intravenous application of contrast medium. Right: [^{123}I]IMT SPET shows a high amino acid accumulation in the glioma while the tumor cannot be delineated in the X-ray CT. (Courtesy K.-J. Langen, Institut für Medizin, KFA Jülich).

the carbon-iodine bond is considerably weaker than the carbon-fluorine bond, and special aromatic or vinylic compounds are needed to avoid in vivo deiodination. The great demand for radioiodinated tracers for in vivo and in vitro applications has stimulated the search for easy no-carrier-added radioiodination procedures. Radiochemists have significantly contributed to this field of chemistry (for reviews see [101, 102]), and simple kit techniques have been developed [102–104]. Several iodine-123 radiopharmaceuticals find application in functional diagnostics (Table 5).

Often the development proceeds from PET to SPET, carbon-11 being the gold standard. For example, the above mentioned [^{11}C]flumazenil was the lead structure for the iodine analog [^{123}I]iomazenil, which is also used for the localization of epileptic foci [105, 106], but due to its higher lipophilicity is also a brain blood flow tracer, a fact which complicates application. Another example is the presynaptic dopamine receptor antagonist IBZM, an iodine analog of the above mentioned benzamide raclopride [107]. More recently an iodinated congener of cocaine, [^{123}I]β-CIT, has been developed, a ligand of the dopamine- and serotonin transporter which is used for the evaluation of neuro-degenerative and psychiatric disorders [105]. Several radioiodinated aromatic amino acid- and fatty acid analogs have been prepared and evaluated as metabolic tracers (for review see [108]). The artificial amino-acid 3-[^{123}I]Iod-alpha-methyl-L-tyrosin (IMT) found application in brain tumor diagnosis [109, 110]. Fig. 15 shows an image of a brain tumor as taken with a modern Three-head-SPET system in comparison to X-ray CT after application of contrast medium. The tumor cannot be delineated on the CT scan while IMT

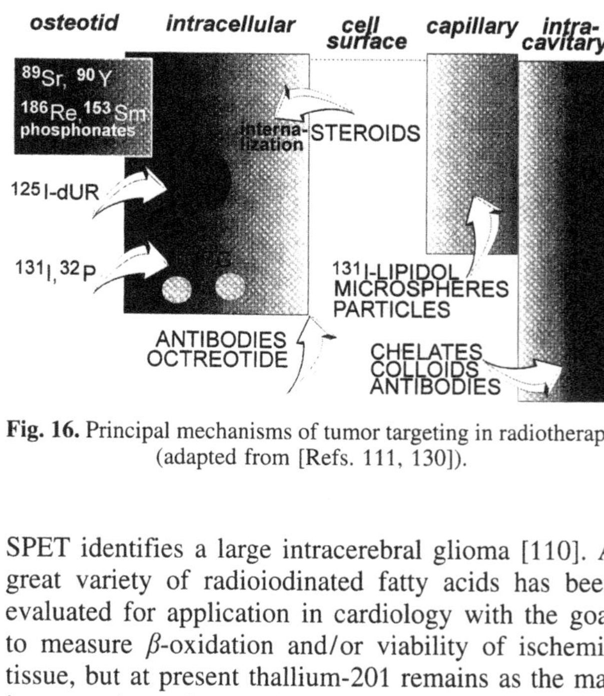

Fig. 16. Principal mechanisms of tumor targeting in radiotherapy (adapted from [Refs. 111, 130]).

SPET identifies a large intracerebral glioma [110]. A great variety of radioiodinated fatty acids has been evaluated for application in cardiology with the goal to measure β-oxidation and/or viability of ischemic tissue, but at present thallium-201 remains as the major tracer in nuclear cardiology with SPET, apart from ^{99m}Tc-compounds (see review by A. Jones).

Radiation therapy

There are several possibilities to make use of radiation for therapeutic purposes. One possibility is to use external radiation like X-rays and high-energy γ-rays, or various energetic particles such as neutrons, protons and heavier ions, the latter utilizing the high selectivity of the Bragg-peak. Another approach is represented by the administration of boron-based compounds and their subsequent irradiation via an external thermal neutron source (boron neutron capture therapy), which uses the emitted α-particles and the ^{7}Li-recoils. Regarding direct administration of therapeutic radionuclides, two methods are used. One route consists of intraarterial or intracavitary introduction of insoluble particles with inserted radionuclides, which remain at the same position within the body. The other route consists of intravenous administration of labeled radiotherapeuticals, which selectively accumulate in the corresponding target tissues according to specific biomolecular interactions. This is what one may best call endoradiotherapy (ERT) and what this section deals with.

There are several principal routes of targeting, such as incorporation into the cell (e.g. [^{131}I]iodide, [^{131}I]MIBG), or onto the cell surface (e.g. monoclonal antibodies (mab), peptides, and other receptor-binding molecules) but also by local intraarterial (microspheres, particles), intracavitary (colloids, mab) and intratumoral administration. These routes are schematically illustrated in Fig. 16 as adapted from [111, 130].

Nuclear properties of therapeutic radionuclides

Table 6 summarizes the radiotherapeuticals available today (at least in Europe), being either routinely used or undergoing clinical trials. In addition to the three therapeutic radionuclides of the early period of ERT, i.e. ^{32}P, ^{89}Sr, ^{131}I, several new radionuclides have been added: ^{80m}Br, ^{67}Cu, ^{90}Y, ^{153}Sm, ^{165}Dy, ^{169}Er, ^{186}Re, ^{198}Au. Moreover, there are many candidates which have not yet undergone clinical trials, but which have been extensively studied in order to reach this status. These are the β^--emitters ^{47}Sc, ^{77}As, ^{111}Ag, ^{105}Rh, ^{109}Pd, ^{165}Dy, ^{166}Ho, ^{188}Re and ^{64}Cu, the Auger-electron emitter ^{117m}Sn as well as the α-emitters ^{211}At and ^{212}Bi. The reader is referred to a number of books and articles, dealing with the suitability of these and further candidates [e.g. 40, 112–117].

This large number of radionuclides raises some questions. What are the criteria for the selection of a given isotope for therapeutic application? And consequently: may a systematic evaluation result in some new, promising candidates? The "ideal" radionuclide delivers its radiation exclusively to the target organ but not to the surrounding healthy tissue. Thus, the first criterion is the mode of decay, i.e. α-, or β-particles or Auger-electrons (as cascades originating from electron capture and/or inner conversion) and the quantity of the accompanying γ-radiation, expressed in terms of branching factor and energy. The type and energy of the emitted particle radiation determine the linear energy transfer (LET) of the radiation in tissues, cf. Fig. 17.

Decay scheme

What dosimetry demands from an "ideal" candidate is in fact a ratio of non-penetrating (particle) to pen-

Table 6. Radiotherapeuticals available today [adapted from Refs. 111, 130]

Isotope	Therapeutical	Indication
^{32}P	phosphate resin, microspheres colloids	bone metastases[g], polycythemia vera[b] liver tumors[i] astrocytoma/cystic craniopharyngioma[j], ALL intratheca therapy[j]; malignant infusions[j]
^{67}Cu	mab	various tumors[f]
^{80m}Br	estrogens	breast carcinoma[c]
^{89}Sr	chloride	bone metastases/ osteosarcoma[g]
^{90}Y	citrate glass microspheres particles mab colloids	bone metastases[g] liver tumors/sarcoma[i] hepatoma[i]; liver tumors/sarcoma[i] malignant infusion[j]; various tumors[j] synovectomy[j]; astrocytoma/cystic craniopharyngioma[j]
^{114m}In	A31 cells	lymphoma[h]
^{125}I	IUdR tamoxifen MIBG	chorioncarcinoma[a] breast carcinoma[c] neural crest tumors[b]
^{131}I	iodide BDP3 MIBG rose bengal lipidol antiferritin SMS-analog mab	differentiated thyroid carcinoma[f], oncocytoma[f] bone metastases[g] neural crest tumors[b] hepatoblastoma[b] liver malignancy[i] HCC/Hodgin's disease[f] neuroendocrine tumors[e] colon/medullary thyroid carcinoma[f], colon/ovarian carcinoma[f]; leukaemia/ lymphoma[f]; melanoma[f]; neuroblastoma[f]; malignant infusions[j]
^{153}Sm	EDTMP	bone metastases/ osteosarcoma[g]
^{165}Dy	FHMA	synovectomy[j]
^{169}Er	citrate	synovectomy[j]
^{186}Re	sulphide, colloid DMSA HEDP	synovectomy[j]; astrocytoma/ cystic craniopharyngiom[a] medullary thyroid carcinoma[d] bone metastases[g]
^{198}Au	colloid	synovectomy[j]; ALL intrathecal therapy[j]

Sites of accumulation and principle of tumor-seeking:
Intracellular: [a] DNA incorporation, [b] metabolic, [c] steroid receptors, [d] non-specific;
Cell surface: [e] hormone receptor, [f] immunologic;
Extracellular: [g] adsorption, [h] cells, [i] intracapillary, [j] intracavitary.

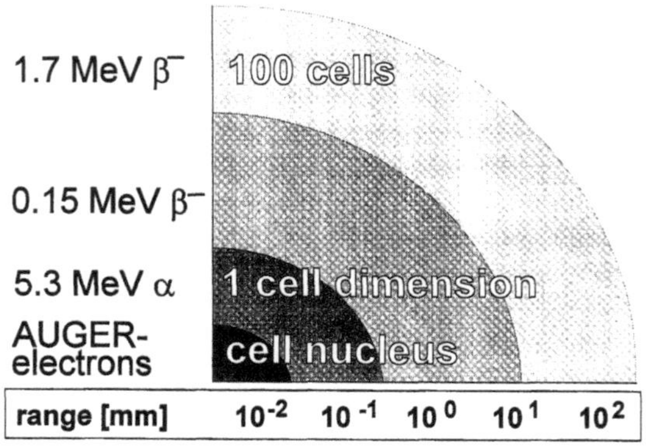

Fig. 17. Correlation between type and energy of particle radiation and the range in tissue (adapted from [Ref. 118]).

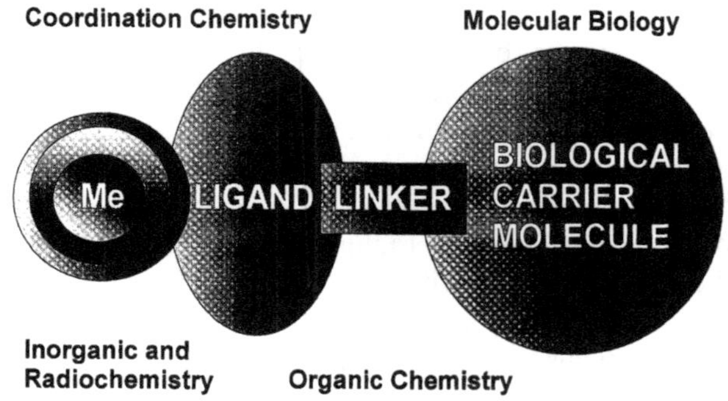

Fig. 18. Multidisciplinary approach to the development of coordinately labeled radiopharmaceuticals.

etrating (γ) radiation as high as possible. Among the β^--emitting candidates, for example, this ratio is high for "pure" β^--emitters such as ^{90}Y ($\approx$1000) but less for radionuclides with some accompanying γ-radiation such as ^{67}Cu ($\approx$100) and ^{111}Ag ($\approx$1).

Half-life

Generally radionuclides having a half-life between some hours and about 70 days are adequate. However, only those candidates are able to reach routine clinical application which can be produced in high yields and may subsequently be distributed to the hospitals. Introduction of short-lived candidates necessitates an effective coordination between the producer (at a reactor or cyclotron), the radiochemical facility, the distributor and the hospitals. In general, this system matches with the "satellite concept" known for the short-lived PET radiopharmaceuticals (see above). Obviously, the generator-based delivery is most convenient. It is, however, limited to a small number of potential generator systems: $^{90}Sr/^{90}Y$, $^{188}W/^{188}Re$, $^{166}Dy/^{166}Ho$ and $^{212}Pb/^{212}Bi$.

Decay products

For an "ideal" candidate, the decay product should be stable. Formation of long-lived decay products having unacceptable radiation characteristics generally prevents the application of the corresponding candidate. Formation of very short-lived decay products ($T_{1/2}$ on the order of a few minutes or less) may be tolerable, if the resulting isotope does not emit particle radiation. If the decay product is not only very short-lived, but in addition emits either low-energy γ-radiation or high-branching positron-emission, it can be used to obtain qualitative images of the therapeutical with SPET or quantitative images with PET. Such pairs are $^{66}Ni/^{66}Cu$ ($T_{1/2}$ = 5.1 s), $^{191}Os/^{191m}Ir$ ($T_{1/2}$ = 4.96 s), $^{140}Nd/^{140}Pr$ ($T_{1/2}$ = 3.39 min) or $^{128}Ba/^{128}Cs$ ($T_{1/2}$ = 3.6 min).

If the decay product itself is a potential candidate for radiotherapy, it may be considered as an in vivo generator, provided the decay of the primary therapeutic radionuclide allows the decay product to stay within the target and/or within the environment of the therapeutical instead of rejecting it due to recoil effects, as studied for the pair ^{212}Bi-DOTA/^{212}Pb-DOTA [119], for example. The most promising example is the pair ^{166}Dy ($T_{1/2}$ = 3.40 d)/^{166}Ho ($T_{1/2}$ = 1.12 d) [120].

Chemistry, tumor-targeting mechanism

Only in a few cases therapeutic radionuclides are used in simple ionic forms, such as $[^{131}I]I^-$ for the treatment of the thyroid cancer or $[^{89}Sr]Sr^{2+}$ for the palliative therapy of bone metastases. Most applications require the synthesis of metal-ligand complexes, the formation of particles or colloids, or most often the radiochemical labeling of biomolecules. This is either radiohalogenation (in the case of ^{131}I, ^{125}I, ^{80m}Br or ^{211}At, for a recent review see [121]) or the coordinate binding of metal cations via bifunctional chelators (for a review see [122]).

Most of the reactor produced radionuclides are not carrier-free. Their specific activity is determined (i.e. limited) by the cross section of the (n,γ) reaction, the isotopic abundance of the target isotope and the neutron flux of the reactor (see above). If, for example, receptor-based interactions are the main source of accumulation of the labeled therapeutical at the tumor surface, a low specific activity, i.e. a relatively high dilution with the stable isotope cannot always be tolerated. For example, the receptor density for many small peptides is only on the order of 10^3-10^4 per cell and thus high specific activities are required.

Radiopharmacology has contributed significantly to ERT. Among the new concepts, the general scheme of radionuclides bound to biologically active peptides and proteins is most promising. Most of the radionuclide candidates are metals, thus bridging coordination chemistry of metals with the pharmacology of biological molecules (Fig. 18). This concept offers several routes for systematic improvements. One is to test various radiometals, depending on their radii and coordination numbers, for the physico-chemical properties as well as in vivo stability of the binding to an appropriate chelator. These ligands have to be bifunctional in order to be conjugated to the biological car-

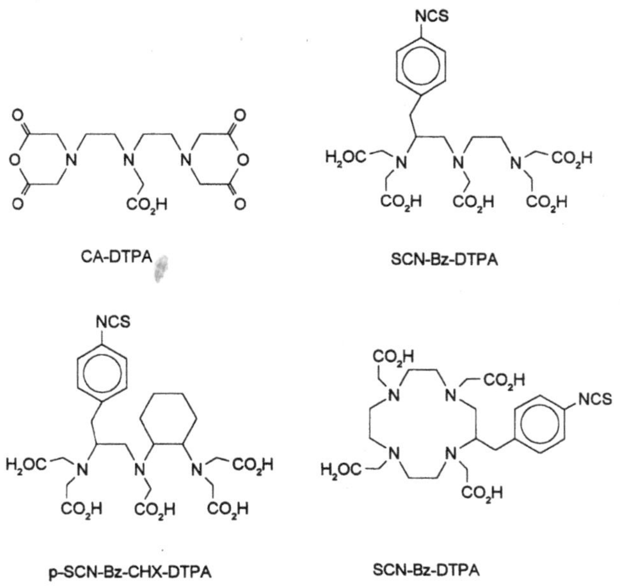

Fig. 19. Structures of some common bifunctional chelating agents.

rier molecule. The chelator itself can be varied not only via the modification of an established structure, but also via the synthesis of new chelators with thermodynamic and kinetic complex stabilities higher than those of any potential competitors in vivo. Figure 19 represents some of these functionalized chelating agents. It illustrates diethylenetriaminepenta-acetic acid (DTPA, shown as its cyclic anhydride CA-DTPA) and the development to backbone substituted derivatives such as p-isocyanatobenzyl-DTPA and substituted p-isocyanatobenzyl-cyclohexyl-DTPA.

These derivatives rely on the basic structure of DTPA but additionally provide sterical hindrance (SCN-Bz-DTPA) and improved rigidity (p-SCN-Bz-CHX-DTPA). Next, the formulation of polyamino macrocycles polyacetic acids such as 1,4,7,10-tetraazacyclododecane-*N,N′,N″,N‴*-tetraacetic acid (DOTA) and its derivatives such as SCN-Bz-DOTA increased the kinetic inertness of the complexes [123–125]. Sometimes linker molecules are introduced. They act either as a steric tool and/or as a biological active module, providing an increased clearance of the activity from the blood or special organs via enzymatic cleavage [126].

Clinical status of endoradiotherapy

Main medical indications are the cancer of the thyroid, the palliative therapy of bone metastases, and the various applications in oncology for primary tumors or melanomas and the synovectomy of joints.

Thyroid carcinoma

About one million patients worldwide have been treated with ^{131}I for differentiated thyroid carcinoma. Doses usually range from 110 to 370 MBq (3–10 mCi) for hyperthyroidism [127, 128], up to 7.4 GBq or even 24 GBq (200 or 650 mCi) for metastatic differentiated thyroid carcinoma. Major series of investigations demonstrated a high efficiency in terms of survival and proved that side-effects and long-term complications do not constitute critical limitations. In 1983 ^{131}I-labeled meta-iodobenzylguanidin [^{131}I]MIBG was introduced [129], which is well established today for the treatment of tumors derived from neural crest. More than one hundred patients with neuroblastoma (children), malignant pheochromocytoma, metastatic carcinoid or medullary thyroid carcinoma have been treated alone at The Netherlands Cancer Institute, receiving a dose of 3.7–7.4 GBq (100–200 mCi) of [^{131}I]MIBG of high specific activity (1.48 GBq/mg) [130].

Bone metastases

Bone therapy refers to the treatment of primary bone tumors such as osteosarcoma [131] as well as the palliative therapy of painful disseminated bone metastases, originating from primary soft tissue tumors, in particular breast and prostate which represent the most common type of cancer. In fact, approximately 80% of patients with prostatic carcinoma develop bone disease and nearly half of them experience bone pain [132]. Accumulation of potential radiotherapeuticals is due to the structure of bone, i.e. the hydroxylapatite $Ca_{10}(PO_4)_6(OH)_2$. Consequently, radiotherapeuticals applied are either PO_4-substituents such as [^{32}P] phosphates or phosphonates or Ca(II) substituents such as [^{89}Sr]Sr^{2+} or [^{90}Y]Y^{3+}. Recently, [^{186}Re]HEDP and [^{153}Sm]EDTMP were introduced, which bind via Sr(II) substitution as well as via the phosphonate structure of the complex ligand. The same mechanism also offers a route to use ^{131}I. Nevertheless, commercially distributed radiotherapeuticals are based on the above mentioned metal radionuclides. In general, the response is >75%. Except for [^{32}P]PO_4 (which has a metabolic pathway to bone marrow cells) side effects are minimum. The bone marrow, however, is the critical target for all the bone-affine radiotherapeuticals (for a recent review see [133]).

Antibodies

The development of the hybridoma technique in 1975 set a historical mark for the clinical use of antibodies. Among others, The John Hopkins Oncology Center Baltimore played a pioneering role, particularly in treating patients for hepatoma and Hodkin's disease with ^{131}I- and ^{90}Y-labeled polyclonal antibodies (^{131}I-antiferritin, ^{90}Y-antiferritin (see [134]). Meanwhile, various other monoclonal antibodies against gastrointestinal, ovarian and colon carcinoma, leukaemia, lymphoma, hepatoma, neuroblastoma and melanoma have been developed. Radioimmuniotherapy is developing rapidly as an additional treatment modality in oncology. This is reflected by a number of recent reviews

Table 7. Potential α-emitting candidates for ERT

Isotope	Half-life	Main γ-emission in keV (abundance in %)	Decay products	Production routes
^{211}At	7.21 h	low	uncritical	$^{209}Bi(\alpha,2n)$; Gen. ^{211}Rn (14.6 h)
^{225}Ac	10.0 d	low	critical	Gen. ^{225}Rn (14.8 d) from ^{229}Th (7340 a)
^{223}Ra	11.4 d	269.4 (13.6)	uncritical?	Gen. ^{227}Th (18.7 d) from ^{235}U (7.108 a)
^{224}Ra	3.66 d	240.8 (3.9)	critical	Gen. ^{228}Th (1.91 a)
^{212}Bi	1.0 h	727.3 (6.65)	uncritical	Gen. ^{212}Pb (10.6 h) from ^{232}Th (1.41 a)
^{213}Bi (^{213}Po)	45.6 min (4.2 µs)	440.3 (16.5)	critical	Gen. ^{225}Ac (10.0 d)
^{255}Fm	20.1 h	low	critical	Gen. ^{255}Es (39.8 d)

on the individual aspects of ERT (see [112, 130, 135, 136]). Recent experience indicates a trend to smaller molecules such as antibody fragments Fab or $F(ab')_2$ down to small peptides, in particular for malignant tumors, which are hormone-dependent in their development and growth. The success of ^{123}I-tyr^3-octreotide or the ^{111}In-labeled somatostatin analogue octreotide (OctreoScan®) (see chapter by Weiner and Thakur) in the diagnosis of somatostatin receptor expressing tumors may lead to the development of similar compounds for ERT, labeled with ^{131}I [137] or β^--emitting metallic radionuclides such as ^{90}Y [138].

Synovectomy

Radiation synovectomy is an attractive alternative to surgical or chemical synovectomy for the treatment of rheumatoid arthritis, for a recent review see [139]. It is based on the careful administration of labeled particles directly into the synovial cavity. The first agents developed were colloidal suspensions. Due to the leakage of these compounds, which are subsequently trapped by the reticuloendothelial system, new compounds of macroaggregate type (insoluble particles such as ferric hydroxide macroaggregate FHMA, calcium oxalate) or biodegradable ones such as liposomes and, more recently, hydroxylapatite [140] were tested. All the radionuclides involved represent β^--emitters. The intercorrelation between maximum β^--energy and the maximum range in tissue was utilized for the stepwise application in synovectomies of different joints (^{90}Y with max. 12 mm range for the knee, ^{186}Re with max. 4.3 mm for medium joints and ^{169}Er with max. 1.0 mm for finger joints). Clinically applied activities range from 0.25 to maximum 370 MBq (10 mCi) for long-lived radionuclides (mainly ^{32}P, ^{90}Y, ^{198}Au), and up to 11 GBq (300 mCi) for short-lived ones (^{165}Dy).

New perspectives of endoradiotherapy

A new generation of therapeutic radionuclides came up. Among the β^--emitters there are 8 candidates without any accompanying γ-emission. Half of them are well established (^{32}P, ^{35}S, ^{90}Y, ^{169}Er), others are rarely considered (^{66}Ni, ^{121}Sn, ^{143}Pr, ^{191}Os). Also some candidates with very low γ-emission should be noted: ^{89}Sr, ^{109}Pd, ^{166}Ho and ^{172}Tm. From the point of view of availability, however, ^{153}Sm and ^{186}Re are superior, although they have poorer dosimetric characteristics. Only ^{177}Lu with about a 10-fold higher formation cross section (2100 barn) may compete [141]. Although these radioisotopes can be produced in very high specific activities of some 10^2 GBq/mg, the radionuclide generator systems ^{90}Sr/Y, ^{166}Dy/Ho and ^{188}W/Re are of special interest (specific activities of $>10^4$ GBq/mg).

Among the Auger-electron emitters, ^{51}Cr, ^{67}Ga, ^{111}In, ^{125}I and ^{201}Tl are well known, but mainly from diagnostic applications, where their accompanying γ-emission is used. So far, ^{125}I is the major Auger-electron emitter of interest in ERT. Candidates without γ-emission, such as ^{71}Ge, ^{140}Nd, ^{160}Er and ^{165}Er, have not yet been seriously considered. Three aspects are common to this group. (i) The isotopes are neutron deficient, being generally produced at cyclotrons and hence, are no-carrier-added. (ii) The EC decay mode dominates for the heavier elements, compared to positron emission for the lighter ones. Consequently, Auger-electron emitters are found mainly among the metal elements. (iii) The LET of the Auger-electrons is on the order of 1 to 10 eV/nm, and their range is a few µm only (cf. Fig. 17). The chemical and biological consequences of the Auger effect are drastic, leading to double strand breaks of the DNA (for a review cf. [142, 143]). It has been shown by double labeling experiments that the pyrimidine base is completely fragmented following the Auger cascade, probably due to charge neutralization and excitation decomposition. Double strand breaks of the DNA occurred not only at the site of decay but also as far as hundred base pairs away [144]. To achieve an effective cell killing, however, the labeled radiotherapeutical must be chemically coupled to the DNA or at least be internalized through

Table 8. Positron-emitting analogs for the bridging of ERT and PET

Therapeutic isotope	β^+-substituent	Half-life	β^+-branching %
^{89}Sr	^{83}Sr	1.35 d	24
^{90}Y	^{86}Y	14.7 h	33
^{153}Sm	$^{142}Sm(Pm)$	72 min (40 s)	5.7 (78)
$^{131}I/^{125}I$	^{124}I	4.18 d	23
^{71}Ge	^{66}Ge	2.26 h	24
	^{69}Ge	1.63 d	24
^{177}Ta	^{173}Ta	3.65 h	23
	^{174}Ta	1.18 h	24

the cell membrane. In the case of ^{125}I this can be accomplished by using labeled DNA precursors. For metals, this is much more difficult. This may inhibit their broad application in ERT. Thus it is a challenge to design suitable Me-labeled carriers.

There are a few promising α-emitters (cf. Table 7). Their potential value is derived from their short-range and high LET radiation of about 100 keV/μm with high relative biological effectiveness. ^{211}At, ^{212}Bi and ^{223}Ra should be preferred because they do not include decay chains where other particle emitting nuclides are involved. Their half-lives are quite different and one may make a selection according to a specific application. ^{212}Bi and ^{223}Ra are available from sophisticated radionuclide generator systems and are used to label monoclonal antibodies [145]. ^{211}At needs to be produced at a nearby cyclotron and offers the useful properties of a halogen for labeling chemistry. The in vivo stability of astatinated biomolecules, however, remains one of the crucial problems for the endotherapeutic application of the corresponding compounds (cf. e.g. [146]).

Many problems still need to be solved in ERT. There is an ongoing need for further investigations on the biochemical base of endotherapeutical actions. New strategies are required. The interactions of individual metal ions with components of the blood serum are to be studied in more detail. Moreover, the optimization of any approach requires the quantitative measurement of many individual parameters – among them the very important value of the in vivo uptake kinetics in individual organs or regions of interest in patients. Some candidates such as ^{153}Sm and ^{186}Re provide a low-energy γ-radiation and thus qualitative imaging of the biodistribution is possible. Another approach to quantify uptake kinetics is the bridging of PET and ERT. A prerequisite is the availability of an isotope of the same element or homolog (a chemically similar element) which emits positrons. Table 8 shows that there are some pairs among the isotopic substituents. Using ^{86}Y and PET, for example, the in vivo uptake kinetics of [$^{86/90}Y$]citrate and [$^{86/90}Y$]EDTMP in patients with bone metastases were measured and the corresponding individual radiation doses were calculated using the MIRD recommendations [147, 148]. This may contribute to the quantitative evaluation of chemically different radiotherapeuticals as well as to the success of the therapy and an individual treatment planning.

In this context, physical and mathematical considerations on the action of radiation have reached a new level, which allows a quantitative estimation of the in vivo dosimetry for various radionuclides [149–152]. It may result in a combined strategy of the treatment of tumors: starting with ^{90}Y-labeled therapeuticals, for example, followed by substitution of ^{153}Sm (max. range 3.0 mm) when tumor regression was achieved [153]. There are even attempts to magnetically enhance the therapeutic effects [154].

Conclusion and outlook

Radioactivity has indeed revolutionized life sciences during this century, and it is still an indispensable tool. Despite an increasing number of non-radioactive tracer methods developed during the last decades, such as luminescent-, fluorescent-, and enzymatic labeling in the field of immuno assay, or functional MRI, which may eventually substitute blood flow measurements presently carried out with emission tomography, or the use of stable isotopes in various fields, the major applications of radioactive tracers will remain or even increase because of their sensitivity combined with the ease of detection from outside a closed system, such as the human body. A further advantage is that only extremely small concentrations in the nanomolar range are needed which do not cause any pharmacodynamic effect. Major emphasis is placed on cyclotron-produced short-lived radionuclides, also in view of environmental aspects. Emission tomography (PET and SPET), which uses such tracers, is a rapidly proliferating technology. It is not only of clinical diagnostic importance but is also a research tool for studying molecular medicine in vivo with the goal to learn how a disease develops on a biochemical level. The fact that this can be done in the living human being is a tremendous advantage, since the complex interactions with the chemistry of the body are taken into account, an information which is not available via in vitro studies. Emission tomography, particularly PET, provides a new window to the chemistry of brain, heart and tumors.

Endotherapy is another area which is receiving renewed interest. Even though there is no ideal radionuclide and endotherapy is a matter of compromise, significant progress has been achieved in tumor targeting and therapy control. Recent advances in molecular biology and the development of synthetic peptides and hormones provide new concepts for the design of potential high affinity biological carriers.

The young field of radiopharmaceutical chemistry is a challenging multidisciplinary research area for a radiochemist interested in synthetic chemistry and biochemistry rather than in nuclear physics.

References

1. Brucer, M.: *A Chronology of Nuclear Medicine,* Mallinckrodt Medical Heritage Publications, Inc., St. Louis, Missouri (1990).
2. Hevesy, G.: Biochem. J. **17**, 439 (1923).
3. Hevesy, G.: Compt. Rend. **178**, 1324 (1924).
4. Christiansen, J. A., Lombholt, S.: Compt. Rend. **179**, 291 (1924).
5. Blumgart, H. L., Weiss, S.: J. Clin. Invest. **4**, 15–31 (1927).
6. Fermi, E.: Nature 1934, 757.
7. Hevesy, G., in: *Adventures in Radiation Research.* The collected papers of George Hevesy. Pergamon Press, Oxford, pp. 11–70 (1962).
8. Chievitz, O., Hevesy, G.: Nature 1935, 754.
9. Evans, R. D.: *Glaser's Medical Physics* (1936), p. 7.
10. Hertz, S., Robert, A., Evans, R. D.: Proc. Soc. Exp. Biol. Med. **38**, 510 (1938).
11. Lawrence, J.: Radium **35**, 51 (1940).
12. Livingood, J. J., Seaborg, G. T.: Phys. Rev. **53**, 1015 (1938).
13. Hamilton, J. G., Soley, M. H.: Am. J. Physiol. **127**, 557 (1939).
14. Pecher, J.: Univ. Calif. Pub. Pharmacol. **11**, 117 (1942).
15. Lauritsen, C. C., Crane, H. R.: Phys. Rev. **46**, 493 (1934).
16. Crane, H. R., Lauritsen, C. C.: Phys. Rev. **45**, 430 (1934).
17. Livingston, M. S., McMillan, E.: Phys. Rev. **46**, 437 (1934).
18. Snell, A. H.: Phys. Rev. **51**, 143 (1937).
19. Volker, J. F., Hodge, H. C., Wilson, H. J.: J. Biol. Chem. **123**, 543 (1940).
20. Tobias, C. A., Lawrence, J., Roughton, F., *et al.*: Am. J. Physiol. **145**, 253 (1945).
21. Weinreich, R., Ritzl, F., Feinendegen, L. E., Schnippering, H. G., Stöcklin, G.: Radiat. Environ. Biophys. **12**, 271 (1975).
22. Kamen, M. D., Ruben, S.: Phys. Rev. **58**, 194 (1940).
23. Berson, S. H., Yalow, R. S., *et al.*: J. Clin. Invest. **35**, 170 (1956).
24. International Atomic Energy Agency, *Isotopes in Day to Day Life,* IAEA, Vienna (1981).
25. Meyers, W. G., Wagner, H. N., Jr., in: *Nuclear Medicine* (Wagner H. N., editor) HP Publishing Co., Inc. Publishers New York, NY, 1975.
26. Tubis, M., Posnich, E., Nordyke, R. A.: Proc. Soc. Exp. Biol. Med. **103**, 497 (1960).
27. Tucker, W. D., Greene, M. W., Weiss, A. J., Murrenhoff, A. P.: Methods of preparation of some carrier-free radioisotopes involving sorption on alumina. USAEC Report BNL-3746, Brookhaven National Laboratory, May 29, 1958.
28. Buckingham, P. D., Forse, G. R.: Int. J. Appl. Radiat. Isot. **14**, 439 (1963).
29. Ter-pogossian, M. M., Wagner, H. N., Nucleonics, Oct. 1966.
30. Lebowitz, E., Greene, M. W., Bradley-Moore, P., Atkins, H., Ansari, A., Richards, P., Belgrave, E.: J. Nucl. Med. **14**, 1421 (1973).
31. Meyers, W. G., Anger, H. O.: J. Nucl. Med. **3**, 183 (1962).
32. Sajjad, M., Lambrecht, R. M., Wolf, A. P.: Radiochim. Acta **39**, 165 (1986).
33. Ruth, T. J., Wolf, A. P.: Radiochim. Acta **26**, 21 (1979).
34. Qaim, S. M., Clark, J. C., Crouzel, C., Guillaume, M., Helmeke, H. J., Nebeling, B., Pike, V. W., Stöcklin, G., in: *Radiopharmaceuticals for Positron Emission Tomography* (G. Stöcklin and V. W. Pike, Editors), Kluwer Academic Publishers, Dordrecht, The Netherlands, pp. 1–42 (1993).
35. Sajjad, M., Lambrecht, R. M., Wolf, A. P.: Radiochim. Acta **36**, 159 (1984).
36. Sixth Workshop on Targetry and Target Chemistry, Vancouver, B. C., Canada, August 1995, Abstracts. See contributions from St. Louis, Turku (IBA-machine), Seattle, etc.
37. Casella, V., Ido, T., Wolf, A. P., Fowler, J. S., MacGregor, R. R., Ruth, T. J.: J. Nucl. Med. **21**, 750 (1980).
38. Roberts, A. D., Oakes, T. R., Nickles, R. J.: Appl. Radiat. Isot. **46**, 87 (1995).
39. Lambrecht, R. M.: Radiochim. Acta **34**, 9 (1983).
40. Knapp, Jr., F. F., Mirzadeh, S.: Eur. J. Nucl. Med. **21**, 1151 (1994).
41. Green, M. A., Mathias, C. J., Welch, M. J., McGuire, A. H., Perry, D., Fernandez-Rubio, F., Perlmutter, J. S., Raichle, M. E., Bergmann, S. R.: J. Nucl. Med. **31**, 1989 (1990).
42. Fujibayashi, Y., Matsumoto, K., Yonekura, Y., Konishi, J., Yokoyama, A.: J. Nucl. Med. **30**, 1838 (1989).
43. Pao, P. J., Silvester, D. J., Waters, S. L.: J. Radioanal. Chem. **64**, 267 (1981).
44. Loc'h, C., Mazière, B., Comar, D., Knipper, R.: Int. J. Appl. Radiat. Isot. **33**, 267 (1982).
45. Grant, P. M., Miller, D. A., Gilmore, J. S., O'Brien, Jr., H. A.: Int. J. Appl. Radiat. Isot. **33**, 415 (1982).
46. *The $^{82}Sr/^{82}Rb$ Generator* (Waters, S. L., Coursey, B. M., Editors) Appl. Radiat. Isot. **38**, 171 (1987). (Special issue).
47. Tárkányi, F., Qaim, S. M., Stöcklin, G.: Appl. Radiat. Isot. **41**, 91 (1990).
48. Mausner, L. F., Prach, T., Srivastava, S. C.: Appl. Radiat. Isot. **38**, 181 (1987).
49. Huszár, I., He Youfeng, Jegge, J., Weinreich, R.: J. Lab. Compd. Radiopharm. **26**, 168 (1989).
50. Qaim, S. M.: Radiochim. Acta **30**, 147 (1982).
51. Plenevaux, A., Guillaume, M., Brihaye, C., Lemaire, C., Cantineau, R.: Appl. Radiat. Isot. **41**, 829 (1990).
52. Rösch, F., Qaim, S. M., Stöcklin, G.: Appl. Radiat. Isot. **44**, 677 (1993).
53. Fessler, A., Alfassi, Z. B., Qaim, S. M.: Radiochim. Acta **65**, 207 (1994).
54. Blessing, G., Lavi, N., Qaim, S. M.: Appl. Radiat. Isot. **43**, 455 (1992).
55. Blessing, G., Weinreich, R., Qaim, S. M., Stöcklin, G.: Int. J. Appl. Radiat. Isot. **33**, 333 (1982).
56. Vaalburg, W., Paans, A. M. J., Terpstra, J. W., Wiegman, T., Dekens, K., Rijskamp, A., Woldring, M. G.: Int. J. Appl. Radiat. Isot. **36**. 961 (1985).
57. Kovács, Z., Blessing, G., Qaim, S. M., Stöcklin, G.: Int. J. Appl. Radiat. Isot. **36**, 635 (1985).
58. Rösch, F., Novgorodov, A. F., Qaim, S. M.: Radiochim. Acta **64**, 113 (1994).
59. Szelecsényi, F., Boothe, T. E., Tavano, E., Plitnikas, M. E., Tárkányi, F.: Appl. Radiat. Isot. **45**, 473 (1994).
60. Tárkányi, F., Szelecsényi, F., Kopecký, P., Molnár, T., Andó, L., Mikecz, P., Tóth, G., Rydl, A.: Appl. Radiat. Isot. **45**, 239 (1994).
61. Qaim, S. M., Stöcklin, G.: Radiochim. Acta **34**, 25 (1983).
62. Scholten, B., Qaim, S. M., Stöcklin, G.: Appl. Radiat. Isot. **40**, 127 (1989).
63. Andó, L., Mikecz, P., Tcheltsov, A. N., Suvorov, I. A., Mahunka, I.: J. Radioanal. Nucl. Chem. Lett. **146**, 169 (1990).
64. Kurenkov, N. V., Malinin, A. B., Sebyakin, A. A., Venikov, N. I.: J. Radioanal. Nucl. Chem. Lett. **135**, 39 (1989).
65. Tárkányi, F., Qaim, S. M., Stöcklin, G., Sajjad, M., Lambrecht, R. M., Schweickert, H.: Appl. Radiat. Isot. **42**, 221 (1991).
66. Firouzbakht, M. L., Schlyer, D. J., Wolf, A. P.: Radiochim. Acta **56**, 167 (1992).
67. Lebowitz, E., Greene, M. W., Fairchild, R., Bradley-Moore, P. R., Atkins, H. L., Ansari, A. N., Richards, P., Belgrave, E.: J. Nucl. Med. **16**, 151 (1975).
68. Blue, J. W., Liu, C. D., Smathers, J. B.: Med. Phys. **5**, 532 (1978).
69. Lagunas-Solar, M. C., Jungerman, J. A., Peek, N. F., Theus, R.: Int. J. Appl. Radiat. Isot. **29**, 159 (1978).

70. Qaim, S. M., Weinreich, R., Ollig, H.: Int. J. Appl. Radiat. Isot. **30**, 85 (1979).
71. Mirzadeh, S., Knapp, Jr., F. F., Callahan, A. P., in *Proc. Int. Conf. on Nuclear Data for Science and Technology*, May 1991, Jülich, Germany (S. M. Qaim, Editor), Springer Verlag, Berlin/Heidelberg, p. 595 (1992).
72. Lambrecht, R. M., Sajjad, M., Qureshi, M. A., Al-Yanbawi, S. J.: J. Radioanal. Nucl. Chem. Lett. **127**, 143 (1988).
73. Firouzbakht, M. L., Schlyer, D. J., Wolf, A. P.: J. Lab. Compd. Radiopharm. **35**, 257 (1994).
74. Scholten, B., Kovács, Z., Tárkányi, F., Qaim, S. M.: Appl. Radiat. Isot. **46**, 255 (1995).
75. Shigeta, N., Matsuoka, H., Osa, A., Kaizumi, M., Izumo, M., Kobayashi, K., Hashimoto, K., Sekine, T.: JAERI Annual Report, Vol. 3, p. 163 (1993).
76. Sokoloff, L., Reivich, M., Kennedy, C., Des Rosiers, M. H., Patlak, C. S., Pettigrew, K. D., Sakurada, O., Shinohara, M.: J. Neurochem. **28**, 897 (1977).
77. Gallagher, B. M., Ansari, A., Atkins, H., Casella, V., Christman, D. R., Fowler, J. S., Ido, T., MacGregor, R. R., Som, P., Wam, C. N., Wolf, A. P., Kuhl, D. E., Reivich, M.: J. Nucl. Med. **18**, 990 (1977).
78. Ido, T., Wan, C. N., Casella, V., Fowler, J. S., Wolf, A. P., Reivich, M., Kuhl, D. E.: J. Lab. Compd. Radiopharm. **14**, 175 (1978).
79. Reivich, M., Kuhl, D., Wolf, A. P., Phelps, M., Ido, T., Casella, V., Fowler, J. S., Hoffman, E., Alavi, A., Som, P., Sokoloff, L.: Civ. Res. **44**, 127 (1979).
80. Ferrieri, R. A., Wolf, A. P.: Radiochim. Acta **34**, 69 (1983).
81. *Radiopharmaceuticals for Positron Emission Tomography. Methodological Aspects* (Stöcklin, G. and Pike, V. W., Editors) Kluwer Academic Publishers, Dordrecht, Boston, London (1993).
82. Kabalka, G. W.: Acc. Chem. Res. **17**, 215 (1984).
83. Kabalka, G. W., Lambrecht, R. M., Sajjad, M., Fowler, J. S., Kunda, S. A., McColum, G. W., McGregor, R.: Int. J. Appl. Radiat. Isot. **36**, 853 (1985).
84. Takahashi, K., Murakami, M., Sasaki, H., Kondo, Y., Mizusawa, S., Nakamich, H., Tida, H., Miura, S., Kanno, I., Uemura, K., Ido, T.: J. Lab. Compd. Radiopharm. **23**, 1111 (1986).
85. Herzog, H., Seitz, R. J., Tellmann, L., Schlaug, G., Kleinschmidt, A., Nebeling, B., Stöcklin, G., Müller-Gärtner, H. W.: Eur. J. Nucl. Med. **21**, 138 (1994).
85a. Långström, B., Dannals, R. F., in: *Principles of Nuclear Medicine*, 2nd edition (Wagner, H. N., Szabo, Z., Buchanan, J. W., eds.), W. B. Saunders Company, London, Toronto, Montreal, Sydney, Tokyo 1995.
86. Jewett, D.: Appl. Radiat. Isot. **43**, 1383 (1992).
87. Holschbach, M., Schüller, M.: Appl. Radiat. Isot. **44**, 887 (1993).
88. Holschbach, M., Schüller, M.: Appl. Radiat. Isot. **44**, 779 (1993).
89. Link, J. M., Clark, J. C., Larsen, P., Krohn, K. A.: J. Lab. Compd. Radiopharm. **37**, 76 (1995).
90. Coenen, H. H., in: Batille, T. A., Jones, J. R. (Eds.), *Synthesis and applications of isotopically labelled compounds.* Elsevier, Amsterdam, 1989, p. 443.
91. Hamacher, K., Coenen, H. H., Stöcklin, G.: J. Nucl. Med. **27**, 235 (1986).
91a. Füchtner, F., Steinbach, J., Mädius, P., Johannsen, B.: J. Lab. Comp. Radiopharm. **37**, 743 (1995).
92. Melon, P., Schwaiger, M.: Eur. J. Nucl. Med. **19**, 453 (1992).
93. Farde, L., Wiesel, F.-A., Halldin, C., Sedvall, G.: Arch. Gen. Psychiatry **45**, 71 (1988).
94. Wagner, H. N., Jr.: Biochem. Pharmacol. **37**, 51 (1988).
95. Smith, M., Wolf, A. P., Brodie, J. D., *et al.*: Biopsychiatry **23**, 653 (1988).
96. Garnett, E. S., Firnau, G., Nahmias, C.: Nature **305**, 137 (1983).
97. Samson, Y., Hantraye, P., Baron, J. C., Soussaline, F., Comar, D., Mazière, M.: Eur. J. Pharmacol. **110**, 274 (1985).
98. Savic, I., Roland, P., Sedvall, G., Persson, A., Panili, S., Widen, L.: Lancet II, 863 (1988).
99. Strauss, L. G., Conti, P. S.: J. Nucl. Med. **32**, 623 (1992).
100. Nunn, A., Linder, K., Strauss, H. W.: Eur. J. Nucl. Med. **22**, 265 (1995).
101. Coenen, H. H., Moerlein, S. M., Stöcklin, G.: Radiochim. Acta **34**, 47 (1983).
102. Dewanjee, M. K.: *Radioiodination: Theory, Practice and Biomedical Applications.* Kluwer Academic Publishers, Boston, Dordrecht, London (1991).
103. Trivedi, M., Cornochan, P.: J. Label. Compd. Radiopharm. **33**, 61 (1993).
104. Krummeich, C., Holschbach, M.: Appl. Radiat. Isot. **46**, 917 (1995).
105. Beer, H. F., Blauenstein, P. A., Hasler, P. H., Delaloye, B., Riccabona, G., Bangerl, I., Hunkeler, W., Bonetti, E. P., Pieri, L., Richards, J. G., Schubiger, P. A.: J. Nucl. Med. **31**, 1007 (1990).
106. Bartenstein, P., Ludolph, A., Schober, O., Lottes, G., Scheidhauer, K., Sciuk, J., Beer, H. F.: Eur. J. Nucl. Med. **18**, 111 (1991).
107. Kung, H. F., Guoy, Z., Billings, J., Xu, Y., Blau, M., Ackerhalt, R. E.: Nucl. Med. Biol. **15**, 195 (1988).
108. Stöcklin, G.: Eur. J. Nucl. Med. **19**, 527 (1992).
109. Biersack, H. J., Coenen, H. H., Stöcklin, G., Reichmann, K., Bockisch, A., Oehr, P., Dashab, M., Rollmann, O.: J. Nucl. Med. **30**, 110 (1989).
110. Langen, K.-J., Roosen, N., Coenen, H. H., Kuikka, J. T., Kuwert, T., Herzog, H., Stöcklin, G.: J. Nucl. Med. **32**, 1225 (1991).
111. E.A.N.M. Task Group Radionuclide Therapy, report 1994: Radionuclide therapy: From palliation to cure.
112. Larson, S. M., Carrasquillo, J. A., Reynolds, J. C.: Cancer Invest. **2**, 363 (1984).
113. Spencer, R. P., Seevers, R. H., Friedman, A. M. (eds.): *Radionuclides in Therapy.* CRC Press, Boca Raton 1987.
114. Srivastava, S. C. (ed.): *Radiolabeled monoclonal antibodies for imaging and therapy.* NATO ASI Series, Plenum Press, New York (1988).
115. Adelstein, S. M., Kassis, A. I., in: Srivastava, S. C. (ed.): *Radiolabeled monoclonal antibodies for imaging and therapy.* NATO ASI Series, Plenum Press, New York 1988, pp. 165–176.
116. Mausner, L. F., Straub, R. F., Srivastava, S. C., in: Srivastava, S. C. (ed.): *Radiolabeled monoclonal antibodies for imaging and therapy.* NATO ASI Series, Plenum Press, New York 1988, pp. 149–164.
117. Volkert, W. A., Goeckeler, W. F., Erhardt, G. J., Ketring, A. R.: J. Nucl. Med. **32**, 174 (1991).
118. Strand, S.-E., Norrgren, K., Ingvar, C., Brodin, T., Hallstadius, L., Ljungberg, M., in: Schubiger, P. A., Hasler, P. H. (eds.), *Radionuclides for therapy.* Proceedings of the 4th Böttstein Colloquium held at Würelingen/Villigen, Switzerland, June 13 and 14, 1986, Editiones ⟨Roche⟩, Basle.
119. Mirzadeh, S., Kumar, K., Gansow, O. A.: Radiochim. Acta **60**, 1 (1993).
120. Smith, S. V., Di Bartolo, N., Mirzadeh, S., Lambrecht, R. M., Knapp, F. F., Hetherington, E. L.: Appl. Radiat. Isot. **46**, 759 (1995).
121. Wilbur, D. S.: Bioconj. Chem. **3**, 433 (1992).
122. Meares, C. F.: *Perspectives in Bioconjugate Chemistry.* The American Chem. Soc., Washington, D.C. (1993).
123. Brechbiel, M. W., Gansow, O. A.: Bioconj. Chem. **2**, 187 (1991).
124. Meares, F. C., Moi, M. K., Diril, H., Kukis, D. L., McCall, M. J., Deshpande, S. V., DeNardo, S. J., Snook, D., Epenetos, A. A.: Br. J. Cancer **62**, 21 (1990). Suppl. X.
125. Yuanfang, L., Chuanchu, W.: Pure Appl. Chem. **63**, 427 (1991).
126. Paik, C. H., Quadri, S. M., Reba, R. C.: Nucl. Med. Biol. **16**, 457 (1989).
127. Maxon, H. R., Thomas, S. R., Wen Chen: Clin. Nucl. Med. **6**, 87 (1981).

128. Becker, D. V., in: Tubiane, M. (ed.), *Thyroid diseases*. Pergamon Press, Oxford (1982), pp. 145–158.
129. Sisson, J. C., Frager, M. S., Valk, T. W., Gross, M. D., Swanson, D. P., Wieland, D. M., Tobes, M. C., Beierwaltes, W. H., Thompson, N. W.: New Engl. J. Med. **305**, 12 (1981).
130. Hoefnagel, C. A.: Eur. J. Nucl. Med. **18**, 408 (1991).
131. Blake, G. M., Zivanovic, M. A., McEwan, A. J., Condon, B. R., Ackery, D. M.: Br. J. Radiol. **60**, 253 (1987).
132. Campa, J. A. III, Payne, R.: Semin. Nucl. Med. **22**, 3 (1992).
133. Lewington, V. J.: Eur. J. Nucl. Med. **20**, 66 (1993).
134. Leichner, P. K., Yang, N.-C., Frenkel, T. L., Loudenslager, D. M., Hawkins, W. G., Klein, J. L., Order, S. E.: Int. J. Rad. Oncol. Biol. Phys. **14**, 1033 (1988).
135. O'Brien, H. A., Jr.: *Radioimmunoimaging and radioimmunotherapy.* Elsevier Science Publ. Co. Inc., New York (1983).
136. DeNardo, G. L., Maddock, S. W., Sguoros, G., Scheibe, P. O., DeNardo, S. J.: J. Nucl. Med. **34**, 1020 (1993).
137. Lamberts, S. W. J., Bakker, W. H., Reubi, J.-C., Krenning, E. P.: New Engl. J. Med. 323, 1246 (1990).
138. Stolz, B., Smith-Jones, P. M., Ruser, G., Albert, R., Briner, U., Mäcke, H., Weckbecker, G., Bruns, C.: Proc. 42nd Ann. Meeting, Soc. Nucl. Med., Minneapolis, June 12–15, 1995; Abstr.; J. Nucl. Med. **36**, 114P (1995).
139. Deutsch, E., Brodack, J. W., Deutsch, K. F.: Eur. J. Nucl. Med. **20**, 1113 (1993).
140. Chinol, M., Vallabhajoshula, S., Goldsmith, S. J., Klein, M. J., Deutsch, K. F., Chinen, L. K., Brodack, J. W., Deutsch, E. A., Watson, B. A., Tofe, A. J.: J. Nucl. Med. **34**, 1536 (1993).
141. Schlom, J., Siler, K., Milenic, D. E., Eggensperger, D., Colcher, D., Miller, L. S., Houchens, D., Cheng, R., Kaplan, D., Goeckeler, W.: Cancer Res. 51, 2889 (1991).
142. Halpern, A., Stöcklin, G.: Radiat. Environ. Biophys. **14**, 167 and 257 (1977).
143. Makrigiorgos, G., Adelstein, S. J., Kassis, A. I.: Radiat. Environ. Biophys. **29**, 75 (1990).
144. Linz, U., Stöcklin, G.: Radiat. Res. **101**, 202 (1985).
145. Ruegg, C. L., Anderson-Berg, W. T., Brechbiel, M. W., Mirzaadeh, S., Gansow, O. A., Strand, M.: Cancer Res. **50**, 4221 (1990).
146. Zalutsky, M. R., in: Amaldi, U., Larsson, B. (eds.): *Hadrontherapy in oncology.* Elsevier (1994).
147. Herzog, H., Rösch, F., Stöcklin, G., Lueders, C., Qaim, S. M., Feinendegen, E. L.: J. Nucl. Med. **34**, 2222 (1993).
148. Rösch, F., Herzog, H., Plag, C., Neumaier, B., Braun, U., Müller-Gärtner, H. W., Stöcklin, G.: Eur. J. Nucl. Med., submitted 1995.
149. Wessels, B. W., Rogus, R. D.: Med. Phys. **11**, 638 (1984).
150. Rao, D. V., Howell, R. W.: J. Nucl. Med. **34**, 1801 (1993).
151. Siegel, J. A., Stabin, M. G.: J. Nucl. Med. **35**, 152 (1994).
152. Leichner, P. K.: J. Nucl. Med. **35**, 1721 (1994).
153. O'Donoghue, J. A., Bardiés, M., Wheldon, T. E.: J. Nucl. Med. **36**, 1902 (1995).
154. Raylman, R. R., Wahl, R. L.: J. Nucl. Med. **35**, 157 (1994).

Radiochimica Acta 70/71, 273–287 (1995)

Metallic Radionuclides: Applications in Diagnostic and Therapeutic Nuclear Medicine

By R. E. Weiner[1] and M. L. Thakur[2]

[1] Department of Nuclear Medicine, University of Connecticut Health Center Farmington, Connecticut 06030-2804, USA
[2] Division of Nuclear Medicine, Thomas Jefferson University Hospital, Philadelphia, PA 19107-5, USA

(Received October 25, 1995)

Cold things are hurtful to the bones, teeth, nerves, brain, spinal marrow, but hot things are good.
Hippocrates

Nuclear Medicine / Radiopharmaceuticals / ^{111}In / ^{67}Ga / ^{201}Tl / ^{89}Sr

Summary

Nuclear Medicine is a medical modality that utilizes radioactivity (radiopharmaceutical) to diagnose and treat disease. Radiopharmaceuticals contain a component which directs the radionuclide to the desire physiological target. For diagnostic applications, these nuclides must emit a γ ray that can penetrate the body and can be detected externally while for therapeutic purposes nuclides are preferred that emit β particles and deliver highly localized tissue damage. ^{67}Ga citrate is employed to detect chronic occult abscesses, Hodgkin's and non-Hodgkin's lymphomas, lung cancer, hepatoma and melanoma and localizes in these tissues utilizing iron-binding proteins. ^{201}Thallous chloride, a potassium analogue, used to diagnosis coronary artery disease, is incorporated in muscle tissue via the Na^+-K^+-ATPase. ^{111}In labeled autologous white blood cells, used for the diagnosis of acute infections and inflammations, takes advantage of the white cell's role in fighting infections. ^{111}In is incorporated in other radiopharmaceuticals e.g. polyclonal IgG, OncoScint CR/OV, OctreoScan and Myoscint by coupling diethylenetriaminepentaacetic acid, a chelate, covalently to these molecules. OncoScint CR/OV and Myoscint localize by antigen-antibody interactions while OctreoScan is taken up by malignant cells in a receptor based process. Polyclonal IgG may share some localization characteristics with ^{67}Ga.·^{89}Sr, a pure β emitter, is used for palliation of bone pain due to metastatic bone lesions. Bone salts [$Ca(PO)_4$] are increased in these lesions and this radionuclide is taken up similarly to Ca^{2+}. ^{186}Re and ^{153}Sm bound to polydentate phosphonate chelates are used similarly and follow the phosphate pathway in lesion incorporation.

Future directions

Despite the availability on many excellent agents used in routine Nuclear Medicine, the development of new or improved radiopharmaceuticals continues unabated. The emphasis is placed those agents which would keep up with the changing times and help establish an essential role for Nuclear Medicine in functional imaging.

Advances in molecular biology begun today will dramatically impact the medicine practiced tomorrow. Nuclear Medicine is based on the tracer theory, which is inherently molecular. We are therefore poised for a pivotal role of molecular probes in Nuclear Medicine and appropriately, the development of a new generation of radiopharmaceuticals such as radiolabeled peptides and molecular recognition units, which has already begun. In the future, such molecular probes will play a vital role in the practice of Nuclear Medicine which will predominantly implicate functional characterization of diseases, rather than just their anatomical locations.

The discovery of the first radioelement merely one hundred years ago has paved the way to the preparation of more than 1800 radionuclides. Although only a few of these possess characteristics suitable for diagnostic and therapeutic applications, they have contributed to the health care management of millions of patients throughout the world and lead investigators to the formation of a well established medical modality, Nuclear Medicine.

Nuclear Medicine stemmed from the pioneering work of George Hevesy who used phosphorus-30 (originally produced by F. Joliot and J. Curie in 1934) in animals in 1935 and of Joseph Hamilton who attempted to treat leukemic patients with sodium-24 in 1936 [1]. The introduction of the cyclotron during this time period rapidly increased the number of radionuclides that were available to investigators for biomedical research. This availability of radionuclides stimulated their investigations even more intensively. While Nuclear Medicine has grown and evolved since this time, the basic theme remains the same. This medical specialty utilizes the nuclear properties of radioactive atoms for non-invasive location of occult lesions and/or physiological evaluation of the human body and provides therapy with unsealed sources. The chemical form in which the radionuclide accomplishes this task is called a radiopharmaceutical. Typically a radiopharmaceutical contains a radionuclide and directs the atom to the desired physiological target for either diagnostic or therapeutic purposes. For diagnostic use, these radionuclides, after localized in a target inside the body, must emit electromagnetic radiation that can penetrate the tissue to be detected externally by present day imaging devices. Gamma rays or X-rays distribute their energy uniformly throughout the tissue and therefore can traverse tissue several centimeters thick. In contrast, for therapeutic use, it is desired to deliver highly localized radiation dose to specific tissue and cause local cellular destruction. This function can be performed by radionuclides that emit β or α particles. Beta particles travel only a few mm (~100 cell diameters) and α particles <100 µm in soft tissue before they expend all their energy, and thereby lessening the chance normal tissue destruction.

For a radionuclide to be used in Nuclear Medicine for diagnostic purposes the emission of γ-rays is essential but there are other factors which also can contribute. The toxicity of the nuclide or radiopharmaceutical is usually not a concern because only minute quantities of the nuclide or radiopharmaceutical are injected into patients. For example, a typical dose of ^{67}Ga citrate is 6 mCi (220 MBq) which is only 8.5 ng of gallium. A concern is unwanted tissue damage. This is caused by excessive ionizing radiation localized in normal organs. An attempt is made to minimize the absorbed radiation dose (the energy deposited by the radiation) to the patient while obtaining the desired diagnostic information. This is mainly accomplished, wherever possible, by using the least amount of radionuclide and radionuclides that have short half-lives ($T_{1/2}$). Nuclides with short $T_{1/2}$'s, 10 seconds to 80 hrs, are preferred in diagnostic applications because higher amounts can be given to the patients without significantly increasing the radiation dose. If the amount of radioactivity is increased, this usually means that more localizes in the target tissue because at these trace quantities, the physiological system is never near saturation. The increased activity in target yields better images and hence improved diagnostic information. Lastly, in an ideal situation one would want to use nuclides that produce solely γ-rays and no particles. This would eliminate high local irradiation. The purpose of this communication is to describe a few radiopharmaceuticals, outline their applications, and describe the mechanisms by which they work.

Diagnostic applications

Nuclide characteristics

Some of the metallic nuclides which are presently extensively used in diagnostic Nuclear Medicine throughout the world are ^{111}In, ^{67}Ga, and ^{201}Tl. They are used in a wide range of applications including the detection of malignant tumors, inflammatory processes, assessment of cardiac viability and inflammation of cardiac muscle. These nuclides may be used as monoatomic element, as in the case with ^{201}Tl or bound to other molecules which chelate ^{111}In or ^{67}Ga.

Table 1 shows the major characteristics of these nuclides. In an aqueous environment, the elements exist as Ga^{3+} and In^{3+}; thallium has two oxidation states, Tl^{3+} and Tl^{+} (the most stable) [2]. Both ^{67}Ga and ^{111}In can undergo extensive hydrolysis at pH 3 and above if not chelated. At physiological pH, the major species for Ga is the gallate, $[Ga(OH)_4]^-$ while for the much less amphoteric In it is the uncharged $In(OH)_3$ [3]. For example, at pH 7.4, gallium forms mostly $[Ga(OH)_4]^-$ (98%) and some $Ga(OH)_3$ (1.6%) with OH^- ligands [4]. However, the solubility of gallate is limited by that of $Ga(OH)_3$ whose minimum value is ~40 nM [3]. The uncharged $Ga(OH)_3$ would precipitate if its concentration is above this value. For ^{67}Ga this limit is not a concern *in vivo* because for the usual initial blood level of ~74 kBq/ml (6 mCi [normal dose]/~3000 ml plasma), the gallium concentration would be 0.050 nM. However, the ^{67}Ga radiopharmaceutical is the citrate complex, a weak chelate K_{ML} = ~10^{10}, to prevent any possibility of hydrolysis and subsequent precipitation [5]. Since the concern of hydrolysis would be greater for ^{111}In, the nuclide is shipped as the chloride salt in a moderate concentration of HCl (~0.05 M). Furthermore, many of presently used radiopharmaceutical for this nuclide incorporate a high affinity polydentate chelate diethylenetriaminepentaacetic acid (DTPA) $K_{ML} = 10^{28}$ [6] in their structure (see later this text).

A variety of factors contribute to the utilization of these nuclides. The first and foremost is the ability of the nuclide or radiopharmaceutical (combination of nuclide and molecule that directs the nuclide to target) to identify some significant clinical parameter. In the late 60's, ^{67}Ga was shown to localize in certain soft tissue tumors and inflammatory processes [7, 8]. Later ^{201}Tl was shown to concentrate in the heart and to provide an indication of blood flow to various myocardial segments as well as the absence of flow which would indicate the extent and position of non-viable tissue [9, 10]. Lastly, Thakur and McAfee [11] devised a technique to label white blood cells (WBC) with ^{111}In so that this radiopharmaceutical could be used as an infection detection agent.

Next important factor is the ability of nuclide which has localized in a particular tissue in a patient to be detected externally by the present day detection devices. These devices, gamma cameras, consist of a lead collimator to direct the γ rays to a spot on a NaI(Tl) crystal. While each of the nuclides have γ rays that can be detected by the camera, there is some physical characteristic that makes these nuclides less than optimal for imaging. Gallium probably has worst characteristics of the three (Table 1). When ^{67}Ga decays, less than 40 γ rays are produced per 100 disintegrations for the 93 keV photon. That is, in the other 60 disintegrations, ^{67}Ga decays directly to the ground state of ^{67}Zn without yielding a photon. The other γ rays of ^{67}Ga are even less efficient. In contrast, the photon efficiencies for both ^{111}In and ^{201}Tl are greater than 90%. Thus, if we compare just photon efficiencies, ^{67}Ga would need to be taken up twice as much by the target tissue to be comparable to the other nuclides. A compensating factor for ^{67}Ga at least compared to ^{111}In is that the efficiency of a camera's crystal to transform the photon energy into light is inversely related to photon energy [12]. The disadvantage of lower energy photons (<100 keV) (this is more of a problem for ^{201}Tl) is increased tissue attenuation and scatter. These photons are more absorbed and scattered by tissue before they emerge to interact with the camera.

Table 1. Important nuclear characteristics of useful nuclear medicine nuclides.

Nuclide	^{67}Ga	^{111}In	^{201}Tl
Characteristic			
Photon energy (keV)	93 188 300	173 247	69–80
%photons/disint	38 20 16	91 94	95
half-life (hr)	78	67.4	73.1
decay	EC to ^{67}Zn	EC to ^{111}Cd	EC to ^{201}Hg
Production method cyclotron	^{68}Zn(p,2n)^{67}Ga	^{112}Cd(p,2n)^{111}In	^{203}Tl(p,3n)^{201}Pb→^{201}Tl
Electron structure	[Ar]$3d^{10}4s^24p$	[Kr]$4s^{10}5s^25p$	[Xe]$4f^{14}5d^{10}6s^26p$

While both clinical usefulness and the ability are important, the cost and ease of distribution is also an essential component for widespread use. Table 1 shows that each of these nuclides are produced by charged particle bombardment in a cyclotron. For example, ^{67}Ga is produced by irradiating an enriched ^{68}Zn (normal abundance for this isotope is 18.6%) target with 12–22 MeV protons [13]. To isolate the ^{67}Ga, the target is dissolved in HCl, the Ga extracted with isopropyl ether, back extracted from the ether into HCl and the solution is evaporated to dryness. Then citric acid is added, the pH adjusted to near neutral and finally the ^{67}Ga citrate solution is sterilized. The need for an enriched zinc target increases the cost because expensive procedures are required for its separation. The energy of the protons is not unusually high and the bombardment can be carried out in moderately sized cyclotron [14]. Indium is produced with a similar energy range of protons but the target does not need to be enriched [15]. This would imply that all things being equal ^{111}In should have a lower per unit production cost. However, at present, primarily due to the low clinical utilization of ^{111}In nationally, ^{111}In costs 4–5 times higher than ^{67}Ga. Indium is purified from cadmium in a similar manner to the Ga purification. There are a number of cycles of extraction with HCl and back extraction into ether to remove iron and other impurities. The final product is dissolved in sterile, pyrogen-free HCl. To produce ^{201}Tl, a somewhat higher proton energy, 31 MeV, is used and the target is pure natural thallium [16]. This target is dissolved in acid and ^{201}Pb produced (Table 1) is isolated by ion-exchange chromatography. The ^{201}Pb is allowed to decay (9.4 hr) to ^{201}Tl which is separated by ion-exchange chromatography as the Cl salt. The other factors that contribute to the cost are: the rate of their production (μCi/μA-hr), the efficiencies of their separation and the ease of the process required to convert them into desired radiopharmaceuticals. The $T_{1/2}$ of these nuclides is relatively long (~3 days) so that the nuclides can be shipped to users radiopharmacies or hospitals without significant losses of radioactivity due to decay. After an order is placed, nuclides can arrive the next day. For these isotopes, the loss due to decay is ~1%/hr. This means to achieve the desired delivered radiopharmaceutical dose the activity must start out about 15–25% greater which is acceptable.

Gallium-67 citrate

In vivo distribution

After it is injected ^{67}Ga binds to the iron transport protein transferrin (TF) in the blood and a small percent of non-protein bound Ga remains in equilibrium. For the normal concentration of unsaturated TF (40 μM) in the blood, the gallate concentration is only 1% of the total injected [17]. Because of the extensive hydrolysis of ^{67}Ga (see above) and the low affinity of TF relative to OH^- [4], ^{67}Ga easily shift between TF and the gallate depending the TF concentration. For example, if the concentration of unsaturated TF is decreased from 40 to 50 μM by an increase in iron-saturated TF. This would increase gallate concentration to 7% and increase the amount of activity that could pass out in the urine [17]. The radioactivity is excreted very slowly. For example, after 7 days, only 26% of the injected activity has lost in the urine and only 9% in stools [18]. In humans, the radioactivity is normally initially distributed in the liver (5% of injected dose), spleen (1%), kidney (2%), and skeleton 24% which remains constant as a function of time [19]. However during this time, a large amount of activity appears in the bowel which tends to obscures lesions in this anatomical region.

Indications for scintigraphy

For many years ^{67}Ga citrate was used for the detection of both acute and chronic abscesses and inflammatory processes. However, with the development of ^{111}In oxine labeled white blood cells (see later) it is now used mostly to identify chronic occult abscess sites and chronic inflammatory lesions in diseases such as Sarcoidosis [20]. This nuclide is also used to detect inflammations in patients where WBC function is impaired such as Pneumocystis Carinii Pneumonia. The major cause of death in AIDS patients. This radiopharmaceutical is the choice for radionuclide imaging of number of soft tissue tumors [21]. These include the detection and staging of Hodgkin's disease, non-Hodgkin's lymphomas, lung cancer, hepatoma, and melanoma. An example of a ^{67}Ga scintigram of a patient with Hodgkin's Lymphoma is shown in Figure 1. Probably more important than just identifying the lesion is the ability of ^{67}Ga to determine

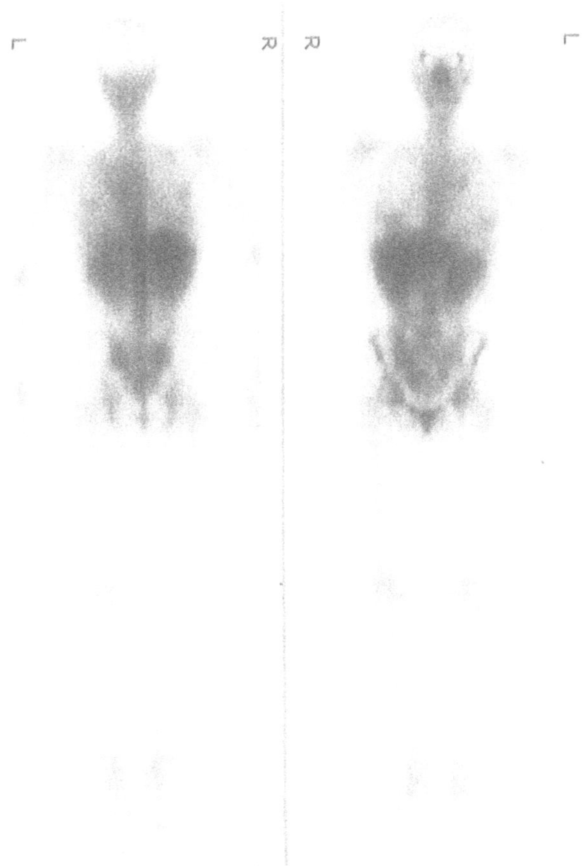

Fig. 1. Posterior (left) and anterior (right) 48-hr images of patient with Hodgkins Lymphoma injected with 10 mCi of ^{67}Ga. Normal uptake in the liver, spleen and skeleton easily seen on the anterior image. Increased activity in mid-chest region and in the lower abdomen on the right side of the patient is also seen on the anterior image.

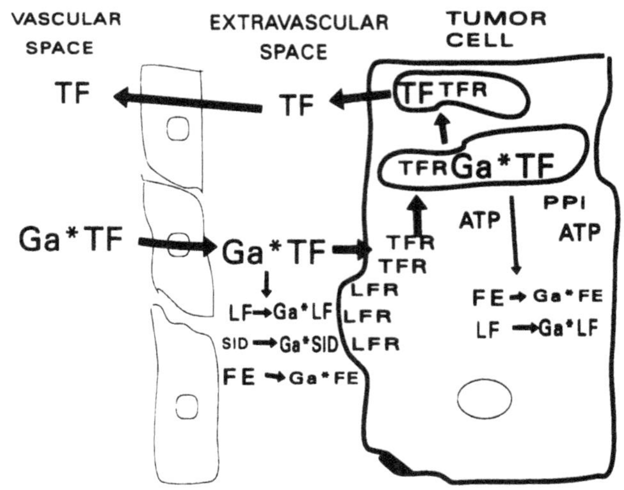

Fig. 2. A proposed model of ^{67}Ga uptake in tumors; ^{67}Ga brought to the tumor by transferrin (TF) where it binds to a transferrin receptor (TFR) and the radionuclide is delivered intracellularly to either proteins, lactoferrin (LF) or ferritin (FE). The proteins, LF and FE or a siderophore-like growth factor (SID) could be present extracellularly to bind ^{67}Ga. The lactoferrin receptor (LFR) and the SID could participate in nuclide transport into the tumor cell.

response to therapeutic intervention. In contrast to radiologic examinations such as computed tomography which detect morphological changes it appears that ^{67}Ga can differentiate recurrent viable tumor compared to scars caused by the chemo- or radiation therapy.

Localization mechanism

The localization mechanism of this nuclide in both tumors and abscess has been studied for more two decades since its clinical utility was identified. Figure 2 shows a proposed model that represents the involvement for a variety of molecules promoting accumulation of ^{67}Ga in tumors. The preponderance of evidence suggests that transferrin (TF) and its receptors on the tumor cell surface are the most critical factor in uptake of ^{67}Ga in tumor cells [22 and references therein]. In the 70's it was shown that TF due to its iron-binding ability stimulated ^{67}Ga incorporation into cultured tumor cells [23]. Other iron-binding molecules; such as desferrioxamine [24], lactoferrin (LF) [25] and ferritin (FE) [26] bind ^{67}Ga with high affinity. Subsequent experiments showed that Ga bound to the iron binding site of TF, LF and a variety of bacterial iron transport molecules called siderophores and caused only slight alterations in the 3-dimensional structures of these molecules. This led investigators to the thesis that ^{67}Ga was handled as an iron analogue i.e., similarly to iron [27]. However, this concept was criticized because animal [28] and human [29] studies showed great disparity in biodistribution between ^{59}Fe and ^{67}Ga. Iron localized in hemopoietic tissue and red blood cells but not in most tumor tissue. In contrast, ^{67}Ga had a great affinity for tumor tissue but low affinity for marrow and remained in liver, spleen and other tissue much longer than ^{59}Fe. Hoffer [27] originally suggested that this pronounced difference was related to the unique oxidation-reduction properties of the two metal ions. For iron to be absorbed in the gastrointestinal tract and eventually incorporated intro hemoglobin, hemopoietic cells in the marrow, and in the cytochrome enzyme system in the mitochondria, requires iron to cycle between its two physiologic stable states; Fe^{2+} and Fe^{3+} [30, 31]. In contrast, Ga^{3+} is the only stable state for gallium under physiological conditions [2]. For example, Ga could not be incorporated into hemoglobin in RBC because it could not be reduced to Ga^{2+} nor could it follow metabolic pathways where reduction was necessary [29]. More recent evidence [4, 32, 33] suggests that the aqueous solubility of the 2 metals are as important as the redox properties in determining their biodistributions (see before). In a physiological environment and at trace levels, ^{67}Ga is soluble whereas iron is not. Therefore iron, not ^{67}Ga, absolutely requires a chelate of some type (e.g., TF) for *in vivo* transport.

Larson and co-workers [34] extended the iron analogue concept to specifically included TF receptors (TFR). Based on both *in vivo* [35, 36] and *in vitro*

[34, 35] model systems they proposed that "a tumor-associated TF receptor is the functional unit responsible for the affinity of gallium for certain neoplasms". Autoradiographic and cell organelle isolation studied demonstrated that ^{67}Ga was incorporated into the lysosomes of viable tumors cells [37]. At normal concentration of Fe*TF (complex of Fe and TF) in the blood, the TFR are saturated with Fe*TF and cells must upregulate the number of TFR to increase iron uptake [30]. Iron is required for several cellular processes but TFR are regulated mainly to meet specific iron needs of DNA synthesis [31]. The specific target for this iron is ribonucleotide reductase, a non-heme iron enzyme which catalyzes the first rate-limiting step in DNA synthesis, conversion of ribonucleotides to deoxyribonucleotides. In rapidly proliferating tumor cells with a high level of DNA synthesis, upregulation of the surface TFR would also lead to increased ^{67}Ga uptake. Increased number of TFR was detected in a variety of tumor cells [34, 38].

TFRs appear to be important in neoplastic cell localization of ^{67}Ga. However, recent experiments with animals that congenitally lack TF still show good tumor uptake [39]. Furthermore, cells in certain circumstances can acquire Fe and possibly ^{67}Ga by TF independent means [40, 41]. This system, however, is regulated quite differently than the TFR system. It is not affected by cellular growth rate, induction of DNA synthesis, cell division or depletion of cellular iron [40]. Thus, this TF-independent system may enhance clearance of excess of iron which may otherwise be toxic. In circumstances where effective TF concentration is reduced or absent, this system, although slower, may be responsible for all the accumulation. However, in normal circumstances TF-dependent system would mainly be responsible for uptake.

The importance of ^{67}Ga scintigraphy as a diagnostic tool lies in its ability to both identify tumor and to differentiate viable tumors from non viable tumor and/or nontumor tissues. There are some data which confirm the connection of cell viability and DNA synthesis to ^{67}Ga uptake. Iosilevsky *et al.* [42] showed that both ^{67}Ga and deoxyglucose uptake in a tumor model had a parallel decline after chemotherapy and radiotherapy. Inhibiting tumor ATP production causes a similar decline in ^{67}Ga incorporation [43] suggesting that ^{67}Ga uptake correlated with cellular metabolism. Most recently, the response to radiotherapy was monitored in a tumor model using metabolic tracers for glucose metabolism, DNA, RNA and protein synthesis and the uptake of these were compared to ^{67}Ga [44]. While ^{67}Ga could not detect early response to treatment, it closely followed the diminution in glucose metabolism. The DNA synthesis indicator and ^{67}Ga could differentiate between tumor with small foci of recurrence and non-viable tumor, whereas the glucose metabolic indicator could not. Non-radioactive gallium has been shown to interfere with DNA synthesis by the specific inhibition of ribonucleotide reductase [45]. The TF-dependent system is intimately connected with

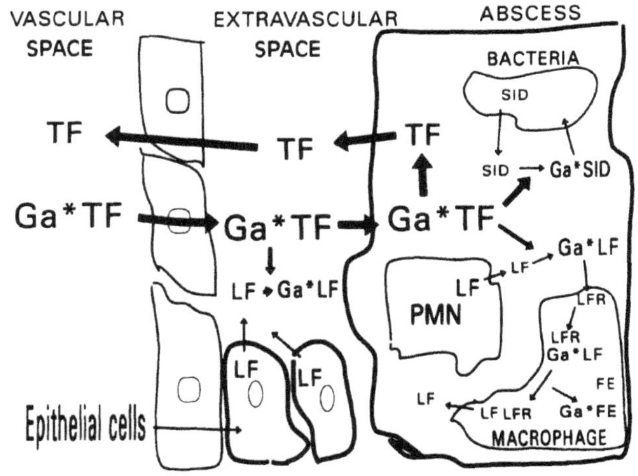

Fig. 3. A model of ^{67}Ga localization in inflammation. The nuclide is delivered to the site by TF and then can translocate to other proteins and to other molecules present in high concentrations. Bacteria could take up ^{67}Ga directly mediated by siderophores (SID). Lactoferrin (LF) deposited by Polymorphonuclear Leukocytes (PMN) or secreted by epithelial cells could bind ^{67}Ga. LF could also deliver its bound nuclide intracellularly to the Macrophage.

actively dividing cells whereas the TF-independent system is not.

Other factors such as LF, FE and Siderophores (SID) could be involved in tumor localization but their role has not been studied extensively. LF, an iron binding protein similar in size and structure to transferrin [30] has an affinity for ^{67}Ga greater than TF [46]. It also forms a complex with ^{67}Ga *in vivo* [47]. Hoffer *et al.* [48] have shown that in two patients with ^{67}Ga-positive tumors (Hodgkin's disease and Burkitt's lymphoma) there was an increased concentration of LF in tumor tissue using immunofluorescence techniques. A LF binding site has been detected on cultured tumor cells [49, 50]. These LF binding sites could be upregulated possibly for iron reutilization [51] or for myelopoiesis regulation [52]. FE is an intracellular iron storage protein [450,000 Da] which has a higher affinity for ^{67}Ga than either TF or LF [32, 53]. This protein could act as a depository for the ^{67}Ga brought into the cell by the TFR. Gallium activity could easily be translocated from TF to FE [53]. High molecular weight proteins from tumor cell lysates contain 8–60% of the total ^{67}Ga activity in these lysates [35, 41]. FE may be secreted from tumor cells and FE binding sites have been found on the surface of hematopoietic cell lines. SID are iron transport molecules that are synthesized by bacteria [see below] [54]. SID have a high affinity for ^{67}Ga and Figure 2 shows how these molecules could participate in the TF-independent pathway [55]. However, there have been only two reports showing that SID have been detected in animal tumor tissue [56, 57].

In an abscess, the mechanism appears to be somewhat different and a proposed mechanism is shown in Figure 3. It is generally believed that TF delivers the ^{67}Ga to the abscess [32, 58, 59]. Once delivered to an

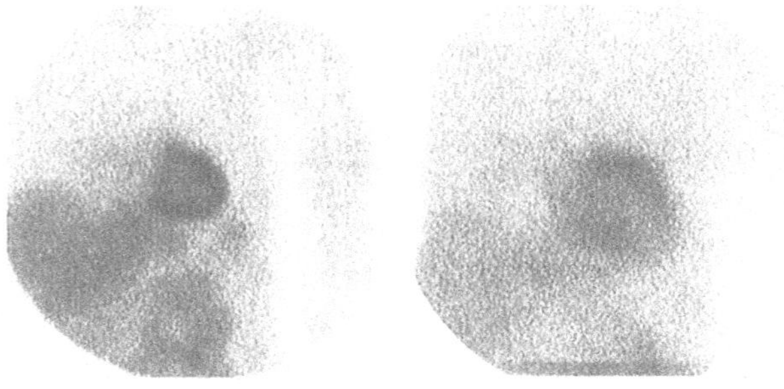

Fig. 4. Anterior images of patients at rest injected with 3 mCi of ^{201}Tl showing normal horseshoe distribution in the myocardium (left) and myocardial infarct in the apical and anteroseptal segments [tip and bottom part of horseshoel] (right). The patient's arm can be seen on the right, liver to the bottom left of the heart and thyroid on the top of the normal image.

abscess site, ^{67}Ga could then be transferred from TF to LF, FE or SID. LF has a high affinity for ^{67}Ga and can remove ^{67}Ga from TF [46]. This protein is present in the secondary granules of polymorphonuclear leukocytes (PMN) [60] and is exocytosed by these cells upon bacterial challenge [61]. Bennett and Skosey [62] have demonstrated the increased presence of LF deposited by PMN in a variety of inflammatory processes and this would provide the abscess exudate with a large quantity of ^{67}Ga avid macromolecules. Active inflammatory lesions are characterized by PMN infiltration [63]. As the lesion age increases a number of monocytes/macrophages appear. LF present at the site could then bind to LF binding sites on monocytes [64] or macrophages [51] and deliver the ^{67}Ga to the intracellular FE. Horse (Ho) FE [32, 53] and Human (Hu) FE have a high affinity for ^{67}Ga [65] and small MW compounds e.g., ATP found intracellularly, can speed the *in vitro* transfer of ^{67}Ga from LF to HoFE and HuFE. Lastly, pathogenic organisms in an purulent abscess are probably the most important factor in ^{67}Ga uptake. Bacteria can take up a significant quantity of ^{67}Ga [58]. Hoffer suggested that SID were involved in ^{67}Ga uptake [27]. These SID are synthesized and secreted by bacteria to scavenge Fe^{3+} and give these organism a survival advantage [54]. These molecules are small molecules (500−1000 Da) and have a extremely high affinity ($K \geq 10^{30}$) for a single Fe^{3+} ion. To deliver the iron, the iron-loaded siderophore passes through a unique membrane protein on the organism's surface. These membrane proteins and SID synthesis are regulated according to iron requirements. SID binds Ga with very high affinity [66] and desferrioxamine, a SID can remove ^{67}Ga from TF [55]. SID can transport ^{67}Ga into organisms in a saturable, energy-requiring manner [66].

Thallous-201 chloride

Indications for scintigraphy

This nuclide is widely used throughout the USA and the world in the management and diagnosis of coronary artery disease [67]. Imaging with ^{201}Tl facilitates a functional assessment of the myocardium which directly reflects the blood flow to myocardial tissue, myocardial tissue viability and cellular membrane integrity. This nuclide provides a physiological assessment of the heart segments which are no longer functioning after a heart "attack" (infarct). In particular, ^{201}Tl scintigraphy is useful in the determination of the presence of viable myocardial tissue which has an inadequate blood supply due to a partial occlusion of coronary arteries (vessel narrowing). In such situations, the ^{201}Tl images (obtained after exercise) usually show a cold spot which indicates the absence of viable tissue. After a period of time, the rest image is obtained. If there are segments in this area which fill with ^{201}Tl, these indicate viable tissue. In contrast, if ^{201}Tl is not refilled in the area in question this indicates dead or fibrotic tissue (Fig. 4). This technique, used in conjunction with coronary artery angiography allows one to determine what vessels could be repaired to improve blood flow with coronary bypass surgery or angioplasty.

Biodistribution

After injection with $^{201}TlCl$, the radioactivity is rapidly taken up by the normal heart and other muscle tissue, achieving maximal concentration of 4−5% of the total injected dose in 10−30 min [9, 10]. The activity in the heart is uniformly distributed and has a moderately slow washout with the $T_{1/2} = 4-7$ hr. This is sufficiently slower than imaging time (~30−45 min). The activity in the blood declines rapidly. In a resting study, 90% of the activity disappears within 20 min for which only 90 seconds are required after exercise. This means that imaging can begin 15 to 20 minutes after injection. The nuclide also concentrates modestly in the kidney (3%) [68] and whole body excretion is slow with $T_{1/2} \approx 10$ d [9].

Localization mechanism

Evidence suggest that Tl is incorporated into the myocardial cell in a similar manner as potassium. Tl^+ has a similar crystal radius (1.5 Å vs 1.38 Å for K^+) and hydrated radii. Thallium uptake into cells mediated in part by the Na^+-K^+-ATPase. Intracellular K^+ levels are maintained in myocardial and other cells by this energy requiring membrane bound enzyme. Thallium uptake is inhibited only 60% by ouabain which normally blocks K^+ exchange [69]. Tl^{1+} is also more tightly bound to the enzyme than K^+ and activates the enzyme [70, 71]. It has a slower efflux rate suggesting possible intracellular binding compared to K^+. In an energy requiring bacterial exchange uptake system, K^+ is a competitive inhibitor of Tl^{1+} and can increase the Tl^{1+} efflux rate [72]. Tl^{1+} myocardial deposition is directly correlated with blood flow. Blood flow is a necessary but not a sufficient condition. The tissue also must be viable with adequate oxygenation [10].

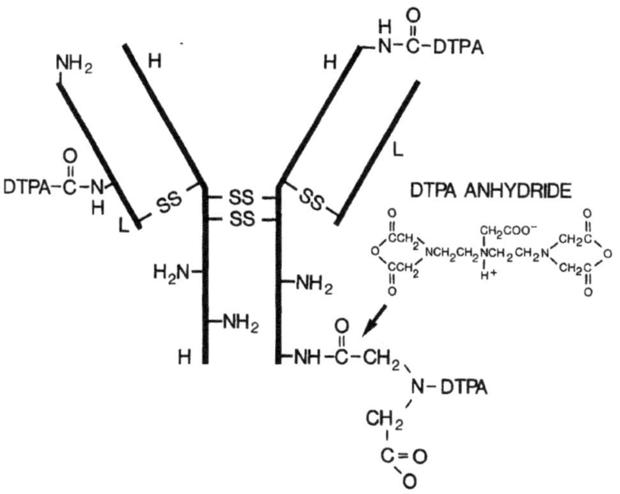

Fig. 5. The structure of IgG with a DTPA attached.

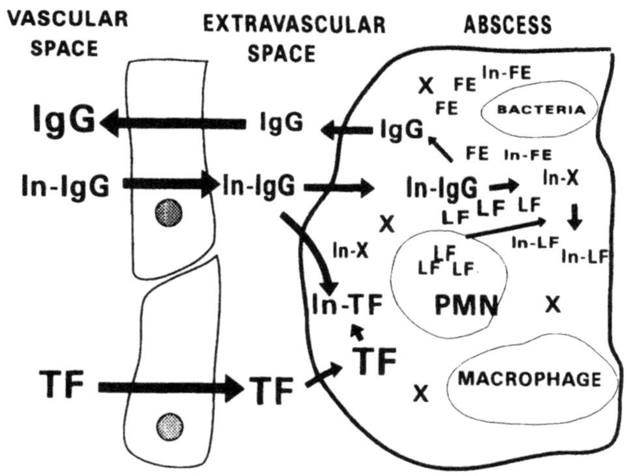

Fig. 6. A model of ^{111}In-IgG localization in inflammation. This includes active transport of IgG across endothelial cells and ^{111}In translocation to; 1, lactoferrin deposited by PMN, 2, unknown low molecular weight molecules, X, that stimulate translocation to proteins, 3, ferritin from disrupted cells and 4, transferrin from blood exudate.

Indium-111 labeled radiopharmaceuticals

Although a large number of ^{111}In radiopharmaceuticals have been prepared and evaluated, only a few prominent compounds are discussed.

Indium-111 labeled polyclonal human immunoglobulin G (^{111}In*IgG)

Indications for scintigraphy

This agent, designed to image infection, is a polyclonal human IgG to which are covalently coupled 2 to 3 DTPA molecules to lysines or arginines on antibody in a random process [73]. Figure 5 shows the most common way to attach a DTPA molecule to an antibody using cyclic anhydride of DTPA [74]. This conjugation is facilitated by formation of an peptide bond, C=OHN, between the protein and DTPA. ^{111}In chloride is then incubated with the DTPA-IgG and injected into the patient. Preliminary studies have been published which indicate that ^{111}In*IgG is an inflammation detection agent that could be a replacement for the present agents, ^{67}Ga or ^{111}In-WBC [75]. It has shown utility for detection of both chronic and acute infections. In a limited number of patients ^{111}In*IgG has shown less background uptake in bone infections compared to ^{67}Ga. At present, this radiopharmaceutical is waiting to be approved by the FDA. It must await for approval and subsequent usage to determine where exactly this agent will fit in the physician's armamentarium for detection of infections and inflammatory processes.

Biodistribution

When injected, ^{111}In*IgG has a long $T_{1/2}$ in the blood [73]. For example, at 12 hr post injection, nearly 60% of 2 min value remains in the blood. Radioactivity is excreted very slowly and no activity in the gastrointestinal tract has been observed. The major organs; liver, spleen, kidney, and bladder are the sites of radionuclide deposition in normal volunteers.

Localization mechanism

Originally it was considered that the F_c portion of the IgG binds to inflammatory cells in the circulation or present at the abscess site was the primary mechanism of the localization process [76]. At present there is little support for this hypothesis [77]. While this mechanism has not been extensively studied, there is evidence that mechanism shares some characteristics with ^{67}Ga abscess localization. A model of the proposed mechanism is shown in Figure 6. While breakdown in the endothelial barrier and increased capillary permeability may explain the initial localization process, it cannot explain the retention over 24 or 48 hr of the clinical time frame. Claessens and co-workers [78] have shown in a rat abscess model that if ^{99m}Tc, ^{111}In, and ^{131}I labeled IgG are compared, the ^{111}In alone remains at the abscess and while the %ID/g of the other nuclides declines as a function of time. It is likely that both the ^{131}I and ^{99m}Tc are stripped off from the protein moiety by the conditions at the abscess site but unlike In^{111}, cannot rebind to other abscess constituents and are removed by diffusion. Most recently these investigators have demonstrated in this model with ^{14}C and ^{111}In labeled IgG that ^{111}In was deposited at the infected site in higher quantities than ^{14}C. This further supported the idea that ^{111}In was released from the IgG in the abscess [79]. Another study in the similar model showed that ^{111}In*IgG uptake increased due to the expansion of the infected space and increased accumulation of ^{111}In into this space [80]. ^{111}In-human serum albumin (HSA), labeled identically, delivered a similar amount of nuclide to the infected site. It was however, a poorer imaging agent because of high uptake in normal tissue. The protein moiety did not appear to influence accumulation in the infected site. High concentrations of TF, LF, FE and low molecular weight mediators e.g. ATP, may permit ^{111}In to be translocated to these proteins and aid in the continual accumulation of radionuclide. TF has an higher effective affinity (1000-fold) for ^{111}In compared to ^{67}Ga [81]. Weiner and co-workers [82–84] have shown *in vitro*, under conditions which mimic those found in abscesses, that ^{111}In can be translocated from IgG-DTPA to both TF and LF. These translocation reactions appear to be effected by pH and are dependent on the presence of phosphoryl and thiol-containing small MW molecules that enhance ligand exchange. The protein moiety had only a modest effect on the translocation over time (transfer from IgG or HSA was also examined). These data suggest that after ^{111}In*DTPA-IgG accumulates into an abscess site via a disrupted endothelial cell barrier, the ^{111}In is translocated to proteins at the site even though ^{111}In has a high affinity for DTPA and thereby permits the ^{111}In to remain at the lesion.

Indium-111 labeled OncoScint

Indications for scintigraphy

This radiopharmaceutical contains a monoclonal antibody originally identified as B72.3 or CYT-103 which is directed ageinst a tumor associated antigen called TAG-72 [85]. A DTPA chelate is coupled specifically via a spacer molecule to the carbohydrate portion of the IgG and has the form, Antibody-Carbo-glycyl-tyrosyl-lysine-DTPA. The ^{111}In is added just before injection. TAG-72, a high MW, mucin-like glycoprotein, is expressed on 94% of colorectal adenocarcinomas and 100% of ovarian cancers tested. OncoScint CR/OV (satumomab pendetide) is the first monoclonal antibody that approved by the FDA (January, 1995) for the diagnosis of the recurrence of ovarian or colorectal cancer. It was developed, is manufactured and distributed by Cytogen. OncoScint, while not as useful for the initial diagnosis of these diseases but may provide an important adjunct in disease management and reoccurrence [86, 87]. One-third of patients originally diagnosed with either of these cancers have recurrence within 18–24 months. A rising concentration of carcinoembryonic antigen (CEA) for colorectal cancer or CA125 for ovarian cancer in the blood is used as indicator. These antigens are readily shed from these tumor types but the blood test is sometimes unreliable. OncoScint scintigraphy is superior to other most used imaging modality, computed tomography, in detecting disease sites in the pelvis, lymph nodes, bone, lung and brain but not in the liver. However, this radiopharmaceutical is not useful if the image is negative since it has a low negative predictive value and the disease still may be present. OncoScint is recommended for only a single use because human anti-mouse antibodies (HAMA) have been detected in 40% of the patients. HAMA causes rapid blood disappearance of OncoScint and interferes with CEA assay. Health related side effects are minimal.

Biodistribution

After injection, OncoScint has a long terminal $T_{1/2}$ in the blood 56 hr [88]. Whole body excretion is very slow with the principle route of excretion via the kidney. In 72 hr, only 10% of the injected activity is eliminated. The normal biodistribution includes; early localization in liver and spleen [89]. Bone marrow activity appears at 24 hr and then persists, as does the activity in blood pool (due to long $T_{1/2}$), in the heart and major blood vessels. Faint kidney or bladder uptake may be seen on later images.

Localization mechanism

Tumor-specific antigenicity was recognized in 1929 and Pressman in the mid 50's attempted to image animal tumors with ^{131}I labeled antibody [90]. This attempt met with limited success. It was thought that the specificity of the polyclonal antibody was the problem. This work was not continued until the mid 70's when the specificity of an antibody preparation was improved by affinity-purification. With the development of monoclonal antibodies large quantities of these antibodies with high specificity could be produced. In early 80's interest was renewed in using monoclonal antibodies to image tumors.

After injection, OncoScint circulates in blood stream and passes through the endothelial cell barrier into extravascular space. The antibody then comes in contact with the tumor cell and binds to the TAG-72 antigen. The incorporation mechanism of antibody-antigen complex on tumor cell surface is not fully clarified and there are a few possibilities. One mechanism is patch and cap, coated pit internalization (typical of receptor-based processes) and catabolism of antibody-antigen complex in lysosomes. Another is modulation where the antibody binds to antigen on the surface and both are expelled from the cell's surface instead of being internalized.

While there was much initial excitement in the Nuclear Medicine community about the superb specificity of the antibody-antigen reaction this has been tempered by persistent problems with this technique. The most prominent problem is the HAMA response which is associated with diminished localization with subsequent injections. Although HAMA does not always prevent tumor localization, this response discourages repeated procedures. Slow clearance from the blood and a long biological $T_{1/2}$, require many days to allow the background to clear. Many studies have shown that antibody uptake is specific but absolute uptake of ^{111}In in the tumor is low (0.01–0.03% injected dose/g of tumor tissue) and this combined with slow clearance give rise to low target to background ratio. Persistently high activity in liver makes liver metastases which commonly occur, difficult to detect. Anti-

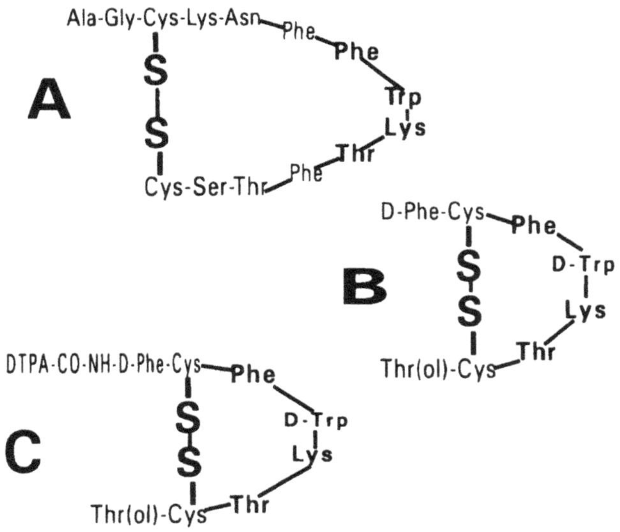

Fig. 7. The amino acid sequence of A, Somatostatin, B, Octreotide, and C, OctreoScan. The amino acids in bold are involved in receptor binding and the DTPA is coupled via a peptide bond. Amino acids are all L-isomers except where noted or an amino alcohol, Thr(ol).

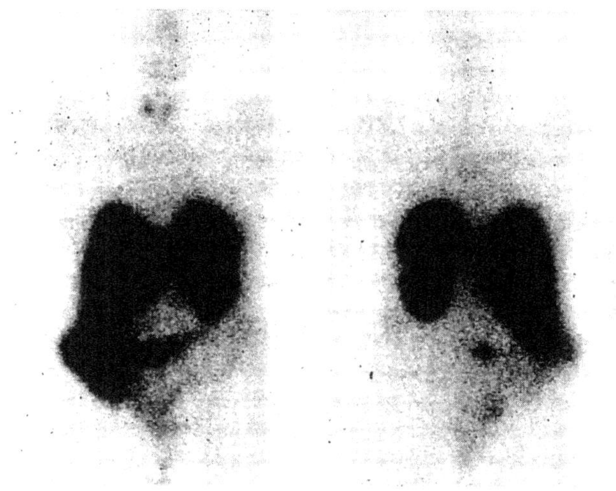

Fig. 8. Anterior (left) and posterior (right) head and torso images of patient injected with 6 mCi of ^{111}In-OctreoScan 24 hrs previously. Focus of activity noted in the anterior lower mid-abdomen which persisted on 96 hr images. Liver, spleen and thyroid gland are visualized on the anterior image and kidneys on the posterior.

body could be prematurely remove from the circulation by binding to circulating levels of antigen (e.g. TAG-72) which have been shed from the tumor cell's surface.

Many approaches have been attempted to overcome these problems and probably the most prominent now is the synthesis of peptides which contain the molecular recognition unit of the antibody. This consists of the amino acids of the antigen combining site. These peptides have many advantages. These molecules are usually non-antigenic, can penetrate the tissue easily and have short blood $T_{1/2}$. While they tend to localize in the target very early their persistence at the target and clinical usefulness will require much further investigation.

Indium-111 labeled OctreoScan (^{111}In*OctreoScan)

Indicates for scintigraphy

OctreoScan is radiolabeled analog of Somatostatin, Octreotide, that has been modified by adding an ^{111}In binding group, DTPA (Fig. 7) [91]. Somatostatin, is an endogenous peptide involved in the inhibition of growth hormones. ^{111}In*OctreoScan is used to localize a wide variety of Neuroendocrine tumors e.g., carcinoids, islet cell tumors, paragangliomas, medullary thyroid cancers, pheochromocytomas, gastrinomas, and anterior pituitary adenomas [92]. Figure 8 shows an image of a patient with an Adrenocorticotropin (anterior pituitary) hormone secreting tumor. Studies are in progress to determine if this radiopharmaceutical can also be used in the detection and staging of Hodgkin's disease and non-Hodgkin's lymphomas. These tumors may contain Somatostatin receptors.

Normal biodistribution

This radiopharmaceutical is rapidly cleared from the blood. For example at 10 min, only 33% of the injected dose remains in the blood and is rapidly excreted into the urine [91]. This is the major excretion route. 25% of the injected radioactivity is present in the urine after 3 hr, 50% after 6 hr and 90% after 48 hr. Hepatobiliary excretion is a minor, <2% appears in the feces after 72 hr. After intravenous administration accumulation of activity is observed in the thyroid gland, pituitary gland, liver, spleen, kidney, and bladder. Gallbladder and intestinal activity are seen at later time points. At 4 hr post injection, 2% of the injected dose is in the spleen, 1.9% in the liver, 7.2% in the kidneys.

Mechanism of localization

Somatostatin is a 14 amino acid, peptide hormone which inhibits the release of growth hormone, insulin, glucagon and gastrin [92]. Somatostatin receptors have been identified on many cells of neuroendocrine origin e.g. anterior pituitary and the islet cells of the pancreas. More importantly, tumors of neuroendocrine origin (see above) have increased expression of these receptors. Somatostatin also has anti-tumor growth effects. To enhance this potential, the hormone was modified to increase its extremely short blood $T_{1/2}$, 2–4 min. Octreotide, an 8 peptide analogue, which includes the receptor binding amino acids, two D-amino acids and an amino alcohol, was the result (Fig. 7). With the longer $T_{1/2}$ (90–120 min), localization and imaging of the radiolabeled octreotide was possible.

There is little metabolism of ^{111}In*OctreoScan. At 4 hr, the ^{111}In is still bound to the intact peptide in the blood and only 10% of the total activity in the urine is nonpeptide bound (package insert). Lastly, there is a strong correlation between a positive image and an *in vitro* assay of somatostatin receptors in excised tumor tissue [92]. This suggests that ^{111}In*OctreoScan is an *in vivo* indicator of somatostatin receptors. Other analogues of somatostatin are being investigated [93].

Indium-111 labeled Myoscint (^{111}In*Myoscint)

Indications for scintigraphy

Myoscint is an anti-myosin antibody, Fab fragment, with an attached DTPA chelate [94]. Myosin is the major intracellular muscle protein and is only exposed when the cell membrane is damaged. Clinical trials have been completed. After the FDA required more clinical information the application was resubmitted in July of 1994 by Centocor. Myoscint is now waiting for approval from the FDA. In contrast to ^{201}Tl, ^{111}In* Myoscint localizes in necrotic myocardial tissue. An infarct would appear as hot spot and normal tissue does not localize this agent. ^{111}In*Myoscint would probably not be used to routinely identify acute myocardial infarctions because of other diagnostic techniques. This agent however may be used to diagnose and evaluate acute myocarditis, acute heart transplant rejection and chemotherapy-induced cardiotoxicity.

Biodistribution

Indium-111*Myoscint remains in the blood with an initial short $T_{1/2}$ of 1.9 hr and slower phase washout of 25.6 hr [95]. At 24 hr, 24% of the initial activity remains in the plasma. Even though this radiopharmaceutical is an Fab fragment, imaging must wait ~24 hr before imaging, in order to let the blood pool radioactivity clear. The ^{111}In activity remaining is usually localized in the liver, spleen, kidney and at later time points, bone marrow. The kidneys receive the greatest concentration of radionuclide. Myoscint does not localize in old infarcted areas and maximum antibody uptake occurs in areas of lowest perfusion [94]. HAMA response has not been observed in patients, even with repeated injections [96, 97].

Mechanism of localization

Myosin, a highly insoluble protein, remains in the myocardial cells long after disruption of the cell membrane [94]. It is not soluble in the salt concentration of normal blood so it is not easily washed away from the sites. A calculation of the number of potential binding sites with a ruptured cell indicates $\sim 2.4 \times 10^{17}$, sites/g of myocardium. This is extremely high and a 100-fold greater than number of sites expected for tumor antigens. While the perfusion of necrotic zone is low, the small size of the Fab fragment, 50 kDa, allows penetration into the infarcted area. Preliminary indications are that fragments of myosin may interfere with uptake.

Indium-111-oxine labeled WBC (^{111}In*WBC)

Indium-111*WBC is prepared by removing the patients blood, isolating the white cell fraction (which normally contains 60% polymorphonuclear leukocytes (PMN) and 30% lymphocytes), incubating it with ^{111}In-oxine to label the WBC and finally re-injecting the labeled cells into the patient. Since its introduction in 1977 [98] ^{111}In*WBC have been successfully used to identify sites of acute inflammation, infection and abscesses [99].

Normal localization

After injection a large fraction of the ^{111}In is distributed equally in the liver and spleen [~30% each] [100]. There is only transient pulmonary uptake. Activity is observed in the bone marrow. A portion of blood activity clears with a $T_{1/2}$ of 2.8–5.5 hr while a small fraction has a longer $T_{1/2}$ of 64 to 116 hr. Very little of the radioactivity is eliminated from the body and release from the liver and spleen is also very slow.

Localization mechanism

8-hydroxyguinoline (oxine) forms a lipophilic, neutrally charged complex with ^{111}In which allows the complex to intercalate into the membrane of the white blood cell [101, 102]. Once inside the cell the ^{111}In in translocation to cytosolic proteins which permits the ^{111}In to remain intracellular and the oxine passively diffuse out of the cell. Inflammations, infections and abscesses are all characterized by a early (usually within 4 hr) PMN influx [63]. An inflammation is the tissue response to some type of irritating stimuli which could be caused by mechanical, chemical thermal trauma. In infections, the initiating stimuli is caused by bacterial, fungal or other organisms and an abscess is an infection where the body has synthesized a barrier surrounding the stimuli to prevent further spread. In the acute phase of each of these cases, the damaged tissues release a variety of chemical factors which increases blood flow, enhances capillary permeability and recruits PMN to the source of stimulation. In this phase there is a relatively rapid turnover of the PMN which allows the ^{111}In*WBC to be recruited to the site. It is also likely that the labeling process activates these cells making them more likely candidates for recruitment [103]. If the irritating stimuli is not eliminated by the PMN then new, more aggressive cells, the macrophage, are recruited to the source. In the next stage, the chronic phase, these cells become the predominate in fighting the irritating source and PMN influx is reduced. Thus, there is much less likelihood that

Table 2. Characteristics of useful therapeutic nuclear medicine nuclides.

Nuclide	^{89}Sr	^{186}Re	^{153}Sm
Characteristic			
Maximum β energy (MeV)	1.46	1.08	0.81
Average β energy (MeV)	0.583	0.346	0.243
half-life (days)	50.5	3.7	1.9
Production method reactor	^{88}Sr(n,γ)^{89}Sr	^{185}Re(n,γ)^{186}Re	^{152}Sm(n,γ)^{153}Sm
Electron structure	[Kr]$5s^2$	[Xe]$4f^{14}5d^5 6s^2$	[Xe]$4f^6 6s^2$

^{111}In*WBC would localize in chronic lesions. This is consistent with clinical studies showing that detection rates decrease for chronic inflammations and abscesses [20].

Indium-111 labeled DTPA (^{111}In-DTPA)

Indications for scintigraphy

This radiopharmaceutical is use to determine the distribution of cerebrospinal fluid (CSF) and more specifically whether there is a leak in the central nervous system (CNS). The CNS which includes the brain and the spinal canal is bathed by this special fluid, CSF. Blood is prevented from entering the CNS by the blood brain barrier (BBB). This fluid is produced in the central part of the brain near the hypothalamus, the ventricles, circulates throughout the brain and spinal chord. The CSF is finally absorbed in an area at the top of the brain near the skull, the arachnoid villi.

Normal localization

After intrathecal injection direct injection into spinal canal space, the ^{111}In-DTPA moves with the flow of CSF down the spinal canal to the head [104]. The nuclide migrates to the base of the brain (basal cisterns), over the top of the cerebral cortex to the parasagittal space (region between top brain and brain membrane) and finally to the arachnoid villi where it exits. The ^{111}In-DTPA does not normally enter the ventricle space. Activity reaches the cisterns in an hour, peaking at 4 hr. Parasagittal radioactivity is first seen at 4 hr, maximizes at 14 to 17 hr, and declines with a $T_{1/2}$ of 10–14 hr. The radionuclide then appears in the blood and is rapidly excreted by the kidneys.

Localization mechanism

The high stability of ^{111}In-DTPA and its small size (~500 Da) maintains this radiopharmaceutical in the CSF until it is absorbed or leaks out through any break in the BBB.

Therapeutic applications

Strontium-89 chloride

Indications for therapy

This nuclide is used for the relief of intractable of pain from widespread metastatic bone lesions [105, 106]. Some of the important characteristic of this nuclide are shown in Table 2. It is a pure β emitter with the maximum range of 8 mm which provides minimal radiation of soft tissue distant from the lesions in which it concentrates. The absence of a γ photon means that localization cannot be verified. The long $T_{1/2}$ means that the lesions will be continually irradiated for an extended period. Strontium-89 provides palliation of bone pain for patients with Prostate or Breast cancer and other malignancies. A common manifestation of these cancers is painful metastatic lesions to the bone. These lesions are usually osteoblastic with an increase in bone turnover and bone deposition. This process allows the increased uptake of bone agents. Chemotherapy may not relieve the pain of these metastases. If there are a few metastases, pain can be controlled by focal external radiation therapy. When there is a great spread of the cancer, whole body or hemi-body beam therapy can be used which is usually associated with significant side effects e.g. nausea, vomiting, diarrhea and bone marrow toxicity. ^{89}Sr provides a less toxic and physically tolerable alternative.

Biodistribution

After injection, ^{89}Sr is rapidly cleared from the blood into the urine [107]. It is primarily excreted through urine (2/3) and feces (1/3) in patients with metastases. Urinary output is greatest in the first two days after injection. Usually 50% of the injected dose is taken up by the skeleton. The skeletal retention after three months ranges between 12 and 90% of the initial uptake [105]. The nuclide remains in metastatic bones lesions longer than normal bone which has a 14 day turnover. The recommended dose is 0.04–0.06 mCi/kg or 4 mCi per patient. Pain relief usually begins 2 weeks after treatment and lasts from 4 to 15 months. The reported effectiveness of ^{89}Sr is from 51% to 86% of treated patients. The main limitation to ^{89}Sr and other bone localizing therapeutical agents is bone marrow toxicity. This radiopharmaceutical causes mild myelosuppression resulting in decreased platelets and leukocytes levels. This decline is usually mild about 15 to 30% of the pre-injection value and recovers in 12 weeks.

Localization mechanism

Ionic strontium is incorporated into bone in a similar manner as calcium. Strontium ion has a similar crystal radius (1.13 Å vs 1.00 Å for Ca^{2+}). ^{85}Sr was first studied in Nuclear Medicine as an imaging replacement for ^{45}Ca [108]. Prior to this period, due to the concern about ^{90}Sr fallout, considerable evidence was generated from studies in animals and in human demonstrating the similarity of Ca^{2+} and Sr^{2+} metabolism [109, 110]. There are, however, quantitative differences in the renal and gastrointestinal handling of these two elements. Calcium is present in the inorganic salts (90%) of cortical bone which also consists of an organic matrix (10%) [111]. To produce cortical bone, this organic matrix, called osteoid, is first laid down by the bone producing cells, osteoblasts. To strengthen the osteoid, Ca^{2+} and PO_4 (P_i) ions are deposited into the osteoid. This provides compressional strength to the bone. The deposition is a two part process with initial material being laid down and then a propagation of mineralization from this site. This initial material is amorphous (noncrystalline) calcium phosphate (ACP). ACP forms foci that grow into crystalline (a regular ordered array) hydroxyapatite (HA). The process may take weeks or months for completion. ACP is a mix of a variety of calcium phosphate salts with various state of hydration. These include $CaHPO_4$, $CaHPO_4 \cdot 2\,H_2O$, $Ca_4H(PO_4)_3$, and $Ca_3(PO_4)_2 \cdot 3\,H_2O$. On the other hand, the formula of HA is mainly $Ca_{10}(OH)_2(PO_4)_6$ but the Ca/P ratio can vary from 1.3 to 2 to 1. HA forms long flat crystals 1−3 nm thick 10−25 nm wide and 45 nm long. This mechanism although slow promotes a rapid incorporation of the small quantities of ionic ^{89}Sr into the bone.

Rhenium-186 hydroxyethylidene diphosphonate (^{186}Re-HEDP)

Indications for therapy

This radiopharmaceutical is chelate of ^{186}Re and HEDP and can be used similarly as ^{89}Sr (Table 2). This nuclide has a relatively shorter $T_{1/2}$, shorter maximum range in tissue, 5 mm but has a γ photon 137 keV (9% abundance) [112]. This allows imaging and quantification of the *in vivo* distribution of the radiopharmaceutical and therefore the calculations of absorbed radiation dose. Phase III clinical trials are presently in progress.

Biodistribution

After injection, ^{186}Re-HEDP is rapidly cleared form the blood into the urine and the skeleton with a blood $T_{1/2}$ of 40 hr [112]. A total of 71% of the dose is excreted in urine in the first 24-hours after injection. The tested dose has been ~35 mCi but optimal dose has not been determined [113, 114]. Pain relief usually begins 2 weeks after treatment and lasts an average of 5 weeks. 80% of the treated patients experienced pain relief. This radiopharmaceutical causes mild myelosuppression from the initial administration which results in decreased platelets and leukocytes counts. This decline is usually mild and averages 30% of the preinjection value.

Localization mechanism

This radiopharmaceutical is probably incorporated into bone in a similar manner as ^{99m}Tc-methylene diphosphonate (MDP) [111]. A complex of phosphate and calcium is present in the inorganic salts of cortical bone. To strengthen the organic matrix of bone, osteoid, both Ca^{2+} and P_i are deposited into the matrix. The present theory of bone formation suggests that a matrix vesicle incorporates the constituents of the inorganic matrix, Ca^{2+} and Phosphate complexes, then organizes this material in to ACP and that this eventually grows into HA. Phosphate complexes are actively (energy-requiring) transported into bone when the ACP is being laid down. Similarly phosphonate metal complexes would be actively transport into these vesicles.

Samarium-153 ethylene diamine tetramethylene phosphonic acid (^{153}Sm-EDTMP)

Indications for therapy

This radiopharmaceutical, a chelate of ^{153}Sm and EDTMP, would also be used similarly as ^{89}Sr (Table 2). This nuclide has a the shortest $T_{1/2}$ of all three agents, shortest maximum range in tissue, 3 mm but also has a γ photon 103 keV (28% abundance) [115]. This allows imaging of the localization. This radiopharmaceutical would be used similarly to ^{89}Sr. Phase III clinical trials have been completed and the data have been submitted to the FDA for approval (Cytogen, June, 1995).

Biodistribution

^{153}Sm-EDTMP rapidly clears from the blood into the urine and only 1% of injected activity remains in the blood after 4 hr [115]. An average of 36% of the dose is excreted in the first 24-hours after injection. Skeletal uptake depends on the number of bony metastases and ranges from ~50% to 80% of the administered dose. The tested dose was 0.5 to 1 mCi/kg which gave good results. The optimal dose has yet not been determined. Pain palliation lasts an average of 4 months with a range of 1 to 11 months. A high fraction (65%) of treated patients experienced pain relief. This radiopharmaceutical, similar to the others, causes mild myelosuppression from the administration which resulted in a decline of the platelet concentration from 43% to 66% of the preinjection value. Leukocytes levels declined in all subjects but only in 2 out of 29 patients fell below the normal range. These counts returned to pretreatment levels by 6 to 8 weeks in all patients.

Localization mechanism

This radiopharmaceutical localizes by an identical manner as ^{99m}Tc-MDP [115]. When ^{99m}Tc-MDP and ^{153}Sm-EDTMP are injected in the same patient complete concordance was observed with each image. Thus, the mechanism is likely to be the same as discussed above for ^{186}Re-HEDP [111].

References

1. Kowalsky, R. J., Perry, J. R.: *Radiopharmaceuticals in nuclear medicine practice*. Appelton & Lange, Norwalk, 1 (1987).
2. Cotton, F. A., Wilkinson, G.: *Advanced inorganic chemistry*, 5th ed. Interscience, New York, 215–230 (1988).
3. Baes Jr., C. F., Mesmer, R. E.: *The hydrolysis of cations*. Wiley-Interscience, New York, 318–319 (1976).
4. Harris, W. R., Pecoraro, V. L.: Thermodynamic binding constants for gallium transferrin. Biochemistry **22**, 292–299 (1983).
5. Harris, W. R., Martell, A. E.: Aqueous complexes of gallium(III). Inorg. Chem. **15**, 713–720 (1976).
6. Welch, M. J., Welch, T. J.: Solution chemistry of carrier free indium. In: Subramanian, G., Rhodes, B. A., Cooper, J. F., *et al.*, eds. *Radiopharmaceuticals*. Society of Nuclear Medicine, New York, 73–79 (1975).
7. Edwards, C. L., Hayes, R. L.: Tumor scanning with ^{67}Ga citrate. J. Nucl. Med. **10**, 103–105 (1969).
8. Lavender, J. P., Lowe, J., Barker, J. R., *et al.*: Gallium-67 citrate scanning in neoplastic and inflammatory lesions. Br. J. Radiol. **44**, 361–366 (1971).
9. Atkins, H. L., Budinger, T. F., Lebowitz, E., *et al.*: Thallium-201 for medical use part 3: human distribution and physical imaging properties. J. Nucl. Med. **18**, 18–25 (1977).
10. Ritchie, J. L., Hamilton, G. W.: Biological properties of thallium. In: Ritchie, J. L., Hamilton, G. W., Wackers, F. J. Th., eds. *Thallium-201 myocardial imaging*. Raven Press, New York, 9–28 (1978).
11. McAfee, J. G., Thakur, M. L.: Survey of radioactive agents for in vitro labeling of phagocytic leukocytes. I. Soluble agents. J. Nucl. Med. **17**, 488–493 (1975).
12. Kowalsky, R. J., Perry, J. R.: *Radiopharmaceuticals in nuclear medicine practice*. Appelton & Lange, Norwalk, 76 (1987).
13. Brown, L. C.: Chemical processing of cyclotron-produced Ga-67. Int. J. Appl. Radiat. Isot. **22**, 710–715 (1971).
14. Kowalsky, R. J., Perry, J. R.: *Radiopharmaceuticals in nuclear medicine practice*. Appelton & Lange, Norwalk, 54–56 (1987).
15. Lamb, J. F.: Commercial production of isotopes for nuclear medicine, 1970–1980. IEEE Transactions on Nuclear Science; NS-28: 1916–1920 (1982).
16. Lebowitz, E., Green, M. W., Fairchild, R., *et al.*: Thallium for medical use I. J. Nucl. Med. **16**, 151–160 (1975).
17. Weiner, R. E., Spencer, R. P., Dambro, T. J., *et al.*: ^{67}Ga distribution in a man with both a decrease in transferrin and hepatic ^{67}Ga concentration. J. Nucl. Med. **33**, 1701–1703 (1992).
18. Package insert. Gallium citrate, ^{67}Ga injection. Mallinckrodt, St. Louis (1992).
19. Nelson, B., Hayes, R. L., Edwards, C. L., *et al.*: Distribution of gallium in human tissues after intravenous administration. J. Nucl. Med. **13**, 92–100 (1972).
20. Alazraki, N. P.: Gallium-67 imaging in infection. In: Early, P. J., Sodee, D. B., eds. *Principles and practice of nuclear medicine*. Mosby, St. Louis, 702–713 (1995).
21. Neumann, R. D., Kemp, J. D., Weiner, R. E.: Gallium-67 imaging for detection of malignant disease. In: Sandler, M. P., Coleman, R. E., Wackers, F., *et al.*, eds. *Diagnostic nuclear medicine*, 3rd ed. Williams & Wilklins, Baltimore, 1243–1260 (1995).
22. Weiner, R. E.: The mechanism of ^{67}Ga localization in malignant disease. Nucl. Med. Biol. In press (1996).
23. Harris, A. W., Sephton, R. G.: Transferrin promotion of ^{67}Ga and ^{59}Fe uptake by cultured mouse myeloma cells. Cancer Res. **37**, 3634–3638 (1977).
24. Larson, S. M., Rasey, J. S., Grunbaum, Z., *et al.*: Pharmacologic enhancement of ^{67}Ga tumors to blood ratios for EMT-6 sarcoma. Radiology **130**, 241–243 (1979).
25. Hoffer, P. B., Huberty, J., Khayam-Bashi, H.: The association of Ga-67 and lactoferrin. J. Nucl. Med. **18**, 713–717 (1977).
26. Hegge, F. N., Mahler, D. J., Larson, S. M.: The incorporation of Ga-67 into the ferritin fraction of rabbit hepatocytes in vivo. J. Nucl. Med. **18**, 937–939 (1977).
27. Hoffer, P. B.: Gallium: mechanisms. J. Nucl. Med. **21**, 282–284 (1980).
28. Sephton, R. G., Hodgson, G. S., DeAbrew, S., *et al.*: Ga-67 and Fe-59 distributions in mice. J. Nucl. Med. **19**, 930–935 (1978).
29. Logan, K. J., Ng, P. K., Turner, C. J., *et al.*: Comparative pharmacokinetics of ^{67}Ga and ^{59}Fe in humans. Int. Nucl. Med. Biol. **8**, 271–276 (1981).
30. Crichton, R. R., Ward, R. J.: Iron metabolism – new perspectives in view. Biochemistry **31**, 11255–11264 (1992).
31. Taetle, R.: The role of transferrin receptors in hemopoietic cells growth. Exp. Hematol. **18**, 360–365 (1990).
32. Weiner, R. E., Schreiber, G. J., Hoffer, P. B., *et al.*: Compounds which mediate Ga-67 transfer from lactoferrin to ferritin. J. Nucl. Med. **26**, 908–916 (1985).
33. Weiner, R. E.: The role of phosphate containing compounds on the transfer of ^{111}In and ^{67}Ga from transferrin to ferritin. J. Nucl. Med. **24**, 608–614 (1989).
34. Larson, S. M., Rasey, J. S., Allen, D. R., *et al.*: Common pathway for tumor cell uptake of gallium-67 and iron-59 via a transferrin receptor. J. Natl. Can. Inst. **64**, 41–53 (1980).
35. Larson, S. M., Rasey, J. S., Nelson, N. J., *et al.*: The kinetics of uptake and macromolecular binding of ^{67}Ga and ^{59}Fe by the EMT-6 sarcoma-like tumor of Balb/c mice. In: *Radiopharmaceuticals II: proceedings of the 2nd international symposium on radiopharmaceuticals*. Society of Nuclear Medicine, New York, 277–308 (1979).
36. Larson, S. M., Rasey, J. S., Allen, D. R., *et al.*: A transferrin-mediated uptake of gallium-67 by EMT-6 sarcoma. II. Studies in vivo (BALB/c mice): concise communication J. Nucl. Med. **20**, 843–846 (1979).
37. Berry, J. P., Escaig, F., Poupon, M. F., *et al.*: Localization of gallium in tumor cells. Electron microscopy, electron probe microanalysis and analytical ion microscopy. Int. J. Nucl. Med. Biol. **10**, 199–204 (1983).
38. DeAbrew, S.: Assays for transferrin and transferrin receptors in tumor and other mouse tissues. Int. J. Nucl. Med. Biol. **8**, 217–221 (1981).
39. Sohn, M.-E., Jones, B. J., Whiting, J. H., *et al.*: Distribution of gallium-67 in normal and hypotransferrinemic tumor-bearing mice. J. Nucl. Med. **34**, 2135–2143 (1993).
40. Kaplan, J., Jordan, I., Sturrock, A.: Regulation of the transferrin-independent iron transport system in cultured cells. J. Biol. Chem. **266**, 2997–3004 (1991).
41. Chitambar, C. R., Zivkovic, Z.: Uptake of gallium-67 by human leukemic cells: demonstration of transferrin receptor-dependent and transferrin-independent mechanisms. Cancer Res. **47**, 3929–3934 (1987).
42. Iosilevsky, G., Front, D., Betman, L., *et al.*: Uptake of gallium-67 citrate and (2 H-3) deoxyglucose in the tumor model following chemotherapy and radiotherapy. J. Nucl. Med. **26**, 278–282 (1985).
43. Higashi, T., Kobayashi, M., Wakao, H., *et al.*: The relationship between ^{67}Ga accumulation and ATP metabolism

in tumor cells *in vitro*. Eur. J. Nucl. Med. **15**, 152–156 (1989).
44. Kubota, K., Ishiwata, K., Kubota, R., *et al.*: Tracer feasibility for monitoring tumor radiotherapy: a quadruple tracer study with fluorine-18-fluorodeoxyglucose or fluorine-18-fluorodeoxyuridine, L-[methyl-^{14}C]methionine, [6-^{3}H]thymidine, and gallium-67. J. Nucl. Med. **32**, 2118–2123 (1991).
45. Chitambar, C. R., Matthaeus, W. G., Antholine, W. E., *et al.*: Inhibition of leukemic HL60 cell growth by transferrin-gallium: effects on ribonucleotide reductase and demonstration of drug synergy with hydroxyurea. Blood **72**, 1930–1936 (1988).
46. Harris, W. R.: Thermodynamics of gallium complexation by human lactoferrin. Biochemistry **25**, 803–808 (1986).
47. Hoffer, P. B., Huberty, J., Samuel, A.: Use of affinity chromatographic method to determine relative binding of ^{67}Ga to lactoferrin and transferrin in tear fluid. J. Nucl. Med. **19**, 732 (1978).
48. Hoffer, P. B., Miller-Catchpole, R., Turner, D. A.: Demonstration of lactoferrin in tumor tissue from two patients with positive gallium scans. J. Nucl. Med. **20**, 424–427 (1979).
49. Ismail, M., Brock, J. H.: Binding of lactoferrin and transferrin to the human promonocytic cell line U937. J. Biol. Chem. **268**, 21618–21625 (1993).
50. Garre, C., Biachi-Scarra, C., Sirito, M., *et al.*: Lactoferrin binding sites and nuclear localization in K562(S) cells. J. Cell. Physiol. **153**, 477–482 (1992).
51. Moguilevsky, N., Masson, P. L., Courtoy, P. J.: Lactoferrin uptake and iron processing into macrophages: a study in familial haemochromatosis. Brit. J. Haematol. **66**, 129–136 (1987).
52. Broxmeyer, H. E.: Iron-binding proteins and the regulation of hematopoietic cell proliferation/differentiation. In: De Souza, M., Brock, J. H., eds. *Iron in immunity, cancer and inflammation*. Wiley, Chichester, 199–222 (1989).
53. Weiner, R. E., Schreiber, G. J., Hoffer, P. B.: *In vitro* transfer of Ga-67 from transferrin to ferritin. J. Nucl. Med. **20**, 608–614 (1983).
54. Neilands, J. B.: Microbial envelope proteins related to iron. Annu. Rev. Microbiol. **36**, 285–309 (1982).
55. Weiner, R. E., Thakur, M. L., Goodman, M., *et al.*: Relative stabilities of In-111 and Ga-67 deferoxamine and transferrin complexes. In: *Radiopharmaceuticals II: proceedings 2nd international symposium radiopharmaceuticals.* Society of Nuclear Medicine, New York, 331–340 (1979).
56. Fernandez-Pol, J. A.: Isolation and characterization of a siderophore-like growth factor from mutants of SV40-transformed cells adapted to picolinic acid. Cell **14**, 489–499 (1978).
57. Herscheid, J. D. M., Knops, G. H. J. N., Boele, S., *et al.*: On the mechanism of gallium accumulation in tumors. Isolation of N-hydroxypeptides. Nucl. Med. Biol. **13**, 63–65 (1986).
58. Tsan, M. F.: Mechanism of gallium-67 accumulation in inflammatory lesions. J. Nucl. Med. **26**, 88–92 (1985).
59. Hayes, R. L., Rafter, J. J., Carlton, J. E., *et al.*: Studies on the in vivo uptake of ^{67}Ga by an experimental abscess: Concise communication. J. Nucl. Med. **23**, 8–14 (1982).
60. Weiner, R. E., Hoffer, P. B., Thakur, M. L.: Lactoferrin: its role as a ^{67}Ga binding protein from polymorphonuclear leukocytes. J. Nucl. Med. **22**, 32–37 (1981).
61. Leffell, M. S., Spitznagle, J. K.: Fate of human lactoferrin and myeloperoxidase in phagocytizing human neutrophils: effect of immunoglobin G subclasses and immune complexes coated on latex beads. Infect. Immun. **12**, 813–820 (1975).
62. Bennett, R. M., Skosey, J. L.: Lactoferrin and lysozyme levels in synovial fluid: differential indices of articular inflammation and degradation. Arthritis Rheum. **20**, 84–90 (1977).
63. Guyton, A. C.: *Textbook of medical physiology*, ed. 8. WB Saunders Co., Philadelphia, 365–373 (1990).
64. Bennett, R. M., Davis, J.: Lactoferrin binding to human peripheral blood cells: an interaction with a B-enriched population of lymphocytes and a subpopulation of adherent mononuclear cells. J. Immun. **127**, 1211–1216 (1981).
65. Weiner, R. E.: The role of human liver ferritin (HuFE) in Ga-67 localization. J. Nucl. Med. **24**, P121 (1984).
66. Emery, T., Hoffer, P. B.: Siderophore-mediated mechanism of gallium uptake demonstrated in the microorganism *Ustilago sphaerogena*. J. Nucl. Med. **21**, 935–939 (1980).
67. Sodee, D. B., Port, S. C., O'Donnell, J. K., *et al.*: Cardiovascular system. In: Early, P. J., Sodee, D. B., eds. *Principles and practice of nuclear medicine*. Misby, St. Louis, 370–442 (1995).
68. Package insert. Thallous Chloride, ^{201}Tl. Mallinckrodt, St. Louis (1990).
69. McGilvery, R. W.: *Biochemistry a functional approach*, 3 ed. W. B. Saunders, Philadelphia, 381–389 (1983).
70. Britton, J., Blank, M.: Thallium activation of Na-K-ATPase of rabbit kidney. Biochim. Biophys. Acta **159**, 160–165 (1968).
71. Skulskii, I. A., Manninen, V., Jarnefeldt, J.: Interaction of thallous ions with the cation transport mechanism in erythrocytes. Biochim. Biophys. Acta **298**, 702–709 (1973).
72. Damper, D. P., Epstein, W., Rosen, B. P., *et al.*: Thallous ion is accumulated by potassium transport systems in *Escherichia coli*. Biochemistry **18**, 4165–4169 (1979).
73. Oyen, W. J. G., Claessens, R. A. M. J., van Horn, J. R., *et al.*: Scintigraphic detection of bone and joint infections with indium-111-labeled nonspecific polyclonal human immunoglobulin G. J. Nucl. Med. **31**, 403–412 (1990).
74. Hnatowich, D. J., Layne, W. W., Childs, R. L., *et al.*: Radioactive labeling of antibody: a simple and efficient method. Science **220**, 613–615 (1982).
75. Oyen, W. J. G., Claessens, R. A. M. J., van der Meer, J. W. M., *et al.*: Detection of subacute infectious foci with indium-111 autologous leukocytes and indium-111 labeled human nonspecific immunoglobulin G: a prospective comparative study. J. Nucl. Med. **32**, 1845–1860 (1991).
76. Rubin, R. H., Fischman, A. J., Nedelman, M., *et al.*: The use of radiolabeled nonspecific polyclonal human immunoglobulin in the detection of focal inflammation by scintigraphy: comparison with gallium-67-citrate and technetium-99m-labeled albumin. J. Nucl. Med. **30**, 385–389 (1989).
77. Fischman, A. J., Fucello, A. J., Pellegrino-Gensey, J. L., *et al.*: Effect of carbohydrate modification on the localization of human polyclonal IgG at focal sites of bacterial infection. J. Nucl. Med. **33**, 1378–1382 (1992).
78. Oyen, W. J. G., Claessens, R. A. M. J., van der Meer, J. W. M., *et al.*: Biodistribution and kinetics of radiolabeled proteins in rats with focal infection. J. Nucl. Med. **33**, 388–394 (1992).
79. Claessens, R. A. M. J., Koenders, E. B., Boerman, O. C., *et al.*: Dissociation of indium-111-labelled diethylene triamine penta-acetic acid conjugated non-specific polyclonal human immunoglobulin G in inflammatory foci. Eur. J. Nucl. Med. **22**, 212–219 (1995).
80. Juweid, M., Strauss, H. W., Yaoita, H., *et al.*: Accumulation of immunoglobulin G at focal sites of inflammation. Eur. J. Nucl. Med. **19**, 159–165 (1992).
81. Harris, W. R., Chen, Y. C., Wein, K.: Equilibrium constants for the binding of indium(III) to human serum transferrin. Inorg. Chem. **33**, 4991–4998 (1994).
82. Rayne, D., Weiner, R. E.: Exchange of In-111 between IgG and lactoferrin, transferrin and ferritin. J. Nucl. Med. **32**, 1099 (1991).
83. Carlson, A. J., Sasso, D., Weiner, R. E.: Transchelation of In-114m from IgG-DTPA and abscess-related proteins. In: Martin-Comin, J., ed. *Radiolabeled blood elements: recent advances in techniques and applications*. Plenum Publishing, New York, 15–18 (1994).

84. Sasso, D., Weiner, R. E.: The effect of protein and pH on exchange of In-114m from IgG and HSA to lactoferrin. J. Nucl. Med. **34**, 245P (1993).
85. Abdel-Nabi, H. H., Doerr, R. J.: Multicenter clinical trials of monoclonal antibody B72.3-GYK-DTPA ^{111}In (^{111}In-CYT-1-03; OncoScint CE103). In: Maguire, R. T., Van Nostrand, D., eds. *Diagnosis of colorectal cancer and ovarian carcinoma: applications of immunoscintigraphy technology*. Marcel Dekker, New York, 73–88 (1992).
86. Collier, B. D., Foley, W. D.: Current imaging strategies for colorectal cancer. J. Nucl. Med. **34**, 537–540 (1993).
87. Krag, D. N.: Clinical utility of immunoscintigraphy in managing ovarian cancer. J. Nucl. Med. **34**, 545–548 (1993).
88. Jusko, W. J., Kung, L.-P., Schmelter, R. F.: Immunopharmacokinetics of ^{111}In-CYT-103 in ovarian cancer patients. In: Maguire, R. T., Van Nostrand, D., eds. *Diagnosis of colorectal cancer and ovarian carcinoma: applications of immunoscintigraphy technology*. Marcel Dekker, New York, 177–190 (1992).
89. Maguire, R. T.: OncoScint image atlas. In: Maguire, R. T., Van Nostrand, D., eds. *Diagnosis of colorectal cancer and ovarian carcinoma: applications of immunoscintigraphy technology*. Marcel Dekker, New York, 141–175 (1992).
90. Gallagher, B. M.: Monoclonal antibodies: the design of appropriate carrier and evaluation systems. In: Lambrecht, R. M., Eckelman, W. C., eds. *Animal models in radiotracer design*. Springer-Verlag, New York, 61–105 (1983).
91. Package insert. OctreoScan. Mallinckrodt, St. Louis (1994).
92. Krenning, E. P., Kwekkeboom, D. J., Bakker, W. H., *et al.*: Somatostatin receptor scintigraphy with [^{111}In-DTPA-D-Phe^{1}] and [^{123}I-Tyr^{3}]-octreotide: the Rotterdam experience with more than 1000 patients. Eur. J. Nucl. Med. **20**, 716–731 (1993).
93. Thakur, M. L.: Radiolabeled peptides: now and the future. Nucl. Med. Commun. **16**, 724–732 (1995).
94. Kaw, B. A., Yasuda, T., Gold, H. K., *et al.*: Acute myocardial infract imaging with indium-111-labeled monoclonal antimyosin Fab. J. Nucl. Med. **28**, 1671–1678 (1987).
95. Johnson, L. L., Seldin, D. W.: The role of antimyosin antibodies in acute myocardial infarction. Sem. Nucl. Med. **19**, 238–246 (1989).
96. Carrio, I., Berna, L., Ballester, M., *et al.*: Indium-111-antimyosin scintigraphy to asses myocardial damage in patients with suspected myocarditis and cardiac rejection. J. Nucl. Med. **29**, 1893–1900 (1988).
97. DeNardo, D., Scibilia, G., Macchiarelli, A. G., *et al.*: The role of indium-111 antimyosin Fab imaging as a noninvasive surveillance method of human heart transplant rejection. J. Heart Transplant. **8**, 407–412 (1989).
98. Thakur, M. L., Lavender, J. P., Arnot, R. N., *et al.*: Indium-111-labeled autologous leukocytes in man. J. Nucl. Med. **18**, 1014–1021 (1977).
99. Preston, D. F.: Indium-111 label in inflammation and neoplasm imaging. In: Early, P. J., Sodee, D. B., *Principles and practice of Nuclear Medicine*. Mosby, St. Louis, 714–724 (1995).
100. Package insert. Indium-111 oxine solution. Arlington Heights, Amersham Healthcare (1986).
101. Thakur, M. L., Segal, A. W., Louis, L., *et al.*: Indium-111-labeled cellular components: mechanism of labeling and intracellular localization. J. Nucl. Med. **18** 1022–1027 (1977).
102. Hwang, K. J.: Modes of interaction of In-111-8-hydroxyquinoline with membrane bilayer. J. Nucl. Med. **19**, 1162–1168 (1978).
103. Sasso, D., Gionfriddo, M., Thrall, R., *et al.*: Biodistribution of ^{111}In labeled antibody directed against intracellular adhesion molecule-1 in normal rats. J. Nucl. Med. In press (1996).
104. Kowalsky, R. J., Perry, J. R.: *Radiopharmaceuticals in nuclear medicine practice*. Appelton & Lange, Norwalk, 157–163 (1987).
105. Laing, A. H., Ackery, D. M., Bayly, J., *et al.*: Strontium-89 chloride for pain palliation in prostatic skeletal malignancy. Br. J. Radiol. **64**, 816–822 (1991).
106. Robinson, R. G., Preston, D. F., Spicer, J. A., *et al.*: Radionuclide therapy of intractable bone pain: emphasis on strontium-89. Sem. Nucl. Med. **22**, 28–32 (1992).
107. Package insert. Metastron, Arlington Heights, Amersham Healthcare (1993).
108. Charkes, N. D., Sklaroff, D. M.: Early diagnosis of metastatic bone cancer by photoscanning with Strontium-85. J. Nucl. Med. **5**, 168–179 (1964).
109. Dow, E. C., Stanbury, J. B.: Strontium and calcium metabolism in metabolic bone diseases. J. Clin. Invest. **39**, 885–892 (1960).
110. Spencer, H., Laszlo, D., Brothers, M.: Strontium-85 and calcium-45 metabolism in man. J. Clin. Invest. **36**, 680–692 (1957).
111. Spencer, R. P., Hosain, F., Weiner, R. E.: Bone imaging radiopharmaceuticals. In: Henkin, R. E., Boles, M. A., Dillehay, G. L., *et al.*, eds. *Nuclear Medicine*, 2nd ed. Mosby Yearbook, St. Louis, 1151–1166 (1996).
112. de Klerk, J. N. H., Zonneberg, B. A., ven het Schip, A. D., *et al.*: Dose escalation study of rhenium-186 hydroxyethylidene diphosphonate in patients with metastatic prostate cancer. Eur. J. Nucl. Med. **21**, 1114–1120 (1994).
113. de Klerk, J. N. H., van Dijk, A., ven het Schip, A. D., *et al.*: Pharmacokinetics of rhenium-186 after administration of rhenium-186 hydroxyethylidene diphosphonate in patients with bone metastases. J. Nucl. Med. **33**, 646–651 (1992).
114. Maxon III, H. R., Thomas, S. R., Hertzberg, V. S., *et al.*: Rhenium-186 hydroxyethylidene diphosphonate for the treatment of painful osseous metastases. Sem. Nucl. Med. **22**, 33–40 (1992).
115. Farhanghi, M., Holmes, R. A., Volkert, W. A., *et al.*: Samarium-153-EDTMP: pharmacokinetics, toxicity and pain response using an escalating dose schedule in treatment of metastatic bone cancer. J. Nucl. Med. **33**, 1451–1458 (1992).

Radiochimica Acta 70/71, 289–297 (1995)

Technetium in Nuclear Medicine

By A. G. Jones

Department of Radiology, Harvard Medical School, Boston, MA 02115, USA

(Received December 18, 1995)

Technetium-99m / Radiopharmaceuticals / Gammy ray imaging / Single photon emission tomography / Nuclear medicine

Abstract

The introduction of the technetium generator and the development of the Anger camera in the early 1960s formed the basis of modern clinical nuclear medicine. Chosen originally purely for its physical decay characteristics, the radionuclide ^{99m}Tc has since become the backbone of routine imaging. In the United States, for example, more than 85% of all studies involve the use in one form or another of this short-lived gamma emitter. Since technetium is an artificial element, which occurs on earth only by virtue of the fission process, this intense medical use of the material seems an oddity. In fact, thanks to the work of a small number of early pioneers, the choice turns out to be eminently logical. This short article outlines some of the history associated with this radionuclide and its radiopharmaceuticals.

Discovery

The discovery of the element was made in Palermo by Segrè and Perrier in 1937 [1], and by far the most complete account of this event is found in a talk given by Segrè in Italy in 1985 [2]. The discovery occurred as the result of a period during 1936 spent by Segrè at the cyclotron of the University of California at Berkeley. He noted that parts of the interior of the machine had been irradiated and replaced, in particular a strongly radioactive deflector lip made of molybdenum that contained long-lived isotopes. Since the bombardments had been made with deuterons, he surmised that there might exist isotopes of the then unknown element 43 in the deflector. Returning to Italy with scraps of the deflector, he worked with Perrier to develop a chemical separation for the new element and to acquire subsequently some knowledge of its chemistry. Although the work announcing the discovery was published in 1937, it would be eleven years before the new element was finally named, by Segrè with the help of a colleague who was a Greek scholar.

In order to study short-lived radionuclides of the element, Segrè returned to Berkeley in 1938 and began working with an instructor in chemistry named Seaborg. The discovery of the radioactivity with a half-life of six hours followed [3], though publication of this finding was delayed because the concept of isomeric states was somewhat controversial at the time. In fact, publication was held up until a paper appeared, authored by Pontecorvo, concerning an isomeric state in rhodium [4]. Subsequent work by others in detecting conversion electrons from the radionuclide confirmed that the activity indeed belonged to element 43 and the discovery of the element itself was corroborated much later when it was isolated on a macroscopic scale from fission products.

Nuclear medicine

Several other isotopes that would become important in nuclear medicine were also discovered at Berkeley during the late 1930s, including ^{131}I [5]. Though this nuclide would be used medically almost immediately, it would be almost twenty years before the potential of ^{90m}Tc was realized. Several things converged to make this possible. The first radionuclide generator system in a convenient form was the ^{132}Te-^{132}I couple developed at Brookhaven National Laboratory in the mid 1950s [6]. This system was loaded with ^{132}Te derived from fission products, and efforts to improve the purity of the eluted iodine led to the detection of a contaminating activity that was determined to be ^{99m}Tc [7]. Subsequent work showed that the molybdenum parent was following the tellurium through the separation process. The technetium generator that resulted was first reported in 1958 [8]. The history of the generator itself has been recounted by Richards who was primarily responsible for promoting the medical applications of the isomeric state during the late 1950s and early 1960s [9, 10].

Another aspect of this period was the early effort by Harper and his colleagues at the University of Chicago to further develop imaging techniques. Based upon estimates by Beck that the optimum detection energy of the sodium iodide crystals being used in scanners lay around 150 keV [11], Harper's interest turned to ^{99m}Tc [12]. Eventually this gave rise to the shipment of a series of technetium generators from Brookhaven to Chicago and to the demonstration of the effectiveness of ^{99m}Tc in imaging the liver, brain, and thyroid [13]. The early generators were eluted with dilute acids: nitric acid originally and then hydrochloric acid in order to be more physiologically compatible. In early 1964 Maynard at the Radiochemical Center (now Amersham International Plc) in England discovered that equivalent yields of pertechnetate could be achieved with physiologic saline rather than acid [14]. At this point the stage was set for establishing commercial generators that gave a sterile, pyrogen-free source of pertechnetate. This was achieved in 1966 when a further convenience was introduced with the evacuated vial technique of elution, which both limited the exposure of the column to water and its radiolysis products and also ensured a fresh supply of sterile eluate on each occasion. While each of these steps individually seems small, taken as a whole they form the essential framework by which a short-lived isotope could be available as required at any site anywhere in the world on a timely basis. This was no small achievement.

It might be interesting at this point to interject a quote from a talk given by Charlton to the Pharmaceutical Society in April 1965 [15] to the effect that the 1963 British Pharmacopœia, together with the 1964 Addendum, listed the following radioactive preparations: cyanocobalamin (^{57}Co), cyanocobalamin (^{58}Co), ferric citrate (^{59}Fe), gold (^{198}Au) injection, iodinated (^{131}I) human serum albumin injection, sodium chromate (^{51}Cr) solution, sodium iodide (^{131}I) solution, sodium phosphate (^{32}P) injection. One year later the abstracts of papers at a meeting of the British Institute of Radiology reflected a sea change in emphasis toward low-energy gamma emitters and technetium [16]. Technetium agents available from Harper's research at this juncture included pertechnetate, sulfur colloid, DTPA iron ascorbate complex, labeled human serum albumin and thiocyanate [17].

The next major contribution to modern nuclear medicine came again from Brookhaven in 1971 with the publication by Atkins *et al.* of a paper describing ^{99m}Tc-DTPA complexes prepared by three different methods [18]. This was the first report of what has since become known as an "instant kit" preparation, using stannous ion as a reducing agent for the pertechnetate precursor [19]. Though the basic chemistry was still poorly understood at this stage, sufficient was known about soft donor ligands and their ability to coordinate the metal to design such kits. Eventually this became a standard model for on-site synthesis of imaging agents, for example, methylene diphosphonate (MDP), pyrophosphate (PYP), glucoheptonate (GH), disofenin and mebrofenin. In each of these instances, the stannous ion serves to reduce the metal prior to complexation and to maintain reducing conditions for a practical portion of the clinical day, usually six hours. The ligand in turn can be seen to perform up to four functions: complexing the technetium present, targeting the product, complexing the tin present in order to avoid the formation of colloidal material, and (certainly in the case of glucoheptonate) helping to maintain the integrity of the complex by mass action. Lyophilization of the vial contents in the presence of an expander such as a sugar allows for long-term storage and instant reconstitution and reaction when the pertechnetate is added. This concept was not only convenient but also introduced simplicity and reliability to the preparation of radiopharmaceuticals as they were needed. These attributes thus brought a range of technetium imaging agents into the mainstream of medicine in the 1970s through approval by governmental regulatory agencies such as the Food and Drug Administration (FDA) in the United States.

For the technetium generator the next major milestone came in 1974 when the first of the fission product columns became routinely available. A carrier-free molybdenum parent allowed the use of a much smaller column, with smaller elution volumes in turn producing a much higher specific activity product. In concert with the standard kit design this also contributed to the routine on-site preparation of radiopharmaceuticals. Molinski has outlined the many difficulties presented in the design of these generator systems [20], not the least of which was the separation of ^{99}Mo from fission products in sufficient purity so that it could be used safely to produce an injectable product. Nevertheless, the radiopharmaceutical industry overcame these problems and the product is now extremely reliable.

Development of new imaging agents

The first phase

Research into the development of new agents over the past twenty-five years can be roughly divided into three different, albeit overlapping, phases. During the first phase in the early 1970s the agents used in the clinic were relatively simple in nature, targeting basic physiologic function, and were essentially commercialized versions of products devised by Harper and other pioneers in the 1960s. The nuclear medicine clinic thus employed colloids and particles to take advantage of capillary blockade in the lungs or the natural function of the liver and spleen in removing debris from the circulation. Labeled albumen served for blood pool scanning and the complex formed with DTPA for glomerular filtration studies of renal function. Figure 1 compares a liver scan made on an early rectilinear scanner with ^{198}Au colloid with a similar study made on a modern Anger camera with ^{99m}Tc-sulfur colloid. This clearly demonstrates the importance of advances in instrumentation as a second key factor in the establishment of modern nuclear medicine.

In this period insufficient basic chemistry was known, but nevertheless some ingenious ideas emerged for new agents. One of the first objectives was to improve upon the ability to monitor the skeletal system and to move away from scanning with ^{85}Sr or, where it might be available, ^{18}F as fluoride ion. Ligands were chosen that had both an avidity for calcium and oxygen donor atoms to bind technetium. Further, the typical aim was that these relatively simple ligands even when complexed would guide the radiolabel to the intended target after injection. Subramanian suggested polyphosphates of particular chain length [21]. Next, technetium pyrophosphate (which later proved to be the active impurity in the particular samples studied by Subramanian) was proposed [22]. Finally came

Fig. 1. Comparison of a liver scan made on a rectilinear scanner with ^{198}Au colloid (left) and one made with ^{99m}Tc-sulfur colloid using an Anger camera (right). Note the absence of splenic uptake in the ^{198}Au scan. (Image courtesy of S. Treves, M. D., The Children's Hospital, Boston).

the organic phosphonates, put forward because they would be more stable than the inorganic phosphates against in-vivo metabolism [23, 24]. One of these (again suggested by Subramanian), ^{99m}Tc-methylene diphosphonate is still the imaging agent of choice and an essential part of routine clinical nuclear medicine practice [25]. Figure 2 shows a comparison of bone scans performed with sodium ^{18}F-fluoride and with ^{99m}Tc-MDP. Though both agents are effective in targeting the tissue of interest, the convenience of the generator–kit system outweighed the logistics of production and the expense of distributing ^{18}F to the point where the latter was eclipsed. For example, the fluorine dose depicted here was generated in Chicago and imaged in Boston. While technetium images such as these are routine fare for those familiar with nuclear medicine, it is still a marvel to this author that the information is being provided by a vanishingly small amount of a coordination complex of an element that does not occur naturally on earth. Furthermore, unlike other modalities the image provides not a still picture but a dynamic one of the state of the bone structure: the density reflects accurately the turnover proceeding in normal areas of bone and the increased reactivity in other areas due to the presence of pathology or trauma.

One of the more entertaining aspects of nuclear medicine is the frequency with which alternative uses – beyond the one originally intended – are found for an agent. ^{99m}Tc-pyrophosphate is a case in point in that it was also used to obtain positive images of the site of myocardial infarction, targeting the crystallites of hydroxyapatite that eventually form after such an event. An accidental observation made early in 1975 by Uren and colleagues in Australia [26] and by others showed that injection of pertechnetate for brain imaging in a patient who had previously had a bone scan with ^{99m}Tc-pyrophosphate led to the labeling of red blood cells. These observations formed the basis for a standard protocol for performing radioventriculograms – functional studies of the left ventricle – using two sequential injections of, respectively, the contents of a pyrophosphate kit and half an hour later a

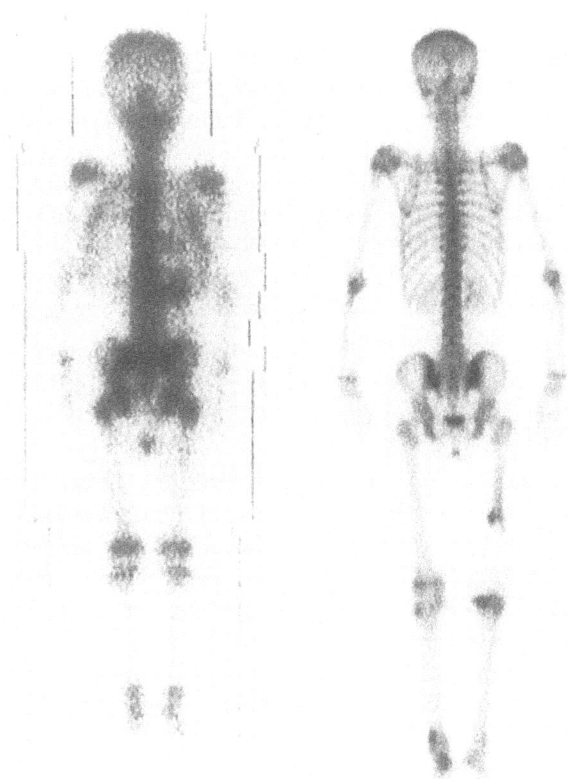

Fig. 2. Anterior views of the skeleton using ^{18}F-sodium fluoride (left) and ^{99m}Tc-MDP (right). Both reflect the dynamic state of bony tissue. Note the increased activity in joints and load-bearing regions. (Image courtesy of S. Treves, M. D., The Children's Hospital, Boston).

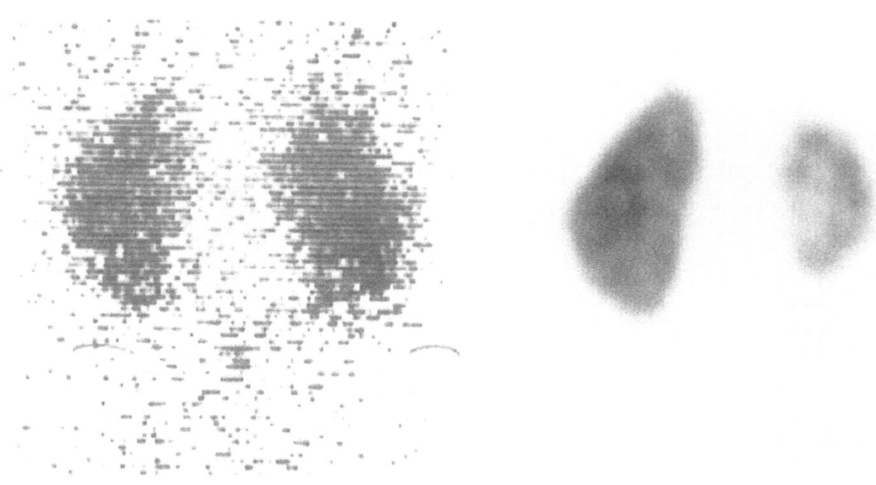

Fig. 3. Comparison of a kidney scan performed with ^{203}Hg-chlormerodrin (left) with a modern example done using ^{99m}Tc-DMSA complex (right). The ^{99m}Tc scan shows a damaged though still partially functioning right kidney. (Image courtesy of S. Treves, M. D., The Children's Hospital, Boston).

dose of $^{99m}TcO_4^-$. This obviated the need for retrieving a sample of the patient's blood for labeling in vitro prior to reinjection. From the standpoint of a chemist there remain questions about this process of adding the reactants together in a different sequence. With this technique roughly half of the radioactivity resided on the erythrocytes while the remainder cleared through the kidneys without targeting bone. This complex has never been studied or compared with the bone-seeking pyrophosphate complex.

The second phase

The second phase of research into new agents was a direct result of the ever increasing amount of fission products being generated by power reactors, and hence of the long-lived nuclide ^{99}Tc. The availability of this isotope in weighable quantities began to make possible conventional studies of technetium chemistry, applying the usual spectroscopic and analytical methodologies of the inorganic chemist. This phase began in the early 1970s and is still continuing. The usual procedure has been to generate new classes of technetium coordination complexes containing ^{99}Tc, to characterize these fully by a variety of means, and then to study the pharmacokinetics of the corresponding ^{99m}Tc material in animal models or in vitro using cellular or tissue preparations. Though there may be an element of serendipity to this approach it has in fact led to significant advances in the design of both radioactive and nonradioactive contrast agents, as well as to the introduction of new tests in routine nuclear medicine.

Much of the original effort was directed toward new complexes that could be formed in high yield in aqueous media, because the starting material in practice is pertechnetate ion in isotonic saline. The first major new class to be discovered was the oxotechnetium complexes in the +5 oxidation state. The first of these was a bis ligand complex with the bidentate mercaptothioacetic acid [27]. This was noteworthy not merely because it was new but also because the technetium selected specifically a ligand that was an impurity in the sample of the intended ligand (thioglycolic acid). This was the first indication of the affinity of Tc(V) for sulfur as a donor and also of the dangers in assuming that at trace levels the chemistry is totally predictable. The second complex containing the $Tc^{V}O^{3+}$ core found by Smith [28] underscored this avidity and also confirmed the fact that these materials can be formed readily and in high yield, an important factor in practice. A study of the relative affinities of different donor sets followed and led to the design of a bisamide-bisthiol (N_2S_2 or DADS) chelate under the assumption that greater in-vivo stability would be gained from the chelate effect than from complexes formed with bidentate ligands [29]. Initially proposed as a renal imaging agent, this system was developed further by Fritzberg into an N_3S chelate [30] that is now commercially available under the name Technescan MAG3™. During these early studies of oxotechnetium complexes the use of mass spectrometry for the identification of coordination compounds was first proposed [31]. This technique has proved so effective for this purpose that it is now a standard method for characterizing metal compounds, not only in radiopharmaceutical chemistry but also in inorganic chemistry as a whole.

An analogous N_2S_2 chelate system containing a bisamine bisthiol (DADT or BAT) was devised by Burns *et al.* [32, 33]. Whereas the amides formed anionic products, this system was chosen to form a

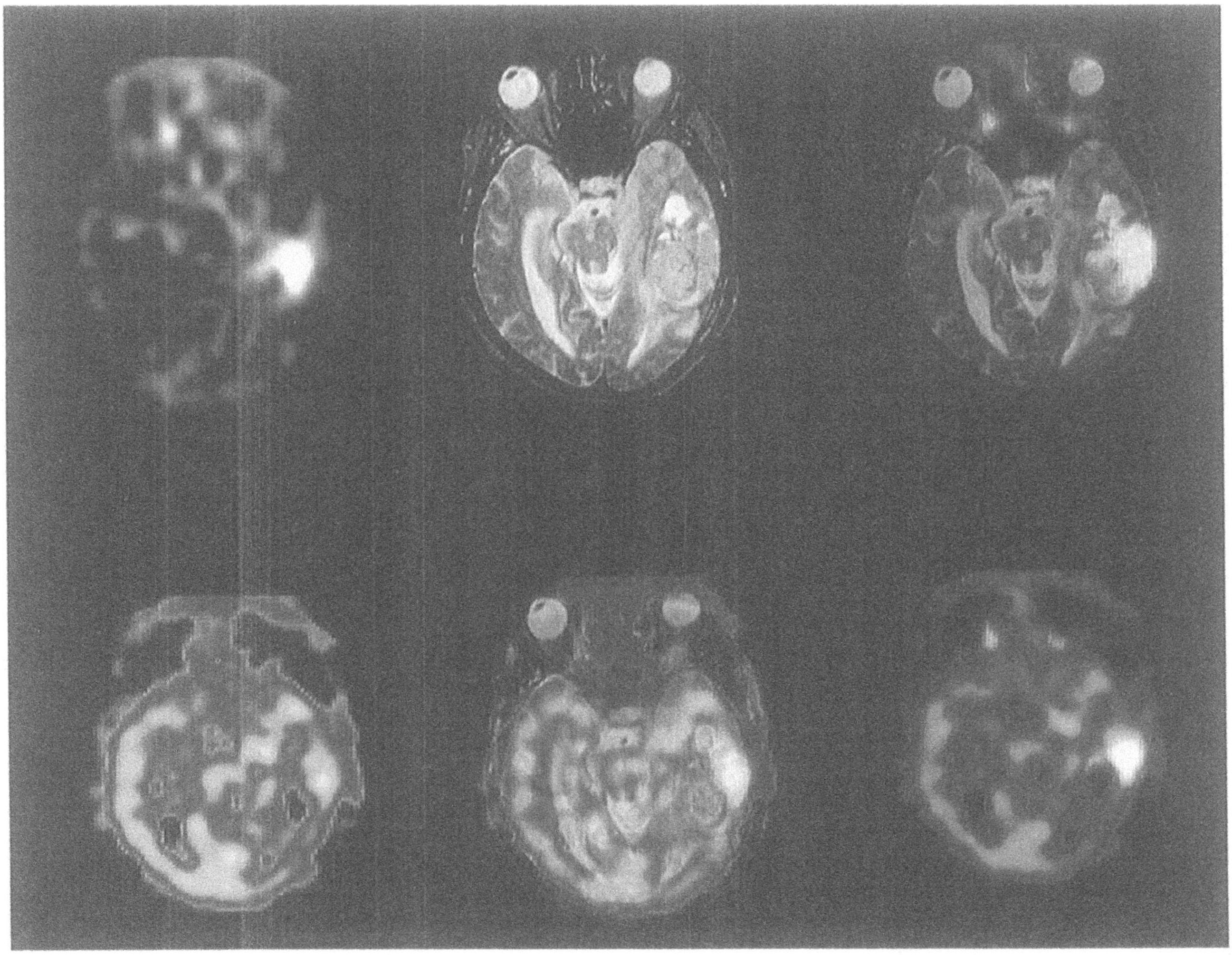

Fig. 4. Images taken of a brain tumor. Top panel: left, ^{201}Tl-thallous chloride scan; center, magnetic resonance imaging (MRI) of tumor bed; right, superpositioning of the nuclear medicine image on the MRI. Bottom panel: left, blood perfusion scan with ^{99m}Tc-Ceretec™; center, co-registration of perfusion and MRI scan; right, superimposed ^{201}Tl and ^{99m}Tc scans. Both agents target the same volume of tumor but the mass is more extensive by MRI. (Image courtesy of B. Leonard Holman, M. D., Brigham and Women's Hospital, Boston.)

neutral species when complexed with the metal, with the objective of penetrating the blood–brain barrier and providing a means of imaging the brain. This class has subsequently been extensively studied by Lever [34] and by Kung [35] for this and a variety of other purposes. For nuclear medicine, the most important result stemming from this work is a brain perfusion agent that is comprised of the bis ester of the original Burns ligand, available in the United States since 1994 and marketed under the name Neurolite™. This material is neutral upon injection, but when it crosses into the brain one only of the two ester moieties becomes hydrolyzed by an enzyme in tissue, thus rendering the overall charge negative and effectively blocking its egress from the cells. This is a symmetrical molecule containing two chiral centers, and only one isomer is active and then only in certain species, including the higher primates and man [36].

At this stage at least seven of the routinely available radiopharmaceuticals are in fact relatively simple oxotechnetium(V) species, including four approved or near approval in the United States over the past ten years. Two of these, complexes formed with glucoheptonic acid and with dimercaptosuccinic acid (DMSA), stem from the era before extensive structural studies were required by the regulatory agencies. Some work has been done on the DMSA system that indicates it is a technetium(V) complex and that isomerism occurs. Not yet explained, however, is the product of an alternative formulation used in Japan for detecting thyroid tumors [37]. Figure 3 shows ^{99m}Tc–DMSA in its more usual role for assessing renal function. This should be compared with the left hand panel which is an image obtained on a rectilinear scanner in the 1960s with ^{203}Hg-chlormerodrin, then commonly used for renal scanning.

A product of the post characterization era among oxotechnetium(V) complexes was first marketed in 1988 in the United States (earlier in Europe) under the name Ceretec™. This complex contains an N_4 or four nitrogen donor set [38] and is designed to assess cerebral blood flow. A typical use for this agent is shown in Figure 4 where several modalities come together, specifically magnetic resonance imaging (MRI) and scanning with two different radiopharmaceuticals, Ceretec™ to monitor blood flow and ^{201}Tl-thallous chloride to delineate a brain tumor. The ligand/complex system on which Ceretec™ is based is another example of isomer formation, with different forms showing different in-vivo properties. This issue of isomerism in technetium complexes and its influence on drug design is a fascinating topic, and the best and most comprehensive account to date has been published recently by Nowotnik [39].

This second phase in radiopharmaceutical development of essentially inorganic complexes also yielded a body of research on myocardial perfusion agents. The first materials to show avidity for cardiac tissue in vivo were cationic phosphine complexes of Tc(III) and Tc(I) [40, 41]. When tested in humans, however, an unexpected species difference emerged that slowed their development as clinical agents until the recent development of tetrofosmin [42] and Q12 [43]. In late 1981, however, a second class of complexes, the hexakis isonitriles of technetium(I), was discovered to have an avidity for cardiac tissue, and three examples of this potentially large group of molecules were successfully tested in man [44–46]. The third of these, ^{99m}Tc-methoxyisobutylisonitrile (MIBI), has been available for clinical use under the name Cardiolite™ in parts of Europe since as early as 1987 and in the United States since 1991. Although designed originally as a myocardial perfusion agent, anecdotal observations by Müller published in 1987 [47] prompted research into the possible use of isonitriles as tumor-seeking agents [48, 49]. Such studies parallel those going on in photoaffinity therapy in which other positively charged lipophilic species with a generally spherical form such as organic dyes are being investigated for ablation of tumor with lasers. During this same period Piwnica-Worms *et al.* discovered the dependence of uptake upon membrane potential, not only that of the plasma membrane but also that of the mitochondrial membrane [50]. Reduction of the membrane potential results in clearance of the technetium complex from the cell. The existence of ^{99}Tc-MIBI in mitochondria of exposed cells has been demonstrated by physical means [51]. Perhaps even more interesting is the recent finding, again by Piwnica-Worms, that the uptake of ^{99m}Tc-MIBI in tumor cells is markedly affected by the presence of P-glycoprotein, a molecule that is expressed on the surface of cells showing one of several forms of multidrug resistance [52]. It is intriguing to note that these properties are no doubt shared by other classes of materials not yet studied in this manner. In keeping with the experience in nuclear medicine with new classes of agents, Cardiolite™ recently underwent clinical trials to assess its usefulness for detection of breast cancer in certain patient populations. Similar results have been reported for ^{99m}Tc-tetrofosmin [53], an agent currently approved for myocardial perfusion studies in Europe and nearing approval in the United States that is another example of an oxotechnetium(V) species.

The third phase

The third phase of radiopharmaceutical development overlaps considerably with the second, briefly outlined above, and represents attempts to capitalize upon known technetium chemistry to link the radionuclide to molecules of biologic interest. It also may reflect the sense that it is difficult or perhaps impossible to adequately screen each and every new class of technetium complex for relevant biologic activity. In the 1990s, therefore, there is an increasing emphasis on modification of molecules, or portions thereof, that may help to detect a particular disease process. There is thus a real dependence upon discoveries made in other fields of research. For example, although N_2S_2 and N_3S chelates have been used for the conjugation and subsequent metallation of monoclonal and polyclonal antibodies for over ten years [54], there is now a concerted effort to investigate the usefulness of technetium-labeled peptides. Instances include the work designed to develop a means of imaging thrombi, using labeled platelet GPIIb/IIIa antagonists, now at or near the stage of clinical testing [55, 56]. More removed from the clinic and yet representative is the attempt by Katzenellenbogen *et al.* to label progestins with ^{99m}Tc in order to develop an agent for assessing therapies in cases of breast cancer [57]. This is a fine instance of a situation in which basic chemistry, in this case the chelation chemistry of the element, is being developed further in order to meet demands set by the required in-vivo properties of the final molecule, including attributes such as lipophilicity, charge and shape.

Even further from routine clinical use is the effort by many groups to develop agents that will target a variety of receptor types and hopefully thereby reflect various aspects of organ activity. The most comprehensively studied technetium agent is ^{99m}Tc-galactosyl-neoglycoalbumin (NGA) targeted to a receptor in the liver [58], though this has seen only investigational use in humans. Much of current research is directed to imaging receptors in the brain, including work by Kung [59, 60], Spies [61], Jurisson [62] and Lever [63]. This is a very difficult exercise for the blood–brain barrier represents a formidable obstacle to the entry of imaging agents. This may be particularly so for those that include a requirement to firmly coordinate a metal radionuclide, i.e., a portion with an extra 300 daltons in weight and the size of a tripeptide. Beyond this, it is necessary to prove that the molecule being studied is indeed specific for one particular receptor type, that the variation in receptor density reflects a given function, and that the progression of a disease state may be followed in a quantitative manner or that the scan is effective in the management of the patient. Practicality also deems that the disease being followed be present in a significant proportion of the patient population; otherwise there will be insufficient incentive to invest the vast sums necessary to bring a drug to the marketplace. These are major hurdles to overcome, but nevertheless it is recognized that for the future of nuclear medicine new windows on the brain and its function must be developed. Figure 4 shows the exquisite structural information that can be provided by MRI. Within the next decade it is predicted that blood perfusion imaging in the heart and the brain will be possible with this technique. Figure 4 also demonstrates, albeit with an agent that shows low specificity for tumor, that the anatomical detail from MRI and the physiologic information available from nuclear medicine could be a very powerful combination.

Summary

The story of ^{99m}Tc is remarkable in that this metastable isotope of an artificial element has exhibited real staying power in a field of medicine. The reasons for this may be summarized as follows.

The radionuclide has a single 140 keV photon and relatively little nonpenetrating radiation, decaying to a long-lived ground state that is a low-energy beta emitter. These properties ensure that the absorbed radiation dose to the patient is kept at acceptable levels, as is proper for a diagnostic test, and the emitted energy meshes well with the efficiency of sodium iodide detectors used in modern gamma cameras. The dosimetry allows the use of higher doses, hence more photons, clearer pictures, and shorter imaging times.

The yield of the precursor nuclide ^{99}Mo from uranium fission by thermal neutrons exceeds 6 percent and the mass 99 decay chain is the only one in the region that results in a radioactive isotope of molybdenum. This molybdenum can be isolated in high purity from other fission products and thus provides a monoisotopic source (from the radioactive standpoint) of the isotope to be administered into humans. This is also invaluable from the standpoint of dosimetry and of imaging without, for example, interfering gamma rays from contaminants.

The daughter isotope when needed can be isolated in high yield (about 80%) from an alumina column with sterile, pyrogen-free physiologic saline at high specific activity. The six hour half-life allows ample time for synthesis using convenient kit formulations and also for quality control procedures. The half-life of the nuclide is reasonably compatible with the duration of the average imaging procedure. The radiological problems associated with all phases of its use are relatively minimal.

Finally, technetium is a second row transition metal with a very rich chemistry, much of which has not yet been exploited. The extent of this chemistry and its applicability to the design of imaging agents are evidenced by the steady flow of new radiopharmaceuticals into the clinic and by the numerous new compounds being investigated in the research laboratory.

References

1. Perrier, C., Segrè, E.: Radioactive isotopes of element 43. Nature **140**, 193–194 (1937).
2. Segrè, E.: The adventurous discovery of technetium. In: *Technetium in Chemistry and Nuclear Medicine 2.* Nicolini, M., Bandoli, G., Mazzi, U. (eds.). Verona, Cortina International (1986), pp. 1–10.
3. Segrè, E., Seaborg, G. T.: Nuclear isomerism in element 43. Phys. Rev. **54**, 772 (1938).
4. Seaborg, G. T.: Private communication (1989).
5. Livingood, J. J., Seaborg, G. T.: Radioactive iodine isotopes. Phys. Rev. **53**, 1015 (1938).
6. Stang, L. G. Jr., Tucker, W. D., Banks, H. O. Jr., Doering, R. F., Mills, T. H.: Production of iodine-132. Nucleonics **12**, 12–24 (1954).
7. Stang, L. G. Jr., Tucker, W. D., Doering, R. F., Weiss, A. J., Greene, M. W., Banks, H. O.: Development of methods for the production of certain short lived radioisotopes. In: Radioisotopes in Scientific Research: Proceedings of the 1st UNESCO International Conference, Paris 1957. London, Pergamon Press (1958), Volume 1, pp. 50–70.
8. Tucker, W. D., Greene, M. W., Weiss, A. J., Murrenhoff, A. P.: Methods of preparation of some carrier-free radioisotopes involving sorption on alumina. USAEC Report BNL-3746, Brookhaven National Laboratory, May 29, 1958.
9. Richards, P.: Nuclide generators. In: *Radioactive Pharmaceuticals.* Andres, G. A., Kniseley, R. M., Wagner, H. N. Jr., Anderson, E. B. (eds.). Oak Ridge, US Atomic Energy Commission (1966), pp. 155–163.
10. Richards, P., Tucker, W. D., Srivastava, S. C.: Technetium-99m: an historical perspective. Int. J. Appl. Radiat. Isot. **33**, 793–799 (1982).
11. Beck, R. N.: A theoretical evaluation of brain scanning systems. J. Nucl. Med. **2**, 314 (1961).
12. Richards, P.: Private communication (1981).
13. Harper, P. V., Beck, R., Charleston, D., Lathrop, K. A.: Optimization of a scanning method using Tc^{99m}. Nucleonics **22**, 50–54 (1964).
14. Charlton, J. C.: Private communication (1982).
15. Charlton, J. C.: Radioactive pharmaceuticals. The Pharmaceutical Journal, July 10, 1965.
16. Abstracts of papers, Isotope Study Section on Radioisotopes in Clinical Research and Diagnosis, British Institute of Radiology, June 8, 1966. J. Radiol. **40**, 76 (1967).
17. Lathrop, K. A.: Preparation and control of ^{99m}Tc radiopharmaceuticals. In: Radiopharmaceuticals From Generator-produced Radionuclides. Vienna, International Atomic Energy Agency (1970), STI/PUB/294, pp. 39–52.
18. Atkins, H. L., Cardinale, K. G., Eckelman, W. C., Hauser, W., Klopper, J. F., Richards, P.: Evaluation of ^{99m}Tc-DTPA prepared by three different methods. Radiology **98**, 674–677 (1971).
19. Eckelman, W. C., Richards, P.: Instant ^{99m}Tc compounds. Nucl. Med. **10**, 251–254 (1971).
20. Molinski, V. J.: A review of ^{99m}Tc generator technology. Int. J. Appl. Radiat. Isot. **33**, 811–819 (1982).
21. Subramanian, G., McAfee, J. G., Bill, E. G.: ^{99m}Tc-labeled polyphosphate as a skeletal imaging agent. Radiology **102**, 701–704 (1972).
22. Hosain, P.: Technetium-99m labelled pyrophosphate: a simple and reproducible bone scanning agent. Radiology **109**, 467 (1973).
23. Castronovo, F. P., Callahan, R. J.: A new bone scanning agent: ^{99m}Tc-labeled 1-hydroxy-ethylidene-1,1-disodium phosphate. J. Nucl. Med. **13**, 947 (1972).
24. Tofe, A. J., Francis, M. D.: Optimization of the ratio of stannous tin:ethane-1-hydroxyl-1,1-diphosphonate for bone

scanning with ^{99m}Tc-pertechnetate. J. Nucl. Med. **15**, 69–74 (1974).
25. Subramanian, G., McAfee, J. G., Blair, R. J., Kallfelz, F. A., Thomas, F. D.: Technetium-99m-methylene diphosphonate – a superior agent for skeletal imaging: comparison with other technetium complexes. J. Nucl. Med. **16**, 744–755 (1975).
26. Uren, R. F.: Private communication (1977).
27. DePamphilis, B. V., Jones, A. G., David, M. A., Davison, A.: Preparation and crystal structure of oxotechnetium bis-(thiomercaptoacetate) and its relationship to radiopharmaceuticals labeled with ^{99m}Tc. J. Am. Chem. Soc. **100**, 5570–5571 (1978).
28. Smith, J. E., Byrne, E. F., Cotton, F. A., Sekutowski, J. C.: A thiol complex of technetium pertinent to radiopharmaceutical use of ^{99m}Tc. J. Am. Chem. Soc. **100**, 5571–5572 (1978).
29. Davison, A., Jones, A. G., Orvig, C., Sohn, M.: A new class of oxtechnetium(5+) chelate complexes containing a $TcON_2S_2$ core. Inorg. Chem. **20**, 1629–1632 (1981).
30. Fritzberg, A. R., Kasina, S., Eshima, D., Johnson, D. L.: Synthesis and biologic evaluation of technetium-99m MAG_3 as a hippuran replacement. J. Nucl. Med. **27**, 111–116 (1986).
31. Wilson, B. W., Costello, C. E., Carr, S. A., Biemann, K., Orvig, C., Davison, A., Jones, A. G.: Field desorption mass spectrometry of complex anions containing rhenium and technetium. Anal. Lett. **12**, 303–311 (1979).
32. Burns, H. D., Dannals, R. F., Dannals, T. E., Kramer, A. V., Marzilli, L. G.: Synthesis of tetradentate aminothiol ligands and their technetium complexes. J. Labeled Comp. Radiopharm. **18**, 54–55 (1981).
33. Kramer, A. V., Epps, L. A., Burns, H. D., Johnson, A. N., Ranganathan, N., Ravert, H. T.: Synthesis and characterization of neutral aminoethanol complexes of technetium. J. Nucl. Med. **24**, P127 (1983).
34. Lever, S. Z., Burns, H. D., Kervitsky, T. M., Goldfarb, H. W., Woo, D. V., Wong, D. F., Epps, L. A., Kramer, A. V., Wagner, H. N. Jr.: Design, preparation, and biodistribution of a technetium-99m triaminedithiol complex to assess regional cerebral blood flow. J. Nucl. Med. **26**, 1287–1294 (1985).
35. Kung, H. F., Molnar, M., Billings, J., Wicks, R., Blau, M.: Synthesis and biodistribution of neutral lipid-soluble Tc-99m complexes that cross the blood-brain barrier. J. Nucl. Med. **25**, 326–332 (1984).
36. Walovitch, R. C., Hill, T. C., Garrity, S. T., Cheesman, E. H., Burgess, B. A., O'Leary, D. H., Watson, A. D., Ganey, M. V., Morgan, R. A., Williams, S. J.: Characterization of technetium-99m-L,L-ECD for brain perfusion imaging, part 1: pharmacology of technetium-99m ECD in nonhuman primates. J. Nucl. Med. **30**, 1892–1901 (1989).
37. Yokoyama, A., Hata, N., Saji, H., Horiuchi, K., Tanaka, H., Morita, R., Torizuka, K.: Chemically designed 99m-Tc radiopharmaceuticals for tumor diagnosis. J. Nucl. Med. **22**, P69 (1981).
38. Leonard, J. P., Nowotnik, D. P., Neirinckx, R. D.: Technetium-99m-d,l,HM-PAO: a new radiopharmaceutical for imaging regional brain perfusion using SPECT – A comparison with iodine-123-HIPDM. J. Nucl. Med. **27**, 1819–1823 (1986).
39. Nowotnik, D. P., Jurisson, S. S.: Structure and stereochemistry in technetium coordination complexes. In: *The Chemistry of Technetium in Medicine*. Nuclear Science Series (NAS-NS-3204). Steigman, J., Eckelman, W. C. (eds.). Washington D.C., National Academy Press (1992), 111–130.
40. Deutsch, E., Bushong, W., Glavan, K. A., Elder, R. C., Sodd, V. J., Scholz, K. L., Fortman, D. L., Lukes, S. J.: Heart imaging with cationic complexes of technetium. Science **214**, 85–86 (1981).
41. Gerson, M. C., Deutsch, E. A., Libson, K. F.: Myocardial perfusion imaging with ^{99m}Tc-tris-DMPE in man. Eur. J. Nucl. Med. **9**, 403–407 (1984).
42. Higley, B., Smith, F. W., Smith, T., Gemmell, H. G., Das Gupta, P., Gvozdanovic, D. V., Graham, D., Hinge, D., Davidson, J., Lahiri, A.: Technetium-99m-1,2-bis[bis(2-ethoxyethyl)phosphino]ethane: human biodistribution, dosimetry and safety of a new myocardial perfusion imaging agent. J. Nucl. Med. **34**, 30–38 (1993).
43. De Rosch, M. A., Brodack, J. W., Grummon, G. D., Marmion, M. E., Nosco, D. L., Deutsch, K. F., Deutsch, E.: Kit development for the Tc-99m myocardial imaging agent Technecard™. J. Nucl. Med. **33**, 850 (1992).
44. Holman, B. L., Jones, A. G., Lister-James, J., Davison, A., Abrams, M. J., Kirshenbaum, J. M., Tumeh, S. S., English, R. J.: A new myocardial imaging agent, Tc-99m-labeled hexakis(t-butylisonitrile)technetium(I)[Tc-99m-TBI]: initial experience in the human. J. Nucl. Med. **25**, 1350–1355 (1984).
45. Holman, B. L., Sporn, V., Jones, A. G., Sia, S.T.B., Perez-Balino, N., Davison, A., Lister-James, J., Kronauge, J. F., Mitta, A.E.A., Camin, L. L., Campbell, S., Williams, S. J., Carpenter, A. T.: Myocardial imaging with technetium CPI: initial experience in the human. J. Nucl. Med. **28**, 13–18 (1987).
46. Taillefer, R., LaFlamme, L., Dupras, G., Picard, M., Phaneuf, D. C., Leveile, J.: Myocardial perfusion imaging with 99m-Tc-methoxyisobutylisonitrile (MIBI): comparison of short and long time intervals between rest and stress injections: preliminary results. Eur. J. Nucl. Med. **13**, 515–522 (1988).
47. Müller, S. T., Guth-Tougelides, B., Creutzig, H.: Imaging of malignant tumors with Tc-99m-MIBI SPECT. J. Nucl. Med. **28**, 562 (1987).
48. Delmon-Moingeon, L. I., Piwnica-Worms, D., Van den Abbeele, A. D., Holman, B. L., Davison, A., Jones, A. G.: Uptake of the cation hexakis(2-methoxyisobutylisonitrile)-technetium-99m by human carcinoma cell lines in vitro. Cancer Res. **50**, 2198–2202 (1990).
49. O'Tuama, L. A., Treves, S. T., Larar, J. N., Packard, A. B., Kwan, A. J., Barnes, P. D., Scott, R. M., Black, P. M., Madsen, J. R., Goumnerova, L. C., Sallan, S. E., Tarbell, N. J.: Thallium-201 versus technetium-MIBI SPECT in evaluation of childhood brain tumors: a within-subject comparison. J. Nucl. Med. **34**, 1045–1051 (1993).
50. Piwnica-Worms, D. P., Kronauge, J. F., Chiu, M. L.: Uptake and retention of hexakis(2-methoxyisobutyl isonitrile) technetium(I) in culture chick myocardial cells. Mitochondrial and plasma membrane potential dependence. Circulation **82**, 1826–1838 (1990).
51. Piwnica-Worms, D., Kronauge, J. F., LeFurgey, A., Backus, M., Hockett, D., Ingram, P., Liebermann, M., Holman, B. L., Jones, A. G., Davison, A.: Mitochondrial localization and characterization of ^{99}Tc-Sestamibi in heart cells by electron probe x-ray microanalysis and ^{99}Tc-NMR spectroscopy. Magn. Reson. Imaging **12**, 641–652 (1994).
52. Piwnica-Worms, D., Chiu, M. L., Budding, M., Kronauge, J. F., Kramer, R. A., Croop, J. L.: Functional imaging of multidrug-resistant P-glycoprotein with an organotechnetium complex. Cancer Res. **53**, 977–984 (1993).
53. Mansi, L., Rambaldi, P. F., La Provitera, A., Di Gregorio, F., Proaccini, E.: Tc-99m Tetrofosmin uptake in breast tumors. J. Nucl. Med. **36**, 83P (1995).
54. Fritzberg, A. R., Abrams, P. G., Beaumier, P. L.: Specific and stable labeling of antibodies with technetium-99m with a diamide dithiolate chelating agent. Proc. Natl. Acad. Sci. USA **5**, 325–334 (1988).
55. Barrett, J., Edwards, S., Harris, T., Rajopdhye, M., Lazewatsky, J., Liu, S., Damphousse, D., Heminway, S., Mazaika, T., Thomas, J., Carroll, T., Smith, J.: The development of a technetium-99m agent for imaging deep vein thrombosis. In: *Technetium and Rhenium in Chemistry and Nuclear Medicine 4*. Nicolini, M., Bandoli, G., Mazzi, U. (eds.). Padova, SGEditoriali 1995, pp. 275–280.
56. Ben-Haim, S., Kahn, D., Weiner, G. J., Madsen, M. T., Waxman, A. D., Williams, C. M., Clarke-Pearson, D. L., Cole-

man, R. E., Macguire, R. T.: The safety and pharmacokinetics in adult subjects of an intravenously administered ^{99m}Tc-labeled 17 amino acid peptide (CYT-379). Nucl. Med. Biol. **21**, 131–142 (1994).

57. Hom, R. K., Chi, D., Katzenellenbogen, J. A.: Stereochemical issues in the synthesis of bis-bidentate $(NS)_2$ amino thiol complexes of oxorhenium(V) and oxotechnetium(V) whose structures mimic those of steroids. Proceedings 11th International Symposium on Radiopharmaceutical Chemistry, Vancouver, August 1995, pp. 441–443.
58. Vera, D. R., Stadalnik, R. C., Trudeau, W. L., Scheibe, P. O., Krohn, K. A.: Measurement of receptor concentration and forward-binding rate constant via radiopharmacokinetic modeling of technetium-99m-galactosyl-neoglycoalbumin. J. Nucl. Med. **32**, 1169–1176 (1991).
59. Kung, H. F., Bradshaw, J. E., Chumpradit, S., Zhuang, Z.-P., Kung, M.-P., Mu, M., Frederick, D.: New TcO(III) and ReO(III) N_2S_2 complexes as potential CNS 5-HT_{1A} receptor imaging agents. In: *Technetium and Rhenium in Chemistry and Nuclear Medicine 4.* Nicolini, M., Bandoli, G., Mazzi, U. (eds.). Padova, SGEditoriali (1995), pp. 293–298.
60. Plossl, K., Zhuang, Z.-P., Meegalla, S. K., Bradshaw, J., Frederick, D., Stevenson, D. A., Kung, M.-P., Kung, H. F.: Tc-99m[MPP] complexes: potential 5-HT_{1A} receptor imaging agents. Proceedings 11th International Symposium on Radiopharmaceutical Chemistry, Vancouver, August 1995, pp. 306–308.
61. Spies, H., Johannsen, B., Pietzch, H.-J., Noll, B., Noll, St., Scheunemann, M., Fietz, T., Berger, R., Brust, P., Leibnitz, P.: Technetium and rhenium complexes as serotonin receptor binding ligands. Proceedings 11th International Symposium on Radiopharmaceutical Chemistry, Vancouver, August 1995, pp. 319–320.
62. Jurisson, S. S., Pirro, J., DiRocco, R. J., Rosenspire, K. C., Jagoda, E., Nanjappan, P., Eckelman, W. C., Nowotnik, D. P., Nunn, A. D.: Boronic acid adducts of technetium dioxime (BATO) complexes derived from quinuclidine benzilate (QNB) boronic acid stereoisomers: syntheses and studies of their binding to muscarinic acetylcholine receptor. Nucl. Med. Biol. **22**, 269–281 (1995).
63. Del Rosario, R. B., Jung, Y.-W., Baidoo, K. E., Lever, S. Z., Wieland, D. M.: Synthesis and in vivo evaluation of a $^{99m/99}$Tc-DADT-benzovesamicol: a potential marker for cholinergic neurons. Nucl. Med. Biol. **21**, 197–204 (1994).

Radiochimica Acta 70/71, 299–304 (1995)

Formation of Heavy Elements in Nature

By P. K. Kuroda
4191 Del Rosa Court, Las Vegas, Nevada 89121 (USA)

(Received May 23, 1995)

Plutonium-244 / Fission xenon / Carbonaceous chondrites / Lunar fines / Age of the solar system

Abstract

A brief historical account is presented of the progress made during the second half of the 20th century in the field of studies on formation of heavy elements in nature. Results from the latest calculations on the amounts of ^{244}Pu fission xenon in carbonaceous chondrites and in lunar fines indicate that the generally accepted view that the solar system formed about 4,500 to 4,600 million years ago is in error.

1. Introduction

In 1896, Becquerel [1] discovered radioactivity and nine years later in 1905, Curie [2] stated: "At the beginning of our investigations we stated, Mme. Curie and I, that the phenomena (of radioactivity) could be explained by two distinct and very general hypotheses which were described by Mme. Curie in 1899 and 1900 (*Revue Générale des Sciences*, January 10, 1899, and *Revue Scientifique*, July 21, 1900): (1) In the first hypothesis it can be supposed that the radioactive substances borrow from an external radiation which they release, and their radiation would then be a secondary radiation. It is not absurd to suppose that space is constantly traversed by very penetrating radiations which certain substances would be capable of capturing in flight. (2) In the second hypothesis it can be supposed that the radioactive substances draw from themselves the energy which they release. The radioactive substances would then be in course of evolution, and they would be transforming themselves progressively and slowly in spite of the apparent invariability of the state of certain of them."

Curie [2] then went on to remark: " – the second hypothesis has shown itself the more fertile in explaining the properties of the radioactive substances properly so called" and, consequently, the first hypothesis became largely forgotten during the first half of the 20th century.

In 1957, Burbidge *et al.* [3] published the so-called B_2FH theory of the synthesis of the elements in stars and it became obvious that the process of formation of heavy elements in stars occurs in such a manner as it was described by Curie [2] in his first hypothesis. Very penetrating radiations which the Curies had envisioned turned out to be the neutrons discovered by Chadwick [4] in 1932.

On the eve of the 100th anniversary of the discovery of radioactivity by Becquerel [1], a brief account of the progress made in the field of studies on formation of heavy elements in nature during the second half of the 20th century will be presented here.

2. The 31 October 1952 thermonuclear weapon's test

On 31 October 1952, the U.S. Government conducted the first thermonuclear weapon's test at the Enewetak (Eniwetok) Atoll in the Pacific Ocean, which was a 10.4-megaton device named "MIKE". It was a remarkable coincidence that the year 1952 was the 100th anniversary of the birth of Becquerel [1]. It is interesting to recall that in the same year 1952, Urey [5] made the following remark: " – Dr. Edward Teller remarked recently the origin of the earth was somewhat like the explosion of the atomic bomb: the physical effects are often temporary, but the chemical effects, such as radioactive and non-radioactive elements, remain. It is possible by a study of these substances to learn much about the bomb, and also about the origin of the earth."

In the fallout samples from the 31 October 1952 thermonuclear explosion, elements 99 (einsteinium, symbol E) and 100 (fermium, symbol Fm) were discovered by a large group of scientists at the University of California, Berkeley, the Argonne National Laboratory and the Los Alamos Scientific Laboratory in 1955 [6].

Recalling the events which took place during the 1950's, Seaborg [7] remarked, in 1992: "Elements 99 and 100 were unexpectedly discovered in debris from the "MIKE" thermonuclear explosion. – Early analysis of the "MIKE" debris by scientists at the Argonne National Laboratory near Chicago and the Los Alamos Laboratory in New Mexico showed the unexpected presence of new isotopes of plutonium, ^{244}Pu and ^{246}Pu. – This observation led to the conclusion that ^{238}U in the device had been subjected to an enormous neutron flux and had successively captured numerous neutrons. (Later calculations showed that an integral neutron fluence of $(1-4)\times10^{24}$ neutrons was delivered in a few nanoseconds – a few *moles* of neutrons!!)."

A year prior to the discovery of elements 99 and 100, Studier *et al.* [8], in 1954, discovered ^{244}Pu in pile-irradiated plutonium samples. They irradiated several ^{239}Pu samples in the high neutron flux of the Materials Testing Reactor (MTR) for progressively longer periods of time. They reported that the samples which they studied received up to an integrated flux of 4×10^{21} neutrons and it was reasonable to assume that most of the ^{244}Pu came from the multiple-order neutron-capture reactions in plutonium: $^{239}Pu(n,\gamma)^{240}Pu(n,\gamma)^{241}Pu(n,\gamma)^{242}Pu(n,\gamma)^{243}Pu(n,\gamma)^{244}Pu$.

It is interesting to note here that Seaborg [9], in 1958, remarked: " – The isotope ^{244}Pu (half-life 7.6×10^7 years) is especially interesting because of its long half life, and offers intriguing possibilities for the ore distant future."

3. The B$_2$FH theory

In 1956, Suess and Urey [10] stated: "On October 26, 1889, Frank Wigglesworth Clarke [11] read a paper before the Philosophical Society of Washington in which he outlined a research program that covered the contents of (our) paper. Clarke's paper was the first of a number of papers by many different authors." They quoted Clarke [11] as saying in 1889: "An attempt was made in the course of this investigation to represent the relative abundances of the elements by a curve, taking their atomic weight for one set of ordinates. It was hoped that some sort of periodicity might be evident, but no such regularity appeared." Since Suess and Urey [10] in 1956 made no mention of the motive behind Clarke's research, the above statement about the attempt made by Clarke [11] in 1889 became largely forgotten during the second half of the 20th century.

It so happened that as early as in 1873, Clarke [11] published an article in a journal called *Popular Science Monthly*, in which he expressed his idea that the chemical elements must have originally developed by a process of evolution from much simpler forms of matter, as is indicated by the progressive chemical complexity observed in passing from the nebulae through the hot stars to the cold planets. Clarke [11] was therefore the first chemist to attempt to study the process of formation of heavy elements in nature in the 19th century.

In 1889, Clarke [11] drew the following conclusion from his studies: "If, despite Mendeléeff's recent demurrer we assume that the elements have been evolved from one primordial form of matter, their relative abundance becomes suggestive. Starting from the original *protyle*, as Crookes has called it, the process of evolution seems to have gone on slowly until oxygen was reached. At that point the process exhibited its maximum energy, and beyond it the elements forming stable oxides were most rapidly developed, and in the largest amounts. On this supposition the scarcity of the elements above iron becomes somewhat intelligible, but the theory does not account for everything and is to be regarded as merely tentative."

After having compiled the most reliable table of the *cosmic* abundances of the elements in 1956, Suess and Urey [10] reported that the nuclear abundance distribution as derived from the discussions supplied a basis for comparison of empirical data with the various theories of the origin of the nuclear species, but they remarked: " – none of the existing theories can account for all the empirical facts, even in a crude way. Hence, it seems hopeless to attempt to explain the finer details presented here by any of these (existing) theories in their present form".

In the same year 1956, Baade *et al.* [12] and Burbidge *et al.* [13] put forward the fascinating hypothesis that the spontaneous fission of ^{254}Cf with a half-life of 55 days may be responsible for the form of the decay of light-curves of supernova of type I, which have an exponential shape with a half-life of 55 days. They described the way in which ^{254}Cf may be synthesized in a supernova outburst, and the reason why the energy released by its decay may dominate all others. They also noted that the presence of Tc in red giant stars and of Cf in Type I supernovae appeared to be observational evidence that neutron capture processes on both the *slow* and a *fast* timescale have been necessary to synthesize the heavy elements in their observed cosmic abundances in stars.

In the following year 1957, Burbidge *et al.* [3] published a 103-page article entitled "Synthesis of the elements in stars" and pointed out that in order to explain all the fine details of the Suess-Urey [10] cosmic abundance curve, at least eight different types of nucleosynthesis processes were demanded, if it is assumed that only hydrogen is primordial. The eight processes required were: (1) Hydrogen burning, (2) helium burning, (3) α-process, (4) e (or equilibrium)-process, (5) s-process (or neutron-capture at a slow rate), (6) r-process (or neutron-capture at a rapid rate), (7) p-process (the process of proton-capture) and finally (8) x-process, which is responsible for the synthesis of the deficient elements (deuterium, lithium, beryllium and boron). It is worthy of note here that the exact nature of the x-process nucleosynthesis has remained as a mystery for many years.

4. The Oklo Phenomenon

In 1956, the writer [14, 15] made the prediction that natural reactors should have existed on the earth about 2 billion years ago. Although this prediction was not taken seriously by scientists of the 1950's, sixteen years later in 1972, French investigators [16, 17] discovered the remnants of natural reactors at the Oklo uranium mines located in the Republic of Gabon, Africa.

Until the middle of the 20th century, scientists believed that chemical elements were synthesized only in stars, but the discovery of the so-called Oklo

Phenomenon has demonstrated that a nuclear *fire* had once existed on our planet earth and formation of heavy elements was occurring in nature. A vast amount of experimental data on the remnants of the Oklo reactors have since been obtained mostly by French investigators during the 1970's and a new knowledge gained from the results of these studies were to play important roles in the studies on lunar samples and meteorites in the field of cosmochemistry.

5. The ^{244}Pu hypothesis

In 1960, newly elected President Kennedy declared that the United States would send Man to the moon before the decade was over and Reynolds [18] at the University of California, Berkeley, made the important discovery that the xenon from the chondritic(stone) meteorite Richardton was heavily enriched in ^{129}Xe. He concluded that this isotope almost certainly was formed from the radioactive decay of ^{129}I with a half-life of 16 million years, now extinct as a natural radioactivity but not so at the time of formation of the meteorite.

In the same year 1960, the writer [19] pointed out that ^{244}Pu with a half-life of 82 million years should have been also present in the early solar system and the experimental evidence for its existence could be secured by searching for the presence in meteorites of excess heavy xenon isotopes, which are produced by the spontaneous fission of ^{244}Pu:

$$^{129}\text{I} \xrightarrow[1.6\times10^{7}\text{ years}]{\beta^-} {}^{129}\text{Xe(stable)}$$

and

$$^{244}\text{Pu} \xrightarrow[6.55\times10^{10}\text{ years}]{\text{Spontaneous fission}} {}^{131-136}\text{Xe(stable)}.$$

The first experimental evidence for the ^{244}Pu hypothesis was obtained in the writer's laboratories at the University of Arkansas in 1965 [20] and the occurrence of ^{244}Pu in the early solar system became firmly established at the beginning of the 1970's [21]. It is important to note here, however, that Reynolds and Turner [22], in 1964, discovered that the carbonaceous chondrite Renazzo contained a strange fission-like xenon component, which they called the carbonaceous chondrite fission (CCF) xenon. This strange xenon component was later found to be present in all the carbonaceous chondrites and became widely known as xenon-*HL*. According to most investigators in the field of cosmochemistry, this strange component is a mixture of the products of *r*- and *p*-process nucleosynthesis and the amounts found in carbonaceous chondrites are at least one order of magnitude greater than that of the ^{244}Pu fission xenon, which is found to occur in chondrites and achondrites.

6. The FGH theory

In 1962, Fowler *et al.* [23] at the California Institute of Technology published a paper entitled "Nucleosynthesis during the Early History of the Solar System" and reported that the synthesis of ^{2}D, ^{6}Li, ^{7}Li, ^{9}Be, ^{10}B and ^{11}B and possibly ^{13}C and ^{15}N occurred during an intermediate stage in the early history of the solar system. According to these investigators, the planetary material had become largely separated, but not completely, from the hydrogen which was the main constituent of primitive solar material during this intermediate stage. Appropriate physical conditions were satisfied by solid planetesimals of dimensions from 1 to 50 meters consisting of silicates and oxides of the metals embedded in an icy matrix. The synthesis occurred through spallation and neutron reactions simultaneously induced in the outer layers of the planetesimals by bombardment of high energy charged particles, mostly protons, accelerated in magnetic flares at the surface of the condensing Sun. The total particle energy was approximately 10^{50} MeV while the average energy was close to 500 MeV per nucleon. Studies of the abundance of Li in young *T Tauri* stars served as the primary astronomical evidence for Fowler *et al.* [23]'s point of view.

Fowler *et al.* [23] went on to state that 10 percent of terrestrial-meteoritic material was irradiated with a thermal neutron flux of 10^7 (n/cm^2 second) for an interval of 10^7 years. It is interesting to note here that the integrated thermal neutron flux, to which 10 percent of terrestrial-meteoritic material was exposed, was approximately equal to 3×10^{21} (n/cm^2), which was about twice the value of 1.5×10^{21} (n/cm^2) of thermal neutrons, to which some of the uranium ore deposits at the site of the Oklo natural reactors were exposed [24].

The most remarkable feature of the FGH theory [23] was that the neutron flux was sufficient to produce the radioactive ^{107}Pd and ^{129}I necessary to account for the radiogenic ^{107}Pd and ^{129}Xe anomalies observed in meteorites. Eight years later in 1970, however, a set of experimental data, which appeared to be a fatal blow to the FGH theory were reported by Eugster *et al.* [25] at the California Institute of Technology. They determined the isotopic composition of Gd in a number of meteorites and reported that the results indicated that they seemed to set an upper limit of 3×10^{15} neutrons per cm^2 on a differential integrated thermal neutron irradiation of the earth and these meteorites, which seemed to leave little room for an intense irradiation of solid bodies in the early solar system, as it was envisioned by Fowler *et al.* [23] in 1962.

It was totally unexpected, but it appeared to the writer at that time that Fowler *et al.* [23] were pleased to accept the conclusion reached by Eugster *et al.* [25] that the early irradiation of terrestrial-meteoritic material did not play a significant role in the early history of the solar system.

7. Is the Moon younger than the meteorites?

The first report on the examination of the samples returned from the Apollo 11 landings in Mare Tranquilitatis on 20 July 1969 appeared in the 19 September 1969 issue of *Science* [26]. The Lunar Sample Preliminary Examination Team (LSPET) stated in this report that the isotopic composition of xenon in the Apollo 11 fines closely resembled that of trapped xenon found in carbonaceous chondrites, except for a small addition of the lighter xenon isotopes, presumably from spallation reactions, and a deficiency of ^{134}Xe and ^{136}Xe. The ratio of ^{129}Xe to ^{132}Xe in the material measured, on the other hand, was essentially the same as that in carbonaceous chondrites.

In the following year 1970, Reynolds *et al.* [27] at the University of California, Berkeley, reported that excess ^{129}Xe was absent in both the Apollo 11 lunar fines 10084,59 and the fine-grained lunar rock 10057,20 and they concluded that the Moon must be younger than the meteorites. They did find a small excess of fission xenon in the lunar rock 10057,20, but they stated: "Fission xenon in rock 57 was indicated, after subtraction of cosmogenic xenon, by higher values of ^{136}Xe/^{132}Xe and ^{134}Xe/^{132}Xe than in lunar fines. The amount of fission xenon observed is approximately what should accumulate in 4.1×10^9 years (see below) of uranium decay in rock 57 (780 ppb uranium)." Reynolds *et al.* [27] thus attributed the excess fission xenon which they found in this sample to the spontaneous fission of ^{238}U and concluded that the ^{244}Pu fission xenon must be absent in lunar samples.

Meanwhile, at the University of Arkansas, the writer [28] reported in 1971 that the difference in the isotopic compositions of xenon found in carbonaceous chondrite Renazzo, lunar fines and in the atmosphere can be attributed to the alterations of the isotopic composition by a combined effect of (a) mass-fractionation, (b) spallation and (c) 10 keV (stellar temperature) neutron-capture reactions. This meant that the isotopic composition of the primordial xenon must have been severely altered by the processes (a), (b) and (c) occurring in the vicinity of the site of a supernova explosion.

At about the same time, the isotopic composition of xenon in the carbonaceous chondrite Allende, which fell two years earlier in 1969 in Southern Chihuahua, Mexico, was being measured in the writer's laboratories at the University of Arkansas. The xenon isotope data obtained were published in 1972 [29] and the results clearly demonstrated that this meteorite also contained a large excess of the so-called carbonaceous chondrite fission (CCF) xenon.

8. The so-called FUN anomalies

In the following year 1973, Clayton *et al.* [30] at the University of Chicago reported to have found that the oxygen of anhydrous, high temperature minerals in the carbonaceous chondrite Allende was strongly depleted in the heavy stable isotopes ^{17}O and ^{18}O. They reported that the effect was the result of nuclear rather than chemical processes and probably resulted from the admixture of a component of almost pure ^{16}O. They went on to state that this component may have predated the solar system and may represent interstellar dust with a separate history of nucleosynthesis.

In 1974, Lee and Papanastassiou [31] and, independently, Gray and Compston [32] made another important discovery that the isotopic composition of magnesium displayed anomalies which could not be attributed to isotopic fractionation. During the period from 1974 to 1982, it was found that virtually every element analyzed in two kinds of mineral inclusions (C-1 and EK-1-4-1) of the Allende meteorite had an anomalous isotopic composition (O, Mg, Ca, Ti, Sr, Ba, Nd and Sm). In 1982, Wasserburg and Papanastassiou [33] published a review article and stated that these samples of mineral inclusions have been called FUN samples to denote mass fractionation (F) and unknown nuclear anomalies (UN).

In the same year 1982, the writer [34] reported in Chapter 7 of his monograph entitled "The Origin of the Chemical Elements and the Oklo Phenomenon" that the origin of these anomalies observed in meteorites is attributable to the *x*-process nucleosynthesis, but the writer's view was not taken seriously by cosmochemists of the 1980's.

Meanwhile, Fowler [35] attached a great significance to the discovery of FUN anomalies and the occurrence of ^{129}I and ^{26}Al in the early solar system in his 1983 Nobel Lecture entitled "The Quest for the Origin of the Elements". He stated: "The fact that the anomalies produced by these short-lived nuclei relative to normal abundances all are of the order of 10^{-4} despite the wide range in their mean lifetime (1.04×10^6 years to 23×10^6 years) indicates that this anomaly range must be the result of inhomogeneous mixing of *exotic* materials with much larger quantities of normal solar system materials over a short time rather than the result of free decay." Fowler [36] then went on to remark: "The anomalies in stable isotope abundances are of the same order as those for short-lived radioactive nuclei and strongly support the view that the solar nebula was inhomogeneous, with regions containing exotic materials up to 10^{-4} or more of normal material."

It was noteworthy, however, that Fowler [35] made no mention of the occurrence of ^{244}Pu in the early solar system in his 1983 Nobel Lecture and it appeared as if he was no longer interested in the existence of this important extinct nuclide in the early solar system.

9. Plutonium-244 in the early history of the solar system

In 1983, the writer became firmly convinced of the fact that the generally accepted view that ^{244}Pu fission xenon is absent in carbonaceous chondrites and lunar fines must be in error. In order to prove this, however, it was necessary to find a new method of interpreting the xenon isotope data, which enabled one to calculate the precise amount of ^{244}Pu fission xenon present in a sample of meteorite or lunar fines in a straightforward manner.

For more than two decades, Reynolds *et al.* [22] were having serious difficulties in their attempt to determine the isotopic composition of the CCF xenon, which became widely known as xenon-*HL*. It appeared to the writer that the main reason for this was they assumed that xenon fractions released from carbonaceous chondrites were binary mixtures of the average carbonaceous chondrite (AVCC) xenon and an additional xenon component, which could be identified as an isotopically *pure* substance.

In 1989, Lewis *et al.* [36] at the University of Chicago made the important discovery that xenon-*HL* is carried by interstellar grains of tiny diamonds (less than 3 nm in diameter). It appeared, however, that these investigators were happy to accept the view that xenon-*HL* was an *exotic* substance and made no further attempt to reexamine the xenon isotope data, which they have obtained. The writer, on the other hand, felt that xenon-*HL* was not likely to be an isotopically *pure* substance, but instead, its isotopic composition could be explained as a mixture of ^{244}Pu fission xenon and a primordial xenon, whose isotopic composition resembles that of the atmospheric xenon, but is severely altered by a combined effect of (a) mass-fractionation, (b) spallation and (c) 10 keV (stellar temperature) neutron-capture reactions, as it was first pointed out by him [28] in 1971.

It is important to recall that Burbidge *et al.* [3], in 1957, stated: "The characteristic of all these (deficient) elements is that they are very unstable at temperatures of stellar interiors, so that it appears probable that they have been produced in regions of low density and temperature. Now, the regions in our Galaxy in which such conditions are present are (i) in stellar atmosphere and (ii) in gaseous nebula." This meant, in the opinion of the writer, that what Burbidge *et al.* [3] had in mind in 1957 was that the *x*-process nucleosynthesis occurred in regions located in the vicinity of the site of a supernova explosion.

Four years later in 1993, the writer [37] reported on a new method of interpretation of the xenon isotope data, which appeared to be universally applicable to all the meteorites and lunar samples, including carbonaceous chondrites and lunar fines. In this new method, alterations of the relative abundances of nine stable xenon isotopes by the processes of (a) mass-fractionation, (b) spallation and (c) 10 keV (stellar temperature) neutron-capture reactions were taken into consideration. The use of this method, which enabled one to calculate the exact amount of ^{244}Pu fission xenon in a straight-forward manner yielded a number of surprising results.

The greatest surprise of them all was that the carbonaceous chondrites contained as much as 20 to 60×10^{-12} (cm^3 STP/g) of ^{244}Pu fission xenon (^{136f}Xe) [38]. Results from the latest calculations showed that the isotopic compositions of the strange xenon components, such as xenon-*HL* and the *s*-type xenon, can be explained in terms of the two-component model in which they are regarded as representing the isotopic composition of a mixture of the ^{244}Pu fission xenon and a trapped xenon component, whose isotopic composition resembles that of the atmospheric xenon, but is often markedly altered by a combined effect of (a) mass-fractionation, (b) spallation and (c) neutron-capture reactions caused primarily by the 10 keV (stellar temperature) neutrons.

Results from these calculations have revealed that the Cl carbonaceous chondrite Orgueil contains as much as 59.8×10^{-12} (cm^3 STP/g) of ^{244}Pu fission xenon (^{136f}Xe) and it began to retain its xenon 5,072 million years ago, when the ^{244}Pu to ^{238}U ratio was as high as 0.511 (atom/atom). The C2 carbonaceous chondrites Cold Bokkeveld, Murray, Murchison and Renazzo contained $(21.6 \text{ to } 46.9) \times 10^{-12}$ (cm^3 STP/g) of ^{244}Pu fission xenon (^{136f}Xe) and they began to retain their xenon 4,908 to 5,008 million years ago, when the ^{244}Pu to ^{238}U ratio was 0.131 to 0.299 (atom/atom), while the C3V carbonaceous chondrites Mokoia and Allende contained $(21.8 \text{ to } 24.7) \times 10^{-12}$ (cm^3 STP/g) of ^{244}Pu fission xenon (^{136f}Xe) and they began to retain their xenon 4,879 to 4,902 million years ago, when the ^{244}Pu to ^{238}U ratio was 0.103 to 0.124 (atom/atom).

It was also found [39] that the lunar fines contained as much as $(71.3 \text{ to } 973) \times 10^{-12}$ (cm^3 STP/g) of ^{244}Pu fission xenon (^{136f}Xe) and they started to retain their xenon 4,568 to 4,857 million years ago, when the ^{244}Pu to ^{238}U ratio was 0.0078 to 0.0855 (atom/atom). These results indicate that the lunar fines started to retain their xenon prior to the time the chondrites and achondrites started to retain their xenon and hence the former is older than the latter.

Results from the latest calculations [40] also demonstrated that it is possible to explain the isotopic anomalies observed for several elements other than xenon, for example, carbon, oxygen, neon, barium and neodymium, as due to the alterations of the isotopic compositions by a combined effect of (a) mass-fractionation, (b) spallation and (c) 10 keV (stellar temperature) neutron-capture reactions, which took place in the *x*-process nucleosynthesis, as it was first pointed out by the writer [34] in 1982.

10. Concluding remarks

In the summer of 1992, the writer sent a preprint of the paper [37], which appeared in this Journal in 1993, to Professor William A. Fowler at the California Institute of Technology. Several days later, the writer received a letter dated August 31, 1992, from him, in which he stated: "Thank you for the interesting reprints which you recently send to me. I will read them in the course of time. However, I no longer make comments on papers on any subjects. Respectfully yours."

Professor Fowler died on March 14, 1995, on the eve of the 100th anniversary of the discovery of radioactivity by Becquerel [1].

References

1. Becquerel, H.: Compt. rend. Acad. Sci. Paris **122**, 501–503 (1896).
2. Curie, P.: *Radioactive Substances, Especially Radium, Nobel Lecture*, June 6, 1905 in *Nobel Lectures, Physics*, 1901–1921 Elsevier Publishing Company, Amsterdam, London, New York, 73–78 (1967).
3. Burbidge, E. M., Burbidge, G. R., Fowler, W. A., Hoyle, F.: Rev. Mod. Phys. **29**, 547–650 (1957).
4. Chadwick, J.: Nature **129**, 312 (1932).
5. Urey, H. C.: *The Planets, Their Origin and Development*, Yale University Press, New Haven, Preface, ix, 1–245 (1952).
6. Ghiorso, A., Thompson, S. G., Higgins, G. H., Seaborg, G. T., Studier, M. H., Fields, P. R., Fried, S. M., Diamond, H., Mech, J. F., Pyle, G. L., Huizenga, J. R., Hirsch, H., Manning, W. M., Browne, C. S., Smith, H. L., Spence, R. W.: Phys. Rev. **99**, 1048–1049 (1955).
7. Seaborg, G. T.: *Transuranium Elements: A Half Century*, edited by L. R. Morss and J. Fuger, American Chemical Society, Washington D.C., 10–49 (1992).
8. Studier, M. H., Fields, P. R., Sellers, P. H., Friedman, A. M., Stevens, C. M., Mech, J. F., Diamond, H., Sedlet, J., Huizenga, J. R.: Phys. Rev. **93**, 1433 (1954).
9. Seaborg, G. T.: *The Transuranium Elements*, Yale University Press, New Haven, page 75 (1958).
10. Suess, H. E., Urey, H. C.: Rev. Mod. Phys. **28**, 53–74 (1956).
11. Clarke, F. W.: Bull. Phil. Soc. Washington **11**, 131–142 (1889).
12. Baade, W., Burbidge, G. R., Hoyle, F., Burbidge, E. M., Christy, R. F., Fowler, W. A.: Pub. Astron. Soc. of the Pacific **68**, no. 403, 296–300 (1956).
13. Burbidge, G. R., Hoyle, F., Burbidge, E. M., Christy, R. F., Fowler, W. A.: Phys. Rev. **103**, 1145–1149 (1956).
14. Kuroda, P. K.: J. Chem. Phys. **25**, 781 (1956).
15. Kuroda, P. K.: J. Chem. Phys. **25**, 1295 (1956).
16. Bodu, R., Bouziques, H., Morin, N., Pfiffelmann, J. P.: C. R. Acad. Sci. Paris **275D**, 1731 (1972).
17. Neuilly, M., Bussac, J., Frejacques, C., Nief, G., Vendryes, G., Ivon, J.: C. R. Acad. Sci. Paris **275D**, 1847–1849 (1972).
18. Reynolds, J. H.: Phys. Rev. Lett. **4**, 8–10 (1960).
19. Kuroda, P. K.: Nature **187**, 36–38 (1960).
20. Rowe, M. W., Kuroda, P. K.: J. Geophys. Res. **70**, 709–714 (1965).
21. Alexander, E. C., Jr., Lewis, R. S., Reynolds, J. H., Michel, M. C.: Science **172**, 837–840 (1971).
22. Reynolds, J. H., Turner, G.: J. Geophys. Res. **69**, 3263–3281 (1964).
23. Fowler, W. A., Greenstein, J. L., Hoyle, F.: Geophys. J. Royal Astron. Soc. **6**, No. 2, 148–220 (1962).
24. Hagemann, R., Devillers, C., Lucas, M., Lecompte, T., Ruffenach, J.-C.: *The Oklo Phenomenon*, IAEA, Vienna, 415–428 (1975).
25. Eugster, O., Tera, F., Burnett, D. S., Wasserburg, G. J.: J. Geophys. Res. **75**, 2753–2768 (1970).
26. The Lunar Sample Preliminary Examination Team (LSPET): Science **165**, 1211–1227 (1969).
27. Reynolds, J. H., Hohenberg, C. M., Lewis, R. S., Davis, P. K., Kaiser, W. A.: Science **167**, 545–548 (1970).
28. Kuroda, P. K.: Nature Phys. Sci. **230**, 40–42 (1971).
29. Manuel, O. K., Wright, R. J., Miller, D. K., Kuroda, P. K.: Geochim. Cosmochim. Acta **36**, 961–983 (1972).
30. Clayton, R. N., Grossman, L., Mayeda, T. K.: Science **182**, 485–488 (1973).
31. Lee, T., Papanastassiou, D. A.: Geophys. Res. Letters **1**, 225–228 (1974).
32. Gray, C. M., Compston, W.: Nature **251**, 495–497 (1974).
33. Wasserburg, G. J., Papanastassiou, D. A.: *Essays in Nuclear Astrophysics*, ed. by C. A. Barnes, D. D. Clayton and D. M. Schramm, Cambridge University Press, pp. 77–140 (1982).
34. Kuroda, P. K.: *The Origin of the Chemical Elements and the Oklo Phenomenon*, Springer Verlag, Berlin, Heidelberg, New York, pp. 115–146 (1982).
35. Fowler, W. A.: *The Quest for the Origin of the Elements*, Nobel Lecture, December 8, 1983, Science **226**, 922–935 (1984).
36. Lewis, R. S., Anders, E., Draine, B. T.: Nature **339**, 117–121 (1989).
37. Kuroda, P. K.: Radiochim. Acta **62**, 27–33 (1993).
38. Kuroda, P. K., Myers, W. A.: Radiochim. Acta **68**, 81–90 (1995).
39. Kuroda, P. K., Myers, W. A.: Radiochim. Acta **64**, 155–165 (1994).
40. Kuroda, P. K., Myers, W. A.: Radiochim. Acta **69**, 31–34 (1995).

Radiochimica Acta 70/71, 305–316 (1995)

Radioactivity: A Tool to Explore the Past

By Hans R. von Gunten
Paul Scherrer Institut, CH-5232 Villigen PSI, Switzerland

(Received October 28, 1994; revised March 23, 1995)

Dating / Age determinations / Time markers / Cosmogenic radionuclides / Lead-210 / Uranium-thorium-lead / Rubidium-strontium / Potassium-argon / Radioactive disequilibrium / Fission tracks / Thermoluminescence

Abstract

The use of the radioactive decay for dating purposes is reviewed. The most important dating methods are discussed. They include radioactive time markers (e.g., ^{137}Cs), cosmogenic radionuclides (e.g., ^{14}C, ^{10}Be), decay products of ^{226}Ra (e.g., ^{222}Rn, ^{210}Pb), uranium-thorium-lead chronometers, rubidium-strontium and potassium-argon nuclide pairs, applications of radioactive disequilibrium, dating with fission tracks, and thermoluminescence. Some important results, for instance the age of the Earth and the Moon, are also presented.

I. Introduction

Priests, astronomers, philosophers and other educated people of the great Oriental, Antique, and other societies, already wanted to known more about creation and past times. In this context, one of the important questions was to establish the origin and age of the Earth and Cosmos. However, these problems remained unsolved or were answered by rough estimates only, until **radioactivity** was discovered in 1896 and enabled an accurate dating. The following historic résumé is extracted by the present author from a more detailed overview by Roth [1] in the comprehensive book about dating methods based on nuclear phenomena by Roth and Poty [2]:

"Until the 18th century, the Bible set the age of the Earth very accurately to 4004 years. But still today, 'creationists' derive from studies of the Old Testament an age between 6000 and 10,000 years. At the end of the 19th century Lord Kelvin used heat balances to estimate the Earth's age to 50 to 100 million years (MY). However, already then, geologists concluded from many observations that this age was too short. After the discovery of radioactivity by Becquerel in 1896, and the demonstration of heat evolution in this process by Curie and Laborde, Lord Rutherford recognized in 1904 that Kelvin's Earth's age was indeed much too short, because he had neglected the important additional heat source in the interior of the Earth. Using for the first time a *decay constant,* Rutherford estimated the Earth's age from the amount of helium in uranium minerals to 500 MY. Assuming, without experimental verification, that lead was the final product in the decay chain of uranium, Boltwood assigned 1907 about the same age value to uranite samples. Based on the first modern geological time scale, Holmes attributed, in 1913, an age of 1600 MY to certain archaic rocks".

Scientifically sound dating involving the radioactive decay started with the work of Nier [3, 4] on lead isotopes, in combination with the decay chains of uranium isotopes and of thorium. These authors demonstrated that different isotopic compositions of the analyzed lead samples resulted from mixtures of a 'primitive' lead and 'radiogenic' lead. Their findings enabled Holmes [5] and Houtermans [6], independently, to calculate an Earth's age which was close to the presently accepted value of 4500 MY.

The discovery of 'cosmogenic' radionuclides (e.g., ^{14}C; Libby [7]), and the use of relatively short-lived nuclides of the natural decay chains (e.g., ^{210}Pb; Goldberg [8]), have made age determinations of archeological artifacts, and estimates of time scales for the natural environmental cycles possible.

The extensive testing of atomic bombs in the atmosphere in the late fifties and early sixties, along with other nuclear activities (and accidents) have left narrow and discrete *radioactive time markers* that can be used to date certain environmental archives very accurately. But more generally, the radioactive dating methods are based on the temporal decrease of the number of radioactive atoms and often on the related ingrowth of radioactive or stable daughter nuclides. These changes are accurately described by the *radioactive decay laws.* Radioactive decay follows first order kinetics which cannot be changed by external influences, thus providing a very accurate clock. Of prime importance are the *decay constants* (or the *half-lives*) of the involved radionuclides.

Besides the radioactive time markers, four general methods of age determination can be distinguished: i) the *present day activity* of a radionuclide in a sample of interest is compared to the known *initial activity* of the same radionuclide; ii) a *radioactive parent* nuclide decays into a *stable daughter* nuclide; iii) a *radioactive daughter* is separated from a *radioactive parent nuclide* by natural chemical or physical processes; and iv) the *cumulative radiation damage* or the *total energy deposition* resulting from the decay of long-lived radionuclides present in or close to a sample.

If the decay constant of the radionuclide in method i) from above is known, the age of the sample can be calculated. In method ii), the age of the sample is calculated from the number of stable daughter atoms formed (which equal the number of decayed parent atoms) during the elapsed time since the formation of the sample (e.g., the time of solidification). This method applies also to the end members of the radioactive decay chains, if radioactive equilibrium is reached among *all* chain members, and *no loss or gain* of either parent or daughter occurs other than by radioactive decay. In method iii), the *decay of the separated daughter* or its *ingrowth* into a sample can be used to calculate the time span since the separation. In this case one speaks of *radioactive disequilibrium* between parent and daughter nuclide.

Nuclear dating methods have been discussed to some extent in most textbooks on nuclear and radiochemistry (e.g., Choppin and Rydberg [9]; Friedlander *et al.* [10]; Ehmann and Vance [11]; Lieser [12]). Comprehensive reviews of the subject have been published by Currie [13], Roth and Poty [2], Durrance [14] and Faure [15]. In the following, a personally biased selection of methods and applications is discussed.

II. Dating methods

1. Radioactive time markers

Fallout from atmospheric testing of very large nuclear devices (e.g., the largest H-bombs developed by the Soviet Union, 1961/62) has left discrete and well-conserved radioactive deposits in the environment. For some radionuclides, and under favorable environmental conditions these deposits remained essentially at the original site and can be used as radioactive time markers. For instance, ^{3}**H**, ^{14}**C**, the fission product ^{137}**Cs** or unreacted ^{239}**Pu** are firmly sorbed onto soils, sediments and other samples (e.g., glacier ice), and buried by soil forming processes, ongoing sedimentation, or precipitation (e.g., snow). These activities mark the year 1962/63 very accurately. More recently, the widespread fallout from the reactor accident at Chernobyl has formed (at least in Europe) a well defined marker of the year 1986. Because ^{137}Cs is easily detected by γ-ray spectroscopy, the two time markers (i.e., fallouts of bombs and from the Chernobyl reactor accident) covering a time interval of about 23 years, can be used in a simple way to determine the recent sedimentation rate in lakes and other aquatic systems, and the accumulation rate of glaciers. Fig. 1 presents a ^{137}Cs peak in sediments of Lake Zürich, Switzerland, together with a ^{210}Pb activity/depth profile (discussed in a later paragraph). The relatively narrow ^{137}Cs peak of the year 1962/63 allows assignment of an absolute time scale to the ^{210}Pb data.

Tritium (^{3}H) resulting from hydrogen bombs, has left a time marker in glacier ice (e.g., Gäggeler *et al.* [16]). Tritium can, furthermore, be used to estimate residence times of groundwater. If no ^{3}H is detected, the groundwater would have to have been isolated from surface waters for more than 30 years, i.e., before 1962/63. The relatively narrow time window for the input of bomb produced ^{14}C can be used in investigations of ocean mixing.

2. Comparison of present and initial radioactivity

In this simple case the radioactive decay law, $\mathbf{A_t} = \mathbf{A_0} \cdot \mathbf{e}^{-\lambda t}$, applies. (1)

$\mathbf{A_t}$, is the present-day activity, $\mathbf{A_0}$, the initial activity, and λ, the decay constant of the radionuclide of interest.

The number **N** of radioactive atoms is related to the activity by $\mathbf{A} = \lambda \cdot \mathbf{N}$. (2)

The range of the applicability of a radionuclide for dating purposes is very roughly 0.1 to 10 half-lives.

2.1 Cosmogenic radionuclides

The simple decay law applies very often to radionuclides that are produced in the atmosphere by interactions of highly energetic cosmic radiation. A selection of 'cosmogenic' radionuclides is presented and discussed in this paragraph.

Carbon-14 [$T_{1/2}$ = 5730 years]

The most important cosmogenic radionuclide is unquestionably ^{14}C. Its application for dating purposes was first proposed by Libby [7] in 1952, and has since found thousands of applications in archeology, geology, hydrology, glaciology and other fields. The radionuclide is produced at a practically constant rate in the natural atmosphere by interactions of cosmic rays with nitrogen, i.e., [^{14}N(n,p)^{14}C]. After production, the 'hot' ^{14}C atoms are rapidly oxidized to form $^{14}CO_2$, which then follows the natural CO_2-cycle in the environment. Like inactive CO_2, it participates in the biocycle, reaching steady state concentrations dictated by the living biomass. After the death of an organism the CO_2-exchange with the environment is interrupted, and ^{14}C starts to decay.

Calibrations of the method (Fig. 2) by dendrochronology (tree-ring counting) established a variation in the ^{14}C production rate of about 10 percent during the past 10,000 years (Damon *et al.* [17]; Stuiver [18, 19]). On a longer time scale, variations of the geomagnetic field (Bard *et al.* [20]) and of the intensity of solar modulation can influence the production rate by about a factor of two. In addition, anthropogenic activities have led to significant changes in the ^{14}C/C ratio: nuclear weapons tests in the atmosphere and nuclear industry have contributed considerably to the ^{14}C reservoir by artificial production of ^{14}C. On the other hand, the accelerated production of CO_2 in this century due to burning of fossil fuels (which contain no ^{14}C) has led to a reduction of the specific activity of ^{14}C. In addition, mass fractionation can occur in the transfer of carbon between different reservoirs. These variations have to be taken into account for accurate age determinations.

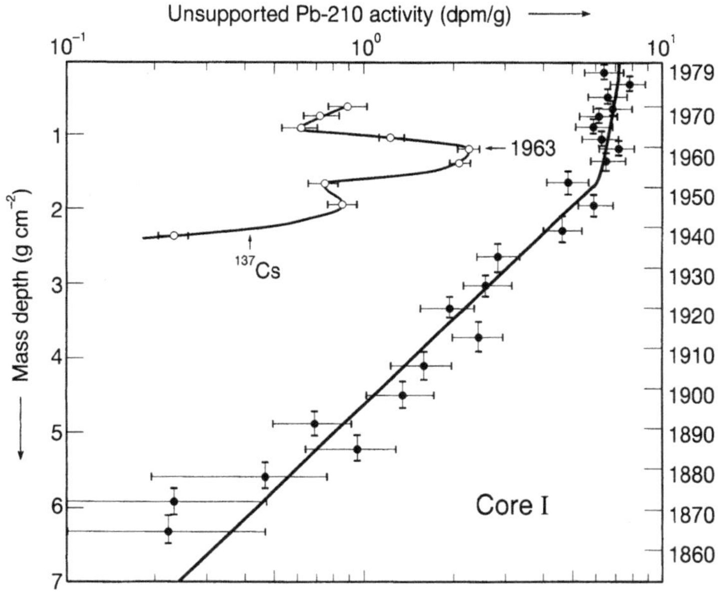

Fig. 1. Activities of 'unsupported' ^{210}Pb (see text) and ^{137}Cs from a sediment core of Lake Zürich, Switzerland (Erten *et al.* [49]). The ^{137}Cs peak of 1962/63 enables an absolute dating with ^{210}Pb. Reprinted by permission of Birkhäuser Verlag, Basel, Switzerland.

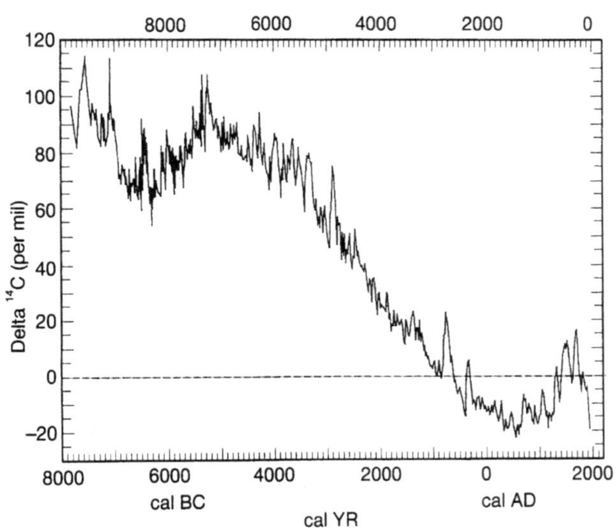

Fig. 2. Atmospheric $\Delta^{14}C$ during the past 10,000 years (Stuiver *et al.* [19]). $\Delta^{14}C$ is the difference (in permil) between the actual ^{14}C specific activity and the ^{14}C specific activity based on a constant production rate. Reprinted by permission of RADIOCARBON.

The sensitivity of the ^{14}C dating method has been increased considerably by the advent of *accelerator mass spectrometry (AMS)*. This method detects ^{14}C atoms directly, after ionization and acceleration in a tandem Van de Graaff (Fig. 3), and enables the analysis of much smaller specimen than by the classic activity measurements of ^{14}C. This technique has recently been reviewed by Finkel and Suter [21].

Important applications of ^{14}C dating are in archeology; Quaternary geology (e.g., lake sediments containing organic matter, Zbinden *et al.* [22]); atmospheric chemistry (Currie *et al.* [23]); glaciology; hydrology (e.g., Davis and Murphy [24]); oceanography; meteoritics (Jull *et al.* [25]); etc. For instance, ^{14}C was used to measure ages of seeds that were found at archeological sites, ages of paper (e.g., the Dead Sea scrolls showing an age of 1900 years), papyrus and textiles (e.g., the shroud of Turin (Damon *et al.* [26]), and to assess ocean mixing (Broecker and Peng [27]). A very important recent application of ^{14}C is the dating of ice cores, which can be used as archives of climatic history and anthropogenic pollution (e.g., Wilson and Donahue [28]).

Beryllium-10 [$T_{1/2} = 1.5 \times 10^6$ years] and beryllium-7 [$T_{1/2}$ = 53 days]

^{7}Be and ^{10}Be are mainly produced by interactions of high energetic cosmic rays with atmospheric nitrogen and oxygen. Beryllium is a very rare element and because of its lithophilic character, is rather immobile in the environment. The very low content of stable beryllium in the environment enables the measurement of radioactive beryllium isotopes at concentrations, that are not accessible for other radionuclides (because they are diluted by their stable isotopes). Whereas ^{7}Be can be used for measurements of short residence times and for mixing investigations in the lower atmosphere or in lakes, the very long half-life of ^{10}Be (Hofmann *et al.* [29]) has made possible long time investigations in the environment. ^{10}Be has found many useful applications which were reviewed, e.g., by Bourlès [30], general; Beer *et al.* [31], polar ice; Anderson *et al.* [32], sediments; Brown [33], soils; Nishiizumi [34, 35], meteorites and lunar rocks.

^{10}Be measurements in ice cores provide an important historical record for cosmic ray and solar activity. It has been shown (e.g., Beer *et al.* [31]) that the concentrations of ^{10}Be in Antarctic ice cores are anticorrelated with sunspot activity. This anticorrelation results from an association of sunspots with the solar magnetic activity which modulates the cosmic ray flux to the Earth's atmosphere. A good correlation has also

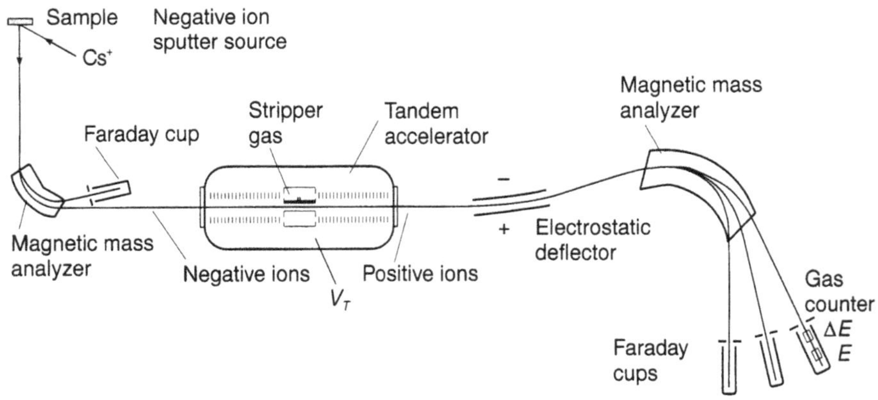

Fig. 3. Schematic diagram of an AMS-facility. Negative ions are produced from the investigated material in a Cs-sputter source. The ions are mass analyzed, and then accelerated in a tandem accelerator. In the 'stripper', the fast ions lose electrons and are converted to a positive charge state. In the second section of the tandem they are further accelerated. They are electrostatically deflected, mass analyzed and finally detected and identified by nuclear counter techniques. Courtesy M. Suter (ETH Zürich).

been observed for variations with a time period of centuries (Suess Wiggles) in the ^{10}Be concentrations and variations of ^{14}C activity in tree rings (Siegenthaler and Beer [36]).

Silicon-32 [$T_{1/2}$ = 133 years]

The production rate of ^{32}Si by cosmic ray spallation of Ar in the atmosphere is very small (~1.6 atoms/cm^2 and s) and its use as a dating nuclide is still under investigation. Its half-life was not well known until recently, but the latest measurements by two different methods (Hofmann *et al.* [37]; Chen *et al.* [38]) agree on a value of 133±9 years. This half-life would allow to date samples in a time span not accessible neither by ^{210}Pb (22 years) nor ^{14}C (5730 years). For dating with ^{32}Si by conventional counting techniques, ^{32}P (in radioactive equilibrium with its parent ^{32}Si) is used. However, ^{32}Si has to be extracted from very large samples (⩾ one ton of material; e.g., Gellerman *et al.* [39]), in order to obtain significant results. This restricts the spacial and time resolution. Presently, the use of accelerator mass spectrometry is under development (Zoppi *et al.* [40]) for ^{32}Si, but the method is still not fully exploited and its sensitivity is at present not better than that attainable using direct radioactive counting. A problem in the use of ^{32}Si as a dating nuclide lies in the ubiquitous occurrence of silicon in the environment leading to very low specific concentrations (i.e., ^{32}Si/Si$_{tot}$ $10^{-15}-10^{-17}$).

Additional important cosmogenic nuclides

Several other radionuclides are produced by the interaction of cosmic rays with constituents of the atmosphere, but the limited length of the present paper allows only a short enumeration.

Aluminium-26 [$T_{1/2}$ = 7.05×10^5 years] has become an important nuclide for investigations on marine archives, erosion studies of rocks, and lunar and meteorite samples. A review of ^{26}Al applications in the geosciences has been provided by Middleton and Klein [41].

Chlorine-36 [$T_{1/2}$ = 3.01×10^5 years] has found a number of applications in hydrology (Bentley *et al.* [42]), atmospheric sciences (e.g., Synal *et al.* [43]), geoscience (Gove [44]; Hallet and Putkonen [45]) and cosmology (e.g., Nishiizumi [34]).

Iodine-129 [$T_{1/2}$ = 1.6×10^7 years] may be used for research in hydrology and in extraterrestrial samples. A summary of the current ^{129}I research in the geosciences has been published by Fabryka-Martin *et al.* [46].

2.2 Decay products of radon

Radon-222 *[$T_{1/2}$ = 3.8 days]*, the direct decay product of ^{226}Ra emanates from the Earth's crust and enters into the atmosphere where it decays through a series of shorter lived products into ^{210}Pb. Several of these decay products may be used for residence time studies in the atmosphere, but the most important member in the ^{222}Rn decay chain is certainly ^{210}Pb.

Lead-210 [$T_{1/2}$ = 22.3 years]

^{210}Pb is deposited as wet or dry fallout onto the Earth's surface and can be used to obtain age information over the last 100 to 150 years. Goldberg [8], in 1963, was the first to use the radioactive decay of 'unsupported' ^{210}Pb to date glacier ice. Several years later, Krishnaswami *et al.* [47], applied this method to lake sediments.

A basic assumption of the ^{210}Pb dating method is a constant flux of ^{210}Pb to the environmental archives. A monitoring experiment, covering a time period of 11 years, has shown that this assumption is accurate, if average fluxes are computed for periods of the order of one year (von Gunten and Moser [48]). For shorter time periods (i.e., ≤ a few months), however, considerable and irregular variations in the ^{210}Pb flux are observed.

^{210}Pb, deposited on glaciers is covered by succeeding precipitations of snow, and that entering lake sediments, by settling particles. Finally, the deposits form a *closed system* (i.e., no exchange with the surrounding environment, e.g., no percolating water) in

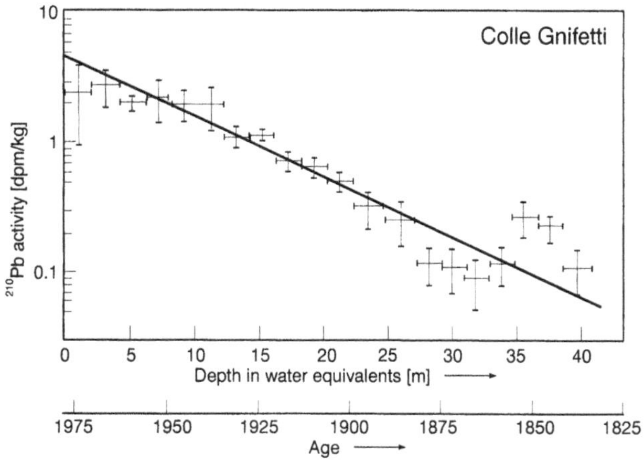

Fig. 4. ^{210}Pb activity versus depth measured in an ice core from Colle Gnifetti (Monte Rosa, 4450 m, Switzerland). The solid curve shows an exponential fit through data points averaged for core sections of 2-m water equivalent.

which ^{210}Pb decays with its half-life. The measured ^{210}Pb activity must be corrected for a background activity resulting from ^{210}Pb ('supported ^{210}Pb') directly produced by the decay of radium in the sample.

The literature on applications of the ^{210}Pb dating method is abundant. Some dating examples have been discussed by von Gunten and Moser [48]. Examples, for ^{210}Pb depth profiles in sediments of Lake Zürich [49] and in an ice core from Colle Gnifetti are presented in Figs. 1 and 4, respectively. Dating of sediments, or glacier ice is important because these materials contain conserved information on the concentrations of pollutants in the past. The half-life of ^{210}Pb allows an excellent archival record from the beginning of the industrial revolution in the past century to the present time.

3. Radioactive parent and stable decay product

In this method, the decay to a stable nuclide is considered. Knowing the present-day activity, the half-life of the radionuclide, and the concentration of the stable nuclide that has built-up by the decay of the radioactive parent nuclide, the age of the sample can be determined. This age corresponds for most of the samples to the time since their formation (e.g., crystallization from a melt or solution, etc.). Important examples for this dating possibility are the decays of ^{40}K into stable ^{40}Ar, and that of ^{87}Rb into stable ^{87}Sr, and the decay chains of ^{238}U, ^{235}U and ^{232}Th leading to the stable lead isotopes ^{206}Pb, ^{207}Pb and ^{208}Pb, respectively.

The number of parent atoms at the time t of the measurement, $\mathbf{P}_t$, is related to the number of parent atoms $\mathbf{P}_0$ at time zero (which is assumed to be the time of the formation of a mineral), by

$$\mathbf{P}_t = \mathbf{P}_0 \cdot \mathbf{e}^{-\lambda t}, \tag{3}$$

$\mathbf{S}_t$, the number of stable daughter atoms at time t is represented by

$$\mathbf{S}_t = \mathbf{S}_0 + (\mathbf{P}_0 - \mathbf{P}_t) = \mathbf{S}_0 + \mathbf{P}_t(\mathbf{e}^{\lambda t} - 1), \tag{4}$$

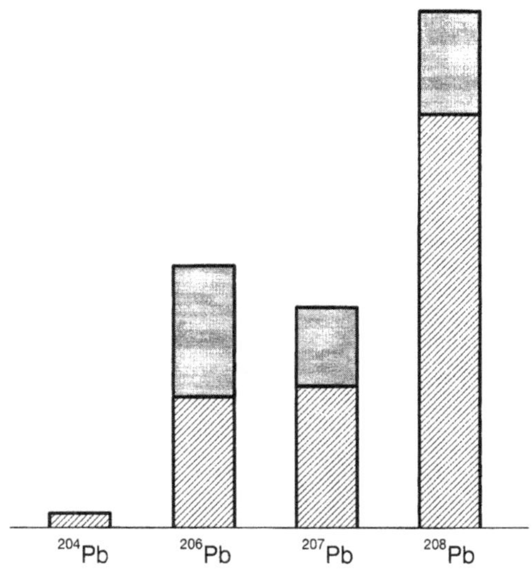

Fig. 5. Mean isotopic composition of present-day 'common' lead (sum of primordial and common lead). ▨ primordial lead, ▩ radiogenic lead. After Albarède *et al.* [50]. Reprinted by permission of Kluwer Academic Publishers.

where $\mathbf{S}_0$ is the number of atoms of a stable daughter present in the sample at $t = 0$.

This equation is only valid if *no loss or addition* (other than by the radioactive transformation) of either parent or daughter nuclide occurred during the time period t. This condition is fulfilled if the sample represents a **closed system** (i.e., is completely isolated from the surrounding environment).

3.1 Uranium-Thorium-Lead Chronometers

The U-Th-Pb decay chains are probably the most accurate possibilities for measurements of geological samples with ages >30 million years. A comprehensive discussion of this method has been given by Albarède *et al.* [50]. The stable end products in the decay chains of ^{238}**U** *($T_{1/2}$ = 4.468 × 10^9 years)*, ^{235}**U** *($T_{1/2}$ = 7.038 × 10^8 years)*, and ^{232}**Th** *($T_{1/2}$ = 1.405 × 10^{10} years)* are the lead isotopes ^{206}**Pb,** ^{207}**Pb** and ^{208}**Pb**, respectively. The abundance of these lead isotopes in a sample can be used to determine the time since the formation of the sample as a closed system. In modern methods, the lead isotopes are always determined by thermal ionisation mass spectrometry, and the amount of the non-radiogenic ^{204}**Pb** can be used to estimate and correct the concentrations of the other lead isotopes that were in the sample at time $t = 0$. The composition of common lead is shown in Fig. 5.

The **age** based on the decay of ^{238}**U** is obtained by transformation of Eq. (4):

$$t = 1/\lambda_{238} \ln\{1 + [(^{206}\mathbf{Pb}_t - {}^{206}\mathbf{Pb}_0)/{}^{238}\mathbf{U}_t]\}\,. \tag{5}$$

Similar equations apply for ^{235}**U**/^{207}**Pb** and ^{232}**Th**/^{208}**Pb**. Comparison of different dating methods can be used for the verification of the closed system condition (i.e., **concordia**, Wetherill [51]). The curve in Fig. 6 represents data that are in agreement for the ^{238}U/^{206}Pb and the ^{235}U/^{207}Pb systems under the concordia for-

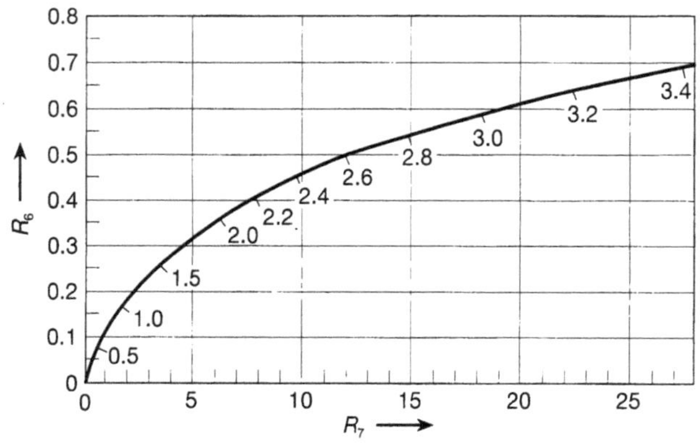

Fig. 6. Concordia diagram for the U/Pb system, where:

$$R_7 = \frac{(^{207}Pb/^{204}Pb)_t - (^{207}Pb/^{204}Pb)_0}{^{235}U/^{204}Pb},$$

$$R_6 = \frac{(^{206}Pb/^{204}Pb)_t - (^{206}Pb/^{204}Pb)_0}{^{238}U/^{204}Pb}.$$

The curve represents data that agree for the $^{238}U/^{206}Pb$ and the $^{235}U/^{207}Pb$ systems, i.e., no losses or gains of lead and uranium occurred since the formation of the samples. The numbers on the concordia curve indicate the age of the samples in 10^9 years. After Friedlander *et al.* [10], p. 485. Copyright 1981. Reprinted by permission of John Wiley and Sons.

malism. Values below the concordia curve indicate a loss of lead or a gain of uranium, and values above the curve, a gain of lead or a loss of uranium.

Ages can also be determined from the ratio of $\mathbf{^{206}Pb/^{207}Pb}$, if the sample contains no non-radiogenic lead (indicated by the absence of ^{204}Pb), using the following equations:

$$\mathbf{^{206}Pb = {}^{238}U(e^{\lambda_{238}t} - 1)}, \tag{6}$$

$$\mathbf{^{207}Pb = {}^{235}U(e^{\lambda_{235}t} - 1)}. \tag{7}$$

Combining Eqs. (6) and (7)

$$\mathbf{^{206}Pb/^{207}Pb = {}^{238}U(e^{\lambda_{238}t} - 1)/{}^{235}U(e^{\lambda_{235}t} - 1)}, \tag{8}$$

where $(\mathbf{^{238}U/^{235}U}) = \mathbf{137.8}$ is the present isotopic ratio.

The variation of the $^{206}Pb/^{207}Pb$ ratio with time is shown in Fig. 7.

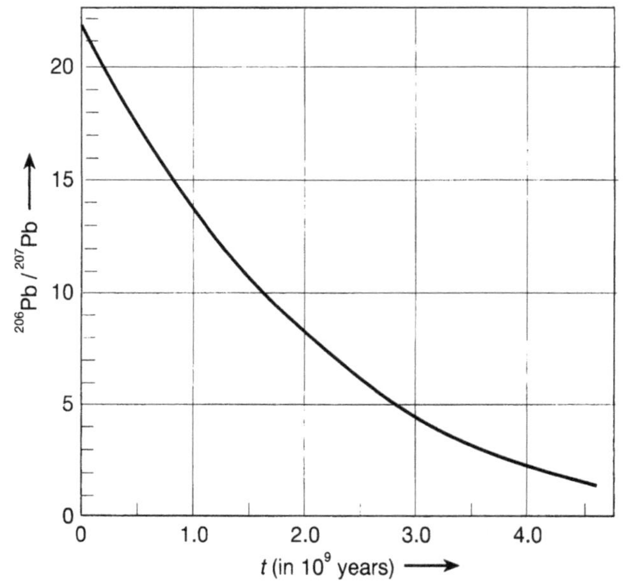

Fig. 7. Variation of radiogenic $^{206}Pb/^{207}Pb$ with age. After Friedlander *et al.* [10], p. 486. Copyright 1981. Reprinted by permission of John Wiley and Sons.

3.2 Rubidium-Strontium dating [$T_{1/2}$ $^{87}Rb = 4.8 \times 10^{10}$ years]

^{87}Rb decays by β-emission into ^{87}Sr. A detailed description of the rubidium-strontium method has been published by Bernard-Griffiths [52]. This very accurate dating method is based on a mass spectrometric measurement of ^{87}Rb and of the $^{87}Sr/^{86}Sr$ ratio, and a conventional analysis of the elemental concentrations of rubidium and strontium in the sample.

The age of a sample can then be calculated from:

$$\mathbf{(^{87}Sr/^{86}Sr)_t = (^{87}Sr/^{86}Sr)_0 + (^{87}Rb/^{86}Sr)_t(e^{\lambda t} - 1)}, \tag{9}$$

where the subscripts t and 0 refer to the present and initial values, respectively. For the stable, nonradiogenic ^{86}Sr, the present and the initial values are equal. The initial $(^{87}Sr/^{86}Sr)_0$ ratio can be obtained from the intercept of the straight line with the ordinate in an *isochrone presentation* of a group of samples (Fig. 8, Moorbath *et al.* [53]). The slope of the straight isochrone line corresponds to the age of a group of samples.

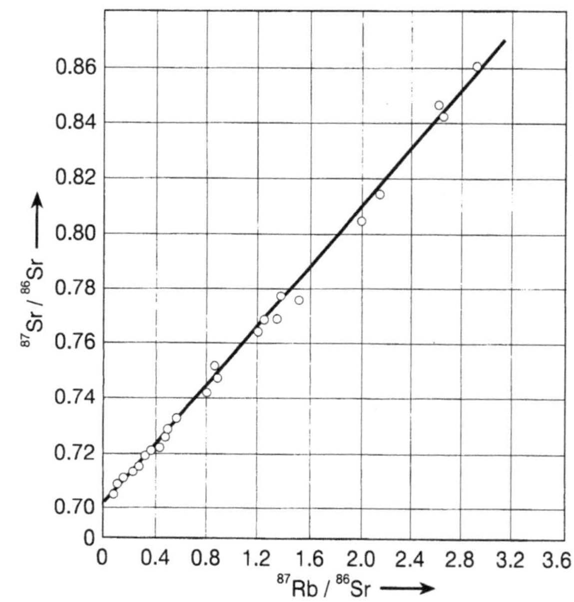

Fig. 8. Rb/Sr isochron for a gneiss sample from Greenland [53]. The slope of the isochron shows an age of 3660±90 MY. The intercept with the ordinate indicates the $^{87}Sr/^{86}Sr$ ratio at t = 0. Reprinted with permission from NATURE. Copyright 1972, MacMillan Magazines Limited, and S. Moorbath.

3.3 Potassium-Argon dating [$T_{1/2}$ $^{40}K = 1.28 \times 10^9$ years]

Potassium-argon dating methods have been reviewed by Montigny [54] and Malusky [55]. The 10.7% decay branch of ^{40}K to stable ^{40}Ar can be used to date geological, meteoritic and lunar samples if the systems are closed with respect to potassium and argon. The following equation is used, in analogy to the rubidium-strontium couple:

$$(^{40}Ar/^{36}Ar)_t = (^{40}Ar/^{36}Ar)_0 + f(^{40}K/^{36}Ar)_t(e^{\lambda t} - 1), \quad (10)$$

where **f** represents the branching ratio (EC/β^-) in the decay of ^{40}K. ^{36}Ar is (like ^{86}Sr) non-radiogenic and its abundance does not change with time. As for rubidium-strontium, isochron plots of $^{40}Ar/^{36}Ar$ versus $^{40}K/^{36}Ar$ will give a straight line for a group of related samples. The slope of the isochron represents the age of the samples, and the intercept with the ordinate the value of the initial $(^{40}Ar/^{36}Ar)_0$ ratio.

In recent years a *modification of the potassium-argon dating method* has been introduced in many laboratories (Malusky [55]). It involves irradiating the sample with a flux of fast neutrons to produce ^{39}Ar from ^{39}K (main K isotope) in a (n,p) reaction. The amount of ^{39}Ar produced is proportional to the potassium concentration in the sample so that one need only measure the $\mathbf{^{40}Ar/^{39}Ar}$ ratio. The main advantage of this modified method is that reliable ages can be obtained even when diffusion losses have occurred. The $^{40}Ar/^{39}Ar$ ratios can be measured with a mass spectrometer during a stepwise heating procedure. The $^{40}Ar/^{39}Ar$ will be low in those fractions of the sample where a loss of Ar has occurred (^{39}Ar is a measure of K). A plateau value for the $^{40}Ar/^{39}Ar$ ratio in the heating procedure indicates a region with no loss of Ar, and, hence, a reliable age of the sample.

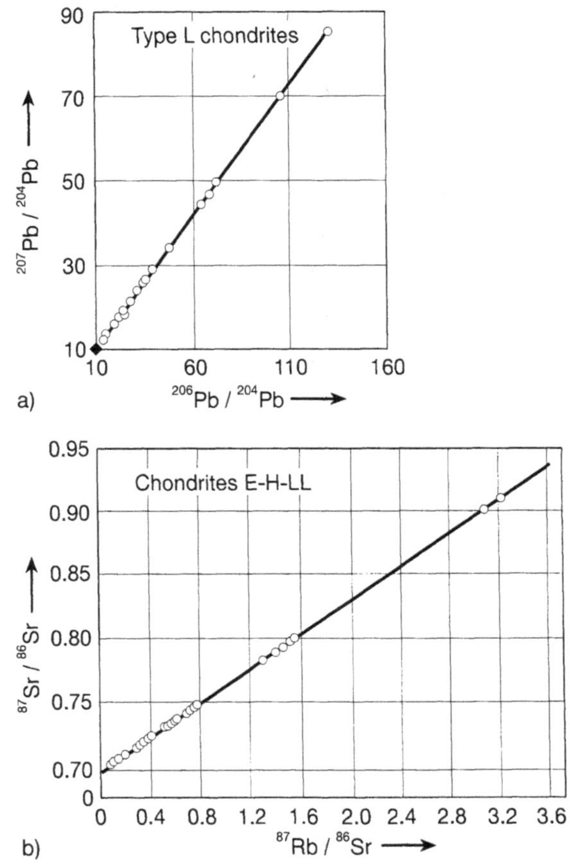

Fig. 9. (a) $^{207}Pb/^{204}Pb$ isochron of meteorites of the L type (○) and a troilite from Canion Diablo (◆) showing an age of $(4.555 \pm 0.014) \times 10^9$ years (after G. Manhes, Thèse, Univ. de Paris VII, 1982). (b) Rb/Sr isochron of E, H, and LL type chondrites with an age of $(4,504 \pm 0.015) \times 10^9$ years and an initial ratio $^{87}Sr/^{86}Sr_0 = 0.69885 \pm 0.00010$ (after J. F. Minster, Thèse, Univ. de Paris VII, 1979). Reprinted by permission of Kluwer Academic Publishers.

3.4 Other dating techniques

Principally, the decay of any long-lived natural radionuclide can be used for dating purposes. Based on the principles described above, the following parent-daughter systems have found many applications and are used extensively by several highly specialized laboratories:

Samarium-neodymium (Bernard-Griffiths and Gruau [56]), *rhenium-osmium* (Luck [57]), and *lutetium-hafnium* (Rees [58]).

3.5 The age of the Earth and of the solar system

The radionuclides discussed in this section, and, when possible, combinations of several methods, enable a determination of the age of the Earth and meteorites and an estimate of the age of the solar system. These dating methods have helped considerably to increase our knowledge of the development and history of geological, lunar and meteoritic materials (e.g., Wetherill [59]; Faure [15]).

The method of isochrones, as discussed above for rubidium-strontium and potassium-argon can be extended to the uranium lead system, according to:

$$\frac{(^{207}Pb/^{204}Pb)_t - (^{207}Pb/^{204}Pb)_0}{(^{206}Pb/^{204}Pb)_t - (^{206}Pb/^{204}Pb)_0} = \frac{^{235}U\,(e^{\lambda_{235}t} - 1)}{^{238}U\,(e^{\lambda_{238}t} - 1)}. \quad (11)$$

Using isochrones of uranium-lead and rubidium-strontium, the age of *meteorites* has been measured very precisely (Fig. 9). These ages fall in a very narrow range of $(4.54 \pm 0.02) \times 10^9$ years (Albarède [60]), strongly supporting the hypothesis that meteorites were formed simultaneously from a homogeneous cloud of gas and dust.

Patterson [61] found that terrestrial lead from ocean sediments with $^{206}Pb/^{204}Pb$ and $^{207}Pb/^{204}Pb$ ratios of 19.0 and 15.8, respectively, representing a well-mixed sample, matched the isochrone of the meteoritic lead samples (e.g., that of Fig. 9a). Again, this strongly supports the hypothesis that the *Earth* was formed as an isolated body at the same time as the *meteorites*, i.e., 4540 MY years ago, or, as suggested by Albarède [60] and others, about 60 MY later. The oldest rocks on Earth have ages between 3500–3700 MY (e.g., Fig. 8, Moorbath [53]). Some Australian zircons may even be older, i.e., 4100–4200 MY (Moorbath [62]).

Ages of lunar samples show a considerable variation, depending on their origin. This variation is caused by the cratering activity during the intensive meteoritic bombardment of the lunar surface in the early history of the Moon. However, the oldest samples, recovered from the lunar highlands by the Apollo-17 astronauts, are about 4500 MY old (e.g., Taylor [63]), and represent the *age* of the *Moon*. Samples from the mare regions, on the other hand, are younger, about 3500 MY old (Dalrymple [64]; Swindle *et al.* [65]). This shows that the *maria* (the dark basins on the Moon, that are visible from Earth) were formed later. Large meteoritic impacts on the Moon probably triggered the outflow of lava streams from the molten interior. A chemical fractionation of parent and daughter nuclides in the lava melt then set back the radioactive clock. An example for a lunar site which was formed by an impact event is the crater Copernicus. Its age was recently redetermined to 800 MY (Bogard *et al.* [66]). Dating of lunar samples was a very important mean for a reconstruction of the history of the Moon.

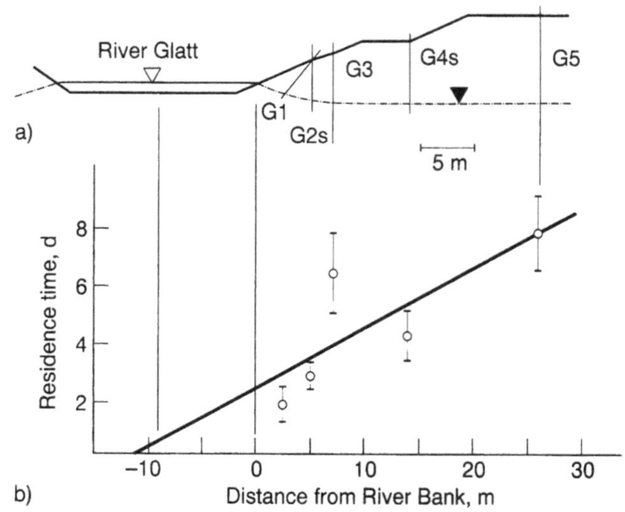

Fig. 10. Dating with radon: (a) schematic cross section of the River Glatt site, Switzerland, and (b) linear regression line through data points corresponding to a mean groundwater flow velocity of 5 m per day, after [69]. Courtesy American Geophysical Union.

4. Radioactive disequilibrium

Many aspects of uranium-series disequilibrium have been discussed in detail in the recent book by Ivanovich and Harmon [67] including the application of radioactive disequilibria for dating purposes (Ivanovich *et al.* [68]). In the natural decay chains of uranium and thorium, the secular radioactive equilibrium is often externally disturbed, i.e., the radioactive decay chain is interrupted at a certain chain member. In this case, the decay of a radionuclide that was separated by chemical or physical means from its precursor (*daughter-excess method*), or the ingrowth of a radioactive daughter product, which was separated before, into a parent sample (*daughter-deficiency method*) can be used for dating purposes. A separation of parent and daughter nuclide can, for instance, occur during sedimentation processes, and, thus, initiate a radioactive disequilibrium. If the separated sediments become a closed system, it is possible to determine the date of the separation event. Important examples of this type are disequilibria between ^{230}**Th** and ^{234}**U**, or between ^{231}**Pa** and ^{235}**U**.

4.1 Dating methods based on the accumulation of a radioactive daughter

Determination of groundwater residence times with radon-222 [$T_{1/2}$ = 3,825 days]

In geological regions containing unconsolidated sediments (e.g., glaciofluvial deposits), surface water may infiltrate and become shallow groundwater. Hoehn and von Gunten [69] and Hoehn *et al.* [70] have used ^{222}Rn to estimate the residence time and the related flow velocity in groundwater reservoirs of Switzerland. In surface waters, the radioactive equilibrium between ^{226}Ra and ^{222}Rn is strongly perturbed, because the water is only in contact with the uppermost layers of sediments and outgases to the atmosphere. This leads to very low radon concentrations in surface waters. When this radon-depleted surface water infiltrates into an aquifer, it picks up more and more radon (emanating from aquifer grains) which, finally, at a certain distance from the point of infiltration, reaches a steady state activity. Measurements of this steady state activity, $\mathbf{A}_\infty$, and activities, $\mathbf{A}_t$, along the flow path, enable a calculation of the groundwater residence time and its related flow velocity, using the radioactive decay law (Fig. 10), e.g.,

$$\mathbf{A}_t = \mathbf{A}_\infty(\mathbf{1} - \mathbf{e}^{-\lambda t}). \quad (12)$$

This method has also been used to obtain information about residence times in artificial bank filtration in the purification process of drinking water (Schulte-Ebbert *et al.* [71]). The described radon method allows, under favorable conditions, to assess groundwater residence times of about 15 days.

The $^{230}Th/^{238}U$, and the $^{231}Pa/^{235}U$ methods [$T_{1/2}$ ^{230}Th = 7.54×10^4 years, $T_{1/2}$ ^{231}Pa = 3.276×10^4 years]

The ^{230}Th/^{234,238}U method is based on the decay of ^{238}U (*α, $T_{1/2}$ = 4.468×10^9 years*) through ^{234}Th (*β⁻, 24.1 days*), ^{234m,234}Pa (*β⁻, 1.2 min and 6.7 h*) and ^{234}U (*α, 2.446×10^5 years*) into ^{230}Th. ^{234}Th and ^{234m,234}Pa always reach secular equilibrium with ^{238}U. If secular radioactive equilibrium also exists between ^{238}U and ^{234}U, the age, ***t***, of a sample is calculated according to:

$$^{230}\mathbf{Th}/^{238}\mathbf{U} = \mathbf{1} - \mathbf{e}^{-\lambda t}. \quad (13)$$

However, the ratio of ^{234}**U**/^{238}**U** is often not unity, and therefore, taking into account this additional disequilibrium, the correct equation becomes more complicated than Eq. (13). The accumulation (ingrowth) of ^{230}Th from ^{238}U can be used to date marine and terrestrial phosphorites (fluorapatite) and carbonates (e.g., fossil corals, shells, speleothems, bones, travertines) and volcanic rocks with ages of $\leq$350,000 years. These applications have been reviewed by Burnett and Veeh [72], and Schwarcz and Blackwell [73]. For instance, sedimentary apatite contains relatively high concentrations of uranium, but generally low contents of thorium. If the apatite-inclusions in phosphorites acquired mainly uranium and, any or very little ^{230}Th (or ^{231}Pa) during mineral formation, and if the time of the mineral formation is short as compared to the half-lives of these radionuclides, and the formed minerals remained as a closed system, the age of formation of the minerals can be determined accurately. These assumptions seem to be valid for marine phosphorites. Marine carbonates (e.g., reef-building corals) contain, at the time of formation, several parts per million (ppm) of uranium, but essentially no ^{230}Th. Samples from drill cores showed that ^{230}Th increased in stratigraphically older corals, and finally reached equilibrium values. Calcium carbonate precipitation occurs usually from saturated solutions that contain a few ppm of uranium as carbonate, phosphate or humic acid complexes, but, due to its low solubility, essentially no thorium. The resulting closed calcite formations contain then a few ppm uranium, but no ^{230}Th, and are ideally suited for dating samples which are related to archeological and or Quaternary geological problems.

The ^{231}Pa/^{235}U method is based on the decay of ^{235}U through ^{231}Th (β^-, *25.5 h*) into ^{231}Pa. Again, the half-life of ^{231}Th is short enough to be always in secular equilibrium with ^{235}U, and the age of a sample can be calculated:

$$^{231}\mathbf{Pa}/^{235}\mathbf{U} = \mathbf{1} - \mathbf{e}^{-\lambda t}. \tag{14}$$

The applications of the ^{231}Pa/^{235}U method are similar as for ^{230}Th/^{238}U, however, the accessible time is shorter, $\leq$150,000 years.

4.2 Dating methods based on the decay of separated (excess) uranium daughter products

In the hydrological cycle a natural chemical fractionation occurs due to different solubilities of uranium-series daughter nuclides from their parents. The insoluble radioactive daughter nuclides (i.e., ^{230}Th and ^{231}Pa) accumulate in bottom sediments of oceans and lakes, whereas uranium remains in solution as stable uranyl carbonate complexes. The excess of the separated radionuclide decays according to the decay constant of the radionuclide. The age or the accumulation rate of a deposit can then be determined by the general radioactive decay law.

A review of these methods, in relation to marine sediments and sedimentation processes, has been published by Huh and Kadko [74]. The half-lives of ^{230}Th and ^{231}Pa are very favorable to date sedimentation rates of deep-sea sediments which accumulate by about 1 to 30 mm in 1000 years. Both nuclides adsorb strongly to particles and accumulate in marine sediments, where they are practically unsupported by their parents. After burial by ongoing sedimentation, the unsupported nuclides decay with depths (time) according to their half-lives. Assumptions for the validity of the method are a constant production in sea water, and a constant sedimentation rate. They were found valid in many investigations.

5. Fission tracks

The method of fission track dating has been reviewed by Carpéna and Mailhé [75]. For details, the reader is referred to this publication. Here, only the basic principles are summarized. The energetic fragments resulting from the spontaneous fission of ^{238}U produce micro-damages in solid materials which can be enlarged by chemical etching to be visible under a light microscope. The density of tracks (number of tracks per surface unity) is a function of the age of the sample (the duration of the spontaneous fission process), and its concentration of uranium, if no physical (e.g., heating) and chemical changes occurred in the minerals during this time period. For an age determination it is necessary to measure the fission track density, and the uranium content. Usually, this is done in two steps: i) the tracks that were naturally produced in minerals and glasses by ^{238}U spontaneous fission are counted; ii) the uranium concentration is determined from the tracks formed by ^{235}U-fission during a reactor irradiation in a well known thermal neutron flux. The age of the sample is calculated from:

$$t = \mathbf{1}/\lambda_{238} \, \mathbf{log}\{\mathbf{1}+(\mathbf{D}_{235}/\mathbf{D}_{238}) \cdot (\lambda_{238}/\lambda_{SF}) \cdot \sigma \cdot \mathbf{I} \cdot \phi\}, \tag{15}$$

where λ_{238} and λ_{SF}, decay constant of ^{238}U and partial decay constant for spontaneous fission of ^{238}U, respectively,
$\mathbf{D}_{238}$ and $\mathbf{D}_{235}$, natural fission tracks (^{238}U) per cm^2 and tracks produced in the thermal-neutron fission of ^{235}U per cm^2, respectively,
I, isotope ratio of natural uranium, i.e., 137.8,
σ, fission cross-section of ^{235}U, and ϕ, reactor neutron flux per cm^2 and s.

There exists still an uncertainty of about 15% among different authors for the value of the decay constant for spontaneous fission of ^{238}U. An other problem inherent to this method is, that the fission tracks may disappear ('anneal') when the samples are subjected to elevated temperatures. The annealing temperatures depend very much on the investigated minerals. For instance, in an autunite the fission tracks disappeared completely already at 60 °C (Fleischer and Price [76]). On the other hand, the annealing of fission tracks may be directly used to obtain informations on the temperature history of a sample, if its age is known from other methods. Sometimes it is also difficult to distinguish fission tracks from other damages in the samples. Furthermore, fission tracks may overlap, be hidden and lost in counting. Due to these problems, the method of fission track dating, is, in the opinion of this author, less reliable and less accurate than other dating methods, like for instance the uranium-lead method. An advantage of the fission track chronometer is that it is less affected than other methods by chemical changes in the minerals. This is due to the fact that the fission fragments produce relatively stable physical damages in the minerals or glasses, and not chemically mobile radionuclides that can migrate out of the samples.

6. Thermoluminescence

Many electrically non-conducting minerals, that were exposed to artificial or natural irradiation, emit light when they are heated. This phenomenon, called thermoluminescence, has been observed already in 1663 by Boyle, when he heated a diamond. A comprehensive review of this method has been given by Lalou and Valladas [77].

In this method, as in the fission track method, neither the decrease of the number of radioactive nuclides with time nor the transformation of a parent nuclide into a daughter product is used to obtain the dating information. The age of a sample is deduced from the effects of the radiation energy that is deposited and has accumulated in the sample since its formation. Due to the long half-lives of the main radionuclides involved, the radiation dose rate can be assumed as constant. In nature, minerals of quartz, feldspar, micas, calcite, etc., are continuously irradiated by α-, β-, and γ-radiation resulting from the decay of members of the natural uranium and thorium decay chains, or from the decay of ^{40}K. Hereby, atoms are ionized and the electrons are trapped in crystal defects, from where they cannot escape without an external excitation. In the thermoluminescence method the external excitation is promoted by heating. Part of the trapped electrons combine with holes, others form luminous centers. With increasing temperature, one may observe several light signals. First, the least stable trapped electrons are released, and at the highest temperature, the most stable ones. Thus, a thermoluminescence curve is obtained which is characteristic for a sample and its irradiation history.

If a sample is heated for a second time, no further light signals are observed. However, if such annealed crystals are exposed to a radiation source, a new thermoluminescence curve is obtained which differs from the first one by showing peaks at lower temperatures (i.e., above 100 °C), in addition to the signals at higher temperature (230−400 °C) which were already observed during the first heating. For applications, it is advisable to use the signals emitted at high temperature. Artificial irradiation experiments with increasing radiation doses have shown that the high temperature thermoluminescence is proportional to the irradiation dose of the material. If the investigated sample remained in the same environment during its whole history, then the irradiation dose rate can be assumed as constant, and it is in principal possible to evaluate the age of the sample from the magnitude of the high temperature thermoluminescence peak. The age of a sample can be calculated from the following relation, if the annual irradiation dose is exactly known:

$$\textbf{age = natural dose/annual dose}\,. \tag{16}$$

The natural dose is the total radioactive decay energy acquired per unit mass since the formation (e.g., crystallization) of a sample. The annual dose is the energy corresponding to the radioactive decay of the nuclides in the sample. This emitted decay energy is not related in a simple way to the energy acquired by the sample. By far the largest amount of energy is contributed by α-decay of the uranium and thorium decay families. But the range of α-particles in solid matter is only 20−50 µm, depending on their energy and the density of the material. On the other hand, the ranges of β-particles in solids are of the order of a mm, and γ-rays travel much farther and have only a few interactions when traversing the solid material. It is, therefore, necessary to establish efficiencies for the energy deposition for each type of radiation and material. Furthermore, corrections for inhomogeneities in the samples have to be made. The range of linearity between the dose and the thermoluminescence signal is limited; at higher doses a saturation effect is observed. For quartz, saturation occurs at 100−500 Gy (10−50 krad), whereas calcite can accept 3000 Gy (300 krads).

Despite these problems (for a more comprehensive review, see the paper by Lalou and Valladas [77]), thermoluminescence has been successfully used to date archaeological (e.g., ceramic artifacts, burned stones used as kitchen utensils, stone tools, etc.), and geological samples (e.g., recent volcanic rocks, young calcite formations in caves, very fine aeolien deposit in which the thermoluminescence was annealed by exposure to bright sunlight when they resided on the surface, etc.).

Thermoluminescence is certainly related to the age of the samples. However, its application is not as straightforward as in many of the other dating methods, and practically each application has its specific and inherent problems. The method should not be applied blindly. But if all the problems are treated properly, it can be used to assess a time-range which falls between that of the applicability of ^{14}C and the lower limit of the potassium-argon method.

Conclusions

The discovery of the radioactivity in 1896, and the laws of the radioactive decay have opened the possibility to date accurately many objects in our environment. Sensitive instruments and methods, and an impressive number of radionuclides enabled to determine the age of archaeological, geological, cosmological and other samples. These methods have greatly helped to improve our knowledge about archeology and historical developments, and our insight into the evolution of the solar system, and among it, of the Earth and the Moon. The radioactive dating methods are in principle straightforward, quite easy to apply, and accurate, but may lead to erroneous results if the physical and chemical background of the investigated system is not thoroughly understood. In order to avoid misinterpretations and systematic errors it is advisable, whenever possible, to verify the age of a sample by more than one method.

Acknowledgement

The author thanks J. P. Adloff and an anonymous reviewer for constructive comments, and U. Krähenbühl and K. Tabor for improvements.

References

1. Roth, E.: Dating methods using natural radioactive phenomena: principles and general aspects. *Nuclear Methods of Dating* (E. Roth and B. Poty, Eds.), pp. 1–34, Kluwer Academic Publishers, 1989.
2. Roth, E., Poty, B. (Eds.): *Nuclear Methods of Dating.* Kluwer Academic Publishers, 1989.
3. Nier, A. O.: Variations in the relative abundances of the isotopes of common lead from various sources. J. Am. Chem. Soc. **60**, 1571–1576 (1938).
4. Nier, A. O., Thompson, R. W., Murphy, B. F.: The isotopic constitution of lead and the measurement of geologic time III. Phys. Rev. **60**, 112–116 (1941).
5. Holmes, A.: An estimate of the age of the Earth. Nature **157**, 680–684 (1946).
6. Houtermans, F. G.: Die Isotopen-Häufigkeiten im natürlichen Blei und das Alter des Urans. Naturwissenschaften **33**, 185–186 (1946).
7. Libby, W. F.: *Radiocarbon Dating,* University of Chicago Press 1952.
8. Goldberg, E. D.: Geochronology with lead-210. *Symp. Radioactive Dating*, pp. 121–131. Internat. Atomic Energy Agency, IAEA, Vienna 1963.
9. Choppin, G. R., Rydberg, J.: *Nuclear Chemistry, Theory and Applications.* Pergamon Press, 1980.
10. Friedlander, G., Kennedy, J. W., Macias, E. S., Miller, J. M.: *Nuclear and Radiochemistry,* 3rd ed. Wiley-Interscience, 1981.
11. Ehmann, D. E., Vance, D. E.: *Radiochemistry and Nuclear Methods of Analysis.* Wiley, 1991.
12. Lieser, K. H.: *Einführung in die Kernchemie.* 3. Aufl. Verlag Chemie, 1991.
13. Currie, L. A. (Ed.): *Nuclear and Chemical Dating Techniques. Interpreting the Environmental Record.* ACS Symposium Series **176**, American Chemical Society, 1982.
14. Durrance, E. M.: *Radioactivity in Geology, Principles and Applications.* Ellis Horwood, 1986.
15. Faure, G.: *Principles of Isotope Geology,* 2nd Ed. Wiley, 1986.
16. Gäggeler, H., von Gunten, H. R., Rössler, E., Oeschger, H., Schotterer, U.: ^{210}Pb-dating of alpine firn/ice cores from Colle Gnifetti, Switzerland. J. Glaciology **29**, 165–177 (1983).
17. Damon, P. E., Lerman, J. C., Long, A.: Temporal functions of ^{14}C: Causal factors and Implications. Ann. Rev. Earth Planet. Sci. **6**, 457–494 (1978).
18. Stuiver, M., Kromer, B., Becker, B., Ferguson, C. W.: Radiocarbon age calibration back to 13,300 years BP and the ^{14}C age matching of the German oak and US bristlecone pine chronologies. Radiocarbon **28**, 969–979 (1986).
19. Stuiver, M., Pearson, G. W., Braziunas, T.: Radiocarbon age calibration of marine samples back to 9000 cal yr BP. Radiocarbon **28**, 980–1021 (1986).
20. Bard, E., Hamelin, B., Fairbanks, R. G., Zindler, A.: Calibration of the ^{14}C time scales over the past 30,000 years using mass spectrometric U-Th ages from Barbados corals. Nature **345**, 405–410 (1990).
21. Finkel, R. C., Suter, M.: AMS in the Earth Sciences: Techniques and Applications. Adv. Anal. Geochem. **1**, 1–114 (1993).
22. Zbinden, H., Andree, M., Oeschger, H., Ammann, B., Lotter, A., Bonani, G., Wölfli, W.: Atmospheric radiocarbon at the end of the last glacial: An estimate based on AMS radiocarbon dates on terrestrial macrofossils from lake sediments. Radiocarbon **31**, 795–804 (1989).
23. Currie, L. A., Fletcher, R. A., Klouda, G. A.: On the identification of carbonaceous aerosols via ^{14}C accelerator mass spectrometry and laser microprobe mass spectrometry. Nucl. Instr. Meth. **B29**, 346–354 (1987).
24. Davis, S. N., Murphy, E.: *Dating Ground Water and the Evaluation of Repositories for Radioactive Waste.* NUREG/CR-4912. U.S. Nuclear Regulatory Commission, 1987.
25. Jull, A. J. T., Donahue, D. J., Linick, T. W.: ^{14}C activities in recently fallen meteorites and Antarctic meteorites. Geochim. Cosmochim. Acta **53**, 2095–2100 (1989).
26. Damon, P. E., Donahue, D. J., Gore, B. H., Hatheway, A. L., Jull, A. J. T., Linick, T. W., Sercel, P. J., Toolin, L. J., Bronk, C. R., Hall, E. T., Hedges, R. E. M., Housley, R., Law, I. A., Perry, C., Bonani, G., Trumbore, S., Wölfli, W., Ambers, J. C., Bowman, S. G. E., Leese, M. N., Tite, M. S.: Radiocarbon dating of the shroud of Turin. Nature **337**, 611–616 (1989).
27. Broecker, W. S., Peng, T.-H.: *Tracers in the Sea.* ELDIGIO Press, Palisades, N.Y., 1982.
28. Wilson, A. T., Donahue, D. J.: AMS carbon-14 dating of ice-progress and future prospects. Nucl. Instr. Meth. **AMS 5** (1990).
29. Hofmann, H. J., Beer, J., Bonani, G., von Gunten, H. R., Raman, S., Suter, M., Walker, R. L., Wölfli, W., Zimmermann, D.: ^{10}Be: Half-life and AMS-standards. Nucl. Instr. Meth. **B29**, 32–36 (1987).
30. Bourlès, D. L.: Beryllium isotopes in the Earth's environment. In: *Encyclopedia of Earth System Science,* pp. 337–352. Academic Press, 1992.
31. Beer, J., Blinov, A., Bonani, G., Finkel, R. C., Hofmann, H. J., Lehmann, B., Oeschger, H., Sigg, A., Schwander, J., Staffelbach, T., Stauffer, B., Suter, M., Wölfli, W.: Use of ^{10}Be in polar ice to trace the 11-year cycle of solar activity. Nature **347**, 164–166 (1990).
32. Anderson, R. F., Lao, Y., Broecker, W. S., Trumbore, S. E., Hofmann, H. J., Wölfli, W.: Boundary scavenging in the Pacific Ocean: a comparison of ^{10}Be and ^{231}Pa. Earth Planet. Sci. Lett. **96**, 287–304 (1990).
33. Brown, L.: ^{10}Be: recent applications in Earth sciences. Philos. Trans. R. Soc. Lond. **A323**, 75–86 (1987).
34. Nishiizumi, K.: ^{53}Mn, ^{26}Al, ^{10}Be and ^{36}Cl in meteorites: data compilation. Nucl. Track Radiation Meas. **13**, 209–237 (1987).
35. Nishiizumi, K., Imamura, M., Kohl, C. P., Nagai, H., Kobayashi, K., Yoshida, K., Yamashita, H., Reedy, R. C., Honda, M., Arnold, J. R.: ^{10}Be profiles in lunar surface rock 68815. Proc. Lunar Planet Sci. Conf. **18th**, 79–85 (1987).
36. Siegenthaler, U., Beer, J.: Model comparison of ^{14}C and ^{10}Be isotope records. In: *Secular Solar and Geomagnetic Variations in the Last 10,000 Years* (F. R. Stephenson and A. W. Wolfendale, Eds.), Kluwer Academic Publishers, **Vol. 236**, 315–328 (1988).

37. Hofmann, H. J., Bonani, G., Suter, M., Wölfli, W., Zimmermann, D., von Gunten, H. R.: A new determination of the half-life of ^{32}Si. Nucl. Instr. Meth. **B 52**, 544–551 (1990).
38. Chen, Y., Kashy, E., Banzin, D., Benenson, W., Morrissey, D. J., Orr, N. A., Sherrill, B. M., Winger, J. A., Young, B., Yurkon, J.: Half-life of ^{32}Si. Phys. Rev. **C 47**, 1462–1465 (1993).
39. Gellerman, R., Börner, J., Franke, T., Fröhlich, K.: Isotopenpraxis **24**, 114–117 (1988).
40. Zoppi, U., Kubik, P. W., Suter, M., Synal, H.-A., von Gunten, H. R., Zimmermann, D.: High Intensity isobar separation at the Zürich AMS facility. Nucl. Instr. Meth. **B 92**, 142–145 (1994).
41. Middleton, R., Klein, J.: ^{26}Al: Measurements and applications. Philos. Trans. R. Soc. Lond. **A 323**, 121–143 (1987).
42. Bentley, H. W., Philipps, F. M., Davis, S. N.: Chlorine-36 in the terrestrial environment. In: *Handbook of Environmental Isotope Geochemistry* (P. Fritz and J. C. Fontes, Eds.), Elsevier Scientific Publishing, **Vol. 2**, pp. 427–480 (1986).
43. Synal, H.-A., Beer, J., Bonani, G., Suter, M., Wölfli, W.: Atmospheric transport of bomb-produced ^{36}Cl. Nucl. Instr. Meth. **B 52**, 483–488 (1990).
44. Gove, H.: Tandem-accelerator mass-spectrometry measurements of ^{36}Cl, ^{129}I and osmium isotopes in diverse natural samples. Philos. Trans. R. Soc. Lond. **A 323**, 103–119 (1987).
45. Hallet, B., Putkonen, J.: Surface dating of dynamic landforms: young boulders on aging moraines. Science **265**, 937–940 (1994).
46. Fabryka-Martin, J., Bentley, H., Elmore, D., Airey, P. L.: ^{129}I as an environmental tracer. Geochim. Cosmochim. Acta **49**, 337–347 (1985).
47. Krishnaswami, S., Lal, D., Martin, J. M., Meybeck, M.: Geochronology of lake sediments. Earth Planet. Sci. Lett. **11**, 407–414 (1971).
48. von Gunten, H. R., Moser, R. N.: How reliable is the ^{210}Pb dating method? Old and new results from Switzerland. J. Paleolimnol. **9**, 161–178 (1993).
49. Erten, H. N., von Gunten, H. R., Rössler, E., Sturm, M.: Dating of sediments from Lake Zürich (Switzerland) with ^{210}Pb and ^{137}Cs. Schweiz. Z. Hydrol. **47**, 5–11 (1985).
50. Albarède, F., Michard, A., Cuney, M.: Uranium-thorium-lead dating. In: *Nuclear Methods of Dating* (E. Roth and B. Poty, Eds.), pp. 111–161, Kluwer Academic Publishers, 1989.
51. Wetherill, G. W.: Discordant uranium-lead ages I. Trans. Am. Geophys. Union **37**, 320–326 (1956).
52. Bernard-Griffiths, J. B.: The rubidium-strontium method. In: *Nuclear Methods of Dating* (E. Roth and B. Poty, Eds.), pp. 73–94, Kluwer Academic Publishers, 1989.
53. Moorbath, S., O'Nions, R. K., Parkhurst, R. J., Gale, N. H., McGregor, V. R.: Further rubidium strontium age determinations on the very early precambrian rocks of the Godthaab district, West Greenland. Nature Phys. Sci. **240**, 78–82 (1972).
54. Montigny, R.: The conventional potassium-argon method. In: *Nuclear Methods of Dating* (E. Roth and B. Poty, Eds.), pp. 295–324, Kluwer Academic Publishers, 1989.
55. Malusky, H.: Argon-39 – argon-40 dating. Principles and applications to minerals from terrestrial rocks. In: *Nuclear Methods of Dating* (E. Roth and B. Poty, Eds.), pp. 325–351, Kluwer Academic Publishers, 1989.
56. Bernard-Griffiths, J., Gruau, G.: The samarium-neodymium method. In: *Nuclear Methods of Dating* (E. Roth and B. Poty, Eds.), pp. 53–72, Kluwer Academic Publishers, 1989.
57. Luck, J.-M.: The rhenium-osmium system. In: *Nuclear Methods of Dating* (E. Roth and B. Poty, Eds.), pp. 95–104, Kluwer Academic Publishers, 1989.
58. Rees, T.: The lutetium-hafnium method. In: *Nuclear Methods of Dating* (E. Roth and B. Poty, Eds.), pp. 105–110, Kluwer Academic Publishers, 1989.
59. Wetherill, G. W.: Radiometric chronology of the early solar system. Ann. Rev. Nucl. Sci. **25**, 283–328 (1975).
60. Albarède, F.: The age of the earth and meteorites, and extinct radioactivities. In: *Nuclear Methods of Dating* (E. Roth and B. Poty, Eds.), pp. 45–52, Kluwer Academic Publishers, 1989.
61. Patterson, C.: Age of meteorites and the Earth. Geochim. Cosmochim. Acta **10**, 230–237 (1956).
62. Moorbath, S.: The most ancient rocks? Nature **307**, 585–586 (1983).
63. Taylor, S. R.: *Lunar Science: A Post-Apollo View.* Pergamon Press, 1975.
64. Dalrymple, G. B.: ^{40}Ar/^{39}Ar ages of six Apollo 15 impact melt rocks by laser stepwise heating. Geophys. Res. Lett. **18**, 1163–1166 (1991).
65. Swindle, T. D., Spudis, P. D., Taylor, G. J., Korotev, R. L., Nichols, R. H.: Searching for Crisium Basin ejecta. Chemistry and ages of Luna 20 impact melts. Proc. Lunar Planet. Sci. Conf. **21**, 167–181 (1990), Houston 1991.
66. Bogard, D. D., Garrison, D. H., Shih, C.-Y., Nyquist, L. E.: ^{39}Ar-^{40}Ar dating of two lunar granites: The age of Copernicus. Geochim. Cosmochim. Acta **58**, 3093–3100 (1994).
67. Ivanovich, M., Harmon, R. S. (Eds.): *Uranium-series Disequilibrium: Applications to Earth, Marine, and Environmental Sciences.* Clarendon Press, 2nd Ed., 1992.
68. Ivanovich, M., Latham, A. G., Ku, T.-L.: Uranium series disequilibrium. Applications in geochronology. In: *Uranium-series Disequilibrium: Applications to Earth, Marine, and Environmental Sciences*, pp. 62–94, Clarendon Press, 2nd Ed., 1992.
69. Hoehn, E., von Gunten, H. R.: Radon in groundwater: A tool to assess infiltration from surface waters to aquifers. Water Resour. Res. **25**, 1795–1803 (1989).
70. Hoehn, E., von Gunten, H. R., Stauffer, F., Dracos, T.: Radon-222 as a groundwater tracer. A laboratory study. Environ. Sci. Technol. **26**, 734–738 (1992).
71. Schulte-Ebbert, U., Hollerung, R., Willme, U., Kaczmarcyk, B., Bahrig, B., Schöttler, U.: Verhalten von anorganischen Spurenstoffen bei wechselnden Redoxverhältnissen im Grundwasser. Veröffentlichungen des Instituts für Wasserforschung GmbH Dortmund und der Dortmunder Stadtwerke AG, **Nr. 43**, p. 128 (1991).
72. Burnett, W. C., Veeh, H. H.: Uranium-series studies of marine phosphates and carbonates. In: *Uranium-series Disequilibrium: Applications to Earth, Marine, and Environmental Sciences*, pp. 487–512, Clarendon Press, 2nd Ed., 1992.
73. Schwarcz, H. P., Blackwell, B. A.: Archaeological applications. In: *Uranium-series Disequilibrium: Applications to Earth, Marine, and Environmental Sciences*, pp. 513–552, Clarendon Press, 2nd Ed., 1992.
74. Huh, C.-A., Kadko, D. C.: Marine sediments and sedimentation processes. In: *Uranium-series Disequilibrium: Applications to Earth, Marine, and Environmental Sciences*, pp. 460–486, Clarendon Press, 2nd Ed., 1992.
75. Carpéna, J., Mailhé, D.: Fission-track dating. In: *Nuclear Methods of Dating* (E. Roth and B. Poty, Eds.), pp. 189–233, Kluwer Academic Publishers, 1989.
76. Fleischer, R. L., Price, P. B.: Techniques for geological dating of minerals by chemical etching of fission fragment tracks. Geochim. Cosmochim. Acta **28**, 1705–1714 (1964).
77. Lalou, C., Valladas, G.: Thermoluminescence dating. In: *Nuclear Methods of Dating* (E. Roth and B. Poty, Eds.), pp. 235–273, Kluwer Academic Publishers, 1989.

Radiochimica Acta 70/71, 317–321 (1995)

Radioactivity and Radiation Chemistry of Water

By Ivan G. Draganić

The Institute of Nuclear Sciences "Vinča", P.O.B. 522, 11000 Beograd, Serbia, Yugoslavia

(Received July 31, 1995)

Radioactivity / Radiolysis of water / Nuclear reactors / Oklo / Early Earth / Cometary nuclei

Abstract

The chemical changes of water induced by radiation, first observed on aqueous solutions of radium salts, were reported less than five years after the discovery of radioactivity. The importance of the radiolytic behaviour of water was progressively recognized in various domains of nuclear science and technology. More recent interest concerns the roles of radioactivity, radiation and water in chemical evolution processes because of their ubiquitous presence on the Earth and in Space.

1. The early days of radioactivity and water radiolysis

Already in 1903, Giesel [1] quoted the observation of Runge and Bodländer that a gas consisting of hydrogen and oxygen is constantly released from an aqueous solution of radium bromide; at the same time the solution becomes brownish because of the generation of bromine. It seems, however, that Ramsay [2] was the first physicist to carry out irradiation experiments reported in his article "The chemical action of radium emanation on distilled water". This work and publications by Cameron and Ramsay [3–4] contained a number of quantitative data, among which those on the generation of H_2 and O_2 under various experimental conditions.

Debierne [5] presented in 1909 the results of a systematic approach with simple experiments. He used a sealed ampoule of radium chloride as the radiation source, placed in a vessel containing nondegassed distilled water. The device was isolated from air and connected to a manometer. The increase in pressure was continuous and proportional to the time of exposure. In the interpretation of his results, Debierne concluded that the decomposition of water was due to the action of penetrating β and γ rays. The first data on the action of α rays on water were reported in 1913 by Duane and Scheuer [6].

Compared to present day irradiation facilities, the radiation sources used in these experiments were very weak and seldom stronger than 100 mg radium. In addition to the low radiation intensities, the methods of chemical analysis available at the time were relatively insensitive. For these reasons, the investigations were limited mainly to qualitative observations of chemical changes or, less often, to semi-quantitative measurements of gross effects. Nevertheless, the early literature of radiation chemistry contains fine observations on the radiolytic behaviour of water and various aqueous solutions like those of Kailan [7–11] at the Radium Institute in Vienna.

No much efforts were made to explain the origin of the observed facts. An exception was Debierne's hypothesis [12] on the formation of ionized water molecules during the passage of radiation. The latter produce H and OH species which react among themselves; molecular products H_2 and H_2O_2 are formed as well as water molecules by recombination of H and OH. With this hypothesis, Debierne was well ahead of the time because it contains the main elements of the presently generally accepted model of water radiolysis.

2. Nuclear reactors: World War II and post-war years

From the very beginning of the nuclear war program, it was obvious that ordinary as well as heavy water, were an important material for the realization of nuclear reactors, both as neutron moderator and as coolant. The question arose as to how water would behave in a field of radiation equivalent to that from thousands of kilograms of radium. Moreover, the water would be exposed to a complex mixture of charged particles, neutrons and γ rays. It was known that the main components were fast neutrons and γ rays, but their action on water was not clear. It was thought that neutrons, through the intermediary of recoil protons which they displace in water, would have an effect similar to that of α particles, and hence would decompose water efficiently. In this connection, A. O. Allen, one of the radiation chemist involved in the research wrote [13]: "Physicists had decided that the best scheme for cooling the reactor would be by water rushing through aluminum pipes. The difficulty was that no one knew whether under the influence of the unprecedently intense radiation in these reactors the water might react with aluminum as if it were calcium or sodium, as indeed thermodynamics suggested".

The secret radiation chemistry program was carried out under the guidance of two physicists, Milton

Burton and James Franck. These scientists based their research on the free radical theory derived from physical concepts. In a first instance, radiation could solely transfer energy to the electrons of the water molecule, leading to its excitation and/or to its ionization. The ionized and excited water molecules participate in various reactions producing free radicals H and OH. Further reactions of these species lead to the formation of molecular products, H_2 and H_2O_2.

At the time, the experiments performed and the reasonings could not be published. The American team had no contact with outside research, even not with the British team including F. S. Dainton and N. Miller, who pursued in Canada similar studies of water radiolysis with the same objective [14]. In the USA, the work was conducted under military secrecy in several laboratories. It was progressively recognized that water decomposes under radiation to yield the molecular products, but also that the radicals H and OH convert the latter back to water molecules. It become apparent that water used in reactors should be extremely pure, as almost any substance dissolved in water increased the radiolytic decomposition of the solvent. These findings were summarized by Allen after the war [15].

In the fifties, it was generally admitted that the induced radioactivity and the decomposition of water under the intense radiation present in reactors could be controlled [16].

For a while, the possibility of using ionizing radiation as a cheap "by-product" of the operation of nuclear power reactors was considered as a promising tool for industrial applications. However, ten years later, it became evident that any use of nuclear reactor, outside the production energy, was complex and costly, particularly when compared with the increasing availability of irradiation facilities provided by radioisotopic sources and modern, high intensity accelerators. Nevertheless, the experience gained in the unconventional radiation chemistry with large doses of mixed radiations was a valuable heritage [17]. It showed how on prolonged irradiations, the initially present compounds decompose while new products are formed by the action of radiation and take part in chemical processes. Unconventional radiation chemistry is particularly relevant to the field of nuclear waste management and long term waste disposal.

Owing to the important role of water in contemporary nuclear energy programs and in the action of radiation on living matter, the radiation chemistry of water [18] is nowadays a pre-eminent topic in the general field of radiation chemistry [19].

3. Natural nuclear reactors and underground waters on the early Earth

Natural nuclear reactors were geological arrangements of uranium and water in configurations such as a fission chain reaction could be triggered. This phenomenon occurred in uranium deposits on the Precambrian Earth, as was the case at the fossil reactor site discovered in Oklo (Gabon, Africa) [20]. The present concentration of ^{235}U ($T_{1/2}$ = 700 My) in natural uranium is 0.72% which precludes the start of a chain fission process even in the richest uranium deposits.

On the early Earth, the concentration of ^{235}U was up to 40 times higher and could have led to a chain reaction, as was suggested by P. K. Kuroda long before the discovery of the Oklo phenomenon [21].

From the point of view of nuclear reactor physics and chemistry, the reactions which occurred at the fossil nuclear reactors site at Oklo (which includes six reactor cores) are well understood [22–23]. The operation time of individual reactors lasted between 0.6 and 0.8 My, while the total duration of the Oklo phenomenon was up to several million years. It has also been concluded that none of the geological and geochemical characteristics at the Oklo deposit are unique, and that the latter could not have been a single occurrence on the Precambrian Earth.

Individual natural reactors in operation constituted efficient localized sources of energy over a large geological time span. In comparison with man-made reactors, the natural ones operated at a very low power, on the average about one kW. Routine procedures for in-pile dose evaluation [24] show that these natural reactors delivered dose rates of the order of tens of gray (Gy) per hour. Such dose rates are similar to those available in small laboratory irradiation units and could have been very convenient for radiation-induced syntheses.

It is possible to make a rough estimate of the amount of uranium which took part in natural nuclear reactors. Based on the slight variations of the isotopic ratio $^{238}U/^{235}U$ in presently accessible uranium ores with 1000 ppm UO_2 or more, it appears that thousands of natural reactors were in operation between 1800 and 4100 My ago. On the other hand, if a one km thick layer of the Earth's crust is taken into account with an average concentration of 3 ppm, then hundred million reactor sites of the Oklo type could have been in activity in the past of our planet [24].

Water had an important role in the making of a natural reactor from a uranium deposit. It acted as neutron moderator and coolant and contributed to the propagation and control of the fission chain. At the same time, water was radiolized and the dissolved substances were chemically altered. The material subjected to irradiation was efficiently circulated by underground waters.

Computer simulation helped to examine the action of mixed nuclear radiations on a model system of sandstone underground water [25]. It showed that considerable amounts of O_2, H_2 and H_2O_2 could have been formed on a geological time scale. Many chemically reactive free radical intermediates could also have been involved in reactions with various inorganic and organic compounds dissolved in waters flowing through or in the vicinity of the reactors. This radiation-induced oxidizing capacity of underground waters is usually neglected in the explanations of the oxidizing processes in primitive underground waters and in the search of the origins of primary oxidized minerals on the early Earth.

It is worth mentioning that the presence of long-lived fission products in the reaction core ensured that the reactor was a place of localized radiations for long periods of time (hundred of years or more) after the chain process was terminated.

On the other hand, some of the fission radionuclides would migrate and pursue the irradiation after fixation in the environment or after uptake in Precambrian living matter. The dose rates were low, but combined with irradiation over large periods of time, they could have been effective in producing evolutionary changes in primitive life. An examination of major events in the early biological evolution, shows that the main evolutionary changes of precariotic life resulted from biochemicals rather than morphological innovations [26]. With the initiation of free radical processes in the early biochemistry, nuclear radiations could have been a very effective mutagenic agent for Precambrian biota.

4. Radioactive potassium-40 in the early ocean waters

The presence of potassium salts in primitive ocean waters and the radioactive decay of ^{40}K ($T_{1/2}$ = 1250 My) have provided an internal source of radiation in Arachaean oceans [27]. 3800 My ago, the natural abundance of ^{40}K was 0.1%. The amount of energy delivered by ^{40}K depended on the concentration of potassium salts in primitive ocean waters. The latter were recycled many times over geological history and may not have retained the information of their past potassium content. Nevertheless, a reliable estimate was possible. It is based on interferences drawn from the nature of ancient sediments and the products of their metamorphism. Various observations suggest that the main components of the lithosphere and the hydrosphere were extant by the time of Issua water-laid sediments of West Greenland [28]. From the average energy released per decay, the calculated absorbed dose rate is 2.95×10^{-4} $Jkg^{-1}y^{-1}$.

Compared to the amounts of energy provided by other sources on the early Earth, the contribution of ^{40}K is extremely small. However, in many respects, ^{40}K was an ideal *in situ* source. The energy was deposited homogeneously throughout the water volume, including in the deep ocean. In addition, owing to the very nature of radioactive decay, the energy deposit occurred undisturbed and uninterrupted so that the total amount of the energy available for radiation chemistry over a geological time-scale was significant.

The chemical effects induced by the radiation were examined by computer simulation [27]. The latter was based on the estimate of the energy deposited and on well established radiation-induced reactions of substances which were assumed to be constituents of sea water. The radiation-induced process was followed over a geological time scale of one hundred million years (100 My). The beginning of the time scale was deliberately fixed at 3800 My ago. This is the age of the oldest known sedimentary rocks and most likely of ocean waters whose content in salt and major chemical elements was not essentially different from the contemporary ones.

The results obtained show that one molecule H_2 is formed per H_2O decomposed. The total amount of molecular hydrogen generated in the sea waters over 100 My is 3.8×10^{18} g, assuming an ocean mass of 1.4×10^{24} g. This amount is similar to that released by volcanic activity on the early Earth over the same period of time [28]. A considerable total amount of radiolytically generated molecular oxygen over 100 My was also calculated: 3.0×10^{19} g (for comparison the content of the contemporary atmosphere is 1.2×10^{21} g).

In addition to H_2 and O_2, hydrogen peroxide is produced in significant amounts. A survey of reaction mechanisms shows the presence of strongly oxidizing hydroxyl radicals (OH) produced by direct decomposition of water. Many other oxidizing free radicals appeared as intermediates, contributing also to the total oxidizing capacity of early ocean water. They may have played a significant role in the chemistry of Earth before the emergence of oxygen due to photosynthetic microorganisms, for example in the origin of primary oxidized iron minerals [31] in the oldest known terrestrial sediments, the 3800 My-old rocks in the Isua Supracrustal Group of West Greenland.

It has been proposed that comets brought vast quantities of water to the primitive Earth, which by this acquired an exogenous ocean of water [32] with a significant delivery of organic compounds [33]. Organic compounds of endogenous and exogenous origins potentially present may have contributed to the radiation-induced synthesis of biogenic molecules of importance for chemical evolution processes.

5. Radiation processing of water in cometary nuclei

Isotopic anomalies found in meteorites, particularly that pointing to the presence of ^{26}Al in the early solar system, suggest that very likely a supernova explosion occurred at the beginning of the accretion process [34]. The nucleosynthesis during the explosion has provided a large number of radionuclides which were built into the comet nuclei like in all other bodies in the solar system. Radiation emitted in their decays, as well as decays from long-lived radionuclides already present in the nebula like isotopes of U and Th, constituted a source of energy available for chemical reactions in the icy interior of a comet's body [35].

The surface layers were constantly exposed to cosmic rays, a mixture of electromagnetic radiation and charged particles. The latter also produced hot atoms and cosmogenic radionuclides, which on their way contributed to radiation chemical changes [36–37].

The intensities of radiations were rather weak, in average about a tenth of gray per year or less, but the time available for irradiation is enormous, 10^9 years. In addition, the environmental conditions, such as low temperature and density of matter, are almost ideal for the accumulation of radiolytic products and preservation of radiation-induced alterations.

The International Halley Watch Program was prepared and carried out with a general belief that the event will be an encounter with a pristine material of a dense interstellar cloud where the solar system and comets were born. Some findings in 1986 indeed confirmed the presence of simple interstellar molecules which are also believed to be the primitive constituents of the presolar nebula. However, for a water-dominated body, an unexpected solid dark-gray surface of the cometary nucleus appeared. It was attributed to some unknown chemical compound(s); the puzzling presence of larger nitrogen-containing organics was found in ejected dust. These and some other findings suggest that the pristine cometary material must have been chemically altered, most likely by the action of radiation.

Simulation experiments performed in the laboratory offer an insight into the radiation chemistry involved. The inventory of radiolytic products found is impressive: about forty compounds including some of primary biological importance, as well as larger ones with molecular weights up to 80000 amu [38]. The experiments suggest that the material in the comet nucleus is more likely a remnant of material in the presolar nebula which has been chemically altered to various degrees by radiations from imbedded radionuclides and cosmic rays.

6. Perspectives

The first century of the nuclear era ends with an impressive inventory of knowledge on radioactivity and radiation, both on the early and on the contemporary Earth and in the interplanetary medium. Nevertheless, various challenging problems are bequeathed to the coming generation [39]. Some of them will be mostly technological, like the decommissioning of hundreds of nuclear power plants, the reprocessing (or safe storage) of hundred thousands tons of spent nuclear fuel elements and the disposal of million tons of intermediate and high-level nuclear waste. Water will be an unavoidable tool in some stages of the handling of huge amounts of radioactive material, often in conditions seldom met in routine laboratory work: materials exposed to large radiation doses and radiogenic heat, sometimes very complex chemical systems. A better understanding of radiation-induced chemical processes at high temperature will be needed to find technological solutions. More data on free radical intermediates and on rate constants of their reactions will also be required.

Other challenging problems will be less urgent, but not less exciting, like those relevant to radiation chemistry of water in cometary nuclei [40] or on the early Earth [41]. Radioactivity is generally considered as a minor contributor to the energy sources available for chemistry in these media. Yet, it deserves more attention than presently given because of the efficiency in producing chemically reactive species (ions, free radicals, radical ions) in conditions hostile for chemistry, such as those in water-dominated commentary icy matrix or in the ocean depths and underground primitive waters. Present achievements in radiation and hot atom chemistry provide a solid foundation for further research.

References

1. Giesel, F.: Ber. Dtsch. Chem. Ges. **36**, 342 (1903).
2. Ramsay, W.: J. Chem. Soc. **91**, 937 (1907).
3. Cameron, A. T., Ramsay, W.: J. Chem. Soc. **92**, 966 (1908).
4. Cameron, A. T., Ramsay, W.: J. Chem. Soc. **92**, 992 (1908).
5. Debierne, A.: C. R. Acad. Sci. Paris **148**, 703 (1909).
6. Duane, W., Scheuer, O.: Radium **10**, 33 (1913).
7. Kailan, A.: Wien. Ber. **120**, 1213 (1911).
8. Kailan, A.: Wien. Ber. **120**, 1373 (1911).
9. Kailan, A.: Monatsh. Chem. **33**, 1329 (1912).
10. Kailan, A.: Monatsh. Chem. **33**, 1361 (1912).
11. Kailan, A.: Monatsh. Chem. **34**, 1269 (1913).
12. Debierne, A.: Ann. Phys. (Paris) **2**, 97 (1914).
13. Allen, A. O.: J. Chem. Educ. **45**, 290 (1968).
14. Allen, A. O.: Chpt. 1 in *Early Developments in Radiation Chemistry* (J. Kroh, Ed.), The Royal Society of Chemistry, Cambridge 1989.
15. Allen, A. O.: J. Phys. Colloid Chem. **52**, 479 (1948).
16. Roth, E.: *Chimie nucléaire appliquée*, Masson and Cie, Paris 1968.
17. Draganić, I. G.: Chpt. 7 in *Early Developments in Radiation Chemistry* (J. Kroh, Ed.), The Royal Society of Chemistry, Cambridge 1989.
18. Draganić, I. G., Draganić, Z. D.: *The Radiation Chemistry of Water*, Academic Press, New York 1971.
19. Spinks, J. W. T., Woods, R. J.: *An Introduction to Radiation Chemistry*, John Wiley and Sons, New York 1990.
20. Neuilly, M., Bussac, J., Frejacques, C., Nief, G., Vendryes, G., Yvon, J.: C. R. Acad. Sci. **275 D**, 1847 (1972).
21. Kuroda, P. K.: J. Chem. Phys. **25**, 781 (1956).
22. *The Oklo Phenomenon*, IAEA, Vienna 1975.
23. *Natural Fission Reactors,* IAEA, Vienna 1978.
24. Draganić, I. G., Draganić, Z. D., Altiparmakov, D.: Precambrian Res. **20**, 283 (1983).
25. Bjergbakke, E., Draganić, Z. D., Sehested, K., Draganić, I. D.: Radiochim. Acta **48**, 73 (1989).
26. Schopf, J. W. (edit.): *Earth's Earliest Biosphere*, Princeton University Press, Princeton 1983.
27. Draganić, I. G., Bjergbakke, E., Draganić, Z. D., Sehested, K.: Precambrian Res. **52**, 337 (1991).
28. Holland, H. D.: *The Chemical Evolution of the Atmosphere and Oceans*. Princeton University Press, Princeton 1984.
29. Weast, R. G. (Ed.): *CRC Handbook of Chemistry and Physics*, CRC Press, Boca Raton 1987.
30. Miller, S. L., Orgel, L. E.: *The Origins of Life on the Earth*, Prentice-Hall 1974.
31. Dymek, F. R., Klein, C.: Precambrian Res. **39**, 247 (1988).

32. Chyba, C. F.: Nature **330**, 129 (1987).
33. Chyba, C. F., Thomas, P. J., Brookshaw, L., Sagan, C.: Science **249**, 366 (1990).
34. Reeves, H.: Astrophys. J. **231**, 229 (1979).
35. Draganić, I. G., Draganić, Z. D., Vujošević, S.: Icarus **60**, 464 (1984).
36. Ryan Jr., M. P., Draganić, I. G.: Astrophys. Space Sci. **125**, 49 (1986).
37. Draganić, I. G., Ryan, M. P., Draganić, Z. D.: Adv. Space Sci. **5**, 13 (1987).
38. Draganić, Z. D., Vujošević, S. I., Negron-Mendoza, A., Azamar, J. A., Draganić, I. G.: J. Mol. Evol. **22**, 175 (1985).
39. Draganić, I. G., Draganić, Z. D., Adloff, J. P.: *RADIATION and RADIOACTIVITY on EARTH and BEYOND*, 2nd edition, CRC Press, Boca Raton 1993.
40. Draganić, I. G., Draganić, Z. D.: J. Chim. Phys. **86**, 55 (1988).
41. Draganić, I. G., Vujošević, S. I.: in *Chemical Evolution: Origin of Life* (C. Ponnamperuma and J. Ch. Flores, Eds.), A. Deepak Pub., Hampton 1993.

Radiochimica Acta 70/71, 323–331 (1995)

Radioactivity and the Nuclear Environment

By A. G. Maddock
The Chemical Laboratories, Lensfield Road, Cambridge CB2 1 EW, UK

(Received July 7, 1995)

Change of half-life / Mössbauer spectroscopy / Perturbed angular correlation

Abstract

Radioactive decay provides the basis of three methods of exploring the nuclear environment. Changes in the decay constant with the state of chemical combination of the disintegrating atom are difficult to measure and only provide very limited information. The time dependence of the angular correlation between the directions of emission of successive photons is more useful, but the method has serious limitations.

The outstanding successful technique of Mössbauer spectroscopy however has proved remarkably versatile in its applications. The quantitative data it yields are often difficult to obtain in any other way.

The study of radioactivity has led to the discovery of three techniques for exploring the environment of the nucleus that suffers decay. One of them is of extraordinary versatility and its importance overwhelms that of the other two.

1. Changes in radioactive decay constant

The first to be discovered has proved to be of limited application. For nearly fifty years it was always stated that radioactive decay could not be influenced by any treatment accessible in a laboratory. However in the explosion of interest in nuclear physics following the Second World War the Daudels [1a, b] in Paris and Segré [2] in California pointed out that radioactive decay, at least under terrestrial conditions, involved the atom and not simply the nucleus. Hence the decay constant for orbital electron capture decay of ^{7}Be, which must depend on the orbital electron density at the nucleus, must be affected by the state of chemical combination of the beryllium.

2. Changes in decay constant for orbital electron capture and isomeric transition

This dependence was soon verified; the fractional change in decay constant, $\delta\lambda/\lambda$, for different pairs of beryllium compounds was found to be as much as 7.4×10^{-4} in some cases [3]. Even the difference in the decay rates of ^{7}Be in the two crystal modifications of beryllium oxide could be measured.

For elements of greater Z the values of $\delta\lambda/\lambda$ were generally smaller and their measurement was very time consuming. Measurements of such small differences in decay constant require close attention to the experimental conditions. They are most commonly made using Rutherford's balanced ionisation chambers method.

An interesting alternative is possible for isotopes where orbital electron capture competes with positron emission. In these systems only the former decay mode should be noticeably sensitive to the environment of the nucleus, so that the ratio of the two decay modes should depend on the chemical state of the affected element [4].

The fractional change in decay constant, $\delta\lambda/\lambda$, like the difference in isomer shift found from Mössbauer spectra, depends on the difference in the electron density at the nuclei in the two compounds. A comparison of the two measurements, where this is possible, is of interest. It can be done for example using ^{71}Ge, which decays by orbital electron capture, and ^{73}Ge which can provide Mössbauer spectra [5].

Analogous changes can be found in the constants for decay by isomeric transition, when there is substantial internal conversion, especially if this takes place mainly in the outer orbitals of the atom containing the decaying nucleus.

Examples of species for which quantitative data are available include ^{99m}Tc, ^{90m}Nb and ^{125m}Te, in the first two cases the excitation of the isomeric state is so small that internal conversion is only energetically possible in the O and N shells [6a, b, c]. The 76 eV excited state of ^{235}U would seem especially favourable for study [7a, b, c]. In these systems $\delta\lambda/\lambda$ values of a few times 10^{-3} have been found.

3. Changes in β decay constant

Decay by β^- emission should also be influenced by the chemical state of the affected atom [8]. The total beta decay probability is the sum of two terms, the probability of decay yielding an electron in the continuum and the probability of decay giving a bound electron – orbital electron creation [9]. It is interesting to observe that the direct detection of this last mode of decay is one of the unsolved problems in radioactivity.

The bound electron creation is not an important mode if the total decay energy is large, but it becomes more important at low energies and its greater sensitivity to the chemical environment of the nucleus magnifies its importance in relation to the change of decay probability with environment.

The available energy from the decay process ${}^{A}Z^{n+} \rightarrow {}^{A}(Z+1)^{(n+1)+}$ depends on the difference in the energies of the initial and final atoms. This difference is dependent on n. Further the product atom may be formed in an electronically excited state or even suffer further ionization due to "shake off". Thus the maximum beta ray energy has a spectrum of values and depends on the state of chemical combination of the decaying atom.

The bound electron creation decay mode also depends on the chemical state of the atom containing the decaying nucleus since it is dependent on the available orbitals for electron creation. The two effects will together determine $\delta\lambda/\lambda$.

Calculations for ^{63}Ni, a very soft beta emitter, show that the two contributions are of opposite sign. Hence although the individual contributions are of the order of 10^{-3} the overall effect to be expected is of the order 10^{-4} [10].

The experimental data on the dependence of the decay constant for beta decay on the state of chemical combination is unsatisfactory and indeed often contradictory. This is at present the case notwithstanding the relevance of such information to some other problems.

Chemical effects should be measurable for the, a priori, favourable case of tritium decay, but present experimental data are only broadly in keeping with calculated effects [11]. A few years ago half life data on ^{241}Pu spanned an unreasonable range of values and the possibility existed that differences in the compounds measured could account for the spread. Calculation showed that this was unlikely and more recent measurements span a narrower range [12].

The decay of ^{187}Re is particularly important because of its significance as a cosmochronometer. It has the remarkably low maximum beta particle energy of 2.7 keV and bound electron creation might make a significant contribution to the overall decay. The constant for decay giving an electron in the continuum can be measured, albeit with difficulty, but the decay by bound electron production and the total decay constant are hard to determine [13].

As a means of exploring the atomic environment changes in beta decay constants are unattractive.

4. Mössbauer spectroscopy

Almost a decade later than the observations on ^{7}Be decay an enormously more powerful technique that could provide information on the magnetic and electrical environment of the disintegrating nucleus was discovered. While investigating the possibility of resonant fluorescent gamma ray scattering Mössbauer found that if some weakly excited nuclei were present in atoms in a solid lattice some proportion of recoiless emission of photons took place [14]. This emission readily gave rise to resonant absorption and scattering. The line width for emission was extremely small, being determined by the mean life time of the excited state. A typical mean life of a suitable excited state was 10^{-7} s so that a line width of about 10^{-8} eV could be obtained.

5. Mössbauer absorption spectroscopy

In order to have a useful source of such radiation it is necessary to have a parent radioactive species whose decay feeds the very short lived excited state. A scatterer or absorber containing the stable, or at least very long lived, isotope produced in the de-excitation then gives resonant absorption. By using the Doppler effect, moving the source towards, or away from, the absorber an extremely small range of photon energies could be spanned by the source. It was quickly realized that this tiny band of photon energies would embrace the changes and splittings of nuclear energy levels produced by the magnetic [15] and electrical [16a, b] fields existing at the nucleus by virtue of its presence in an atom, perhaps also combined in a molecule, in the solid.

A monochromatic source could be obtained if the Mössbauer atom was in a cubic environment in the solid and had a full, or empty valence shell. Thence modifying the energy of the emission using the Doppler effect the absorption spectrum of a sample could be investigated. Alternatively, using a single line absorber the spectrum emitted by the source could be examined.

6. Parameters obtainable from the spectra

Spectra obtained in this way enable one to compare the electron density at the nuclei in two absorbers, using the isomer shift; to obtain the quantities specifying the electric field gradient at the nucleus, the quadrupole splitting and the asymmetry parameter and, using single crystal absorbers, to determine the orientation of the electric field gradient in relation to the crystal axes. In magnetically ordered solids the magnitude of the field at the nucleus and its orientation could be determined.

The changes in the spectra with the temperature of measurement provided more information, especially regarding the smaller crystal field and spin orbit splittings of the electronic energy levels of the Mössbauer atom in the absorber.

Very different information was supplied by the change in the fraction of recoiless absorption events with temperature. The change in the areas under the absorption lines yielded similar information, regarding the phonon spectrum of the absorber, more simply.

Altogether the technique has led to a much deeper insight into the nature of the solid state.

7. Simple applications

The extraordinary variety of the applications of Mössbauer spectroscopy is illustrated in the examples that follow. The combination of the isomer shift and quadrupole splitting is usually characteristic of the oxidation state of the Mössbauer species in the absorber. At first sight the application of the spectra for this purpose would seem trivial, but there are compounds where such a spectral identification is very useful.

One of the first applications of Mössbauer spectroscopy in chemistry, the demonstration of the identity of Prussian and Turnbull's blue, was of this kind [17a, b].

In a more recent study an iron(III) tetraphenyl porphyrin complex was oxidized to yield an apparently iron(IV) complex. FeTPPCl gave [FeTPPCl]$SbCl_6$, but the isomer shift and quadrupole splitting of the product was not much different from the parent iron(III) compound; the oxidation predominantly involves the ligand. By contrast $(FeTPP)_2N \cdot ClO_4$ gives a much lower isomer shift as is appropriate for iron(IV) [18].

The Mössbauer spectra generally show whether there is more than one coordination environment of the Mössbauer atoms in the absorber by giving spectra composed of two or more component spectra. Discontinuities in the spectral parameters, isomer shift and quadrupole splittings, for spectra taken at different temperatures indicated phase changes.

But the more interesting applications are those where alternative methods of investigation are much less satisfactory, if indeed available.

8. Mixed valence compounds and valence delocalization

Many transition element compounds are known in which the element is present in more than one oxidation state. If the coordination environments of the two kinds of atoms are not very different electron hopping between the atoms, or valence delocalization, may take place. When the mean life time between gaining and losing an electron is much greater than the mean life of the Mössbauer excited state the spectrum arising from these atoms will be the addition of the spectra due to the two oxidation states of the atom; the valences are localized. But if the electron hopping is much faster than this a single spectrum with parameters determined by the weighted mean of the parameters from the above separate spectra will be obtained. The weighting allows for the proportions of the atoms in the different oxidation states.

When the mean life time of the atom in a given oxidation state becomes comparable to the mean life of the Mössbauer excited state line broadening occurs and a detailed analysis of the spectrum yields the frequency of hopping [19].

Iron chemistry provides many instances of mixed valence compounds. In many the valences are localized and the spectra are a superposition of spectra due to the two oxidation states. This is the case for the hydrates of Fe_2F_5, some mixed valence iron phosphates and the aforementioned Prussian and Turnbull's blue. But in others such as $L_2Fe_2(\mu OH)_3(ClO_4)_2(CH_3OH)_2$, where L = NN′N″trimethyl triazacyclononane, a single spectrum with an isomer shift of 0.74 mms^{-1} is found even at 4.2 K. This value is about the mean of the isomer shifts expected for Fe^{2+} and Fe^{3+} in such a complex [20].

Particular interest attaches to those compounds that undergo a transition from localized to delocalized valences at some temperature. Such transitions may or may not be accompanied by a structural change. One of the earliest examples of such a transition, identified before Mössbauer spectroscopy was discovered, was the Verwey transition in Fe_3O_4, magnetite. The compound suffers a structural phase change at about 119 K, at the same time there occur substantial changes in electrical and magnetic properties. This oxide is an inverse spinel and it was suggested that this change was due to electron hopping between the Fe^{2+} and Fe^{3+} on the octahedral *B* lattice above 119 K [21].

The Mössbauer spectrum provides excellent confirmation of this suggestion. Above 119 K the spectrum comprises two magnetically split sextets, one due to the Fe^{3+} on the tetrahedral sites of the *A* lattice, giving the characteristic isomer shift of 0.27 mms^{-1}. The other arising from the *B* lattice gives an isomer shift of 0.67 a value effectively for $Fe^{2.5+}$ due to rapid electron hopping from Fe^{2+} to Fe^{3+} [22a, b].

It was some time before similar temperature dependent valence delocalization was identified in other iron compounds. However the series of compounds that can be represented by the formula $Fe_3O(O_2C \cdot R)_6L_3 \cdot S$ nearly all show temperature dependent delocalization of valences. The complexes consist of planar triangular Fe_3O units stacked on a threefold axis with solvent molecules, S, between planes. The L are generally nitrogen bases such as 3 methyl pyridine and the S benzene, toluene, CH_3CN or other solvent. The temperatures of onset of delocalization vary considerably with R, S and L. In most cases the transition is not sharp but extends over a range of temperatures in which both localized and delocalized spectra can be seen. Some of the effects observed may be due crystal size and density of defects; macro-crystalline and finely powder material from the same preparation behave differently. Hardly any single crystal data are available [23].

It is interesting to note that the comparatively slow Mössbauer spectrum measurement, effective time of measurement about 10^{-7} s is advantageous in these measurements. If the much faster technique of infra red spectroscopy, 10^{-13} s, is used all the iron atoms appear to have localized valences.

Another group of compounds that have been extensively investigated are the one electron oxidation products of the biferrocenes. For these a somewhat different approach is necessary, the isomer shifts for low spin iron(II) and iron(III) are too close for mean values to be identified but the similar averaging of the larger quadrupole splittings can be sought. With these complexes the role of conjugation in facilitating the electron hopping can be shown. In compound shown in the figure, with $R=CH \cdot CH_3$ delocalization is incomplete at 300 K, but with $R=-CH=CH-$ delocalization is essentially complete at ≈78 K [24].

These studies have shown that the localized – delocalized transition depends not only on the chemical identity of the substance but also on the properties of the solid it forms. Cooperative effects in the solid are involved.

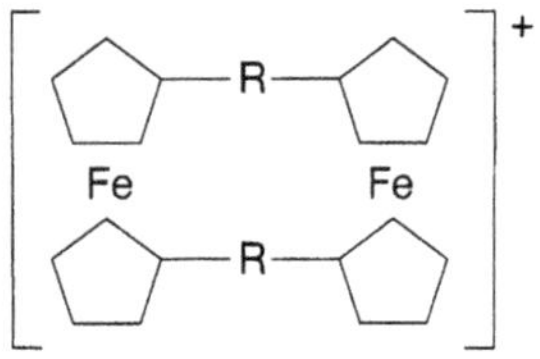

9. Spin cross over systems

Another area of investigation by Mössbauer spectroscopy, sharing some of the unusual features of the mixed valence solids, is the study of spin cross over. Although compounds which change from high spin to low spin configurations on lowering the temperature were discovered many years ago, using magnetic measurements, such behaviour is much more conveniently investigated quantitatively by Mössbauer spectroscopy. With iron(II) complexes the high and low spin forms give spectra with substantially different isomer shifts and quadrupole splittings and the areas under the absorption lines can be used to measure the proportions of the two forms present at the temperature of measurement. In some compounds the transition from high to low spin forms takes place like a first order phase change; in others the transition extends over a substantial temperature range within which spectra due to both forms are seen. In some compounds only partial conversion of one form to the other takes place [25a, b].

Like the delocalization in mixed valence compounds the spin cross over transition is sensitive to the characteristics of the solid involved; different solvates generally behave differently, even to the extent of the cross over failing to take place. Deuteration of the solvent produces easily measurable changes in the transition process. Macrocrystalline and powder samples often behave quite differently.

These two examples show how Mössbauer spectroscopy is yielding a detailed picture of processes taking place in solids.

10. Solid state decompositions

The technique is invaluable in the investigation of reactions, such as thermal decompositions, occurring in solid substances. The sample is not destroyed by the measurement and the emergence of new phases can be detected. It must be remarked, however, that the technique is not very sensitive to small amounts. In most systems it gives information about the major components of the solid [26].

The temperature and condition of the firing, whether oxidizing or reducing, can be inferred from the Mössbauer spectra of pottery. Spectra of shards of ancient pottery have been used to try to identify the kiln used in its manufacture [27].

11. Changes in bond character

The spectra often contain information about the bonding of the Mössbauer atom that is difficult to obtain in other ways. Analysis of the spectrum gives the electric field gradient at the nucleus of the Mössbauer atom. Examination of such data for tin(IV) compounds, for which there is no free ion contribution to the electric field gradient, shows that, to a reasonable approximation, each ligand attached to the tin makes its individual contribution to the quadrupole splitting [28]. This point charge model had already been developed by nuclear quadrupole spectroscopists.

With some compounds however this model failed rather badly. For example it predicts that the quadrupole splittings for the tetrahedral compounds SnR_3X and $SnRX_3$ should be of equal magnitude but opposite sign. The experimental data for the compounds with $R=C_6H_5$ and $X=Cl$ gave splittings of 1.39 mms^{-1} and 1.19 mms^{-1} respectively. The discrepancy was not due to the tin acquiring a higher coordination number than four in the solid since quickly frozen solutions gave practically the same values. This and related anomalies could be accounted for by supposing that the four bonds in such compounds are not equivalent s–p hybrids [29a, b]. As proposed in Bent's rule the bond to the least electronegative ligand has the higher s content. Similar conclusions about changes in hybridization were reached in studies of complexes of gold(I) of the type LAuX [30].

12. Calculations of isomer shifts and electric field gradients

It is possible to calculate both isomer shifts and electric field gradients. The difficulties in calculating the former quantities are familiar. The calculations must use wave functions referring not to the isolated Mössbauer atom but to the cluster in which it is present, a simple example is FeF_6^{3-}. Relativistic wave functions are essential. The measured isomer shifts give the difference in electron density at the nuclei for two compounds; it is the difference of two large quantities, the greater part of the electron density being due to the sub valence shell electrons. Screening effects of the *p* and other orbitals are difficult to calculate. Calculations avoiding serious approximations are very time consuming.

As might be expected in these circumstances the calculated value is very often not too close to the observed value. The observed isomer shifts provide an interesting test of the validity of such calculations.

Calculations of electric field gradients are generally rather more successful, but they also demand considerable computer time.

13. Magnetic studies

Mössbauer spectroscopy has proved to be of great value in the study of the magnetic properties of solids. The most extensive work has concerned iron compounds, especially the mixed oxides such as the spinels and perovskites.

The onset of magnetic ordering of the iron atoms leads to a persistent hyperfine magnetic field at the nuclei and magnetic splitting of the spectrum ensues. The magnitude of the field and its orientation in relation to the principal axis of the electric field gradient can be deduced from the spectrum.

Using an externally applied field and single crystal samples the nature of the ordering, antiferromagnetic, ferrimagnetic or ferromagnetic, can be determined and the sign and orientation of the hyperfine field in relation to the crystal axes obtained. Magnetic phase changes are readily detected. Using an applied field normal to the Mössbauer photon beam the spin flop transition is clearly revealed [31].

By substituting diamagnetic ions, such as Ga^{3+}, for the iron information about the magnetic coupling between the lattices can be explored. In the substituted samples next nearest neighbour effects, as described in section 18, are observed [32].

In conjunction with other techniques Mössbauer spectroscopy has helped disentangle the complicated magnetic behaviour of chemically simple substances like magnetite, Fe_3O_4.

Magnetic properties, since they arise from cooperative interactions in the solid, depend on the particle size in the sample measured. Very small particles behave paramagnetically at temperatures below the Néel temperature, this behaviour is called superparamagnetism. This has provided the basis of a technique in which the Mössbauer spectra at different temperatures can be used to deduce something about the particle size in the sample [33].

14. Conversion electron Mössbauer spectroscopy

The absorption events regenerate the excited state of the Mössbauer nucleus in the absorber. De-excitation leads to re-emission of Mössbauer photons in all directions, so that besides measuring the reduction in the intensity of the photon beam from the source along the beam axis the scattered photons can be measured in other directions. In such a measurement the velocity of movement of the source leading to greatest reduction in beam intensity will give the highest intensity of the scattered photons [34].

With most Mössbauer nuclei however internal conversion dominates the decay and a high proportion of the de-excitation takes place with electron and X-ray emission. Hence it is more satisfactory to measure these scattered radiations. Both the electrons and the X-rays are generally of low energy, 7.3 keV and 6.45 keV respectively in the case of ^{57}Fe spectra, and if the scatterer, that is absorber, is incorporated directly in a proportional counter the spectrum recorded will be determined by the superficial and near surface layers of the scatterer. This is because the conversion electrons have a very limited range in a solid sample.

Much greater specificity as to the depth below the surface of the sample from which the spectrum emanates can be obtained using more elaborate apparatus. The scatterer and conversion electron detector are housed in a vacuum chamber. The detector is a channel electron multiplier and conversion electrons of selected energy are recorded so that the spectra are determined by the material at selected depths below the surface of the scatterer [35].

These techniques, often called CEMS, have been used extensively for the study of the corrosion of metals, especially iron [36]. They are also convenient for the study of thin films. The way in which the magnetic ordering in iron metal films changes with the thickness has been explored with this technique [37]. Not surprisingly the various treatments that have been used to modify the properties of iron, including nitriding and phosphate treatments to reduce corrosion, have been studied in this way [38]. An example of industrial interest is the way the spectrum from steel ballbearings changes as the bearings age [39].

15. Applications in mineralogy and petrology

A high proportion of minerals contain iron, an especially favourable element for Mössbauer spectroscopy. In the numerous silicate minerals the iron is most frequently present as Fe^{2+}. The isomer shift helps to determine whether the iron cations are 4, 5, 6 or 8 coordinate. It also distinguishes Fe^{2+} from Fe^{3+}. A very wide variety of quadrupole splittings are found for the Fe^{2+} ions even with the same coordination number. The environments are always distorted from the ideal tetrahedral and octahedral arrangements for four and six coordination.

Since the free ion and ligand contributions to the quadrupole splitting are normally of opposite sign the splitting is generally greatest for the least distorted environment. As a result the Mössbauer spectrum of a mineral is a unique property and can be used for identification [40a, b].

The nondestructive measurement of the Mössbauer spectrum was particularly useful for the identification of minerals in precious samples such as those collected on the moon [41]. Terrestrial pyroxenes and plagioclase usually show some Fe^{3+} in their spectra but none could be seen in the spectra of moon rock samples. However the Mössbauer spectrum is not very sensitive to the presence of Fe^{3+} in the presence of much larger amounts of Fe^{2+}, less than 1% of the iron in the oxidized form will not normally be seen. The moon "soil" contains ilmenite and in the impact craters glassy materials. No Fe^{3+} could be detected in the spectra of either of these, although more sensitive methods of detection showed tiny amounts of Fe^{3+} in superficial material [42].

In many metamorphosed rocks both Fe^{2+} and Fe^{3+} are present and the determination of their proportions by classical analysis is not always reliable; the solution of the sample for analysis often leads to oxidation of some Fe^{2+}. The area of the absorption peaks due the iron in a given valence and lattice site is proportional to the amount of iron in that form in the absorber. It is necessary that the spectrum be measured with a thin sample, that there is minimal overlap of component spectra and that the Mössbauer fractions for the iron in the different sites are about the same. The latter condition is likely to be fulfilled if spectra are measured at low temperature and the sites have the same coordination number.

Although the precision of this method of analysis is limited the results are at least as reliable as those from chemical analysis [43].

16. Cation ordering in minerals

Many minerals have more than one kind of cationic site which can be occupied by Fe^{2+} or other divalent cations such as Mg^{2+}, Ca^{2+} etc. Often a range of minerals are available in which various proportions of the sites are occupied by Fe^{2+} and the question arises does the iron prefer to occupy one kind of site more than another. The number of kinds of site and their local configuration is generally known from diffraction studies. In many cases this will permit the spectrum arising from iron in a given site to be identified and the area under the absorption lines gives the amount of iron in that kind of site. With two kinds of site a random occupation will lead to half the total iron in each kind of site. More generally the areal measurements will give the mole fraction of iron in each site and, supposing sites not occupied by iron are occupied by magnesium, the data will yield an equilibrium constant for the reaction $Fe^{2+}(1) + Mg^{2+}(2) \geq Fe^{2+}(2) + Mg^{2+}(1)$ the 1 and 2 referring to the two kinds of site. A refinement is to introduce activity coefficients, for example on the basis the theory of regular solutions.

At very high temperatures the equilibrium constant will always approach unity, but at lower temperatures it may differ very substantially from one. Measurement of the Fe^{2+} occupation of the sites may then supply information about the thermal history of the mineral [44].

17. Valence delocalization in minerals

Some minerals, usually giving very complex Mössbauer spectra, have component spectra with anomalous isomer shifts and, usually, very temperature dependent spectra. The anomalous isomer shifts are too high for four or six coordinate Fe^{3+} but too small for even four coordinate Fe^{2+}. These characteristics are typical of delocalization between Fe^{2+} and Fe^{3+}. The minerals include Deerite, Howeite, Ilvaite, Augite, Schorlomite and others. No other technique of investigation can reveal this delocalization so clearly [45].

18. Next nearest neighbour effects

The electric field gradient at the iron atom in minerals is affected not only by the coordination geometry of the oxide ion ligands but also by the next nearest neighbour ions. This is because these ions are also interacting with the ligand oxide ions. If there are two or more cationic components in the mineral and they are randomly distributed over the cation sites there will be a distribution of the proportions of the two cations as the next nearest neighbours to different iron atoms. Thus there will be a distribution of electric field gradients and the lines in the spectrum will broaden. By contrast with one cationic species or an ordered occupation of cationic sites, leading to a constant next nearest neighbour environment, sharp line spectra will ensue [46].

19. Soil analysis

The determination of the identity and proportions of the various iron oxides and hydroxides present in soils and earthy deposits is a matter of considerable difficulty. Combined with X-ray diffraction studies Mössbauer spectra supply the best solution to the problem. However rather extensive measurements are needed, including recording of spectra at temperatures down to liquid helium temperature. Long measurements ensuring a low statistical error on each point on the spectrum are required. Conclusions from one set of spectra should then be used to calculate expected values for other spectra and an analysis of the quality of the fit made [47].

20. Dating

Several possibilities of using Mössbauer spectra for dating minerals have been suggested. They include a slow cis-trans change in minerals with octahedral Fe^{2+} with two OH ligands and a slow ordering of next nearest neighbours if, these are initially disordered. As yet none of the procedures can be regarded as established.

21. An example of potential industrial application

Since good quality spectra generally take a day to accumulate and expert interpretation is usually necessary Mössbauer spectra are not attractive for industrial application. However there are exceptions. In a number of coals their sulphur content is largely present as pyrites. A two point measurement can be made, one at a velocity giving an absorption maximum for pyrites, the other at a velocity far off resonance. The difference in counting rate at the two velocity will be a measure of the pyrites content of the coal. A measurement of this kind can be made in less than an hour [48].

22. Application to iron containing natural products

Mössbauer spectra have proved of considerable value in helping to elucidate the structure and mode of operation of the numerous biologically important iron complexes. Most of these iron-protein compounds are very complex and no single method of investigation gives a complete picture of their nature and properties. In many cases their iron content is very low, perhaps one iron atom in a molecule of weight 10^4 to 10^6. The abundance of ^{57}Fe in natural iron is only about 2% so that it is best to use samples produced from organisms which have been supplied with ^{57}Fe enriched iron. Good quality spectra with small statistical errors are important for these studies and a very detailed computer analysis of the spectra is essential.

Perhaps the simplest of these generally complicated compounds are the iron transport complexes, these include transferrin, lactoferrin and the siderochromes. The last of these may have molecular weights as low as 1000. They are formed in the simpler organisms. These compounds behave as strong high spin iron(III) complexes, over a wide range of temperatures they give quadrupole split paramagnetic spectra but at very low temperatures magnetic splitting ensues. The iron atoms in the solid complex are well separated and there is no magnetic coupling. An analysis of the spectra yields details of the symmetry at the iron site and the ligand field splittings. In higher organisms the iron transport compounds appear to contain two iron atoms per molecule and the spectra suggests they may occupy different sites.

23. The iron storage compounds, such as ferritin and hemosiderin, are also amongst the simpler of these compounds

They have a relatively high iron content and give somewhat simpler spectra than the other compounds that will be considered. They consist of a core of iron(III) hydroxide, usually containing some phosphate, surrounded by a protein sheath. At low temperatures they give magnetically split spectra but over some temperature range below room temperature the magnetic spectrum collapses to give a quadrupole split doublet. This suggests superparamagnetic behaviour (See section 13). The size of the iron hydroxide core can be estimated on this basis. Ferritins from different organisms, animals, plants or fungi, differ substantially in core size [49a, b, c].

24. Not surprisingly extensive work has been done on the Heme group of iron proteins

In these compounds the iron is in a quasi-octahedral environment, surrounded in the equatorial positions by four nitrogens in a porphyrin type ligand. The fifth axial position binds to a nitrogen in a histidine unit in the protein aminoacid chains. The sixth position can be occupied by oxygen, water, cerbon monoxide or other ligands, or be vacant. Hemoglobin has four such units in the molecule, myoglobin only one. Mössbauer spectroscopy has confirmed and amplified conclusions from earlier work on these compounds and uncovered new information.

In deoxyhemoglobin, where the iron is not in the plane of the four nitrogens, the spectra confirm that it contains high spin Fe^{2+}. A quadrupole split spectrum is obtained even at low temperatures, but magnetic splitting takes place in an external applied magnetic field. From the temperature dependence of these spectra information about the environment of the iron can be obtained.

Another complication is revealed by these and other data. Model porphyrin complexes of iron have shown that abnormal spin states, $S=1$ and $S=3/2$ for iron(II) and iron(III) respectively, may occur. Ferricytochrome *c* seems to contain a heme unit of this kind [50]. The deoxyhemoglobin data suggest a low lying state of spin unity. A further complication is the possibility of spin cross over with these compounds.

In oxyhemoglobin the iron is low spin and a quadrupole split spectrum is found even at very low temperatures. The unusual temperature dependence of quadrupole splitting provides information about the orientation and rotation of the oxygen around the iron oxygen bond. The O–O axis is neither parallel nor normal to the plane of the four nitrogens.

Other heme complexes can be shown to contain high spin or low spin iron(III) and there is even evidence for iron(IV) complexes [51].

25. Another rather numerous group of natural iron compounds are the iron sulphur proteins

These contain iron interacting with four sulphur atoms in a tetrahedral arrangement. In the rubredoxins all the sulphur atoms are in cysteinyl units of the protein giving units with a single iron atom. In the ferridoxins there are two bridging sulphide ligands giving a unit containing two iron atoms. There are also ferridoxins with three iron atoms with the same proportions of sulphide and cysteinic sulphur ligands and a ring . Indeed Mössbauer spectroscopy played an important role in the identification of this group [52]. Finally a very important group contain a quasi-cubic, or cubane, unit with four sulphidic atoms and four iron atoms each bonded to three sulphidic and one cysteinic sulphurs [53].

These compounds are often involved in natural redox reactions. Mössbauer spectroscopy plays an important part in the investigation of these compounds, very detailed studies are essential. In the compounds containing two or more iron atoms the question of the oxidation state of the iron atoms is complicated by valence delocalization [54].

Numerous iron containing enzymes as well as the iron molybdenum species nitrogenase involved in nitrogen fixation, have been studied by Mössbauer spectroscopy [55].

26. Differential perturbed angular correlation (DPAC)

The third technique in which radioactive decay yields information about the environment of the radioactive atom depends on the angular correlation between the directions of emission of successive photons emitted by the nucleus. In these circumstances the emission of the first photon coincides with spin polarisation of the nucleus, different spin orientations are no longer equally populated, therefore the emission of the second photon is not isotropic in relation to the direction of the first.

In the absence of any perturbing fields the relative probability of the second photon being emitted in a direction making an angle θ with the direction of the first photon is given by the sum of a number of products composed of a coefficient and a Legendre polynomial. Only the even Legendre functions are involved. The coefficients are functions of the spin quantum numbers for the nuclear states involved and the multipolarity of the emission.

Considering now a differential measurement in which the relative probability is measured a time t after the emission of the first photon, the result will be the above mentioned sum corrected for the decay of the intermediate state during the interval t. This correction term will be the only time dependent term.

If however the nucleus is perturbed by a magnetic field or an electric field gradient the spin polarisation of the intermediate state will be affected and the relative probability for emission at an angle θ becomes time dependent. This time dependent term contains the components of the electric field and the magnetic field at the nucleus, the same quantities as are obtained from the Mössbauer spectrum.

The theory of perturbed angular correlation was developed in the early 1950's [56a, b, c] but experimental data only began to appear about the time Mössbauer spectroscopy developed.

27. Experimental considerations

The nuclear state emitting the first photon, or its parent, should have a reasonably long half life, preferably longer than an hour to permit measurements over a long enough period for good statistics. The intermediate state should have a half life appreciably longer than the magnetic or quadrupole precession times and the time interval δt set by the circuit detecting coincidence of first and second photons. A convenient species for DPAC is the nuclear isomer ^{111m}Cd, the first photon has an energy 150 keV and a half life of 48.6 min, rather shorter than desirable. The second photon of 247 keV is emitted by a state with a half life of 84 ns. With a coincidence resolution time of 2 ns the time dependence can be followed over some hundreds of nanoseconds. Such measurements on a cadmium compound labelled with the nuclear isomer will yield the same information as a hypothetical Mössbauer spectrum of the compound, except that nothing corresponding to the isomer shift will be obtained.

Unfortunately there are not very many suitable nuclear isomers available [57]. Several more systems become available if one produces the state emitting the initiating photon by another radioactive decay processes. But this introduces a complicating factor; how far has the decay of the parent species affected the chemical state of the species emitting the first photon?

The recoil following soft β particle emission is generally too small to effect bond rupture but both β decay and orbital electron capture lead to considerable electronic disturbance. In metals and conductors such disturbance is quickly repaired and PDAC is entirely suitable for measuring electric field gradients and magnetic fields. In insulators however the measurements may be more useful for what they tell one about the chemical effects of the decay of the parent species.

Sometimes both DPAC and Mössbauer data are available for a compound. Agreement of the electric field gradient values obtained by the two methods is satisfactory. Discrepancies can be attributed to the effects of the decay. DPAC measurements are possible producing excited states of ^{111}Cd by β decay of ^{111}Ag and by orbital electron capture in ^{111}In. In both cases comparison with nuclear isomer data shows that both decay processes lead to some ^{111}Cd in abnormal environments.

An interesting simple result is to find that a closed shell parent almost always leads to a closed shell decay product [58].

DPAC is especially well suited to exploring relaxation effects taking place faster than the life time of the intermediate state, such as a fluctuating electric field gradient. Unlike Mössbauer spectroscopy it can be used with solutions or gases. In solutions it provides a means of measuring the rotation period of large molecules after labelling with a suitable radioactive species [58].

References

1. (a) Bouchez, R., Daudel, R., Daudel, P., Muxaet, R.: J. Phys. Radium **8**, 336 (1947).
(b) Daudel, R.: Rev. Sci. **87**, 162 (1947).
2. Segré, E.: Phys. Rev. **71**, 274 (1947).
3. Johlige, H. W., Aumen, D. C., Born, H. J.: Phys. Rev. **C2**, 1616 (1970).
4. Remeikis, V., Dragunas, A., Makariunas, K., Ruzece, B.: Izv. Akad. Nauk S.S.R. Ser. Fiz. **52**, 2 (1988).

5. Makariunas, K., Dragunas, A., Makariuniene, E.: Hyperfine Interact. **36**, 211 (1987).
6. (a) Geresimov, V. N., Zelekov, A. G., Kulakov, V. M., Pchelin, V. A., Sokolovskaya, M. V., Soldavac, A. A., Chistyakov, L. V.: Zh. Eksp. Teor. Fiz. **86**, 1169 (1984).
(b) Bainbridge, K. T., Goldharber, M., Wilson, E.: Phys. Rev. **90**, 430 (1953).
(c) Makariunas, K., Makariuniene, E., Dragunas, A.: Liet. Fiz. Rinkinys **22**, 114 (1982). [Chem. Abstr. **98**, 186880].
7. (a) Mazaki, H., Shimizu, S.: Phys. Rev. **148**, 1161 (1966).
(b) Nève de Mevergnies: Phys. Lett. **32B**, 482 (1970).
(c) Koltsov, V. V.: Izv. Akad. Nauk S.S.R. Ser. Fiz. **57**, 100 (1993).
8. de Benedetti, S., de S. Barros, F., Hoy, G. R.: Ann. Rev. Nucl. Sci. **16**, 31 (1966).
9. Pyper, N. C., Harston, M. R.: Proc. Roy. Soc. Lond. **420A**, 277 (1988).
10. Harston, M. R., Pyper, N. C.: J. Phys. **17B**, L839 (1984).
11. Harston, M. R., Pyper, N. C.: Phys. Rev. **48A**, 268 (1993).
12. Harston, M. R., Pyper, N. C.: Phys. Rev. Lett. **56**, 1790 (1986).
13. (a) v.e.g. Zhongua Chen, Rosenberg, L., Spruch, L.: Adv. At. Mol. Opt. Phys. **26**, 297 (1990).
(b) Janecke, J. W., Becchetti, F. D., Roberts, D. A.: Phys. Rev. **47C**, 2954 (1993).
14. Mössbauer, E.: Z. Physik **151**, 124 (1958).
15. Hanna, S. S., Heberle, J., Littlejohn, C., Perlow, G. J., Preston, R. S., Vincent, D. H.: Phys. Rev. Lett. **4**, 177 (1960).
16. (a) Kistner, O. C., Sunyar, A. W.: Phys. Rev. Lett. **4**, 412 (1960).
(b) de Benedetti, S., Lang, G., Ingalls, R.: Phys. Rev. Lett. **6**, 60 (1961).
17. (a) Kerler, W., Neuwirth, W., Fluck, E., Kuhn, P., Zimmermann, B.: Z. Physik **173**, 321 (1963).
(b) Duncan, J. F., Wigley, P. W. R.: J. Chem. Soc. 1963, 1120.
18. Ercolani, C., Gardini, M., Pennesi, G., Rossi, G., Russo, V.: Inorg. Chem. **27**, 422 (1988).
19. v.e.g. Dattagupta, S.: in *Mössbauer Spectroscopy*, Eds. D. P. E. Dickson and F. J. Berry, Camb. Univ. Press (1986).
20. Xiao Q. Ding, Bominaar, E. L., Eckhard, B., Heiner, W., Trautwein, A. X., Druece, S., Chaudhuri Phalgun, Winkler, K.: J. Chem. Phys. **92**, 178 (1990).
21. Verwey, F. J. W.: Nature **144**, 327 (1939).
22. (a) Bauminger, R., Cohen, S. G., Marinov, A., Ofer, S., Segal, E.: Phys. Rev. **122**, 1447 (1961).
(b) Banerjee, S. K., O'Reilly, W., Johnson, C. E.: J. Appl. Phys. **38**, 1289 (1967).
23. v.e.g. Wohler, S. E., Williams, R. J., Oh, S. M., Hendrickson, D. N., Inniss, D., Strause, C. E.: J. Am. Chem. Soc. **108**, 2938 (1986).
24. Kramer, J. A., Hendrickson, R. N.: Inorg. Chem. **19**, 3330 (1980).
25. For review see (a) Gütlich, P.: Struct. Bonding **44**, 83 (1981).
(b) Konig, E., Ritter, G., Kulshreshtha, S. K.: Chem. Rev. **55**, 219 (1985).
26. v.e.g. Bancroft, G. M., Dharmarwardena, K. G., Maddock, A. G.: J. Chem. Soc. A, 1969, 2914.
27. v.e.g. Chevalier, R., Coey, J. M. D., Bouchez, R.: J. Physique **37, C6**, 861 (1976).
28. Clark, M. G., Maddock, A. G., Platt, R. H.: J. Chem. Soc. Dalton, 1972, 281.
29. (a) Parish, R. V.: in *Mössbauer Spectroscopy applied to Chemistry, Vol. 1*, Ed. G. J. Long, Plenum Press, 1984.
(b) Marks, A. P., Drago, R. S., Herber, R. H., Potasek, M. J.: Inorg. Chem. **15**, 257 (1976).
30. Jones, P. G., Maddock, A. G., Mays, M. J., Muir, M. M., Williams, A. F.: J. Chem. Soc. Dalton 1977, 1434.
31. For review see Thomas, M. F., Johnson, C. E.: Chapter 4 in *Mössbauer Spectroscopy*, Eds. D. P. E. Dickson and F. J. Berry, Camb. Univ. Press, 1986.
32. v.e.g. Kostikas, A., Simopoulos, A., Gangas, N. N. J.: Phys. Status Solidi **42**, 705 (1970).
33. Mørup, S., Dumesic, J. A. Topsoe, H.: in *Applications of Mössbauer Spectroscopy*, Ed. R. L. Cohen, Academic Press N.Y., 1980.
34. For review see Berry, F. J.: Trans. Met. Chem. **4**, 209 (1979).
35. Liljquist, D., Bodlund-Ringstrom, B.: Nucl. Instrum. Methods **160**, 131 (1979).
36. (a) Sete Camara, A., Keune, W.: Corrosion Sci. **15**, 441 (1973).
(b) Simmons, G. W., Kellerman, E., Leidheiser, H.: Corrosion **29**, 227 (1973).
37. Przybyski, M., Koreki, J., Gradmann, V.: Appl. Phys. **A52**, 33 (1981).
38. Berry, F. J., Maddock, A. G.: Chem. Commun. 1978, 308.
39. Spijkerman, J. J.: in *Mössbauer Methodology* **7**, 86 (1971).
40. (a) Pollak, H., de Coster, M., Amelinkx, S.: Phys. Status Solidi **2**, 1653 (1962).
(b) Bancroft, G. M.: *Introduction to Mössbauer Spectroscopy for Inorganic Chemists and Geochemists*, Van Nostrand, 1973.
41. See Proc. 1st., 2nd., 3rd. and 4th. Lunar Sci. Confs. Published in Supplements to Geochim. Cosmochim. Acta 1970–1973.
42. Niebuhr, H. H., Zeira, S., Hafner, S. S.: in Proc. 4th. Lunar Sci. Conf. 1973.
43. Bancroft, G. M.: Chem. Geol. **5**, 255 (1970).
44. Bancroft, G. M., Burns, R. G., Maddock, A. G.: Am. Mineral. **52**, 1009 (1967).
45. Evans, B. J., Amthauer, G.: J. Phys. Chem. Solids **41**, 985 (1980).
46. Bancroft, G. M., Osbourne, M. B., Fleet, M. E.: Solid State Commun. **47**, 623 (1983).
47. For review see Bowen, L. H., Webb, S. B.: in *Mössbauer Spectroscopy and its chemical applications*, Advances in Chemistry Series, ACS No 194, 247 (1981).
48. Williamson, D. L., Guettinger, T. W., Dickerhoof, D. W.: Advances in Chemistry, ACS 194, 177 (1981).
49. (a) Boas, J. F., Window, B.: Aust. J. Phys. **19**, 573 (1966).
(b) Oosterhuis, W. T., Spartalian, K.: in *Applications of Mössbauer Spectroscopy*, Ed. R. L. Cohen, Academic Press, Vol. 1, 1976.
(c) Bauminger, E. R., Harrison, P. M., Hechel, D., Nowik, I., Treffry, A.: Biochem. Biophys. Acta **48**, 1118 (1991).
50. Maltempo, M. M., Moss, T. H., Spartalian, K.: J. Chem. Phys. **73**, 2100 (1980).
51. For review see Spartalian, K., Long, G.: in *Applications of Mössbauer Spectroscopy*, Ed. R. L. Cohen, Academic Press, Vol. 2, 1980.
52. Huynh, B. H., Moura, J. J. G., Moura, I., Kent, T. A., Le Gall, J., Xavier, A. V., Munck, E.: J. Biol. Chem. **225**, 3242 (1980).
53. For review see Trautwein, A. X., Bill, E., Bominaar, E. L., Winkler, H.: Struct. Bonding **78**, 1 (1991).
54. Dickinson, D. P. E., Johnson, C. E.: in *Structural and Resonance Techniques in Biological Research*, Ed. D. L. Rousseau, Academic Press 1984.
55. Newton, W. E., Chelier, S. F., Sands, R. H., Durham, W. R.: Biochem. Biophys. Res. Commun. **162**, 882 (1989).
56. (a) Goertzel, G.: Phys. Rev. **70**, 897 (1946).
(b) Alder, K., Albers-Schonberg, H., Heer, E., Novey, T. B.: Helv. Phys. Acta **26**, 761 (1953).
(c) Abragam, A., Pound, R. V.: Phys. Rev. **92**, 943 (1953).
57. Shirley, D. A., Haas, H.: J. Chem. Phys. **58**, 3339 (1973).
58. For review see Boyer, P., Baudry, A.: in *Hot Atom Chemistry*, Ed. Tatsuo Matsuura, Kodansha Press, 1984.

Radiochimica Acta 70/71, 333–341 (1995)

X-Rays and Radiochemistry

By H. K. Yoshihara

Japan Isotope Data Institute, Nishikatsuyama 25-20, Aobaku, Sendai 981, Japan

(Received May 29, 1995; revised June 19, 1995)

X-Rays / Historical review / Analysis of radioactive materials / PIXE / TXRF / Chemical effects

Abstract

The contribution of X-rays to the development of radiochemistry is reviewed. Historical accounts are presented together with the present status of the applications of X-rays, including PIXE and synchroton radiation, to structural and analytical chemistry. Finally, the chemical effects on X-ray emission and their application to chemical speciation are discussed.

1. Introduction

X-rays were discovered by Roentgen in November 1895, less than 4 months before Becquerel's discovery of radioactivity. The succession of both events appears logical, because X-rays are generated in the inner electron shells of the atom, while radioactivity is a property of the atomic nucleus itself. The high penetrating power of the X-rays was surprising to scientists who were familiar only with visible, IR and UV radiations. Radioactivity was accompanied by the emission of even more energetic and penetrating radiations.

The recognition of X-rays not only imposed a revolutionary change in the concept of the atom proposed by Dalton and his contemporaries, but opened the way to the discovery of radioactivity on grounds of the relation between X-rays and fluorescence. The famous French mathematician Henri Poincaré suggested that since X-rays seem to emerge from the fluorescence of the walls of a Crookes tube, any other fluorescence, whatever its origin, should also be a source of X-rays. This observation prompted Henri Becquerel to verify whether the phosphorecence of uranium salts, a phenomenon already investigated by his father, was accompanied by the emission of X-rays. As is well known, Becquerel did not observe X-rays but rather a new kind of radiation which was able to blacken a photographic plate. The origin of this emission was the radioactivity of uranium, which was confirmed later by the Curies.

The chemical action of X-rays was gradually elucidated in the years following 1910 by S. Lind. This aspect is relevant to radiation chemistry and will not be covered in this review.

In 1914, Moseley established his law which relates the wavelength of X-rays to the atomic number of the associated element. This law was applied successfully by G. von Hevesy, the pioneer of the use of X-rays in chemistry and 1944 Nobel Laureate, to the identification in 1923 of a new element with atomic number 72 (hafnium).

At that time, Hevesy, a Hungarian of noble birth, was working at the Niels Bohr Institute in Copenhagen and was helped by Coster, a Dutch specialist on X-ray spectroscopy. The fruitful collaboration between the chemist and the physicist was stimulated by the international and interdisciplinary atmosphere which prevailed at the Institute. The new element discovered by Hevesy and Coster resembled Ti and Zr, contrary to the belief that it should belong to the group of rare earths. The discovery of hafnium supported the atomic theory of Bohr, who heard about the discovery during his stay in Stockholm on the occasion of his Nobel Prize lecture delivered in 1922.

A young Japanese scientist, attracted by Bohr's reputation, joined the Institute in Copenhagen. This was Y. Nishina, who was to become famous for his quantum mechanical calculation, together with Klein, of the cross section of Compton scattering, and later also for the discovery of the symmetric fission of uranium induced by fast neutrons. In 1925, Nishina and Coster successfully performed the first quantitative chemical analysis based on X-rays. They used standards of Ta and Lu to determine the amount of Hf, a practice which was later extensively applied in radioactivation analysis, the major radioanalytical method introduced by Hevesy in 1936.

Coster was also interested in the chemical effects of the X-ray absorption edge, which had been observed in the early 1920's, by Bergenbren, Lindh and others. He suggested that the chemical shift of the absorption wavelength was closely related to the deformation of the outer electron shells caused by the chemical state of the atom investigated.

Nishina extended his stay in Copenhagen and showed in 1927 that the chemical shift of the X-ray absorption in Ca, Cl and S could be explained quantitatively in terms of the lattice energy and of the character of chemical bonding. This work was achieved in cooperation with two other Japanese visitors in Copenhagen, Aoyama and Kimura. The latter is well known for the discovery of the first nuclide in the neptunium family, the β^- emitter ^{237}U.

Inspired by the spirit of the Bohr Institute, Nishina was to become the most influential physicist for the

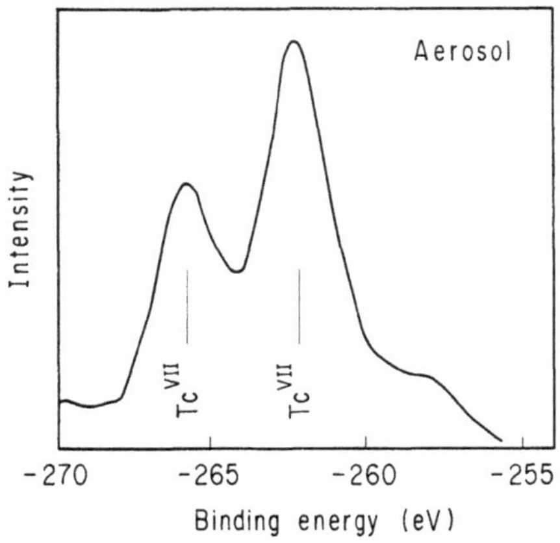

Fig. 1. The XPS spectrum of aerosol from high level wastes [3].

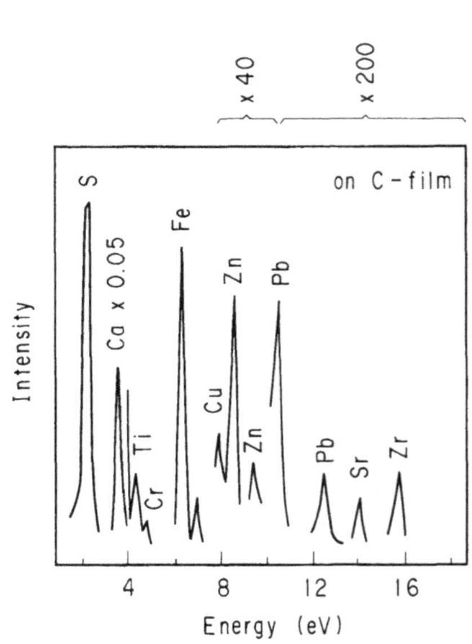

Fig. 2. PIXE analysis of atmospheric deposits on a carbon film [5].

following generation of Japanese nuclear scientists, as pointed out by Segré in his book "From X-rays to Quarks" published in 1976. This role has been acknowledged by Yukawa, 1949 Nobel Laureate for his prediction of the existence of the meson.

Nishina's impact was not limited to the physicists' circles, since he also stimulated interest on nuclear chemistry. Back in Japan, Kimura, who had participated in Hevesy's work, undertook the search for transuranium elements at the Physical and Chemical Research Institute of Tokyo shortly after McMillan's discovery of neptunium in 1940. One of his students, N. Saito, was for a time Director at the International Atomic Energy Agency. Kimura's laboratory was well equipped with a Siegbahn X-Ray spectrometer for analytical purposes. The present author was one of Kimura's youngest students.

The importance of X-rays in radiochemistry is well established today. Striking examples are X-ray measurements in particle-induced X-ray emission (PIXE), X-ray diffraction studies of Tc compounds and actinide elements, structural analysis by XAFS (including EXAFS and XANES) and X-ray fluorescence (XRF) analysis, including total reflection X-ray fluorescence (TXRF) of synchrotron radiation (SR).

However, X-rays have further potentialities. The chemical effects on X-ray characteristics are being elucidated step by step. Progress in the field is expected with the advent of instruments with higher resolution. Speciation of elements based on chemical effects has a promising future and recent studies in this field are included in the review.

2. X-rays in the analysis of radioactive materials

For a time, X-ray measurements were considered relatively inconvenient when compared to the detection of the radiations emitted by radioactive nuclides. The demerits were overcome with the advent of semiconductor detectors in the 1960's. Si(Li) detectors are now extensively used in energy-dispersive X-ray spectrometers (EDX). Compared to the wavelength-dispersive systems (WDX), which resolve the X-rays with an analyzing crystal coupled to a goniometer, they offer the advantages of a higher counting efficiency, a shorter measurement time and easier data treatment. However, the resolution of the EDX systems is definitely below that of the other type of spectrometer.

EDX spectrometers with X-ray tubes are widely used for the multielemental and fast analysis of numerous materials. They are convenient for the determination of U and Th, but the sensitivity is relatively poor. A more sophisticated analysis is performed on small samples with an electron probe X-ray microanalyzer (EPMA). With this technique, Bradbury *et al.* [1] first demonstrated that the fission products Ru, Mo, Ce and Ba existed as solid solutions in slices of irradiated UO_2 pellets. Interestingly, metallic inclusions of noble metals such as Ru, Rh and Pd were also found as a consequence of the redistribution of oxygen in irradiated fuels. Cladding interactions in neutron-irradiated MOX fuels (UO_2-PuO_2) were studied by Johnson *et al.* [2]. With EPMA, these authors showed that intergranular attack of the stainless steel cladding had occurred at the fuel-cladding interface; in particular, fission-produced iodine played an important role in the corrosion of the cladding.

An energy-dispersive spectrometer combined with a scanning electron microscope (SEM-EDX) is a powerful tool for drawing an elemental map of the material to be measured. Baumgärtner and coworkers [3] investigated the chemical behaviour of technetium in aerosols generated from high level wastes by means of SEM-EDX, Auger electron spectroscopy (AES) and X-ray photoelectron spectroscopy (XPS or ESCA). All these methods rely on X-rays, either as the excitation source or as the radiation emitted in relaxation processes. The authors found that the behaviour of Tc in the map was similar to that of Cs measured by SEM-EDX and AES. Since the XPS spectrum revealed Tc as pertechnetate, the authors concluded that $CsTcO_4$ was present in the aerosol (Figure 1). This observation

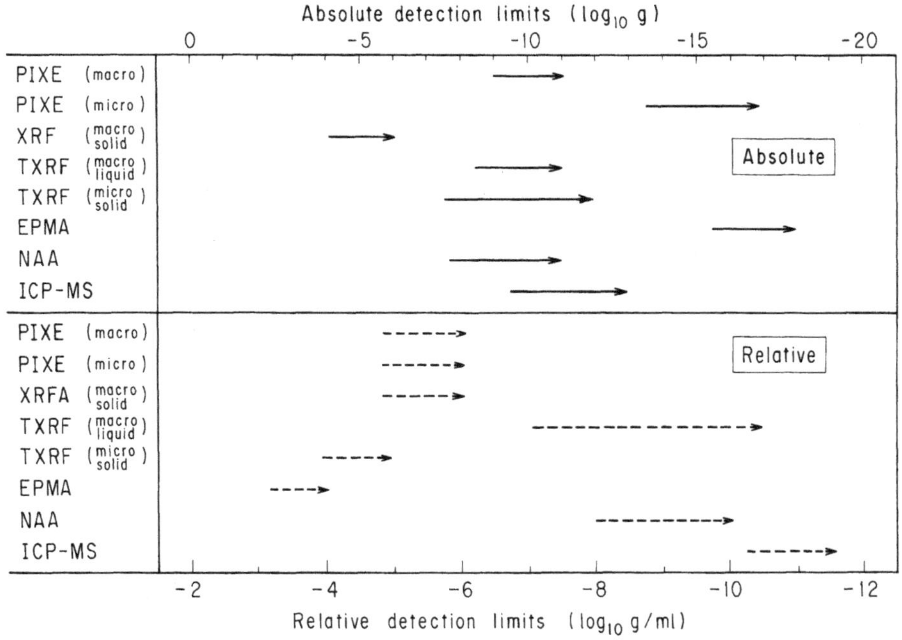

Fig. 3. Comparison of the detection limit of PIXE with those of other methods. XRF: X-ray fluorescence, TXRF: total reflection X-ray fluorescence, EPMA: electron probe X-ray microanalysis, NAA: neutron activation analysis, and ICP-MS: inductively coupled plasma mass spectrometry.

is compatible with the known behaviour of fission iodine in spent nuclear fuels that combines with Cs as CsI [4].

3. Analysis of environmental and biological samples by PIXE

Protons, deuterons, helium ions and heavy ions delivered by accelerators have large interaction cross sections with atomic electrons, followed by the emission of X-rays, thus enabling a highly sensitive and multi-elemental analysis of various materials. This is the essence of the PIXE method (particle-induced X-ray emission), which has been applied successfully to the analysis of trace elements since its introduction in 1970 by Johansson and coworkers [5]. Before Johansson's decease in 1992, the PIXE method had been recognized as a standard technique in analytical chemistry with a wide range of applications in physics, chemistry, biology and medicine, as well as in technology.

The first example of PIXE analysis presented by Johansson *et al.* is shown in Figure 2 [5]. A carbon film was exposed outdoors overnight and the X-ray spectrum of the sample bombarded for one hour with 5 MeV protons was measured with a Si(Li) detector. The characteristic X-rays of the elements S, Ca, Ti, V, Cr, Fe, Cu, Zn, Sr, Zr and Pb were clearly identified.

The sensitivities of PIXE and of other methods are compared in Figure 3. The absolute sensitivity is high for PIXE, but the method is less efficient on a relative sensitivity scale. It can be improved with preconcentration procedures and detection of trace elements in water at the ng/l level has been reported [6]. The detection limit of PIXE, as well as that of other analytical methods, is greatly enhanced by the use of microbeams (μ PIXE).

PIXE is very commonly applied to environmental samples and is particularly useful for the analysis of aerosols. Cahill suggested the establishment of an international network for the control of fine sulfate aerosols which arise largely from human activities [7]. Data useful for meteorology, based on PIXE measurements of aerosols in Europe, have recently been obtained by Koltay [8]. A model has been established to estimate the relative fractions of anthropogenic aerosols arising from local emissions and from sources in neighbouring countries.

Environmental pollution caused by studded car tires has been demonstrated by a Japanese group using the PIXE method [9]. It was found that at the beginning and end of the snow-fall season in the northern districts of Japan (Sapporo, Sendai), the erosion of asphalt roads by studded tires released dusts. The profile of elements found in dusts, shown in Figure 4, is in good agreement with that of asphalt, i.e., rich in Ca and quite different from that in cities (e.g. Nagoya) where studded tires are not in use. Legal regulation for the use of studded tires was established in consequence.

Biology and medicine are also important fields for application of PIXE. Human blood serum and hair samples have been analysed by Valković [10]. The uptake of environmental pollutants by hair has obvious implications in public health.

Yumoto *et al.* have analysed the content of Al in brain [11]. The element is suspected to contribute to Alzheimer's disease. Upon administration (oral or by injection) of Al in rats, the metal accumulates in the

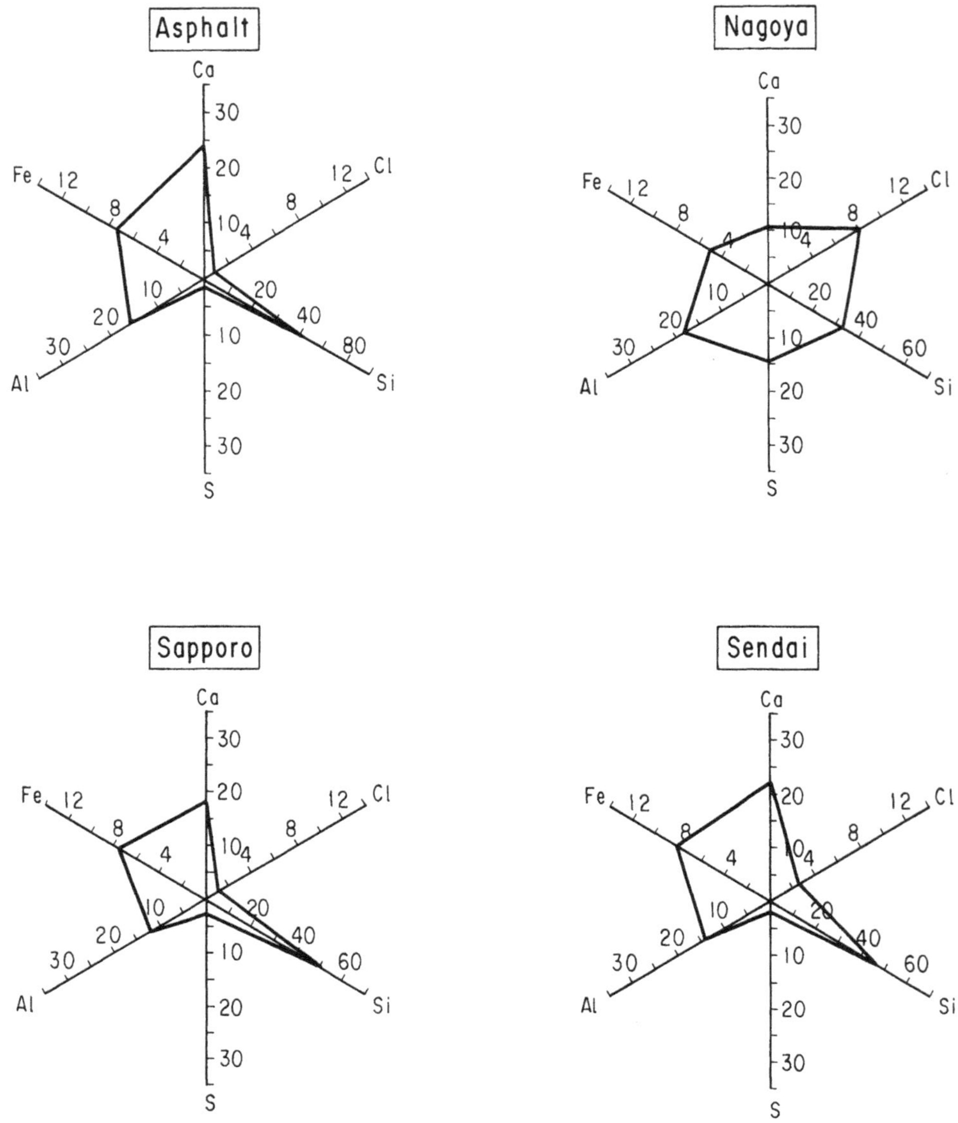

Fig. 4. The elemental concentration of six main elements for asphalt, and for airborne particulates in Sapporo, Sendai and Nagoya [9].

brain and morphological changes occur simultaneously. The authors concluded that Alzheimer's disease is caused by irreversible accumulation of Al in the brain, as well as in the nuclei of brain cells.

For the examination of fine art and paintings, the object is bombarded with a proton beam extracted from the accelerator window. The non-destructive PIXE technique with an external beam is convenient for surface analysis of precious samples of very small sizes. Works of art from the Louvre and other museums have been analysed [12]. The external PIXE milliprobe was used in the investigation of the 42-line Bible printed by Gutenberg in the 1450's. Different lots of papers used for the document could be distinguished on the basis of their Fe/Ca and Mn/Ca ratios [13].

With the development of a focussing device for accelerated ions, a powerful method for scanning microprobes became available. μ PIXE using a scanning proton, microbeam has been applied by Vis *et al.* to biomedical samples (serum, tissues and hair) [14]. An elemental map drawn with the same technique has provided useful information on the formation of aerosols, Al uptake in beech plant roots [15] and in sphagnum leaves [16]. The usefulness of μ PIXE in life sciences, geosciences, environmental sciences, archaeometry, material science and in industry has been demonstrated with the scanning proton microprobe at the University of Oxford [17].

In the practice of PIXE analysis, pile-up and charge buildup must be avoided. Peisach *et al.* have noted an abnormal X-ray yield for insulating fluoride compounds [18]; in the case of CuF_2 the enhancement exceeded 1000 for the Cu K X-rays with respect to metallic Cu [19]. This anomaly was ascribed to charge buildup followed by a discharge which could generate extra X-rays. It has been pointed out that the phenomenon might find applications in the production of monoenergetic ultra-soft X-rays in ESCA studies [20].

Applications of PIXE are found in the proceedings of the conferences "Particle induced X-ray Emission and its Analytical Applications" (in Nuclear Instruments and Methods in Physics Research B: Special issues) and in the International Journal of PIXE.

4. Synchrotron radiation (SR) in structural and analytical chemistry

Synchrotron radiation is emitted when relativistic charged particles move along a curved path in a magnetic field. Many synchrotron facilities are in operation in the world, and provide a wide range of spectra with wavelengths extending from the infrared to the X-ray region. In this section, applications of SR to structural and analytical problems related in some way to radiochemistry are described.

4.1 EXAFS using synchrotron radiation

Extended X-ray absorption fine structure (EXAFS) studies of Tc compounds have been performed by the Cincinnati group [21]. The oligomer/polymer structure in aqueous solution of the bone-seeking agent Tc-MDP (MDP = methylene diphosphonate) was elucidated. An Australian group has been concerned with the determination of the μ-oxo or bis(μ-oxo) dimer of Tc compounds [22]. The two varieties yield different EXAFS pattern, thus enabling confirmation of the presence of the bis(μ-oxo) dimer in the nitridotechnetium complex.

4.2 TXRF using synchrotron radiation

Total reflection X-ray fluorescence spectrometry (TXRF) was pioneered by Yoneda and applied to spectrochemical analysis [23]. The technique is widely used for surface analysis. Superficial concentrations of elements and impurities on Si wafers have been measured with a detection limit down to 10^9 atoms cm^{-2} [24]. The analysis is rapid, multielemental and non destructive.

The method has been combined with intense SR sources. The very high sensitivity of SR-TXRF has been demonstrated for the elements K to Zn [25, 26]. The detection limit is enhanced with polarized SR beams, with a limit of 3 pg for Co [27].

5. Chemical effects on X-rays characteristics

The chemical shift of X-ray wavelengths is known since the 1920's, but detailed features of chemical effects on the properties of X-rays have been elucidated only within the last two or three decades. In the simple case of X-ray emission, four different types of chemical effects are observed (Figure 5):

(a) energy (wavelength) shift; (b) change in the emission probability (intensity); (c) appearance of satellite lines as a result of multiple electron vacancies and electron pick-up from central atoms; (d) change of the peak profile (broadening) [28]. Although these effects may sometimes perturb X-ray measurements, they are very useful for chemical state analysis and chemical speciation [29].

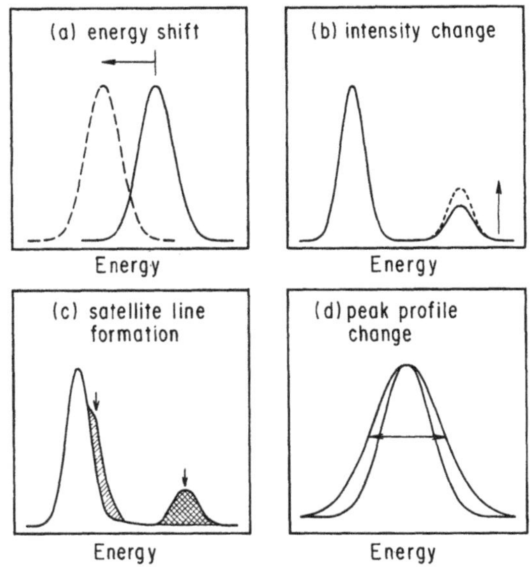

Fig. 5. Chemical effects on X-ray spectra.

5.1 X-ray energy shift

X-ray energy shifts due to chemical effects have been reported for various elements. Assignment to specific chemical states has been attempted for S [30−33], Al [34−36], Si [31], Sn [31] and other elements. For example, Gohshi determined the distribution of sulfur species in coke (or coal) with a high resolution X-ray spectrometer equipped with two crystals [37]. Figure 6 shows the spectra of thiosulfate, which serves as standard, and of coke. The coke spectrum was deconvoluted into two components corresponding to S^{6+} and S^{2-}. Typical analyses of coal samples are given in Table 1. Kawai *et al.* have found a linear relationship between the chemical shift of the Sc $K\alpha$ line and the electronegativity of the counter atom [38].

Recently, Iihara and coworkers have reported chemical shifts in iodine and antimony compounds with equipment similar to that of Gohshi [39]. The energy shifts for the $L\gamma_4$, $L\gamma_1$ and $L\beta_{2,15}$ iodine lines are shown in Table 2. The most important shift is observed for $L\gamma_4$ owing to the location of iodine valence electrons in the corresponding level.

Sugiura measured the X-ray emission from the valence band in metal complexes [40−42]. The $L\beta_{2,15}$ lines of Pd and Pt and the chlorine $K\beta$ line have been recorded for the square planar complexes K_2PdCl_4, $(NH_4)_2PdCl_4$, trans-$[Pd(NH_3)_2Cl_2]$, $[Pd(NH_3)_2Cl_2] \cdot H_2O$, K_2PtCl_4, cis- and trans-$[Pd(NH_3)_2Cl_2]$. The spectra of geometrical isomers are different.

5.2 Change in the X-ray emission probability

A change in the intensity of X-ray emission with the chemical state has been observed at Tohoku University

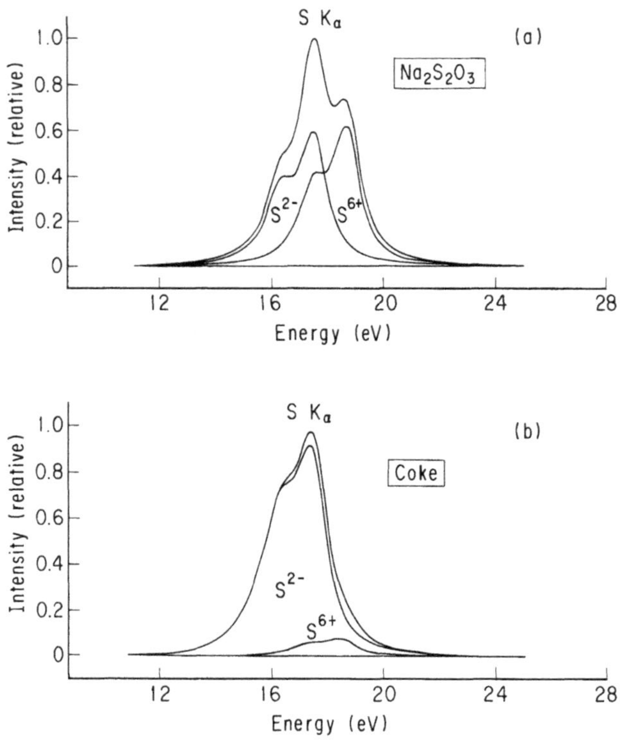

Fig. 6. Sulfur $K\alpha$ spectrum and its components; (a) sodium thiosulfate and (b) coke [31].

Table 1. Analysis of sulfur composition by X-ray fluorescence (weight%)

Sample	Total	S^{6+}	S^{2-}
A	0.31	0.02	0.29
B	0.99	0.21	0.78
C	0.36	0.06	0.30
D	0.70	0.25	0.45
E	0.51	0.04	0.47
F	0.22	0.01	0.21
G	0.37	0.03	0.34
H	2.71	0.37	2.34

Table 2. X-ray energy shift of iodine compounds

Compound	Energy shift (eV)		
	$L\gamma_4$	$L\gamma_1$	$L\beta_{2,15}$
NaI	0.94	−0.36	−0.44
KI	0.82	−0.51	−0.41
CuI	1.35	−0.52	−0.57
SnI_2	0.60	−0.68	−0.60
HIO_3	−0.11	−0.37	−0.46
KIO_3	−0.33	0.06	0.01
KIO_4	0.00	0.00	0.00

in Sendai in the electron capture decay of ^{51}Cr [43] and in the isomeric transition of ^{99m}Tc [44]. The $K\beta_2/K\alpha$ intensity ratio for ^{99m}Tc has been compared to that of ^{95m}Tc and ^{97m}Tc. Mukoyama *et al.* [45] used the simplified model of Brunner [46] to correlate successfully the change with the effective charge on Tc. The changes are small, being 10−20% for chromium and at maximum 4% for technetium, because the electronic transitions at the origin of the emission lines occur between the subvalence shells $3p$ for Cr and $4p$ for Tc.

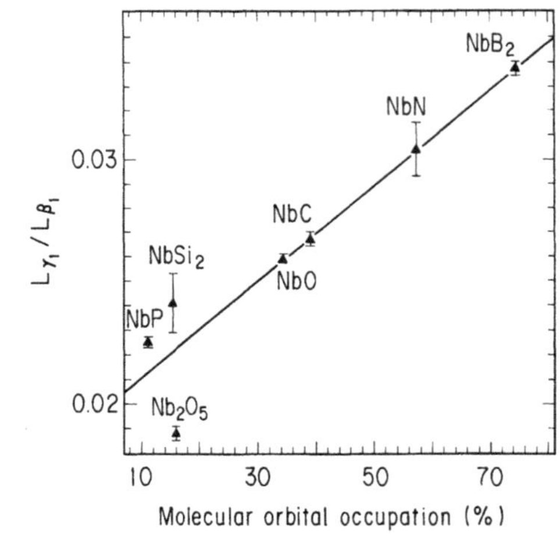

Fig. 7. The $L\gamma_1/L\beta_1$ X-ray intensity ratios for various niobium compounds bombarded with protons as a function of molecular orbital occupation [47].

More significant effects on the X-ray intensity ratio have been observed in the PIXE $L\gamma_1$ lines of niobium and molybdenum. Iihara *et al.* have measured the $L\gamma_1/L\beta_1$ intensity ratio in proton-bombarded Nb and Mo compounds [47]. As shown in Figure 7, the intensity ratio in Nb increases with the molecular orbital occupation calculated by the GAMESS-MO method. The maximum change is as high as 60%, demonstrating the effectiveness of direct transitions starting with the $4d$ valence shell. The linear relationship seen in the figure could be explained by the change in the electron density in the orbital concerned. The emission probability T is

$$T = A\, N_i E_x^3 D_{i,j}^2 \qquad (1)$$

where A is a constant, N_i the electron density in the orbital i, E_x the X-ray transition energy and $D_{i,j}$ the dipole matrix element. The subscripts i and j denote respectively the initial and final orbitals in the transition. The terms E_x and $D_{i,j}$ are only slightly dependent on the chemical state. Therefore the intensity is essentially determined by the population of the molecular orbital.

The Tohoku group has also measured X-ray intensity ratios in iodine and antimony. The chemical effects are more pronounced for the $L\gamma_4/L\beta_3$ ratio (with a maximum of 60 to 70%) than for $L\gamma_1/L\beta_1$ or $L\beta_{2,15}$ because the $L\gamma_4$ X-ray line corresponds to transitions involving valence shells, while the latter lines arise from subvalence shells.

In general, electronic transitions from the valence shell are more sensitive to the chemical environment than those occurring in subvalence shells. This trend has been verified by comparing in the same chromium compounds the $L\beta/L\alpha$ ratio to the $K\beta/K\alpha$ ratio in the Cr X-ray spectrum [48].

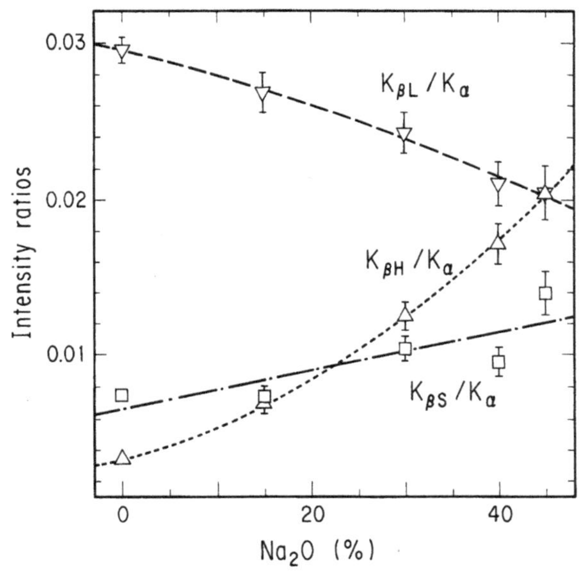

Fig. 8. The Si $K\beta/K\alpha$ intensity ratios against the content of Na_2O in glasses [49].

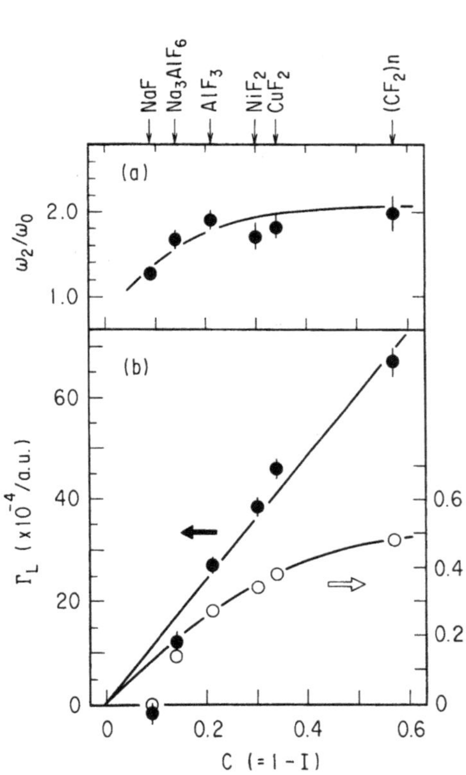

Fig. 9. The fluorescence yield ratio ω_2/ω_0, the L-shell refilling probability f and the L-shell width Γ_L plotted as a function of covalency C [52].

A large variation of the $K\beta/K\alpha$ ratio has been observed in Na_2O-SiO_2 glasses as a function of the Na_2O content [49]. The intensity ratios of fine structure components ($K\beta_S$, $K\beta_L$ and $K\beta_H$) of the $K\beta$ line to $K\alpha$ change remarkably with the Na_2O content. The intensity of $K\beta_L$ decreases while the intensities of $K\beta_H$ and $K\beta_S$ increase with Na_2O content, as shown in Figure 8. This behaviour has been explained theoretically by assuming direct interactions between Si and Na.

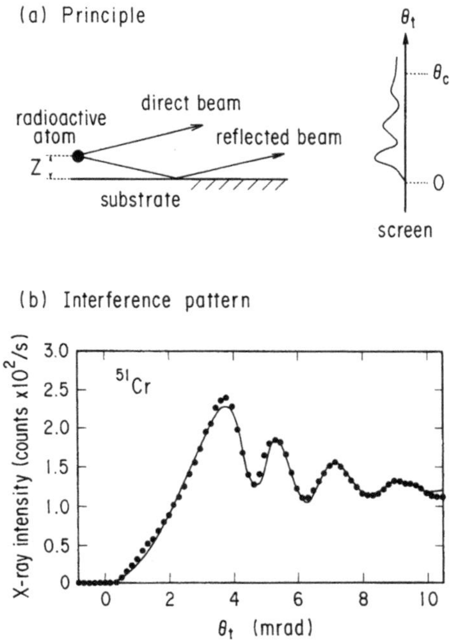

Fig. 10. The interference pattern from radioisotope-originated X-rays: (a) principle and (b) a result for ^{61}Cr-labeled arachidate [57].

5.3 Satellite lines

Satellite lines in X-ray spectra have various origins, such as multivacancy satellites and charge transfer satellites. In particular, satellite peaks appear in PIXE spectroscopy with heavy ions. The intensity of the KL^n peak follows a binomial probability distribution

$$I(KL^n) = \binom{8}{n} P_L^n (1 - P_L^n)^{8-n} \tag{2}$$

where P_L is the probability of ionization of the L shell and n the number of L vacancies. As pointed out by Watson *et al.*, the KL^n satellite line is closely related to the electronegativity of the neighbouring atoms of the X-ray emitter [50]. Results obtained by Benka *et al.* indicated a strong influence of the chemical environment on the second and third peaks in the spectra [51]. Uda *et al.* found that various parameters were related to the covalency term C (or $1-I$, where I is the ionicity): the fluorescence yield ratio ω_2/ω_0 of the first to the third peak, the probability f of refilling the L shell, and the L shell width Γ_L, as shown in Figure 9 [52]. The authors suggested that chemical speciation could be achieved with these relations. Cr $K\alpha$ satellite lines emitted in the bombardment of four chromium targets with 14 MeV/u O^{4+} were clearly resolved by Folkmann with a 20×50 mm analyzing crystal [53]. As in the previous cases [51, 52], chemical effects appear in the second and third peaks.

Charge transfer satellites in Be K X-ray spectra formed by pick-up of electrons from the L shell in oxygen have been observed in the bombardment of BeO with nitrogen or oxygen ions [54].

Quantum mechanical calculations have been performed by Kawai *et al.* to interpret the formation of charge transfer satellites [55]. The $K\beta''$ satellite in the

Sc X-ray spectrum is strong in ScF_3, but weak in Sc_2O_3. This pattern was assigned to electron transitions from an antibonding MO involving Sc 3 p and ligand 2 s electrons to the 1 s^{-1} hole. Strong $K\alpha''$ and $K\beta''$ satellites appear in metallic Sc owing to a better screening of the core hole by electrons in the conduction band than by 3 p^{-1} holes. The $K\beta''$ satellite was assigned to the hole transition from 1 s^{-1} to an antibonding MO formed between Sc 3 p electrons and electrons of neighbouring atoms. Satellite lines due to splitting of molecular orbitals have been reported [55].

5.4 Peak profile change

The profile of X-rays peaks may change by broadening or deformation. The broadening of the Nb $L\gamma_1$ peak in amorphous Nb-Ni alloy with respect to that of metallic Nb has been observed by Iihara and Yoshihara [56], and attributed to mixed band formation involving Nb 4 d and Ni 3 d electrons.

Chemical effects are also known for more energetic muonic and pionic X-rays, but they are not yet fully explored [29].

6. Other applications

Interfering effects of X-rays following electron capture decay of ^{51}Cr have been reported recently [57]. When the radioactive atom is located at a short distance over the surface of a substrate, X-rays from glancing reflections interfere with the direct beam (Figure 10). The interference pattern recorded on an imaging plate depends on the distance z between the radionuclide and the surface of the substrate and on the take-off angle θ_t. The X-ray yield Y is determined by the relations

$$I(\theta_t, z) = |E_d + E_r|^2 \quad (3)$$

$$Y(\theta_t) = \int N(z)(\theta_t, z)\, dz \quad (4)$$

where $I(\theta_t, z)$ is the intensity of the characteristic X-ray, E_d and E_r are the field plane waves E for the direct and reflected emissions and N(z) is the distribution of atoms at the distance z. From the experiment performed by Sasaki *et al.* [57] with ^{51}Cr-arachidate LB films, a distance z shorter than the theoretically expected value has been determined. The authors argued that the surface monolayer was unstable in air owing to its hydrophilic nature and may have undergone a reversal of orientation known as overtuning.

7. Concluding remarks

In this review, the importance and role of X-rays in radiochemistry have been described, together with the fields of application. X-rays are indispensable as analytical tools. Chemical effects on X-ray properties are not only of interest for basic science, but bear promising applications to chemical analysis. However, the acquisition of further and refined data is essential.

References

1. Bradbury, B. T., Demant, J. T., Martin, P. M., Poole, D. M.: J. Nucl. Mater. **17**, 227 (1965).
2. Johnson, I., Johnson, C. E., Crouthamel, C. E., Seils, C. A.: J. Nucl. Mater. **48**, 21 (1973).
3. Henkelmann, R., Baumgärtner, F.: Fresenius Z. Anal. Chem. **319**, 795 (1984).
4. Cronenberg, A. W., Langer, S.: Nucl. Technol. **87**, 234 (1989).
5. Johansson, T. B., Akselsson, R., Johansson, S. A. E.: Nucl. Instrum. Methods **84**, 151 (1970).
6. Johansson, E. M., Johansson, S. A. E.: Nucl. Instrum. Methods **B3**, 154 (1984).
7. Cahill, T. A.: Nucl. Instrum. Methods **B75**, 217 (1993).
8. Koltay, E.: Nucl. Instrum. Methods **B85**, 75 (1994).
9. Amemiya, S., Tsuriya, Y., Masuda, T., Asawa, A., Tanaka, K., Katoh, T., Mohri, M., Yamashita, T.: Nucl. Instrum. Methods **B3**, 516 (1984).
10. Valković, V.: Nucl. Instrum. Methods **142**, 151 (1977).
11. Yumoto, S., Ohashi, H., Nagai, H., Ogawa, Y., Iwata, Y., Ishii, K.: Intern. J. PIXE **2**, 493 (1992); Yumoto, S., Ohashi, H., Nagai, H., Kakimi, S., Ishikawa, A., Kobayashi, K., Ogawa, Y.: Nucl. Instrum. Methods **B75**, 188 (1993).
12. Menu, M.: Nucl. Instrum. Methods **B75**, 469 (1993).
13. Eldred, A., Kusko, B. H., Cahill, T. A.: Nucl. Instrum. Methods **B3**, 579 (1984); Kusko, B. H, Cahill, T. A., Eldred, R. A., Schwab, R. N.: ibid 689 (1984).
14. Vis, R. D., Van der Stap, C. C. A. H., Bos, A. J. J.: Nucl. Instrum. Methods **B3**, 319 (1984).
15. Hult, M., Bengtsson, B., Larsson, N. P.-O., Yang, C.: Scanning Microsc. **6**/2, 581 (1992).
16. Watkinson, S. C., Watt, F.: Biologist **95**, 181 (1992).
17. Grim, G. W., Watt, F.: Nucl. Instrum. Methods **B75**, 495 (1993).
18. Peisach, M.: J. Radioanal. Nucl. Chem. **110**, 461 (1987).
19. Peisach, M.: Nucl. Instrum. Methods **B85**, 100 (1994).
20. Pillay, A. E., Peisach, M.: J. Radioanal. Nucl. Chem. **189**, 283 (1995).
21. Martin Jr., J. L., Yuan, J., Lunte, C. E., Elder, R. C., Heineman, W. R., Deutsch, E.: Inorg. Chem. **28**, 2899 (1989).
22. Williams, G. A., Martin, L. J.: Private communication.
23. Yoneda, Y.: Phys. Rev. **131**, 2010 (1963).
24. Prange, A.: Spectrochim. Acta **48B**, 107 (1993).
25. Horowitz, R., Howell, J.: Science **178**, 608 (1972).
26. Iida, A., Yoshinaga, A., Sakurai, K., Gohshi, Y.: Anal. Chem. **58**, 394 (1986).
27. Wobrauschek, P., Kregsamer, P., Ladisich, W., Rieder, R., Streli, C.: Spectrochim. Acta **48B**, 143 (1993).
28. Yoshihara, K., Iihara, J.: Intern. J. PIXE **2**, 93 (1992).
29. Yoshihara, K.: Top. Current Chem. **157**, 1 (1990).
30. Hurley, R. G., White, E. W.: Anal. Chem. **46**, 2234 (1976).
31. Gohshi, Y., Hirao, O., Suzuki, I.: Adv. X-Ray Anal. **18**, 406 (1975).
32. Watanabe, T., Hashiguchi, Y., Sato, H.: Trans. Iron Steel Inst. Jap. **21**, 173 (1981).
33. Ban, Y., Sugiura, T., Furuya, K., Gohshi, Y.: Bunseki Kagaku **34**, 238 (1985).
34. Gel'mukhanov, F. K., Mazalov, L. N., Kondratenko, A. V.: Chem. Phys. Lett. **46**, 133 (1977).
35. Gohshi, Y.: Spectrochim. Acta **36B**, 763 (1981).
36. Grimaldi, R., Paris, V., Mariani, E.: X-Ray Spectrometry **10**, 163 (1981).
37. Gohshi, Y., Fukao, Y., Hori, K.: Spectrochim. Acta **27B**, 135 (1972); Gohshi, Y., Kamada, H., Kohra, K., Utaka, T., Arai, T.: Appl. Spectrosc. **36**, 171 (1982).
38. Kawai, J., Nakamura, E., Nihei, Y., Fujisawa, K., Gohshi, Y.: Spectrochim. Acta **45B**, 463 (1990).
39. Iihara, J., Fukuhara, K., Yagi, M., Omori, T., Yoshihara, K.: Hyperfine Interactions **84**, 433 (1994).
40. Sugiura, C.: J. Phys. Soc. Jap. **59**, 2134 (1990).
41. Sugiura, C.: Jap. J. Appl. Phys. **29**, 1116 (1990).
42. Sugiura, C.: Jap. J. Appl. Phys. **29**, 2426 (1990).

43. Tamaki, Y., Omori, T., Shiokawa, T.: Radiochem. Radioanal. Lett. **20**, 255 (1975).
44. Yamoto, I., Kaji, H., Yoshihara, K.: J. Chem. Phys. **84**, 522 (1986).
45. Mukoyama, T., Kaji, H., Yoshihara, K.: Phys. Lett. **A118**, 44 (1986).
46. Brunner, G., Nagel, M., Hartmann, E., Arndt, E.: J. Phys. **B15**, 4517 (1982).
47. Iihara, J., Izawa, G., Omori, T., Yoshihara, K.: Nucl. Instrum. Methods **A299**, 394 (1990).
48. Iihara, J., Omori, T., Yoshihara, K., Ishii, K.: Nucl. Instrum. Methods **B75**, 32 (1993).
49. Iihara, J., Kawai, J., Sekine, T., Yoshihara, K.: Intern. J. PIXE **3**, 177 (1993).
50. Watson, R. L., Leeper, A. K., Sonobe, B. I., Chiao, T., Jensen, F. E.: Phys. Rev. **A15**, 914 (1977).
51. Benka, O., Watson, R. L., Bandong, B.: Phys. Rev. **A28**, 3334 (1983).
52. Uda, M., Benka, O., Fuwa, K., Maeda, K., Sasa, Y.: Nucl. Instrum. Methods **B22**, 5 (1987).
53. Folkmann, F.: Nucl. Instrum. Methods **B75**, 9 (1993).
54. Kawatsura, K., Ozawa, K., Fujimoto, F., Terasawa, M.: Phys. Lett. **60A**, 327 (1977).
55. Kawai, J.: Nucl. Instrum. Methods **B75**, 3 (1993).
56. Iihara, J., Yoshihara, K.: Intern. J. PIXE **3**, 215 (1993).
57. Sasaki, Y. C., Suzuki, Y., Tomioka, Y., Ishibashi, T., Satoh, I., Hirokawa, K.: Phys. Rev. **B50**, 15516 (1994).

E. Radionuclides in the Environment and Nuclear Energy

Radiochimica Acta 70/71, 345–353 (1995)

Radioactivity in the Atmosphere

By H. W. Gäggeler

Institut für anorganische, analytische and physikalische Chemie, Labor für Radio- und Umweltchemie, Universität Bern, 3012 Bern, Switzerland
and Labor für Radio- und Umweltchemie, Paul Scherrer Institut, 5232 Villigen PSI, Switzerland

(Received August 24, 1995)

Radioactivity / Atmosphere / Radon / Thoron / Cosmogenic radionuclides

Abstract

A short review is made on atmospheric radionuclides with emphasis on their activity concentrations in the troposphere, mostly for surface air over Europe. Discussed are those species which have activity concentrations above about 1 mBq/m^3. For this atmospheric layer highest average activity concentrations ($\geq$1 Bq/m^3) have i) ^{222}Rn ("radon") and its short-lived decay products, (ii) ^{220}Rn ("thoron"), however, only for the lowest meter above ground, and iii) ^{85}Kr. In the range 1 mBq/m^3 to 1 Bq/m^3 are T(^{3}H), ^{14}C, ^{37}Ar and ^{7}Be as well as the "thoron" decay product ^{212}Pb. Average activity concentrations slightly below 1 mBq/m^3 has ^{210}Pb, the only long-lived decay product of "radon". The sources of these radionuclides are discussed. Not included in this review are radionuclides contained in the atmosphere at very low activity concentrations or from single emissions such as e.g. from the Tschernobyl accident in 1986.

1. Introduction

Environmental radioactive nuclides have three different sources; they are either primordial, cosmogenic or anthropogenic.

Primordial radionuclides are those which were produced during element evolution and which have (partly) survived since then due to their long half-lives. Our solar system has an age of about 4.5 · 10^9 years; this means that nuclides with half-lives longer than about 10^8 years can still be found in our environment. With respect to the atmosphere, progenies of the primordial nuclides ^{238}U and ^{232}Th with half-lives of 4.5 · 10^9 and 1.4 · 10^{10} years, respectively, are of special importance. These actinides are mainly contained in the lithosphere and decay via several nuclides to different isotopes of lead. In both decay chains one intermediate nuclide is an isotope of the noble gas radon, ^{222}Rn (called "radon") in case of ^{238}U and ^{220}Rb ("thoron") in case of ^{232}Th which have half-lives of 3.8 days and 56 s, respectively. These gases can escape from the lithosphere into the atmospheres. Of less importance – with respect to the atmosphere – is ^{235}U with a half-life of 7.0 · 10^8 years, caused by its low isotopic abundance of 0.7 % in natural uranium and the fact that the corresponding radon isotope ^{219}Rn ("actinon") has a very short half-life of 4 s.

The second reservoir of radionuclides is the upper atmosphere. The earth is permanently bombarded by cosmic ray particles. These particles are of solar or galactic origin and are predominantly protons. Mainly the high energy galactic protons (typical energies 100 MeV – 3 GeV with a flux of about 3 cm^{-2}s^{-1} [1]) interact with the atmospheric species in nuclear reactions and form a variety of radionuclides. Also secondary particles such as neutrons and myons have high enough energies to produce several nuclides. The beam flux of galactic cosmic ray particles onto the earth surface is not constant in time; it is modulated by the solar activity which has a well known 11 year cycle. In time periods with high solar activity (i.e. high sun spot number) the flux of galactic particles bombarding the earth surface is low and vice versa. This is due to a magnetic shielding of the earth caused by the solar activity. In addition, the solar activity has low-frequency variabilities e.g. with a time constant of several hundred years. This leads to time periods with extremely low solar activities such as e.g. during the Maunder minimum between 1645–1715. During this time period the production of cosmogenic nuclides was increased.

Finally, with the beginning of nuclear technology, humans have started to emit radioactive species into the environment. This was for the first time a serious problem during the atmospheric nuclear weapons testing programs of USA and the former Soviet Union in the late 50th and early 60th. Most pronounced were the ejections into the atmosphere in 1962 which led to a test-stop agreement for overground explosions on 5 August 1963 in Geneva. Even now, the radioactive debris from these nuclear weapons testing period can be found in archives such as lake or sea sediments or in glacier ice cores. During the following period of underground testing only radioactive gases such as ^{85}Kr or ^{37}Ar (a secondary product from the ^{40}Ca(n,α)^{37}Ar reaction) were released into the atmosphere.

Nuclear power plants and nuclear reprocessing plants emit radionuclides into the atmosphere too. Of special importance was the Tschernobyl accident: on April 26, 1986, due to human failure, a 1 GW reactor exploded and caused a high contamination of the local environment. By long-range atmospheric transport Tschnerobyl fallout could be detected nearly everywhere in the northern hemisphere.

Also the non-nuclear industry releases some radionuclides into the atmosphere such as T, ^{14}C or several isotopes of iodine from tracer applications in industry, chemistry or nuclear medicine.

In the following, the main radionuclides contained in the different atmospheric compartments will be discussed, with main emphasis on their activity concentrations in the planetary boundary layer, i.e. in the lower troposphere. No general overview on all existing radionuclides will be given but rather a discussion according to decreasing average activity levels, mainly over Europe. As a lower limit, nuclides with about $\geq 1\ mBq/m^3$ average activity concentration will be discussed. Therefore, some radionuclides of great importance to environmental studies but of low average activity concentration will not be included into this review. Examples of this latter category are ^{10}Be which can be used to reconstruct the history of the solar activity, ^{81}Kr as a nuclide of great potential to study very old aquatic systems, ^{129}I as a tracer to reconstruct the total emission history from nuclear industry or ^{32}Si to be used to date environmental archives over time periods between about 10^2 and 10^3 years.

2. ^{222}Rn and its decy products

2.1 Radon-222

^{222}Rn and its decay products are important atmospheric radionuclides because they are responsible for a sizable fraction of the total public exposure to ionizing radiation. Typical exhalation rates of ^{222}Rn from the soil are on the order of one atom/cm^2 s or $2 \cdot 10^{-4}$ Bq/cm^{2}s [2]. ^{222}Rn can highly accumulate in houses, especially in cases with low ventilation rates. According to estimates [3], indoor ^{222}Rn exposure may be responsible for more than 10% of the US incidence of lung cancer. Indoor activity concentrations are typically between 10 and 10^2 Bq/m^3 [3–6], for e.g. Swiss houses on the average 60 Bq/m^3 [6].

For outdoor conditions, emanating ^{222}Rn and its decay products distribute in the troposphere mainly via turbulent mixing which is characterized by the eddy diffusivity. Eddy diffusion is strongly depending on micrometeorological conditions. This may cause pronounced diurnal variations of the ^{222}Rn activity concentration at low altitudes (1–10 m above ground) [2, 5, 7]. Average diurnal variations for clear weather condition at e.g. 1.5 m above ground in a river valley in Germany for the time period May/July 1989 showed minimum activity concentrations of about 5 Bq/m^3 in the afternoon and maximum values of about 40 Bq/m^3 at 4 am [7, 8].

Exhalation rates of ^{222}Rn are about 100 times higher over continents if compared to the sea [9]. This leads to a gradient of the average activity concentrations of ^{222}Rn over the continent. Average values (annual means for 1980–1983) were about 0.8 Bq/m^3 on the research platform Nordsee, 3.3 Bq/m^3 at Waldhof, 5 Bq/m^3 at Heidelberg and 6.5 Bq/m^3 at Krakau [10]. The surface activity concentrations of ^{222}Rn are nearly equal over the different continents, except for

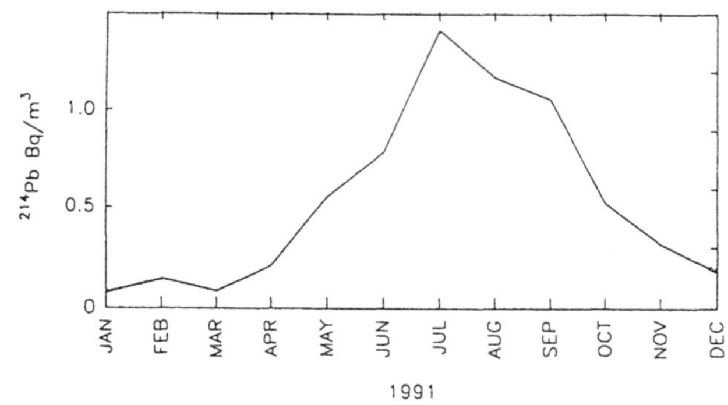

Fig. 1. Monthly mean activity concentrations of the short-lived ^{222}Rn ("radon") decay product ^{214}Pb at Jungfraujoch (3450 m asl), Switzerland, for the year 1991 (from [15]).

arctic areas [11]. For Antarctica, typical values are $(1-3) \cdot 10^{-2}$ Bq/m^3 [12]. Over the sea, at distances far from land masses, ^{222}Rn activity concentrations are typically 0.1 Bq/m^3 [13].

This well defined source term of ^{222}Rn makes this radionuclide a good marker for air parcels of continental or maritime origin [11, 14, 15]. ^{222}Rn measurements can therefore be used as an additional help to identify source regions of air masses. Examples are measurements at Jungfraujoch (3450 m asl), Switzerland. This is a high-alpine site, where high radon activities could be correlated with air trajectories arriving to this site from southerly (Africa) or easterly (continental) direction, but not from westerly direction (maritime origin) [15]. In Antarctica, high ^{222}Rn activities were connected to air mass transport from South America [12]. Another example is Mauna Loa, Hawaii, where high ^{222}Rn activity concentrations could be attributed to long-range transport from Asia [16].

The vertical concentration profile of ^{222}Rn within the planetary boundary layer is mainly determined by the atmospheric stability. The activity decreases exponentially, as predicted theoretically [17] and found experimentally [18, 19]. Within the lowest 4 km it decreases by about a factor of four to ten [19, 20]. For ground-based measurements at elevated sites, meteorological conditions strongly influence the local ^{222}Rn activity concentrations. Especially during the summer period convective vertical transport from nearby valleys cause an increase of the local activity concentrations (Fig. 1) [15, 21].

^{222}Rn activity concentrations reach values of about 10 to 100 mBq/m^3 (STP) in the tropopause and lower stratosphere [20, 22, 23]. However, it was found that advective transport of radon-rich air from the boundary layer can reach the upper troposphere [24] or even the stratosphere [22] which can lead to bimodal distributions of the ^{222}Rn activity concentration [24].

2.2 Short-lived decay products of ^{222}Rn

^{222}Rn decays by a chain of α- and β-emissions via ^{218}Po ($T_{1/2}$ = 3.1 min), ^{214}Pb ($T_{1/2}$ = 26.8 min), ^{214}Bi ($T_{1/2}$ = 19.9 min) and ^{214}Po ($T_{1/2}$ = 164 μs) to the long-lived

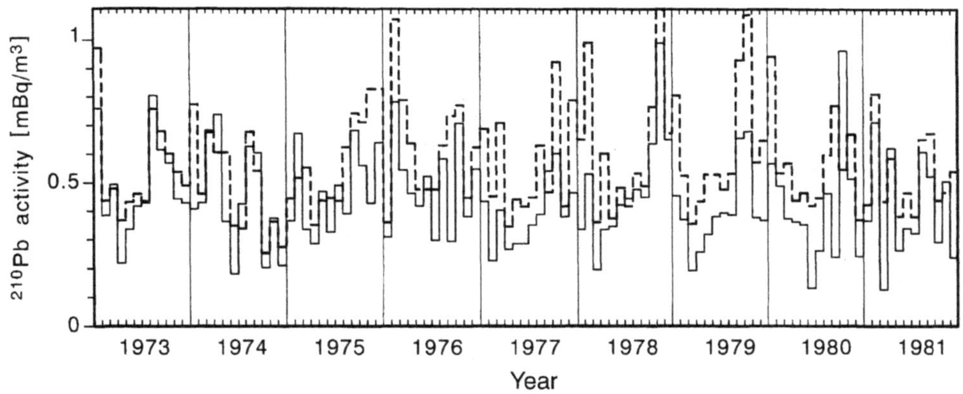

Fig. 2. Monthly mean activity concentrations of ^{210}Pb in surface air of Fribourg, Switzerland [37] (solid line) and Neuherberg, Munich, Germany for 1973–1981 [36] (dashed line). Both sites are located on the northern side of the Alps at a distance of 360 km.

nuclide ^{210}Pb ($T_{1/2}$ = 22 years). Since these progenies of radon are non-gaseous species, they adsorb to aerosol particles. It is assumed that this process proceeds in two steps. First, after thermalization of the primary nuclides in the gas, the atoms or positive ions form clusters of 0.5 nm to 5 nm size – depending on parameters such as temperature and humidity – by attachment of H_2O molecules [2, 25]. These clusters are called "unattached" radon progenies. Second, these clusters attach to aerosol particles. Typical attachment rates for outdoor conditions (aerosol particle concentration $(2-7) \cdot 10^4$ cm^{-3}) were found to be 20–100 h^{-1} [26], corresponding to average attachment times of 40 s to 3 min. For an aerosol particle concentration of only 100 cm^{-3}, an attachment time of about 1 hour was observed [27]. Of special relevance is the fact, that due to a 112 keV nuclear recoil, the first progeny of the α-emitting nuclide ^{218}Po, the daughter nuclide ^{214}Pb, escapes with about 80% probability [28, 29] from the particles which have typical median aerodynamic diameters of 50 to 200 nm [28]. In addition, for indoor condition or outdoor situations close to ground, prior to an attachment of the short-lived ^{222}Rn progenies to aerosol particles, these nuclides may plate-out to nearby-surfaces. Since the diffusion coefficients of the nm-size clusters are much higher compared to those of aerosol particles, the plate-out effect may deplete the progenies of ^{222}Rn considerably in the lowest few meters above ground [9, 17]. This is of special importance if aerosol particle collection techniques such as air-filter measurements are used to determine the ^{222}Rn activity concentrations.

At altitudes higher than about 10 m above ground ^{222}Rn is generally in secular equilibrium with its short-lived decay products [17].

2.3 Lead-210

The only long-lived ($\geq$ years) decay product of ^{222}Rn is ^{210}Pb. This nuclides is attached to aerosol particles in the sub-micron range and therefore considered to be a good tracer for an aerosol produced by gas-to-particle conversion [30]. Its residence time in the atmosphere is determined by the average residence time of sub-micron aerosol particles from the accumulation mode which is typically one to two weeks [31, 32]. Removal from the atmosphere occurs mainly by wet and dry deposition of the carrier aerosol particles. Estimated continental deposition rates are about 20 mBq/cm² y [33, 34]. Yearly average activity concentrations in surface air over Europe are about 0.2–0.7 mBq/m³, except for arctic regions with lower values, e.g. 15 µBq/m³ at Franz Joseph Island [35]. Similar to the European values are those in America, Asia and Australia [35]. Lower values are found in Antarctica, e.g. 13 µBq/m³ at the South Pole [35].

Monthly mean activity concentrations exhibit no pronounced seasonal variation, except for some maxima during long-periods with clear weather [36–38]. Fig. 2 shows monthly mean activity concentrations for ^{210}Pb from Fribourg, Switzerland [37] and Neuherberg, Germany [36] for the time period 1973–1981. Both locations are on the northern side of the Alps at a distance of 360 km. The very similar concentration patterns at both locations indicates that ^{210}Pb is well mixed in the planetary boundary layer over large areas.

Due to the long half-life of ^{210}Pb it can be used to date environmental archives such as lake sediments or glacier ice [34]. Nuclear dating is, however, only possible if the input function is known over the time period to be determined. Surprisingly, recent ^{210}Pb measurements in ice cores from Greenland indicated that the deposition rates changed significantly during the last 150 years [39, 40]. Also for high-alpine ice cores recent ^{210}Pb measurements give evidence for variations in the deposition rates over the same time period [41, 42]. Presently there is no explanation for this observation.

The vertical profile of the ^{210}Pb activity concentration is expected to be rather constant up to an altitude of about 10 km [17]. Experimental data from e.g. United States show, however, that the activity concentration of ^{210}Pb decreases by about a factor of five within this atmospheric layer [20], presumably due to lower residence times of aerosol particles than assumed in [17]. Also data from Jungfraujoch (3450 m asl) yield annual mean values of about 0.2 mBq/m^3 [15] which is a factor of two lower compared to the values at the nearby low-altitude site Fribourg [38].

Decay products of ^{210}Pb

^{210}Pb decays via ^{210}Bi ($T_{1/2}$ = 5.0 d) to ^{210}Po ($T_{1/2}$ = 138.38 d). Average activity concentrations of ^{210}Po in air at e.g. Neuherberg measured during 1976–1982 were found to be a factor of 10 to 20 lower than those of ^{210}Pb [36].

Activity ratios of the decay products of ^{210}Pb are commonly used to determine residence times of aerosol particles in the troposphere. In general, values deduced from measurements of ^{210}Bi and ^{210}Pb are typically one week [43], in good agreement with theoretical estimates for the lower troposphere. On the other hand, residence times obtained from measurements of ^{210}Po and ^{210}Pb are usually significantly higher, mostly about one month [44, 45]. Such high values are discussed controversially in the literature: most authors assume additional sources of ^{210}Po besides radioactive decay of ^{210}Pb [20] such as the soil [46, 47] or pollution sources of volcanic origin or from fossil fuel burning [47].

3. ^{220}Rn and its decay products

Due to the short half-life of about one minute, ^{220}Rn ("thoron") escapes only from a small surface soil layer into the atmosphere. For the same reason the gradient of the activity concentration in the lowest atmospheric layer is very steep, typically decreasing by more than a factor of five within the first meter above ground [9]. Since the average thorium concentration in surface soil is generally higher than that of uranium, in the atmospheric layer below 1.5 m above ground, ^{220}Rn activity concentrations are higher than those of ^{222}Rn. Typical values measured for outdoor conditions at 1.5 m height in New Mexico were 22 Bq/m^3 for ^{220}Rn and 5.5 Bq/m^3 for ^{222}Rn [47]. In a flat river valley close to Göttingen strong diurnal variations of the activity concentrations of ^{220}Rn were found during a summer month, similar to those of ^{222}Rn, with up to a factor of 50 higher values during night compared to the noon minima [48]. Average values at this site at the 1.5 m level were about 80 Bq/m^3 [7]. During late fall with strongly reduced vertical advective mixing, no diurnal variation of the activity concentrations for the same level above ground were observed [2].

With respect to radiation dose it is presently unclear whether the thoron decay products ^{212}Pb ($T_{1/2}$ = 10.6 h), ^{212}Bi ($T_{1/2}$ = 60.6 min) and ^{212}Po ($T_{1/2}$ = 0.3 μs) are of relevance. For indoor conditions it is estimated that these radionuclides may add up to 20% relative to that of the ^{222}Rn decay products [49].

Atmospheric measurements of ^{220}Rn decay products are scarce. In Malaga, Spain, close to the sea shore, during a 6 month measuring campaign of air collected 2 m above ground, ^{212}Pb activity concentrations of about 0.1 Bq/m^3 and ^{222}Rn activity concentrations of about 2 Bq/m^3 were determined [50]. Continuous measurements of ^{212}Pb over 3.5 months in a river valley close to Göttingen resulted in average activity concentration of 0.4 Bq/m^3, with large variations of the single measurements by nearly two orders of magnitude [48]. At Mauna Loa, ^{212}Pb activity concentrations were strongly depending on meteorological conditions with up to 4 mBq/m^3 for up-slope wind conditions and much lower values for non-up-slope wind conditions [51]. At Jungfraujoch, ^{212}Pb monthly mean activity concentrations showed large variations with minimum values of 1 mBq/m^3 in winter and maximum values of 8 mBq/m^3 in summer [15].

4. Krypton-85

^{85}Kr is a noble gas isotope with a half-life of 10.76 years. It is produced in the thermal neutron induced fission of ^{235}U with a cumulative yield of 0.28%, which is the product of the isobaric chain yield of 1.32% and the yield of the internal transition of ^{85m}Kr of 21.4%. This radionuclide is emitted into the atmosphere during nuclear weapons testing (both above and underground), from nuclear power plants as well as from nuclear reprocessing plants. The average activity concentrations of ^{85}Kr in the northern hemisphere is presently slightly higher than 1 Bq/m^3 [52]. Its activity concentration is steadily increasing with a doubling time of about 20 years [53]. At collection sites within the planetary boundary layer, activity peaks are frequently observed, presumably due to well defined releases from reprocessing plants [53, 54]. At Jungfraujoch, which is in the free troposphere, the average activity concentration of ^{85}Kr is also about 1 Bq/m^3 but with much less pronounced fluctuations with time [52]. This means that this radionuclide is well mixed in the troposphere of the northern hemisphere [54]. In antarctica, the activity concentration of ^{85}Kr is about 20% lower compared to the values in the northern hemisphere [53]. This is caused by the average transportation time from the emission sources, all being in the northern hemisphere, to Antarctica, which is on the order of a few years.

Even though ^{85}Kr is the artificial radionuclide with the highest average activity concentration in the troposphere, its radiological importance is negligible, because – unlike ^{222}Rn – it has no non-gaseous radioactive decay products which can be accumulated in the human body.

5. Tritium

Tritium (^{3}H=T) is a weak β-emitter (E_β=18 keV) with a half-life of 12.3 years. Tritium is both of natural and anthropogenic origin. The natural production occurs via interaction of cosmic rays with the atmosphere. The anthropogenic production has several

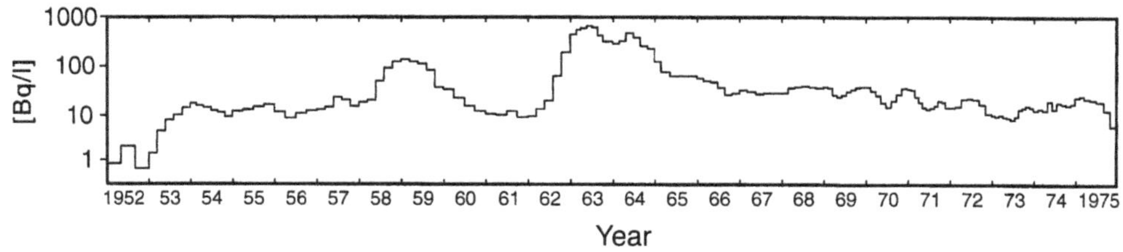

Fig. 3. Tritium activity concentration (corrected for decay) from an ice core from Colle Gnifetti (4450 m asl), Monte Rosa massif (from [37]). The dating is based on the observation of well documented stratigraphic layers (Saharan dust horizons etc.).

sources. First, tritium was released to the atmosphere by thermonuclear explosions. Second, a release into the atmosphere occurs from nuclear power plants or reprocessing plants as well as from industry. Due to its widespread applications, the actual tropospheric activity concentration of tritium is significantly increased compared to the natural level from prenuclear times. The activity concentration of tritium is often given in T.U. (Tritium Units), with 1 TU = 1 T atom per 10^{18} H atoms. 1 T.U. corresponds to a specific tritium activity of 0.12 Bq/kg water.

In the atmosphere – except for the upper stratosphere – the dominating species containing tritium is HTO [55, 56]. Most of the tritium is deposited on the earth surface by precipitation. However, depending on humidity and temperature, in the atmosphere a large part of tritium is in form HTO vapor. Due to isotopic enrichment, the T/H ratio in liquid HTO is 1.1 higher compared to that in gaseous HTO [57]. Some minor part of tritium is also found in form of HT.

5.1 Natural tritium

As a cosmogenic radionuclide, tritium is produced in the stratosphere in spallation reactions of galactic cosmic rays, mainly protons, with oxygen and nitrogen. In addition, tritium is also produced by secondary particles, high-energy ($E \geq 4.4$ MeV) neutrons in the $^{14}N(n,T)^{12}C$ reaction. As for all cosmogenic radionuclides, the rate of production depends on the solar activity and varies between $(1.5-3) \cdot 10^{17}$ Bq/year [56]. Maximum solar flares correspond to minimum production and vice versa.

The specific natural tritium concentration in precipitation depends on the latitude, with about 5 Bq/l at the poles and 0.06 Bq/l at the equator [57]. For a latitude of 45 °N or S, the natural specific tritium activity concentration is about 0.6 Bq/l. This behavior is caused by the stratosphere-troposphere exchange of air masses [57, 58]. In addition, increased downward air mass flow is observed in late spring [55, 59]. This leads to seasonal fluctuations of the tropospheric activity concentration with maxima in late spring [57].

There exists a continental effect of the specific tritium activity concentration with lower values for sites close to the sea shore compared to those located deep inside the continents [57]. For sites close to the coast the isotopic composition of HTO corresponds to that of oceanic surface water whereas for inland sites the HTO composition is increasingly determined by the tropospheric values at this latitude and by evaporated water from the land surface.

For tritium in form of HT, the specific activity is much higher. However, due to the low abundance of hydrogen in air, the total amount of natural tritium in form of HT is only about 10^{-4} compared to that in form of HTO [55].

5.2 Anthropogenic tritium

Begemann and Libby [60] estimated, that about $4 \cdot 10^{17}$ Bq of tritium was released for each megaton thermonuclear explosion. The total amount of tritium released during the overground test series performed by USA and USSR between 1954 and 1963 is estimated to be about $1.1 \cdot 10^{20}$ Bq [55].

French and Chinese tests between 1968 and 1977 have added another $(7-10) \cdot 10^{18}$ Bq [55]. These nuclear tests have, of course, strongly increased the HTO concentration in the atmosphere. As an example, the measured tritium concentrations in precipitation, as determined from an Alpine glacier ice core from Colle Gnifetti (4450 m asl) near Zermatt, Switzerland, for the time period 1952–1975 is shown in Fig. 3 [37, 61]. From a level of about 1 Bq/l in 1952 the tritium activity concentration increased to 740 Bq/l in 1963 and then decreased to about 10 Bq/l in 1975. Since the snow deposition at this high-alpine wind exposed site occurs not very regularly [62], the seasonal fluctuation of the tritium deposition is not well preserved. Between 1975 and present, the tritium activity concentration decreased further. Typical tritium activity concentrations in precipitation, e.g. 1993 in Switzerland at different locations were between 1.5 and 5.5 Bq/l [6, 63]. The reasons for a still increased level of tritium compared to the natural background level are industrial applications, such as e.g. from the production of luminous dials.

5.3 Atmospheric activity concentration of tritium

If we assume, as an example for 1992, an average relative humidity of 70% and an average winter and summer temperature of 2 °C and 20 °C, respectively (values for Zürich, Switzerland [64, 65]), the measured tritium activity concentrations in precipitation of about

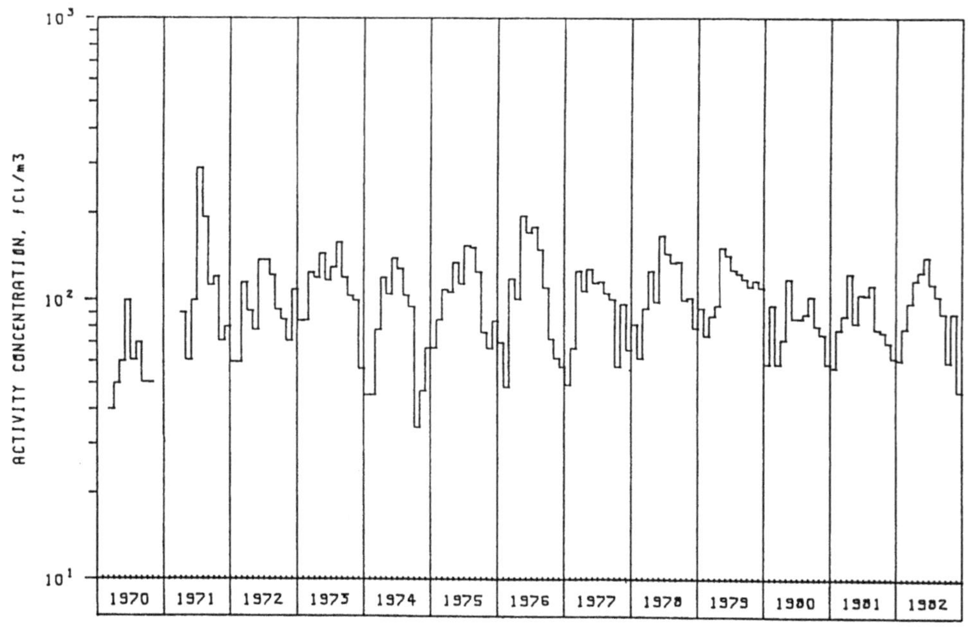

Fig. 4. Monthly mean activity concentrations of ^{7}Be in surface air of Neuherberg, Germany (from [36]). Activities are given in femtocurie/m^3. 100 fCi/m^3 = 3.7 mBq/m^3.

2 Bq/l corresponds to an average activity concentrations of gaseous HTO of 8 to 24 mBq/m^3 for winter and summer, respectively. However, for the same year, in the neighborhood of a company using tritium for industrial applications, much higher gaseous HTO activity concentrations of about 5 – 25 Bq/m^3 have been measured [52].

6. Beryllium-7

^{7}Be is a cosmogenic radionuclide with a half-life of 53.3 d. It is produced both by primary (mostly protons) and secondary (neutrons) cosmic ray particles with nitrogen and oxygen according to the following reactions: $^{14}N(p,2\alpha)^{7}Be$, $^{16}O(p,^{10}B)^{7}Be$, $^{14}N(n,^{8}Li)^{7}Be$, and $^{16}O(n,^{10}Be)^{7}Be$.

About 67% of ^{7}Be is produced in the stratosphere and 33% in the troposphere [66]. The peak production is approximately in the middle of the atmosphere (in mass units) [67]. After production, ^{7}Be attaches to submicron-sized aerosol particles [68]. Cascade impactor measurements showed that in ambient aerosol particles most of the ^{7}Be was found in the 0.7 to 1.1 μm range [69]. Due to its main γ-line at 478 keV this nuclide can be easily measured in air filter samples. The surface activity concentration of ^{7}Be in air depends on four processes [70], i) stratosphere-troposphere exchange, ii) downward transport in the troposphere, iii) wet scavenging, and iv) horizontal transfer from mid latitudes to higher and lower latitudes. Hence, the annual cycle of ^{7}Be in surface air may be quite different at different sites. For mid latitudes in the northern hemisphere, typical annual cycles with maximum activity concentrations in early summer are found (Fig. 4), caused by tropopause folding during this time period. In polar regions or at subtropical locations the annual cycles of ^{7}Be activity concentrations are more complex as shown in a very thorough review on ^{7}Be in the atmosphere [71].

The question whether the surface activity concentration of ^{7}Be depends on the solar activity is discussed controversially. In Ref. [71] no such clear correlation was found in contrast to Ref. [72], where for the years 1985 – 1990 at 8 stations from 71 °N to 53°S a pronounced decrease of the annual mean activity concentration was correlated to an increase of the solar activity (sun spot number) for the same time period [73].

The annual mean surface activity concentrations are higher at low latitudes compared to high latitudes, which is somewhat different to expectations based on pure vertical stratosphere-troposphere exchange. This might be a result of the complex transport processes mentioned above. As an example, the annual mean activity concentrations for 1987 were (in mBq/m^3): 1.4 in Barrow, Alaska (71°N), 2.9 in Moosonee, Canada (51°N), 4.5 in Chester, USA (41°N), 5.4 in Miami, USA (26°N), 4.5 in Lima, Peru (12°S), 4.8 in Perth, Australia (32°S), 3.5 in Cape Grim, Tasmania, Australia (41°S), and 1.7 in Punta Arenas, Chile (53°S), respectively [72]. In Europe (e.g. Neuherberg, Germany, or Fribourg, Switzerland), annual man values in surface air are about 3 mBq/m^3 [36, 37].

The activity concentrations are increasing with altitude. For northern mid latitudes typical values are 18 mBq/m^3 at the tropopause and 155 mBq/m^3 in the lower stratosphere [74].

7. Carbon-14

^{14}C ($T_{1/2}$ = 5730 years) is a pure β-emitter with a maximum energy of 0.2 MeV. Similar to tritium, also modern atmospheric ^{14}C is of natural and anthropogenic origin. Natural ^{14}C is mainly produced in interactions of secondary cosmic ray particles (neutrons) with nitrogen, according to the reaction $^{14}N(n,p)^{14}C$. About 55% of ^{14}C is produced in the stratosphere and 45% in the upper troposphere [75]. Primary ^{14}C oxidizes immediately in the atmosphere to $^{14}CO_2$. The average natural atmospheric activity concentration of ^{14}C is about 40 mBq/m³ air, based on a measured specific activity of 0.219 Bq ^{14}C/g total carbon [76] and an atmospheric volume concentration of 0.033% CO_2. This corresponds to a $^{14}C/^{12}C$ ratio of $1.2 \cdot 10^{-12}$. The largest reservoir of ^{14}C is the ocean (94%), the residual part being distributed in the atmosphere (1.5%), biosphere (2.0%) and the humus (2.5%) [77].

^{14}C is incorporated into living material via CO_2 uptake. This nuclide has therefore found widespread applications for dating purposes of dead organic material with an age up to 32,000 years [77]. By comparing "true" ages from dendrochronology of tree rings with the ^{14}C-age of the carbon from the corresponding annual layers it was found that the past atmospheric ^{14}C activity concentration was not constant. The longest period of the observed activity fluctuations has a period of about 11,000 years and is attributed to geomagnetic dipole fluctuations of the earth [78]. The shortest fluctuation is caused by the well known 11-year cycle of the solar activity. The amplitudes of these activity fluctuations are usually $\leq 2\%$ [79].

With the beginning of the nuclear age, anthropogenic ^{14}C was introduced into the atmosphere. In the years 1962/63 with maximum overground testing of thermonuclear weapons, the ^{14}C activity concentration in the northern hemisphere increased by nearly a factor of two [80]. In the southern hemisphere the increase was much smaller [81]. After the stop of overground nuclear explosions in 1963 the anthropogenic ^{14}G activity concentration decreased significantly, every 17 years be a factor of two [81]. Actual ^{14}C activity concentrations are still increased by about 10% above the natural level, mainly as residual contribution from the nuclear explosions but also due to emissions from nuclear power plants, reprocessing plants and from ^{14}C tracer applications in chemistry. This means that the actual ^{14}C activity concentration in the troposphere is about 45 mBq/m³.

Besides the anthropogenic increase of the ^{14}C activity concentration there exists also an anthropogenic decrease which is caused by human fossil fuel burning. In 1950 it was realized by Suess that ^{14}C activities in modern wood samples were too low by about 2% [82]. This so-called Suess effect was later masked by the much larger increase of the ^{14}C activity concentration from the nuclear weapons testing.

8. Argon-37

^{37}Ar decays via electron capture and has a half-life of 35 d. It is produced in cosmic ray induced reactions with atmospheric argon. In underground nuclear weapons tests it is formed in the $^{40}Ca(n,\alpha)^{37}Ar$ reaction. Some release of this nuclide occurs also from nuclear facilities [6]. In surface air the natural activity concentration from cosmic ray production is about 0.5 mBq/m³ [6]. During the early 70th with large underground test series, the ^{37}Ar activity concentration e.g. in Bern, Switzerland, was increased shortly after every explosion by about a factor of 100 [37]. Presently, the activity concentration at the same location is between 0.6 and 3.8 mBq/m³ [6].

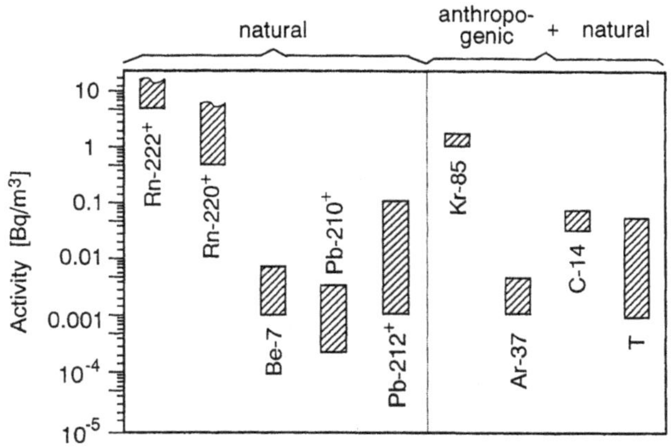

Fig. 5. Average activity concentrations of those radionuclides which have values above about 1 mBq/m³ in surface air over continents – mainly over Europe. The sign ⁺ means that also short-lived decay products of these nuclides have similar activity concentrations.

9. Summary

Typical ranges of actual surface air activity concentrations of different radionuclides over continents are shown in Fig. 5. Only those nuclides are depicted which presently have average activity concentrations above about 1 mBq/m³. Highest values have ^{222}Rn ("radon") and ^{220}Rn ("thoron") including their short-lived decay products. Over the sea, the highest activity concentration has the anthropogenic radionuclide ^{85}Kr, due to the much reduced emanation rates of radon and thoron from the oceans.

Fig. 5 also shows that the highest tropospheric activity concentration from all cosmogenic radionuclides has ^{14}C, even though about 10% of its present-day activity is caused by anthropogenic sources.

The gaseous radionuclides shown in Fig. 5 are ^{222}Rn, ^{220}Rn, ^{85}Kr, ^{37}Ar, ^{14}C (in form of CO_2), and T (in form of water vapor). They are of minor importance with respect to public exposure to ionizing radiation. This is different for those radionuclides which are attached to aerosol particles such as the decay products of ^{222}Rn and ^{220}Rn but also for ^{7}Be because they can be accumulated in the human respiratory organ.

Not included in this review are those radionuclides which were released during the Tschernobyl accident on April 26, 1986 because they caused a contaminations of tropospheric air masses over a few days only, e.g. in Switzerland between about April 30 and Mai 3, 1986 [83].

References

1. Reedy, R. C.: Science **219** 127 (1983).
2. Porstendörfer, J.: J. Aerosol Sci. **25**, 219 (1994), and references therein.
3. Bruno, R. C.: J. Air Pollut. Control. Ass. **33**, 105 (1983).
4. Burkart, W.: Nucl. Technol. ***60***, 114 (1983).
5. Jacobi, W.: Health Phys. **22**, 441 (1972).
6. *Environmental Radioactivity and Radiation Exposure in Switzerland 1993*, Bundesamt für Gesundheitswesen, Bern, ISBN 3-905235-15 (1994).
7. Butterweck, G., Reineking, A., Kesten, J., Porstendörfer, J.: Atmos. Environ. **28**, 1963 (1994).
8. Porstendörfer, J., Butterweck, G., Reineking, A.: Health Phys., in press.
9. Jacobi, W.: Die natürliche Radioaktivität der Atmosphäre und ihre Bedeutung für die Strahlenbelastung des Menschen, Hahn Meitner Institut, Berlin, Report HMJ-B21 (1962), in german.
10. Volpp, H. J.: Untersuchung des großräumigen atmosphärischen Transports in Mitteleuropa mit Hilfe von ^{222}Rn. Doctoral thesis, Universität Heidelberg (1984), unpublished.
11. Feichter, J., Crutzen, P. J.: Tellus **42B**, 46 (1990).
12. Pereira, E. B.: Tellus **42B**, 39 (1990).
13. Lambert, G., Ardouin, B., Sanak, J.: Tellus **42B**, 76 (1990).
14. Heimann, M., Monfray, P., Polian, G.: Tellus **42B**, 83 (1990).
15. Gäggeler, H. W., Jost, D. T., Baltensperger, U., Schwikowski, M., Seibert, P.: Atmos. Environ. **29**, 607 (1995).
16. Whittlestone, S., Robinson, E., Ryan, S.: Atmos. Environ. **26A**, 251 (1992).
17. Jacobi, W., André, K.: J. Geophys. Res. **68**, 3799 (1963).
18. Nazarov, L. E., Kuzenkov, F., Malakhov, S. G., Volokitina, L. A., Gaziyev, Ya. I., Vasil'yev, A. S.: J. Geophys. Res. **75**, 3575 (1970).
19. Lambert, G., Polian, G., Sanak, J., Ardouin, B., Buisson, A., Jegou, A., Le Roulley, J. C.: Ann. Géophys. **38**, 497 (1982).
20. Moore, H. E., Poet, S. E., Martell, E. A.: J. Geophys. Res. **78**, 7065 (1973).
21. Reiter, R., Munzert, K.: Arch. Met. Geoph. Biokl., Ser. B, **31**, 39 (1982).
22. Lambert, G., Le Roulley, J.-C., Kritz, M.: Tellus **42B**, 135 (1990).
23. Liu, S. C., McAfee, J. R., Cicerone, R. J.: J. Geophys. Res. **89**, 7291 (1984).
24. Kritz, M. A., LeRoulley, J. C., Danielsen, E. F.: Tellus **42B**, 46 (1990).
25. Rogak, S. N., Baltensperger, U., Flagan, R. C.: Aerosol Sci. and Techn. **14**, 447 (1991).
26. Porstendörfer, J., Mercer, T. T.: Atmos. Environ. **12**, 2223 (1978).
27. Whittlestone, S.: J. Atmos. Chem. **11**, 27 (1990).
28. Mercer, T. T.: Health Phys. **31**, 173 (1976).
29. Mercer, T. T., Stowe, W. A.: Radioactive Aerosols produced by radon in room air, in: *Inhaled Particles III* (Edited by W. H. Walton, Unwin, Old Woking (1971).
30. Sanak, J., Gaudry, A., Lambert, G.: Geophys. Res. Lett. **8**, 1067 (1981).
31. Poet, S. E., Moore, H. E., Martell, E. A.: J. Geophys. Res. **77**, 6515 (1972).
32. Francis, C. W., Chesters, G., Haskin, L. A.: Environ. Sci. Technol. **4**, 587 (1970).
33. El-Dahoushy, F.: Proc. First Int. Summer School on Low-Level Measurements and Their Applications to Environmental Radioactivity, La Rabida, Andalusia, Spain 1987, Published by World Scientific, M. Garcia-Leon and G. Madurga (Eds.), p. 224–273 (1988).
34. von Gunten, H. R., Moser, R. N.: J. Paleolimnol. **9**, 161 (1993).
35. Rangarajan, C., Madhavan, R., Gopalakrishnan, S. S.: J. Environ. Radioact. **3**, 23 (1986).
36. Report Gesellschaft für Strahlen- und Umweltforschung mbH München, GSF, S-956, ISSN-0721-1694 (1983), in german.
37. Report "25 Jahre Radioaktivitätsüberwachung in der Schweiz", KUER, c/o Physics Institute, University of Fribourg, Switzerland (1982), unpublished.
38. Gäggeler, H., von Gunten, H. R., Nyffeler, U.: Earth Plan. Sci. Lett. **33**, 119 (1976).
39. Nijampurkar, V. N., Clausen, H. B.: Tellus **42B**, 29 (1990).
40. Dibb, J. E.: Tellus **44B**, 72 (1992).
41. Gäggeler, H., von Gunten, H. R., Rössler, E., Oeschger, H., Schotterer, U.: J. Glaciol. **29**, 165 (1983).
42. Gäggeler, H. W., von Gunten, H. R., Füeg, B.: unpublished data (1995).
43. Rangarajan, C., Eapen, C. D.: Tellus **42B**, 142 (1990) and references therein.
44. Anand, S. J. S., Rangarajan, C.: J. Environ. Radioact. **11**, 235 (1990).
45. Kuroda, P. K., Daniel, P. Y., Nevissi, A., Beck, J. N., Meason, J. L.: J. Radioanal. Chem. **43**, 443 (1978).
46. Moore, H. E., Martell, E. A., Poet, S. E.: Environ. Sci. Technol. **10**, 586 (1976).
47. Zarcone, M. J., Schery, S. D., Wilkening, M. H., McNamee, E.: Atmos. Environ. **20**, 1273 (1986).
48. Porstendörfer, J., Butterweck, G.: unpublished data referred to in Ref. [2] (1991).
49. Schery, S. D.: J. Air Waste Manage. Assoc. **40**, 493 (1990).
50. Duenas, C., Fernandez, M. C., Senciales, M.: Atmos. Environ. **24A**, 1255 (1990).
51. Whittlestone, S., Schery, S. D., Yanxia, Li: Climate Monitoring and Diagnostocs Laboratory, National Oceanic and Atmospheric Administration, U.S. Depart of Commerce, CMDL Summary Report 1992, Boulder, Colorado (1993).
52. *Umweltradioaktivität und Strahlendosen in der Schweiz 1992,* Bundesamt für Gesundheitswesen, Abteilung Strahlenschutz, Chemin du Musée 3, CH-1700 Fribourg, ISBN 3-905235-09-9 (1993), in german.
53. Weiss, W., presented at: Seminar über Umweltbelastung durch langlebige künstlich erzeugte Radionuklide, Verkehrshaus, Luzern, Switzerland, 29. Juni 1995, KUER/KSA Organizers, unpublished.
54. Wilhelmova, L., Tomasek, M., Stukheil, K.: J. Radioanal. Nucl. Chem. Lett. **144**, 125 (1990).
55. see e.g. in: A. C. Chamberlin, *Radioactive aerosols, Cambridge Environmental Chemistry Series,* Cambridge University Press, ISBN 0-521-40121-6 (1991).
56. *Tritium in the environment.* National Council on Radiation Protection and Measurements (NCRP), Washington DC (1979).
57. Hebert, D.: *Tritium in der Atmosphäre – Quellen, Verteilung, Perspektive –*, Freiberger Forschungshefte, C 443 Geowissenschaften, ISSN 0071-9404, VEB Deutscher Verlag für Grundstoffindustrie, Leipzig (1990), in german.
58. Douglass, A. R., Stanford, J. L.: J. Geophys. Res. **87**, 5001 (1982).
59. Ehalt, D. H.: J. Geophys. Res. **76**, 7351 (1971).
60. Begemann, F., Libby, W. F.: Geochim. Cosmochim. Acta **12**, 277 (1957).
61. Oeschger, H., Schotterer, U., Stauffer, B., Haeberli, W., Röthlisberger, H., Gäggeler, H. W.: Z. Gletscherk. Glazialgeol. **13**, 193 (1977).
62. Schotterer, U., Oeschger, H., Wagenbach, D., Münnich, K. O.: Z. Gletscherk. Glazialgeol. **21**, 379 (1985).
63. Schotterer, U.: priv. comm. (1995).
64. Lugauer, M.: priv. comm. (1995).
65. Schriftreihe Umwelt Nr. 207 Luft. NABEL Luftbelastung 1992, Bundesamt für Umwelt, Wald und Landschaft (Eds.), Bern, August (1993), in german.
66. Lal, D., in: *Earth Science and Meteoritics,* compiled by J. Geiss and E. D. Goldberg 312 pp., North Holland, New York (1963).

67. Junge, C. E.: *Air Chemistry and Radioactivity,* 382 pp., Academic, San Diego, Calif. (1963).
68. Bondietti, E. A., Hoffman, F. O., Larsen, I. L.: J. Environ. Radioact. **1**, 5 (1984).
69. Papastefanou, C., Ioannidou, A.: J. Environ. Radioact. **26**, 273 (1995).
70. Feely, H. W., Larsen, R. J., Sanderson, C. G.: Annual Report of the surface air sampling program, July 1988, Rep. EML-497, Environ. Meas. Lab., U.S. Dept. of Energ., New York (1988).
71. Brost, R. A., Feichter, J., Heimann, M.: J. Geophys. Res. **D12**, 22'423 (1991).
72. Larsen, R. J.: J. Environ. Radioact. **18**, 85 (1993).
73. Barry, R. G., Chorley, R. J.: *The sun spot number between 1610 and 1990, Atmosphere, Weather and Climate,* 6th ed., Rontledge, London (1992).
74. Dutkiewicz, V. A., Husain, L.: Geophys. Res. Lett. **6**, 171 (1979).
75. Damon, P. E., Sternberg, R. E.: Radiocarbon **31(3)**, 697 (1989).
76. Hagemann, F. T., Gray Jr., J., Machta, L., Turkevich, A.: Science **130**, 542 (1959).
77. Stuiver, M.: Nature **273**, 271 (1978).
78. see e.g.: *Méthodes de datation par les phénomènes nucléaires naturels application,* E. Roth, B. Poty (Eds.), Masson, Paris, New York, Barcelone, Milan, Mexico, Sao Paulo (1985).
79. Hauenstein, J.: Herstellung von mikrogramm-großen Kohlenstoffproben für die Bestimmung des ^{14}C Gehaltes sowie Untersuchung von fossilen Pollen und polarem Gletschereis, Dissertation, Universität Bern, Switzerland (1994), unpublished (in german).
80. Levin, I., Bösinger, R., Bonani, G., Francey, R. J., Kromer, B., Münnich, K. O., Suter, M., Trivet, N. E. A., Wölfli, W., in: *Radiocarbon after four decades*, eds. R. E. Taylor, A. Long, R. S. Kra, Springer Verlag, pp 503–518 (1992).
81. Kaimei, D., Youneng, Q., Fan, C. Y.: Radiocarbon **34(3)**, 753 (1992).
82. Suess, H. E.: Natural radiocarbon and the rate of exchange of carbon dioxide between the atmosphere and the sea, Proc. Williams Bay Conf. Sept. 1953, NAS-NSF Publ. 52 (1954).
83. Jost, D. T., Gäggeler, H. W., Baltensperger, U., Zinder, B., Haller, P.: Nature **324**, 22 (1986).

Radiochimica Acta 70/71, 355–375 (1995)

Radionuclides in the Geosphere: Sources, Mobility, Reactions in Natural Waters and Interactions with Solids

By K. H. Lieser

Fachbereich Chemie, Zintl-Institut, Technische Hochschule, D-64289 Darmstadt, Germany

(Received December 16, 1994; revised February 7, 1995)

Radionuclides / Geosphere / Speciation / Solubility / Natural waters / Sorption / Ion exchange

Summary

The various sources of radioactivity are considered. Natural and anthropogenic radionuclides are selected with regard to their importance in the geosphere and listed according to their chemical properties. The importance of the solubility of the radionuclides is emphasized and in view of their mobility five compartments are distinguished: surface of the earth, atmosphere, near-surface layers (with connection to the surface), nuclear installations and deep layers without connection to the surface.

In a following chapter, the reactions of radionuclides with the components of natural waters are discussed. The importance of the knowledge of the chemical forms (speciation) is stressed and illustrated by examples. The influences of Eh and pH, inorganic salts, colloids, suspended matter, organic compounds and microorganisms as well as that of precipitation and coprecipitation is brought out. Then, the interaction of radionuclides with solid components of the geosphere are dealt with. After description of general aspects, interactions with consolidated rocks, unconsolidated rocks, various minerals, sediments and soils are discussed. Isotopic and non-isotopic exhange reactions of radionuclides on minerals are also considered. Finally, the problems of modelling are discussed.

1. Introduction

Radioactivity of uranium has been discovered hundred years ago (1896) by Becquerel [1]. In the following years the natural radionuclides (mainly members of the transmutation series of U and Th) have been identified, and after the discovery of the neutron by Chadwick (1932), the discovery of nuclear fission by Hahn and Strassmann (1938) and the construction of accelerators a great number of artificial radionuclides and new artificial elements have been produced [2]. Today, the investigation of the behaviour of radionuclides in the geosphere, in particular in the environment, has become one of the most important fields of radiochemistry [3].

In this paper the attempt is made to give a survey on the sources and the mobility of the most important radionuclides, to order their numerous interactions with the various components of the geosphere and to elucidate the diverse chemical aspects of these interactions. Emphasis is laid on the chemical properties of the relevant radionuclides. Because detailed description of all kinds of interactions is not possible within the limited scope of this article, references are given for further reading. From these references it is obvious that besides radiochemistry itself various other fields are involved: chemistry of the elements and their complexes, physical chemistry, organic chemistry, colloid chemistry, geochemistry, chemistry of water, sediments and soils.

In chapter 2 a survey is given on the various natural and man-made sources of radioactivity on the earth. In chapter 3 dispersal, dissolution and mobility of radionuclides are discussed. Chapter 4 is devoted to the reactions of radioactive species with the components of aquatic systems and chapter 5 to their interactions with solid components of the geosphere.

2. Sources of radioactivity in the geosphere

Two kinds of sources of radioactivity on the earth can be distinguished: natural sources and anthropogenic sources. Natural sources, like ^{40}K, ^{232}Th, ^{235}U and ^{238}U, are present on the earth from the beginning. They have been formed in the course of nucleogenesis and are called primordial radionuclides. Further natural sources of radioactivity, like ^{3}H and ^{14}C, are produced continuously by the interaction of cosmic rays with the atmosphere.

By mining of ores and minerals appreciable amounts of natural radionuclides, like ^{232}Th, ^{235}U, ^{238}U and their radioactive daughters are brought up to the surface of the earth and contribute to the radioactivity in the environment. In nuclear power stations artificial radionuclides, mainly fission products and transuranium elements are produced and great care is taken in handling the resulting radioactive waste safely and to localize it at certain places. The optimal conditions of final storage of the high-active waste (HAW) are still a subject of discussion. By application of nuclear weapons, by nuclear weapon tests and nuclear accidents, like the accident at Chernobyl, considerable amounts of fission products and radioelements have been set free and distributed via the atmosphere as radioactive fall-out over large areas, in particular in the northern hemisphere.

Radionuclides of special importance in the geosphere are listed in Table 1. In compiling this list the following criteria have been applied: Half-life > 1 d (in case of activation products of materials used in nuclear reactors > 1 y); natural radionuclides of ex-

Table 1. Radionuclides of importance in geomedia
(Half-lives and fission yields σ_f from Refs. [4] and [5] – radionuclides with half-lives >10 y and present in relatively high concentrations are underlined)

Group of the Periodic Table	Element	Radionuclide	Source
	H C	^{3}H (12.3 y) ^{14}C (5730 y)	atmosphere; nuclear fission
I	K	^{40}K (1.28×10^9 y)	potassium salts
	Rb	^{87}Rb (4.80×10^{10} y; σ_f = 2.56%)	rubidium salts; nuclear fission
	Cs	^{134}Cs (2.06 y); ^{135}Cs (3.0×10^6 y; σ_f=6.54%); ^{137}Cs (30.0 y; σ_f=6.18%)	nuclear fission
II	Sr	^{89}Sr (50.5 d; σ_f=4.82%); ^{90}Sr (28.5 y; σ_f=5.77%)	nuclear fission
	Ba	^{140}Ba (12.75 d; σ_f=6.21%)	nuclear fission
	Ra	^{223}Ra (11.43 d); ^{224}Ra (3.66 d); ^{225}Ra (14.8 d); ^{226}Ra (1600 y); ^{228}Ra (5.75 y)	ores/minerals
IV	Sn	^{121}Sn (1.127 d; σ_f(121 + 121 m) = 0.013% ^{121m}Sn (55 y); ^{123}Sn (129.2 d; σ_f=0.016%); ^{125}Sn (9.64 d; σ_f=0.029%); ^{126}Sn ($\approx 10^5$ y; σ_f=0.054%)	nuclear fission
	Pb	^{210}Pb (22.3 y)	ores/minerals
V	Sb	^{125}Sb (2.73 y; σ_f=0.029%); ^{126}Sb (12.4 d; σ_f=0.054%); ^{127}Sb (3.85 d; σ_f=0.125%)	nuclear fission
	Bi	^{210}Bi (5.013 d)	ores/minerals
VI	Se	^{79}Se ($<6.4 \cdot 10^5$ y; σ_f=0.044%)	nuclear fission
	Te	^{127m}Te (109 d; σ_f(127+127m)=0.125%) ^{129m}Te (33.6 d; σ_f(129+129m)=0.757%) ^{131m}Te (1.2 d; σ_f(131+131m)=2.89%) ^{132}Te (3.62 d; σ_f=4.31%)	nuclear fission
	Po	^{210}Po (138.38 d)	ores/minerals
VII	I	^{129}I (1.57×10^7 y; σ_f=0.757%) ^{131}I (8.04 d; σ_f=2.89%)	nuclear fission
0	Kr	^{85}Kr (10.72 y; σ_f=1.32%)	nuclear fission
		^{133}Xe (5.24 d; σ_f(133+133m)=6.70%) ^{133m}Xe (2.19 d)	nuclear fission
	Rn	^{222}Rn (3.825 d)	ores/minerals
I A	Ag	^{111}Ag (7.45 d; σ_f=0.019%)	nuclear fission
II A	Cd	^{109}Cd (1.267 y)	activation
		^{113m}Cd (13.7 y; σ_f(113+113m)=0.0152%)	fission; activation
		^{115}Cd (2.228 d; σ_f(115+115m)=0.0106%) ^{115m}Cd (44.6 d)	nuclear fission
III A	Y	^{90}Y (2.671 d; σ_f=5.77%); ^{91}Y (58.5 d; σ_f=5.93%)	nuclear fission
(Ln)	La	^{140}La (1.678 d; σ_f=6.21%)	nuclear fission
	Ce	^{141}Ce (32.5 d; σ_f=5.79%); ^{143}Ce (1.38 d; σ_f=5.96%) ^{144}Ce (284.9 d; σ_f=5.49%)	nuclear fission
	Pr	^{143}Pr (13.58 d; σ_f=5.96%)	nuclear fission
	Nd	^{147}Nd (10.98 d); σ_f=2.27%)	nuclear fission
	Pm	^{147}Pm (2.623 y; σ_f=2.27%) ^{149}Pm (2.212 d; σ_f=1.08%) ^{151}Pm (1.183 d; σ_f=0.420%)	nuclear fission
	Sm	^{151}Sm (90 y; σ_f=0.420%) ^{153}Sm (1.946 d; σ_f=0.162%)	nuclear fission
	Eu	^{155}Eu (4.96 y; σ_f=0.0320%) ^{156}Eu (15.2 d; σ_f=0.0148%)	nuclear fission
(An)	Ac	^{225}Ac (10.0 d)	decay of ^{237}Np
		^{227}Ac (21.77 y)	ores/minerals
	Th	^{227}Th (18.72 d); ^{228}Th (1.913 y)	ores minerals
		^{229}Th (7340 y)	decay of ^{237}Np
		^{230}Th (7.54×10^4 y) ^{232}Th (1.41×10^{10} y); ^{234}Th (24,10 d)	ores/minerals

Table 1. (Continued)

Group of the Periodic Table	Element	Radionuclide	Source
	Pa	^{231}Pa (3.276×10^4 y)	ores/minerals
		^{233}Pa (27.0 d)	decay of ^{237}Np
	U	^{233}U (1.592×10^5 y)	decay of ^{237}Np
		^{234}U (2.45×10^5 y); ^{235}U (7.04×10^8 y)	ores/minerals
		^{236}U (2.342×10^7 y); ^{237}U (6.75 d)	nuclear reactors
		^{238}U (4.468×10^9 y)	ores/minerals
	Np	^{236}Np (1.55×10^5 y)	nuclear reactors
		^{237}Np (2.140×10^6 y); ^{238}Np (2.117 d)	nuclear reactors
	Pu	^{239}Pu (2.411×10^4 y); ^{240}Pu (6560 y); ^{241}Pu (14.4 y); ^{242}Pu (3.76×10^5 y)	nuclear reactors
	Am	^{241}Am (432.7 y); ^{242m}Am (141 y): ^{243}Am (7380 y)	nuclear reactors
	Cm	^{242}Cm (162.9 d); ^{243}Cm (28.5 d): ^{244}Cm (18.11 y); ^{245}Cm (8500 y); ^{246}Cm (4730 y); ^{247}Cm (1.56×10^7 y) ^{248}Cm (3.40×10^5 y)	nuclear reactors
	Bk	^{249}Bk (320 d)	nuclear reactors
	Cf	^{249}Cf (351 y; ^{250}Cf (13.1 y) ^{251}Cf (898 y); ^{252}Cf (2.64 y)	nuclear reactors
IV A	Zr	^{93}Zr (1.5×10^6 y; $\sigma_f = 6.38\%$)	fission+activation
		^{95}Zr (64.02 d; $\sigma_f = 6.52\%$)	nuclear fission
V A	Nb	^{94}Nb (2.0×10^4 y)	activation
		^{95}Nb (34.97 d; $\sigma_f(95+95m) = 6.52\%$) ^{95m}Nb (3.61 d)	nuclear fission
VI A	Mo	^{99}Mo (2.748 d; $\sigma_f = 6.07\%$)	nuclear fission
VII A	Tc	^{99}Tc (2.13×10^5 y; $\sigma_f = 6.07\%$)	nuclear fission
	Re	^{187}Re (5×10^{10} y)	rhenium compounds
VIII A	Fe	^{55}Fe (2.73 y)	activation
	Co	^{60}Co (5.271 y)	activation
	Ni	^{59}Ni (7.5×10^4 y); ^{63}Ni (100 y)	activation
	Ru	^{103}Ru (39.25 d; $\sigma_f = 3.03\%$); ^{106}Ru (1.020 y; $\sigma_f = 0.402\%$)	nuclear fission
	Rh	^{105}Rh (1.473 d; $\sigma_f = 0.969\%$)	nuclear fission
	Pd	^{107}Pd (6.5×10^6 y; $\sigma_f = 0.147\%$)	nuclear fission

tremely long half-lifes (> 10^{11} y) and radionuclides with fission yields < 0.01% are not taken into account; radioisotopes of the elements 81–92 not being members of the decay series of ^{232}Th, ^{235}U and ^{238}U and radionuclides produced for medical or technical purposes by activation are not considered. The radionuclides are arranged according to the position of the elements in the Periodic Table, in order to facilitate the discussion of their chemical behaviour. Radionuclides with half-lives > 10 y that are present in relatively high concentrations are underlined, because their behaviour over long periods of time, in particular their migration behaviour during storage in nuclear waste repositories, is of great importance.

With respect to radiation doses and possible hazards, the local concentrations and the radiotoxicities of the radionuclides have to be taken into account. Local concentrations of fission products and transuranium elements are high in anthropogenic sources, like nuclear reactors, reprocessing plants and high-active waste (HAW). Local concentrations of radionuclides are also high in natural sources, like ores of U or Th. On the other hand, local concentrations are, in general, low in case of fall-out (with the exception of the region of the Chernobyl accident) and off-gas and effluents of nuclear installations. They are also low for dispersed naturally occurring radionuclides, like ^{3}H, ^{14}C, and the other wide-spread natural sources containing K, U or Th and daughters in low concentrations. The radiotoxicity of ^{3}H, ^{14}C and ^{40}K is also low, whereas it is high for many fission products and for the actinides.

3. Mobility of radionuclides

3.1 General aspects

Mobility of radionuclides is the most important aspect in discussing the behaviour of radionuclides in the geosphere. Gaseous species, aerosols and species dissolved in aquifers are mobile and easily transported with air or water. Solids, however, must be dissolved to become mobile. Therefore, if solid forms of radionuclides are taken into account, mobility depends on solubility and dissolution (leaching). Dissolution and dispersal, respectively, are prerequisites of mobility.

Solubility depends on the nature and the properties of the solid compound containing the radionuclide and on the properties of the solvent, in general water in which various components are dissolved. In the solid, the radionuclide may be present as a macrocomponent or as a microcomponent, and the solubility may vary considerably with the composition. Besides, solubility increases with decreasing size and increasing degree of disorder of the solid particles. Eh, pH and complexing agents may be of significant influence on solubility, and the effects of radiation can not be neglected. Consequently, many investigations are devoted to the problems of solubility and leaching, respectively [3, 6–11]. It should also be mentioned that even for pure and well-defined compounds, like $SrSO_4$, $PbSO_4$ or AgI, solubility is only in a restricted range given by the solubility product and solubility of neutral particles and of complexes has to be taken into account [12–17]. Furthermore, in case of solid solutions (mixed crystals) the solubility of the components is different from that of the pure compounds [18, 19].

With respect to mobility of radionuclides in the geosphere, five compartments may be distinguished, as illustrated schematically in Fig. 1.

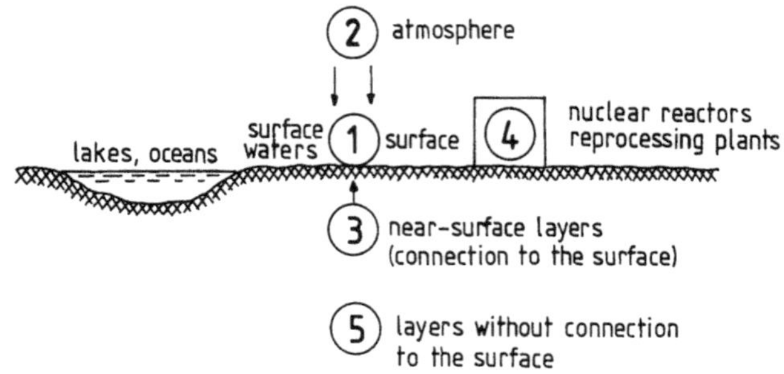

Fig. 1. Compartments of radionuclides in the geosphere.

All radionuclides on the surface of the earth (compartment 1) and in the atmosphere (compartment 2) have easy access to the biosphere. From surface waters as well as from near-surface groundwaters, from soils and from the air radionuclides may be taken up by microorganisms, plants, fish and other animals, thus also entering the various pathways of the food chain. As a measure of the relative amounts of radionuclides transferred to living things and to the food chain transfer factors are used. They depend on the chemical properties and the chemical form of the radionuclide considered, the kind of microorganism, plant or animal and their metabolism and may differ by several orders of magnitude for the various chemical and biological species. Some radionuclides are enriched in certain plants and organs of animals and man, whereas others are released again rather early. The study of the behaviour of radionuclides in the biosphere, in particular their transfer to the human body and their metabolism in man, is the field of radioecology and will not be considered in this article.

3.2 Surface of the earth (compartment 1)

Compartment 1 comprises the surface of the earth, surface waters, including lakes and oceans. The inventory of radionuclides in this compartment is mobile, provided that the species are soluble in water. The relevant radionuclides are mainly of natural origin [20–30]. ^{40}K is widely distributed in nature and easily soluble in form of K^+-ions, which are enriched in clay minerals by sorption. Th present as a minor or major component in minerals is immobile, because the chemical species of Th(IV) are very sparingly soluble in natural waters. However, some of the transmutation products, like ^{228}Ra, ^{224}Ra and ^{220}Rn, are mobile. In contrast to Th(IV), U(IV) is oxidized by air to U(VI) and easily soluble in natural waters containing carbonate or hydrogen carbonate, respectively. The triscarbonato complex $[UO_2(CO_3)_3]^{4-}$ is found in all rivers, lakes and oceans in concentrations of the order of 10^{-6} to 10^{-5} g/l. The daughters of ^{238}U, the long-lived ^{226}Ra and ^{222}Rn are also mobile. ^{226}Ra is found in relatively high concentrations at salt water spas, and ^{222}Rn contributes considerably to the radioactivity in the air. ^{222}Rn and the daughters ^{218}Po, ^{214}Pb, ^{214}Bi, ^{214}Po, ^{210}Pb, ^{210}Bi and ^{210}Po are the major sources of the radioactive dose received by man under normal conditions.

Additional radionuclides are brought into compartment 1 from the compartments 2, 3 and 4, and their mobility depends on the chemical form (i.e. the species) and the solubility. Radionuclides set free on the earth's surface from nuclear installations, by nuclear explosions and by nuclear weapon tests will be considered in more detail in section 3.5.

3.3 Atmosphere (compartment 2)

In the atmosphere the radionuclides ^3H (T) and ^{14}C are generated continuously by nuclear reactions of the neutron component of cosmic radiation with nitrogen in the upper layers of the atmosphere [31–35]. The most important reactions are $^{14}N(n,^3H)^{12}C$ and $^{14}N(n,p)^{14}C$. T is also produced by spallation of heavier nuclides. Both radionuclides are mobile. T is transformed into tritiated water (HTO) and enters the water cycle [36], whereas ^{14}C is predominantly transformed into $^{14}CO_2$ and takes part in the CO_2 cycle. The natural concentration of T in the air is about 1.8×10^{-3} Bq m^{-3} and that of ^{14}C about 5×10^{-4} Bq m^{-3}.

Other radionuclides are emitted from the earth's surface into the atmosphere. The most important radionuclide is ^{222}Rn which will be discussed in more detail in section 3.4. The decay products of ^{222}Rn as well as radionuclides released from nuclear installations into the air are mainly found in form of radioactive aerosols [37].

3.4 Near-surface layers (compartment 3)

From the layers with connection to the surface of the earth (compartment 3) radionuclides are brought to the surface (compartment 1) by natural processes and by human activities. From a few places ^{222}Rn is able to escape through crevises. Ions, like Ra^{2+} or UO_2^{2+} are leached from ores or minerals by groundwater and may enter the surface. Volcanic activities are also bringing up radionuclides to the surface, where they may be leached and enter the water cycle.

Mining of uranium ores and of other ores and minerals (e.g. phosphates) has brought up appreciable amounts of U, Th and the members of their transmutation series to the surface [21] and initiated mobilization of relevant radionuclides. Large amounts of Rn are released into the air. The residues of mining and processing operations are stockpiled and the slag heaps as well as the waste waters cause considerable problems. The isotopes of Ra continue to emit Rn. Thus, about 1 GBq of ^{222}Rn are released per ton of ore containing 1% U_3O_8. Ra and its daughters are migrating to natural oil and gas reservoirs and constitute the major radioactive contaminants of crude oil. The global activity of Ra isotopes brought up to the surface by the production of oil is of the order of 10 TBq per year. Mining of potassium salt deposits for use of K as fertilizer brings additional amounts of ^{40}K to the surface of the earth, where it is leached and enters the surface waters.

Radionuclides are also set free by the burning of coal in thermal power stations [38]. Depending on the origin, coal contains various amounts of U and Th, and these as well as the daughters are released by combustion. Volatile species, in particular Rn, are emitted with the waste gas, ^{210}Pb and ^{210}Po are emitted with the fly ash, and the rest, including U and Th, is found in the ash. The global release of Rn is of the order of 10^{14} Bq per year.

Other activities of man have led to the distribution of appreciable amounts of radionuclides in the atmosphere and on the earth's surface. In the first place nuclear explosions and nuclear weapon tests have to be mentioned, by which Pu and fission products have been deposited, either directly in the ground or via the atmosphere in the form of fall-out. The amount of Pu released by nuclear weapon tests between 1958 and 1981 is estimated at 4.2 tons, of which 2.8 tons were dispersed in the atmosphere and 1.4 tons disposed locally. By underground nuclear explosions about 1.5 tons of Pu have been set free.

3.5 Nuclear installations (compartment 4)

Nuclear reactors and reprocessing plants (compartment 4) are constructed in such a way that the radioactive inventory is confined to certain places and carefully shielded. Only limited amounts of radionuclides are allowed to enter the environment [39–44].

T and ^{14}C are produced in nuclear reactors by the reactions $^2H(n,\gamma)^3H$, $^{10}B(n,2\alpha)^3H$, $^{17}O(n,\alpha)^{14}C$, $^{14}N(n,p)^{14}C$ and $^{13}C(n,\gamma)^{14}C$, respectively. The amounts vary with the type of the reactors: between about 10^{12} and 10^{13} Bq of T and about 10^{12} Bq of ^{14}C per GWe per year. T is released as HTO and about one third of the ^{14}C is released in form of $^{14}CO_2$.

Under normal operation conditions, very small amounts of fission products are set free from nuclear reactors and reprocessing plants, and strong regulations have been established to ensure that the safety limits are not exceeded. Of greatest importance are the actinides and the long-lived fission products, such as ^{90}Sr, ^{99}Tc, ^{129}I, and ^{137}Cs. These radionuclides are released in small amounts, mainly from low-active waste (LAW) streams (e.g. $^{99}TcO_4^-$) or as gases (e.g. organic compounds of ^{129}I and ^{131}I, respectively, from nuclear reactors or reprocessing plants, and $^{99}TcF_6$ in the course of $^{235}UF_6/^{238}UF_6$ isotope separation). ^{99}Tc is found in practically all stages of reprocessing [45, 46] and released in small, but measurable amounts into the sea. ^{129}I is separated from the off-gas of reprocessing plants, mainly in form of $Ag^{129}I$. From this compound it is liberated very slowly by radiolytic decomposition [47].

Despite the safety regulations, accidents occurred with nuclear reactors and reprocessing plants, primarily due to mistakes of the operators. By these accidents parts of the inventory of compartment 4 have entered compartment 1, either directly or via the atmosphere (compartment 2). Mainly gaseous fission products and aerosols have been emitted, but solutions have also been given off. The most severe accident was that at Chernobyl in 1986 [48–54]. Gaseous fission products and aerosols have been transported through the air over great distances [55]. Even molten particles from the reactor core ("hot particles") have been carried with the air over distances of several hundred kilometers.

The behaviour of the radionuclides on the surface of the earth depends primarily on their chemical and physico-chemical form (species) and the solubility, as already mentioned. The degree of dispersal is different (compact solids, coarse particles, fine particles, molecules), and the solubilities vary considerably. In water, the radionuclides may be present in form of suspended matter, colloids, free ions or molecules. Therefore, chemical speciation is of great importance [3, 56, 57]. The formation of radiocolloids has found particular interest [3, 58–66]. General statements with respect to the mobility can be made on the basis of the properties of the elements [67–70]:

Alkali and alkaline earth ions, such as $^{137}Cs^+$ or $^{90}Sr^{2+}$ are easily dissolved in water, independent of pH, and considered to be mobile species. However, their solubility is restricted and their mobility is limited, if they are incorporated into ceramics or glass or bound in clay minerals. The solubility of the alkaline earth sulfates decreases in the order $SrSO_4 > BaSO_4 > RaSO_4$, but the solubility product is, in general, not reached. ^{129}I forms rather mobile species and reacts easily with organic substances. ^{85}Kr and ^{133}Xe remain predominantly in the air.

The lanthanides, e.g. ^{144}Ce, ^{147}Pm, ^{151}Sm, are not soluble in water under normal conditions, because of hydrolysis of the Ln^{3+} cations. However, colloids may be formed, either intrinsic colloids with lanthanides as main components, if their concentration is high enough, or carrier colloids with natural colloids as main components on which the lanthanides are sorbed. Complexes with organic substances may also be formed, in particular with humic substances.

The solubility of the actinides in the oxidation states +3 and +4 is similar to that of the lanthanides, and hydrolysis is more pronounced in case of An(IV). On the other hand, the dioxocations AnO_2^+ and AnO_2^{2+} exhibit relatively high solubility in water in presence of carbonate or hydrogen carbonate, as already mentioned for UO_2^{2+}. Consequently, the actinides may show rather high mobility, provided that they are able to form dioxocations, that the redox potential is high enough (aerobic conditions), and that the water contains HCO_3^- ions in sufficiently high concentrations.

The behaviour of Zr(IV), Nb(IV) and Tc(IV) is comparable to that of An(IV). The influence of the redox potential is very pronounced in case of Tc: Whereas Tc(IV) is not dissolved in water and immobile, Tc(VII) is easily dissolved as TcO_4^- and very mobile.

3.6 Layers without connection to the surface (compartment 5)

Compartment 5 (Fig. 1) comprises all radionuclides in deposits deep under the earth's surface, inaccessible to water and without connection to the surface. Large amounts of ores and minerals with considerable contents of U, Th and their daughters are buried in this compartment. The radionuclides can be considered to be immobile as long as they are not brought up to the surface by geological activities or by man and as long as the contact with water is excluded [43].

The conception with respect to the radioactive waste produced in the shielded compartment 4 (nuclear reactors and reprocessing plants, respectively) is to bring this waste down into compartment 5 in order to confine it there safely for long periods of time. Safety discussions are based on the scenario that water may enter the waste deposits, that the waste containers corrode, that the waste is leached and the radionuclides are set free [43]. The most important step of this scenario is the dissolution of the high-active waste (HAW) by groundwater. In addition, some radionuclides, e.g. ^{129}I, may be liberated in form of volatile species [47].

As far as the mobility of radionuclides in the geosphere over long periods of time is concerned, investigations of the natural deposits of U and Th and of the natural reactors at Oklo are of great importance [3, 43, 71]. The reactors at Oklo have been in operation about two billion years ago [72, 73]. Due to the fission of ^{235}U, the relative concentration of this isotope of U is lower than at other places, whereas the relative concentrations of the stable end products of the fission chains are higher, and spreading of the nuclides from their original location in the course of time gives information about their migration behaviour.

4. Reactions of radionuclides with the components of natural waters

4.1 General aspects

In aqueous solution, the vast majority of the radionuclides listed in Table 1 is present in cationic forms, and the following chemical reactions have to be taken into account:

a) Hydration: By ion-dipole interaction aquo complexes are formed: $M^{n+}\text{---}\overset{(\delta-)}{O}\langle^{H^{(\delta+)}}_{H^{(\delta+)}}$. The energy of hydration increases with decreasing ionic radius and with increasing positive charge of the cations. Thus, the energy of hydration is low for Cs^+ and high for Th^{4+}.

b) Hydrolysis: If the charge of the cations is high ($n > 2$), the repulsion of the protons by the cations may lead to the formation of hydroxo complexes:

$$M^{n+}\text{---}O\langle^{H}_{H} \rightleftharpoons (M{-}OH)^{(n-1)+} + H^+. \quad (1)$$

(Additional water molecules are omitted.) As it is evident from Eq. (1), hydrolysis depends strongly on pH. The repulsion and, consequently, hydrolysis increase with the charge of the cations, and hydrolysis is very pronounced in case of Pa^{5+}. Hydrolysis increases also with decreasing ionic radius of the cations.

c) Condensation: Polynuclear complexes are formed by condensation reactions, like

$$2\,(M{-}OH)^{(n-1)+} \rightleftharpoons (M{-}O{-}M)^{2(n-1)+} + H_2O \quad (2)$$

and subsequent condensation reactions. (Additional water molecules are omitted.) The driving force of this kind of reaction is the formation of the predominantly covalent M−O−M bonds and of one molecule of water. This type of reaction is not only observed with hydroxo complexes of the same metal M, but it takes also place with mononuclear or polynuclear hydroxo complexes of other elements (e.g. polynuclear iron(III) hydroxide), with silicic or polysilicic acid and with colloids carrying hydroxyl groups:

$$(M(1))_m{-}OH + HO{-}M(2) \rightleftharpoons (M(1))_m{-}O{-}M(2) + H_2O. \quad (3)$$

(m indicates the polynuclear nature of the hydroxo complex; charges and additional water molecules are omitted.)

Anions compete with the formation of aquo and hydroxo complexes, depending on their concentration:

$$M^{n+}(H_2O)_m + X^- \rightleftharpoons MX^{(n-1)+}(H_2O)_{m-1} + H_2O \quad (4)$$

and

$$(M{-}OH)^{(n-1)+} + X^- \rightleftharpoons MX^{(n-1)+} + OH^-. \quad (5)$$

(Water molecules are omitted in Eq. (5).)

In case of cations of transition elements (groups IA to VIIIA in Table 1) the possibility of formation of relatively strong covalent bonds with ligands containing donor atoms has to be considered.

With anionic forms of radionuclides, such as $^{129}I^-$ or $^{99}TcO_4^-$, these kinds of reactions do not apply.

Various species of radionuclides are formed by interaction with the components of natural waters. Groundwaters, rivers, lakes and the oceans contain a great variety of substances. Besides the main component water, other inorganic components have to be considered:

- Dissolved gases, like oxygen and carbon dioxide, which influence the redox potential Eh and the pH of the water;
- salts, like NaCl, $NaHCO_3$ and many others which affect pH and complexation and are responsible for the ionic strength;
- inorganic colloids, like polysilicic acid, iron oxides/hydroxides and finely dispersed clay minerals;
- inorganic suspended matter (i.e. coarse particles, only found in agitated water).

Organic components comprise:

- Compounds of low molecular mass, like organic acids, amines, amino acids and other metabolites;
- compounds of high molecular mass, like humic and fulvic acids, and colloids formed by these substances or by other degradation products of organic matter;
- suspended coarse particles of organic matter;
- microorganisms.

The concentrations of these compounds in natural waters vary over a wide range. Coarse particles are observed only in agitated waters and settle down as soon as agitation decreases or stops. Even though many of the components listed above may be present in rather low concentrations (microcomponents), their concentration is often many orders of magnitude higher than that of the radionuclides in question, and they can therefore not be neglected.

The aspect of very low concentration is of special importance for short-lived isotopes of radioelements, the concentration of which is, in general, extremely low, and the behaviour of single atoms has to be considered [74–76]. In case of radioisotopes of stable elements, on the other hand, the ubiquitous presence of these elements leads to measurable concentrations of the elements with the consequence that these radioisotopes show the normal behaviour of trace elements.

Due to the large number of components, natural waters are rather complex systems. The relative concentrations of many components, as well as pH and Eh, are controlled by chemical equilibria. However, there are also components, in particular the colloids and the microorganisms, for which thermodynamic equilibrium conditions are not applicable. The complexity of the chemistry in natural waters and the non-applicability of thermodynamics to various components are the main reasons for the fact that calculations are very difficult and problematic. The same holds for laboratory experiments with model waters. Results obtained for a special kind of water are, in general, not applicable for other natural waters of different origin.

The species of the radionuclides present in natural waters can be characterized by the chemical form and the physico-chemical state (e.g. free ions, complexes, intrinsic colloids ("Eigenkolloide"), carrier colloids ("Fremdkolloide") and speciation of radionuclides has become an important branch of radiochemistry [3, 57, 77–95].

The species of the actinides are of special interest. General aspects of speciation and results obtained for trace elements in aqueous systems are also applicable [96–116]. Eh-pH-diagrams [117, 118] give information about the ranges of stability of oxidation states and hydroxo complexes.

A great number of data have been calculated with respect to the chemical form of radionuclides in aquatic systems [3], but the general applicability of these data is limited, due to the reasons mentioned above.

4.2 Redox potential and pH

The redox potential Eh (aerobic or anaerobic conditions) is of great influence on the behaviour of radionuclides in geomedia, if different oxidation states have to be taken into account. In this context, the concentrations of oxygen and hydrogen sulfide are significant. The presence of oxygen in natural waters is due to the access of air, whereas that of hydrogen sulfide may be due to weathering of sulfidic minerals or to decomposition of organic compounds. The redox potential Eh is of special importance for the behaviour of I, U, Np, Pu, and Tc [83–85, 90, 119–121]. Elemental iodine is volatile and reacts with organic compounds, in contrast to I^- ions. U(IV) does not form soluble species in natural waters, in contrast to U(VI). Therefore, U(IV) is dissolved under aerobic conditions after oxidation to UO_2^{2+}. The latter is reduced again, if the water enters a reducing zone, and redeposited as UO_2. The most important feature of Np in water is the great range of stability of Np(V) in form of NpO_2^+, and in this respect Np differs markedly from the neighbouring elements U and Pu. Under reducing conditions, Np(IV) is stable which resembles U(IV) and Pu(IV). The chemical form of Tc depends also on the redox potential. Under anaerobic conditions, the stable oxidation state is +4 (TcO_2, $TcO(OH)_2$), and under aerobic conditions Tc is easily oxidized to TcO_4^-. This anionic form of Tc shows very different behaviour and has a high tendency to react with proteins [122]. In presence of H_2S, sparingly soluble sulfides may be formed with elements of groups IV (Pb), V (Sb, Bi), IA (Ag), IIA (Cd), VIA (Mo), VIIA (Tc), and VIIIA (Fe, Co, Ni).

In natural waters pH varies between about 6 and 8 and influences the chemical behaviour of elements that are sensible to hydrolysis, i.e. elements of groups III, IIIA, IV, IVA, V, VA and VIIIA of the Periodic Table [123–125]. Relevant radionuclides belong mainly to the groups of the lanthanides and the actinides. The tendency of the actinides to hydrolyze increases in the order $MO_2^+ < M^{3+} < MO_2^{2+} < M^{4+}$. As already mentioned in section 4.1, formation of complexes with various inorganic and organic ligands present in natural waters competes with the formation of hydroxo complexes, with the consequence that those complexes are formed that exhibit the highest stability under the given conditions. Whether polynuclear hydroxo complexes are formed by condensation of mononuclear hydroxo complexes of the same kind according to Eq. (2) or by condensation with reactive species of other origin according to Eq. (3) depends on the relative concentrations of the mononuclear hydroxo complexes of the element considered and that of reactive species of other origin. Accordingly, either intrinsic colloids ("Eigenkolloide") or carrier colloids ("Fremdkolloide") may be formed [65, 114]. Reactive species with respect to formation of carrier colloids have already been mentioned in section 4.1: polysilicic acid, iron(III) oxide/hydroxide and finely dispersed (colloidal) clay particles. Due to hydrolysis and interaction of hydroxo complexes with other components, actinides are not found as monomeric species in natural waters in absence of complexing agents, whereas in presence of carbonate or hydrogen carbonate, respectively, monomeric carbonato complexes prevail.

4.3 Inorganic salts

Inorganic salts affect the behaviour of radionuclides in natural waters in various ways. At high salinity (high ionic strength) formation of colloids is hindered and colloids already present are coagulated, if salt water enters the system. Thus precipitation of colloids carried by rivers occurs in a large scale in the estuaries. Further, dissolved salts influence pH, hydrolysis and complexation [126]. They may act as buffers, e.g. in seawater, where pH is kept constant at about 8.2 by the presence of $NaHCO_3$. Finally, the anions in natural waters form ion pairs and stable complexes with cationic radionuclides and affect solubility, colloid formation and sorption behaviour. Mobility may be enhanced by complexation [127]. Cl^- ions are weak complexing agents, but they are able to substitute OH^- ions in hydroxo complexes and to suppress hydrolysis. HCO_3^- ions are also found in significant quantities in natural waters, particularly in seawater, and form

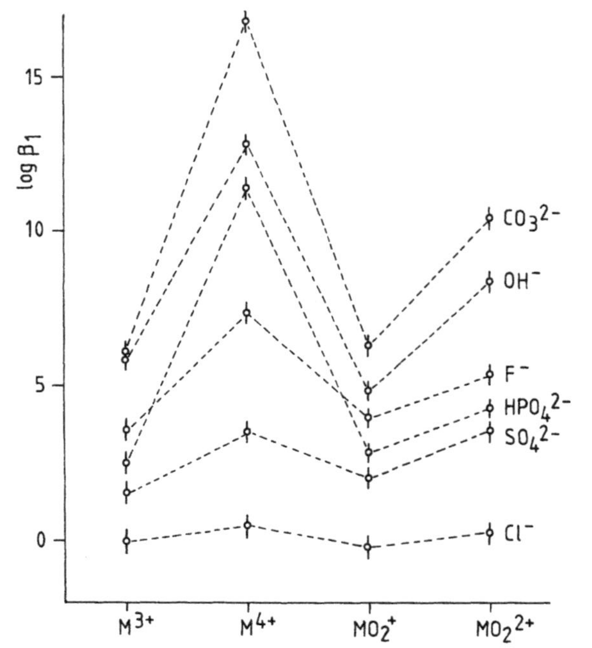

Fig. 2. Logarithms of the stability constants β_1 for the formation of 1:1 complexes of the actinide ions M^{3+}, M^{4+}, MO_2^+ and MO_2^{2+} with various anions [67, 68].

rather stable carbonato and hydrogen carbonato complexes, as already mentioned for UO_2^{2+} ions. Complexation constants are known for the alkaline earth ions, lanthanides and actinides [3, 128, 129]. Dissolution of $CaCO_3$ in groundwater by formation of the hydrogen carbonato complex $Ca(HCO_3)_2$ in presence of CO_2 is responsible for the hardness of water and the transport of Ca^{2+}. Sr^{2+} ions show similar behaviour. The carbonato complexes of the lanthanides and the actinides are more stable than the complexes formed with sulfate and fluoride. Carbonato complexes may also stabilize oxidation states. The logarithms of the stability constants for the formation of 1:1 complexes of the actinide ions M^{3+}, M^{4+}, MO_2^+ and MO_2^{2+} with various inorganic ligands are plotted in Fig. 2.

Quite a number of carbonato, hydrogen carbonato and mixed hydroxo/carbonato complexes are reported in the literature for lanthanides and actinides [129]. Examples for the oxidation state +3 are: $CeCO_3^+$, $Ce(CO_3)_2^-$, $Ce(CO_3)_3^{3-}$, $AmHCO_3^{2+}$, $Am(HCO_3)_2^+$, $AmCO_3^+$, $Am(CO_3)_2^-$, $Am(CO_3)_3^{3-}$, $AmOH(CO_3)_2^{2-}$; for the oxidation state +4: $PuCO_3^{2+}$, $Pu(CO_3)_3^{2-}$, $Pu(CO_3)_4^{4-}$, $U(CO_3)_5^{6-}$, $Pu(CO_3)_5^{6-}$, $PuOHCO_3^+$; for the oxidation state (V): $NpO_2CO_3^-$, $NpO_2(CO_3)_2^{3-}$, $NpO_2(CO_3)_3^{5-}$; for the oxidation state (VI): $UO_2(CO_3)_2^{2-}$, $UO_2(CO_3)_3^{4-}$.

4.4 Inorganic colloids

The interest in the migration behaviour of radionuclides in the geosphere led to a revival of colloid chemistry of inorganic compounds [3, 65, 130–136]. Formation of intrinsic colloids ("Eigenkolloide") in natural waters can be excluded for radioisotopes of elements of groups 0, I and VII, and the probability that they may be formed is small for radioisotopes of elements of other groups as long as the concentration of the elements is low. Therefore, only carrier colloids ("Fremdkolloide") are considered in this section. Radiocolloids of this kind are formed by interaction of radionuclides with colloids present in natural waters. Clay particles have a high affinity for the heavy alkali ions that are bound by ion exchange in the order $Cs^+ > Rb^+ > K^+ > Na^+$ [137]. This leads to the formation of carrier colloids with ^{137}Cs [65, 138]. Alkaline earth ions are also bound by ion exchange on clay minerals, actually in the order $Ra^{2+} > Ba^{2+} > Sr^{2+} > Ca^{2+}$ [137]. Due to the difference in affinity for Cs^+ and Sr^{2+} ions and the competition of Ca^{2+} with Sr^{2+} ions for the sorption sites, fixation of ^{90}Sr on clay particles is not as strong as that of ^{137}Cs [65]. Hydroxo complexes of elements of groups III, IIIA, IV, IVA, V, VA and VIIIA, on the other hand, interact with hydroxyl groups of colloidal polysilicic acid, iron(III) hydroxide, hydrated iron(III) oxide and the hydroxyl groups at the surface of colloidal clay particles in the way mentioned in section 4.1 (Eq. (3)) and described in the literature by the term hydrolytic adsorption [139–142].

It should be emphasized that the formation of colloids is not only a matter of the size, but it depends primarily on the surface properties of the particles. They may carry charges (e.g. adsorbed ions), a shell of water molecules, or hydrophobic substances on the surface which lead to mutual repulsion and prevent coagulation [135, 143]. Without these surface properties, the particles could not exist as colloids. The behaviour of radionuclides bound on natural colloids by ion exchange or sorption follows that of the colloids which serve as carriers.

4.5 Suspended inorganic matter

With suspended inorganic matter (coarse particles of quartz, silicates, including clay, iron oxide and other inorganic substances) radionuclides may also interact by sorption or ion exchange [144, 145]. However, these particles are only observed in flowing or agitated natural waters (e.g. in rivers). After transportation over certain distances, that depend on the mass and the density of the particles, the latter are settling down in form of sediments which contain the radionuclides bound on these particles.

4.6 Organic compounds of low molecular mass

Organic compounds are primarily found in surface waters, but also in groundwaters, if these are or have been in contact with organic substances [146]. Compounds of low molecular mass may be of natural origin (e.g. metabolites, such as organic acids, amines or amino acids) or anthropogenic (e.g. detergents, aromatic sulfonic acids used in manufacturing textiles, as colour-brighteners or for fabrication of dyestuffs). Many of these organic compounds are strong complexing agents and well soluble in water. They are able to form stable complexes with elements of groups III, IV, V, and in particular with transition elements (groups IA to VIIIA) [128]. If they are not separated off, they may carry trace elements including radionuclides over large distances [127, 147].

4.7 Organic compounds of high molecular mass, organic colloids and suspended organic matter

Organic compounds of high molecular mass play an important role in natural waters [3, 102, 115, 127, 146–159] and the knowledge about the behaviour of trace elements in aquatic systems [96, 103, 106, 108, 160–164] can also be applied to radionuclides. The most important examples are humic and fulvic acids, both degradation products of organic matter. They are polyelectrolytes and contain carboxylic and phenolic hydroxyl groups which make these compounds hydrophylic and enable them to form rather stable complexes. Other compounds of high molecular mass are proteins, lipids and carbohydrates. The concentration of dissolved organic matter in natural waters varies considerably. It may be as low as 0.1 mg/l DOC (dissolved organic carbon) in deep groundwaters, ranges from 0.5 to 1.2 mg/l DOC in oceans and may go up to about 50 mg/l DOC in swamp waters. The stability constants of complexes formed between cations and humic or fulvic acids can be expressed as a function of the degree of ionization of the acids. For the actinides rather high stability constants have been measured [146].

With increasing mass of the organic compounds colloidal properties may prevail and a pronounced difference between non-colloidal and colloidal species will not exist any more [135]. Radiocolloids containing organic compounds may be formed in two different ways: By sorption or ion exchange of radionuclides on macromolecular organic compounds or by sorption of organic complexes of the radionuclides on inorganic colloids. These organic complexes may be of low or high molecular mass; they form a coating on the inorganic material. In all cases, the surface properties of the resulting colloids are decisive for their behaviour.

With increasing size of colloidal particles filtration effects become important, in particular, if the water passes through layers or zones with small pores, by which particles of larger size are filtered off [65].

Interaction of radionuclides with suspended coarse particles of organic matter depends on the nature of the substances and is, in principle, comparable to the interaction with colloidal macromolecular compounds. The suspended particles may stay on the surface of rivers or lakes, if their density is low, or they may settle down together with other coarse particles in form of sediments, if the agitation decreases [96, 103, 106, 108, 160–164].

4.8 Microorganisms

Microorganisms may also influence the fate of radionuclides in natural waters [109, 165–167], in particular in surface waters and near-surface groundwaters. Depending on the metabolism of the microorganisms and their preference for certain elements, the radionuclides may be incorporated and migrate with the microorganisms. Incorporation of radionuclides can be studied by microautoradiography [165, 166]. With the uptake by microorganisms, radionuclides may enter the food chain.

4.9 Precipitation and coprecipitation

If a certain supersaturation of a component is reached, precipitation occurs, starting with nucleation and crystal growth [103, 168, 169]. Supersaturation may be caused by entering of those species into the system that form sparingly soluble compounds with the components already present, by change of pH or temperature, redox reactions, loss of volatile components, like CO_2, shift of complexation equilibria or evaporation of water. Examples often observed in natural waters are the precipitation of $CaCO_3$ due to the loss of CO_2 and the shift of the equilibrium between $CaCO_3$ and $Ca(HCO_3)_2$, and the precipitation of Fe(III) hydroxide due to the oxidation of Fe(II) by the access of air. Less common examples are precipitation of sparingly soluble sulfides by reaction of group VIIIA elements with H_2S or of sulfates by reaction of group II elements with SO_4^{2-} ions or precipitation of U(IV) hydroxide by reduction of UO_2^{2+} ions. Depending on the conditions and the properties of the compounds, the process of precipitation may stop at an intermediate stage with the formation of a colloid. Examples are sols of colloidal Fe(III) hydroxide in water. Formation of colloids is also possible for other sparingly soluble compounds, if further growth of the particles is hindered by protective coatings. Precipitation of radionuclides may be stopped by adsorption of the intermediates (e.g. polynuclear complexes) on the surface of natural colloids. On the other hand, colloids may be precipitated by increase of salt concentration, as has been mentioned in section 4.3.

In the first place those elements are involved in these precipitation reactions that are present in sufficiently high concentrations in natural waters, so that the solubility limits are exceeded. For trace elements this condition is in most cases not fulfilled. However, solubility limits may be exceeded even for trace elements, if compounds of extremely low solubility, e.g. hydroxides, sulfides or phosphates are formed. A practical example is the deposition of U(IV) in zones of low redox potential [170]. Precipitation of sparingly soluble hydroxides has been studied as function of pH by use of paper chromatography for Th hydroxide and Fe(III) hydroxide [92, 171].

If low concentrations of radionuclides are considered, coprecipitation is, in general, the most important process. Two mechanisms of coprecipitation of radionuclides have been proposed by Hahn [172], isomorphous substitution or adsorption. In the first case solid solutions (mixed crystals) are formed in which the microcomponents replace the macrocomponent in its normal lattice by syncrystallization (cocrystallization). In the second case, the microcomponents are incorporated as impurities during formation of amorphous precipitates or crystal growth, respectively. They may be given off again, at least partly, in the course of recrystallization of crystalline compounds or in the course of ageing.

Coprecipitation with sparingly soluble hydroxides, preferably Fe(III) hydroxide, is frequently used in analytical chemistry for separation of trace elements [173]. The amorphous intermediates that are formed by condensation of mononuclear to polynuclear hydroxo complexes offer a high specific surface area with hydroxyl groups as reactive sorption sites for many species of trace elements or radionuclides, respectively, in particular cationic hydroxo complexes of metals in the oxidation states +3 and +4, like MOH^{2+}, MOH^{3+}, $M(OH)_2^+$, $M(OH)_2^{2+}$ and others, and anions, like HPO_4^{2-}, MoO_4^{2-} and others. Coprecipitation of trace elements with Fe(III) hydroxide or Fe(III) phosphate, respectively, is often used as an effective procedure in the preparation of drinking water [174].

Actinides(IV) and actinides(III) are very effectively coprecipitated with $BaSO_4$ [175–177], although they do not form compounds that are isomorphous with $BaSO_4$. Detailed investigation showed that by coprecipitation of Th and U(IV) with $BaSO_4$ anomalous solid solutions are formed in which in the lattice of $BaSO_4$ $3Ba^{2+}$ ions are substituted by $2K^+$ and $1Th^{4+}$ or $1U^{4+}$, respectively [178, 179]. Coprecipitation of actinides(IV) with $SrSO_4$ is taking place in a similar way by formation of anomalous solid solutions [180]. The results show that even if the microcomponent and the macrocomponent have different stoichiometry and do not form isomorphous crystals, if considered separately, the microcomponent may be coprecipitated very effectively by lattice substitution. Generally, coprecipitation of microcomponents is to be expected, if these would also be precipitated under the given conditions, provided they would be present in higher concentrations. This rule must also be extended in the case of coprecipitation of actinides(IV) with $BaSO_4$ or $SrSO_4$, because the sulfates of actinides(IV) do not exhibit low, but medium solubility. Furthermore, so far formation of solid solutions has only be observed, if the ionic radii of the substituting and the substituted ions do not differ by more than 15%. But this limit is appreciably exceeded in case of formation of anomalous solid solutions of Th^{4+} with $BaSO_4$, where the ionic radius of Th^{4+} is about 24% smaller than that of Ba^{2+}.

As a result it can be concluded that coprecipitation may considerably influence the fate of radionuclides in natural waters, in particular that of actinides.

5. Interaction of radionuclides with solid components of the geosphere

5.1 General aspects

The migration behaviour of radionuclides in the geosphere is strongly influenced by the interaction of the species present in the aqueous system with the surrounding solids. This is of special interest in case of long-lived radionuclides [3, 147, 181–183] the transport of which may be appreciably retarded or even stopped, if the interaction is strong, particularly if the radionuclides are incorporated into the solids. Therefore, sorption of radionuclides on solids has been investigated extensively for materials in the neighbourhood of planned high-active waste repositories [3, 184].

Sorption of radionuclides on solids may be due to various kinds of interactions [185–187]: Fixation by predominantly ionic bonds (ion exchange), by mainly covalent bonds (chemisorption), by weak (van der Waals) bonds (physisorption); sorption on the outer surface of the solids (substantial quantity: outer surface area), sorption on inner surfaces (substantial quantity: inner surface area), sorption by ion exchange at positions within the solids, sorption by precipitation, coprecipitation, or incorporation (one directional processes). In all cases the surface and exchange properties of the solids, in particular the quality of the sorption and exchange sites, are important, as well as the knowledge of the species of the radionuclides in the aqueous system [137, 185, 188–197], in order to understand the migration behaviour of radionuclides.

The following main groups of interactions of radionuclides with solids can be distinguished:

a) Reactions of hydroxo complexes or anionic forms of radionuclides with hydroxyl groups at the surface of solids ($\geqslant$SiOH, $>$AlOH or $>$FeOH), e.g.

$$\geqslant \mathrm{SiOH(s)} + \mathrm{HO{-}M^{(n-1)+}(aq)} \rightleftharpoons \geqslant \mathrm{Si{-}O{-}M^{(n-1)+}(s)} + \mathrm{H_2O}. \qquad (6)$$

(Additional water molecules are omitted.) The complex of M in solution is converted to a complex at the surface of the sorbent ("surface complex") with a partly changed coordination sphere. Reaction (6) is equivalent to reaction (3) in section 4.1. It has first been described in detail by Kautsky [140] who studied the interaction of Th^{4+} with hydrous silicon dioxide and the subsequent reaction with phosphate. As already mentioned in section 4.4, this type of reaction has been characterized by the term "hydrolytic adsorption" in the literature [139–142]. It is very common and has been used in analytical chemistry for selective separations [141, 142]. By this group of interactions, bonds of predominantly covalent nature are formed. Reaction (6) is often described as an ion exchange

$$\geqslant \mathrm{SiOH(s)} + \mathrm{M^{n+}(aq)} \rightleftharpoons \geqslant \mathrm{Si{-}O{-}M^{(n-1)+}(s)} + \mathrm{H^+}, \qquad (7)$$

although the presence of hydroxo complexes of M is a prerequisite of the reaction and contribution of ionic bonding in $\geqslant$SiOH is extremely small. Sorption of the anion $^{99}MoO_4^{2-}$ on hydrous aluminium oxide is applied in the commercial ^{99}Mo/^{99m}Tc radionuclide generators.

b) Exchange of non-hydrolyzed cationic species of radionuclides at the outer surface or within the layer structure of solids. This type of reactions plays an important role in case of the non-hydrolyzing ions $^{137}Cs^+$, $^{90}Sr^{2+}$ and $^{226}Ra^{2+}$ and is very effective, if clay minerals are present that exhibit a high exchange capacity [137]. The predominant type of interaction is ion exchange. Exchange of cationic or anionic species of radionuclides on the surface of ionic compounds may also contribute to sorption, in particular, if the radionuclides are incorporated in the course of recrystallization.

c) Adsorption of complexes of radionuclides with various inorganic ligands (besides hydroxo complexes) or with organic ligands, and adsorption of colloidal species of radionuclides. This group of reactions comprises a great variety of radionuclide species and of sorption equilibria. Adsorption of organic complexes of radionuclides, particularly complexes with humic acids, and adsorption of colloids are important examples. The predominant kind of interaction is physical adsorption with formation of relatively weak (van der Waals) bonds.

The investigation of radioactive disequilibria in natural systems [3, 198] gives valuable information about solid/solution interactions and their influence on the migration behaviour of radionuclides.

The presence of colloids may influence the sorption ratios observed in natural systems appreciably [199, 200], and the large variations of sorption ratios that have been measured in the same systems could be explained by the formation of colloids.

The sorption behaviour of radionuclides in various natural systems have been studied in great detail [e.g. 3, 84, 85, 90, 93, 138, 184, 186, 187, 201–215], and it has been found in all investigations that sorption and migration may vary considerably. They depend on the properties and the dispersal of the radionuclide species present in solution, the nature of the sorption sites, the surface area of the solids and the interference and the competition, respectively, of other species. Again, knowledge about the behaviour of trace elements in aquatic systems [96, 103, 106, 108, 160–163, 216] is of great practical importance.

The solids in the geosphere are of very different kind and composition. As already mentioned, the surface properties are of special interest with respect to the interaction with radioactive species. Keeping this aspect in mind, the following main components of the geosphere can be distinguished:

- Consolidated rocks (magmatic rocks, like basalt, granite, feldspar, quartz, olivines, plagioclases, pyroxenes, and sedimentary rocks, like sandstone, limestone or dolomite);
- unconsolidated rocks (more or less loose-packed material, consisting mainly of glacial deposits of gravel, sand and clay;
- sediments in rivers, lakes and oceans;
- soils (mainly sand, clay, humus with plant residues, little animals and plenty of microorganisms).

Besides the main components, many other minerals have to be taken into account:

- Oxides and hydroxides (e.g. hydrargillite, diaspor, corundum, spinels, hematite, magnetite, perowskites);
- halides (e.g. rock salt, cryolite, carnallite);
- sulfides (e.g. pyrite);
- sulfates (e.g. gypsum, anhydrite, alum);
- phosphates (e.g. apatites, monazites – some phosphates contain exchangeable protons);
- carbon and carboniferous material.

Of special significance with respect to their properties as sorbents are the clay minerals (e.g. kaolinite, montmorillonite, illite, chlorite), mainly due to their high exchange capacity. They exhibit the three main groups of interactions a), b) and c) mentioned above. In the case of clay minerals with layer structure, sorption by ion exchange proceeds into the inner parts of the particles, and the term "adsorption" is not appropriate.

5.2 Consolidated rocks

The surfaces of silicate rocks are altering in the course of time. Cations, like Na^+, K^+, Mg^{2+}, Ca^{2+}, that are integrated parts of the silicates, are leached out and hydroxyl groups are formed at the surface which may add or give off protons, depending on the pH. The resulting sorption sites are of different nature and quality. $\equiv$SiOH, $>$AlOH, $>$FeOH groups are found most frequently. They are binding hydroxo complexes or divalent oxoanions by formation of surface complexes, as described in section 5.1. A review on anion adsorption is given in the literature [217].

Most investigations with consolidated rocks have been made with granites, because these are possible host rocks for high-active waste repositories [3, 44, 218–228]. Qualitatively, the interactions of radionuclides with various natural materials are comparable with the interactions in chromatographic columns, filled with $SiO_2 \cdot xH_2O$, $Al_2O_3 \cdot xH_2O$, or $Fe_2O_3 \cdot xH_2O$. The specific surface area of silicate rocks, however, is relatively low. In neutral media preferably those radionuclides are sorbed that are present as hydroxo complexes or as divalent anions. Other stable complexes with inorganic or organic ligands as well as colloids show only weak interactions. The competition with other components of the aqueous system for the sorption sites must also be considered in discussion the sorption equilibria.

With sedimentary rocks radionuclides may also interact in various ways. They are sorbed on sandstone, limestone and other sedimentary rocks [3, 229–232]. In addition to reactions with hydroxyl groups, heterogeneous exchange reactions may contribute to the sorption [233, 234]. Examples are: Exchange of $H^{14}CO_3^-(aq)$ for $^{12}CO_3^{2-}(s)$ or exchange of $^{90}Sr^{2+}(aq)$ for $Ca^{2+}(s)$ at the surface of calcite, limestone or dolomite. These exchange reactions have a certain energy of activation, but they may lead to a considerable decrease of the concentration of radionuclides, like ^{14}C, ^{90}Sr, ^{210}Pb or ^{226}Ra, while the water is moving slowly along the surfaces of these kinds of rocks. In addition, incorporation into the inner parts of the crystals due to recrystallization and formation of solid solutions may take place.

Interaction of radionuclides with consolidated volcanic tuffs has been investigated intensively [3, 235–239], because the deposits at the Yucca Mountains, USA, have been found to be suitable as repositories for high-active waste.

5.3 Unconsolidated rocks

The loose-packed material of unconsolidated rocks consists mainly of sand and clay. Clay minerals are the most important components, because of their high sorption capacity that has already been mentioned in section 5.1. Compared with clay minerals, sand is a rather poor sorbent, although particles of crystalline hydrated silica exhibit also exchange properties [240], due to the presence of $\equiv$SiOH groups.

The exchange properties of clay minerals have been investigated in great detail [137, 195, 241–249]. On the one hand, sorption and exchange sites, respectively, are available at the outer surfaces, similar to those present at the surface of alumosilicates ($\equiv$SiOH and $>$AlOH groups). On the other hand, clay minerals with layer structure (e.g. montmorillonite, vermiculite) are able to exchange cations (mainly Na^+, K^+, Mg^{2+}, Ca^{2+}) in positions between the layers. Due to the high specific surface area and the large number of exchange sites, the exchange capacity goes up to values of about 1 to 2 meq/g clay. The ion exchange equilibria depend on the charge density of the clays and the valence of the exchangeable ions. Clay minerals with high charge densities, like illite and micas, exhibit a high preference for monovalent cations ($Cs^+ > Rb^+ > K^+$) which are bound more or less irreversibly. Divalent cations ($Ra^{2+} > Ba^{2+} > Sr^{2+}$) are also firmly bound.

These highly charged, mica-type clay minerals play an important role in nature, because they are present in all fertile soils and capable of binding mono- and divalent cations. They exhibit a high tendency to form colloids and affect strongly the behaviour of radionuclides in the geosphere. ^{137}Cs, ^{90}Sr and ^{226}Ra are fixed in positions between the layers in partly irreversible form. They are set free again only in presence of relatively high concentrations of competing ions. The hydroxo complexes of lanthanides and actinides are mainly bound at the outer surfaces by interaction with ⋗SiOH and ⋗AlOH groups. In addition, complexes, in particular those with organic ligands, are bound at the outer surfaces rather effectively, due to the large specific surface area of the clay minerals.

Often compounds present in relatively low concentration play a decisive role for the sorption behaviour of natural solids. Examples are: Small amounts of clay besides large amounts of sand in case of sorption of ^{137}Cs [138] and small amounts of carboniferous materials besides large amounts of sand and clay in case of sorption of ^{129}I [85].

As the layers above the salt dome at Gorleben, Germany, consist of unconsolidated rocks, they have been investigated in great detail with respect to the migration behaviour of radionuclides [3]. In other countries, clay is discussed as host for high-active waste repositories, because of the favorable sorption properties of clay minerals.

5.4 Isotopic and non-isotopic exchange on minerals

Heterogeneous exchange on carbonates has already been mentioned in the previous section. Such exchange reactions are also taking place at the surface of other sparingly soluble compounds and minerals (e.g. halides, sulfates, phosphates). They may lead to rather selective separations [196, 233, 234, 250–261]. Following the exchange at the surface, ions may be incorporated into the solids in the course of recrystallization which is a very slow but continuous process. The rate of recrystallization increases with decreasing size and increasing imperfection of the crystals. The uptake of foreign ions is not only observed, if these have the same charge, but anomalous solid solutions with radionuclides of different charge may also be formed [180].

5.5 Sediments in rivers, lakes and oceans

The sediments in rivers, lakes and in the oceans are worth special consideration, because sediments are present in all surface waters and radionuclides disposed at the surface of the earth (compartment 1, section 3.2) will sooner or later come into contact with these sediments.

Aquatic sediments are formed by settling down of coarse and fine inorganic and organic particles. Radionuclides may enter river, lake or ocean sediments in different ways: By sorption of molecular-dispersed species (ions, molecules) on the sediments, by precipitation or coprecipitation of originally molecular-dispersed species, by coagulation and sedimentation of colloids, in particular carrier colloids (e.g. by increase of the ionic strength in estuaries) or by sedimentation of coarse particles (suspended matter). By desorption, the radionuclides may by remobilized and released again into the water.

The main components of these sediments are similar to the sediments that have been formed in former geological times: sand and clay minerals. However, river and lake sediments contain also relatively great amounts of organic material and of living things, in particular microorganisms.

Due to the importance of river, lake and ocean sediments with respect to the migration behaviour of radionuclides, many authors have investigated the interactions of radionuclides or trace elements with these sediments [3, 262–278]. In many cases, appreciable fixation and retention, respectively, of radionuclides have been found. It is difficult, however, to distinguish the influences of the various processes, mentioned above, and to correlate the fixation in the sediments to certain components. As far as the inorganic components are concerned, the interaction with clay minerals plays the most important role.

5.6 Soils

With respect to the components and the chemistry, soils are even more complex than river or lake sediments. On the other hand, large areas of the continents are covered with soils of various composition, and therefore the interest in the behaviour of trace elements or radionuclides, respectively, in soils is justified [3, 279–306]. Above that, radionuclides are easily transferred from soils to plants and animals, and in this way they may enter the biosphere and the food chain.

The main components of soils are sand, clay and humus. Whereas interaction of radionuclides with sand is rather weak, as in the case of sediments, sorption by clay minerals and reactions with the organic compounds in humus are most important for the migration behaviour of radionuclides. The following statements can be made for some relevant radionuclides: $^{137}Cs^{+}$ ions are rather strongly bound on clay particles, as already mentioned in section 5.3. $^{90}Sr^{2+}$, $^{226}Ra^{2+}$ and $^{210}Pb^{2+}$ are also retained by clay particles or bound on chalky soil by precipitation or ion exchange. $^{129}I^{-}$ is oxidized and reacts easily with organic compounds, and these take part in the metabolism of microorganisms. Lanthanides and actinides are either present as hydroxo complexes or as organic complexes. These species are rather strongly bound on the components of the soils, but the organic complexes may also stay in solution, possibly in form of colloids. $^{99}TcO_4^-$ reacts easily with proteins [122] and ^{99}Tc may thus be incorporated into organic matter.

Besides the composition of the soils, other factors are also of major influence on the migration of radionuclides: rainfall, thickness of the soil layers and their permeability to water, and nature of the layers underneath. For instance, ^{137}Cs will be washed down quickly through layers of sand, but stay in layers of clay. ^{239}Pu will be sorbed by clay more strongly than by sand and may stay in soils for rather long times, if it is not dissolved by complexation or displaced by other compounds.

5.7 Modelling

Because of the great practical importance of the interactions between radionuclides and solids, particularly with respect to safety aspects of high-active waste repositories, attempts have been made to model these interactions with the aim to describe the influence of sorption quantitatively by a theoretical concept and characteristic data.

The first approach was the "K_d-concept", i.e. the determination of distribution coefficients K_d for the calculation of sorption and retardation of radionuclides in an aquatic solid/solution system [307]. In this concept it is assumed that an equilibrium distribution of a certain radionuclide X exists between solution and solid: X(solution) $\rightleftharpoons$ X(solid). It was found, however, that the K_d values or sorption ratios R_s, respectively, may vary in a range of several orders of magnitude in the same solid/solution systems [83, 84, 138, 187, 207]. Every measured sorption ratio is valid only for a certain species and for a special set of conditions, including the kind and the composition of the solid, Eh, pH, temperature, concentrations of all possible complexing agents, concentration of other components in the water, particularly those competing for the sorption sites, colloid formation, volume to mass ratio, agitation, time of contact between water and solid, filtration effects. Furthermore, equilibrium conditions are generally not given, e.g. if colloids are involved. Sorption data without detailed description of the species and the conditions for which they have been measured, are not useable for quantitative statements. If the influence of only one parameter has been overlooked, the results may be useless. Therefore, the concept that sorption of radionuclides in natural multicomponent systems may be described by one (or possibly two) distribution coefficients ("K_d concept") must fail, because of the complexity of the systems and because equilibrium conditions are, in general, not fulfilled.

Another concept is the surface complexation model [3, 193, 308–316]. In this concept, it is assumed that the species of the radionuclides react with certain groups at the surface of solids (primarily hydroxyl groups, e.g. $\geqslant$SiOH, $>$AlOH, or $>$FeOH groups) by formation of surface complexes, as indicated in Eq. (6). Various approaches with respect to the formulation of surface complexation are described in the literature [309–316], but the common feature is the reaction of metal ions M^{n+} with hydroxyl groups SOH at the surface of solids

$$SOH + M^{n+} \rightleftharpoons SOM^{(n-1)+} + H^{+}. \qquad (8)$$

(equilibrium constant K_M)

This equation is equivalent to Eq. (7) in section 5.1.

By the surface complexation model reactions of type a) in section 5.1 are taken into account, and in application of this model the same general problems are encountered as in the case of the "K_d-concept". If one process of sorption (i.e. one species of M and one kind of sorption sites) is dominating, if the concentration of the sorption sites is known, and if the sorption properties are not changed by the sorption of other cations or anions, sorption may be described by equilibrium constants, such as K_M in Eq. (8). If several radionuclide species and/or several kinds of sorption sites are involved, however, several equilibria have to be taken into account.

Sorption by ion exchange is not considered in Eq. (8), and uptake of radionuclides by precipitation, coprecipitation or recrystallization, or adsorption of colloids are not taken into consideration in the surface complexation model. Therefore, it is unavoidable to investigate all possible interactions in every given system very carefully.

References

1. Becquerel, H.: Sur les Radiations Invisibles Émises par les Corps Phosphorescents, C. R. Acad. Sci. Paris **122**, 501 (1896).
2. See textbooks on Nuclear and Radiochemistry, e.g. Friedlander, G., Kennedy, G. W., Macias, E. S., Miller, J. M.: *Nuclear and Radiochemistry,* 3rd. ed., Wiley, New York 1981; Lieser, K. H.: *Einführung in die Kernchemie*, 3rd. ed., VCH-Verlag, Weinheim 1991; Choppin, G. R., Liljenzin, J.-O., Rydberg, J.: *Radiochemistry and Nuclear Chemistry,* 2nd ed. Butterworth-Heinemann Ltd., Oxford 1995; Adloff, J.-P., Guillaumont, R.: *Fundamentals of Radiochemistry,* CRC Press, Boca Raton, FLA, 1993.
3. Chemistry and Migration Behaviour of Actinides and Fission Products in the Geosphere, Proceedings of the International Conference Munich 1987, Radiochim. Acta **44/45** (1988); Proceedings of the Second International Conference Monterey, CAL., 1989, Radiochim. Acta **52/53** (1990); Proceedings of the Third International Conference, Jerez de la Frontera, Spain, 1991, Radiochim. Acta **58/59** (1992); Proceedings of the Fourth International Conference Charleston, SC, 1993, Radiochim. Acta **66/67** (1994).
4. Browne, E., Firestone, R. B. (Ed. Shirley, V. S.), *Table of Radioactive Isotopes*, Wiley, New York 1986.
5. Seelmann-Eggebert, W., Pfennig, G., Münzel, H., Klewe-Nebenius, H.: *Chart of the Nuclides,* Kernforschungszentrum Karlsruhe, 5th ed. 1981.
6. Allard, B., Torstenfeld, B.: *Actinide Solubilities and Speciation in a Repository Environment,* NAGRA-NTB-85-18, Nagra, Baden, Switzerland 1985.
7. Nitsche, H., Edelstein, N. M.: Solubilities and Speciation of Selected Transuranium Ions, Radiochim. Acta **39**, 23 (1985).
8. Felmy, A. R., Rai, Dhanpat, Schramke, J. A., Ryab, R. L.: The Solubility of Plutonium Hydroxide in Dilute Solution and in High-Ionic Strength Chloride Brines, Radiochim. Acta **48**, 29 (1989).
9. Felmy, A. R., Rai, Dhanpat, Fulton, R. W.: The Solubility of $AmOHCO_3$ (c) and the Aqueous Thermodynamics of the System $Na^+-Am^{3+}-HCO_3^--OH^--H_2O$, Radiochim. Acta **50**, 193 (1990).
10. Nitsche, H., Müller, A., Standifer, E. M., Deinhammer, R. S., Becraft, K., Prussin, T., Gatti, R. C.: Dependence of Actinide Solubility and Speciation on Carbonate Concen-

tration and Ionic Strength in Groundwater, Radiochim. Acta **58/59**, 27 (1992).
11. Lemire, R. J., Garisto, F.: The Effect of Ionic Strength, Groundwater Composition and Temperature on Calculated Radionuclide Solubilities, Radiochim. Acta **58/59**, 37 (1992).
12. Lieser, K. H.: Radiochemische Messung der Löslichkeit von Silberhalogeniden in Wasser und in Natriumhalogenidlösungen und die Komplexbildung der Silberhalogenide mit Halogenidionen, Z. Anorg. Allg. Chem. **292**, 97 (1957).
13. Lieser, K. H.: Radiochemische Bestimmung der Löslichkeit von Erdalkalisulfaten in Wasser und in Natriumsulfatlösungen, Z. Anorg. Allg. Chem. **335**, 225 (1965).
14. Lieser, K. H., Beyer, G., Lakatos, E.: Radiochemische Messung der Löslichkeit von Bleisulfat, Z. Anorg. Allg. Chem. **339**, 208 (1965).
15. Coetzee, P. P., Förster, M., Lieser, K. H.: Solubility of AgI in Saturated NaCl Solution, Radiochim. Acta **37**, 159 (1984).
16. Lieser, K. H., Mühlenweg, U., Sipos-Galiba, I.: Dissolution of Neptunium Dioxide in Aqueous Solutions under Various Conditions, Radiochim. Acta **39**, 35 (1985).
17. Lieser, K. H., Bauscher, Ch., Nakashima, T.: Dissolution of TcO_2 in Aqueous Solutions under Various Conditions, Radiochim. Acta **42**, 191 (1987).
18. Oeser, H.: Radiochemische Messung der partiellen Löslichkeit von Strontiumsulfat und Bariumsulfat im System $SrSO_4/BaSO_4$ und der Löslichkeit von Strontiumsulfat in Abhängigkeit von der Sulfatkonzentration, Diplomarbeit T. H. Darmstadt 1961.
19. Lieser, K. H.: unpublished results.
20. Eisenbund, M.: *Environmental Radioactivity,* Academic Press, New York 1973.
21. Kathren, R. L.: *Radioactivity in the Environment; Sources, Distribution and Surveillance,* Harwood Acad. Publ., Chur 1984.
22. Durrance, E. M.: *Radioactivity in Geology, Principles and Applications,* Ellis Horwood, Chichester 1986.
23. Stein, L.: The Chemistry of Radon, Radiochim. Acta **32**, 163 (1983).
24. Hopke, P. K. (Ed.): *Radon and its Decay Products: Occurrence, Properties and Health Effects,* Amer. Chem. Soc., Washington D.C. 1987.
25. Crameri, R., Burkart, W.: The Radon Problem, Radiat. Phys. Chem. **34**, 251 (1989).
26. Semkow, T. M.: Recoil Emanation Theory Applied to Radon Release from Mineral Grains, Geochim. Cosmochim. Acta **54**, 425 (1989).
27. Tieh, T. T., Ledger, E. B., Rowe, M. W.: Release of Uranium from Granitic Rocks during in Situ Weathering and Initial Erosion (Central Texas), Chem. Geol. **29**, 227 (1980).
28. Benes, P., Obdrzalek, M., Cejchanova, M.: The Physicochemical Forms of Traces of Radium in Aqueous Solutions Containing Chlorides, Sulphates and Carbonates, Radiochem. Radioanal. Lett. **50**, 227 (1982).
29. Benes, P., Sebesta, F., Sedlacek, J., John, J., Sandrik, R.: Particulate Forms of Radium and Barium in Uranium Mine Waste Waters and Receiving River Waters, Water Res. **17**, 619 (1983).
30. *The Environmental Behaviour of Radium,* International Atomic Energy, Vienna 1990.
31. Grosse, A. V., Johnston, W. M., Wolfgang, R. L., Libby, W. F.: Science **113**, 1 (1951).
32. Fireman, E. L.: Phys. Rev. **91**, 922 (1953).
33. Fireman, E. L., Schwarze, D.: Phys. Rev. **94**, 385 (1954).
34. Kaufman, S., Libby, W. F.: Phys. Rev. **93**, 1337 (1954).
35. Andersen, E. C., Libby, W. F., Weinhouse, S., Reid, A. F., Kirschenbaum, A. D., Grosse, A. V.: Science **105**, 576 (1947); Phys. Rev. **72**, 931 (1947).
36. Momoshima, N., Okai, T., Kaji, T., Takashima, Y.: Distribution and Transformation of Various Chemical Forms of Tritium in the Environment, Radiochim. Acta **54**, 129 (1991).
37. Chamberlain, A. C.: *Radioactive Aerosols, Cambridge Environmental Chemistry Series 3,* Cambridge University Press, Cambridge 1991.
38. Tadmor, J.: Radioactivity from Coal-fired Power Plants, J. Environ. Radioact. **4**, 177 (1986).
39. Pigford, T. H.: Environmental Aspects of Nuclear Energy Production, Ann. Rev. Nucl. Sci. **24**, 515 (1974).
40. Aurand, K. (Ed.): *Kernenergie und Umwelt,* Erich Schmidt Verlag, Berlin 1976.
41. Farmer, F. R.: *Nuclear Reactor Safety,* Academic Press, New York 1977.
42. The Environmental Impacts of Production and Use of Energy, UNEP Energy Report Series 2–79, Nairobi 1979.
43. Brookins, D. G.: *Geochemical Aspects of Radioactive Waste Disposal,* Springer Verlag, New York, Berlin, Heidelberg, Tokyo 1984.
44. Barney, G. S., Navratil, J. D., Schulz, W. W. (Eds.): *Geochemical Behaviour of Disposed Radioactive Waste,* Amer. Chem. Soc. Symp. Series 246, Seattle 1984.
45. Lieser, K. H.: Technetium und Neptunium im Kernbrennstoff-Kreislauf, Chem. Ztg. **110**, 215 (1986).
46. Lieser, K. H.: Technetium in the Nuclear Fuel Cycle, in Medicine and in the Environment, Radiochim. Acta **63**, 5 (1993).
47. Lieser, K. H., Coetzee, P. P., Förster, M.: Liberation of Iodine from AgI during Radiolysis, Radiochim. Acta **38**, 33 (1985).
48. Summary Report on the Post-Accident Review Meeting on the Chernobyl Accident, Safety Series No. 75 – INSAG 1, IAEA, Vienna 1986.
49. WHO Assesses Chernobyl Radiation in Europe, Nucl. Europe **6**, 29 (1986).
50. Symposium Chemical Phenomena Associated with Radioactivity Releases During Severe Nuclear Plant Accidents, Amer. Chem. Soc., New York 1986.
51. Radionuclides from the Chernobyl Accident, Special Issue, Radiochim. Acta **41** (no. 4), 131–201 (1987).
52. Rogers, J. T. (Ed.): *Fission Product Transport Processes in Reactor Accidents,* Hemisphere Publishing 1990.
53. Proceedings Seminar on Comparative Assessment of the Environmental Impact of Radionuclides Released during three Major Nuclear Accidents: Kyshtym, Windscale, Chernobyl, EUR 13574, Commission of the European Communities, Luxembourg 1990.
54. Sandalls, F. J., Segal, M. G., Victorova, N.: Hot Particles from Chernobyl: A Review, J. Environ. Radioact. **18**, 5 (1993).
55. Cains, P. W., Yewer, K. C., Waring, S.: Volatilization of Ruthenium, Caesium and Technetium from Nitrate Systems in Nuclear Fuel Processing and Waste Solidification, Radiochim. Acta **56**, 99 (1992).
56. Ure, A. M., Davidson, C. M. (Eds.): *Chemical Speciation in the Environment,* Chapman and Hall, London 1994.
57. von Gunten, H. R., Benes, P.: Speciation of Radionuclides in the Environment, PSI-Bericht Nr. 94-03, Februar 1994, ISSN 1019-0643.
58. Paneth, F. A.: Akad. Wiss. Wien, Abt. IIa **121**, 2193 (1912); **122**, 1079 (1913).
59. Zsigmondy, R.: Kolloid-Z. **6**, 304 (1913).
60. Hahn, O., Imre, L.: Z. Phys. Chem. A **144**, 161 (1929).
61. Haissinsky, M.: *Les Radiocolloides*, Hermann, Paris 1934.
62. Straik, I.: Z. Phys. Chem. A **157**, 269 (1951).
63. Ramsay, J. D. F.: The Role of Colloids in the Release of Radionuclides from Nuclear Waste, Radiochim. Acta **44/45**, 165 (1988).
64. Maiti, C. T., Smith, M. R., Saul, J. C.: Colloid Formation Study of U, Th, Ra, Pb, Sr, Rb and Cs in Briny Groundwaters, Nucl. Technol. **84**, 82 (1989).
65. Lieser, K. H., Ament, A., Hill, R., Singh, R. N., Stingl, U., Thybusch, B.: Colloids in Groundwater and Their Influence on Migration of Trace Elements and Radionuclides, Radiochim. Acta **49**, 83 (1990).

66. Kim, J. I.: Actinide Colloid Generation in Groundwater, Radiochim. Acta **52/53**, 71 (1991).
67. See Textbooks and Handbooks on Inorganic Chemistry.
68. Keller, C.: *The Chemistry of the Transuranium Elements,* Verlag Chemie, Weinheim 1971.
69. Katz, J. J., Seaborg, G. T., Morss, L. R. (Eds.): *The Chemistry of Actinide Elements*, 2nd ed., Chapman and Hall, London 1986.
70. Cleveland, J. M.: *The Chemistry of Plutonium*, Gordon and Breach, New York 1971.
71. Hagemann, R., Roth, E.: Relevance of the Studies of the Oklo Natural Reactors to the Storage of Radioactive Wastes, Radiochim. Acta **25**, 241 (1978).
72. The Oklo Phenomenon, IAEA Symposium, Vienna 1975.
73. Natural Fission Reactors, IAEA Tech. Commun., Vienna 1978.
74. Guillaumont, R., Adloff, J. P., Peneloux, A., Delamoye, P.: Sub-tracer Scale Behaviour of Radionuclides. Application to Actinide Chemistry, Radiochim. Acta **54**, 1 (1991).
75. Guillaumont, R., Adloff, J. P.: Behaviour of Environmental Plutonium at very Low Concentration, Radiochim. Acta **58/59**, 53 (1992).
76. Adloff, J. P.: Chemical Behaviour of Matter at very Low Concentration, to appear in Pure Appl. Chem.
77. Allard, B., Kipatsi, H., Liljenzin, J. O.: Expected Species of Uranium, Neptunium and Plutonium in Neutral Aqueous Solutions, J. Inorg. Nucl. Chem. **42**, 1015 (1980).
78. Edelstein, N. M. (Ed.): *Actinides in Perspective,* Pergamon Press, Oxford 1982.
79. Carnall, W. T., Choppin, G. R. (Eds.): *Plutonium Chemistry,* Amer. Chem. Soc. Symp. Series 216, Washington, D.C., 1983.
80. Choppin, G. R.: Solution Chemistry of the Actinides, Radiochim. Acta **32**, 43 (1983).
81. Bulman, R. A., Cooper, J. R. (Eds.): *Speciation of Fission and Activation Products in the Environment,* Elsevier, Amsterdam 1985.
82. Kim, J. I.: Chemical Behaviour of Transuranic Elements in Natural Aquatic Systems, in: *Handbook on the Physics and Chemistry of Actinides,* Vol. 4 (Freeman, A. J., Keller, C., (Eds.), p. 413, Elsevier, Amsterdam 1986.
83. Lieser, K. H., Bauscher, Ch.: Technetium in the Hydrosphere and in the Geosphere, I. Chemistry of Technetium and Iron in Natural Waters and Influence of the Redox Potential on the Sorption of Technetium, Radiochim. Acta **42**, 213 (1987).
84. Lieser, K. H., Mühlenweg, U.: Neptunium in the Hydrosphere and in the Geosphere. I. Chemistry of Neptunium in the Hydrosphere and Sorption of Neptunium from Groundwater on Sediments under Aerobic Conditions, Radiochim. Acta **43**, 27 (1988).
85. Lieser, K. H., Steinkopff, Th.: Chemistry of Radioactive Iodine in the Hydrosphere and in the Geosphere, Radiochim. Acta **46**, 49 (1988).
86. Nelson, D. M., Orlandini, K. A., Penrose, W. R.: Oxidation States of Plutonium in Carbonate-rich Natural Waters, J. Environ. Radioact. **9**, 189 (1989).
87. Caceci, M. S.: Environmental Chemistry of the Actinide Elements, in: *Metal Speciation in the Environment* (Broekart, J. A. C., Gücer, S., Adams, F., Eds.) NATO ASI Series, Vol. G 23, Springer Verlag, Berlin, Heidelberg, New York 1990.
88. Guillaumont, R., Peneloux, A.: Radiochemistry and Actinide Chemistry, J. Radioanal. Nucl. Chem. **143**, 275 (1990).
89. Ahrland, S.: Hydrolysis of the Actinide Ions, in: *Handbook on the Physics and Chemistry of the Actinides,* Vol. 9 (Eds. Freeman, A., Keller, C.), p. 471, Elsevier, Amsterdam 1991.
90. Lieser, K. H., Hill, R., Mühlenweg, U., Singh, R. N., Tu Shude, Steinkopff, Th.: Actinides in the Environment, J. Radioanal. Nucl. Chem. **147**, 117 (1991).
91. Dozol, M., Hagemann, R.: Radionuclide Migration in Groundwaters: Review of the Behaviour of Actinides, Pure Appl. Chem. **65**, 1081 (1993).
92. Lieser, K. H., Hill, R.: Hydrolysis and Colloid Formation of Thorium in Water and Consequences for its Migration Behaviour – Comparison with Uranium, Radiochim. Acta **56**, 37 (1992).
93. Lieser, K. H., Hill, R.: Chemistry of Thorium in the Hydrosphere and in the Geosphere, Radiochim. Acta **56**, 141 (1992).
94. Proc. Intern. Symp. on Radioecology: Chemical Speciation – Hot Particles, Commission European Communities, Znojmo 1992.
95. Lieser, K. H., Ament, A.: Radiochemical Investigation of the Behaviour of Lead in Groundwaters, Radiochim. Acta **57**, 45 (1992).
96. Förstner, U., Müller, G.: *Schwermetalle in Flüssen und Seen,* Springer Verlag, Berlin, Heidelberg 1974.
97. Stumm, W., Brauner, P. A.: Chemical Speciation, in: *Chemical Oceanography,* 2nd Ed. (Riley, J. P., Skirow, G., Eds.), p. 173, Academic Press, New York 1975.
98. Benes, P., Steinnes, E.: On the Use of Radionuclides in the Study of Behaviour and Physico-chemical State of Trace Elements in Natural Waters, Int. J. Environ. Anal. Chem. **4**, 263 (1976).
99. Lerman, A. (Ed.): *Lakes Chemistry, Geology, Physics.* Marcel Dekker, New York 1978.
100. Bowen, H. J. M., Page, E., Valente, I., Wade, R. J.: Radiotracer Methods for Studying Speciation in Natural Waters, J. Radioanal. Chem. **48**, 9 (1979).
101. Benes, P., Majer, V.: *Trace Chemistry in Aqueous Solutions,* Elsevier, Amsterdam 1980.
102. Florence, T. M., Batley, G. E.: Chemical Speciation in Natural Waters, CRC Crit. Rev. Anal. Chem. **9**, 219 (1980).
103. Stumm, W., Morgan, J. J.: *Aquatic Chemistry,* 2nd Ed., Wiley, New York 1981.
104. Wong, C. S., Boyle, E., Bruland, K. W., Burton, J. D., Goldberg, E. D. (Eds.): *Trace Metals in Sea Water,* Plenum Press, New York 1983.
105. Leppard, G. G. (Ed.): *Trace Element Speciation in Surface Waters and Its Ecological Implications,* Plenum Press, New York 1983.
106. Salomons, W., Förstner, U.: *Metals in the Hydrocycle,* Springer Verlag, Berlin 1984.
107. Bernhard, M., Brinckman, F. E., Sadler, P. J. (Eds.): *The Importance of Chemical Speciation in Environmental Processes,* Springer Verlag, Berlin 1986.
108. Stumm, W.: *Aquatic Surface Chemistry,* Wiley, New York 1987.
109. Glück-Macholdt, C., Lieser, K. H.: Untersuchung des Verhaltens von Zn, Cd. Hg and Pb in Wasser ohne und mit Zusatz von Schwebstoffen und Komplexbildnern sowie in Rheinwasser mit Hilfe der Radionuklidtechnik, Vom Wasser **69**, 183 (1987).
110. Mizuike, A.: Recent Developments in Trace Metal Speciation in Fresh Waters, Pure Appl. Chem. **59**, 555 (1987).
111. L. Landner (Ed.): *Speciation of Metals in Water, Sediment and Soil Systems,* Springer Verlag, Berlin 1987.
112. Kramer, J. R., Allen, H. E. (Eds.): *Metal Speciation, Theory, Analysis and Application,* Lewis Publishers 1988.
113. Batley, G. E. (Ed.): *Trace Element Speciation: Analytical Methods and Problems,* CRC Press, Boca Raton, FLA, 1989.
114. Lieser, K. H.: Speziation: Eine Herausforderung für die Analytische Chemie, GIT Fachz. Lab. **33**, 293 (1992).
115. Salbu, B., Steinnes, E. (Eds.): *Trace Elements in Natural Waters,* CRC, in press.
116. Bidoglio, G., Stumm, W. (Eds.): *Chemistry of Aquatic Systems: Local and Global Perspectives,* Kluwer Academic Publishers, 1994.
117. Pourbaix, M.: *Atlas of Electrochemical Equilibrium Aqueous Solutions,* Pergamon Press, Oxford 1966.
118. Brookins, D. C.: *Eh-pH Diagrams for Geochemistry,* Springer Verlag, Berlin, Heidelberg 1988.

119. Beaucaire, C., Poalhoat, P.: Redox Chemistry of Uranium and Iron, Radium Geochemistry, and Uranium Isotopes in the Groundwaters of the Lodève Basin, Massif Central, France, Appl. Geochem. **2**, 417 (1987).
120. Büppelmann, K., Kim, J. I., Lierse, Ch.: The Redox Behaviour of Plutonium in Saline Solutions under Radiolysis Effects, Radiochim. Acta **44/45**, 65 (1988).
121. Grenthe, I., Stumm, W., Laaksuharju, M., Nilsson, A.-C., Wikberg, P.: Redox Potentials and Redox Reactions in Deep Groundwater Systems, Chem. Geol. **98**, 131 (1992).
122. Desmet, G., Myttenaere, C. (Eds.): *Technetium in the Environment,* Elsevier, London, New York 1986.
123. Baes Jr., C. F., Mesmer, R.: *The Hydrolysis of Cations,* Wiley, New York 1976.
124. Smith, R. M., Martell, A. E.: *Critical Stability Constants,* Plenum Press, New York 1976.
125. Pashalidis, I., Kim, J. I., Lierse, Ch., Sullivan, J. C.: The Hydrolysis of Pu(VI) and Time-dependent Nucleation Reactions, Radiochim. Acta **61**, 29 (1993).
126. Buffle, J.: *Complexation Reactions in Aquatic Systems: An Analytical Approach,* Ellis Horwood 1988.
127. Means, J. L., Crerar, D. A., Duguid, J. O.: Migration of Radioactive Wastes: Radionuclide Mobilization by Complexing Agents, Science **200**, 1477 (1978).
128. Smith, R. M., Martell, A. E.: *Critical Stability Constants,* Plenum Press, Oxford 1976.
129. Newton, T. W., Sullivan, J. C.: Actinide Carbonate Complexes in Aqueous Solution, in: *Handbook on the Physics and Chemistry of the Actinides, Vol. 3* (Eds. Freeman, A. J., Keller, C.), p. 387, Elsevier, Amsterdam 1985.
130. Ho, C. H., Miller, N. H.: Chemistry of Uranium Colloids in Groundwater Environment, in: Proc. 2. Intern. Conf. on Radioactive Waste Management, Canadian Nucl. Soc., Toronto 1986.
131. Buddemeier, R. W., Hunt, J. R.: Transport of Colloidal Contaminants in Groundwater: Radionuclide Migration at the Nevada Test Site, Appl. Geochem. **3**, 535 (1988).
132. von Gunten, H. R., Krähenbühl, U.: The Reactor Accident at Chernobyl: A Possibility to test Colloid-controlled Transport of Radionuclides in a Shallow Aquifer, J. Contaminant Hydrol. **2**, 237 (1988).
133. Ramsay, J. D. F.: The Role of Colloids in the Release of Radionuclides from Nuclear Waste, Radiochim. Acta **44/45**, 165 (1988).
134. Dearlove, J. P. L., Longworth, G., Ivanovich, M., Kim, J. I., Delakowitz, B., Zeh, P.: Study of Groundwater Colloids and their Geochemical Interactions with Natural Radionuclides in Gorleben Aquifer Systems, Radiochim. Acta **52/53**, 83 (1991).
135. Lieser, K. H.: Kolloide in der Analytischen Chemie, GIT Fachz. Lab. **33**, 583 (1991).
136. Kim, J. I., Zeh, P., Delakowitz, B.: Chemical Interactions of Actinide Ions with Groundwater Colloids in Gorleben Aquifer Systems, Radiochim. Acta **58/59**, 147 (1992).
137. Amphlett, C. B.: *Inorganic Ion Exchangers,* Elsevier, Amsterdam 1964.
138. Lieser, K. H., Steinkopff, Th.: Chemistry of Radioactive Cesium in the Hydrosphere and in the Geosphere, Radiochim. Acta **46**, 39 (1988).
139. Fricke, R., Schmäh, H.: Z. Anorg. Chem. **255**, 253 (1948).
140. Kautsky, H., Weßlau, H.: Z. Naturforsch. **9b**, 569 (1954).
141. Kohlschütter, H. W., Getrost, H., Miedtank, S.: Z. Anorg. Allg. Chem. **308**, 190 (1961).
142. Kohlschütter, H. W., Miedtank, S., Getrost, H.: Fresenius Z. Anal. Chem. **192**, 381 (1963).
143. See Textbooks on Colloid Chemistry.
144. Calmano, W., Lieser, K. H.: Austauschvorgänge von Spurenelementen an Schwebstoffen, Naturwissenschaften **68**, 264 (1981).
145. Calmano, W., Lieser, K. H.: Untersuchung der Austauschvorgänge von Spurenelementen an Schwebstoffen mit Hilfe der Radionuklidtechnik, Fresenius Z. Anal. Chem. **307**, 356 (1981).
146. Choppin, G. R., Allard, B.: Complexes of Actinides with Naturally Occurring Organic Compounds, in: *Handbook on the Physics and Chemistry of the Actinides, Vol. 3* (Eds. Freeman, A. J., Keller, C.), p. 407, Elsevier, Amsterdam 1985.
147. Environmental Migration of Long-lived Radionuclides, IAEA, Vienna 1982.
148. Wallace, G. T.: The Association of Copper, Mercury and Lead with Surface-active Organic Matter in Coastal Seawater, Mar. Chem. **11**, 379 (1982).
149. Boggs, S. Jr., Livermore, D., Seitz, M. G.: Humic Substances in Natural Waters and their Complexation with Trace Metals and Radionuclides: A Review, Argonne National Laboratory ANL-84-78, Argonne, ILL, 1985.
150. Giesy, J. P., Geiger, R. A., Kevern, N. R., Alberts, J. J.: UO_2^{2+}-Humate Interactions in Soft, Acid, Humate-rich Waters, J. Environ. Radioact. **4**, 39 (1986).
151. Sparkes, S. T., Sanders, T. W., Long, S. E.: A Literature Survey of the Organic Speciation of Radionuclides, AERE R-12487, Harwell 1987.
152. Choppin, G. R.: Humics and Radionuclide Migration, Radiochim. Acta **44/45**, 23 (1988).
153. Choppin, G. R.: Chemistry of Actinides in the Environment, Radiochim. Acta **43**, 82 (1988).
154. Bidoglio, G., De Plano, A.: Interactions and Transport of Plutonium-Humic acid Particles in Groundwater Environments, in: *Proc. Conf. Scientific Basis for Nuclear Waste Management XII,* p. 823, Materials Research Society, 1989.
155. Kim, J. I., Buckau, G., Bryant, E., Klenze, R.: Complexation of Americium(III) with Humic Acid, Radiochim. Acta **48**, 135 (1989).
156. Berry, J. A., Baston, G. M., Bond, K. A., Linklater, C., Pilkington, N. J.: Studies on the Effects of Degradation Products on the Sorption of Tin and Radium, Mat. Res. Soc. Symp. Proc. **212**, 577 (1991).
157. Choppin, G. R.: J. Radioanal. Nucl. Chem. **147**, 109 (1991).
158. Kim, J. I., Wimmer, H., Klenze, R.: A Study of Curium(III) Humate Complexation by Time-resolved Laser Fluorescence, Spectroscopy (TRLFS), Radiochim. Acta **54**, 35 (1991).
159. Choppin, G. R.: The Role of Natural Organics in Radionuclide Migration in Natural Aquifer Systems, Radiochim. Acta **58/59**, 113 (1992).
160. Burton, J. D., Liss, P. S. (Eds.): *Estuarine Chemistry,* Academic Press, London, New York 1976.
161. Kavanough, M. C., Leckie, J. O. (Eds.): *Particulates in Water,* Amer. Chem. Soc., Advances in Chemistry Series **189**, Washington, D.C., 1980.
162. Förstner, U., Wittmann, G. T. W.: *Metal Pollution in the Aquatic Environment, 2nd Ed.,* Springer Verlag, Berlin, Heidelberg, New York 1981.
163. Morel, F. M. N.: *Principles of Aquatic Chemistry,* Wiley, New York 1983.
164. Duursma, E. K., Dawson, H. J.: *Marine Organic Chemistry,* Elsevier, Amsterdam 1981.
165. Glück, C., Lieser, K. H.: Microautoradiography of Trace Elements in Suspended Matter, Naturwissenschaften **69**, 391 (1982).
166. Glück-Macholdt, C., Lieser, K. H.: Untersuchung der Sorption von Schwermetallen an Schwebstoffen des Rheins mit Hilfe der Mikroautoradiographie, Vom Wasser **70**, 255 (1988).
167. Pedersen, K., Albinsson, Y.: Possible Effects of Bacteria on Trace Element Migration in Crystalline Bed-rock, Radiochim. Acta **58/59**, 365 (1992).
168. Nielsen, A. E.: *Kinetics of Precipitation,* Pergamon Press, Oxford 1964.
169. Lieser, K. H.: Steps in Precipitation Reactions, Angew. Chem., Int. Ed. **8**, 188 (1969).
170. Deutsch, W. J., Serne, R. J.: Uranium Mobility in the Natural Environment. Evidence from Sedimentary Roll-Front

Deposits, in: *Geochemical Behaviour of Disposed Radioactive Waste* (Eds. Barney, G. S., Navratil, J. D., Schulz, W. W.), ACS Symposium Series 246, Amer. Chem. Soc., Washington, D.C., 1984.
171. Fabrikanos, A., Lieser, K. H.: Papierchromatographische Trennung während der Hydroxidfällung, Z. Phys. Chem. Neue Folge **51**, 130 (1966).
172. Hahn, O.: *Applied Radiochemistry,* Cornell University Press, Ithaca, N. Y., 1936.
173. See textbooks on Analytical Chemistry.
174. Dyck, W., Lieser, K. H.: Mitfällung von Cu, Zn, As, Ag, Cd und Pb bei der Eisenhydroxid- und der Eisenphosphatfällung, Vom Wasser **56**, 183 (1981).
175. Sill, C. W., Willis, C. P.: Anal. Chem. **36**, 622 (1964).
176. Sill, C. W., Willis, C. P.: Anal. Chem. **38**, 97 (1966).
177. Sill, C. W., Williams, R. L.: Anal. Chem. **41**, 1624 (1969).
178. Ambe, S., Lieser, K. H.: Coprecipitation of Thorium with Barium Sulfate, Radiochim. Acta **25**, 93 (1978).
179. Lieser, K. H., Bachtanian, S.: Coprecipitation of U(IV) with Barium Sulfate, Radiochim. Acta **68**, 121 (1995).
180. Lieser, K. H., Medert, K.: Coprecipitation of Th with $SrSO_4$ and Uptake of Th by $SrSO_4$, Radiochim. Acta **68**, 191–195 (1995).
181. Paul, A. C., Pillai, K. C.: Transfer and Uptake of Radium in a Natural and Technologically Modified Radiation Environment, J. Environ. Radioact. **3**, 55 (1986).
182. Coughtrey, P. J., Thorne, M. C.: *Radionuclide Distribution and Transport in Terrestrial and Aquatic Ecosystems. A Critical Review of Data, Vol. 1–4,* Balkema 1984.
183. Desmet, G., Nassimbeni, P., Belli, M.: *Transfer of Radionuclides in Natural and Semi-Natural Environments,* Elsevier, Amsterdam 1990.
184. Proc. Conf. Scientific Basis for Nuclear Waste Management, North Holland, Amsterdam (e.g. VI, 1982, VI, 1983, VII, 1984).
185. Lieser, K. H.: Sorption Mechanisms, in: *Sorption and Filtration for Gas and Water Purification* (Ed. Bonnevie-Svendsen, M.), NATO Advanced Study Institutes Series, Noordhoff-Leyden 1975.
186. Lieser, K. H., Gleitsmann, B., Steinkopff, Th.: Sorption of Trace Elements or Radionuclides in Natural Systems Containing Groundwater and Sediments, Radiochim. Acta **40**, 33 (1986).
187. Lieser, K. H., Steinkopff, Th.: Sorption Equilibria of Radionuclides or Trace Elements in Multicomponent Systems, Radiochim. Acta **47**, 55 (1989).
188. Gregg, S. J., Sing, K. S. W.: *Adsorption, Surface Area and Porosity,* Academic Press, London, New York 1965.
189. Gould, R. F.: *Adsorption from Aqueous Solutions, Advances in Chemistry Series No. 79,* Amer. Chem. Soc., Washington, D.C., 1968.
190. Hauffe, K., Morrison, S. R.: *Adsorption,* Walter de Gruyter, Berlin, New York 1974.
191. Lieser, K. H., Quandt, S., Gleitsmann, B.: Abtrennung von Schwermetallen aus natürlichen Wässern an anorganischen Sorbentien (TiO_2, ZrO_2, Al_2O_3), Fresenius Z. Anal. Chem. **298**, 378 (1979).
192. Sipos-Galiba, I., Lieser, K. H.: Comparison of the Properties of Inorganic Ion Exchangers with Respect to the Separation of the Fission Products Cesium and Strontium, Radiochem. Radioanal. Lett. **42**, 329 (1980).
193. Anderson, M. A., Rubin, A. J. (Eds.): *Adsorption of Inorganics at Solid-Liquid Interfaces,* Ann Arbor Sciences, Ann Arbor 1981.
194. Clearfield, A.: *Inorganic Ion Exchange Materials*, CRC Press, Boca Raton, FLA, 1982.
195. Weiß, A., Sextl, E.: Clay Minerals as Ion Exchangers, in: *Ion Exchangers* (Ed. Dorfner, K.), p. 492, Walter de Gruyter, Berlin, New York 1991.
196. Lieser, K. H.: Non-Siliceous Inorganic Ion Exchangers, in: *Ion Exchangers* (Ed. Dorfner, K.), p. 519, Walter de Gruyter, Berlin, New York 1991.
197. Lieser, K. H., Quandt-Klenk, S., Thybusch, B.: Sorption of Uranyl Ions on Hydrous Silicon Dioxide, Radiochim. Acta **57**, 45 (1992).
198. Ivanovich, M., Harmon, R. S.: *Uranium Series Disequilibrium: Applications to Earth, Marine, and Environmental, Sciences,* Oxford Sciences Publ., Oxford 1992.
199. Lieser, K. H., Peschke, S., Gleitsmann, B.: Sorption Equilibria in Colloid Systems and their Influence on Distribution Ratios of Trace Elements, Fresenius Z. Anal. Chem. **321**, 119 (1985).
200. Lieser, K. H., Gleitsmann, B., Peschke, S., Steinkopff, Th.: Colloid Formation and Sorption of Radionuclides in Natural Systems, Radiochim. Acta **40**, 39 (1986).
201. Langmuir, D., Herman, J. S.: The Mobility of Thorium in Natural Waters at Low Temperatures, Geochim. Cosmochim. Acta **44**, 1753 (1980).
202. Lieser, K. H., Peschke, S.: The Geochemistry of Fission Products (Cs, I, Tc, Sr, Zr, Sm), OECD Workshop on Geochemistry and Radioactive Waste Disposal, Paris 1982.
203. Ames, L. L., McGarrah, J. E., Walker, B. A.: Sorption of Uranium and Radium by Biotite, Muscovite and Phlogopite, Clays Clay Miner. **31**, 343 (1983).
204. Grütter, A., von Gunten, H. R., Rössler, E.: Sorption, Desorption and Isotope Exchange of Caesium ($10^{-9}-10^{-3}$ M) on Chlorite, Clays Clay Miner. **34**, 677 (1986).
205. Lieser, K. H., Bauscher, Ch.: Technetium in the Hydrosphere and in the Geosphere, II. Influence of pH, of Complexing Agents and of Some Minerals on the Sorption of Technetium, Radiochim. Acta **44/45**, 125 (1988).
206. Lieser, K. H., Mühlenweg, U.: Neptunium in the Hydrosphere and in the Geosphere, II. Column Experiments with Neptunium, Radiochim. Acta **44/45**, 129 (1988).
207. Lieser, K. H., Steinkopff, Th.: Chemistry of Radioactive Cerium in the Hydrosphere and in the Geosphere, Radiochim. Acta **48**, 79 (1989).
208. Mahara, Y., Matsuzuru, H.: Mobile and Immobile Plutonium in Groundwater Environment, Water Res. **23**, 43 (1989).
209. Grütter, A., von Gunten, H. R., Kohler, M., Rössler, E.: Sorption, Desorption and Exchange of Caesium on Glaciofluvial Deposits, Radiochim. Acta **50**, 177 (1990).
210. Hobart, D. E.: Actinides in the Environment, in: Proc. of Robert A. Welch Foundation Conf. on Chemical Research XXXIV, Fifty Years with Transuranium Elements, Houston, TX, 1990, p. 379.
211. Sheppard, S. C., Sheppard, M. I., Evenden, W. G.: A Novel Method Used to examine Variation in Tc Sorption among 34 Soils, Aerated and Anoxic, J. Environ. Radioact. **11**, 215 (1990).
212. Ivanovich, M., Froehlich, K., Hendry, M.: Uranium-Series Radionuclides in Fluids and Solids, Milk River Aquifer, Alberta, Canada, Appl. Geochem. **6**, 405 (1991).
213. Silva, R. J.: Mechanism for the Retardation of Uranium-(VI) Migration, Mat. Res. Soc. Symp. Proc. **257**, 323 (1992).
214. Baston, G. M. N., Berry, J. A., Linklater, C. M.: Factors Influencing the Sorption of Radium onto Geological Materials, Anal. Proc. **30**, 194 (1993).
215. Ticknor, K. V.: Uranium Sorption on Geological Materials, Radiochim. Acta **64**, 229 (1994).
216. Schindler, P.: Surface Complexes at Oxide-Water Interfaces, in: *Adsorption of Inorganics at Solid-Liquid Interfaces* (Eds. Anderson, M. A., Rubin, A. J.), Ann Arbor Science, Ann Arbor 1981.
217. Hingston, F. J.: A Review of Anion Adsorption, in: *Adsorption of Inorganics at Solid-Liquid Interfaces* (Eds. Anderson, M. A., Rubin, A. J.) Ann Arbor Science, Ann Arbor 1981.
218. Ames, L. L., McGarrah, J. E., Walker, B. A.: Sorption of Trace Constituents from Aqueous Solutions onto Secondary Minerals, I. Uranium, Clays Clay Miner. **31**, 321 (1983); II. Radium, Clays Clay Miner. **31**, 335 (1983).

219. Gascoyne, M., Cramer, J. J.: History of Actinide and Minor Element Mobility in an Archean Granitic Batholith in Manitoba, Canada, Appl. Geochem. **2**, 37 (1987).
220. Torstenfeld, B., Rundberg, R. S.: Actinide Sorption on Granites and Minerals as a Function of pH and Colloids/Pseudocolloids, Radiochim. Acta **44/45**, 111 (1988).
221. Hochella Jr., M. F., White, A. F. (Eds.): *Mineral-Water Interface Geochemistry, Reviews in Mineralogy, Vol. 23*, Mineralogical Society of America, Washington, D.C., 1990.
222. Gauthrie, V. A., Hart, K. P., Levins, D. M., Seatonberry, B. W.: Geochemical Interactions between Leachants from Actinide-Doped Synroc and three Australian Granites, Radiochim. Acta **52/53**, 159 (1990).
223. Gutierrez, M. G., Bidoglio, G., Avogadro, A., Mingarro, E., d'Alessandro, M.: Experimental Investigations of Radionuclide Transport through Cored Granite Samples, Radiochim. Acta **52/53**, 213 (1990).
224. Mecherri, O. M., Budiman-Sastrowardoyo, P., Kamal, L., Rouchaud, J.-C., Fedoroff, M.: Study of the Sorption of Neodymium on Granite Minerals, Radiochim. Acta **58/59**, 267 (1992).
225. Gutierrez, M. G., Bidoglio, G., Avogadro, A., Yllera de Llano, A.: Studies on Hydro-Geochemical Controls of Neptunium and Selenium Migration in Granite Columns, Radiochim. Acta **58/59**, 277 (1992).
226. Gascoyne, M., Barber, J. H.: The Mobility of U, Ra and Rn in a Granite Batholith on the Canadian Shield, Radiochim. Acta **58/59**, 277 (1992).
227. Hölttä, P., Hautojärvi, A., Hakanen, M.: Transport and Retardation on Non-Sorbing Radionuclides in Crystalline Rock Fractures, Radiochim. Acta **58**/U5, 285 (1992).
228. Griffault, L. Y., Gascoyne, M., Kaminenir, C., Kerrich, R., Vandergraaf, T. T.: Actinide and Rare Earth Element Characteristics of Deep Fracture Zones in the Lac du Bonnet Granitic Batholith, Manitoba, Canada, Geochim. Cosmochim. Acta **57**, 1181 (1993).
229. Berry, J. A., Bourke, P. J., Coates, H. A., Green, A., Jefferies, N. L., Littleboy, A. K., Hooper, A. J.: Sorption of Radionuclides on Sandstones and Mudstones, Radiochim. Acta **44/45**, 135 (1988).
230. Liu, Wen-Chung, Lo, Jiunn-Guang, Tsai, Chao-Ming: Sorption of Cs, Sr and Co on Andesite and Coral Limestone, Radiochim. Acta **52/53**, 169 (1990).
231. Carroll, S. A., Bruno, J., Petit, J.-C., Dran, J.-C.: Interaction of U(VI), Nd and Th(IV) at the Calcite-Solution Interface, Radiochim. Acta **58/59**, 245 (1992).
232. Winkler, A., Brühl, H., Trapp, Ch., Bock, W.-D.: Mobility of Technetium in Various Rocks and Defined Combinations of Natural Minerals, Radiochim. Acta **44/45**, 183 (1988).
233. Lieser, K. H., Hild, W.: Principles and Applications of Heterogeneous Isotopic and Non-Isotopic Ion Exchange on the Surface of Ionic Crystals, Radioisotopes in the Physical Sciences and Industry, IAEA, Vienna 1962, p. 338.
234. Lieser, K. H.: Radiochemische Messung des heterogenen Austausches von isotopen und nicht-isotopen Ionen an der Oberfläche von Erdalkalicarbonaten, Radiochim. Acta **3**, 93 (1964).
235. Triay, I. R., Meijer, A., Cisneros, M. R., Miller, G. G., Mitchell, A. J., Ott, M. A., Hobart, D. E., Palmer, P. D., Perrin, R. E., Aguilar, R. D.: Sorption of Americium in Tuff and Pure Minerals Using Synthetic and Natural Groundwaters, Radiochim. Acta **52/53**, 139 (1990).
236. Rundberg, R. S., Triay, I. R., Ott, M. A., Mitchell, A. J.: Observation of Time Dependent Dispersion in Laboratory Experiments with Intact Tuff, Radiochim. Acta **52/53**, 219 (1990).
237. Thompson, J. L., Gilmore, J. S.: Migration of Fission Products at the Nevada Test Site: Detection with an Isotopic Tracer, Radiochim. Acta **52/53**, 229 (1990).
238. Buddemeier, R. W., Finkel, R. C., Marsh, K. V., Ruggieri, M. R., Rego, J. H., Silva, R. J.: Hydrology and Radionuclide Migration at the Nevada Test Site, Radiochim. Acta **52/53**, 275 (1990).
239. Thompson, J. L.: Studies of Fission Product Movement in Tuffaceous Media, Radiochim. Acta **58/59**, 273 (1992).
240. Iler, R. K.: Ion Exchange Properties of Crystalline Hydrated Silica, J. Colloid Sci. **19**, 648 (1964).
241. Hofmann, U., Giese, K.: Cation Exchange by Clay Minerals, Kolloid-Z. **87**, 21 (1939).
242. Weiß, A., Hofmann, U.: Batavite, Z. Naturforsch. **6b**, 405 (1951).
243. Weiß, A.: Cation Exchange Ability of Clay Minerals, I. Comparison of Investigations, Z. Anorg. Allg. Chem. **297**, 232 (1958); II. Cation Exchange in Minerals of the Mica, Vermiculite and Montmorillonite Groups, Z. Anorg. Allg. Chem. **297**, 257 (1958).
244. Weiß, A., Mehler, A., Koch, G., Hofmann, U.: Anion Exchange in Clays, Z. Anorg. Allg. Chem. **284**, 247 (1956).
245. Boehm, H.-P., Lieser, K. H.: Radiochemische Bestimmung des Kationenaustauschvermögens von Tonmineralen, Z. Anorg. Allg. Chem. **304**, 207 (1960).
246. Halevy, E.: Exchangeability of Hydroxyl Groups in Kaolinite, Geochim. Cosmochim. Acta **28**, 1139 (1964).
247. Weiß, A.: in: *Ullmanns Enzyklopädie der Technischen Chemie* (Ed. W. Foerst), Urban und Schwarzenberg, München, Berlin, Wien 1966, p. 583; 1964, p. 697.
248. von Olphen, H.: *An Introduction to Clay Colloid Chemistry*, Wiley, New York 1977.
249. Hafez, M. B., Abdelrasoul, A. A., Algasmi, R.: Fixation of Radioelements on Illite and Bentonite, Isotopenpraxis **18**, 321 (1982).
250. Lieser, K. H., Hild, W.: Zur Abtrennung von Strontium-90 aus Lösungen, Naturwissenschaften **46**, 599 (1959).
251. Lieser, K. H., Hild, W.: Selektive Abtrennung von radioaktivem Strontium, Naturwissenschaften **47**, 494 (1960).
252. Lieser, K. H., Hild, W.: Heterogener Austausch isotoper und nicht-isotoper Ionen an Erdalkalisulfaten, I. Untersuchungsmethode und Isotopenaustausch mit Strontium-90, Z. Anorg. Allg. Chem. **320**, 117 (1963); II. Austauschgleichgewichte an der Oberfläche der Erdalkalisulfate, Z. Anorg. Allg. Chem. **331**, 133 (1964).
253. Lieser, K. H., Gütlich, Ph., Rosenbaum, I.: Kinetics and Mechanism of Heterogeneous Exchange Reactions on the Surface of Ionic Crystals, Proc. Exchange Reactions, IAEA, Vienna 1965, p. 375.
254. Lieser, K. H., Gütlich, Ph., Hild, W., Hecker, A., Rosenbaum, I.: Exchange Equilibria on the Surface of Ionic Crystals, Proc. Exchange Reactions, IAEA, Vienna 1965, p. 385.
255. Lieser, K. H.: Kinetik heterogener Isotopenaustauschreaktionen, Monatsh. Chem. **101**, 1630 (1970).
256. Mager, H., Lieser, K. H.: Rekristallisation und Alterung von Strontiumsulfat, I. Z. Phys. Chem. Neue Folge **79**, 288 (1972); II. Z. Phys. Chem. Neue Folge **81**, 81 (1972); III. Z. Phys. Chem. Neue Folge **81**, 89 (1972).
257. Lang, G., Lieser, K. H.: Heterogener Isotopenaustausch, Kristallwachstum und Auflösung von Ionenkristallen am Beispiel Bleisulfat, Z. Phys. Chem. Neue Folge **86**, 143 (1973).
258. Lieser, K. H., Braun, H., Mager, H.: Selective Separations via Recrystallization, J. Radioanal. Chem. **32**, 367 (1976).
259. Lieser, K. H., Pallikaris, G.: Investigation of the Mechanism of Recrystallization, Faraday Discuss, Chem. Soc. **61**, 149 (1976).
260. Lieser, K. H., Mager, H., Pallikaris, G.: Mechanism of Recrystallization of Ionic Crystals under Solution, Z. Phys. Chem. Neue Folge **105**, 35 (1977).
261. Lieser, K. H.: Dynamic Behaviour of Solid Surfaces, in: *Reactivity of Solids* (Eds. Wood, J., Lindqvist, O., Helgesson, C., Vannerberg, N.-G.), Plenum Press, New York 1977.
262. Shearer Jr., S. D., Lee, G. F.: Leachability of Radium-226 from Uranium Mill Solids and River Sediments, Health Phys. **10**, 217 (1964).

263. Edgington, D. N., Alberts, J. J., Wahlgren, N. A., Karttunen, J. O., Reeve, C. A.: Plutonium and Americium in Lake Michigan Sediments, in: *Transuranic Nuclides in the Environment,* p. 493, IAEA, Vienna 1981.
264. Aston, S. R., Assinder, D. J., Stanners, D. A., Rae, J. E.: Plutonium Occurrence and Phase Distribution in Sediments of the Wyre Estuary, Northwest England, Mar. Pollut. Bull. **12**, 308 (1981).
265. Mahara, Y., Kudo, A.: Fixation and Mobilization of ^{60}Co on Sediments in Coastal Environments, Health Phys. **41**, 645 (1981).
266. Foerstner, U., Schoer, J.: Some Typical Examples of the Importance of the Role of Sediments in the Propagation and Accumulation of Pollutants, in: *Sediments and Pollution in Waterways. General Considerations,* p. 137, Techn. Doc. 302, IAEA, Vienna 1982.
267. Pimpl, M., Schüttelkopf, H.: The Plutonium Contamination of the Altrhein River. The Plutonium Contamination as a Function of Sediment Particle Diameter (in German), KfK-3074, Kernforschungszentrum Karlsruhe 1982.
268. Evans, D. W., Alberts, J. J., Clark, R. A.: Reversible Ion-Exchange of Cesium-137 Leading to Mobilization from Reservoir Sediments, Geochim. Cosmochim. Acta **47**, 1041 (1983).
269. Azevedo, H. L., Schüttelkopf, H.: Studies on the Retention of Ra-226 and Pb-210 by Sediments from the Receiving River of the Waste Waters of an Uranium Prospection in Menzenschwand, West Germany, Kernforschungszentrum Karlsruhe 1984.
270. Foerstner, U., Schoer, J.: Diagenesis of Chemical Associations of ^{137}Cs and other Artificial Radionuclides in River Sediments, Environ. Technol. Lett. **5**, 295 (1984).
271. Joshi, L. U., Zingde, M. D., Desai, B. N.: Studies on ^{226}Ra and ^{210}Pb Activities and the Concentration Factors of ^{226}Ra in the Surface Organic Layers of the Estuarine Sediments of Mindola and Purna Rivers in India, J. Radioanal. Nucl. Chem. **82**, 369 (1984).
272. Schoer, J., Foerstner, U.: Chemical Forms of Artificial Radionuclides and their Stable Counterparts in Sediments, in: Environmental Contamination, Internat. Conf., London, July 1984, p. 738, CEP Consultants, Edinburgh 1984.
273. McKay, W. A., Baxter, M. S.: The Partitioning of Sellafield-Derived Radio-Caesium in Scottish Coastal Sediments, J. Environ. Radioact. **2**, 93 (1985).
274. Benes, P., Strejc, P.: Interaction of Radium with Freshwater Sediments and their Mineral Components. IV. Waste Water and Riverbed Sediments, J. Radioanal. Nucl. Chem. **99**, 407 (1986).
275. Terada, K., Hamano, M., Natsumoto, K.: The Distribution and Characterization of ^{137}Cs in Shallow Water Sediments, Bull. Chem. Soc. Japan **59**, 1863 (1986).
276. Azevedo, H. L., Amaral, E. C. S., Godoy, J. M.: Evaluation of the ^{226}Ra Transport by River Sediments Surrounding the Brasilian Uranium Mining and Milling Facilities, Environ. Pollut. **51**, 259 (1988).
277. Benes, P., Lam Ramos, P., Poliak, R.: Factors Affecting Interaction of Radiocesium with Freshwater Solids, I. pH, Composition of Water and Solids, J. Radioanal. Nucl. Chem. **133**, 359 (1989).
278. Milton, G. M., Cornett, R. J., Kramers, S. J.: The Transfer of Iodine and Technetium from Surface Waters to Sediments, Radiochim. Acta **58/59**, 291 (1992).
279. Raja, M. E., Babcock, K. L.: On the Soil Chemistry of Radio-Iodine, Soil Sci. **91**, 1 (1961).
280. Squire, H. M., Middleton, L. J.: Behaviour of ^{137}Cs in Soils and Pastures. A Long Term Experiment, Radiat. Bot. **6**, 413 (1966).
281. Wada, K., Harward, M. E.: Amorphous Clay Constituents of Soils. Adv. Agron. **26**, 211 (1974).
282. Whitehead, D. C.: The Sorption of Iodide by Soil Components, J. Sci. Food Agric. **25**, 73 (1974).
283. Nriagu, J. O. (Ed.): *The Biochemistry of Lead in the Environment,* Elsevier, Amsterdam 1978.
284. Means, J. L., Crerar, D. A., Borcsik, M. P., Duguid, J. O.: Adsorption of Co and Selected Actinides by Mn and Fe Oxides in Soils and Sediments, Geochim. Cosmochim. Acta **42**, 1763 (1978).
285. Bohn, H. L., McNeal, B. L., O'Connor, G. A.: *Soil Chemistry,* Wiley, New York 1979.
286. Lindsay, W. L.: *Chemical Equilibria in Soils,* Wiley, New York 1979.
287. Benes, J.: Concentration Dependence of Iodine Ion Sorption by Soils, Jad. Energ. **25**, 344 (1979).
288. Davies, B. E. (Ed.): *Applied Soil Trace Elements,* Wiley, New York 1980.
289. Yamamoto, M., Yamamori, S., Komura, K., Sakanoue, K.: Behaviour of Plutonium and Americium in Soils, J. Radiat. Res. **21**, 204 (1980).
290. Hanson, W. C. (Ed.): Transuranic Elements in the Environment, US Department of Energy, Technical Information Center, Washington, D.C., 1980.
291. Fowler, E. B., Essington, E. H., Polzer, W. L.: The Interaction of Low Level Liquid Radioactive Waste with Soils: I. Behaviour of Radionuclides in Soil-Waste Systems, Soil Sci. **132**, 2 (1981).
292. Yamamoto, M., Tanii, T., Sakanoue, M.: Characteristics of Fall-out Plutonium in Soil, J. Radiat. Res. **22**, 134 (1981).
293. Bachhuber, H., Bunzl, K., Schimmack, W., Gaus, I.: The Migration of ^{137}Cs and ^{90}Sr in Multilayered Soils: Results from Batch, Column and Fallout Investigations, Nucl. Technol. **59**, 291 (1982).
294. Cook, G. T., Baxter, M. S., Duncan, H. J., Malcolmson, R.: Geochemical Associations of Plutonium and Gamma-emitting Radionuclides in Caithness Soils and Marine Particulates, J. Environ. Radioact. **1**, 119 (1984).
295. Sposito, G.: *The Surface Chemistry of Soils,* Oxford University Press, Oxford 1984.
296. Spalding, B. P.: Extraction of Radiostrontium Coprecipitated with Calcium Carbonate in Soil, Soil Sci. Soc. Am. J. **49**, 69 (1985).
297. Bors, J., Martens, B.: Studies on the Significance of Organic Soil Components on the Immobilization of Radioiodine, in: Proc. 4th Internat. Symp. Radioecology. The Impact of Nuclear Origin Accidents on Environment, p. D142, Cadarache 1988.
298. Livens, F. R., Rimmer, D. L.: Physico-chemical Controls of Artificial Radionuclides in Soil, Soil Use Manag. **4**, 63 (1988).
299. Livens, F. R., Baxter, M. S.: Particle Size and Radionuclide Levels in Some West Cumbrian Soils, Sci. Total Environ. **70**, 1 (1988).
300. Robens, E., Hauschild, J., Aumann, D. C.: Iodine-129 in the Environment of a Nuclear Fuel Reprocessing Plant: IV. ^{129}I and ^{127}I in Undisturbed Surface Soils, J. Environ. Radioact. **9**, 17 (1989).
301. Sposito, G.: *The Chemistry of Soils,* Oxford 1989.
302. Rauret, G., Llaurado, M., Vidal, M.: Solid Speciation of Radiocaesium in Soils, in: *Proc. Workshop Transfer of Radionuclides in Natural and Semi-Natural Environments* (Eds. Desmet, P., Nassimbeni, P., Belli, M.), p. 538, Elsevier, Amsterdam 1990.
303. Bors, J., Erten, H., Martens, R.: Sorption Studies of Radioiodine in Soils with Special References to Soil Microbial Biomass, Radiochim. Acta **52/53**, 317 (1991).
304. Assimakopoulos, P., Ioannides, K. G., Karamanis, D. T., Pakou, A. A., Stamoulis, K. C., Vayonakis, A., Veltsos, E.: Radiocaesium Aging in Surface Soils, in: *Proc. Internat. Symp. Radioecology, Znojmo,* Commission of European Communities 1992.
305. Oughton, D. H., Salbu, B., Riise, G., Lien, H., Oestby, G., Noeren, A.: Radionuclide Mobility and Bioavailability in Norwegian and Soviet Soils, Analyst **117**, 481 (1992).
306. Arapis, G., Voutsinas, M., Martinez, A., Millan, R., Gutierrez, J.: Behaviour of Cesium in Contaminated Soils with and without Agricultural Practices, Radiochim. Acta **58/59**, 309 (1992).

307. Cremers, A., Maes, A.: Radionuclide Partitioning in Environmental Systems: A Critical Analysis, in: *Application of Distribution Coefficients in Radiological Assessment Models* (Eds. Sibley, T. H., Myttenaere, C.), Elsevier Applied Science, London 1985.

308. Stumm, W. (Ed.): Equilibria Concepts in Natural Water Systems, Amer. Chem. Soc., Adv. Chem. Series 67, Washington, D.C., 1967.

309. Schindler, P. W., Fürst, B., Dick, R., Wolf, P. U.: Ligand Properties of Surface Silanol Groups: Surface Complex Formation with Fe^{3+}, Cu^{2+}, Cd^{2+} and Pb^{2+}, J. Colloid Interface Sci. **55**, 469 (1976).

310. Davis, J. A., Leckie, J. O.: Effect of Adsorbed Complexing Ligands on Trace Metal Uptake by Hydroxides, Environ. Sci. Technol. **12**, 1309 (1978).

311. Davis, J. A., James, R. O., Leckie, J. O.: Surface Ionization and Complexation at the Oxide-Water Interface. I. Computation of Electrical Double Layer Properties in Simple Electrolytes, J. Colloid Interface Sci. **63**, 480 (1978).

312. Davis, J. A., Leckie, J. O.: Surface Ionization and Complexation at the Oxide-Water Interface. II. Surface Properties of Amorphous Iron Oxyhydroxide and Adsorption on Metal Ions, J. Colloid Interface Sci. **67**, 90 (1978).

313. Jenne, E. A.: *Chemical Modelling in Aqueous Systems,* Amer. Chem. Soc. Symp. Series 93, Washington, D.C., 1979.

314. Stumm, W., Kummert, R., Sigg, L.: A Ligand Exchange Model for the Adsorption of Inorganic and Organic Ligands at Hydrous Oxide Surfaces, Croat. Chim. Acta **530**, 291 (1980).

315. Westall, J., Hohl, H.: A Comparison of Electrostatic Models for the Oxide-Solution Interface, Adv. Colloid Interface Sci. **12**, 265 (1980).

316. Schindler, P. W.: Surface Complexes at Oxide-Water Interfaces, in: *Adsorption of Inorganics at Solid-Liquid Interfaces* (Eds. Anderson, M. A., Rubin, A.), Ann Arbor Science, Ann Arbor 1981.

Radiochimica Acta 70/71, 377–396 (1995)

Actinide Environmental Chemistry

By R. J. Silva and H. Nitsche[1,]*

San Jose State University, San Jose, CA 95102, USA

[1] Institut für Radiochemie, Forschungszentrum Rossendorf e.V., P. O. Box 510119, D-01314 Dresden, Germany

(Received September 18, 1995)

Actinides / Chemistry / Environment / Migration

Summary

In order to predict release and transport rates, as well as design cleanup and containment methods, it is essential to understand the chemical reactions and forms of the actinides under aqueous environmental conditions. Four important processes that can occur with the actinide cations are: precipitation, complexation, sorption and colloid formation. Precipitation of a solid phase will limit the amount of actinide in solution near the solid phase and have a retarding effect on release and transport rates. Complexation increases the amount of actinide in solution and tends to increase release and migration rates. Actinides can sorb on to mineral or rock surfaces which tends to retard migration. Actinide ions can form or become associated with colloidal sized particles which can, depending on the nature of the colloid and the solution conditions, enhance or retard migration of the actinide. The degree to which these four processes progress is strongly dependent on the oxidation state of the actinide and tends to be similar for actinides in the same oxidation state.

In order to obtain information on the speciation of actinides in solution, i.e., oxidation state, complexation form, dissolved or colloidal forms, the use of absorption spectroscopy has become a method of choice. The advent of the ultrasensitive, laser induced photothermal and fluorescence spectroscopies has made possible the detection and study of actinide ions at the parts per billion level. With the availability of third generation synchrotrons and the development of new fluorescence detectors, X-ray absorption spectroscopy (XAS) is becoming a powerful technique to study the speciation of actinides in the environment, particularly for reactions at the solid/solution interfaces.

1. Introduction

Due to nuclear testing, reentry and disintegration of nuclear powered satellites, nuclear reactor accidents, uranium mining and nuclear weapons production, actinides have been introduced into the environment. Proposed disposal of high-level radioactive waste in mined repositories in geologic formations as well as the storage and disposal of uranium and plutonium from weapons dismantlement are further possible sources for actinide releases to the environment. In order to design methods for the cleanup of contaminated sites, predict the transport behavior in the environment, perform safety assessment studies to determine the ability of repositories to adequately contain them, and design ways to retard their release and migration rates, it is essential to understand the chemical behavior and forms of actinides under environmental conditions. Because aqueous transport is the most likely scenario for wide-spread actinide intrusion into the biosphere, one needs to know the aqueous chemistry of actinides under environmental conditions and how that chemistry controls the interactions of the actinides with the surrounding geologic media and thus the actinide transport behavior.

The amounts and half-lives of the actinides produced in nuclear explosions and spent reactor fuel need to be considered to decide which actinide isotopes are of most concern from a radiological hazards points of view. The half-lives of selected long-lived isotopes of the actinides are given in Table 1 [1]. The actinides that have long-lived isotopes with large production amounts, and thus are the major contributors to the present contamination and possible future contaminations, are U, Np, Pu and Am. For example, a few hundred years after enriched uranium spent fuel has been removed from boiling and pressurized water reactors, the actinides U through Am account for more than 98 percent of the radioactivity in the fuel [2, 3]. In Figure 1, the distribution of curies per metric ton of heavy metal is shown as a function of time after removal from a typical BWR with 30,000 megawatt-day burnup as calculated using the ORIGEN computer code [2, 3].

2. Reactions in the environment

If one wants to model or predict the release or transport rates of actinides under aqueous environmental conditions, several processes that control the amounts and forms of the actinide in solution should be considered. Four important processes that can occur with actinide cations are: (1) precipitation, (2) complexation, (3) sorption and (4) colloid formation.

Precipitation can occur if there is sufficient concentration of the actinide in solution to exceed the solubility product constant for the formation of a solid phase. This effect will limit the amount of the actinide in solution in the vicinity of the solid phase at any given time and thus will tend to have a retarding effect on release and migration rates.

A number of inorganic and organic ligands that can form complexes with the actinide ions in solution may be present in groundwaters. The charge of the solution complex can vary widely. Complexation increases the

* Corresponding author.

Table 1. Nuclear properties of selected actinide isotopes [1]

Nuclide	Mode of decay	Half-life	Availability	Source
Th-227	α	18.7 d	trace	nature
Th-228	α	1.9 y	trace	nature
Th-229	α	7.3×10^3 y	g	^{233}U daughter
Th-230	α	8.0×10^4 y	g	nature
Th-232	α	1.41×10^{10} y	kg	nature
	SF	$>1 \times 10^{21}$ y		
Th-234	β^-	24.10 d	trace	nature
Pa-231	α	3.28×10^4 y	g	nature
Pa-233	β^-	27.0 d	g	^{233}Th daughter
				^{237}Np daughter
Pa-234	β^-	6.75 h	trace	nature
U-232	α	68.9 y	g	^{232}Th daughter
	SF	8×10^{13} y		
U-233	α	1.59×10^5 y	kg	^{233}Pa daughter
	SF	1.2×10^{17} y		
U-234	α	2.45×10^5 y	g	nature
	SF	2×10^{16} y		
U-235	α	7.0×10^8 y	kg	nature
	SF	3.5×10^{17} y		
U-238	α	4.47×10^9 y	kg	nature
	SF	8.19×10^{15} y		
Np-237	α	2.14×10^6 y	kg	nature(?)
	SF	$>1 \times 10^{18}$ y		^{237}U daughter
				^{241}Am daughter
Np-238	β^-	2.1 d	trace	^{237}Np(n,γ)
Np-239	β^-	2.35 d	μg	nature(?)
				^{243}Am daughter
				^{239}U daughter
Pu-238	α	87.7 y	kg	^{242}Cm daughter
	SF	4.8×10^{10} y		^{238}Np daughter
Pu-239	α	2.41×10^4 y	kg	nature(?)
	SF	5.5×10^{15} y		^{239}Np daughter
				n capture
Pu-240	α	6.56×10^3 y	g	multi n capture
	SF	1.34×10^{11} y	g	
Pu-241	β^- (>99%)	14.4 y	g	multi n capture
	α (<0.002%)			
Pu-242	α	3.76×10^5 y	g	multi n capture
	SF	6.8×10^{10} y		
Pu-244	α	8.26×10^7 y	g	nature(?)
	SF	6.6×10^{10} y		multi n capture
Am-241	α	432.7 y	kg	^{241}Pu daughter
	SF	1.15×10^{14} y		multi n capture
Am-242	β^- (82.7%)	16.01 h	g	^{241}Am(n,γ)
	EC (17.3%)			
Am-242m	IT (99.5%)	141 y	g	^{241}Am(n,γ)
	α (0.48%)			
	SF (0.02%)	9.5×10^{11} y		
Am-243	α	7.38×10^3 y	kg	multi n capture
	SF	2.0×10^{14} y		
Cm-242	α	162.9 d	g	^{239}Pu(α,n)
	SF	6.1×10^6 y		^{242}Am daughter
Cm-244	α	18.11 y	kg	multi n capture
	SF	1.35×10^7 y		^{244}Am daughter
Cm-245	α	8.5×10^3 y	g	multi n capture
Cm-246	α	4.73×10^3 y	g	multi n capture
	SF	1.8×10^7 y		
Cm-247	α	1.56×10^7 y	g	multi n capture
Cm-248	α (91.74%)	3.4×10^5 y	g	multi n capture
	SF (8.26%)			
Bk-249	β^- (>99%)	320 d	mg	multi n capture
	α (<0.002%)			
Bk-250	β^-	3.217 h	trace	^{254}Es daughter
Cf-249	α	351 y	mg	^{249}Bk daughter
	SF	6.9×1010 y		
Cf-252	α (96.91%)	2.645 y	mg	multi n capture
	SF (3.09%)			
Es-253	α	20.47 d	μg	multi n capture
	SF	6.3×10^5 y		
Es-245g	α	275.7 d	μg	multi n capture
	SF	$>2.5 \times 10^7$ y		
Es-254m	β^- (99.6%)	39.3 h	μg	^{253}Es(n,γ)
	SF (0.33%)	$>1 \times 10^5$ y		
Es-255	β^- (92.0%)	39.8 d	trace	multi n capture
	α (8.0%)			
	SF (4×10^{-3}%)			

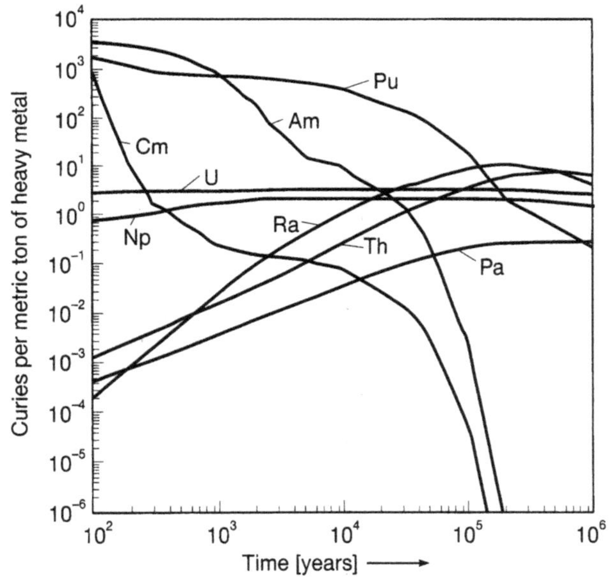

Fig. 1. The distribution of actinide curies per metric ton of heavy metal fuel as a function of time after removal from a typical Boiling Water Reactor with 30,000 megawatt-day burnup as calculated using the ORIGEN code.

amount of the actinide in solution and thus tends to increase release and migrations rates.

The actinide ions may attach themselves to mineral or rock surfaces in contact with the aqueous phase. This process is analogous to the precipitation process since it removes actinide ions from solution. Thus, this process tends to reduce the amount of actinide ion in solution and produce a retarding effect on the migration process.

Finally, the actinide ions can form or become associated with colloidal sized particles. Actinides in this form will not behave like a dissolved species and may exhibit a considerably different migration behavior than the dissolved species. Depending on the nature of the colloid and the solution conditions, this process may enhance or retard migration of the actinide.

All four processes must be considered if one wants to understand and predict the chemical behavior and transport properties of actinides in the environment. Detailed knowledge of the actinide species and charges are essential because the nature and extent of precipitation, complexation, sorption and colloid formation are strongly dependent on them. In addition, because these processes are not independent of one another, they need to be considered simultaneously. These four processes for the actinides U, Np, Pu and Am will be discussed.

3. General actinide chemistry

The actinide series of elements, actinium through lawrencium, results from the filling of the $5f$ electron shell. The chemical similarity to the $4f$ lanthanide series was recognized nearly 50 years ago [4, 5] and provided the frame work for the design of chemical methods for the isolation and identification of the heavier actinides [6]. Table 2 gives the electronic configurations of the lanthanide and actinide atoms and ions. While the trivalent oxidation state prevails throughout the lanthanide series, higher oxidations states are found to be the most stable in the early members of the actinide series [7]. The conversion from one oxidation state to another in aqueous solution is determined by the magnitude of the energy required to remove an electron from the gaseous ion (ionization energy) and the energy released when the gaseous ion combines with water to form the aqueous ion (hydration energy). There exists a large binding energy difference between the $4f$ electrons and the $5d$, $6s$ electrons for the lanthanide elements and the trivalent state owes its stability to a fortuitous combination of ionization and hydration energies [8]. There is not sufficient hydration energy gain over the ionization energy for the $4f$ electrons to produce oxidation states above 3+ as the most stable in aqueous solution. However, other oxidations states can be produced under oxidizing [Ce(IV), Tb(IV)] and reducing conditions [Eu(II), Yb(II)] due to the special stability of the $4f^0$, $4f^7$ and the $4f^{14}$ electronic configurations [8]. In contrast, the $5f$ electrons are shielded to a larger extent from the nucleus of the atom than are the $4f$ electrons. This results in a smaller energy difference between the $5f$, $6d$ and $7s$ electrons than in the lanthanide series [7] and a wider range of oxidation states are possible. The oxidation states that have been observed are given in Table 3. Not all of these oxidation states are possible under environmental conditions as will be discussed later.

While the charge remains 3+, the ionic radii of the lanthanides decrease slightly across the series (the lanthanide contraction) [8]. Many chemical reactions of an ionic nature are correlated with the ratio of the charge to volume of the ions, the charge density [9]. The change of the charge density across the series is sufficiently small that the chemical properties are nearly identical. The lanthanide contraction is essentially parallel among the di, tri and tetravalent ions [10]. The trivalent $5f$ elements exhibit a similar decrease in ionic radii across the series (the actinide contraction) and actinides in the 3+ state tend also to exhibit similar chemical properties [7, 11]. Further, trivalent actinides tend to exhibit similar chemical properties to trivalent lanthanides of the same ionic radius [7], e.g., Nd^{3+} and Am^{3+} form isostructural oxides of similar solubilities [12] and elute in similar positions from ion-exchange resin columns using α-hydroxyisobutyrate as eluant [13]. Also trivalent actinides and trivalent lanthanides have similar hydrolysis constants [14]. There are many other such examples. The radii of the tetravalent actinides exhibit the same regular contraction with atomic number [11]. Because of the similarities and the fact that actinide ions in the same oxidation state have essentially the same core structure, actinides in the same oxidation state will tend to have similar chemical properties. This has certainly been found to be true

Table 2. Electronic configurations of *f*-block atoms and ions[a] [7]

Lanthanide series			Actinide series					
Element	Gaseous atom	$M^{3+}(g)$	Element	Gaseous atom	$M^{+}(g)$	$M^{2+}(g)$	$M^{3+}(g)$	$M^{4+}(g)$
La	$5d6s^2$		Ac	$6d7s^2$	$7s^2$	$7s$		
Ce	$4f5d6s^2$	$4f$	Th	$6d^27s^2$	$6d7s^2$	$5f6d$	$5f$	
Pr	$4f^36s^2$	$4f^2$	Pa	$5f^26d7s^2$	$5f^27s^2$	$5f^26d$	$5f^2$	$5f$
Nd	$4f^46s^2$	$4f^3$	U	$5f^36d7s^2$	$5f^37s^2$	$5f^36d$?	$5f^3$	$5f^2$
Pm	$4f^56s^2$	$4f^4$	Np	$5f^46d7s^2$	$5f^57s$?	$5f^5$?	$5f^4$	$5f^3$
Sm	$4f^66s^2$	$4f^5$	Pu	$5f^67s^2$	$5f^67s$	$5f^6$	$5f^5$	$5f^4$
Eu	$4f^76s^2$	$4f^6$	Am	$5f^77s^2$	$5f^77s$	$5f^7$	$5f^6$	$5f^5$
Gd	$4f^75d6s^2$	$4f^7$	Cm	$5f^76d7s^2$	$5f^77s^2$	$5f^8$	$5f^7$	$5f^6$
Tb	$4f^96s^2$	$4f^8$	Bk	$5f^97s^2$	$5f^97s$	$5f^9$	$5f^8$	$5f^7$
Dy	$4f^{10}6s^2$	$4f^9$	Cf	$5f^{10}7s^2$	$5f^{10}7s$	$5f^{10}$	$5f^9$	$5f^8$
Ho	$4f^{11}6s^2$	$4f^{10}$	Es	$5f^{11}7s^2$	$5f^{11}7s$	$5f^{11}$	$5f^{10}$	$(5f^9)$
Er	$4f^{12}6s^2$	$4f^{11}$	Fm	$5f^{12}7s^2$	$(5f^{12}7s)$	$(5f^{12})$	$(5f^{11})$	$(5f^{10})$
Tm	$4f^{13}6s^2$	$4f^{12}$	Md	$(5f^{13}7s^2)$	$(5f^{13}7s)$	$(5f^{13})$	$(5f^{12})$	$(5f^{11})$
Yb	$4f^{14}6s^2$	$4f^{13}$	No	$(5f^{14}7s^2)$	$(5f^{14}7s)$	$(5f^{14})$	$(5f^{13})$	$(5f^{12})$
Lu	$4f^{14}5d6s^2$	$4f^{14}$	Lr	$(5f^{14}6d7s^2$ or $5f^{14}7s^27p)$	$(5f^{14}7s^2)$	$(5f^{14}7s)$	$(5f^{14})$	$(5f^{13})$
			Rf	$(5f^{14}6d^27s^2)$				$(5f^{14})$

[a] Predicted configurations in parentheses.

Table 3. The oxidation states of the actinide elements [7]

Atomic number Element	89 Ac	90 Th	91 Pa	92 U	93 Np	94 Pu	95 Am	96 Cm	97 Bk	98 Cf	99 Es	100 Fm	101 Md	102 No	103 Lr
Oxidation states													1?		
							(2)			(2)	(2)	2	2	**2**	
	3	(3)	(3)	3	3	3	**3**	**3**	**3**	**3**	**3**	**3**	**3**	3	**3**
		4	4	4	4	**4**	4	4	4	(4)	4?				
			5	5	**5**	5	5	5?		5?					
				6	6	6	6	6?							
					7	(7)	7?								

Bold type = most stable; ( ) = unstable; ? = claimed but not substantiated.

for the solubilities of many compounds and for the formation constants for many complexes.

Actinide ions in the trivalent and tetravalent states in acidic solutions are in the form of the simple hydrated ions An^{3+} and An^{4+}. Actinides in higher oxidation states form oxygenated species in solution called actinyl ions [7]. They have a symmetric and nearly linear structure. The actinyl ions AnO_2^+ and AnO_2^{2+} are extremely stable and act as a unit in chemical reactions. However, this structure decreases the effective charge on the central actinide ion, e.g., values of +3.2 and +2.2 have been reported for the effective charge on the Pu atom in the PuO_2^{2+} and PuO_2^+ ions, respectively [15].

4. Environmental actinide chemistry

4.1 Oxidation states

The single most important property of an actinide ion is its oxidation state because precipitation, complexation, sorption and colloid formation behavior differ considerably from one oxidation state to another. Water has an important oxidation-reduction chemistry that puts limits on the oxidation states that actinides can have in aqueous solutions. The oxidizing and reducing reactions of water (relative to the standard hydrogen electrode) are [16]:

In acid solution

$$H^+ + e^- = 1/2\ H_2 \qquad E_f = 0.00\ V. \tag{1}$$

$$1/4\ O_2 + H^+ + e^- = 1/2\ H_2O \qquad E_f = 1.23\ V. \tag{2}$$

In basic solution

$$H_2O + e^- = 1/2\ H_2 + OH^- \tag{3}$$

$$E_f^b = -0.826\ V.$$

$$1/4\ O_2 + 1/2\ H_2O + e^- = OH^- \tag{4}$$

$$E_f^b = 0.404\ V.$$

The formal electrode potentials, E_f, refer to standard conditions of an ideal one molal solution of ions and one bar gaseous pressure. The way in which these po-

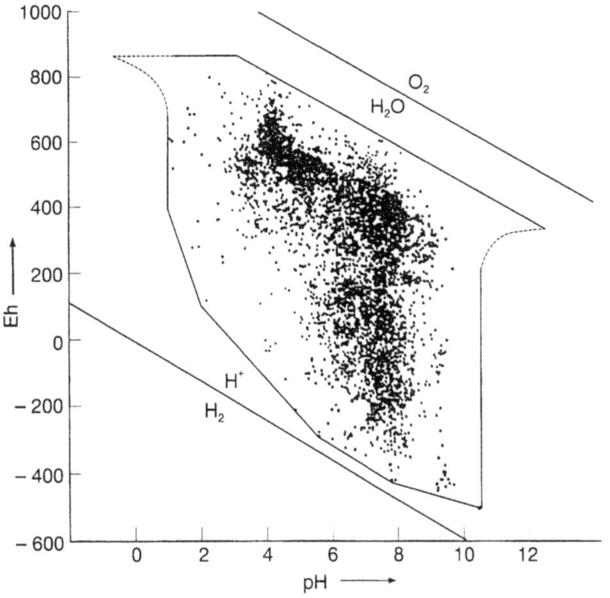

Fig. 2. Eh-pH range found in waters in the natural environment. Heavy solid line represent the stability of water (reprinted with permission from [17], Copyright 1960, The University of Chicago Press).

tentials vary with pH is given by the solid lines in Figure 2. The points shown in Figure 2 are values obtained from Eh and pH measurements on natural waters from hundreds of different sources around the world [17] and illustrate the wide range in Eh and pH values. The points in the upper half of the diagram represent samples taken from environments in contact with the atmosphere. They range from around pH 4 for mine waters, to pH 6 for rain and streams, to pH 8 for normal sea water and 10 for aerated saline residues. The points in the lower half of the Eh-pH diagram are for waters that were isolated from the atmosphere. They range from about pH 5 for water-logged soils to pH 10 for organic saline waters. Any half-reaction with a potential between the two lines of H_2O is stable with respect to oxidation or reduction by water. Of the many oxidation states that have been found for the actinides, only those with potentials that lie between the limits of water stability will persist in natural aquatic environments. In addition to oxygen, a few other elements (C, N, O, S, Fe, Mn) participate in redox reactions that are responsible for establishing the Eh of groundwater [16].

Figure 3 shows the reported redox potentials for the actinide ions at pH = 0 and pH = 14 [7]. Comparisons of these values with the limits set by the stability of water given in reactions 1 through 4 above indicate that U (4+, 5+, 6+), Np (3+, 4+, 5+, 6+), Pu (3+, 4+, 5+, 6+) and Am (3+) would be expected to be stable in natural aqueous environments.

The potential for other than standard conditions is given by the Nernst equation [18]:

$$E = E^0 + (RT/nF)\ \ln([\mathrm{ox}]/[\mathrm{red}]) \qquad (5)$$

$$= E^0 + (0.059/n)\ \log([\mathrm{ox}]/[\mathrm{red}])$$

where R = the gas constant, T = degrees Kelvin, n = number of electrons involved, F = the Faraday constant, [ox] = the concentration of the oxidized species, and [red] = the concentration of the reduced species. While the redox reaction between the 3+ and 4+ states as well as the 5+ and 6+ states are pH independent, oxidation or reduction reactions between either of the two lower oxidation states and either of the two upper oxidation states will be pH dependent because of the formation of the actinyl ion. For example:

$$Pu^{3+} = Pu^{4+} + e^- \qquad (6)$$

$$Pu^{4+} + 2\,H_2O = PuO_2^+ + 4\,H^+ + e^- \qquad (7)$$

$$PuO_2^+ = PuO_2^{2+} + e^- . \qquad (8)$$

Low pH values tend to favor the lower oxidation states while the higher oxidation states become more accessible as the pH is increased, e.g., in acid solutions (pH < 1) in contact with air, Pu^{4+} is the most stable oxidation state while PuO_2^+ is the more stable at neutral pH [19].

Strong complexation of one member of a redox pair can significantly shift the potential in favor of its formation. Therefore, in order to calculate the exact Eh value where oxidation state conversions occurs for U, Np, Pu and Am, all of the significant reactions of the actinide ions in solution would need to be included. For a solution of low ionic strength (0.01 M NaCl), exposed to atmospheric CO_2 and at a pH of 7, the concentrations of different oxidations states for U, Np, Pu and Am were calculated as a function of Eh using the thermodynamic data base for the formation of actinide complexes with hydroxide and carbonate given in references [20], [21] and [22] and the reduction potentials given in Figure 3. The calculation was performed using the speciation computer code HYDRAQL [23]. The results are shown in Figure 4. In a mined repository open to the air with such a dilute groundwater, the oxidation states in solution one would expect to be dominant are: UO_2^+, NpO_2^+, PuO_2^+ and PuO_2^{2+}, and Am^{3+}. However, the oxidation of Am^{3+} to AmO_2^+ in brine solutions with pH > 8 and NaCl concentrations > 3 M has been observed [24]. This is due to the formation of oxidizing forms of chlorine produced by the alpha radiation of the Am.

4.2 Precipitation

Precipitation can limit the amount of actinide in solution and thus can be the first step in the retardation process. The common anions in groundwaters are hydroxide, carbonate, sulfate, phosphate, fluoride, chloride, nitrate and silicate. In Table 4, the composition of water from a number of sources are shown [1]. Hydroxide, carbonate, sulfate, phosphate and fluoride form insoluble compounds with the actinides while chloride and nitrate compounds are quite soluble [25]. The silicate compounds have not been investigated to any great extent, however, uranium forms very insoluble compounds with silica, e.g., uranophane, sodium

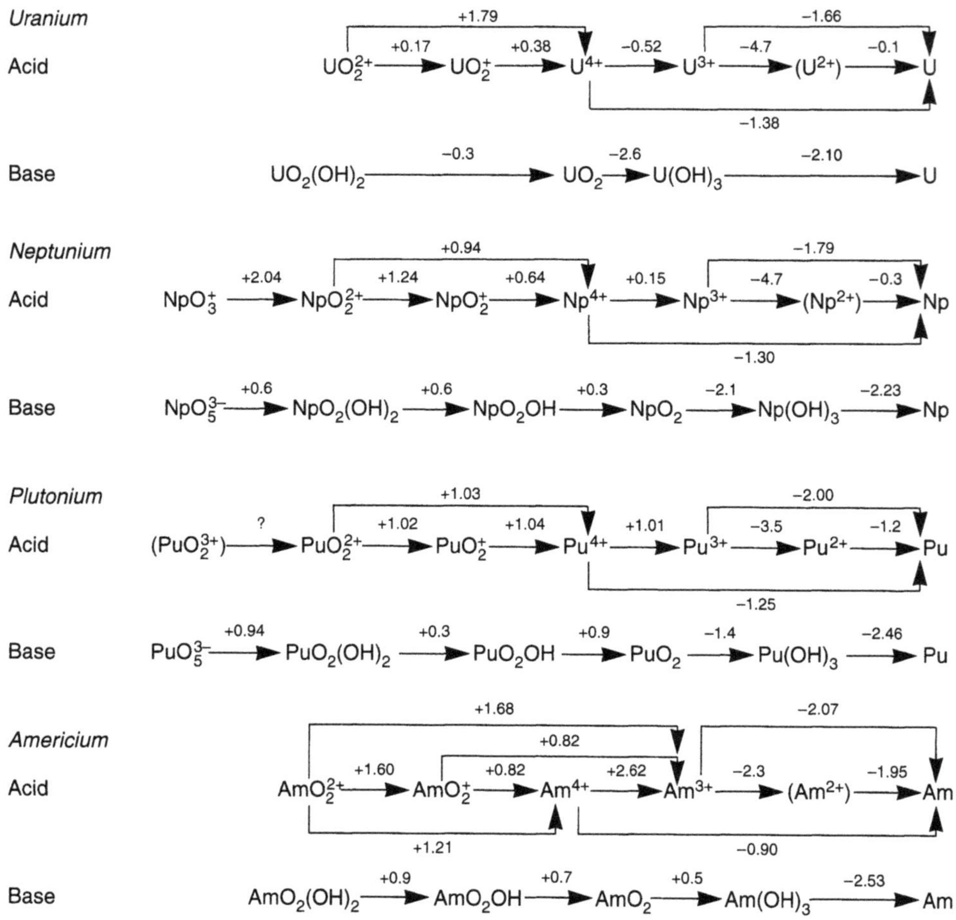

Fig. 3. Standard reduction potentials (volts vs. standard hydrogen electrode) of uranium, neptunium, plutonium and americium in 3znacidic (pH 0) and basic (pH 14) aqueous solution. (From Ref. [7].)

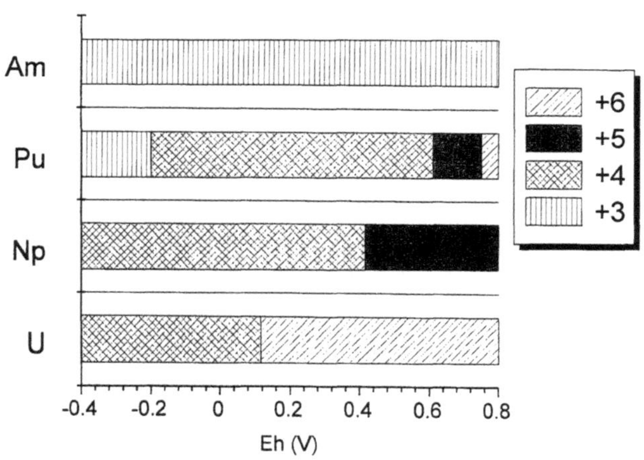

Fig. 4. The calculated distribution in oxidation states of uranium, neptunium, plutonium and americium as function of the Eh of an aqueous solution of 0.01 M NaCl exposed to atmospheric CO_2 at a pH of 7.

boltwoodite, coffinite, soddyite sodium weeksite [26]. The solubilities and solid phases exhibited by U, Np, Pu and Am in all the oxidation states possible under environmental conditions have not been measured. Therefore, we will assume that different actinide elements in the same oxidation state will exhibit similar chemical properties and consider existing thermodynamic data in a general way. Hydroxide and carbonate are the dominant anions in dilute ground waters [25]. Solubility and speciation experiments with U(VI), Np(V), Pu(IV) and Am(III) in dilute ground waters or simulated ground waters have indicated that oxide, hydroxide, carbonate or mixed hydroxycarbonate compounds are the dominant solid phases [21, 22, 25, 27–30]. Possible reactions for the 3+, 4+, 5+ and 6+ oxidation states are shown in Table 5. The nature and solubility of solid phases that form in a solution open to air at a pH of 7 for the four different oxidation states, represented by data for Am^{3+}, Pu^{4+}, NpO_2^+ and UO_2^{2+} are given in Figure 5 [21, 22, 25, 27–30].

4.3 Complexation

4.3.1 Inorganic complexes

The inorganic ligands in ground water responsible for the complexation of the actinides are the same ones discussed under precipitation, i.e., hydroxide, carbonate, sulfate, phosphate, chloride, fluoride, nitrate and silicate. Complexation can occur with more than one ligand or actinide ion so several species may be associated with a given ligand. The expression that describes the complexation reaction is:

$$a\,\mathrm{An}^x + b\,\mathrm{L}^y = (\mathrm{An}_a\mathrm{L}_b)^{ax+by} \quad (9)$$

Table 4. Characteristics of natural waters [1]

Analysis	Rainwater	Surface water (a)	Bedrock water (b)	YMP (f) J-13 water	YMP (f) UE-25p#1 water	Ocean	Mono lake brine
Depth (meters)	0	0.50	≤500			0−20	
Age (years)	0	≤10	≥100				
pH	4−6	7.3−8.4(7.9)	7−10	6.9	6.7	8.1	
Eh (Volts)	0.9	0.0−0.3	−0.05	0.1	0.36	0.8	
O_2 (mg/liter)(c)	10	1−10	<0.1	5.7		≤9	
Na^+	0.3−20	1−20(12)	10−100(d)	45	171	10,766	22,000
K^+	0.1−4	0.3−8(4)	1−5	5.3	13.4	399	
Ca^{2+}	0.5−5	2−100(25)	20−60	11.5	87.8	413	
Mg^{2+}	0.1−0.5	3−30(10)	15−30	1.76	31.9	1292	
Fe (total)		0.1−1	5−30	0.04	<0.1	<0.02	1
F^-		≤0.1	0.5−2	2.1	3.5	1.4	48
Cl^-	0.1−20	0.5−90(10)	5−50(d)	6.4	37	19.353	
Br^-						67	
CO_3^{2-} (total) (d)	≤1	60−200	5−400	118−143	960	~140	28,300
NO_3^-	0.1−4	≤10	<1	10.1	<0.1	<0.7	
PO_4^{3-} (total)	0	≤0.1	≤0.1			≤0.1	20
SO_4^{2-} (total)	≤20(1−5)	3−300(20)	1−15	18.1	129	2712	8300
SiO_2 (total)	0	3−15	5−30	66	66	0.01−7	
SH^- (e)	0	0	<1				
NH_3 (e)	<0.5	<0.1	<0.5			<0.05	
Organic carbon		1−50	<1	0.15		~1	

(a) Lakes, rivers, and shallow wells; typical values within parenthesis, (b) Swedish granite, (c) all concentrations in mg/liter, (d) Mainly as HCO_3^-, (e) SH^- and NH_3 only occur in reducing waters (Eh < 0), except for surface waters in industrial areas, (f) Yucca Mountain, Nevada wells.

Table 5. Solids

3+

$An(OH_3)(s) + 3\,H^+ = An^{3+} + 3\,H_2O$

$AnOHCO_3(s) + 2\,H^+ = An^{3+} + HCO_3^- + H_2O$

$An_2CO_3(s) + 3\,H^+ = 2\,An^{3+} + 3\,HCO_3^-$

4+

$AnO_2(S) + 4\,H^+ = An^{4+} + 2\,H_2O$

5+

$AnO_2OH(s) + H^+ = AnO_2^+ + H_2O$

$NaAnO_2CO_3(s) + H^+ = Na^+ + AnO_2^+ + HCO_3^-$

6+

$AnO_2(OH)_2 \cdot H_2O(s) + 2\,H^+ = AnO_2^{2+} + 3\,H_2O$

$Na_2AnO_4(s) + 4\,H^+ = 2\,Na^+ + AnO_2^{2+} + 2\,H_2O$

$AnO_2CO_3(s) + H^+ = AnO_2^{2+} + HCO_3^-$

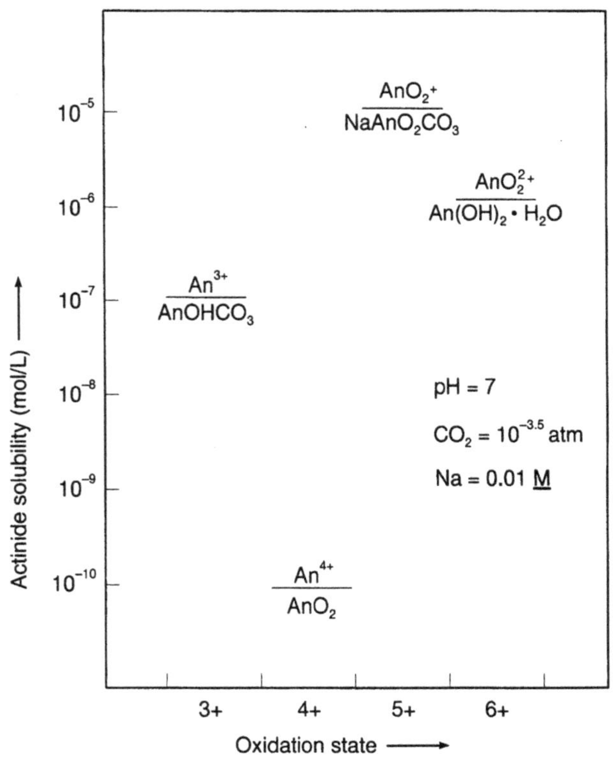

Fig. 5. Solid phases and solubilities of actinides expected for a 0.01 M solution exposed to atmospheric CO_2 at a pH of 7.

and the thermodynamic constant governing the reaction:

$$\beta_{a,b} = [An_aL_b]/[An]^a[L]^b \qquad (10)$$

where the quantities in brackets, [ ], represents concentrations for equilibrium quotients or activities for equilibrium constants. One can compare the relative strengths of different ligands for complexing actinides in different oxidation states by comparing the complex formation constants [31]. This comparison is made in Figure 6 which shows a plot of the formation constants for the different ligands, $\beta_{1,1}$, as a function of oxidation state number. The values for the constants were taken from references [20], [21] and [22]. When a complex formation constant was known for more than one actinide in the given oxidation state, an average value was

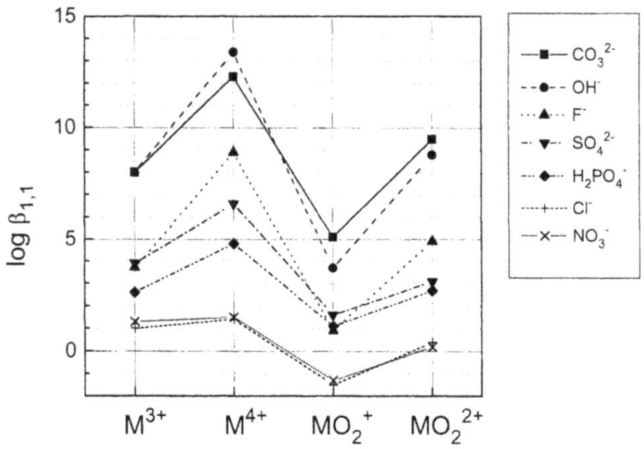

Fig. 6. Average complex formation constants of groundwater ligands versus the oxidation state of the actinide.

Table 6. Calculated concentrations (M)* of ligand required to produce more than 10% complexation

Actinide	Cl^-	NO_3^-	F^-	SO_4^{2-}
UO_2^{2+}	>1	>1	0.1	>1
NpO_2^+	>1	>1	0.1	0.1
Pu^{4+}	>1	>1	0.1	1
Am^{3+}	1	>1	0.01	0.01

* Aqueous solution of pH 8 in contact with atmospheric CO_2.

used in the plots. The values of the constants tend to fall into three groups for a given oxidation state. The trend in strengths of complexation of the various ligands are:

$$OH^-, CO_3^{2-} > F^-, HPO_4^{2-}, SO_4^{2-} > Cl^-, NO_3^- .$$

Another obvious trend is in the strength of complexation of a given ligand for actinides in different oxidation states. This trend is:

$$4+ > 3+ \approx 6+ > 5+.$$

In order to compare the complexation of Cl^-, NO_3^-, F^- and SO_4^{2-} with carbonate complexation and hydrolysis, calculation were made varying the amounts, individually, of the four ligands in a solution of pH = 8 in contact with atmospheric CO_2. The concentrations of the four ligands that were necessary to produce better than 10% complexation by that ligand are given in Table 6 for UO_2^{2+}, NpO_2^+, Pu^{4+} and Am^{3+}.

In Figures 7, 8, 9 and 10 are shown the calculated speciation of UO_2^{2+}, NpO_2^+, Pu^{4+} and Am^{3+} in a dilute sodium chloride solution in contact with atmospheric CO_2 as a function of pH using HYDRAQL. At the low pH values, hydrolysis dominates the speciation but above neutral pH there is sufficient carbonate in solution to make CO_3^{2-} the dominant ligand. The development of negatively charged species, carbonate and hydroxide, is responsible for the increase in solubility of these elements as the pH increases above the neutral region, thus giving rise to the frequently observed U-shaped solubility curves.

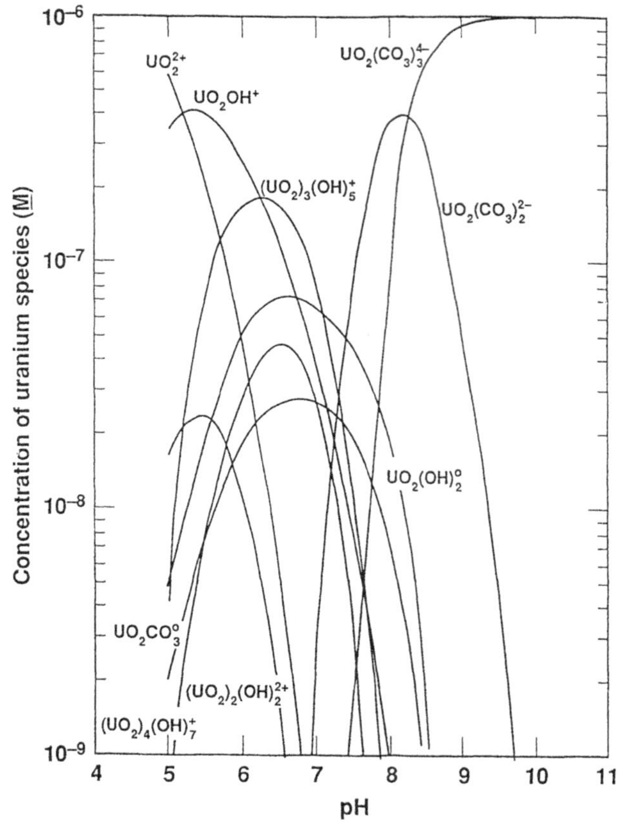

Fig. 7. Calculated speciation of UO_2^{2+} as a function of pH for a 10^{-6} M solution of U in 0.01 M NaCl exposed to atmospheric CO_2.

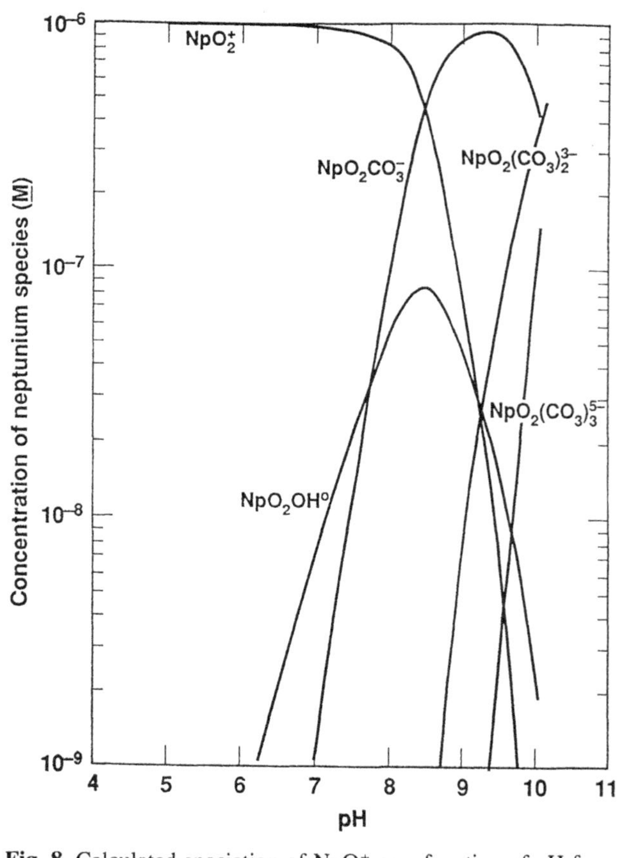

Fig. 8. Calculated speciation of NpO_2^+ as a function of pH for a 10^{-6} M solution of Np in 0.01 M NaCl exposed to atmospheric CO_2.

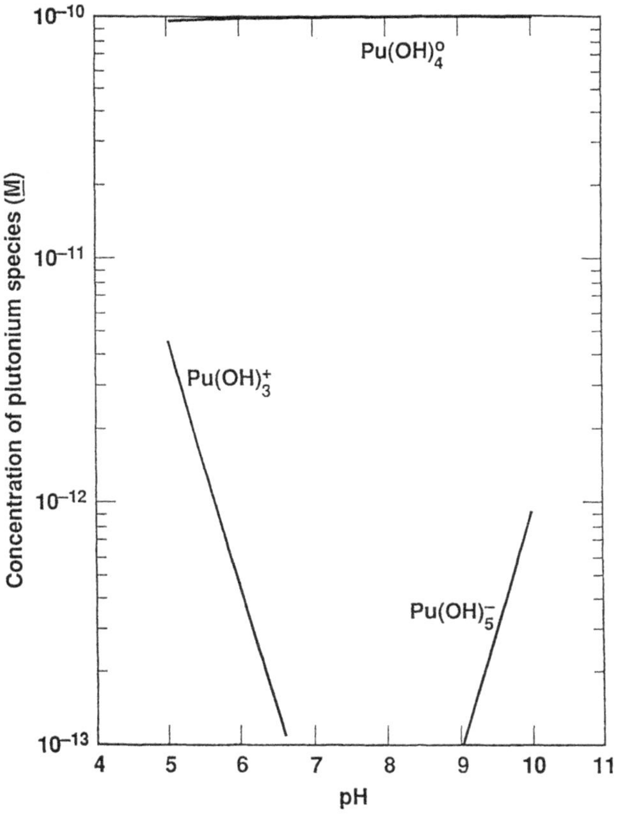

Fig. 9. Calculated speciation of Pu^{4+} as a function of pH for a 10^{-6} M solution of Pu in 0.01 M NaCl exposed to atmospheric CO_2.

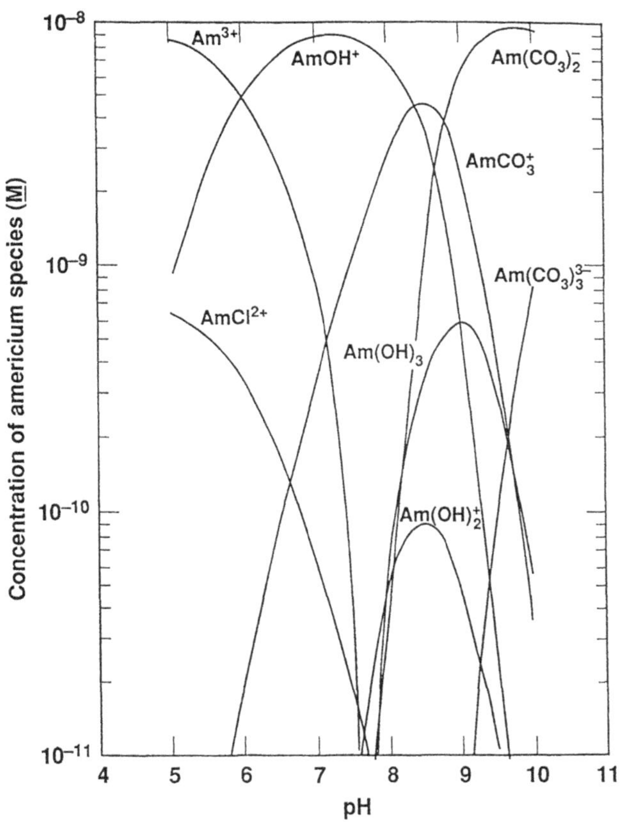

Fig. 10. Calculated speciation of Am^{3+} as a function of pH for a 10^{-6} M solution of Am in 0.01 M NaCl exposed to atmospheric CO_2.

4.3.2 Organic complexes

In addition to the inorganic ligands in ground waters, there are naturally occurring organic ligands that can complex the actinides rather strongly and could effect actinide transport. The most important natural ligands are humic and fulvic acids [32]. These polyelectrolytes tend to be ubiquitous in natural waters, giving rise to a brownish color at sufficient concentration, but vary in concentration and composition depending on their source. It is, therefore, not possible to give a precise value for the actinide complexation constants since they depends on the exact structure. A proposed general structure is shown in Figure 11 [33]. In addition to the phenolic, amine and alcoholic hydroxyl groups that form rather weak complexes, the carbonate groups are numerous and probably account for the strong actinide complex formation.

The humic material in ground waters comes from the decay of natural organic products. Average surface water (no color) is about 5 ppm, the surface of the ocean is about 1 ppm, and the dark swamp waters are about 50 ppm [34]. The names humic and fluvic arise from their method of isolation [32]. Dissolution of the humic substance in NaOH is the first step. The remaining material that is not soluble in the aqueous solution is termed humin. Acidification of the alkaline solution to a pH of 2 with HCl, by definition, precipitates the humic acid fraction. That which remains in solution is the fulvic acid fraction. Fulvic acid molecular weights range from 500 to 2000 amu while the humic acids can be from 1000 to more than 10,000 amu [35] and can attain colloidal dimensions [36].

In addition to the source, the values of the actinide complexation constants depend on the degree of ionization of the humic and fulvic acids and, thus, on pH. The constants tend to increase with pH until about pH = 7 where the acids are completely ionized. Figure 12 [34] shows the variation of the humic acid complex formation constants for Th^{4+}, Am^{3+} and UO_2^{2+} as a function of pH and degree of ionization, α. The logarithmic values of humic acid complex formations constants for Am^{3+}, Pu^{4+}, NpO_2^+ and UO_2^{2+} are estimated to be 12, 16, 5 and 8, respectively, at a pH of 7 using the equations given in reference [34]. These values are about 100 times larger than the carbonate values shown in Figure 6 and humic acid could be an important source of complex formation at concentrations even less than one ppm.

4.4 Adsorption

4.4.1 Adsorption processes

Actinide or other metal ions in ground waters can attach themselves reversibly or irreversibly onto rock or mineral surfaces. According to Stumm and Morgan [37], this adsorption process may result from short-range chemical forces, e.g., covalent bonding, hydrophobic bonding and hydrogen bridges. These chemi-

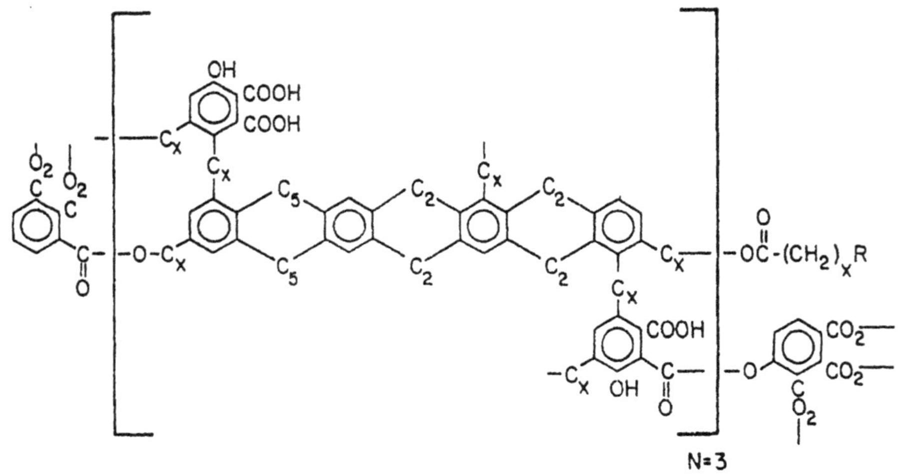

Fig. 11. Hypothetical structural relationships in aqueous humic material based on degradation products of permanganate (reprinted with permission from [32], Copyright 1989, American Chemical Society).

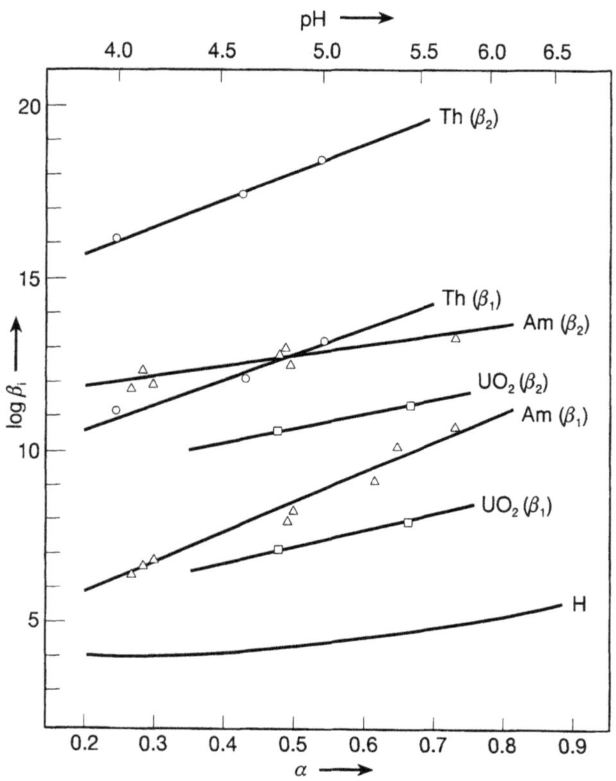

Fig. 12. Complex formation constants of several actinides with humic acid as a function of the degree of ionization (α) and pH (reprinted with permission from [34], Copyright 1985, Elsevier Science Publishers B. V.).

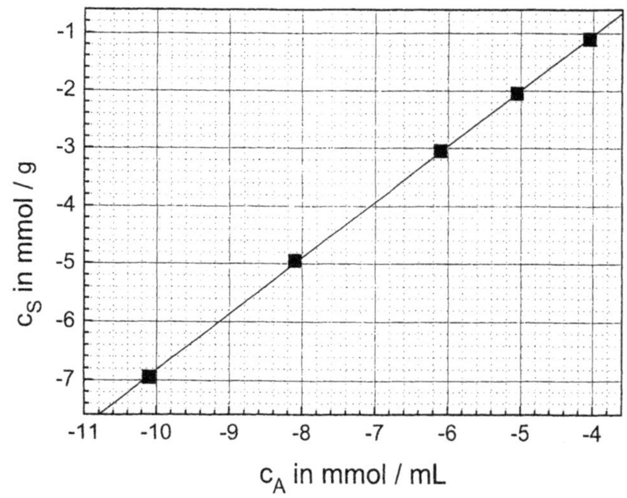

Fig. 13. Isotherm of the adsorption of Cs onto Belle Fourche clay.

sorption processes are frequently irreversible. Sorption may also result from long-range forces, i.e., electrostatic and van der Waals attraction. Electrostatic adsorption is frequently rapid and reversible, e.g., ion-exchange reactions.

The adsorption of a hydrated ion, $M(H_2O)_n$, onto a hydrated surface, $S(H_2O)_m$, can be thought of as a multistep process that involves the dehydration of M, the adsorption of dehydrated M on the hydrated surface and the hydration of the adsorption product [37]. The over all reaction is:

$$S(H_2O)_m + M(H_2O)_n = SM(H_2O)_p + (m+n-p)\, H_2O. \tag{11}$$

Adsorption is usually described quantitatively by adsorption isotherms [37], i.e., by plots of the concentration of the solute in the solid phase versus the concentration of solute in the aqueous phase. An isotherm describing the measured adsorption of cesium on to Belle Fourche clay (montmorillonite) from a solution of pH 7 and 0.01 M NaCl is given in Figure 13 [38]. Several models, e.g., Gibbs, Freundlich and Langmuir have been used to describe the distribution of a solute between solid and liquid phases [37, 39]. A good empirical description of single-solute equilibrium is provided by the Freundlich model [39]. The equation is:

$$K_d^{1/n} = C_s/C_a \tag{12}$$

where C_s and C_a are the equilibrium solid-phase and liquid-phase solute concentrations. The concentrations are usually expressed in units directly connected to the amounts of solute (e.g., moles, disintegrations/minute for radioactive isotopes, etc.) per gram of solid and amounts of solute per ml of solution. Thus, the distribution coefficient, K_d, has the units ml/g. The thermodynamic equilibrium distribution coefficient is given by:

$$K_d^0 = a_s/a_c = \gamma_s C_s/\gamma_a C_a. \quad (13)$$

Because the activities, particularly for the solute in the solid phase, are difficult to obtain or estimate, measured concentrations are used in expression 12 and the constant is frequently refered to as the distribution ratio, R_d. Some average distribution ratios for 12 different minerals and 4 rock types for actinides in the 4 different oxidation states [40, 41] are in the ratio $An^{4+}-500:An^{3+}-50:AnO_2^{2+}-5:AnO_2^{+}-1$. Again, as with precipitation and complexation, there are substantial differences between oxidation states. The trend in R_d values is $4+ > 3+ > 6+ > 5+$.

4.4.2 Source of surface charge

The surface of solids in natural water are electrically charged and, depending on their nature, they may be either positively or negatively charged. For electrical neutrality, there is an equivalent opposite charge of counter ions diffusely distributed in the liquid near the surface. The fixed charge of the surface and the diffuse charge of the liquid are refered to as the electrical double layer [42]. In the simplest model, the charged surface consists of two layers, one on the surface and the other in the solution (Helmholtz model). If the charges in solution are allowed to move due to thermal forces, a balance between the thermal and electrostatic forces is attained and a diffuse layer in solution is formed (Gouy-Chapman diffuse charge model). Stern divided the near surface region into two parts, a layer of ions adsorbed from solution onto the surface (Stern layer) and a diffuse layer (Gouy layer). Integration over the surface layers sum to zero charge. The various models are discussed in detail in reference [42].

There are three main ways for the surface charge to originate [42]:

(1) Ionizable functional groups, e.g., $-OH$, $-COOH$, $-OPO_3H_2$, etc., that depend on the mineral phase. The charge depends on the degree of ionization of the groups and therefore on the pH. The behavior is much like a weak acid.
(2) Lattice imperfections at the solid surface or isomorphous replacement within the lattice, e.g., replacement of Si^{4+} ions in the SiO_2 tetrahedra with Al^{3+} ions generates negatively charged sites.
(3) Adsorption of surfactants via van der Waals interactions, or by hydrogen or hydrophobic bonding, e.g., adsorption of an organic coating onto an inorganic surface.

4.4.3 Surface complexation model

Most metal, metal oxide and metal hydroxide surfaces, e.g., Si, Al, Fe, exhibit an amphoteric behavior [42–44], i.e., the surface hydroxyl group may undergo protonation and deprotonation as follows:

$$S-OH_2^+ \overset{k_1}{=} S-OH \overset{k_2}{=} S-O^-$$

Where S represents the surface, and k_1 and k_2 are the intrinsic acid constants for the two deprotonation reactions. The positively charged species dominate at low pH while the negatively charged species dominate at high pH. The surface charge will go from positive to zero to negative as the pH changes from low to high values. At some pH, which varies from one type of surface to another, the charge integrated over the surface is zero, denoted the point of zero charge (PZC). The $pH_{pzc} = 1/2\,[pk_1 + pk_2]$ [42]. These constants are obtained from acid-base titrations of aqueous suspensions of the finely divided material [42–44]. Because the surface charge is generated by H^+ and OH^-, they are referred to as the potential determining ions. For other than oxide surfaces, in addition to H^+ and OH^-, other ions in solution may contribute to determining the potential, e.g., for carbonate solids, H_2CO_3, HCO_3^- and CO_3^{2-} may be potential determining as well [42]. In these cases, the solution composition must be specified along with the pH_{pzc}. In Table 7, the PZC's for a number of geologic materials are given [42].

Table 7. Zero point of charge (reprinted with permission from [42], Copyright 1981, John Wiley & Sons, Inc.)

Material	pH_{zpt}
α-Al_2O_3	9.1
α-$Al(OH)_3$	5.0
γ-AlOOH	8.2
CuO	9.5
Fe_3O_4	6.5
α-FeOOH	7.8
γ-Fe_2O_3	6.7
"$Fe(OH)_3$"(amorph)	8.5
MgO	12.4
δ-MnO_2	2.8
β-MnO_2	7.2
SiO_2	2.0
$ZrSiO_4$	5
Feldspars	2–2.4
Kaolinite	4.6
Montmorillonite	2.5
Albite	2.0
Chrysotile	>12

The surface groups can interact with solute ions. For example, depending on the surface site speciation, the ions of NaCl can interact as follows [45]:

$$S-OH_2^+ + Cl^- = S-OH_2Cl$$

$$S-O^- + Na^+ = S-ONa$$

negatively charged solution species being adsorbed at the lower pH values and positively charge solution species at higher pH values. In the double layer, one type of counter ion can be exchanged for another, e.g., the ion exchange reaction

$$S-ONa + Mg^{2+} = S-OMg^+ + Na^+$$

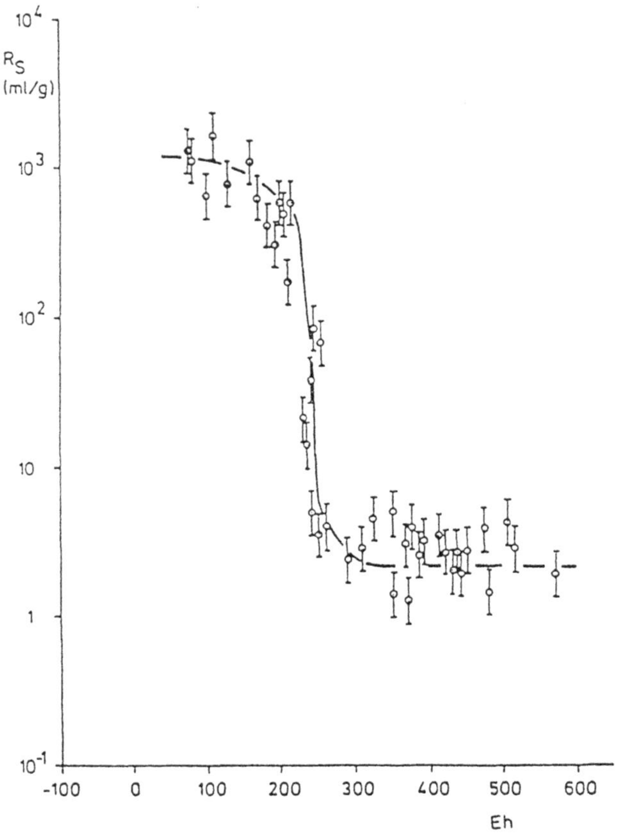

Fig. 14. Neptunium adsorption ratios measured in various systems at pH 7 as a function of the redox potential (reprinted with permission from [31], Copyright 1988, Oldenbourg Verlag GmbH).

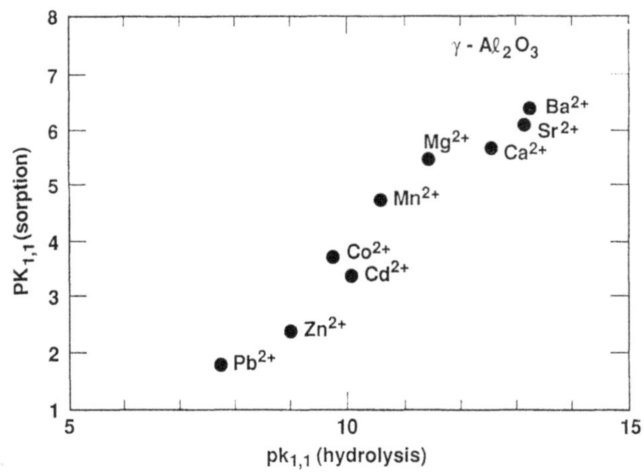

Fig. 15. Correlation of the reaction constants for the adsorption of several metals on to aluminum oxide vs. their first hydrolysis constants.

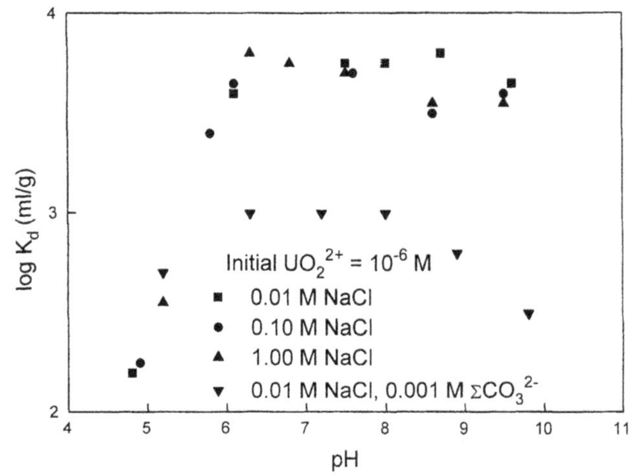

Fig. 16. Distribution coefficients for UO_2^{2+} adsorption on to silica in the presents of carbonate and without carbonate.

or

$$2\,S{-}ONa + Mg^{2+} = 2\,SO{-}Mg + 2\,Na^+.$$

Other solution complexes may also be exchanged, e.g.,

$$S{-}ONa + Mg(OH)^+ = S{-}OMg(OH) + Na^+.$$

The double-layer theory predicts [45] that higher valence ion are generally prefered over lower valence ions, the trend observed for the actinides. The dependence of the adsorption coefficient with oxidation state can be seen dramatically in the measured R_d for Np as a function of solution Eh reported by Lieser *et al.* [31] and shown in Figure 14. The distribution ratio decreases dramatically at the Eh corresponding to the oxidation of Np^{4+} to NpO_2^+.

The R_d approach ignores the chemical properties of absorbent and absorbate, and is valid for a single solution composition and adsorbate concentration. An alternative to the R_d approach is the so-called surface complexation or site-binding model [43, 44, 46, 47]. In this model, the adsorption reactions of aqueous solute species, i.e., hydrated ions, hydrolysis products or other complexed species, with the solid surface are considered as surface complexation reactions. The various solution and surface reactions are allowed to compete for the solute ions. Because the surface adsorption reactions are described in the same form as the solution reactions, the modelling calculations can consider all reactions simultaneously. This model has been used to predict reasonably well the adsorption of many metals on a variety of surfaces [43, 48–50]. The disadvantage of the surface complexation model is the difficulty in the experimental determination of the model parameters and surface reaction constants. However, it has been suggested [42, 51] that there may be correlations between the tendency of given ligands to form complexes in solution and the tendency of that functional group to form surface complexes, e.g., hydrolysis constants of an ion in solution and complexation by surface hydroxide groups. Such a correlation is shown for adsorption of various cations onto aluminum oxide in Figure 15. Possible surface complexation reactions with several actinides and their surface reaction constants are given in references [30] and [43].

In Figure 16, the results of the competition between the two complexation reactions, surface and solution, on the measured adsorption of UO_2^{2+} is readily seen. The figure shows the results of measurements of the adsorption of UO_2^{2+} as a function of pH by silica in NaCl solutions with and without the presence of carbonate [34]. Due to the formation of carbonate complexes of UO_2^{2+} that are not adsorbed or adsorbed to a much lesser extent than the hydrated ion, the distribution ratio is reduced by a factor of about 500 when changing from solutions of NaCl without carbonate present to solutions that contain millimolar amounts of carbonate. The addition of strong actinide complexing agents normally decreases actinide adsorption.

4.5 Colloids

The role of colloids in facilitating actinide transport is far from clear, however, greatly enhanced transport of americium and plutonium associated with colloids over that predicted for ionic species has been observed at a liquid seepage site at Los Alamos. At another site at Los Alamos, plutonium and americium were detected in monitoring wells over a mile away from a liquid waste source and found by ultracentrifugation to be present as colloids [52].

Depending on their size and charge relative to the surrounding porous media, colloids can move more rapidly or more slowly than the average groundwater velocity [53, 54]. The sign and magnitude of the electrical charge of a colloid, the zeta potential, generally is a function of the pH of the solution and there is a specific pH where they are uncharged. When uncharged particles pass through a porous media or thin fracture, they are usually transported without retention by convection and diffusion due to Brownian motion. The water velocity distribution is generally parabolic, the maximum velocity at the center being about two times the average velocity of the water [53, 54]. Particles will randomly sample this velocity variation but will not reach the walls due to their size. Therefore, their average velocity will be greater than the average groundwater velocity; the larger colloids will travel more rapidly than the smaller ones. If the colloids have the same charge as the surrounding media, repulsion will tend to increase the colloids relative velocity even more as they are kept even further away from the walls and will tend to inhibit wall adsorption. In addition, colloid particles that are larger than some pore diameters will be excluded from passage through them and experience a shorter overall path length (size exclusion) and, again, an apparent higher velocity than the average water velocity. If the particles have a charge that is opposite to the surrounding surfaces, retention mechanisms (wall absorption) will decrease the particle velocity relative to that of the ground water. Failure to account for colloid formation and transport could lead to serious underestimations of the rate of transport of actinide contaminants.

4.5.1 Nature of colloids

Colloid systems are usually defined as systems containing particles whose diameters range from one nanometer to one micrometer [54–57]. Colloid systems in aqueous media are divided into two types, lyophilic and lyophobic. The term lyophilic is used to describe colloids produced by the dissolution of macromolecules in the solvent while lyophobic describes colloids formed by dispersion of solid particles in the solvent. A characteristic feature of colloidal dispersions is the large area-to-volume ratio of the particles. The colloids formed by inorganic cations, like the actinide elements, are classified as lyophobic.

Two mechanisms have been proposed to explain the formation of colloids containing actinides in ground waters [36, 54, 57, 58]. The first describes the particles of the colloid as being composed primarily of an actinide and formed by condensation of actinide molecules or ions by a hydrolytic or precipitation process. This type of colloid might be expected to exhibit chemical properties similar to the assumed compound in macroamounts [54]. Such colloids have been called at various times "real", "pure", "eigen", "true" or "intrinsic" colloids.

The other mechanism pictures the formation of actinide containing colloids as the result of the adsorption of trace amounts of actinides onto colloids of impurity or other foreign materials in the aqueous phase. The chemical properties of these colloids would be similar to that of the impurity rather than the actinide. These types of colloids have been called "pseudo", "fremd", or "associative" colloids. There is ample evidence for the formation of both of these types of actinide bearing colloids [15, 36, 58–60].

Many of the important properties of colloids, such as stability and coagulation, are determined by the electric charge on the particles, the zeta potential [60]. The larger the surface charge, the more the particles tend to repell each other, the less likely they are to coagulate, and the more stable the colloid system [61]. For oxides and hydroxides, as the pH changes from acidic to basic conditions, the surface charge on oxide surfaces changes from highly positive to neutral to negative, gradually [61]. The further the pH of the solution is from the pH_{pzc} of the colloid, the larger the surface charge and the more stable is the colloidal system. For example, the pzc for Pu(IV) oxide is around pH 8 [62]. Thus, it is not surprising that the Pu(IV) colloid is very stable at low pH values, i.e., 1–4, but tends to coagulate and precipitate at near neutral pH values [63].

The main effect of increasing electrolyte concentration is to reduce the double layer potential [64]. Therefore, the stability of a colloidal system normally decreases with increasing electrolyte concentration, and coagulation and precipitation are enhanced. The critical electrolyte concentration needed for coagulation varies as $(1/z)^6$ [64]. Thus, the higher the charge on the electrolyte ions, the more effective that electrolyte is in inducing coagulation. Adsorption and surface charge neutralization by species of opposite charge to the colloid promote coagulation at low concentrations but can restabilize colloids at higher concentrations due to reversal of the charge of the colloidal particles [65]. Such behavior has been observed with Pu(IV) colloid in solutions of varying nitrate concentrations [66].

4.5.2 Actinide intrinsic colloids

Kim has given the order in the tendency of actinides to form intrinsic colloids as $M^{4+} > MO_2^{2+} > M^{3+} > MO_2^{+}$ [58]. This is the order of decreasing charge on the central ion. The mechanisms for the nucleation and growth of intrinsic colloids are thought to be similar to those of precipitation [54, 55, 67, 68]. Thus, the process assumes that the concentration of the actinide is sufficiently large so as to exceed the solubility product constant for the formation of a solid phase [54, 68, 69]. Because actinide oxides and hydroxides are very insoluble, they are frequently the source of intrinsic colloids. The route is generally believed to proceed by a gradual transition from ions to hydrolyzed ions to hydrolytic polymers to colloidal-size solid-phase particles. However, other very insoluble compounds of actinides could conceivably be the basis for colloids in natural systems, e.g., carbonates, hydroxycarbonates or phosphates. From this picture, one would not expect intrinsic colloids to form unless the actinide ion solution concentration exceeded the solubility product for the formation of a solid phase.

The colloid of Pu(IV) is the only actinide intrinsic colloid that has been studied in any great detail. The colloid can be easily produced by neutralization of acid solutions of Pu(IV) ion, can remain stable for years, and can be purified by ion-exchange column methods [70]. It exhibits an absorption spectrum distinctly different from those of the other Pu species [70]. The colloid can be precipitated from solution by a variety of anions and it resembles the precipitation of slightly soluble metal salts [66]. While Pu(IV) colloids were once thought to be due to polymer formation from hydrolysis products, Lloyd and Haire have shown that the colloid consists of very discrete particles, which can be amorphous or crystalline, of hydrated plutonium dioxide [71]. The particles become more crystalline with aging. The possible formation of a hydrolytic polymer would be only a step in the colloid formation process.

Silver has estimated the free energy of formation of Pu(IV) colloid [72]. From this number, he has calculated a value of 2.5×10^{-56} for the solubility product constant. This value is close to that measured for $Pu(OH)_4$ [73, 74]. The solubility of the colloid appears to be between that of amorphous $Pu(OH)_4$ and PuO_2, however, the exact value depends on the degree of aging, i.e., crystallinity [75]. Interestingly, the predominant solution species in equilibrium with the colloid under oxidizing conditions was found to be Pu(V) [76]. Rai and Serne found that, if the Pu solution concentration at a given pH falls above the $Pu(OH)_4$ solubility line, Pu(IV) colloid can form. If the Pu concentration falls below this line, the colloid or $Pu(OH)_4$ will

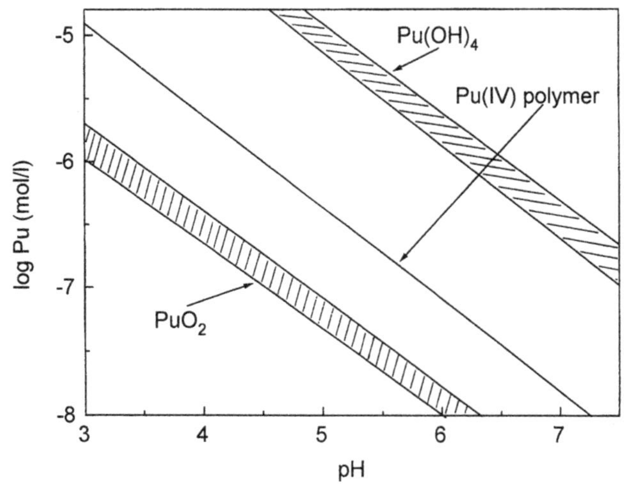

Fig. 17. Solubility curves for $Pu(OH)_4$, PuO_2 and Pu(IV) colloids as a function of pH [76].

not form [77]. The pH dependent solubility curves for the oxide, hydroxide and colloid are shown in Figure 17 [77]. From these curves, it should be possible to estimate the concentration of Pu required for intrinsic colloid formation.

The size of the Pu(IV) intrinsic colloids have been studied by ultracentrifugation [78] and autocorrelation photon spectroscopy [79, 80]. They can vary in size from one to several hundred nanometers. The distribution is not uniform and depends on aging. The fraction of large particles increases with time. Rundberg, *et al.*, found the density to be near but slightly lower than the estimated density of crystalline plutonium dioxide [80].

Plutonium(IV) colloid can form in solutions of Pu(V) and Pu(VI) following reduction of the latter by the effects of alpha radiation [81]. Conversely, the Pu(IV) in the intrinsic colloid is electroactive and the colloid can be broken down and oxidized to soluble Pu(VI) by cerium(IV) [82]. Under oxidizing environmental conditions, the major solution species to form in equilibrium with the colloid is Pu(V) [76, 83], i.e., Pu(IV) colloid will dissolve to produce Pu(V) in solution. Strong complexing agents may inhibit formation or enhance the dissolution of the Pu(IV) colloid [84].

There is indiction that U(IV) forms colloids similar to Pu(IV) [85], however, it has not been studied to any extent. It would be expected to form under reducing conditions where the U(IV) ion is stable. The solubility curve for U(IV), i.e. concentration versus pH, would be the best way to predict the conditions for possible colloid formation.

There is some evidence for the formation of Am(III) [36, 58, 59, 67, 86] and U(VI) [58, 59, 67, 85] intrinsic colloids but not for Np(V) [58, 59, 67]. By analogy with Pu(IV), an estimate of the solution concentration necessary for the formation of intrinsic colloids of actinides in the III, V, and VI oxidation states could be made from their solubility curves. The solubility of $Am(OH)_3$ [12], $UO_2(OH)_2$ [29] and $NpO_2(OH)$ [87] have been reported in the literature.

However, in groundwater, carbonate or other complexation of the solution species can change these solubilities. Compared to Pu(IV), it would seem less likely that one would exceed the solubility products of Am(III) and U(VI) at low pH values where the colloids would be the most stable. Np(V) is very soluble at the low pH values and does not reach a minimum until a pH close to 11 [87]. Conditions for Np(V) colloid formation in environmental waters would usually not be achieved.

4.5.3 Actinide association colloids

Natural colloids are in all groundwaters to some extent. The amount and nature will depend on the associated geologic structure. Colloid particles are commonly composed of materials produced by the disintegration and weathering of rocks and minerals, e.g., clays, silica, aluminum oxide minerals, as well as hydroxides or hydrated oxides of metal, e.g., iron [36, 52, 54, 88–90]. Organic colloids (humic substances) and inorganic particles that are covered with organics may also be present. Groundwater concentrations may vary from as few as $<10^6$ particles per liter at the Nevada Test Site to 10^{17} particles per liter at the Gorleben Site in Germany. The adsorption processes for the adsorption of actinides is similar to those discussed for mineral surfaces in section 4.4 and many of the techniques and conclusions are the same.

5. Optical absorption spectroscopy

Because the oxidation state, degree of complexation and solution form (dissolved or colloidal) have a strong influence on the chemical behavior of an actinides in the environment, the determination of the solution speciation of actinides is nearly as important as a measurement of the amount for predicting migration rates or planning clean-up techniques. While nuclear counting methods provide a sensitive method for determining the amounts, it gives no information about the nature of the solution species. Absorption spectroscopy methods are considered to be one of the most reliable techniques for the detection and characterization of solution species [91, 92] and have been widely used for the characterization of actinides in solutions [93, 94]. The optical absorption spectra of actinides (and lanthanides) show peaks due to the excitation of *f–f* electron transitions [93]. These peaks are quite narrow, usually less than a few nanometers wide (full-width-at-half-maximum height), and the wave length of the absorption peaks are uniquely characteristic of the actinide element and its oxidation state. For example, spectra are shown in Figure 18 for Pu^{3+}, Pu^{4+}, PuO_2^+ and PuO_2^{2+} [93]. The major absorption peak wave lengths for U, Np, Pu and Am in the oxidation states of environmental importance are given in Table 8. In addition to the identification, the quantity of the solution species can be obtained from the measured absorption peaks through the use of Beer's law [95]:

$$A = \varepsilon \times C \times L \tag{14}$$

where A is the measured absorbance (peak height or peak area), ε is the molar absorptivity constant, C is the concentration and L is the light path length in the solution. The reported molar absorptivities of the major absorption peaks of U, Np, Pu and Am in different oxidation states are given in Table 8.

Complexation or colloid formation produces changes in the absorption spectra that can frequently be used to identify the complex species or detect colloid formation. In Figure 19, the absorption spectra of Pu^{4+}, both uncomplexed and carbonate complexed, are shown [96]. The absorption wave length of the uncomplexed hydrated ion is shifted from 470 nm to 486 nm due to formation of a carbonate complexes. Because the concentrations of both complexed and uncomplexed species can frequently be obtained from an analysis of the absorption spectra, this method has been used to measure the complex formation constants for a number of actinide complexes [97–99]. The spectra of dissolved Am^{3+} and of Am^{3+} absorbed onto colloid particles are shown in Figure 20 [87]. The formation of colloids degrades and nearly eliminates the absorption peak of Am at 508 nm.

Unfortunately, conventional absorption spectrometers have detection limits of between 0.01 and 1 millimolar for the actinides [100], depending on their molar absorptivities, while the solubilities of actinides in aqueous solutions of near neutral pH are generally two or more orders of magnitude less (see section 4.2). With the advent of high powered, pulsed lasers in recent years, new laser based techniques have been developed for measuring very weak sample absorbance [101]. These techniques, photoacoustic (PAS) [102], photothermal deflection (PDS) [103] and thermal lensing (TLS) [104], are referred to as photothermal spectroscopies and are potentially three orders of magnitude more sensitive than absorption spectroscopy using conventional spectrometers while supplying the same information [105]. Laser induced fluorescence spectroscopy can be even more sensitive for those actinides that deexcite by the fluorescence process, e.g., U, Cm, and to a lesser extent, Am [106–108]. Unlike conventional spectroscopy, where the absorption is measured as the difference between the intensity of light passing in and out of the absorption cell, photothermal spectroscopies measure directly the amount of energy absorbed by the solute ions in the path of the laser light. The detection limits obtained by PAS for U, Np, Pu and Am in the oxidation states of environmental importance are given in Table 8. In principle, it is possible to increase the sensitivity a factor of ten using TLS [104], however, the experimental arrangement is difficult to set-up and maintain. The use of photothermal spectroscopies coupled with optical fibers for remote measurements of actinides in aqueous

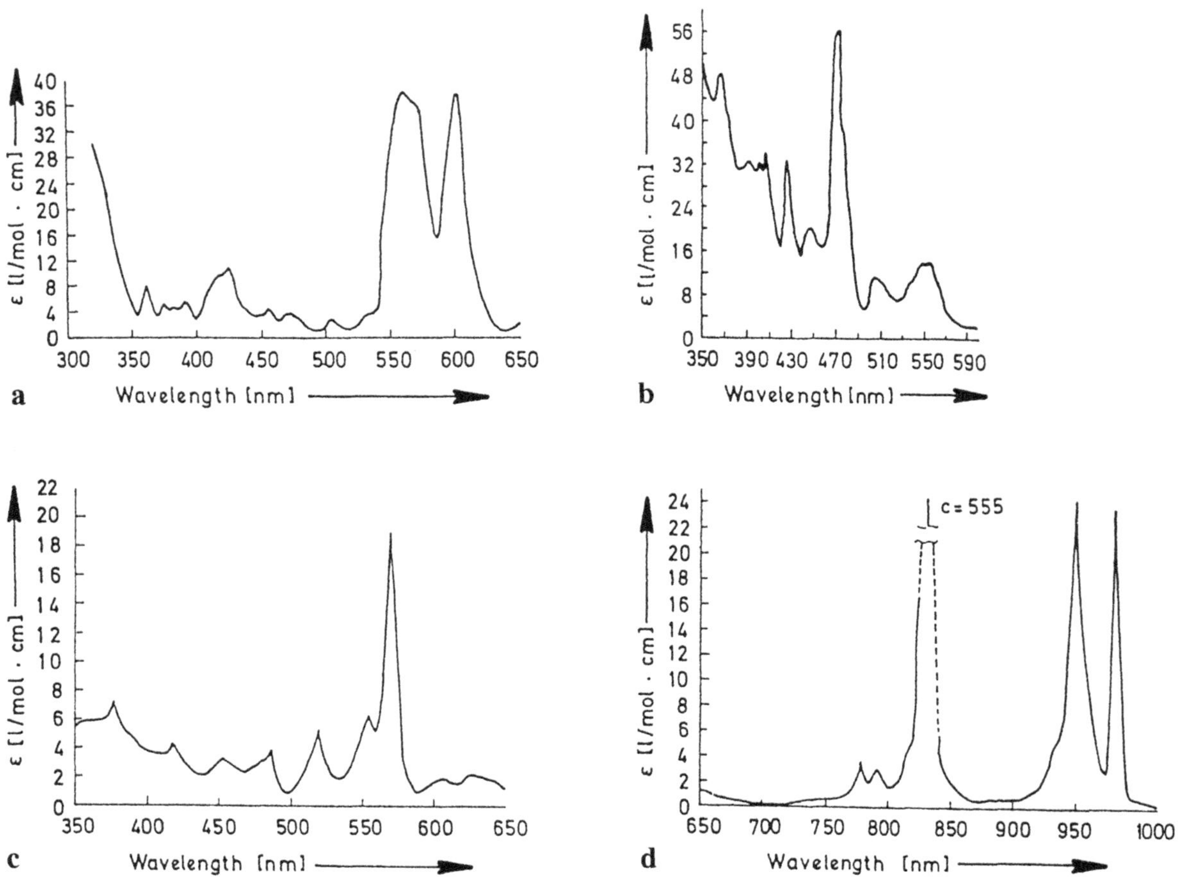

Fig. 18. Absorption spectra for Pu^{3+} (a), Pu^{4+} (b), PuO_2^+ (c), and PuO_2^{2+} (d) [93].

Table 8. Photoacoustic spectroscopy

Ion	Abs. wavelength nm	Molar Abs. $M^{-1}\,cm^{-1}$	Conc. sensitivity M	Conc. sensitivity ppm
Am(3+)	503	380	1×10^{-8}	0.003
Pu(3+)	600	38	1×10^{-7}	0.03
Pu(4+)	470	55	1×10^{-7}	0.03
Pu(5+)	568	19	5×10^{-7}	0.10
Pu(6+)	830	550	1×10^{-7}	0.03
Np(5+)	980	395	1×10^{-7}	0.03
U(4+)	650	58	1×10^{-7}	0.03
U(6+)	415	8	1×10^{-6}	0.20

solution at the parts per billion level [110] has increased rapidly over the last few years. Present systems are usually composed of a Nd-YAG or EXIMER pumped dye lasers [110, 111]. With this system, it is necessary to change dyes several times in order to cover the visible and near infrared region. However, the very recent development of optical parametric oscillator systems using ultrapure barium borate crystals in place of the dye lasers has made continuous spectrum scanning from ultraviolet to the near infrared regions possible [112].

6. X-ray absorption spectroscopy

With the availability of third generation synchrotrons [113] and the development of new fluorescence detectors [114] that can process high count rates, X-ray absorption spectroscopy (XAS) is becoming a powerful technique to study the speciation of actinides in the environment [115, 116]. XAS is particularly useful for investigating non-crystalline and polymeric actinide compounds that can not be identified by X-ray diffraction analysis.

X-ray Absorption Near Edge Spectroscopy (XANES), usually within 40 eV of the adsorption edge, can provide information about the local structure and oxidation state of an actinide in solution, solid or at a solution/solid interface. For example, uranium L_{III} XANES spectra of U(IV) in UO_2, U(VI) in UO_3 and a 50/50 mixture are given in Figure 20 [116]. The peak shift in the spectra with oxidation state is easily detected. This technique has recently been used to show that U(VI) was reduced to U(IV) by cultures of the bacterium Clostridium sp. and that uranium in wastes can be stabilized by the action of this anaerobic bacteria [117]. This technique has also been used recently to investigate uranium speciation in soils and sediments from two former U.S. Department of Energy uranium processing facilities [116].

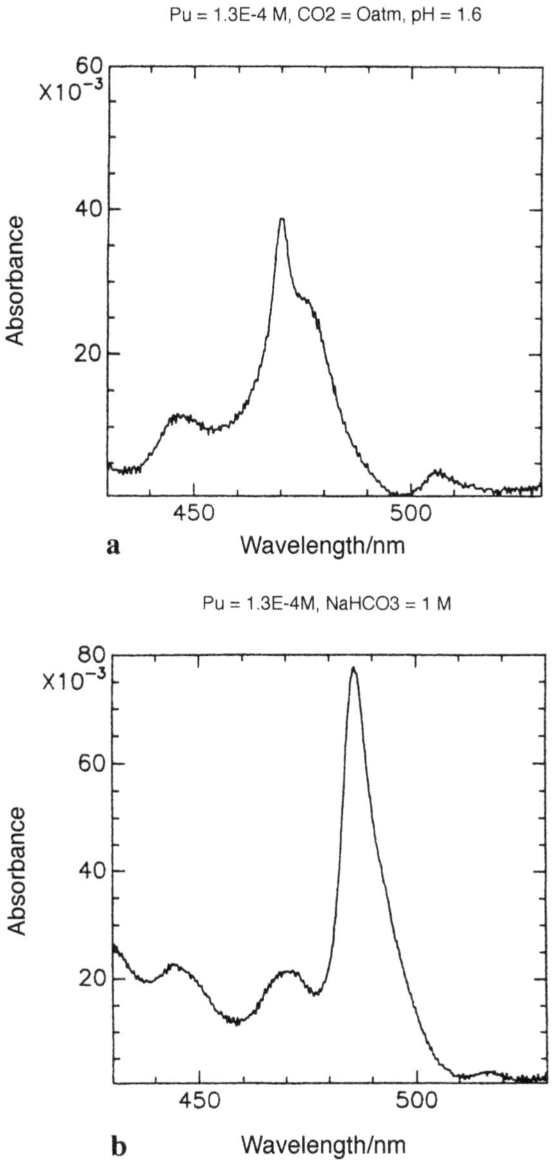

Fig. 19. Absorption spectra for Pu^{4+} (a) in perchloric acid and Pu^{4+} (b) in 1 M Na_2CO_3.

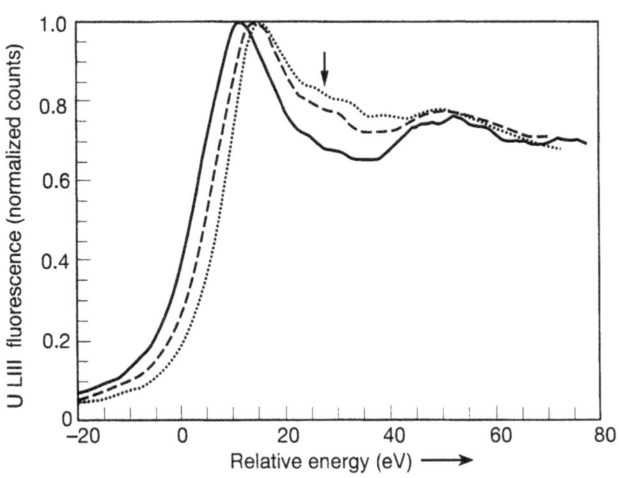

Fig. 20. Uranium L_{III} XANES spectra of pure $U(IV)O_2$ (solid line), $U(VI)O_3$ (dotted line) and a 50/50 mixture of the two (dashed line), (reprinted with permission from [116], Copyright 1994, American Chemical Society).

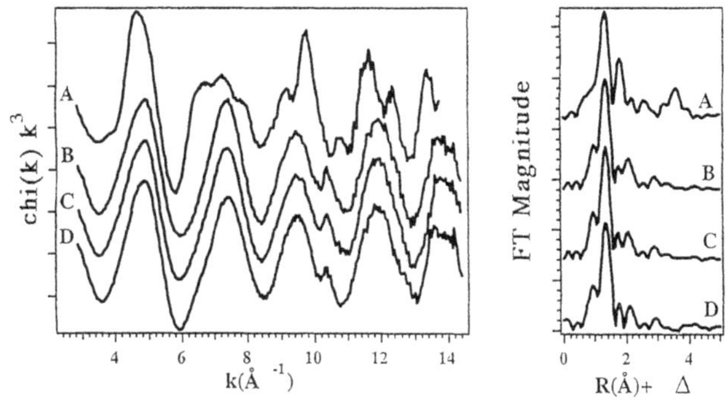

Fig. 21. Uranium L_{III} edge EXAFS spectra of uranyl orthosilicate (A), uranyl adsorbed from a 0.01 M uranyl solution onto silicic acid (B), uranyl adsorbed from a 0.05 M uranyl solution onto silicic acid (C), and uranyl adsorbed from a 0.05 M uranyl solution onto silica gel (D). Right panel shows the Fourier-transformed EXAFS of samples A through C.

Extended X-ray Absorption Fine Structure Spectroscopy (EXAFS) gives additional knowledge on coordination numbers and bond lengths to first, second and even more distant neighbor atoms [115]. This information is very useful for understanding the chemical state and bonding of actinide related to (a) speciation and complexation studies in aqueous and non-aqueous solutions, (b) adsorption processes at solid-solution interfaces involving minerals, mineral assemblies, rocks and soils and (c) adsorption and incorporation of radionuclides in microorganisms and other biological materials. The fine structure in absorption spectra above the L_{III} edge comes from interferences between outgoing photoelectron waves and backscattered waves from neighboring atoms. When EXAFS spectra are converted to momentum space (**k**), the fine structure is directly related to the sum of sinusoidal oscillations for each shell of neighboring atoms. The amplitude of the EXAFS oscillation depends on the number and type of near-neighbor atoms. For example, EXAFS spectra of uranyl orthosilicate and uranium(VI) adsorbed onto precipitated silicic acid and silica gel are shown in Figure 21 [118]. The Fourier-transformed EXAFS of these samples are included in the right panel. From fitting the EXAFS spectrum of the uranium orthosilicate, the following distances for the first four coordination shells of uranium were determined: $U-O_{axial} = 1.79$ Å, $U-O_{equatorial} = 2.39$ Å, $U-Si = 3.16$ Å, and $U-U = 3.88$ Å. The EXAFS spectra of uranyl species adsorbed onto silicic acid and silica gel at pH 4 are identical indicating similar uranyl coordination. The main characteristic of the surface species are two well-separated oxygen coordination shells in the equatorial uranyl plane at 2.27 and 2.50 Å. The results of the analysis favored the interpretation of the uranyl surface species as an innersphere, mononuclear, bidentate complex. A number of papers describing investigations of uranyl(VI) and neptunyl(V) adsorption using EXAFS have appeared recently in the literature [119–122].

The handling of actinides at large synchrotron facilities is very difficult and highly restricted because of the toxicity and radiation properties of the actinides. Therefore, only a few studies have been conducted using elements of higher atomic number than uranium and neptunium. However, proposed facilities of special design for the handling and irradiation of plutonium and beyond at high concentration levels are being constructed at the European Synchrotron Radiation Facility at Grenoble, France [115].

References

1. Hobart, D. E.: Actinides in the Environment, Proc. of the Robert A. Welch Foundation Conference on Chemical Research: XXXIV Fifty Years with Transuranium Elements, Houston, TX, 1990, pp. 379–436.
2. Bell, M. J.: Origen – The ORNL Isotope Generation and Depletion Code, ORNL Report-4628, Oak Ridge National Laboratory, Oak Ridge, TN, 1973.
3. Croff, A. G.: Origen2-A, Revised and Updated Version of the Oak Ridge Isotope Generation and Depletion Code, ORNL Report-5621, Oak Ridge National Laboratory, Oak Ridge, TN, 1980.
4. Seaborg, G. T., Katz, J. J.: *The Actinide Elements,* Natl. Nucl. Energy Ser., Div. IV, 14A, McGraw-Hill, New York 1954.
5. Seaborg, G. T., Katz, J. J., Manning, W. M.: *The Transuranium Elements: Research Papers*, Natl. Nucl. Energy Ser., Div. IV, 14B, McGraw-Hill, New York 1949.
6. Seaborg, G. T.: Elements Beyond 100, Present Status and Future Prospects, Ann. Rev. Nucl. Sci. **18**, 59 (1968).
7. Katz, J. J., Seaborg, G. T., Morss, L. R.: *The Chemistry of the Actinide Elements,* 2nd ed., vol. 2, Chapman and Hall, New York 1986, pp. 1133–1146.
8. Durrant, P. J., Durrant, B.: *Introduction to Advanced Inorganic Chemistry,* Chapter 28, Longman Grp. Ltd., London 1970, pp. 1197–1212.
9. Larsen, E. M.: *Transitional elements,* W. A. Benjamin, Inc., New York 1965, p. 104.
10. Moeller, T.: *The Chemistry of the Lanthanides*, Reinhold Publ. Corp., New York 1963, p. 21.
11. Katz, J. J., Seaborg, G. T., Morss, L. R.: *The Chemistry of the Actinide Elements*, 2nd ed., vol. 2, Chapman and Hall, New York 1986, pp. 1164–1165.
12. Silva, R. J.: Thermodynamic Properties of Chemical Species in Nuclear Waste: The Solubilities of Crystalline Neodymium and Americium Trioxides, LBL Report-15055, Lawrence Berkeley Laboratory, Berkeley, CA, 1982.
13. Katz, J. J., Seaborg, G. T., Morss, L. R.: *The Chemistry of the Actinide Elements*, 2nd ed., vol. 2, Chapman and Hall, New York 1986, pp. 1131–1132.
14. Baes, C. F., Mesmer, R. E.: *The Hydrolysis of Cations,* Krieger Publ. Co., John Wiley and Son, New York 1976.
15. Choppin, G. R.: Chemistry of Actinides in the Environment, Radiochim. Acta **43**, 82–83 (1988).
16. Stumm, W., Morgan, J. J.: *Aquatic Chemistry,* 2nd. ed., John Wiley and Sons, New York 1981, pp. 418–428.
17. Baas Becking, L. G. M., Kaplan, I. R., Moore, D.: Limits of the Natural Environment in Terms of pH and Oxidation-reduction Potentials, J. Geol. **68**, 243–284 (1960).
18. Stumm, W., Morgan, J. J.: *Aquatic Chemistry,* 2nd ed., Wiley-Interscience, New York 1981, p. 436.
19. Choppin, G. R., in: *Plutonium Chemistry* (W. T. Carnall and G. R. Choppin, eds.), ACS Symposium Series 216, Am. Chem. Soc., Washington, D. C., 1983, p. 214.
20. Grenthe, I., chairman: *Chemical Thermodynamics of Uranium,* North-Holland, Amsterdam 1992.
21. Nitsche, H.: Basic Research for Assessment of Geologic Nuclear Waste Repositories: What Solubility and Speciation Studies of Transuranium Elements Can Tell Us, Mat. Res. Soc. Symp. Proc. **212**, 517–529 (1991).
22. Silva, R. J., chairman: *Chemical Thermodynamics of Americium,* North Holland, Amsterdam 1995 (in press).
23. Papelis, C., Hayes, K. F., Leckie, J. O.: HYDRAQL: Aprogram for the Computation of Chemical Equilibrium Composition of Aqueous Batch Systems Including Surface-Complexation Modeling of Ion Absorption at the Oxide/Solution Interface, Tech. Rpt. No. 306, Environmental Engineering and Science, Dept. of Civil Engineering, Stanford University, Stanford, CA 1988.
24. Kim, J. I., Apostolidis, Ch., Büppelmann, K., Kanellakopulos, B., Klenze, R., Lierse, Ch., Magirius, S.: Chemisches Verhalten von Np, Pu und Am in Salzlösungen: Radiolyse, Löslichkeit, Hydrolyse, Chloridkomplexierung und Speziation, TUM Rpt-RCM 00187, Institut für Radiochemie, Technische Universität München, Garching, Germany 1987, p. 15.
25. Allard, B.: Solubilities of Actinides in Neutral or Basic Solutions, Proc. of the Actinide-1989 Conf., Pacific Grove, CA 1981, pp. 553–580.
26. Nguyen, S. N., Silva, R. J., Weed, H. C., Andrews, J. E.: Standard Gibbs Free Energies of Formation at the Temperature 303.15 K of Four Uranyl Silicates: Soddyite, Uranophane, Sodium Boltwoodite, and Sodium Weeksite, J. Chem. Thermodyn. **24**, 359–376 (1992).
27. Nitsche, H.: Solubility Studies of Transuranium Elements for Nuclear Waste Disposal: Principles and Overview, Radiochim. Acta **52/53**, 3–8 (1991).
28. Silva, R. J.: The Behavior of Americium in Aqueous Carbonate Systems, Mat. Res. Soc. Symp. Proc. **26**, 875–881 (1984).
29. Meinrath, G., Kim, J. I.: Solubility Products of Different Am(III) and Nd(III) Carbonates, Eur. J. Solid State Inorg. Chem. **28**, 383–388 (1991).
30. Silva, R. J.: Mechanisms for the Retardation of Uranium-(VI) Migration, Mat. Res. Soc. Symp. Proc. **257**, 323–330 (1992).
31. Lieser, K. H., Mühlenweg, U.: Neptunium in the Hydrosphere and in the Geosphere, Radiochim. Acta **43**, 27–35 (1988).
32. McCarthy, P., in: *Aquatic Humic Substances,* Adv. Chem. Ser. 219, Am. Chem. Soc., Washington, D. C., 1989, pp. xvii–xxx.
33. Sonnenberg, L. B., Johnson, J. D., Christman, R. F., in: *Aquatic Humic Substances,* Adv. Chem. Ser. 219, Am. Chem. Soc., Washington, D. C., 1989, pp. 3–23.
34. Choppin, G. R., Allard, B.: *Handbook on the Physics and Chemistry of the Actinides,* A. J. Freeman, C. Keller, eds., Elsevier Sci. Publ., Amsterdam 1985 pp. 407–429.
35. Kaastrup, E., Halmo, T. M., in: *Aquatic Humic Substances,* Adv. Chem. Ser. 219, Am. Chem. Soc., Washington, D. C., 1989, p. 698.
36. Kim, J. I., in: *Handbook on the Physics and Chemistry of the Actinides,* A. J. Freeman and C. Keller, eds., Elsevier Science Publ., Amsterdam 1985, pp. 413–455.
37. Stumm, W., Morgan, J. J.: *Aquatic Chemistry,* 2nd ed., John Wiley and Sons, New York 1981, pp. 599–647.
38. Silva, R. J., Yee, A. W., White, A. X.: Waste Isolation Safety Assessment Program: Annual Progress Report October 1, 1979–September 30, 1980, LBL Rpt. 9945, Lawrence Berkeley Laboratory, Berkeley, CA 1980.
39. Massoud, P., Ravindran, V., Wong, S.-P., Stevens, M. R., in: *Aquatic Humic Substances*, Adv. Chem. Ser. 219, Am. Chem. Soc., Washington, D. C., 1989, p. 558.
40. Serne, R. J., Relyea, J. F.: The Status of Radionuclide Sorption-Desorption Studies Performed by the WRIT Program, PNL Rpt.-SA-9787, Battelle Pacific Northwest Laboratory, Richland, WA 1981.
41. Moody, J. B.: Radionuclide Migration/Retardation: Research and Development Technology Status Report, ONWI-321, Office of Nuclear Waste Isolation, Battelle Memorial Institute, Columbus, OH 1982.
42. Stumm, W., Morgan, J. J.: *Aquatic Chemistry,* 2nd. ed., John Wiley and Sons, New York 1981, pp. 610–640.

43. Kent, D. B., Tripathi, V. S., Ball, N. B., Leckie, J. O.: Surface-complexation Modeling of Radionuclide Adsorption in Sub-surface Environments, Tech. Rpt. No. 294, Environmental Engineering and Science, Department of Civil Engineering, Stanford University, Stanford, CA 1986.
44. Hayes, K. F., Redden, G. Ela, W., Leckie, J. O.: Application of Surface Complexation Models for Radionuclide Adsorption, PNL Rpt.-7239, Pacific Northwest Laboratory, Richland, WA 1989.
45. Stumm, W., Morgan, J. J.: *Aquatic Chemistry*, 2nd. ed., John Wiley and Sons, New York 1981, pp. 640–647.
46. Davis, J. A., James, R. O., Leckie, J. O.: Surface Ionization and Complexation of the Oxide/Water Interface, J. Colloid Interface Sci. **63**(3), 480–499 (1978).
47. Yates, D. E., Levine, S., Healy, T. W.: Site Binding Model of the Electrical Double Layer at the Oxide/Water Interface, J. Chem. Soc. Faraday Trans. I, **170**, 1807–1818 (1974).
48. Davis, J. A., Leckie, J. O.: Surface Ionization and Complexation at the Oxide/Water Interface. II. Surface properties of Amorphous Iron Oxyhydroxide and Adsorption of Metal Ions, J. Colloid Interface Sci. **67**, 90–107 (1978).
49. James, R. O., Healy, T. W.: Adsorption of Hydrolyzable Metal Ions at the Oxide-Water Interface, J. Colloid Interface Sci. **40**, 42–52 (1972).
50. Silva, R. J., Benson, L. V., Yee, A. W., Parks, G. A.: Waste Isolation Safety Assessment Program. Task 4: Collection and Generation of Transport Data Theoretical and Experimental Evaluation of Waste Transport in Selected Rocks, LBL Rpt.-9945, Lawrence Berkeley Laboratory, Berkeley, CA 1979.
51. Silva, R. J.: An Integrated View of Solubilities, Speciation, Sorption and Migration from a Chemist's Point of View, 3rd, Inter. Conf. on the Chemistry and Migration Behavior of Actinides and Fission Products in the Geosphere, Jerez de la Frontera, October 21–25, 1991.
52. McCarthy, J. F., Zachara, J. M.: Subsurface Transport of Contaminants, Environ. Sci. Technol. **23**(5), 496–502 (1989).
53. Avogadro, A., DeMarsily, G.: The Role of Colloids in Nuclear Waste Disposal, Mat. Res. Soc. Symp. Proc. **26**, 495–505 (1984).
54. Ramsay, D. A.: The Role of Colloids in the Release of Radionuclides from Nuclear Waste, Radiochim. Acta **44/45**, 165–170 (1988).
55. Sheludko, A.: *Colloid Chemistry*, Elsevier Publ. Co., New York 1966, pp. 1–2.
56. Shaw, D. J.: *Colloid and Surface Chemistry*, 4th Ed., Butterworth-Heinemann Ltd., Oxford 1992, pp. 1–7.
57. Bênes, P., Majer, V.: *Trace Chemistry of Aqueous Solutions*, Elsevier Sci. Publ. Co., New York 1980, p. 62.
58. Kim, J. I.: Actinide Colloid Generation in Groundwater, Radiochim. Acta **52/53**, 71–81 (1991).
59. Thompson, J. L.: Actinide Behavior on Crushed Rock Columns, J. Radioanal. Nucl. Chem. **130**, 353–364 (1989).
60. Lieser, K. M., Gleitsmann, B., Peschke, S., Steinkopff, T.: Colloid Formation and Sorption of Radionuclides in Natural Systems, Radiochim. Acta **40**, 39–47 (1988).
61. Yariv, S., Cross, H.: *Geochemistry of Colloidal Systems*, Springer-Verlag, Berlin 1979, pp. 343–349.
62. Bênes, P., Majer, V.: *Trace Chemistry of Aqueous Solutions*, Elsevier Sci. Publ. Co., New York 1980, p. 101.
63. Rai, D., Swanson, J. L.: Properties of Plutonium(IV) Polymer of Environmental Importance, Nucl. Technol. **54**, 107–112 (1981).
64. Stumm, W., Morgan, J. J.: *Aquatic Chemistry*, 2nd ed., John Wiley and Sons, New York 1981, pp. 647–664.
65. Kaastrup, E., Halmo, T. M., in: *Aquatic Humic Substances*, Adv. Chem. Ser. 219, Am. Chem. Soc., Washington, D. C., 1989, pp. 393–395.
66. Bell, J. T., Coleman, C. F., Costanzo, D. A., Biggers, R. E.: *Plutonium Polymerization – III. The Nitrate Precipitation of Pu(IV) Polymer*, J. Inorg. Nucl. Chem. **35**, 629–632 (1973).
67. Olofsson, U., Allard, B., Anderson, K., Torstenfelt, B.: Formation and Properties of Actinide Colloids, SKBF/KBS Tech. Rpt.-TR 83-08, Swedish Nucl. Fuel and Waste Management Co., Stockholm 1983.
68. Kepak, F.: Adsorption and Colloidal Properties of Radioactive Elements in Trace Concentrations, Chem. Rev. **71**(4), 357–369 (1971).
69. Silver, G. L.: Formation of Plutonium(IV) Colloid from Plutonyl Cations, J. Radioanal. Nucl. Chem. Lett. **126**, 429–433 (1988).
70. Costanzo, D. A., Biggers, R. E., Bell, J. T.: Plutonium Polymerization-I. A Spectrophotometric study of the Polymerization of Plutonium(IV), J. Inorg. Nucl. Chem. **34**, 609–622 (1973).
71. Lloyd, M. H., Haire, R. G.: The Chemistry of Plutonium in Sol-Gel Processes, Radiochim. Acta **25**, 139–148 (1978).
72. Silver, G. L.: Free Energy of Plutonium Polymer Formation, J. Less-Common Met. **91**, 317–320 (1983).
73. Rai, D.: Solubility Product of Pu(IV) Hydrous Oxide and Equilibrium Constants of Pu(IV)/Pu(V), Pu(IV)/Pu(VI) and Pu(V)/Pu(VI) Couples, Radiochim. Acta **35**, 97–106 (1984).
74. Kim, J. I., Kanellakopulos, B.: Solubility products of Plutonium(IV) Oxide and Hydroxide, Radiochim. Acta **48**, 145–150 (1989).
75. Rai, D., Ryan, J. L.: Crystallinity and Solubility of Pu(IV) Oxide and Hydrous Oxide in Aged Aqueous Suspensions, Radiochim. Acta **30**, 213–216 (1982).
76. Rai, D., Serne, R. J., Swanson, J. L.: Solution Species of Plutonium in the Environment, J. Environ. Qual. **9**, 417–420 (1980).
77. Rai, D., Serne, R. J.: Solution Species of Pu-239 in Oxidizing Environments: I. Polymeric Pu(IV), PNL Rpt.-SA-6994, Pacific Northwest Laboratory, Hanford, WA, 1979, pp. 1–9.
78. Ichikawa, F., Sato, T.: On the Particle Size Distribution of Hydrolyzed Plutonium(IV) Polymer, J. Radioanal. Nucl. Chem. **84**, 269–275 (1984).
79. Triay, I. R., Hobart, D. E., Mitchell, A. J., Newton, T. W., Ott, M. A., Palmer, P. D., Rundberg, R. S., Thompson, J. L.: Size Determinations of Plutonium Colloids Using Autocorrelation Photon Spectroscopy, Radiochim. Acta **52/53**, 127–131 (1991).
80. Rundberg, R. S., Mitchell, A. J., Triay, I. R., Torstenfelt, J. B.: Size and Density of a Pu-242 Colloid, Mat. Res. Soc. Proc. **112**, 269–275 (1988).
81. Newton, T. W., Hobart, D. E., Palmer, P. D.: The Formation of Pu(IV)-Colloid by the Alpha-Reduction of Pu(V) or Pu(VI) in Aqueous Solutions, Radiochim. Acta **39**, 139–147 (1986).
82. Savage, D. J., Kyffin, T. W.: Oxidation Breakdown of Polymeric Plutonium(IV) Hydroxide by Cerium(IV), Polyhedron **5**(3), 743–752 (1986).
83. Silver, G. L.: The Ratio Pu(V)/Pu(IV) in Natural Water, J. Radioanal. Nucl. Chem. Lett. **137**, 75–78 (1989).
84. Cleveland, J. M. in: Geochemical Modeling in Aqueous Systems, ACS Sym. Ser., American Chemical Society, Washington, D.C., 1979, pp. 321–338.
85. Kraus, K. A.: *Hydrolytic Behavior of the Heavy Elements*, Proc. Intern. Conf. on the Peaceful Uses of Atomic Energy, Geneva, Switzerland, **7**, 245–257 (1956).
86. Buckau, G., Stumpe, R., Kim, J. I.: Americium Colloid Generation in Groundwaters and Its Speciation by Laser-Induced Photoacoustic Spectroscopy, J. Less-Common Met. **122**, 555–562 (1986).
87. Lierse, Ch., Treiber, W., Kim, J. I.: Hydrolysis Reactions of Neptunium(V), Radiochim. Acta **38**, 27–28 (1985).
88. Kim, J. I.: Actinide Colloid Generation in Groundwater, 2nd Int. Conf. on the Chemistry and Migration Behavior of Actinides and Fission Products in the Geosphere, Monterey, CA 1989.
89. Triay, I., Simmons, A., Levy, S., Nelson, S., Nuttall, H., Robinson, B., Steinkampf, W., Viani, B.: Colloid-Facili-

tated Radionuclide Transport at Yucca Mountain, LA Rpt.-12779-MS, Los Alamos National Laboratory, Los Alamos, NM 1995.
90. Yariv, S., Cross, H.: *Geochemistry of Colloidal Systems*, Springer-Verlag, Berlin 1979, p. 23.
91. Rossotti, F. J. C., Rossotti, H. S.: *The Determination of Stability Constants*, McGraw-Hill, New York 1961.
92. Baumann, R. P.: *Absorption Spectroscopy*, John Wiley and Sons, New York 1962, pp. 19–20.
93. Keller, C.: *The Chemistry of the Transuranium Elements*, Verlag Chemie Weinheim, Germany 1971, pp. 94–115.
94. Katz, J. J, Seaborg, G. T., Morss, L. R.: *The Chemistry of the Actinide Elements*, 2nd ed., vol. 2, Chapman and Hall, New York 1986, pp. 1257–1277.
95. Bauman, R. P.: *Adsorption Spectroscopy*, John Wiley and Sons, New York 1962, pp. 368–369.
96. Nitsche, H., Standifer, E. M., Silva, R. J.: Neptunium(V) Complexation with Carbonate, Lanth. and Act. Res. **3**, 203–211 (1990).
97. Nitsche, H., Standifer, E. M., Silva, R. J.: Am(III) Carbonate Complexation in Aqueous Perchlorate Solutions, Radiochim. Acta **46**, 185–189 (1989).
98. Bennett, D. A., Hoffman, D., Nitsche, H., Russo, R. E., Torres, R. A., Baisden, P. A., Andrews, J. E., Palmer, C. E. A., Silva, R. J.: Hydrolysis and Carbonate Complexation of Dioxoplutonium(V), Radiochim. Acta **56**, 15–19 (1992).
100. Cetorelli, J. J., Winefordner, J. D.: Estimation of the Minimum Detectable Concentration and Optimum Instrumental Conditions in Precision Absorption Spectrophotometry of Substances in the Condensed Phase, Talanta **14**, 705–713 (1967).
101. Swafford, R. L., in: *Lasers in Chemical Analysis*, G. M. Hieftje, J. C. Travis, F. E. Lytle, eds., Humana Press, West Lafayette, Indiana 1981, pp. 143–156.
102. Tam, A. C.: Applications of Photoacoustic Sensing Techniques, Rev. Mod. Phys. **58**, 381–486 (1986).
103. Jackson, W. B., Amer, N. M., Boccara, A. C., Fournier, D.: Photothermal Deflection Spectroscopy and Detection, Appl. Opt. **20**, 1333–1344 (1981).
104. Fang, H. L., Swafford, R. L., in: *Ultrasensitive Laser Spectroscopy*, D. S. Kliger, Ed., Academic Press, New York 1983, pp. 175–180.
105. Torres, R. A., Palmer, C. A., Baisden, P. A., Russo, R. E., Silva, R. J.: Comparison of Photoacoustic Spectroscopy, Conventional Absorption Spectroscopy, and Potentiometry as Probes of Lanthanide Speciation, Anal. Chem. **62**(3), 298–303 (1990).
106. Klenze, R., Kim, J. I., Wimmer, H.: Speciation of Aquatic Actinide Ions by Pulsed Laser Spectroscopy, Radiochim. Acta **52/53**, 97–103 (1991).
107. Beitz, J. V., Bowers, D. L., Doxtader, M. M., Maroni, V. A., Reed, D. T.: Detection and Speciation of Transuranium Elements in Synthetic Groundwater via Pulsed-Laser Excitation, Radiochim. Acta **44/45**, 87–93 (1988).
108. Moulin, C., Decambox, P., Mauchien, P., Moulin, V., Theyssier, M.: On the Use of Laser-Induced Time-Resolved Spectrofluorometry for Interaction Studies between Organic Matter and Actinides: Application to Curium, Radiochim. Acta **52/53**, 119–125 (1991).
109. Russo, R. E., Rojas, D., Robouch, P., Silva, R. J.: Remote Photoacoustic Measurements in Aqueous Solutions Using an Optical Fiber, Rev. Sci. Instrum **61**(12), 3729–3732 (1990).
110. Klenze, R., Kim, J. I.: Direct Speciation of Transuranium Element in Natural Aquatic Systems by Laser-Induced Photoacoustic Spectroscopy, Radiochim. Acta **44/45**, 77–85 (1988).
111. European Synchrotron Radiation Facility (ESRF): Beam Line Handbook, Grenoble, France (1994).
112. Mollenauer, L. F., White, J. C., Pollock, C. R., eds.: *Tunable Lasers, Topics in Applied Physics*, Springer-Verlag, New York 1993.
113. Konigsberger, D. C., Prins, R.: *X-ray Absorption, Principles, Application, Techniques of EXAFS, SEXAFS and XANES*, Wiley and Sons, New York 1988.
114. Bucher, J. J., Allen, P. G., Edelstein, N. M., Shuh, D. K., Madden, N., Pehl, D., Cork, C., Malone, D.: A Quad-pixel Monolithic Ge Fluorescence Detector. Stanford Synchrotron Radiation Laboratory (SSRL), Stanford, CA, USA, Annual Users Meeting, Poster Session (1994).
115. Nitsche, H.: Synchrotron X-ray Absorption Spectroscopy: A New Tool for Actinide and Lanthanide Speciation in Solids and Solution, J. Alloys and Compounds **223**, 274–279 (1995).
116. Bertsch, P. M., Hunter, D. B., Sutton, S. R., Bajt, S., Rivers, M. L.: In Situ Chemical Speciation of Uranium in Soils and Sediments by Micro X-ray Absorption Spectroscopy. Environ. Sci. Technol. **28**, 980–984 (1994).
117. Francis, A. J., Dodge, C. J., Lu, F., Halada, G. P., Clayton, C. R.: *XPS* and *XANES* Studies of Uranium Reduction by Clostridium sp., Environ. Sci. Technol. **28**, 636–639 (1994).
118. Reich, T., Moll, H., Denecke, M. A., Geipel, G., Bernhard, G., Nitsche, H., Allen, P. G., Kaltsoyannis, N., Bucher, J. J., Edelstein, N. M., Shuh, D. K.: Characterization of Hydrous Uranyl Silicate by EXAFS Analysis. Fifth International Conference on Migration of Actinides and Fission Products in the Geosphere, Saint-Malo (1995).
119. Chisholm-Brause, C., Conradson, S. D., Buscher, C. T., Eller, P. G., Morris, D. E.: Speciation of Uranyl Sorbed at Multiple Binding Sites on Montmorillonite. Geochim. Cosmochim. Acta **58**(17), 3625–3631 (1994).
120. Chisholm-Brause, C., Conradson, S. D., Eller, P. G., Morris, D. E.: Changes in U(VI) Speciation Upon Sorption onto Montmorillonite from Aqueous and Organic Solutions. MRS Sci. Basis for Nuclear Waste Management, XV, 315–322 (1992).
121. Dent, A. J., Ramsay, J. D. F., Swanton, S. W.: An EXAFS Study of Uranyl Ion in Solution and Sorbed onto Silica and Montmorillonite Clay Colloids. J. Colloid Interface Sci. **150**, No. 1, 45–60 (1992).
122. Combes, J., Chisholm-Brause, D., Brown, Jr., G. E., Parks, G. A., Conradson, S. D., Eller, P. G., Triay, I. R., Hobart, D. E., Meijer, A.: EXAFS Spectroscopic Study of Neptunium(V) Sorption at the α-FeOOH/Water Interface. Environ. Sci. Technol. **26**, No. 2, 376–383 (1992).

Radiochimica Acta 70/71, 397–402 (1995)

Radioactivity and Fission Energy

By L. Koch

European Commission, Joint Research Centre, European Institute for Transuranium Elements, P.O. Box 2340, 76125 Karlsruhe, Federal Republic of Germany

(Received August 23, 1995)

Radioacitvity / Nuclear energy / Transmutation / Fission

Abstract

Anthropogenic radioactivity stems mainly from nuclear energy generation. The contents are given for fission and radioactivation products in a typical LWR and for global accumulation including transuranium elements. The transmutation of radiotoxicity in fast reactors is described.

Introduction

Similar to the CO_2 emissions from the burning of fossile fuels enhancing the greenhouse effect, fission energy generates also an unwanted by-product: artificial radioactivity. Nevertheless fissiogenic radioactivity occurs in nature, too. In 1972, H. Bouzigues observed in natural uranium an odd isotopic abundance for ^{235}U of $0.711 \pm 0.0007\%$, which led to the discovery of a precambrian natural fission energy source in Oklo, Gabon. The uranium ore that Becquerel held in his hands also must have contained radioactivity that stemmed from fission – spontaneous or induced by cosmic rays. During early times on earth, when the ^{235}U abundance was above 2.1%, natural light water boiling reactors were still possible so that the Oklo phenomenon was not a singularity. The Oklo reactors, however, operated only with a power of 1–10 kW during about 1 million years. So, present nuclear power stations cannot simply be regarded as atavisms, although they generate about the same energy. Their power density and hence the accumulation of short-lived fission products is larger by 10^6. A high degree of safety technology is required to contain the accumulated radioactivity in the core of the reactor. In case of a sudden release, as during the Chernobyl catastrophe, a devastating contamination of the environment is the unavoidable consequence. Already before this event, when fission energy became a military weapon in 1945, it turned the public against fission energy use. The following atomic bomb tests in the atmosphere led to a wide-spread contamination by long-living radioactive fission products of which some accumulated in the food chain, which further sensibilized the public, so that now the final disposal of long-living radiotoxic nuclides has become the major acceptance problem of fission energy use.

Nevertheless, it is more or less common belief that mankind cannot do without the energy generated by fission as long as no reasonable substitute is found. Consequently, it ought to be the greatest challenge to the scientific and technical community to develop feasible solutions that will make fission energy acceptable in reducing and containing the artificial radioactivity to such an extent that accidental releases to the environment cannot pose a major hazard to life.

Formation of radioactivity by fission energy

The radioactivity building up during fission energy generation can be described by:

$$\frac{dN}{dt} = \sum (f\sigma^* \phi + z\lambda^*)N^* - (\sigma\phi + \lambda)N. \qquad (1)$$

With N as the radioactive nuclide considered that is formed by Σ routes from different nuclides N^* with the reaction constants $f\sigma^*\phi$ (neutron induced) and $z\lambda^*$ of possible decay modes (ϕ is the neutron flux). The second term of the equation describes the nuclide's depletion by the same type of reactions.

The primary nuclear reaction is the **neutron induced fission** of ^{235}U, which is the only fissile nuclide in accessible quantities on earth to sustain a nuclear chain reaction.

$$^{235}U + n = fp + \nu n + E \qquad (2)$$

For the fission of one ^{235}U atom induced by one thermal neutron n, there are released fission products fp, of varying masses – as described below – together with an average yield of $\nu = 2.423$ neutrons and of average released energy $E = 193.6$ [MeV].

Radioactivity by fission

Following Eq. (1), the formation of a fission product mass N is described by:

$$f\sigma\phi \equiv fy\frac{\sigma_f}{\sigma_a}\phi \qquad (3)$$

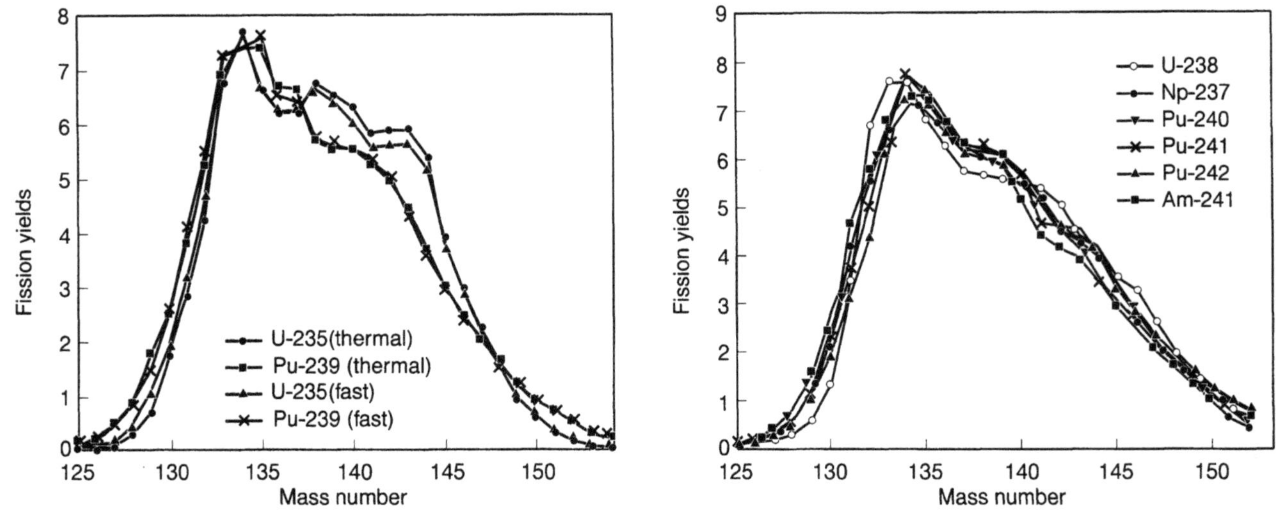

Fig. 1. Isobaric yields of heavier fission products: influence of neutron energy for ^{235}U and ^{239}Pu (left curve), influence of higher masses of fission sources (^{237}Np, ^{238}U, ^{240}Pu, ^{241}Pu, ^{241}Am, ^{242}Pu) [1, 2].

where the fission yield fy, as well as the relative fission cross section σ_f/σ_a varies with the energy spectrum of the neutron flux ϕ. The lighter masses distribution of the asymmetric fission changes little. The heavier fission product distribution reflects the higher masses of the fission sources. With increasing energy of the neutron inducing fission, the symmetric fission increases only slightly (Fig. 1), but the σ_f/σ_a varies strongly, due to so-called resonances in the epithermal neutron energy spectrum (σ_a is the total absorption cross section $\sigma_a = \sigma_f + \sigma_c$).

The (independently by fission) formed nuclide N decays by β^- with the reaction constant λ into another radioactive or stable nuclide. During reactor operation, the short-lived radionuclides determine the total radioactivity which is about 10^{20} Bq for a typical LWR at the time of a reactor shut down. After ten thousand years, the remaining radioactivity is governed by a few radionuclides.

Radioactivity by neutron capture

The neutrons (Eq. (2)) in excess $(\nu - 1)$ to maintain the chain reaction are captured in reactor materials. According to Eq. (1), the build-up is governed by

$$f\sigma^* \phi \equiv \frac{\sigma_c}{\sigma_a} \phi \tag{4}$$

with σ_c the cross section for neutron capture ($\sigma_a = \sigma_f + \sigma_c$). The radiative neutron capture reaction (n, γ) dominates other possible reactions (n, 2n), (n, 3n), (n, p), (n, α). Although for absorbers of excess reactivity and for structural materials elements are chosen such as B, Gd, Zr with small radioactivation, the radioactivity of reactor material is considerable (Table 1).

Of more relevance is the built up radioactivity by activation in the fuel itself. Certain fission products limit the fuel use because of parasitic neutron capture which leads in some cases to the build-up of new radionuclides that are not formed by fission and subsequent β^- decay, because they are shielded in the decay chain by stable nuclides. (Among them are: ^{110}Ag, ^{134}Cs and ^{154}Eu). The neutron capture reactions change the fission product distribution in Light Water Reactors for highly burnt fuels considerably, so that it deviates from the original mass distribution.

The reactivity of a reactor fuel is prolonged by the formation of TU nuclides which are partially fissile, but all are highly radiotoxic. The main reaction is the (multiple) radiative neutron capture (n, γ) followed by β^-- and α-decay which determine the pattern of the reaction path (Fig. 2) leading to the build-up of actinides in measurable quantities up to californium. In Table 2 is given the concentration for the actinide nuclides contained in 1 t HM of spent fuel at the time of discharge.

Although the $^{238}Pu(n, 3n)^{236}Pu$ and ^{234}U $(n, 3n)^{232}U$ reactions have a low probability, they nevertheless contribute substantially, e.g. at discharge of PWR fuel to the formation of the ^{236}Pu (10^{10} Bq/t HM) and ^{232}U (10^8 Bq/t HM), which fall into the natural 4n chain of ^{232}Th [5]. It is the build-up of ^{208}Tl, which emits a 2.6 MeV hard γ-radiation that limits the reuse of recovered uranium from reprocessed LWR fuel, because – during the re-enrichment process with present shielding in the facilities – the acceptable occupational radiation dose would be soon exceeded.

The $^{238}U(n, 2n)^{237}U \xrightarrow{\beta^-} {}^{237}Np$ reaction resurrects the extinct $(4n+1)$ decay chain (Fig. 3). Since ^{237}Np is being produced in the ton scale, it will testify far

Table 1. Radioactivity of fission products (Bq/t HM) from typical PWR fuel and radioactivation in reactor material at discharge and 10000 a [3]

Fission product radioactivity [Bq]		
Radionuclide	Discharge	10000 a
^{3}H	$2 \cdot 10^{13}$	–
^{79}Se	$1 \cdot 10^{10}$	$1 \cdot 10^{10}$
^{85}Kr	$3 \cdot 10^{14}$	–
^{86}Rb	$7 \cdot 10^{13}$	–
^{89}Sr	$3 \cdot 10^{16}$	–
^{90}Sr	$3 \cdot 10^{15}$	–
^{90}Y	$3 \cdot 10^{15}$	–
^{91}Y	$4 \cdot 10^{16}$	–
^{93}Zr	$7 \cdot 10^{10}$	$7 \cdot 10^{10}$
^{95}Zr	$5 \cdot 10^{16}$	–
^{93m}Nb	$5 \cdot 10^{9}$	$6 \cdot 10^{10}$
^{95}Nb	$5 \cdot 10^{16}$	–
^{95m}Nb	$4 \cdot 10^{14}$	–
^{99}Tc	$5 \cdot 10^{11}$	$5 \cdot 10^{11}$
^{103}Ru	$6 \cdot 10^{16}$	–
^{106}Ru	$2 \cdot 10^{16}$	–
^{103m}Rh	$5 \cdot 10^{16}$	–
^{106}Rh	$2 \cdot 10^{16}$	–
^{107}Pd	$4 \cdot 10^{9}$	$4 \cdot 10^{9}$
^{110}Ag	$5 \cdot 10^{15}$	–
^{110m}Ag	$1 \cdot 10^{14}$	–
^{113m}Cd	$1 \cdot 10^{12}$	–
^{115m}Cd	$3 \cdot 10^{13}$	–
^{114}In	$9 \cdot 10^{10}$	–
^{114m}In	$7 \cdot 10^{10}$	–
^{119m}Sn	$4 \cdot 10^{12}$	–
^{121m}Sn	$6 \cdot 10^{9}$	–
^{123}Sn	$7 \cdot 10^{13}$	–
^{126}Sn	$2 \cdot 10^{10}$	$2 \cdot 10^{10}$
^{124}Sb	$2 \cdot 10^{13}$	–
^{125}Sb	$3 \cdot 10^{14}$	–
^{126}Sb	$3 \cdot 10^{13}$	$3 \cdot 10^{9}$
^{126m}Sb	$2 \cdot 10^{13}$	$2 \cdot 10^{10}$
^{123m}Te	$3 \cdot 10^{11}$	–
^{125m}Te	$7 \cdot 10^{13}$	–
^{127}Te	$3 \cdot 10^{15}$	–
^{127m}Te	$4 \cdot 10^{14}$	–
^{129}Te	$1 \cdot 10^{16}$	–
^{129m}Te	$2 \cdot 10^{15}$	–
^{129}I	$1 \cdot 10^{9}$	$1 \cdot 10^{9}$
^{134}Cs	$6 \cdot 10^{15}$	–
^{135}Cs	$1 \cdot 10^{10}$	$1 \cdot 10^{10}$
^{137}Cs	$4 \cdot 10^{15}$	–
^{137m}Ba	$4 \cdot 10^{15}$	–
^{140}Ba	$6 \cdot 10^{16}$	–
^{140}La	$6 \cdot 10^{16}$	–
^{141}Ce	$5 \cdot 10^{16}$	–
^{144}Ce	$4 \cdot 10^{16}$	–
^{143}Pr	$5 \cdot 10^{16}$	–
^{144}Pr	$4 \cdot 10^{16}$	–
^{144m}Pr	$5 \cdot 10^{14}$	–
^{147}Pm	$6 \cdot 10^{15}$	–
^{148}Pm	$6 \cdot 10^{15}$	–
^{148m}Pm	$1 \cdot 10^{15}$	–
^{151}Sm	$1 \cdot 10^{13}$	–
^{152}Eu	$2 \cdot 10^{11}$	–
^{154}Eu	$5 \cdot 10^{14}$	–
^{155}Eu	$3 \cdot 10^{14}$	–
^{156}Eu	$8 \cdot 10^{15}$	–
^{153}Gd	$7 \cdot 10^{11}$	–
^{160}Tb	$3 \cdot 10^{13}$	–
Total	$7 \cdot 10^{17}$	$7 \cdot 10^{11}$

Table 1. Continuation.

Radioactivation products [Bq]		
Radionuclide	Discharge	10000 a
^{3}H	$6 \cdot 10^{7}$	–
^{14}C	$5 \cdot 10^{9}$	$2 \cdot 10^{9}$
^{33}P	$2 \cdot 10^{13}$	–
^{35}S	$3 \cdot 10^{11}$	–
^{45}Ca	$8 \cdot 10^{9}$	–
^{46}Sc	$1 \cdot 10^{12}$	–
^{51}Cr	$7 \cdot 10^{15}$	–
^{54}Mn	$3 \cdot 10^{14}$	–
^{55}Fe	$2 \cdot 10^{15}$	–
^{59}Fe	$2 \cdot 10^{14}$	–
^{58}Co	$1 \cdot 10^{15}$	–
^{60}Co	$8 \cdot 10^{14}$	–
^{59}Ni	$6 \cdot 10^{11}$	$6 \cdot 10^{11}$
^{63}Ni	$9 \cdot 10^{13}$	–
^{65}Zn	$9 \cdot 10^{9}$	–
^{89}Sr	$3 \cdot 10^{12}$	–
^{90}Sr	$8 \cdot 10^{7}$	–
^{90}Y	$1 \cdot 10^{14}$	–
^{91}Y	$7 \cdot 10^{12}$	–
^{93}Zr	$7 \cdot 10^{9}$	$7 \cdot 10^{9}$
^{95}Zr	$2 \cdot 10^{15}$	–
^{93m}Nb	$4 \cdot 10^{8}$	$6 \cdot 10^{9}$
^{94}Nb	$4 \cdot 10^{10}$	$3 \cdot 10^{10}$
^{95}Nb	$2 \cdot 10^{15}$	–
^{93}Mo	$3 \cdot 10^{9}$	$4 \cdot 10^{8}$
^{99}Tc	$4 \cdot 10^{8}$	$4 \cdot 10^{8}$
^{119m}Sn	$4 \cdot 10^{14}$	–
^{121m}Sn	$3 \cdot 10^{10}$	–
^{123}Sn	$3 \cdot 10^{10}$	–
^{124}Sb	$5 \cdot 10^{11}$	–
^{125}Sb	$3 \cdot 10^{12}$	–
^{125m}Te	$6 \cdot 10^{11}$	–
^{182}Ta	$7 \cdot 10^{15}$	–
^{185}W	$2 \cdot 10^{10}$	–
Total	$2 \cdot 10^{16}$	$6 \cdot 10^{11}$

into the future that nuclear energy was once used on earth.

The reactions ^{17}O(n,α) and ^{14}N(n,p) also with low probability lead to the build-up of significant ^{14}C radioactivity (Table 1).

Table 1 specifies the content of fission and radioactivation products and Table 2 the contribution of actinides in one ton Heavy Metal, HM, of a typical LWR.

Accumulation of radioactivity

There exist several computer programmes that calculate according Eq. (1) for a given operation time the radioactivity accumulated in a nuclear reactor [3, 6, 7]. The abundance of a radionuclide in a spent nuclear fuel depends on the neutron fluence and neutron spectrum. The latter is determined by the moderator/coolant, as can be read from Table 2, where the long-living transuranium nuclides from a typical spent PWR fuel (33 GWd/t) are compared to a heavy water moderated power station fuelled with natural uranium (6 GWd/t).

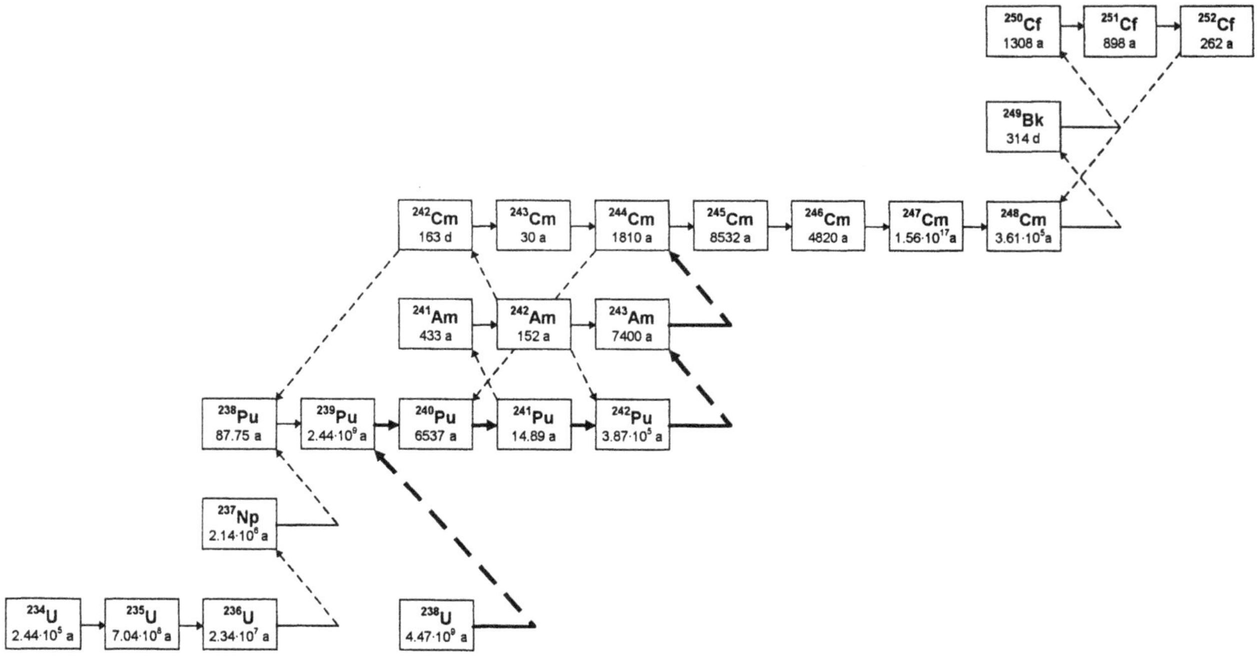

Fig. 2. Build-up chain of long-lived actinides in nuclear fuels with major (bold arrows) and minor (thin arrows) pathways (dashed lines symbolise short-lived radioactive decay) [4].

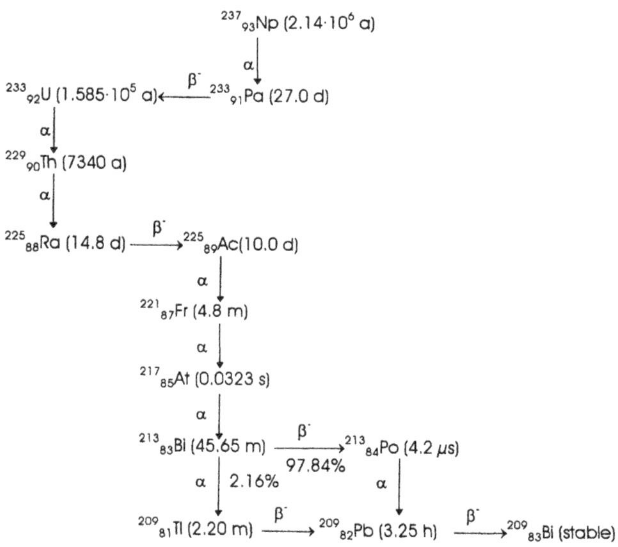

Fig. 3. Extinct decay chain: 4n + 1.

Table 2. Radioactivity [Bq] of built up actinides in 1 t of HM of a typical PWR fuel and HWR fuel at discharge and 10000 a

Radionuclide	Discharge PWR	Discharge HWR	10000 a PWR	10000 a HWR
^{236}U	$1 \cdot 10^{10}$	$2 \cdot 10^{9}$	$1 \cdot 10^{10}$	$3 \cdot 10^{9}$
^{237}Np	$1 \cdot 10^{10}$	$2 \cdot 10^{8}$	$5 \cdot 10^{10}$	$4 \cdot 10^{9}$
^{238}Pu	$8 \cdot 10^{13}$	$5 \cdot 10^{11}$	–	–
^{239}Pu	$1 \cdot 10^{13}$	$6 \cdot 10^{12}$	$1 \cdot 10^{13}$	$4 \cdot 10^{12}$
^{240}Pu	$2 \cdot 10^{13}$	$7 \cdot 10^{12}$	$7 \cdot 10^{12}$	$2 \cdot 10^{12}$
^{241}Pu	$5 \cdot 10^{15}$	$5 \cdot 10^{14}$	$3 \cdot 10^{9}$	$1 \cdot 10^{6}$
^{242}Pu	$7 \cdot 10^{10}$	$4 \cdot 10^{9}$	$8 \cdot 10^{10}$	$4 \cdot 10^{9}$
^{241}Am	$5 \cdot 10^{12}$	$3 \cdot 10^{11}$	$3 \cdot 10^{9}$	$3 \cdot 10^{6}$
^{242m}Am	$1 \cdot 10^{11}$	$6 \cdot 10^{9}$	–	–
^{243}Am	$6 \cdot 10^{11}$	$5 \cdot 10^{9}$	$3 \cdot 10^{11}$	$2 \cdot 10^{9}$
^{242}Cm	$2 \cdot 10^{15}$	$2 \cdot 10^{13}$	–	–
^{243}Cm	$6 \cdot 10^{11}$	$4 \cdot 10^{9}$	–	–
^{244}Cm	$7 \cdot 10^{13}$	$1 \cdot 10^{11}$	–	–
Total	$6 \cdot 10^{15}$	$6 \cdot 10^{14}$	$2 \cdot 10^{13}$	$7 \cdot 10^{12}$

Table 3. Present accumulated actinide radioactivity [Bq] for major abundant, long-living nuclides

Nuclide	^{237}Np	^{239}Pu	^{240}Pu	^{241}Am	^{243}Am
Radioactivity	$6 \cdot 10^{14}$	$2 \cdot 10^{18}$	$1 \cdot 10^{18}$	$4 \cdot 10^{17}$	$2 \cdot 10^{16}$

There is a variety of reactor types in operation world-wide (Fig. 4), from which the so far accumulated artificial radioactivity can be estimated. To do so, either the produced energy or the discharged fuel with its declared burn-up has to be known. Unfortunately the published figures are not detailed enough. From the total generated electricity, the produced plutonium has now been estimated to be about 750 t (to which about 200 t Pu from the military sector have to be added). The average Pu concentration contained in the 165000 t of spent fuel produced so far amounts to 4.5 kg/t HM, which is a rather low value compared to that in present spent fuel, but in accordance with the estimated average burn-up of 15 – 20 MWd/t HM.

The accumulated radioactivity by fission products – including the military sector – follows directly from the produced energy or the mass of plutonium and amounts to about $5 \cdot 10^{16}$ Bq after a cooling time of 1000 a (i.e. mainly ^{99}Tc). The radioactivity of transuranium nuclides – as described above – increases exponentially with discharge burn-up of the fuel. The

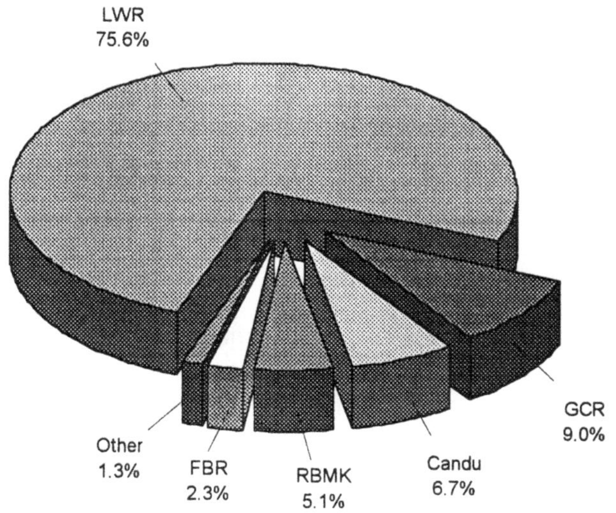

Fig. 4. Share of reactor types in worldwide fission energy generation (1992), 343 GWe (433 units) [8].

latter value is often not accessible, therefore the estimates of the worldwide accumulation of radioactivity can be uncertain by a factor of 2.

Disposal of radioactivity

The option to send the radioactive waste into outer space is presently not considered as a realistic possibility. Hence, all radioactivity has to be stored in the country where the energy was produced. Due to fission energy generation, about 10000 t HM of spent fuel are discharged annually worldwide giving an estimated cumulated total of over 165000 t HM. Of this, about 105000 t HM are currently being stored. The present reprocessing capacity of 4000 t per year is projected to increase in ten years to 7700 t/a [9]. Consequently, the arisings of spent fuel exceed their possible reprocessing.

Several countries with major nuclear energy generating capacities, like Canada, Sweden and USA have decided to dispose of the fuel directly without reprocessing into a geological repository. Others follow both the routes: Finland, Germany, Russian Federation, Spain, Ukraine.

The total amount of commercial spent fuel reprocessed up to 1994 was about 47300 t HM with 34300 t HM stemming from GCR and only 13000 t HM discharged from LWR. From the extracted Pu, about 98 t are stored. Between 2 and 7 t Pu have been used annually in LWRs and FBRs. 40 t HM of the fuel from the latter reactor type have also been reprocessed [9]. As mentioned above, about 200 t Pu have been produced in the military sector.

In the high level waste (HLW) resulting from fuel reprocessing are contained most of the fission products and actinides other than uranium and plutonium. Little waste has been vitrified and most is still stored in liquid form. The radioactivity of the spent fuel and high level waste generate in the first years too much

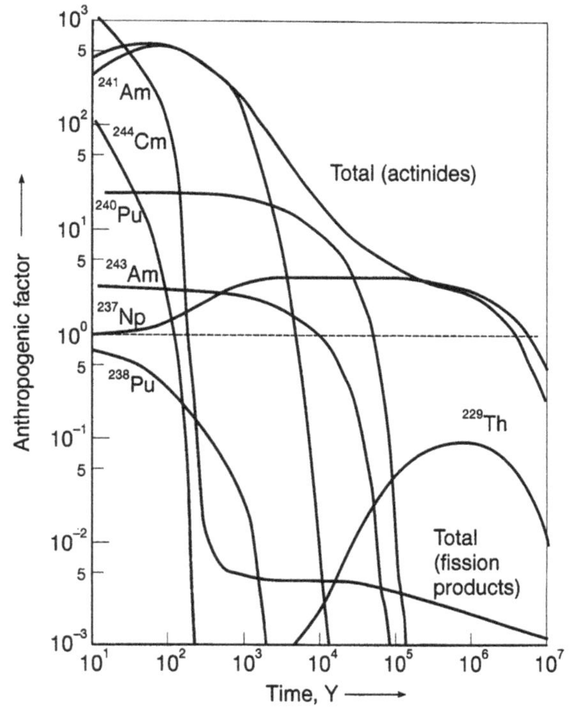

Fig. 5. Time-dependence of radiotoxicity in a spent LWR fuel of 33 GWd/t normalized to the radiotoxicity of the uranium ore (dashed line) mined to produce the fuel [4].

heat (1.5 kW/t HM), to be brought underground. Therefore, an intermediate storage for tens of years is needed to have the radioactivity decayed to acceptable heat levels. During this period, the radiotoxic nuclides still remain accessible for further treatment.

The time dependence of the **radiological risk** of the radioactivity by fission energy in a waste repository is for a given nuclide N according to Eq. (5)

$$\frac{dN}{dt} = z\lambda^* N^* - \lambda N \tag{5}$$

with N^* being its parent nuclide with $z\lambda^*$ decay modes. This has to be expressed in terms of its radiotoxicity as assessed by the ICRP and expressed in annual limits of intake of radionuclides [10]. In order to make the radiological risk comprehensible, one should normalize the radiotoxicity of a certain nuclide to that of the uranium ore which is needed to fabricate the fuel. After use of the fuel, the induced enhanced radioactivity is illustrated in Figure 5 as a function of time.

The radiotoxicity of fission products decreases within a few hundred years to below that of the uranium ore mined to produce the fresh fuel. The accumulated transuranium elements will decay over much longer periods. Hence, their transmutation into short-lived or stable nuclides would reduce the radiotoxic hazard for future generations. This view looks only at the source of the radiotoxic hazard as it is posed by the spent fuel. If stored in a geological repository, the leaching of the fuel and the eventual migration of nuclides back to the biosphere is a more relevant description of the imminent danger. Also fission product nu-

Table 4. List of long-lived radionuclides considered as candidates for nuclear transmutation [11]

Nuclide	$T_{1/2}$ [a]
^{14}C	$5.7 \cdot 10^3$
^{36}Cl	$3.0 \cdot 10^5$
^{129}I	$1.6 \cdot 10^7$
^{135}Cs	$2.0 \cdot 10^7$
^{79}Se	$6.5 \cdot 10^4$
^{93}Zr	$1.5 \cdot 10^6$
^{90}Sr	$2.9 \cdot 10^1$
^{121}Sn	$5.0 \cdot 10^1$
^{126}Sn	$1.0 \cdot 10^5$
^{137}Cs	$3.0 \cdot 10^1$
^{99}Tc	$2.1 \cdot 10^5$

Table 5. Half-lifes [a] of actinides by natural decay and transmutation in a fast reactor [12]

Actinide	Natural decay	Decay in FR
^{237}Np	$2 \cdot 10^6$	2.5
^{238}Pu	88	2.0
^{239}Pu	$2 \cdot 10^4$	1.5
^{240}Pu	$6 \cdot 10^3$	3.1
^{241}Pu	15	1.2
^{242}Pu	$4 \cdot 10^5$	3.4
^{241}Am	433	2.1
^{242m}Am	152	1.2
^{243}Am	$7 \cdot 10^3$	2.9
^{242}Cm	0.45	0.36
^{243}Cm	30	1.3
^{244}Cm	18	2.5
^{245}Cm	$8 \cdot 10^3$	1.5

clides, even if being of lower quantities, could pose a problem if they accumulate in the biosphere.

So far, no common criteria have been accepted which would allow the identification of hazardous nuclides and the extent to which they ought to be partitioned and transmuted. The IAEA has started a co-ordinated research programme (CRP on Evaluation of the Safety, Environmental and Non-Proliferation Aspects of Partitioning and Transmutation of Actinides and Fission Products) aimed at the definition of criteria to determine the radiological hazard posed by the numerous anthropogenic radionuclides (Table 4). Not yet knowing the outcome of this co-ordinated research programme, one can assume that the largest hazard is posed by the transuranium nuclides as well as ^{99}Tc which have no natural diluent, unlike e.g. ^{135}Cs, which will be diluted by natural caesium.

Such radionuclides could be destroyed by nuclear reactions faster than by decay, if in Eq. (1)

$$\sigma a \phi > \lambda. \quad (6)$$

Several possibilities exist: The transmutation by photon-induced reactions or by charged particles seem to have technical difficulties. Therefore, the most promising would be the transmutation by neutron capture or neutron induced fission. For this purpose, existing power stations could be used, but also newly-developed dedicated burner reactors which could be driven by accelerators.

The transmutation of transuranium elements was successfully demonstrated in an irradiation experiment in PHENIX fast reactor in 1985 [11]. Since then, several research programmes have started implying the use of existing nuclear reactors or advanced burner reactors in Russia, Japan, USA and Europe.

With increasing energy of the neutron, the σ_f/σ_c ratio is growing. Consequently, in LWR the build-up of higher actinides is substantially larger than in FR, which not only complicates the fuel handling, but increases the void coefficient of the LWR such that, for reactor safety reasons, a repeated recycling in LWR is excluded. In the fast reactor PHENIX, the transmutation half-life $\ln 2/\sigma a \phi$ is about two years for all actinides which compares favourably, e.g. for ^{237}Np with its natural decay half-life of $2 \cdot 10^6$ a (Table 5).

To achieve a convincingly 100-fold reduction of the radiotoxicity (including also plutonium for warheads), the partitioning and transmutation losses have to be as low as 0.3%, because of the repetitive recycling.

Since worldwide research on partitioning and transmutation has just begun, it would be premature to draw conclusions on technical feasibility and cost/benefit of the different concepts. Moreover, the development status of the various approaches does not yet allow a meaningful comparison. It should be pointed out that spent nuclear fuel has to be stored for tens of years. Also decades will be needed to disassemble nuclear warheads and the recovered Pu has to be stored intermediately, because of the lack of facilities in the US and Russia to feed the Pu into the commercial fuel cycle. Therefore, partitioning and transmutation – if feasible – could offer to the next generation an option to reduce the radioactivity generated by fission energy.

References

1. Cunninghame, J. G.: IAEA-213, vol. 1, p. 351 (1977).
2. Koch, L., *et al.*: EUR 6738e (1981).
3. Fischer, U., Wiese, H. W.: KfK 3014 (1983).
4. Koch, L.: *Handbook on the Physics and Chemistry of the Actinides,* Vol. 4, p. 457–490 (1986).
5. Koch, L., Molinet, R., Wellum, R.: Tagungsbericht Jahrestagung Kerntechnik, p. 472 (1983).
6. Bell, M. J.: Report ORNL-4628 (Oak Ridge National Laboratory) (1973).
7. Croff, A. G., McAdoo, J. W., Bjerke, M. A.: ORNL/TM-7176/R1 (Oak Ridge National Laboratory) (1983).
8. *World Nuclear Industry Handbook 1993,* ISSN 0029-5507, p. 10 (1993).
9. IAEA Yearbook 1994, ISBN 92-0-102394-4 (1994).
10. Annual Limits of Intake of Radionuclides by Workers Based on the 1990 Recommendations, Annals of the ICRP (1990).
11. Koch, L.: Status of Transmutation, IAEA Specialists' Meeting on Use of FBRs for Actinide Transmutation, Obninsk (1992).
12. Koch, L., Cottone, G., Cricchio, A.: Transplutonium 1975, Proceedings of the 4th Transplutonium Element Symposium, p. 461 (1976).

Radiochimica Acta 70/71, 403–412 (1995)

Radioactivity and Fusion Energy

By Hiroshi Kudo

Department of Chemistry, Graduate School of Science, Tohoku University, Sendai 980-77, Japan

(Received June 8, 1995)

Radioactivity / Tritium / Thermonuclear fusion / Muon catalyzed fusion / Fusion neutrons

Summary

Nuclear fusion is expected to give an ultimate solution to energy problems over the long term. From recent progress in developing technology for fusion reactors, we can anticipate a prototype fusion reactor by 2030. This review article describes the present status of nuclear fusion research, including muon catalyzed fusion (μCF) which attracts quite new physical interest. Tritium is an essential component of fusion reactors, because the first-stage fusion reactors will utilize a mixture of deuterium and tritium as their fuel. The knowledge about tritium as well as the fusion-neutron induced radioactivity is summarized in terms of nuclear fusion research.

1. Introduction

It has passed one hundred years, since the discovery of radioactivity by Becquerel in 1896. In the past century, we have experienced the great progress in science and technology. Big changes are also seen in our daily life. In these days, we consume huge amounts of energy in food production, transportation, communication, heating and cooling buildings, materials processing and manufacturing, and virtually all aspects of modern life.

The world total of power demands was about 8 TW in 1975, and estimates of the growth of populations and per-capita energy consumption rates indicates that the world energy demands in 2000 and 2025 will be around 18 TW and 50 TW, respectively [1]. Renewable energy sources, such as solar, geothermal, biomass, hydroelectric, wind, wave, and tidal power, are limited by the usable power they provide. Although the solar and geothermal power flows are large, the useful fractions are small. An additional problem to be solved is restriction of fossil fuel consumption to prevent the increasing atmospheric CO_2 concentrations. Nonrenewable fossil and nuclear fuel are limited by the total amount they can provide.

Utilization of nuclear fusion energy is envisaged to give an ultimate solution to energy problems over the long term. The enormous power is radiated from the sun and stars as the results of nuclear fusion of hydrogen atoms. The most important sequence is the PPI chain [2]:

$$H + H \rightarrow D + e^{+} + \nu \quad (1)$$

$$D + H \rightarrow {}^{3}He + \gamma \quad (2)$$

$${}^{3}He + {}^{3}He \rightarrow {}^{4}He + 2\,H\,. \quad (3)$$

Since middle of the 1950's, continuous efforts have been made to develop technology for nuclear fusion reactors or controlled thermonuclear reactors (CTR). First-stage fusion reactors, which will be constructed in the early 2000's, likely utilize the deuterium and tritium as fuel. One gram of fusion fuel – a mixture of deuterium and tritium gas – produces the same amount of energy as eight tons of oil. The energy in one gram of fusion fuel is five times greater than the energy liberated from nuclear fission of one gram of ^{235}U, which is the fuel of current nuclear reactors. Although the natural abundance of deuterium (0.0153%) would ensure a sufficient supply of this isotope for the fueling of D–T fusion reactors, tritium with negligible abundance (some 10^{-18} of natural hydrogen) is to be artificially produced and a practical fusion reactor should breed at least as much tritium as it consumes. The amount of tritium needed for a fusion reactor is estimated to be more than 3 kg or $\sim 10^{18}$ Bq [3], and precautions are needed to minimize the release of radioactive tritium into the environments.

The 14-MeV neutrons emitted from the D–T reaction would cause severe radiation damage and make the reactor materials radioactive. Nevertheless, a fusion reactor is envisaged to be a clean nuclear energy source, because a fusion reactor does not seem to generate so large amounts of long-lived radioactive waste as current fission reactors do.

This review article describes the present status of nuclear fusion research as well as the radioactivity related to fusion energy. Tritium is an essential component of a fusion reactor, and in developing a clean fusion reactor, it is important to understand a variety of problems concerned with tritium and the fusion-neutron induced radioactivity.

2. Nuclear fusion energy

Our experience of nuclear fusion energy on earth started with thermonuclear weapons in the early 1950's. A series of nuclear tests scattered large amounts of radioactive materials as well as tritium in the environments. The incidents were undesirable, but we realized that really enormous power was released from nuclear fusion.

Fusion research for peaceful use began in middle of the 1950's, with hopes of developing a fusion reactor or controlled thermonuclear reactor. During the first international conference on the peaceful use of

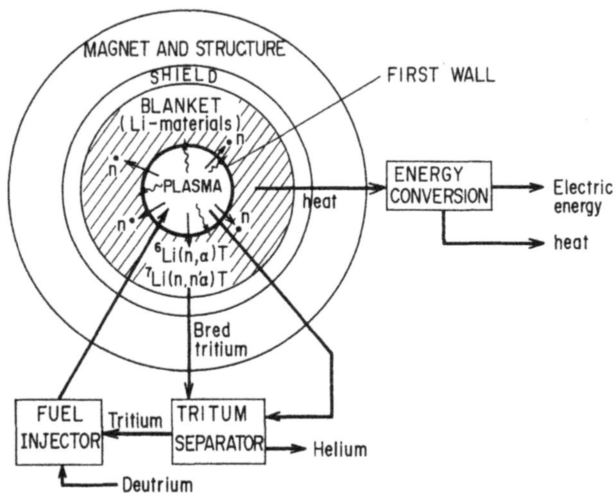

Fig. 1. A schematic drawing of D−T tokamak fusion reactor.

atomic energy, held in Geneva in 1955, the chairman reportedly said that fusion technology would be developed within 20 years. However, nobody at that time understood the difficulty of confining high temperature plasmas which would be the key to nuclear fusion. The difficulty of fusion lies in joining two nuclei. Coulomb repulsion between nuclei with positive charges prevents them from coming close enough to each other. If the temperature in a fusion reactor is raised to 100 million degrees, fuel nuclei become plasmas in which nuclei and electrons move independently at a speed of 1000 km s^{-1}, overcoming the repulsive forces of nuclei.

Nuclear fusion experiments fall into two general categories; magnetic confinement and inertia confinement [1]. Magnetic confinement employs strong magnetic fields to provide thermal insulation between the plasma and the chamber walls. Inertia confinement allows free plasma expansion and cooling, but relies on an extremely high density to attain the reaction conditions in the short expansion time (typically a few ns). The high density is attained by compressing a solid fuel pellet to over 1000 times of its initial density, using laser beams or ion beams. Major experimental programs in inertia confinement fusion (ICF) were initiated in the 1970's.

In the early 1970's many nations shifted experimental emphasis to tokamak-type plasma confinement devices, following the Soviet experimental success in the 1960's. The name "tokamak" stands for Russian words meaning "Toroidal Chamber with Magnet Coils". Tokamaks consist of a doughnut-shaped toroidal magnetic field produced by external coils, a poloidal field produced by a strong plasma current, and a weak vertical field component needed for confining plasma. The plasma must be confined inside a magnetic field so that its particles and energy will not be lost. Figure 1 illustrates a schematic drawing of a magnetic confinement fusion power plant.

In a D−T tokamak fusion reactor, the thermonuclear reaction takes place in the plasma confined in the vacuum vessel, as shown in Fig. 1. The energy produced by the D−T reaction is partly transported by the 14-MeV neutron. In addition, electromagnetic radiation and particles (neutrals and ions) with kinetic energies up to the keV range leave the plasma zone. The "first wall", which separates the vacuum zone from a blanket, would receive a heavy energy load of the 14-MeV neutrons ($10^{14}-10^{15}$ $cm^2\,s^{-1}$) and energetic particles ($\sim 10^{16}$ $cm^2\,s^{-1}$). The blanket is designed to play the following important roles; (1) heat generation, (2) tritium breeding and (3) radiation shields for superconducting magnets outermost. The fast neutrons reaching the blanket lose their energy and produce new tritium through the ^{7}Li(n,n'α)T and ^{6}Li(n,α)T reactions. The generated heat will be removed and eventually converted into electric power.

The progress of plasma confinement technology attained in the past decade with large tokamak devices such as JET (EU), TFTR (US) and JT-60 (Japan) is shown in Fig. 2. The D−T plasma burning experiment using small amounts of tritium (0.2 g) was performed in the Joint European Torus (JET) for the first time in November 1991 [4]. Since December 1993, the high power D−T experimental program with 20−30 g of tritium has been continued on the Tokamak Fusion Test Reactor (TFTR) [5, 6]. Although remarkable progress has been made in creating high temperature plasmas, further progress is required not only in plasma physics, but also in materials science as well as fuel-cycle technology.

To accelerate fusion research and development, the US, Russia, the European Union (EU) and Japan jointly set up in 1988 the International Thermonuclear Experimental Reactor (ITER) project under the auspices of the International Atomic Energy Agency (IAEA). Since 1992, the four parties have been jointly engaged in both conceptual designing for ITER and experiments to collect technical data needed for fully integrating its engineering design. They are aiming at constructing one prototype fusion reactor by 2030. A commercial fusion reactor will hopefully be constructed by around 2050.

3. Nuclear fusion reactions and the fuel cycle

Although there are many possible nuclear fusion reactions as listed in Table 1 [7], the nuclear reactions of prime interest for man-made controlled thermonuclear reactors (CTR) are as follows:

$$D + T \rightarrow {}^4He + n + 17.58\ MeV \quad (4)$$

$$D + D \rightarrow {}^3He + n + 3.27\ MeV \quad (5)$$

$$D + D \rightarrow T + H + 4.04\ MeV \quad (6)$$

$$D + {}^3He \rightarrow {}^4He + H + 18.34\ MeV \quad (7)$$

$$H + {}^6Li \rightarrow {}^4He + {}^3He + 4.02\ MeV\,. \quad (8)$$

$$H + {}^{11}B \rightarrow 3{}^4He + 8.68\ MeV \quad (9)$$

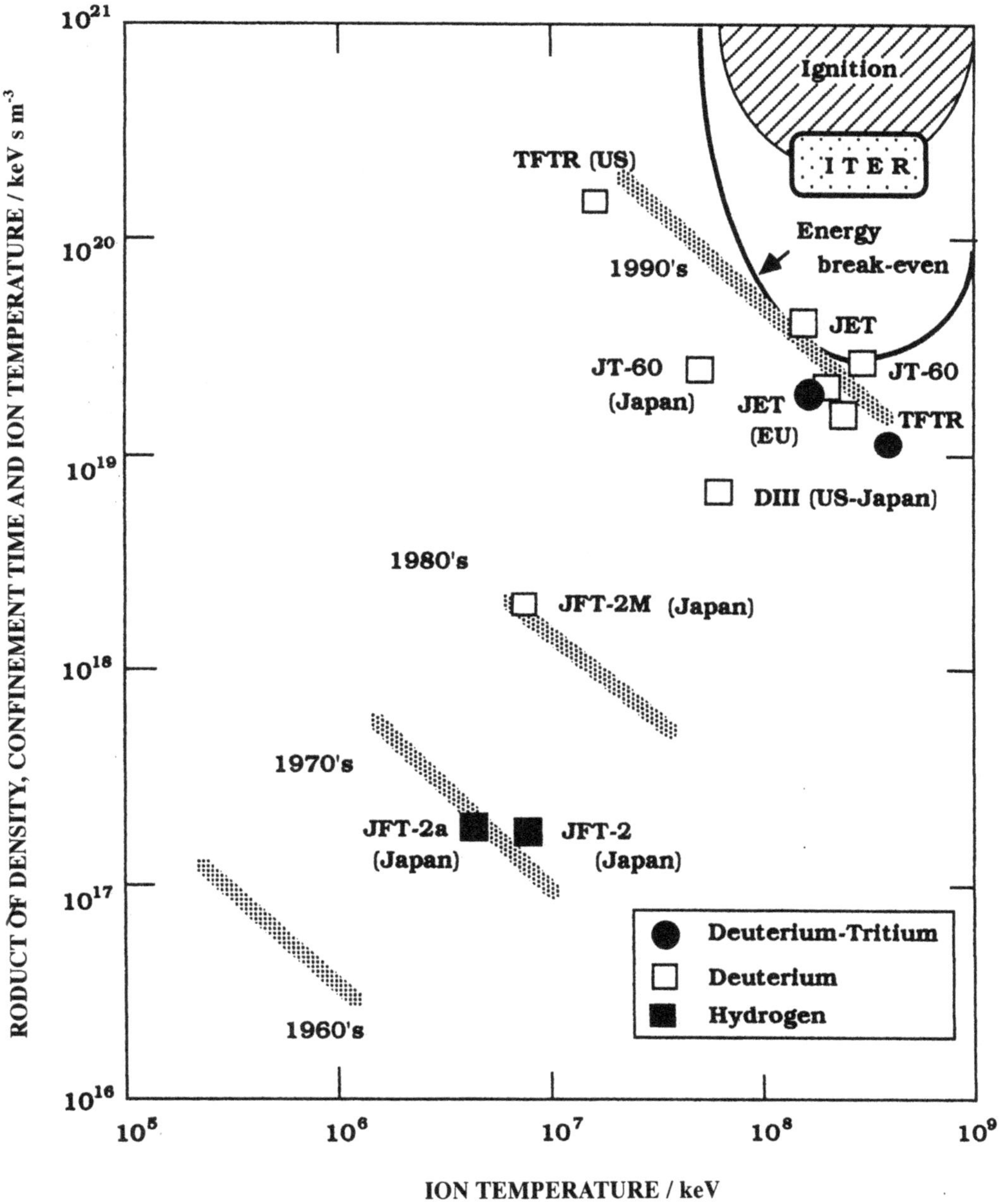

Fig. 2. Progress of tokamak plasma experiments in the world.

Figure 3 shows reaction rate parameters as a function of the plasma temperature [8]. The parameter $\langle \sigma v \rangle$ is the product of the cross section and the relative velocity, averaged over the respective Maxwellian velocity distributions of the plasma.

Of these fusion reactions, deuterium-tritium (D–T) reaction which gives rise to 14.06-MeV neutrons and 3.52-MeV α particle is the most plausible for practical CTR to be constructed in the near future, because a self-sustaining plasma may be realized at the lowest plasma temperatures $T \geq 5$ keV. The optimum plasma temperature is about 15 keV [9, 10]. In this case, it is necessary to breed tritium by bombarding lithium with the fusion neutrons, because tritium does not occur in natural sources. This type is referred to as D–T–Li fuel cycle. Concern for the environmental impacts of D–T reactors is based on the radioactive nature of the primary fuel constituent, the copious activation and radiation damage of materials by 14-MeV neutrons. In addition, a D–T–Li tokamak reactor may need several hundred tons of lithium in the blanket.

The D–D fusion reaction of Eq. (5) is attractive because the primary fuel is only naturally abundant deuterium. It requires no fuel breeding. However, a higher temperature $T \geq 20$ keV is required for a self-sustaining deuterium plasma to overcome bremsstrahlung and synchrotron radiation; the optimum temperature is about 40 keV. Practically, the D–D reaction of Eq. (5) occurs in nearly the same probability as Eq. (6), which produces tritium as the reaction product. We cannot avoid tritium problems, but the amounts of tritium would largely be reduced. A plasma temperature may be lowered to around 35 keV if the fusion products, T and ^{3}He, are recirculated; a scenario referred to as catalyzed D–D cycle. In this cycle, unburned tritium and ^{3}He are recycled into the plasma

Table 1. Thermonuclear reactions in which Z is smaller than about 10 and the cross section is more than one millibarns at energies less than 10 MeV [7]

Reaction	Q value MeV	σ_{max} mb	ε_{max} MeV
D(p,pn)H	−2.2	>400	>5.5
D(d,n)^{3}He	3.27	105	1.9
D(d,p)T	4.03	90	2.0
T(p,n)^{3}He	−0.76	500	3.0
T(d,np)T	−2.2	>700	>6.0
T(d,2n)^{3}He	−3.0	?>1200	>6.5
T(d,n)^{4}He	17.6	5000	0.108
T(t,n)^{5}He(n)^{4}He			
T(t,2n)^{4}He	11.4		
T(t,n)^{5}He*(n)^{4}He			
^{3}He(d,np)^{3}He	−2.2	>70	>1.0
^{3}He(d,p)^{4}He	18.4	700	0.4
^{3}He(t,d)^{4}He	14.3		
^{3}He(t,p)^{5}He(n)^{4}He	11.3+1.0	~50	~1.0
^{3}He(t,np)^{4}He	12.1		
^{3}He(t,n)^{5}Li(p)^{4}He	10.3+1.8		
^{3}He(^{3}He,p)^{5}Li(p)^{4}He	11.0+1.8	>30	>0.8
^{3}He(^{3}He,2p)^{4}He	12.8		
^{4}He(d,np)^{4}He	−2.2	>240	>6
^{6}Li(p,^{3}He)^{4}He	4.02	~200	1.8
^{6}Li(d,n)^{3}He + ^{4}He	1.72	≥600	>5.0
^{6}Li(d,n)^{7}Be(e$_K$)^{7}Li	3.34		
^{6}Li(d,p)^{7}Li	5.02	100	1.2
^{6}Li(d,p′)^{7}Li*(γ)^{7}Li	4.54+0.45		
^{6}Li(d,t)^{5}Li(p)^{4}He	0.9+1.6	>300	>4.0
^{6}Li(d,α)^{4}He	22.4	30	3.7
^{6}Li(t,d)^{7}Li	0.995		
^{6}Li(t,d′)^{7}Li*(γ)^{7}Li	0.509+0.45	>320	>2.1
^{6}Li(t,p)^{8}Li(β)^{8}Be(α)^{4}He	0.800		
^{6}Li(t,n)^{8}Be* or 2^{4}He	16.0		
^{6}Li(^{3}He,p)^{8}Be(α)^{4}He	16.8	30	~5
^{6}Li(^{3}He,p)^{8}Be*($\alpha\gamma$)^{4}He	13.9+2.9	>60	>5
^{7}Li(p,n)^{7}Be	−1.63	>800	>7
^{7}Li(p,α)^{4}He	17.5	65	3
^{7}Li(d,n)^{8}Be(α)^{4}He	15.0	>1000	>5
^{7}Li(d,p)^{8}Li(β)^{8}Be(α)^{4}He	−0.26+16.0	?160	3
^{7}Li(d,t)^{6}Li	−0.995	>150	>4
^{7}Li(t,2nα)^{4}He	8.88		
^{7}Li(t,n)^{9}Be	10.52		
^{7}Li(t,2n)^{8}Be(α)^{4}He	8.83	>1300	>2.1
^{7}Li(t,2nα)^{4}He	8.85		
^{7}Li(t,nα)^{5}He(n)^{4}He	8.08+1.0		
^{7}Li(t,α)^{6}He,^{6}He*	9.83	>40	>1.8
^{7}Li(^{3}He,p)^{9}B(p)^{8}Be(α)^{4}He	9.3+0.3		
^{7}Li(^{3}He,np)^{8}Be(α)^{4}He	9.5+0.1	>600	>1.4
^{7}Li(^{3}He,p)^{9}B	11.2		
^{7}Li(^{3}He,d)^{8}Be(α)^{4}He	11.7+0.1		

σ_{max}: maximum value of the cross section.
ε_{max}: energy at which σ_{max} is obtained.
*: excited nucleus.

such that the burnup rate of these fuels is equal to their production rate [11]. The fusion energy yield comes principally from the D−T, D−D and D−^{3}He reactions. The absence of a breeding blanket is an advantage.

The fusion reaction of Eq. (7) also takes place at higher temperatures, $T \geq 30$ keV, but this reaction provides a relatively "clean" sequence, because little tritium and few neutrons are produced. This D−^{3}He fuel cycle can achieve a reduced level of activation in reactor structures, yielding a low inventory of radioactivity

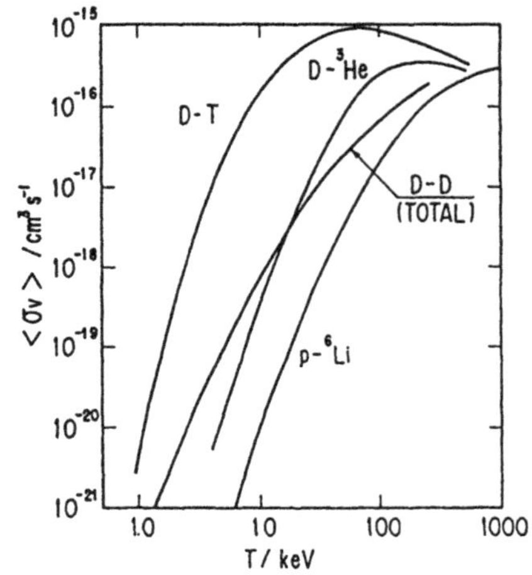

Fig. 3. Plot of fusion reaction rate parameter ⟨σn⟩ as a function of the plasma temperature [8].

in the fuel system [12]. A principal disadvantage is the lack of a ^{3}He source on earth. Apparently it is abundant on the moon, and a proposal has been made to mine the moon to collect ^{3}He [13].

The D−^{6}Li fuel cycle is extremely complicated, having two D−D branches, five D−^{6}Li branches and six ^{6}Li−^{6}Li branches, all of which are exothermic [12]. Although there are slight advantages of D−^{6}Li over D−D reactors, D−^{6}Li is less reactive than D−T, D−D and D−^{3}He.

The p−^{11}B fuel cycle is the only proton-based fuel cycle with adequate reactivity for serious consideration. Although neutron production is the lowest in this fuel cycle of the four cycles considered here, condensable and radioactive ash deposition could introduce both technological and radiological problems in the vacuum systems. The ignition prospects appear good for bremsstrahlung losses and α removal [14], but no steady-state burn has yet been projected for this fuel cycle.

4. Muon catalyzed fusion (μCF)

Muon catalyzed fusion (μCF) is an intramolecular nuclear fusion which was first predicted by Sakharov in 1948, several years before any concepts of thermonuclear fusion were suggested. The idea of energy applications was discarded for a long time because no ways were recognized of efficient use of muon within its lifetime (2.2 μs), but a discovery of the resonance mechanism of muonic ddμ and dtμ molecule formation has shed light on the μCF study again. In recent experiments at LAMPF (USA), PSI (Switzerland) and UT-MSL/KEK (Japan), fusion yields exceeding 100 per muon were observed for the liquid deuterium-tritium system. The μCF research has attracted quite new physical interest and is rapidly developing in theories, experiments and future applications [15].

Nuclear fusion reactions occur when two nuclei are within a few fm (10^{-15} m). Under moderate conditions, however, nuclei with positive charge hardly approach

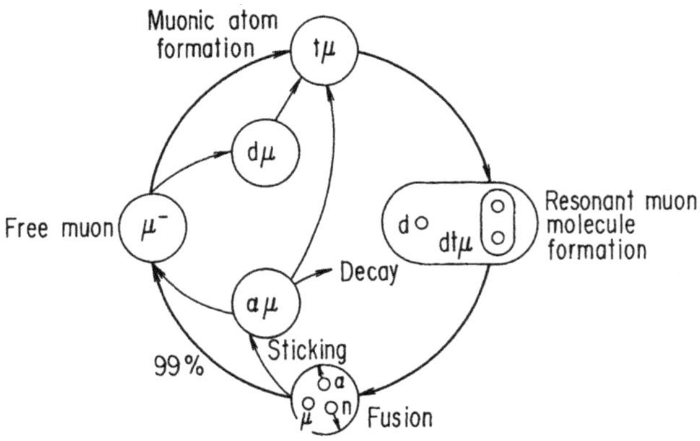

Fig. 4. Schematic expression of a muon catalyzed fusion (μCF) cycle in D−T mixtures.

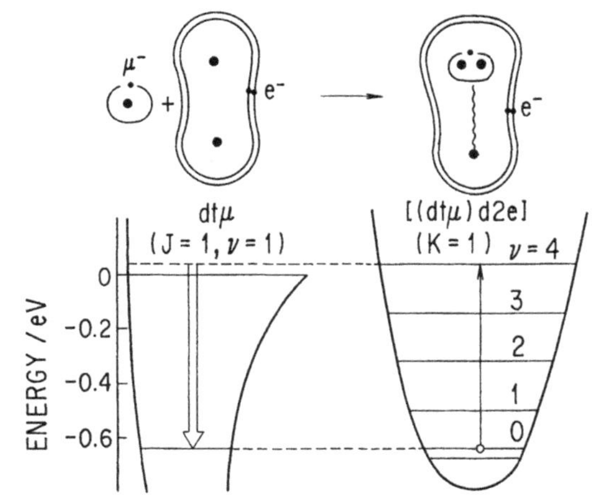

Fig. 5. Level schemes of resonant dtμ formation.

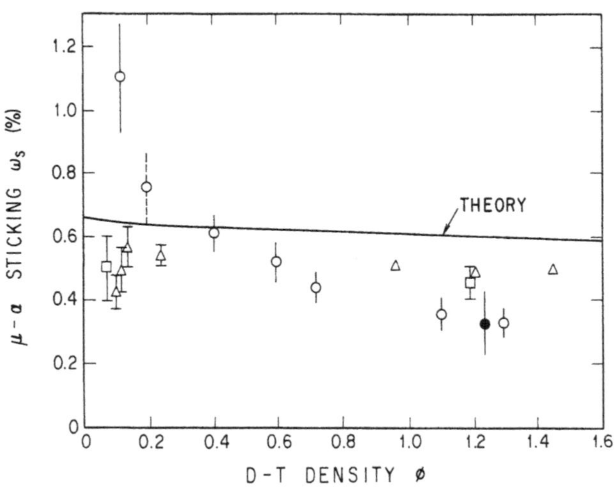

Fig. 6. Observed α-sticking probabilities (ω_s) plotted against the density of D−T mixtures: ○ LAMPF [16, 17], □ and △ PSI [18, 19], and ● UT-MSL/KEK [20]. The curve is derived from theories.

each other because of the strong Coulomb repulsion. In D_2 molecules, for example, the equilibrium internuclear distance is 74 pm, and in this case the fusion rate is as low as 10^{-70} to $10^{-60}\ s^{-1}$. As described in the previous section, we usually use high temperature plasmas to make the nuclear fusion occur. When a negative muon (μ^-) is replaced for the electron in the deuterium atom, the diameter of the neutralized muonic atom is reduced to about 1/200 of normal deuterium atom, because the mass of muon is 207 times as heavy as electrons. In the muonic DT molecule (dtμ) with the internuclear distance of about 500 fm, the fusion rate is enhanced to approximately $10^{12}\ s^{-1}$. Although the lifetime of muon is 2.2 μs, the fusion yields exceeding 100 per one muon have been experimentally observed for liquid deuterium-tritium systems in several laboratories [16−20].

The muon is one of the cosmic rays. For laboratory use, we can obtain intense muon beams from high-energy accelerators. Namely, bombardment of protons or heavy ions with kinetic energy of 300 MeV or more on targets such as beryllium and graphite generates pions ($\pi^{\pm}$). The pions with the lifetime of 26 ns decay to muons ($\mu^{\pm}$). The negative muon (μ^-) is obtained only through the decay of negative pion (π^-) in a vacuum. Figure 4 shows a μCF cycle in a deuterium-tritium mixture. A negative muon from an accelerator thermalizes in the D−T mixture and quickly forms a ground-state muonic dμ atom within 10^{-10} s. The muonic atom is tiny and neutral. Since the ground-state energy level of tμ is deeper by 48 eV than that of dμ, the dμ transfers to tμ at the rate of $10^{-8}\ s^{-1}$, colliding with the surrounding DT and T_2 molecules.

The neutral tμ easily penetrates into D_2 or DT molecules and interacts with the deuterium nucleus to form a muonic dtμ molecule. Binding energies of muonic hydrogen atoms are typically of several 100 eV, so that the normal process of dtμ formation proceeds via the Auger mechanism. Unfortunately, however, the rate of Auger process ($\sim 5\times10^4\ s^{-1}$) is rather small compared with the decay rate of muon ($\lambda_0 = 4.5\times10^5\ s^{-1}$). In the late 1950's this led to a disappointing disbelief of μCF for energy production. In 1967, however, Vesman suggested a resonance mechanism which could greatly enhance the dtμ formation rate [21, 22]. The resonance mechanism is possible due to a weakly bound energy level (−0.64 eV) of the dtμ molecule as depicted in Fig. 5. The excess of energy released by the dtμ formation is given to the electronic excitation of the resonant molecule [(dtμ)d2e]. The rate of this process, called "Vesman mechanism", is a few 10 to 100 times higher than λ_0.

Once the muonic dtμ molecule is formed, the intramolecular D−T fusion occurs at the rate of $10^{12}\ s^{-1}$, giving rise to a neutron and an α particle. The negative muon μ^- is liberated from the molecule in the fusion reaction, and then forms the tμ atom again. Thus, the μCF cycle continues within the lifetime of the muon (2.2 μs). The number of μCF cycle depends mainly upon the μ^- sticking on α particles. Determination of the α-sticking probability (ω_s) is essential in evaluating the energy production possibility with the muon catalyzed fusion process. A number of theoretical and experimental studies have been carried out on this subject [16−27]. Figure 6 shows the α-sticking probabilities observed by several research groups. Re-

Table 2. Physical properties of tritium

Mass	3.016049 u
Nuclear spin	1/2
Magnetic moment	$1.50461 \cdot 10^{-26}$ Am^2
Ionization energy	13.55 eV
Dissociation energy ($T_2 \rightarrow 2$ T)	4.59 eV
β^- decay	100%
Half-life	12.3 y
Maximum β^- energy	18.6 keV
Average β^- energy	5.7 keV
Decay heat	91.1 aW/Bq
Specific activity of T_2	255 TBq/g
Specific volume of T_2 at STP	10.3 fL/Bq

Table 3. Tritium producing reactions with relatively high cross sections

Reaction	Q-value MeV	Energy region of incident particles eV	Cross section 10^{-24} cm^2
D(n,γ)T	6.258	<0.625	5×10^{-4}
^{3}He(n,p)T	0.765	~0.025	5.33×10^3
^{6}Li(n,α)T	4.794	<0.625	9.53×10^2
^{7}Li(n,n'α)T	−2.465	$(3-6)\times10^6$	8.6×10^{-2}
^{10}Be(n,2α)T	0.88	$(1-5)\times10^6$	3.2×10^{-2}
^{6}Li(d,^{5}Li)T	0.59	2.4×10^6	9.5×10^{-2}
^{7}Li(d,^{6}Li)T	−1.00	4.1×10^6	1.6×10^{-1}
^{9}Be(d,^{8}Be)T	4.59	$(1.5-7.7)\times10^6$	2.3×10^{-2}
^{12}C(d,^{11}C)T	−1.25	20×10^6	1.0×10^{-2}
^{19}F(d,^{18}F)T	−4.17	8.7×10^6	3.9×10^{-3}

cently, Nagamine's group has reported an updated experimental result that the ω_s value is around 0.4% or less in a high density D−T system. The result revealed that the μCF cycle per muon was more than 200 and that the energy production would be essentially possible [20].

5. Tritium and its production

Tritium is a radioactive hydrogen isotope. It was discovered in 1938. Following the discovery of deuterium in 1932 by Urey *et al.* [28], efforts were made to discover hydrogen of mass three by two groups; one in Cambridge University and the other in Princeton University. In 1934, the Cambridge group led by Rutherford reported the production of hydrogen-3 by bombarding a deuterium target with fast deuterons [29]. In the same year, the Princeton group reported positive evidence for the existence of hydrogen-3 in natural sources [30]. However, both groups thought that the newly discovered hydrogen isotope was stable [31, 32]. In 1938, tritium was suggested to be radioactive by Bonner's group from observing the formation of an excited ^{3}He in the product of the D(d,p)T reaction [33]. In 1939, Alvarez and Cornog demonstrated that tritium produced by the D(d,p)T reaction was definitely radioactive and had a long half-life [34]. Current determination accords the half-life of tritium 12.3232±0.043 mean solar years where there are 365.25 days/year [35, 36].

Natural tritium is produced by cosmic radiation in the upper atmosphere, where fast neutrons, protons and deuterons collide with molecules to induce tritium; e.g. $^{14}N + n \rightarrow T + {}^{12}C$. The natural occurrence of tritium originated from cosmic radiation was 0.5 to 5 tritium units (1 tritium unit, denoted 1 T.U., equals a ratio of 1 tritium atom to 10^{18} atoms of hydrogen) before March 1954 [37, 38]. Since this time of the beginning of thermonuclear-weapon testing, samples of rain water have contained more than 500 T.U. of tritium [39]. Natural tritium was first detected in the atmospheric hydrogen by Faltings and Harteck [40] and was later shown to be in rain water by Libby's group [41].

Tritium decays to ^{3}He by the reaction,

$$^3H \rightarrow {}^3He + \beta^- + \bar{\nu}, \tag{10}$$

emitting low-energy pure β-radiation. The average energy of the β^- particle is 5.685±0.008 keV [42, 43] and the maximum energy is 18.54±0.1 keV [44]. The β^- particle is the most conspicuous missile shot forth from the decaying tritium nucleus. The β^- particles bounce off the molecular electron clouds and leave a bit of energy in each collision. This goes on until the electron is thermalized. Eventually it finds a positive charge with which to combine [43].

Tritium is unusual in not emitting a γ ray in addition to the β^- particle. No electromagnetic radiation comes from the nucleus. However, the beta particle produces radiation as it plows through the nearby matter. There are two kinds of radiation: bremsstrahlung and fluorescence [45]. Bremsstrahlung is X-ray radiation given off by the β^- particle as it decelerates near a molecule. Its spectrum is continuous, as is the β^--particle spectrum. The bremsstrahlung yield increases with the atomic number of the adsorbing material. The fluorescence radiation is produced from particular atomic transitions. The beta particle knocks out an electron from an atom. The other electrons then cascade down through the energy levels to fill the hole left by the collision. As the electrons drop to lower energy levels, they emit X rays. We should note that, once produced, X rays are more penetrating than β^- particles [43]. The physical properties of tritium are summarized in Table 2.

The environmental impact of the released tritium has been the subject of many years of investigation. The tritium is much more hazardous in its water form (HTO, DTO, T_2O) than in its hydrogen gas form (HT, DT, T_2). The potential hazard from inhalation is 25000 times greater for HTO than for HT [46]. However, the low energy β^- particles and its short biological half-life in the body (6−12 days) lead to low radiotoxity. The tritium gas oxidation rate at ambient temperatures is very low (half-life of 6.5 days). The kinetics of the conversion depend on the tritium concentration in air.

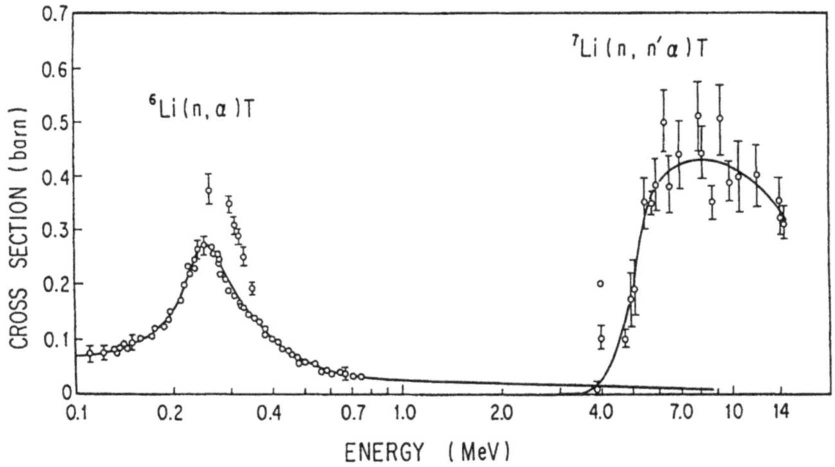

Fig. 7. Excitation functions of the $^{6}Li(n,\alpha)T$ and $^{7}Li(n,n'\alpha)T$ reactions [7]. The circles represent experimental points from several sources.

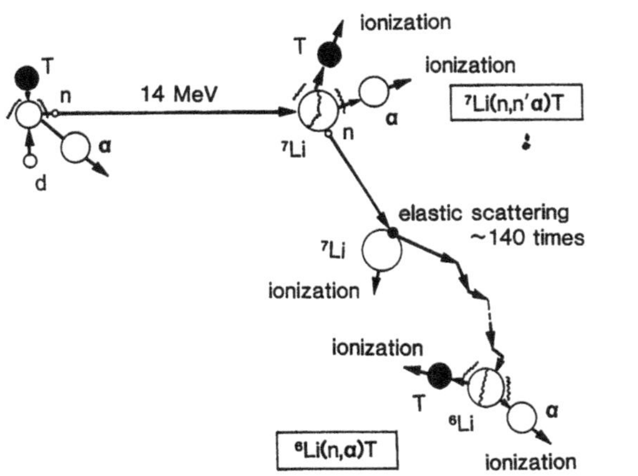

Fig. 8. Pictorial expression of fundamental tritium breeding mechanism.

Table 4. Properties of tritium breeding materials of current interest

Type of blanket	Material	Properties		
		Melting point K	Density g cm^{-3}	Li-atom density 10^{22} cm^{-3}
Liquid blanket	Li	459	0.53	4.4
	$Li_{17}Pb_{83}$	508	9.5	0.56
Solid-state blanket	Li_2O	1711	2.02	8.16
	$LiAlO_2$	1883	2.61	2.38
	Li_5AlO_4	1320	2.22	5.34
	Li_2SiO_3	1474	2.52	3.38
	Li_4SiO_4	1529	2.39	4.81
	Li_2TiO_3	1820	3.46	3.80
	Li_2ZrO_3	1873	4.15	3.27
	Li_8ZrO_6	1568	2.99	5.93

The presence of a gamma radiation field may inhibit the conversion, and fields of 80% neutrons and 20% gamma have been observed to lead to a substantial limitation of the water formation rate in air [47].

Once a D−T fusion reactor is operated successfully, the fuel tritium is produced or bred in a blanket, but we have to produce large amounts of tritium for the first inventory beforehand. Nuclear reactions to produce tritium are summarized in Table 3. Of these reactions, the following reactions would be practically applied to the large scale production:

$$^{3}He + n \rightarrow T + p \tag{10}$$

$$^{6}Li + n \rightarrow T + {}^{4}He \tag{11}$$

$$D + n \rightarrow T + \gamma \tag{12}$$

$$^{235}U + n \rightarrow FP \tag{13}$$

$$^{7}Li + n \rightarrow T + {}^{4}He + n' . \tag{14}$$

Reactions (10)−(13) are exothermic and induced by thermal neutrons. Reaction (10) has a relatively high cross section against thermal neutrons (5330 b), and is frequently used for producing tritium in basic research. However, this reaction should not be applied to the tritium production for fusion reactor fuel, because the natural abundance of ^{3}He is only 1.3×10^{-4}% or less. We usually utilize reaction (11) for producing fusion fuel tritium, irradiating lithium-containing materials such as ^{6}Li-Al alloys in a nuclear reactor [48, 49]. In the past, several kilograms per year of tritium have been produced in this manner for the weapon use. This reaction is plausible to produce large amounts of tritium fuel for the first inventory of fusion reactors.

Although the cross section of reaction (12) is rather small, large amounts of tritium are accumulated in a heavy water reactor for a long period of operation. In these days, 2.5 kg per year of tritium is recovered from CANDU reactors in Canada [50]. Nuclear fission of reaction (13) accompanies the ternary fission which produce tritium. Although the probability is very low, significant amounts of tritium would be accumulated in a core of current fission reactors. The tritium recovery from nuclear fuel reprocessing is anticipated to be a way of ensuring the fuel tritium.

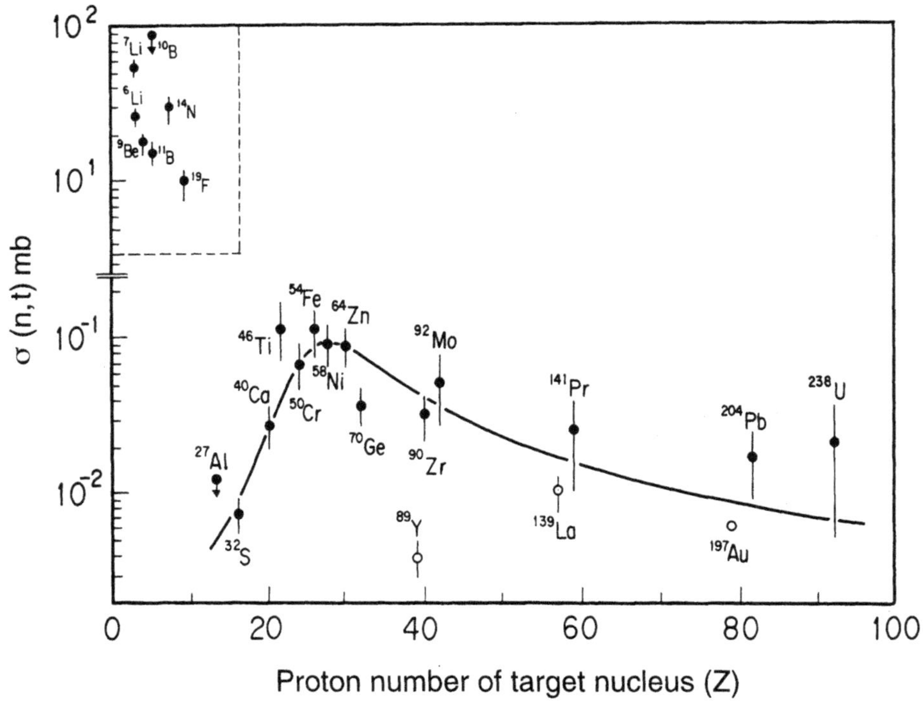

Fig. 9. Systematics in cross section of the (n,t) reactions at 14.6 MeV [68]; the open circles are partial cross sections.

Reaction (14) is endothermic ($Q = -2.47$ MeV) and is effective only for fast neutrons having kinetic energies higher than 2.16 MeV. In Fig. 7 the excitation functions for the ^{6}Li(n,α)T and ^{7}Li(n,n$'\alpha$)T reactions are compared [4]. The ^{7}Li(n,n$'\alpha$)T reaction greatly contributes to the tritium breeding in a blanket of fusion reactors. The fundamental tritium breeding mechanism is depicted in Fig. 8.

Of many lithium-bearing materials, only a few types of materials would be the most promising for practical use. Table 4 lists the candidate breeding materials of current interest. Lithium metal and $Li_{17}Pb_{83}$ will be used in the liquid state (liquid blanket) and the oxidic materials with a higher melting point such as Li_2O are the candidate for solid-state blankets [3, 51]. The high lithium-atom density is favorable for the blanket materials. Although the lithium-atom density is relatively low, $Li_{17}Pb_{83}$ can give a viable tritium breeding rate because of an excellent neutron multiplication by the (n,2n) reaction in the constituent lead [52]. The tritium recovery from breeding materials is a key factor in establishing the viability of blanket concept, and a number of fundamental as well as technological studies have been carried out on the behavior of tritium produced in lithium containing materials [53–63].

Tritium is produced also in the first-wall materials, such as graphite, B_4C and SiC, by the (d,t) reactions listed in Table 3. The use of low-Z materials has been suggested for structural component which are in direct contact with the plasma; e.g., limiters in tokamak device [64, 65]. Graphite is an important candidate for first wall components and, in fact, graphite and carbide coatings are most widely used in today's fusion machines [66, 67]. In addition, 14-MeV fusion neutron produce tritium by the (n,t) reactions in structural materials. Figure 9 shows systematics of the (n,t) cross sections at 14.6 MeV compiled by Qaim and Stöcklin [68].

6. Activation products of fusion reactors

During the operation of a D–T fusion reactor, there will inevitably develop an inventory of induced radioactivity, or activation products, in the first wall, blanket and other components. Neutron-activation products are formed when 14-MeV fusion neutrons strike the main constituents and impurities in the reactor materials. The amounts and assortment of radioactivity are dependent on the nature of the neutron fluence and, more importantly, on the structural materials. Between 20 and 30% of radioactivity in the structural materials is associated with the first wall; and over 95% of it is in the first-wall/blanket components [69]. The relative concentration of radioactivity in various candidate structural materials is shown in Figs. 10 and 11 [70]. It is clear that several materials provide a lower inventory of radioactivity than stainless steel; most notably the V-20% Ti alloy and the SiC ceramic which can be termed a low-activation material. Aluminum is a material of practical interest. Furthermore, it has been proposed that isotopic tailoring may be employed in the long term to reduce the activation in the steel [71].

The biological hazard potential (BHP) of radioactivity induced in the structural material, represented by ^{54}Mn, ^{58}Co and ^{60}Co, is about 100 times as high as that of tritium. However, the radiotoxity of the struc-

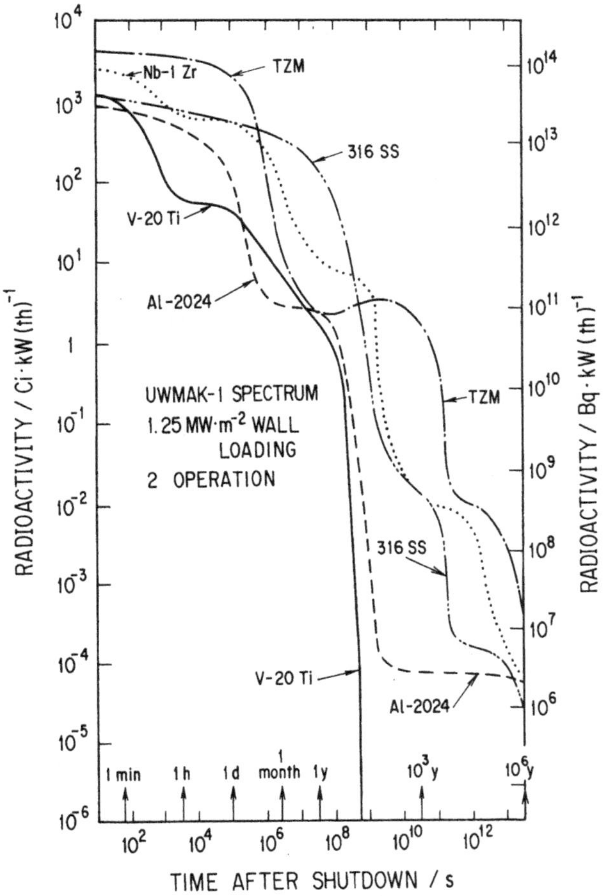

Fig. 10. Blanket radioactivity levels after shutdown for different structural alloys; the neutron spectrum from fusion reactor UWMAK-I [70].

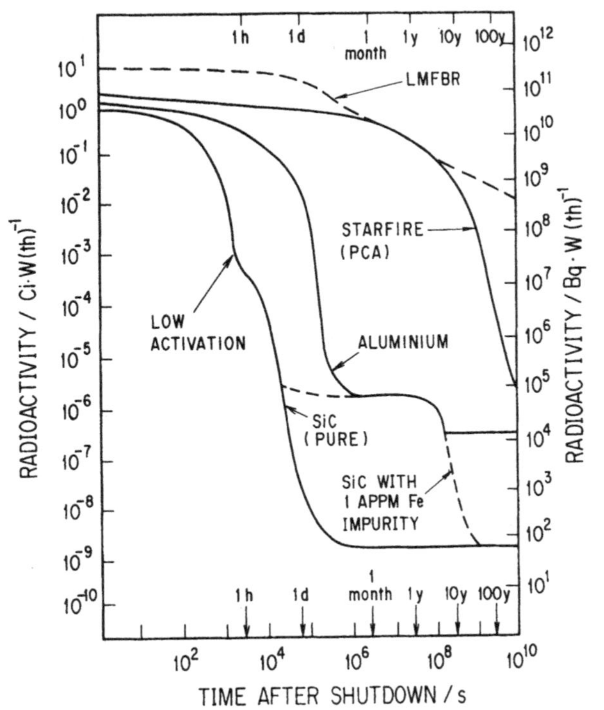

Fig. 11. Radioactivity for reactors after two years of typical operation [70]. STARFIRE is a fusion reactor; the first wall and blanket structures from aluminum, steel and SiC. LMFBR is a fission reactor.

tural materials is several order of magnitudes less hazardous than that of ^{242}Pu and ^{238}U, present in fission reactor waste [72]. From radiological point of view, we should also pay attention to activation products such as ^{13}N, ^{16}N, ^{40}Cl, ^{41}Ar and ^{37}S in the atmosphere around the fusion reactor, although the toxity of these radionuclides is very low because of their short lifetime.

A major difference between fusion and fission is that most of waste from nuclear fusion is produced on the power plant site. This is not the case for fission where substantial amounts of waste are generated during fuel acquisition and processing. A fusion reactor would be easier to control because the fuel is continuously supplied from outside. In case of an emergency, the fusion reaction could be stopped quickly by turning off the fuel supply switch. A fusion reactor does not pose the danger of a meltdown, as a fission reactor does.

Conclusively, the safe handling of huge amounts of tritium must be one of the prime technological subjects of developing D–T fusion reactors. The clue to solving a variety of problems will be found in the plasma experiments on the TFTR and JET devices fueled with tritium [73–75].

References

1. Dolan, T. J.: *Fusion Research*, Pergamon Press, New York (1982).
2. Cox, P. A.: *The Elements. Their origin, abundance and distribution*, Oxford Univ. Press, Oxford (1989).
3. INTOR group: *International Tokamak Reactor*, Phase Two A, Part III, IAEA, Vienna (1988).
4. JET Team: Nucl. Fusion **32**, 187 (1992).
5. Strachan, J. D., *et al.*: Phys. Rev. Lett. **72**, 3526 (1994).
6. Hawryluk, R. J., *et al.*: Phys. Rev. Lett. **72**, 3530 (1994).
7. Crocker, V. S., Blow, S., Watson, C. J. H.: CLM-P240 (1970).
8. Post, R. F.: Ann. Rev. Nucl. Sci. **20**, 509 (1970).
9. Sheffield, J.: Rev. Mod. Phys. **66**, 1015 (1994).
10. McNally, J. R., Jr.: Nucl. Technol./Fusion **2**, 9 (1981).
11. Mills, R. G.: PPL-TM-74191, Princeton Plasma Physics Laboratory (1971).
12. McNally, J. R., Jr.: Nucl. Fusion **18**, 133 (1978).
13. Kulcinski, G. L., Emmet, G. A., Blanchard, J. P., El-Guebaly, L. A., Khater, H. Y., Satarius, J. F., Sawan, M. E., Sviatoslavsky, I. N., Wittenberg, L. J., Witt, R, J.: Fusion Technol. **15**, 1233 (1989).
14. Shuy, G. W., Conn, R. W.: PPG-522, University of California, Los Angeles (1980).
15. Petitjean, C.: Fusion Eng. Des. **11**, 255 (1989).
16. Jones, S. E.: Muon Cat. Fusion **1**, 21 (1987).
17. Caffrey, A. J., Anderson, A. N., van Sichen, C. D., Watts, K. D.: Muon Cat. Fusion **1**, 53 (1987).
18. Petitjean, C., Breunlich, W. H., Cargnelli, M., Kammel, P., Marton, J., Nägele, N., Pawlek, P., Scrinzi, A., Werner, J., Zmeskal, J.: Muon Cat. Fusion **1**, 89 (1987).
19. Petitjean, C., *et al.*: Muon Cat. Fusion **2**, 37 (1988).
20. Nagamine, K., Matsuzaki, T., Ishida, K., Hirata, Y., Watanabe, Y., Kadono, R., Miyake, Y., Nishiyama, K., Jones, S. E., Maltrud, H. R.: Muon Cat. Fusion **1**, 137 (1987).
21. Vesman, E. A.: JETP Lett. **5**, 9 (1967).
22. Gerstein, S. S., Ponomarev, L. I.: Phys. Lett. **72B**, 80 (1977).
23. Nagamine, K., Matsuzaki, T., Ishida, K., Watanabe, Y., Sakamoto, S., Iwasaki, M., Miyake, Y., Nishiyama, K., Tori-

kai, E., Kurihara, H., Kudo, H., Tanase, M., Kato, M., Kurosawa, K., Sugai, H., Fujie, M., Umezawa, H.: Muon Cat. Fusion **5/6**, 289 (1991).
24. Kamimura, M.: Proc. μCF-88 Workshop, Sanibel Island, USA (1988).
25. Bogdanova, L. N., Markushin, V. E., Melezhik, V. S., Ponomarev, L. I.: Sov. J. Nucl. Phys. **50**, 848 (1989).
26. Markushin, V. E.: Muon Cat. Fusion **3**, 395 (1988).
27. Cohen, J. S.: Muon Cat. Fusion **3**, 395 (1988).
28. Urey, H. C., Brickwedde, F. G., Murphy, G. M.: Phys. Rev. **39**, 164 (1932).
29. Oliphant, M. L. E., Harteck, P., Rutherford, L.: Proc. Roy. Soc. **A 141**, 259 (1933).
30. Breit, G., McIntosh, J. S.: Phys. Rev. **83**, 1254 (1951).
31. Nefkens, B. M. K.: Phys. Rev. Lett. **10**, 55 (1963).
32. Alvarez, L. W., Cornog, R.: Phys. Rev. **57**, 248 (1940).
33. Bonner, T. W.: Phys. Rev. **53**, 711 (1938).
34. Alvarez, L. W., Cornog, R.: Phys. Rev. **56**, 613 (1939).
35. Rudy, C. R., Jordan, K. C.: *Tritium Half-life*, Monsanto Research Corporation Report MLM-2458, Mound Laboratory, Maiamisburg, Ohio (1977).
36. Jordan, K. C., Blanke, B. C., Dudley, W. A.: J. Inorg. Nucl. Chem. **29**, 2129 (1967).
37. Kaufmann, S., Libby, W. F.: Phys. Rev. **93**, 1337 (1954).
38. Lujanas, V.: Liet. TSR Mokslu Akad. Darb. Ser. B. **1**, 21 (1963).
39. Evans, E. A.: *Tritium and its Compounds,* 2nd. ed., Butterworths, London (1974).
40. Faltings, V., Harteck, P.: Z. Naturforsch. **5a**, 438 (1950).
41. Grosse, A. V., Johnston, W. H., Wolfgang, R. L., Libby, W. F.: Science, N.Y. **113**, 1 (1951).
42. Pillinger, W. L., Hentges, J. J., Blair, J. A.: Phys. Rev. **121**, 232 (1961).
43. Souers, P. C.: *Hydrogen Properties for Fusion Energy,* Univ. of California Press, Berkeley (1986).
44. Lewis, V. E.: Nucl. Phys. **151A**, 120 (1970).
45. Souers, P. C., Tsugawa, R. T., Stone, R. R.: Rev. Sci. Instrum. **46**, 682 (1975).
46. International Commission on Radiological Protection: ICPR Publication No. 30, Pergamon Press, Oxford (1979).
47. Easterly, C. E., Bennett, M. R.: Nucl. Technol./Fusion **42**, 16 (1983).
48. Rhinehammer, T. B., Wittenberg, L. J.: MLM-2419 (1978).
49. Hugony, P., Sauvage, H., Roth, E.: B.I.S.T., Commissariat à l'Energie Atomique, No. 178, 3 (1973).
50. Sood, S. K., Sissingh, R.A.P., Kveton, O. K.: Fusion Technol. **8**, 2478 (1985).
51. Sako, K., Ohta, M., Yamato, H., Hiraoka, T., Tanaka, K., Asami, N., Mori, S.: KAERI-M 5502 (1973).
52. Meier, W. R.: UCRL-87499 (1982).
53. Kudo, H.: J. Nucl. Mater. **87**, 185 (1979).
54. Kudo, H., Okuno, K.: J. Nucl. Mater. **101**, 38 (1981), (1983).
55. Okuno, K., Kudo, H.: J. Nucl. Mater. **116**, 82 (1983).
56. Kudo, H., Okuno, K.: Radiochim. Acta **33**, 223 (1983).
57. Kudo, H., Okuno, K.: J. Nucl. Mater. **133/134**, 192 (1985).
58. Okuno, K., Kudo, H.: J. Nucl. Mater. **138**, 31 (1986).
59. Okuno, K., Kudo, H.: J. Nucl. Mater. **138**, 210 (1986).
60. Kudo, K., Okuno, K.: J. Nucl. Mater. **155–157**, 524 (1988).
61. Okuno, K., Kudo, H.: Fusion Eng. Des. **8**, 355 (1989).
62. Kudo, H.: Radiochim. Acta **50**, 71 (1990).
63. Kurasawa, T., Watanabe, H., Roth, E., Vollath, D.: J. Nucl. Mater. **155–157**, 544 (1988).
64. Kulcinski, G. L., Conn, W. R., Lang, G.: Nucl. Fusion **15**, 327 (1975).
65. Conn, W. R., Kesner, J.: Nucl. Fusion **15**, 775 (1975).
66. Winter, J.: J. Nucl. Mater. **145–147**, 131 (1987).
67. Dietz, K. J.: J. Nucl. Mater. **155–157**, 8 (1988).
68. Qaim, S. M., Stöcklin, G.: J. Inorg. Nucl. Chem. **35**, 19 (1973).
69. Holdren, J. P.: Ann. Rev. Energy Environ. **16**, 235 (1991).
70. Kazimi, M. S.: Nucl. Fusion **24**, 1461 (1984).
71. Conn, R. W., Okula, K., Johnson, W.: Nucl. Technol. **41**, 289 (1978).
72. Smith, C. F., Kastenberg, W. E.: Nucl. Eng. Des. **39**, 293 (1976).
73. Anderson, J. L., Gentile, C., Hosea, J., Kalish, M., Kamperschroer, J., Kozub, T., LaMarche, P., Murray, H., Nagy, A., Raftopoulis, S., Rossmassler, R., Sissingh, R., Swanson, J., Viola, M., Voorhees, D., VonHalle, A., Walters, R. T.: IAEA-CN/F-2-II-1 (1994).
74. Andrew, P., Cadwell-Nichols, C. J., Coad, J. P., Dietz, K. J., Ehrenberg, J., Goodall, D. H. J., Hemmerich, J., Horton, L., How, J., Lasser, R., Lomas, P., Loughlin, M., McCracken, G. M., Peacock, A., Saibene, G., Sartori, R., Thomas, P., Winkel, T.: J. Nucl. Mater. **196–198**, 143 (1992).
75. Goodall, D. H. J., Andrew, P., Ehrenberg, J., McCracken, G. M., Peacock, A. T., Saibene, G., Sartori, R., van Belle, P.: J. Nucl. Mater. **196–198**, 1002 (1992).

Author Index

www.ingramcontent.com/pod-product-compliance
Lightning Source LLC
Chambersburg PA
CBHW080939240926
51849CB00004BA/502